Point Groups, Space Groups,
Crystals, Molecules

by R. Mirman

Group Theory: An Intuitive Approach
(Singapore: World Scientific Publishing Co., 1995)

Group Theoretical Foundations of Quantum Mechanics
(Commack, NY: Nova Science Publishers, Inc., 1995)

Massless Representations of the Poincaré Group
electromagnetism, gravitation, quantum mechanics, geometry
(Commack, NY: Nova Science Publishers, Inc., 1995)

Point Groups, Space Groups, Crystals, Molecules
(Singapore: World Scientific Publishing Co., 1999)

Point Groups, Space Groups,
Crystals, Molecules

Problem T-1: Finish the diagram.
How is this space group related to that of the cover ?

R. Mirman

World Scientific
Singapore • New Jersey • London • Hong Kong

Published by

World Scientific Publishing Co. Pte. Ltd.
5 Toh Tuck Link, Singapore 596224
USA office: 27 Warren Street, Suite 401-402, Hackensack, NJ 07601
UK office: 57 Shelton Street, Covent Garden, London WC2H 9HE

British Library Cataloguing-in-Publication Data
A catalogue record for this book is available from the British Library.

POINT GROUPS, SPACE GROUPS, CRYSTALS, MOLECULES

ISBN-13 978-981-02-3732-5
ISBN-10 981-02-3732-4

Preface

GROUPS, MATHEMATICS, PHYSICS, ART

Group theory is a diverse, extensive subject, playing roles, often fundamental, in numerous areas — there are many different ways in which it is useful, enlightening, often necessary. This book considers thus only a small part of a broad, substantial topic. While a paramount goal is, of course, teaching the theory, certainly providing the basis for application, its deeper rationale is development of (part of) the foundation for the study of group theory more generally, and of mathematics, physics, chemistry, crystallography — and beyond. The hope is to teach group theory, explaining how it is used, not merely that part that it covers — and to nurture the skills to develop and apply it, helping and stimulating the reader to gain thorough understanding.

Group theory is important, essential, for students in many different fields, for many different types of researchers. The book tries to achieve the impossible by appealing to all. It can be used by those only slightly interested, to get a quick definition, a brief introduction to a concept, a slight understanding of what is being read elsewhere, even of only vocabulary. And it can, hopefully, be studied for deeper and deeper understanding, as profound as wished. For those who really study it, who make serious attempts at many of the problems, who think about questions raised, it is designed to develop not only insightful, profound knowledge of groups, and applications, but more generally of mathematics, physics, chemistry, even philosophy. For such readers the desire is to aid also in the development of those skills and habits that lead to success more generally.

How difficult is this book? It is written to be easy to read and understand, for those who want an introduction to the subject, a reading knowledge, fundamental concepts, vocabulary. And it should not be more difficult for readers who wish to go deeper, although of course that requires more work. But it also attempts to provide the material, the questions, the problems, the challenge, the irritation, the skepticism, so that those who wish a very deep understanding, who wish potent skills, and not only in this subject, will have the stimulation and exercises to help obtain them.

Crystals and molecules play a preeminent role in physics and chemistry (and technology and economics), and their groups of transformations are basic in determining their properties and those of objects that compose them. But beyond such applications, they provide subjects quite suited for the development and illustration of many general concepts of group theory, geometry, mathematics. It is thus surprising that there are have been so few books devoted to the theory of these groups. There have been (in the past) books available, but there seems to never have been a comprehensive treatment of this quite vast subject. This book is certainly not that, and perhaps such is impossible. But it attempts to at least introduce the major parts of the subject, to show how they are related, and what their importance is — and to provide guidance (certainly references) to those who wish to learn more, to go deeper.

Because of the extent of the subject it is impossible to cover more than a small part. In many cases material is merely mentioned, often in problems, to inform the reader that it exists, to at least provide definitions, and also to indicate where further information can be found. Thus, of regrettable necessity, many topics had to be only outlined, but with references to other discussions, perhaps in more depth. And if the reader should find all these discussions are also only outlines, the conclusions to be drawn would be obvious. Often a topic is referred to only to be able to provide references. Unfortunately these are in too many cases to books no longer in print, frequently long out-of-print. Yet the material that they contain, even just the approach, is often not found anywhere else. As they disappear, the knowledge within them, typically gained with so much difficulty and so useful, disappears also. Thus one hope is to at least keep alive the memory of books long gone, but whose disappearance is a substantial loss. A book at least differs from a person — it can be brought back from the dead. Perhaps if memory can be kept alive, some books themselves might be restored to life.

This book is, of course, aimed at providing knowledge of group theory, to enable its readers (to apply and develop it), specifically the part that deals with discrete, particularly crystallographic, groups, and to use these applications to help in mastering groups, and the systems to which they can be applied. But more, what is emphasized is understanding, intuition, not only about groups, crystals, molecules, but about mathematics, physics, chemistry, crystallography, and related fields, and about understanding. It is designed to push the reader to understand the theory, the reasons for it, for the properties of (these) groups, for what we have decided should be the properties, and why. So it aims to stimulate, to goad, the reader to think about, not only the material, but its foundations, its reasons, about what is being read. Those in education know that too often we do not educate our students, we program them. Like a good computer program they are able to deal well with material for which they have been programmed. But too often, just as a computer program, they do not understand what they are doing, or why. They do not understand the reasons the particular software has been read into them. They are unable to think about what they are doing, to extend their capabilities to areas other than those for which they have been programmed. Obviously researchers should be able to deal with novelty, but too often they are able to merely apply routinely their built-in software to subjects at most slightly different then those taught to them. And educators also must be able to think innovatively. How else can they teach their students to do so? This is part of the underlying philosophy of the book and the way it was written (like the first book of this series [Mirman (1995a)]). Is the approach correct? Can it achieve what it tries to? Can any approach? This is an experiment. Hopefully it will not only teach the material, and its understanding, to enable readers to deal with it, but also it will teach those who educate a little about how, and about how not, to do so — or at least stimulate (or disturb) them enough so that they will consider these questions. A fundamental rationale is to raise such issues, to prod the reader to think — and to provoke those who use and teach from this book, to also think, not only about group theory, mathematics, physics, but about education. It aims to stimulate others to try to devise ways of better educating, their students and themselves, and hopefully to spread what they have learned, so others may learn, may be sensitized, aroused, so that they too will look — and find — ways of better, not merely programing, but educating.

What is learned in (good) formal education is never sufficient for a successful career. Rather it provides the foundation for necessary further learning. A successful education imparts not only knowledge — necessary for the acquisition of further knowledge — but skills, to go deeper into subjects, to utilize and extend knowledge and skills, to gain further knowledge. It provides not only depth, and the experience to penetrate further, but breadth, flexibility, abilities to go into other fields, to deal with other forms of knowledge, to acquire and use them; it provides the skills necessary to learn other skills.

This is a major aspect of the philosophy of these books. Not only to teach the subjects and the skills to use them, but also those needed to learn more, about these fields, about others, and the capacity to deal with whatever problems the reader encounters, in whatever fields, however they may arise. For success, especially during uncertainty, breadth and adaptability are essential.

One skill needed by a researcher is judgment — what problems should be attempted? Many problems here are partially trivial, partially research, partially impossible. Deciding what can, and what cannot be done, what is worth trying, is valuable exercise.

Although the question of rigor has been discussed previously [Mirman (1995a), sec. P.5, p. xv], it is worth emphasizing again, for this also is part of the underlying philosophy, and here quite relevant. An exposition is rigorous if it is written in a way so that it can be widely, and easily, understood, thus easily checked. Having been checked by many people, we can be more confident that there are no errors, that all assumptions have been stated. By this standard much of present mathematics and physics is highly nonrigorous. Obscurantism is not rigor, despite the much too popular belief. It is just the opposite.

What is so disturbing is not merely the way much material is so often presented, or written so as not to be presented, but that there is no attempt to explain or understand it. Science becomes a religion; these are the accepted beliefs, there are no reasons for them, there is no need to explain them. Read not only many, but (at least, close to) all, books on these topics. Consider the number of subjects presented, not discussed, without explanation, without justification, without reasons, without understanding, without even the hint of why they are true (if they are), where they came from, or where further discussion or justification can be found (for they cannot). This is not education, it does not lead to understanding. Here then we attempt to fill in (unfortunately, necessarily only some of) these gaps, or at least point them out — and to emphasize how often books on mathematics and science present not explanations, rationales, proofs, but merely beliefs. And very often what everyone believes, while it may be the truth, is not the whole truth. Someone learning the subject should understand this, should know what is true, what is not true, what is not fully true.

For the insightful comprehension we hope the reader will gain, it is important, even essential, to have different views of every (or at least the most important) aspects, and to see how they are related. Thus there are topics covered in great depth. This also allows flexibility. The reader, or the instructor, has available much material from which the choice most fitting for a particular case can be made. There are fewer restrictions because of choices the writer has made.

And to achieve understanding it is necessary to stimulate the reader to think, so to raise questions, doubts even paradoxes, mysteries, contradictions. This underlies many of problems.

While an attempt has been made to explain and justify much material very fully, it is impossible to do this for everything. Much had to be just outlined, with the material taken from standard sources (which are referenced so the reader can find, hopefully, more comprehensive discussions), and many of these took their material from other standard sources. No claim is made that all material taken from such sources has been carefully checked, and perhaps the standard sources did not check everything taken from previous standard sources. It is not unheard-of for errors to be propagated because "facts" become well-known and established, so never checked. Readers should always be very careful. Teaching this is another hope.

Attitudes of researchers, physicists, mathematicians — any researchers — are essential, for success and the production of useful and meaningful work. Of great importance is open-minded skepticism, being willing to examine with care and understanding new ideas, approaches, results no matter how disturbing or contrary to previous beliefs, but also care and skepticism, not only with these but also with accepted beliefs, establishment views. Too many people have instead a closed-minded credulity, closed to anything different from what they are told to believe, but credulous for the sanctioned, the authorized, no matter weakly supported, even absurd, that may be. Too often we accept, uncritically, what we are told, what is on a piece of paper. So the book tries to instill an attitude of open-minded skepticism, to break the connection between appearance of a statement on paper or in a book and acceptance of it. Of course, only slight progress can be made, but it is hoped that at least this is possible — and that others will be stimulated to also try to develop, the so essential, open-minded skepticism. Thus (perhaps too) occasionally, doubts are raised about the statements being presented so to accustom the reader to skepticism, to care, to the habit of checking. Open-mindedness is harder, but perhaps once thought becomes a habit, useful doubt also will.

The philosophy underlying this book is the same as the first one [Mirman (1995a)]. To help the reader understand what is being attempted, an explanation of the approach is attached.

ACKNOWLEDGEMENTS: For much help, this is to thank the Department of Natural Sciences of Baruch College (City University of New York) and World Scientific Publishing Company.

GROUP THEORY:
AN INTUITIVE APPROACH
R. Mirman

Level:

First-year graduate, but accessible to many juniors. The right text for a course in group theory (even for undergraduates), as well as a supplementary text for other courses, and for self-study. Valuable also to those working in group theory, or using it in their research. Versatile: provides a quick understanding of a definition, an introduction, or a thorough knowledge and understanding, as desired by the user.

Eminently suitable, as a text, as a developer of skills (including teaching skills), as a reference, as a source of material not available elsewhere, and as a stimulant to thought (pb. VII.7.e-3, p. 213).

Unusual as text; it is designed to teach — to give competitive edge in these difficult times: Its goal is not merely learning, but understanding — group theory and more broadly, physics, mathematics and the application of mathematics to physics. It is written in a style that makes it easy to read, understand, and master. The approach is concrete, coherent, physical, emphasizing motivation (see especially beginning of most chapters), explanation, vocabulary and definitions, and examples (sec. I-6, p. 25; chap. II, p. 29; sec. III-2, p. 68). Problems are stressed, obviously. All the relevant standard ones are here, in one convenient place, usually expanded and elaborated on, plus a vast number of other, original, problems. Valuable to instructors as a source of problems, and for ideas for projects or research.

The book has an educational philosophy (which is unusual); it is based on not only knowledge of group theory and its use in physics gained during many years of research, but also an understanding of the learning process obtained from long experience in teaching. Clearly, the philosophy and approach are original.

Explanatory depth

Explains not only what is happening, but why (sec. I.3.f, p. 10; sec. IV.5, p. 119; sec. X.6.c, p. 292; sec. X.7.e, p. 299). It not merely derives results, but explains why they hold (sec. VI.1.b, p. 171) and how they arise from the assumptions and definitions about groups and representations (sec. VII.5.b, p. 195). What properties of groups, and the objects to which they are applied, give the results being derived (sec. XI.4.a.ii, p. 317; sec. XII.5.b.iii, p. 361)? Why should these hold (sec. IV.5, p. 119)? What do they imply about the mathematics underlying the material, and mathematics in general, and about physics (sec. XI.6, p. 337)? Why are these properties important (sec. VII.3, p. 186)? Illustrations, examples (throughout the book), even counter-examples (sec. I.3.h, p. 12; pb. III.2.i-4, p. 81; sec. XII.5.e, p. 366), help the reader see what the definitions mean, and understand, remember, and apply them.

Comprehensive: learning different parts of a subject reinforces the rest. The book allows the reader to easily study any topic, strengthening the grasp of all material; and the different parts are explicitly related to each other.

An (unusually) extensive table of contents serves as an outline which organizes material, shows relationships, helps learning, understanding. The style (see, say, the beginning of chapters), also does this.

Definitions and vocabulary

Vocabulary is essential. The book stresses the meaning and significance of terms and concepts (pb. III.4.b-5, p. 87; sec.III.4.e, p. 91; sec.III.4.f, p. 97), making these not only as clear, but as familiar, and easy to remember, as possible. Not only are the terms defined, but often (group-theoretical) reasons for the choice of words are indicated.

Problems improve derivations:

Compare derivations, often uniquely stated, with standard ones (which are also included). Those in other books leave out many steps. While more is put in here, it is neither possible nor desirable to include everything. So many parts are given as problems, reminding the reader that steps are left out and stimulating students to think, and put them in. This is less confusing and makes it easier to understand, learn, and remember the material. And discussions here provide more explanations. Also this helps develop general skills — after working through the book a student will not only know, and understand, group theory, but will be better trained as a mathematician or physicist.

Fruitful as a reference:

Easy to use as a reference, to quickly look up and understand material — without having to read the entire book.

Designed to make finding definitions, explanations, formulas, proofs easy. Very thorough index and table of contents. Extensive and unusually complete cross-references, on almost all pages. Copious citations to other works to provide different views, and indicate where to find material not covered — so assisting readers to expand their capabilities. This is part of its philosophy of not merely presenting material, but educating, helping and encouraging the reader to learn.

Trains (successful) scholars:

The book, perhaps uniquely, is designed to train research mathematicians and physicists, and teachers (pb. I.5.a-7, p. 22; sec. IV.4, p. 116; pb. VII.3.c-1, 2, 3, p. 190; pb. XI.6-4, p. 338). One way is to stimulate them to think (sec. III.6, p. 103), which the entire book tries to do, especially think like a mathematician and physicist (pb. VI.1.b-2, p. 172; p. 181, top; pb. VIII.4.c.ii-1,2, p. 230; pb. XI.4.a.ii-5, p. 318; pb. XII.5-2, p. 358). It is important for a researcher, and especially a teacher, to be able to explain a subject in words (sec. IV.8.c.iii, p. 142). Another aspect of learning mathematics is "art appreciation", to see (and marvel at) the beauty and unity of mathematics, and its application to the physical world (sec. V.5.d, p. 167; sec. IX.2.c, p. 253), and so the beauty of nature. Problems are designed to be thought provoking, to stimulate and teach readers to imagine. And different views of the same subject, the same derivation, not only aid understanding, but teach comparison and evaluation, thus judgment.

Stimulus to research:

Many problems raise questions whose (full) answers are unknown (pb. IV.3.d.iv-9, 10, p. 116; pb. IV.4-18, 19, p. 118; pb. XI.1.a-2, p. 306). It makes easily available important material otherwise difficult to find and organize, and understand, so valuable to those learning, using, and working in group theory, and any of the fields it touches, even people thoroughly familiar with these.

Original:

There is material in the book that appears nowhere else (sec. VI.4, p. 176; sec X.4.b, p. 281; sec. X.6, p.287). Much is pedagogical, but some includes proofs or explanations that may not otherwise exist.

The book is unusual, starting with the cover, in ways that make it exceptionally and widely useful, and provocative.

The examples mentioned are only a small sample, as can be seen by looking at almost any page. It is easy to compare the treatment of this book with the corresponding ones of others, where they exist, since complete references are given.

Standards listed are those on which the book should be judged. Compare these to the standards on which other books ask to be judged. People never live up to their standards, but those that they set tell much about them.

World Scientific Publishing Co.

Table of Contents

List of Tables

List of Figures

Chapter I

Transformations With a Point Fixed: Point Groups

I.1 SYMMETRIES OF SPACE AND OF OBJECTS

Groups play a major role in physics and chemistry because they describe, and control, properties of material objects [Mirman (1995a), sec. II.5, p. 60]. Physical substances, what they are, what they are like, are governed by the nature of space, by geometry, and (so) the laws of physics. Thus their symmetries, and those of physical entities, strongly affect behaviors and characters, of space, laws and objects [Mirman (1995b), sec. I.7, p. 22; Rosen (1981)]. The transformations leaving physical systems invariant are controlled by those of space — and (?) of physical laws, though those of systems are in some ways less, in some ways more, than symmetries of empty space. These greatly affect what objects are and how they behave — and so what we are and how we behave [Mirman (1995c), p. v]. Here we start to study actual materials, crystals and molecules, and try to understand how — and why — their properties are thus limited and determined, and what these properties are determined as. And we use these bodies and their properties to study mathematical systems — the groups that describe the symmetries — for the mathematics is relevant far beyond crystals and molecules.

Physical entities are largely, certainly not completely, geometrical. Crystals and molecules consist of points; that they can be interpreted as physical is often, partially or mostly, irrelevant. Their physical properties are constrained, even largely fixed, by their geometrical ones. Indeed physics itself is largely, even completely, constrained and determined by geometry [Mirman (1995b,c)]. So it is geometry that we study. But geometry is a vast field — much can be interpreted as geometrical. The tools on whose development we concentrate are therefore broadly applicable, to physics, not only to crystals and molecules, and to geometry, to group theory — and in fields that cannot be thought of as any of these.

That a simple system, a hydrogen atom say, is quite restricted by rotational invariance, for example, is not surprising. But an atom in a complicated molecule or solid appears not able to see the symmetries of space so might seem less affected by these — its complex environment appears to make it difficult to find such symmetries as those given by the rotation group. Yet the properties of space are fundamental. How do these limit even complicated substances? And even if we restrict an object, perhaps require that the atoms be arranged in a regular manner, say form a crystal, because a solid is complicated we might expect almost complete freedom to choose such arrangements. Yet the types of crystals are limited, and small in number. Likewise molecules with symmetry are also relatively few (perhaps better, the symmetry types are few, but it is possible to construct many different, quite complicated, molecules for each type). Why? And in what sense are these types the same, and different?

We organize our experience, and objects we deal with, into a small number of categories, else we would be overwhelmed by them. But we cannot require nature to be limited by our simplifications, and often it is not. There are only a few crystal types, but materials do not have to fit all requirements that we impose to define a crystal (as quasicrystals emphasize) — although (interestingly) a very large number do (almost). Thus we want to understand what requirements we have placed, why these, and how they restrict the subjects, mathematical and physical, we study — and why nature so often accepts our rules. Then we will be more able to understand, use, and create such objects, to understand how nature might evade constraints, so be equipped to search for, discover, and use objects and phenomena that do not fit into our, perhaps oversimplified, framework. And we will be better able to use the tools that we must develop for such studies.

The properties of these materials are, in part, consequences of our geometry. Here then we start with geometry, in particular the rotation group [Mirman (1995a), chap. X, p. 269] and its finite subgroups — the objects we consider consist of discrete sets of points — and the relationship of these subgroups to (the relevant subgroups of) the translation group [Mirman (1995a), sec. II.3.f, p. 44].

I.2 WHY STRUCTURES OF CRYSTALS AND MOLECULES ARE LIMITED

To be useful the descriptions of objects must be limited, or they would provide no information. One reason that we (here) study crystals, rather than arbitrary solids, is that by definition they have invariance groups and these are small in number, hence restricted, with definitive, often idiosyncratic, properties, so also making them interesting for physics, technology, economics, and group theory. What in their definition so limits what they are, and how their characteristics are expressed?

The symmetries of crystals and molecules come, in part, from invariance under rotations and reflections, plus also translations. But the points at which objects (atoms, ions, say, or sets of these) are placed form a discrete set, rather than a continuous one as for transformations of space, so the symmetries are of finite subgroups of the rotation and translation groups (thus of the inhomogeneous rotation group [Mirman (1995a), pb. IV.8.c.i-9, p. 140]). These subgroups are quite limited — so the types of crystals and (symmetrical) molecules are. Analysis of such materials requires therefore the determination of the (appropriate) rotation subgroups.

Beyond this, simply saying that the atoms of a solid form a crystal, or that a molecule has symmetry, constrains their configurations. Why, and how can we find these — the reasons and the structures?

I.2.a What are the essential properties of these objects?

The objects studied here are quite unusual (in reality, nonexistent): (almost) perfect crystals and symmetrical molecules are rare among solids, liquids and molecules (though interestingly, they are often excellent approximations to real objects). They are studied because they are, in many ways, easier than less "perfect" substances — and because (being often so close to reality, and so common) they are important. What distinguishes them from other matter — fluids, amorphous solids, and so on?

First is atomicity. Materials really are made of atoms, and this underlies our whole discussion. The analysis then does not apply to continuous fluids (but there are liquid crystals — which are not really continuous), nor to the motion of a particle whose path has to be studied in other ways, being a continuous set of points. The objects we consider consist of discrete points (separated by finite distances). Obvious, but central. *And they are regular, they have symmetry.*

Thus the basic tool is the theory of finite groups [Mirman (1995a)].

And because crystals and molecules can be visualized and are intuitive they provide tools for the development of the theory of finite groups.

Moreover both the types of objects at the points and the set of distances between neighboring ones are few in number. Crystals and molecules consist of one, or a few, kinds of (sets of) atoms (ions, molecules), at least to a reasonable approximation. Of course, if we consider ones of different energy states as different, and at times we should, the system becomes more complicated. So what we do here is merely a start, but it turns out — amusingly — to be a quite good one — for many, many materials.

Atomicity holds for any solid or molecule. But here there is a further condition on the materials — they have symmetry, there are regularities in their structures. What is perhaps surprising is that this sentence alone, though it may appear close to vacuous, almost determines (the organization of) the systems. Of the perhaps infinite number of solids and molecules, this requirement reduces the types being investigated to a (relatively) few (types — but surprisingly (?) many actual objects).

I.2.b There are only a few categories of crystals

The uniformities of our objects result from invariances under interchange of points, and rotations and reflections, and for crystals (distinguishing them from molecules) in addition one other invariance. Translational invariance is imposed (by definition, not by nature, though its concurrence is essential in making definitions reasonable and useful). Moving from any point in a crystal in any direction a finite (direction-dependent) distance, or any integer multiple of it, leaves all physical (and geometrical) quantities unchanged. Crystals are infinite in all directions — generally a good approximation (except near a boundary).

It is largely rotations, and these combined with translations, that provide limitations on the structures; there are only a small number of kinds because there are only a few sets of finite subgroups of the rotation group. So we start with the rotation group. Rotations (of what we have to see) — through a discrete set of angles — keep lattices, so crystals, and many molecules unchanged. There are two reasons. The object has its points arranged so that interchanges (by rotations) leaves the point set invariant. And space is rotationally invariant — and this is needed. But what is a lattice, and what does it have to do with crystals? And what is a crystal?

Problem I.2.b-1: Consider a rod with two identical positive charges at its end. Interchanging them — a physical process — does not change the system. Put the rod in an electric field, with the entire system being the field and rod (there is no external observer to perform operations) and see if a method can be found for performing this rotation while leaving the system unchanged. If space is not rotationally invariant, as it might seem with a field present, is the system invariant?

I.3 LATTICES

A crystal is a set of objects, (collections of) atoms, say, placed to give a lattice. A lattice is an infinite array of (a discrete set of) points that is translationally invariant — a translation in any direction by an integer multiple of a minimum (direction-dependent) distance goes to a point whose surroundings are indistinguishable from those of the initial one [Borchardt-Ott (1993), p. 8; Falicov (1966), p. 139; Sands (1993), p. 4; Senechal (1990), p. 39]. Physically, if we do an experiment at the two points, the results are identical. Mathematically, the surroundings, the set of all other points of the lattice, that is their positions relative to that on which we are standing, are the same no matter at which point we stand; any lattice point can be taken as the origin for there will be no difference in any mathematical or physical properties or relations. Crystals are determined — but only in part — by their lattices. Thus we start by considering not a physical object, a crystal, but a geometrical one, a lattice.

A lattice is given by lattice vectors, in n-dimensional space, any n non-coplanar (linearly-independent) vectors from one point to n identical points — ones (geometrically, for a crystal also physically) indistinguishable from the first. Lattice vectors give the translations from any point in the lattice to any equivalent one. There are clearly such translations of smallest length (lattices consist of discrete points), the vectors giving these are the primitive lattice vectors. Any other lattice vector is a sum of these primitive ones — with integer coefficients, as in

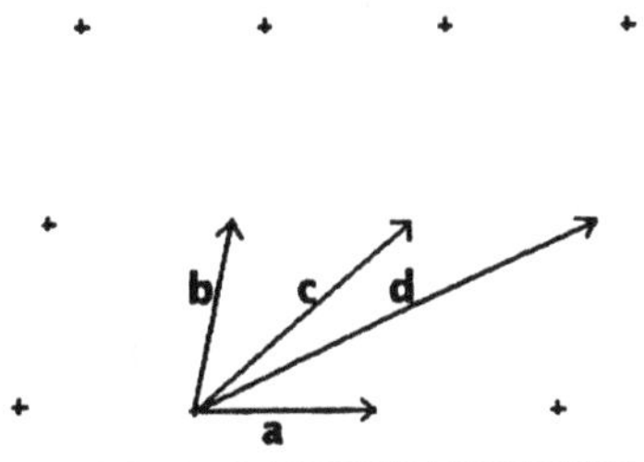

Figure I.3-1: LATTICE VECTORS.

Points here do not look like points. The lattice points are at their centers; we can interpret this as a diagram of a crystal, for which there can be extended objects going with the lattice points.

What are the conditions on these (geometrical) lattice-defining vectors? Are there others that we wish to, or can, impose besides translational invariance? How many different lattices are possible (in two or three dimensions, all we consider), what are they, and why are lattices both possible and limited? And in what ways are lattices different? Why?

Problem I.3-1: Why do the coefficients consist of all integers?

Problem I.3-2: If the smallest displacements under which a lattice is unchanged are $\underline{a}$, $\underline{b}$, $\underline{c}$ — three non-mutually-coplanar vectors — then an arbitrary translation that leaves the lattice invariant is

$$\underline{t} = n_a\underline{a} + n_b\underline{b} + n_c\underline{c}, \tag{I.3-1}$$

where the n's are (all) integers. One might think that for an arbitrary direction it would be impossible to go from one point on the lattice to another by adding these three vectors using only integer coefficients. But show that it is always possible (remembering that a lattice is defined as a set of points invariant under integer multiples of three given translations). For example, show that $\underline{d}$ can be written as a sum, with integral coefficients, of the three lattice vectors $\underline{a}$, $\underline{b}$, $\underline{c}$.

I.3.a Lattices and crystals

A crystal is a lattice, but it is more. At each point of the lattice we place a set of objects, atoms, molecules, ions, or several of these; this set of objects at a lattice point is a basis. Thus a crystal is a lattice with each lattice point having attached a structure (perhaps extended over space — the objects are close to, rather than necessarily at, the point) [Altmann (1994), p. 49; Borchardt-Ott (1993), p. 20; Burns and Glazer (1990), p. 19; Joshua (1991), p. 42]; each lattice point has placed on it a physical object consisting of sets of particles, and the sets for all lattice points are identical, and they are all identically placed. The structure may have symmetry other than that of the lattice, thus a crystal can have lesser, or more complicated, symmetry than its lattice.

Vectors $\underline{a}$, $\underline{b}$, $\underline{c}$ then define the lattice, but not the crystal — that consists of sets of objects with the positions of the sets determined by the lattice. Not all atoms need be the same, nor need they be points — they usually have, say, angular momentum (and magnetic moments), giving a direction. Also each crystal "point" may consist of several (atoms or molecules); each set can have lower symmetry than the lattice (so a rotation, for example, of the lattice may leave it invariant, but not necessarily the objects at each of its points). The symmetry group of a crystal consists of those transformations that leave invariant the lattice plus the structure at each point, the set of rotations and reflections (and translations) taking it into another indistinguishable from it. The nature of the points is irrelevant here. Our interest is only in the group of the object, this called a crystal, though it can be any such object, physical or not.

A crystal may be invariant under a different group than that of the lattice — and how these can be related we have to see.

I.3.b Unit cells

A crystal having translational symmetry is a repetition of some smallest unit. This smallest unit may hold several lattice points, each with several different types of entities, as with NaCl, for example. However for an object to be a crystal it must be a (translationally-invariant) repetition of a smallest unit. It is this smallest part of the crystal giving all its features — the unit cell — that we study [Sands (1993), p. 7; Senechal (1990), p. 42]. Our discussion is for the symmetry given by its unit cell, which carries over to the symmetry of the crystal as a whole, with the meanings of "given by" and "carries over" perhaps more subtle than might appear [Borchardt-Ott (1993), p. 26; Wood (1977), p. 69].

The unit cell often contains only a single lattice point. If we take unit cells such that each corner lies on a point there are eight in a cell, but each is shared by eight cells, again averaging one per cell. There are various ways of picking unit cells — various sets of points that we can choose to specify these; each of the several common types has its own advantages and disadvantages. We study below unit cells, and how they relate to, and show, the symmetry of the crystal (sec. IV.3, p. 196).

Problem I.3.b-1: A unit cell holds a set of neighboring atoms and is specified by its three sides (and three interior angles). Of course it has more than three of each. Is there a maximum, minimum, number? Why are only three non-coplanar lines needed?

Problem I.3.b-2: All unit cells have the same volume [Burns and Glazer (1990), p. 32; Megaw (1973), p. 114; Sands (1993), p. 9; Yale (1988), p. 194] with sides a, b, c and angles α, β, γ, it is obviously (?)

$$V = abc(1 - \cos^2\alpha - \cos^2\beta - \cos^2\gamma + 2\cos\alpha\cos\beta\cos\gamma)^{\frac{1}{2}}. \quad \text{(I.3.b-1)}$$

Check this, and check that it reduces to the correct values for various simple angles.

I.4 THE ROTATION GROUP AND ITS FINITE SUBGROUPS

Lattices, crystals, and symmetrical molecules are invariant under rotations, but only through discrete sets of angles, thus under — finite — subgroups of the rotation group. What are these subgroups? It may seem that there are many, infinitely many — just divide 2π by any integer, no matter how large, or even any rational or real number — but these are groups, so we cannot do arbitrary things with them [Armstrong (1988), p. 37; Burn (1991), p. 57; Coxeter (1973), p. 5, 15; Ghyka (1977), p. 40; Martin (1987), p. 198; Murnaghan (1963), p. 328; Schwarzenberger (1980), p. 43; Simon (1996), p. 11].

I.4.a The finite subgroups of the two-dimensional rotation group

The two-dimensional rotation group, SO(2), is easiest to consider [Armstrong (1988), p. 104].

Problem I.4.a-1: With $R(n)$ a rotation of $\frac{2\pi}{n}$, prove that the only finite subgroups of SO(2) are the sets

$$R(n), R(n)^2, \ldots, R(n)^k, \ldots, \quad k = 0, 1, 2, \ldots, n-1, \qquad \text{(I.4.a-1)}$$

where n is an integer, and there is one set for each integer n. These are the cyclic groups [Mirman (1995a), sec. II.4.b, p. 49], and all. For O(2) [Mirman (1995a), sec. XII.4.b, p. 355], there is added to each one other element, the reflection; the subgroups of O(2) consist thus of all cyclic and dihedral groups, the groups of rotations of objects with two surfaces — dihedral — so having the symmetry of rotations about a central axis, plus an interchange of the two sides by a rotation of π about an axis parallel to, and halfway between, the sides; some dihedral groups contain an inversion interchanging the two sides [Mirman (1995a), sec. II.4.c, p. 50]. Construct such an object, of paper for example. An ordinary blank sheet is an example (with what symmetry group?). With a scissors other dihedral groups can be realized. Show the effect of these transformations on coordinate axes. Also find the effect of a reflection in (a plane perpendicular to) an axis, giving the realization [Mirman (1995a), sec. V.3.c, p. 157] more relevant to O(2).

I.4.b The finite subgroups for three-dimensions

The two-dimensional rotation group is quite simple and we would not expect many subgroups from it. The three-dimensional group has much

more structure, but surprisingly not many more subgroups [Armstrong (1988), p. 105; Neumann, Stoy and Thompson (1994), p. 174]. This is important enough to state as a
THEOREM: A finite subgroup of SO(3) — reflections are not included — is (isomorphic to) either a cyclic group, a dihedral group or the rotational symmetry group of one of the regular polyhedra of which there are only five. Thus besides the cyclic and dihedral groups the only finite subgroups of SO(3) are the symmetric group S_4, its subgroup A_4, and alternating group A_5 [Mirman (1995a), sec. II.4.d.ii, p. 54].

That this places strong limitations on objects with symmetry can already be seen. Since crystals consist of cells with symmetry we have greatly limited the types possible — the types of crystals that we are permitting nature to have (?). Of course there may be further limitations from other conditions, and we also have to show that there are physical crystals corresponding to each allowed mathematical one — that the requirements are not inconsistent (with physics).

Why should the finite subgroups of the rotation group be so restricted? Since SO(2) is a subgroup of SO(3) its subgroups are subgroups of the latter, hence all the cyclic groups are. A cyclic group plus a rotation around an axis perpendicular to that of the group (an element of SO(3)) gives a dihedral group, so these are SO(3) subgroups. Thus we have to consider those finite subgroups that contain rotations about different axes and show that these consist only of A_5, S_4 and their subgroup A_4; not many. The objects invariant under these groups are called the Platonic solids (or polyhedra) [Holden (1991), p. 1]. Why do these groups give the Platonic polyhedra — and only them — and why, in two cases, does a single group give two polyhedra (and no more)?

Problem I.4.b-1: Cyclic (rotation) groups are Abelian, so each element is in a class by itself. Show that for dihedral groups, D_n (sec. I.7.a.i, p. 52; sec. V.2.c, p. 249), the number of classes is $\frac{1}{2}(n+6)$, for n even, and $\frac{1}{2}(n+3)$, for n odd, and that each element in the same class as its inverse [Dixon (1973), p. 5, 75; Hamermesh (1962), p. 43; Ledermann (1987), p. 65, 209; Lomont (1961), p. 32, 78; Wilson, Decius and Cross (1980), p. 316].

Problem I.4.b-2: In the list of rotation-group finite subgroups, subgroups of S_4 and A_5 are not listed, except for A_4. Are all others cyclic or dihedral, or direct products [Mirman (1995a), pb. IV.8.b-3, p. 138]; if not why are they not rotation-group subgroups [Lomont (1961), p. 34, 134; Streitwolf (1971), p. 62]?

I.4.c Regular polyhedra

There are thus three-dimensional figures — polyhedra — that have a special relationship to the rotation group, the regular (Platonic) polyhedra. A regular polyhedron [Armstrong (1988), p. 37; Lines (1965), p. 134; Yale (1988), p. 89] is one invariant under rotational interchange of any two vertices — the polyhedron does not change if any vertex is rotated into any other — so all its angles, and the lengths of all sides, are equal. A polyhedron — many sides — is a three-dimensional closed object. In two dimensions the analogous figure is a polygon — many angles. The generalization to arbitrary dimensions can be called a polytope [Coxeter (1973), p. vi, 126, 289; Fejes Toth (1964), p. 124, 132] — implying many places, like many vertices, from the Greek word that gives the name topology, and also utopia.

How many regular polyhedra are there? Only five [Coxeter (1973), p. 5; Holden (1991), p. 1; Ghyka (1977), p. 40; Rosen (1977), p. 52]. These are the tetrahedron [Armstrong (1988), p. 1] (with rotational symmetry group A_4), the cube and the octahedron, for both the rotational symmetry group is S_4, the dodecahedron and the icosahedron, both with group A_5. The tetrahedron (four surfaces) has four triangular faces, the cube six square ones, the octahedron (eight surfaces) has eight triangular faces, the dodecahedron has twelve pentagonal faces, the icosahedron twenty triangular ones. The dodecahedron and the icosahedron are dual figures — having the same group — so the dodecahedron (dozen surfaces) has twelve faces and twenty vertices, the icosahedron, twenty faces and twelve vertices. A diagram of the three most unfamiliar of these has been given previously [Mirman (1995a), fig. III.2.g-1, p. 78].

Problem I.4.c-1: The cube and octahedron are also dual. Check that the numbers of vertices and faces are correct.

Problem I.4.c-2: Coordinates of the vertices of the regular polyhedra are easily found [Coxeter (1973), p. 52].

I.4.c.i *Determination of the regular polyhedra*

There are various ways of proving that these are the only regular polyhedra. For them, angles between edges are almost fully determined, allowing very few such objects.

Problem I.4.c.i-1: Consider vertex V in the next diagram (which we interpret as the corner of a solid) with q lines, which we can take as rods (edges of the solid) [Coxeter (1973), p. 5], shown as the q (here three) heavy solid ones. Each pair of (heavy solid) lines defines a plane, giving for each pair an angle in their plane. What is the sum, $\angle(A + B + C)$, of the angles between the lines? To give an experimental proof take a set of rods and place their ends on a sheet of paper (containing the thin lines, the projections of the heavy ones), with V above the paper, and draw from point O directly under V (as shown by the dotted line) the thin lines to each point at which a rod touches the paper, as in

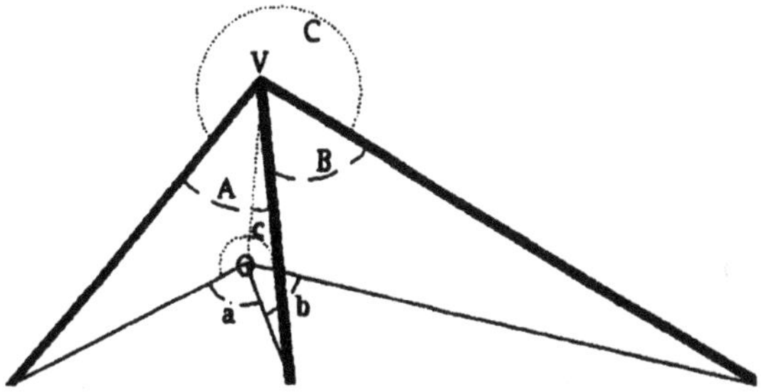

Figure I.4.c.i-1: PROJECTED ANGLES.

It is clear that angle A between two lines from V is less than its projected angle a from point O in the plane under V to the rods (and similarly for B and C). The sum of the angles between the lines from O is of course 2π. If the rods were collapsible and we moved vertex V toward the paper the angles between them would increase (and decrease moving away, becoming 0 for the vertex at infinity), while the angles in the plane would be unchanged, as we see, noting that $\angle A' < \angle A$,

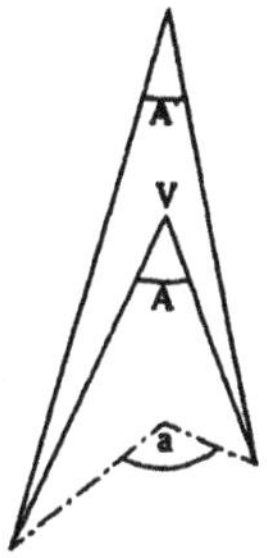

Figure I.4.c.i-2: VERTEX ANGLES.

Thus $\angle (A + B + C)$ is less than the sum of the projected angles on the paper,

$$\angle (A + B + C) < \angle (a + b + c) = 2\pi. \qquad \text{(I.4.c.i-1)}$$

Prove analytically, and with a drawing and also with a computer program that in general, for a regular polyhedron, the sum of the q (equal) angles θ between the edges at each vertex is less than 2π, so for each

$$\theta < \frac{2\pi}{q}; \qquad \text{(I.4.c.i-2)}$$

this is one reason the number of regular polyhedra is limited. Each face of a polyhedron has p edges, the sum of exterior angles is 2π, each then being $\frac{2\pi}{p}$, and the interior angle (the supplement) is

$$\theta = (1 - \frac{2}{p})\pi. \quad \text{(I.4.c.i-3)}$$

Draw these and check. Hence,

$$(1 - \frac{2}{p})\pi < \frac{2\pi}{q}, \quad \text{(I.4.c.i-4)}$$

giving an inequality relating the number of lines bounding each face and the number of lines leaving each vertex. This should give

$$\frac{1}{p} + \frac{1}{q} > \frac{1}{2}. \quad \text{(I.4.c.i-5)}$$

Note that edges and angles enter symmetrically. The number of faces and angles between edges are strongly bounded. Explain why

$$3 \leq p, q \leq 5, \quad \text{(I.4.c.i-6)}$$

and both p and q are integers. It would not be surprising if there were no regular polyhedra, thus no (other) finite subgroups of the rotation group, but (fortunately) there are.

Problem I.4.c.i-2: The reason then that the number of regular polyhedra is limited is that as the number of edges of a face increases the exterior angle decreases, and as the number of lines leaving a vertex increases so does the number of angles they make, and the sum of these is bounded; but there is one question. Explain what the number of sides of a face has to do with the number of lines at a vertex. What does this have to do with a polyhedron being regular? How would the argument break down otherwise? (What does otherwise mean here?) It is clear from this that the polyhedra are paired (dual) — the number of edges of a face of one equals the number of lines emerging from a vertex of the other. Explain why this shows that the number of faces of one equals the number of vertices of the other. Why do they have the same symmetry group?

Problem I.4.c.i-3: Check that, with V the number of vertices, F the number of faces, E the number of edges, p the number of edges per face, q the number of faces per vertex, and N the order of the symmetry group, the allowed values [Cotton (1990), p. 46; Elliott and Dawber (1987), p. 188; Ford (1972), p. 128; Gasson (1989), p. 231; Hargittai and Hargittai (1987), p. 66; Holden (1991), p. 8; Martin (1987), p. 200; Neumann, Stoy and Thompson (1994), p. 175] are

polyhedron	p	q	V	F	E	group	N
tetrahedron	3	3	4	4	6	A_4	12
hexahedron (cube)	4	3	8	6	12	S_4	24
octahedron	3	4	6	8	12	S_4	24
dodecahedron	5	3	20	12	30	A_5	60
icosahedron	3	5	12	20	30	A_5	60

Table I.4.c.i-1: REGULAR POLYHEDRA.

This gives the number of edges for each face, so their shapes. The tetrahedron is self-conjugate, the others are conjugate in pairs. Explain the other name for the cube. Check that "icosa" (from the Greek) means twenty.

I.4.c.ii *Other ways of determining these polyhedra*

It is not unusual in mathematics for there to be different proofs, often seeming to have nothing to do with each other, giving (fortunately) the same truth. Here we consider some other approaches, which besides providing useful comparisons, also provide other useful information.

Problem I.4.c.ii-1: Euler's topological equation

$$V - E + F = 2. \qquad \text{(I.4.c.ii-1)}$$

gives another proof [Coxeter (1973), p. 9, 165; Elliott and Dawber (1987), p. 186; Ford (1972), p. 127; Gasson (1989), p. 229; Hamermesh (1962), p. 48; Hilton and Pederson (1996); Lines (1965), p. 135; Martin (1987), p. 198; Meserve (1983), p. 301; Senechal (1989), p. 34; Sternberg (1994), p. 45]. Prove it. For a regular polyhedron

$$pF = 2E = qV. \qquad \text{(I.4.c.ii-2)}$$

Why? Use this to verify the table.

Problem I.4.c.ii-2: A tetrahedron has four vertices, the smallest possible (in three dimensions), a cube eight. Try to construct (say with sticks) a regular polyhedron with six vertices. Explain what goes wrong. Repeat with seven and nine.

Problem I.4.c.ii-3: Show that these polyhedra are invariant under the groups stated, and no larger ones. One way of doing this is to construct three-dimensional models and find all symmetries. Another is to use coordinates. Also it might be possible to write a computer program to do it (visually).

Problem I.4.c.ii-4: For the tetrahedron, label each vertex and verify that the permutations form S_4. However only half of these can be achieved by rigid-body rotations — without taking the object apart. Which are they? Why? What do the others do? Repeat this for the other regular polyhedra and verify that the stated symmetry groups are correct.

Problem I.4.c.ii-5: Coordinates of the vertices can be found from three-dimensional representations of their symmetry groups by having representation matrices act on a (column) vector (from which point?) giving the (arbitrary) coordinates of a single vertex. Compare with known coordinates [Coxeter (1973), p. 52].

Problem I.4.c.ii-6: Explain why the symmetry group of the cube [Mirman (1995a), sec. II.2.h, p. 40] has odd permutations, like transpositions, but that of the tetrahedron does not. Also construct a tetrahedron and perform transpositions on it. Repeat for a cube. How about the other regular polyhedra?

Problem I.4.c.ii-7: That these polyhedra actually exist is clear but it is useful to construct models of them to check. It should be obvious (why?) that an object invariant under a subgroup of the rotation group is a regular solid, and conversely. What objects are invariant under the (various) cyclic and dihedral subgroups, and why is there a difference (in that they are — presumably — not regular) between objects invariant under these groups? Verify that the relevant subgroups of the symmetric groups do give the correct objects. This requires an extra step beyond showing that the groups are correct.

Problem I.4.c.ii-8: For each regular polyhedron find the numbers of two-fold, three-fold, four-fold and five-fold symmetry axes and the numbers and positions of all mirrors (planes of reflection symmetry) [Holden (1991), p. 13].

Problem I.4.c.ii-9: A regular polyhedron is one for which there is a set of finite rotations about axes through every corner (all corners are equivalent) that leave it unchanged. So its sides are invariant under rotations through a set of discrete angles — the same for all sides. Thus all edges are the same length and all angles are equal. For each such polyhedron there is a finite subgroup of SO(3). Since the latter are limited the regular polyhedra are. Thus it is impossible to take, say, a figure of sixty equal sides and rotate it so that each side generates another similar figure. Rotating a regular polygon (out of its plane) so that each side generates a similar polygon is hard and there are only three cases for which it can be done (which three?). A regular figure cannot be obtained using a six-sided polygon. Rotate a regular hexagon. What goes wrong? Repeat for other polygons; notice, and explain, the differences between ones that give regular polyhedra, and those that do not.

Problem I.4.c.ii-10: Would there be a difference if instead of SO(3), we considered O(3) [Mirman (1995a), sec. XII.4.b, p. 355]? There is a slight ambiguity here. Why? What? Does it matter?

I.4.c.iii *Dual polyhedra*

Regular polyhedra, as we see, come in pairs dual to each other (the tetrahedron is self-dual); thus there are only three of these finite subgroups, though there are five regular polyhedra. This can be shown pictorially by drawing lines connecting the centers of the faces of a solid; these lines form the edges of faces of the dual solid [Gasson (1989), p. 247; Ghyka (1977), p. 43; Holden (1991), p. 4]. The dual of the dual is the polyhedron. A polyhedron and its dual clearly have the same symmetry axes and planes, thus the same symmetry group.

The dual of a cube is an octahedron; perpendiculars through the centers of its eight faces pass through the corners of the cube. Its six diagonals go through the centers of the six faces of the cube. While the axes of maximum symmetry of a cube go through its faces, for a octahedron they lie along the diagonals.

Problem I.4.c.iii-1: A tetrahedron, with four faces and four vertices, is self-dual. Its symmetry group is a subgroup of S_4, but it has fewer symmetries than a cube, so it is proper subgroup — it is neither the group itself, nor the identity. However it is not Abelian (why?), so it must be (?) A_4. In a way, a tetrahedron is half a cube, so it has only half the symmetry operations. Is the last sentence correct?

Problem I.4.c.iii-2: Draw (or construct) a cube and an octahedron and check that they are duals [Armstrong (1988), p. 39] — either can be placed in the other with (all) vertices of one touching the centers of (all) faces of the other — so they have the same symmetry group. Why?

Problem I.4.c.iii-3: The other dual polyhedra, the dodecahedron with twelve faces and twenty vertices, and the icosahedron, with twenty faces and twelve vertices, have sixty rotational symmetries, which is the order of A_5. Rotations through vertices of the dodecahedron produce all the three-cycles of S_5 (why?), and these generate A_5 [Mirman (1995a), pb. VIII.2.c-5, p. 218], so this is the symmetry group of the two objects. Check (diagrammatically) that these are dual. Why is the symmetry group A_5 rather than S_5; why are the other permutations not symmetries?

Problem I.4.c.iii-4: Obviously the dual of the dual is the original polyhedron. However an analytic proof would be interesting.

I.4.c.iv *Extensions and generalizations*

Three dimensions, and regular figures, are familiar. To what extent do these considerations depend on such limitations?

Problem I.4.c.iv-1: There are many polyhedra with different numbers of faces and vertices, but these are not regular, their angles are not all equal and their faces not all congruent. It might be interesting, say with balls and sticks, to try various possibilities, and see why they do not work. A computer graphics program that tries to draw them, in color, can prove enlightening.

Problem I.4.c.iv-2: Regular polyhedra can be inscribed in a sphere (obviously — why?). In fact, they can be inscribed within each other [Bishop (1993), p. 40; Ghyka (1977), p. 43; Holden (1991), p. 29]. Why? However these are not the only ones that can be inscribed in a sphere. There are others, the semi-regular (Archimedean) polyhedra (sec. IV.3.b, p. 199). For these there are sets of faces, all identical with all sides of a face equal, but unlike the Platonic polyhedra, they have more than one set of faces [Coxeter (1973), p. 30; Gasson (1989), p. 228; Ghyka (1977), p. 50; Hargittai and Hargittai (1987), p. 70; Holden (1991), p. 46; Lines (1965), p. 159; Martin (1987), p. 206]. They are semi-regular because the number of faces meeting at each vertex is the same, although these faces need not be the same (in what way?). It would be interesting to derive these, finding the number, and types of faces, and the number of vertices, with the angle(s) at each. Are there relationships between the regular and semi-regular polyhedra? What (if any) are the symmetry groups of the Archimedean polyhedra? How are they related to the rotation-group subgroups? Why? To those of the Platonic polyhedra? Could such relationships be generalized to get other comparable figures (and with what properties)? Why?

Problem I.4.c.iv-3: Euler's equation can be generalized to any dimension giving Schläfli's equation [Coxeter (1973), p. 118; Ghyka (1977), p. 69]. Use this to find the number of regular polyhedra, so the finite subgroups of the rotation group, for each dimension, in particular for four dimensions [Ghyka (1977), p. 68]. Consider various projections of these into three-space [Coxeter (1973), p. 236; Senechal (1995), p. 54]. Do they give anything interesting? Could materials be constructed with these as unit cells? Might they have (economically) useful properties?

Problem I.4.c.iv-4: These arguments are firmly rooted in Euclidean geometry, emphasizing the close connection between a geometry, the objects in it, and the properties of the transformation group under which it is (and they are) invariant. Thus could there be other types of geometries, giving different sets of (analogs to) regular polyhedra? What would their transformation groups be? What are the finite subgroups (these presumably still being relevant)? It might be expected that such geometries are limited — there are only a few (sets of semisimple) Lie groups [Mirman (1995a), sec. XIII.5, p. 383]. Or are there ways of avoiding this problem, if it is relevant?

Problem I.4.c.iv-5: What postulates [Blumenthal (1980), p. 50, 149; Fejes Toth (1964), p. 124; Meserve (1983), p. 185; Mirman (1995a), sec. P.5.b, p. xviii; Neumann, Stoy and Thompson (1994), p. 116; Nikulin and Shafarevich (1987), p. 45] of Euclidean geometry lead to there being five regular solids, only five, these five? Do the same postulates give the Archimedean polyhedra? Which postulates, if any, can be changed to give other results? What would these be? What would the resultant geometries be like?

I.4.d How the rotation group determines its finite subgroups

The preceding determination of the finite rotation groups uses objects; it is interesting because, among other reasons, objects are what physics studies. There are other proofs [Armstrong (1988), p. 104; Coxeter (1973), p. 53; Fejes Toth (1964), p. 58; Heine (1993), p. 132; Lyubarskii (1960), p. 18; Sternberg (1994), p. 27; Yale (1988), p. 89], some perhaps more clearly related to SO(3). The result is important enough to make these interesting, and helpful in the understanding, so application, of crystals and molecules — and groups. And it is always possible that some objects are different from what the proof expects — all the conditions may be needed to obtain a class of materials that we have suitably limited, but may not be for a material that nature likes. Understanding the result, including different proofs, could help in finding and mastering such objects. Thus, and because of its intrinsic interest for group theory, including the rotation group and its generalizations, we consider a different proof in depth [Weyl (1989), p. 77, 149].

I.4.d.i *Orbits and poles and why there are different types*

A rotation leaves fixed all points on a line, its axis (Euler's theorem [Goldstein (1953), p. 118, 132; Jansen and Boon (1967), p. 335.]). These points include two called poles, at which it pierces the surface of a (say, unit) sphere [Murnaghan (1963), p. 329]. Some rotations leave the poles fixed, others move them, causing a pole to trace a curve on the sphere's surface — the orbit of the pole (for a finite group the orbit consists of discrete points). Each pole is obtained from any other one by a transformation of the point group, R. And each transformation takes that given one into a different pole, except for p transformations forming the cyclic group of rotations around the pole, C_P. The transformations that take a pole into a different one form a group, a subgroup of finite group R, specifically a factor group of R with respect to the cyclic subgroup, R/C_P. There is a one-to-one correspondence between poles and transformations of this subgroup.

The groups of symmetry of polyhedra have rotations that form disjoint subsets, determined by the different types of axes, each giving a different orbit [Inui, Tanabe and Onodera (1990), p. 179; Janssen (1973), p. 74; Miller (1972), p. 27; Schwarzenberger (1980), p. 45; Senechal (1990), p. 28; Sternberg (1994), p. 16; Weyl (1989), p. 149; Yale (1988), p. 91]. So the set of poles splits into subsets — orbits — with the members of each set intermixed by the rotations of C_P, but not mixed with members of other sets. How many sets — orbits — are there? The limitation on this number is a reason for that on the number of finite subgroups of the rotation group.

Problem I.4.d.i-1: For a three-sided pyramid (not a regular tetrahedron, but with a regular base so one axis of symmetry), there are two orbits; each contains just one of the two points at the opposite ends of this axis. Draw and check this — inscribe the pyramid in a sphere; the poles are the points at which it is pierced by the axis. Find and describe the orbits. Note that no symmetry rotation of the object takes one pole into the other.

Problem I.4.d.i-2: A cube has three types of rotations [Ford (1972), p. 124; Inui, Tanabe, and Onodera (1967), p. 172; Mirman (1995a), sec. II.2.h, p. 40], with axes through opposite vertices, through midpoints of opposite faces, and through the midpoints of edges diagonally opposite each other (drawn more heavily in the rightmost figure below). We draw these (for guidance with extra lines and different widths — destroying the symmetry) with one pole, P, marked; there is another at the second point at which the axis pierces the (undrawn) sphere. The lines going into each other under a rotation around the labeled $\frac{\pi}{3}$ axis are emphasized. Also the other three equivalent axes are shown as dashed lines. The rotations in each set are equivalent — they are obtained from each other by a rotation. Check the correctness of these axes,

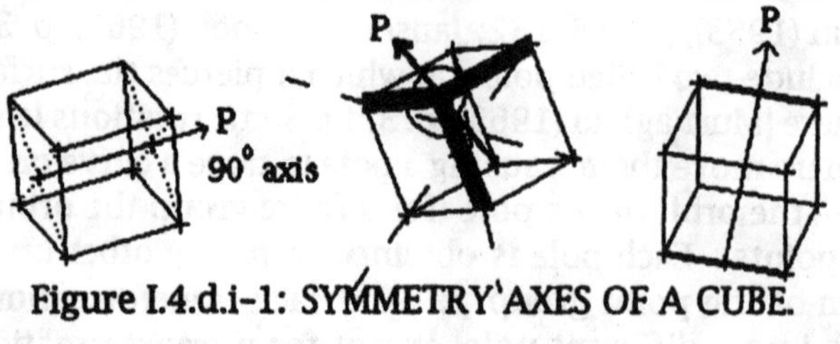

Figure I.4.d.i-1: SYMMETRY AXES OF A CUBE.

To emphasize the meaning of different types of axes, in the next diagram all three equivalent axes of the same type are labeled, and the cube, clearly invariant, is drawn before and after rotations of $\frac{2\pi}{3}$ about the heavy dashed line,

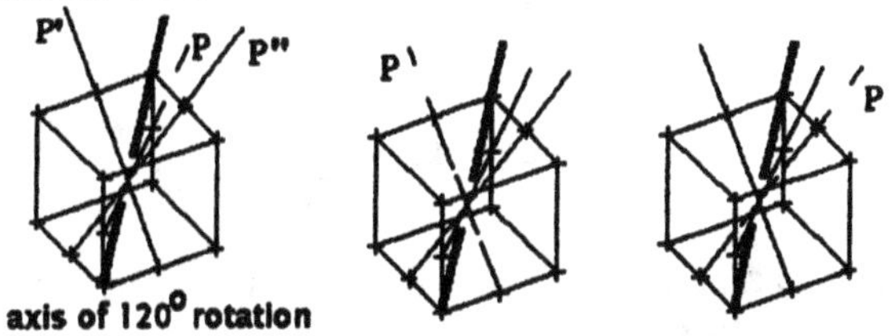

Figure I.4.d.i-2: ROTATION OF A CUBE BY $\frac{2\pi}{3}$.

The poles are given by P, P' and P"; these three points (at which the lines pierce a sphere) form the orbit. Finish the labeling for the two incomplete cases. Find the orbit of the light dashed line when the cube is rotated about each of the solid lines. Are the orbits the same? Rotations in different sets are not equivalent (a $\frac{\pi}{2}$ axis cannot be converted into a $\frac{2\pi}{3}$ one; these also pierce the sphere at different points — different poles). There are three types of orbits. Each contains only a finite number of points, and their numbers of points are different. Check that these are correct, and also the equivalence, and inequivalence, of the sets of rotations. Why do inequivalent rotations give different types of orbits? Take the poles given by an axis through the midpoints of the faces and apply to this axis, and so the poles, all rotations about the other two sets of axes. Find the orbits so generated. Compare with those for the other two types of axes. How do they differ? How many points are in each? Find each set of poles by drawing (as exemplified in the figure). It can be seen that there are three sets, the members taken into those of the same set by rotations, but the sets are not mixed. This should be clear, but checked using this as an example.

I.4.d.ii *The number of orbits and the finite subgroups*

In these examples there are two or three sets of orbits, and this is true in general — a finite subgroup of the (three-dimensional) rotation group can have only two or three types of orbits: for a nontrivial, finite group of rotations (with n elements) and a sphere centered at the point the group leaves fixed, the number k of orbits of poles is either 2 or 3 [Senechal (1990), p. 33]. Why should this be?

To find the number of inequivalent orbits we let α_i be the number of rotations leaving pole i fixed, and β_i the number of poles in orbit i (labeled by any pole i on it), the number of poles each is taken into by the group transformations. Then, with n the order of the group,

$$n = \alpha_i \beta_i, \qquad \text{(I.4.d.ii-1)}$$

for each orbit; a rotation either takes a pole to another of the same orbit, there are β_i of these, or leaves it fixed, there being α_i that do. This, if not obvious, can be shown rigorously [Weyl (1989), p. 150]. Summing over the k orbits gives

$$\sum_{i=1}^{k} \beta_i(\alpha_i - 1) = \sum_{i=1}^{k} (n - \beta_i). \qquad \text{(I.4.d.ii-2)}$$

Each element is a rotation so there are on the sphere two poles that it does not move. Thus (not including the identity)

$$2(n-1) = \sum_{i=1}^{k} (n - \beta_i); \qquad \text{(I.4.d.ii-3)}$$

dividing the equation by n and using eq. I.4.d.ii-1, gives

$$2 - \frac{2}{n} = \sum_{i=1}^{k} (1 - \frac{1}{\alpha_i}). \qquad \text{(I.4.d.ii-4)}$$

As $n \geq 2$,

$$1 \leq 2 - \frac{2}{n} < 2. \qquad \text{(I.4.d.ii-5)}$$

Also $\alpha_i \geq 2$ so

$$\frac{1}{2} \leq 1 - \frac{1}{\alpha_i} < 1, \qquad \text{(I.4.d.ii-6)}$$

for each i. Summing over the orbits we get

$$\frac{k}{2} \leq 2 - \frac{2}{n} \leq k. \qquad \text{(I.4.d.ii-7)}$$

Problem I.4.d.ii-1: Verify that these two inequalities are consistent only for $k = 2$ or 3. Carry out the algebra explicitly for both the pyramid (pb. I.4.d.i-1, p. 18) and the cube.

Problem I.4.d.ii-2: Since the poles in each pair are not mixed, if $k = 2$ there are two poles, thus a single axis for all rotations, and the finite subgroup is isomorphic to a finite subgroup of O(2). The number of poles fixed by the identity equals the number of poles — the number obtained from one by using all group transformations — and here that equals 2. Why? This is the case discussed above (sec. I.4.a, p. 8). More interesting is $k = 3$. There are four cases. In the first the finite subgroup is a cyclic group of order n with rotations through angle $\frac{2\pi}{n}$, $n = 1, 2, 3, \ldots$. This is C_n. Show that this also gives dihedral groups D_n, $n = 2, 3, \ldots$. Why is $n = 1$ not included? For the second case each orbit consists of four points, and rotations preserve distance so any three are at the same distance from the fourth and are vertices of an equilateral triangle. Taking all sets of three-at-a-time shows that the four points are the vertices of a (regular) tetrahedron (a four-surfaced figure, with each surface an equilateral triangle). Check that this is the group of rotations of the tetrahedron, and that it has the right order. Verify that for case 3 there are six points on an orbit and the same argument gives a (regular) octahedron (an eight-surfaced figure). Show that this is also the symmetry group of a cube. How are the cube and octahedron related geometrically? The last case has twelve points on an orbit. The polyhedron is a (regular) icosahedron (with twenty surfaces). This is also the group of the dodecahedron. Why? What is the geometrical relationship between the dodecahedron and the icosahedron? Also show that these are the only possibilities, so we get [Neumann, Stoy and Thompson (1994), p. 179; Yale (1988), p. 93],

case	α_1	α_2	α_3	β_1	β_2	β_3	n	group symbol
							n	C_n
1	2	2	q	q	q	2	$2q$	D_n
2	2	3	3	6	4	4	12	T
3	2	3	4	12	8	6	24	O
4	2	3	5	30	20	12	60	I

Table I.4.d.ii-1: FINITE SUBGROUPS OF SO(3).

Problem I.4.d.ii-3: Of course no pair of these groups should be isomorphic.

Problem I.4.d.ii-4: There is a slightly different version of the algebra [Armstrong (1988), p. 106] that uses the counting theorem [Armstrong (1988), p. 98]: the number of distinct orbits equals the average number of points left fixed by a group element. Prove this. Since the elements are rotations, each leaves two points fixed, except the identity leaves more (all). The average number of points must be at least 2, and it cannot be much more than 2. We might expect the counting theorem even without a rigorous proof. Suppose that each element left fixed the same number of points. How many poles are there? The answer is the number of poles generated from some set (the orbit) by the transformations of the group times the number of orbits; this is the number of orbits times the number on each, which equals the number generated by an element. However this is not the same for each element, so the number of orbits equals the average number so generated. Each rotation leaves two poles fixed, except that the identity leaves all. From this show that $k = 2$ or 3.

Problem I.4.d.ii-5: The important point is that the number of poles is finite, since the number of rotations is, so the number of orbits is.

This strictly limits the number of orbits. It is not surprising that it is limited to 3 since there are only three different sets of rotations, the rotations around three different axes, for the three-dimensional rotation group. We might expect the number of distinct orbits to be determined by and likely equal to this (as it is) so restricting the number of finite subgroups. But does the number of distinct orbits equal the number of axes in other dimensions?

I.4.d.iii *Implications and extensions*

These results are of interest in themselves, and as a foundation for deeper thought. Limitations on the number of poles can be viewed in another way. The set of rotations leaving a pole invariant forms a subgroup of finite subgroup G of SO(3). Thus the points on the orbit correspond to equivalent subgroups, and the rotations taking one orbit point to another, a subgroup, give the cosets of G [Mirman (1995a), sec. IV.6, p. 128]. Different orbits result from cosets based on different subgroups. The number of orbits equals the number of inequivalent subgroups of G, and this is either 2 or 3.

Problem I.4.d.iii–1: For the pyramid (pb. I.4.d.i-1, p. 18), and likewise the cube (pb. I.4.d.i-2, p. 18), what are the rotation groups that leave them invariant? What are the inequivalent subgroups? Find the cosets for all, and check that they give the poles in each orbit.

Problem I.4.d.iii–2: There are several derivations of the finite subgroups of the rotation group. How are they related? Can one be obtained from others? Do they all use the same assumptions? Which is most easily generalized to higher dimensions [Brown, *et al*, (1978)]? How do these proofs and results come from the postulates of Euclidean geometry (which proofs, which results, which postulates)? Can they be generalized to other geometries? If a material is found that does not fit into the present scheme (it is not quite a crystal) which proof best shows the conditions that such material could (or must) violate?

Problem I.4.d.iii–3: It may not be intuitively obvious that there are only a small number of finite subgroups of SO(3). After all, we can partition the surface of a sphere into a set of identical regions and perform a — finite — rotation from the center of one to the center of another. And there are only a finite number (but one as large as we wish) of such regions. Start with a ring (any great circle of such a partitioned sphere), divided into equal strips. Rotate it through some angle about an axis through the center of the sphere. Notice that the distance between the center of a strip and the point to which it goes under the rotation differs for a strip near a pole (the point at which the axis pierces the sphere) and for one near the equator, thus so does the partition of the surface these strips generate. Is it possible to find a finite rotation taking one partition of a spherical surface, partitioned into identical regions, into another? How about a set of rotations about different axes?

Problem I.4.d.iii-4: Crystals are much limited. There are not many types. Of course, we cannot prevent nature from producing solids that still have symmetry, including especially either translational or rotational symmetry. The result just would not be as esthetically enjoyable — although this does not say anything about the scientific, technological or economic value. Are there such solids? Are there limitations on them? Do they have value? How do the violations of these requirements affect their value?

Problem I.4.d.iii-5: It is interesting to extend this [Coxeter (1973); Fejes Toth (1964), p. 124; Schwarzenberger (1980), p. 49; Senechal (1990), p. 103], and find the finite subgroups of the n-dimensional rotation group (and even the finite subgroups of the pseudo-rotation groups). For $n = 4$, the group, and its Lie algebra, are not simple, only semisimple [Mirman (1995a), pb. X.7.e-4, p. 300; (1995b), sec. 7.1.2, p. 124]. Does this make a difference? Are there more, or fewer, subgroups in higher dimensions? Why? What happens to these subgroups upon restriction to three dimensions? Do these provide any (useful) information about three-dimensional space, or the symmetrical objects that can exist within it? What are the (generalizations of the) regular polyhedra in higher dimensions? Find their (various) three-dimensional projections. Are these regular (in any way)? Might solids be constructed with these as unit cells? Would they be (technologically) useful? Why?

I.5 HOW TRANSLATIONS LIMIT ROTATION GROUPS OF CRYSTALS

The finite subgroups of SO(3), so crystals and molecules with symmetry, are limited. But a lattice has an additional symmetry, invariance under translations. How does this effect which rotation groups can be crystallographic symmetry groups — symmetry groups of crystals? A rotation giving a symmetry of a crystal (but not of a molecule) is limited to orders 1,2,3,4 or 6, only, that is only with angle (less than 2π),

$$\theta = 0, \frac{2\pi}{2}, \frac{2\pi}{3}, \frac{2\pi}{4}, \frac{2\pi}{6}. \qquad \text{(I.5-1)}$$

(The order of element r is the power n for which

$$r^n = \text{the identity}, \qquad \text{(I.5-2)}$$

so for a rotation through $\frac{2\pi}{n}$ it is n.) This is the crystallographic restriction [Senechal (1990), p. 17]. Why should this be? Why should translations determine which rotations are allowed?

A lattice is simultaneously invariant under both rotations and translations — and both are discrete. Atomicity is essential for the result. Because the translations are discrete, and limited, so must the rotations be. There are several ways of showing this, and since the result is so fundamental to crystallography (and is not true for quasicrystals or other systems) it is worth studying, and comparing, different proofs. It is also interesting because it is an example of symmetries interacting — the presence of one type affects the other. Also crystallographic symmetry groups are subgroups of the inhomogeneous rotation group, and this emphasizes that all subgroups of a subgroup (here finite rotations, of every integral order, form subgroups of the rotation subgroup) need not be subgroups of the full group; in finding the representations of a subgroup, it is important to remember that it is a subgroup [Mirman (1995a), sec. X.6.b.ii, p. 290; (1995c), sec. 2.3.3, p. 18; sec. 3.2.3, p. 35].

I.5.a Translations limit the trace of the rotation matrix

One proof uses the trace of a matrix [Armstrong (1988), p. 151; Burns and Glazer (1990), p. 22; Fässler and Stiefel (1992), p. 219; Heine (1993), p. 131; Jansen and Boon (1967), p. 353; Lomont (1961), p. 133; Senechal (1995), p. 50; Streitwolf (1971), p. 61; Yale (1988), p. 152]. A rotation, $R(\theta)$ through angle θ using any standard matrix realization [Mirman (1995a), eq. X.4.a.ii-4, p. 280], has trace

$$\sum R(\theta)_{ii} = 1 + 2\cos\theta. \qquad \text{(I.5.a-1)}$$

This matrix acting on a set of lattice vectors sends the set into another such set — it is a symmetry of the lattice. These vectors can be written as sums of those of the first set (it is a complete set, for the lattice), and with integer coefficients since the lattice is translationally invariant only under discrete translations. Thus, because of translational invariance, the rotation matrix has integer entries, so its trace is an integer. Translations limit rotations since vectors obtained with rotations must also be obtainable with translations. There are two conditions on θ, this requirement of integral trace, and that 2π be an integral multiple of θ. Thus what is surprising is not that there are so few crystallographic rotation groups, but that lattices and crystals can exist at all (in any dimension? — why?).

Problem I.5.a-1: Check the trace. Verify that the only values of θ that satisfy are those given.

I.5.b Discreteness is essential

How geometry and atomicity limit the angles can be seen pictorially [Bravais (1969), p. 50; Burns and Glazer (1990), p. 20; Falicov (1966), p. 79; Farmer (1996), p. 52; Hargittai and Hargittai (1987), p. 382; Ladd (1989), p. 139; Senechal (1995), p. 6; Wherrett (1967), p. 227; Yale (1988), p. 104]. Take two nearest-neighbor lattice points, A and B; these lie a finite distance d apart. If a rotation by θ around an axis through A perpendicular to the plane of the paper is a symmetry, it brings B to lattice point C. This rotation about B moves A to D. The key is that by translational invariance a translation from C to D is a symmetry operation, and the distance between two points is an integer multiple of a smallest distance, here d. Thus the distance from C to D is md, where m is an integer, which is

$$CD = d + 2d\cos(\pi - \theta) = md. \tag{I.5.b-1}$$

So

$$\cos\theta = \frac{1-m}{2}, \text{ and } |\cos\theta| \le 1; \tag{I.5.b-2}$$

this gives the constraint on θ. It is seen that rotations of both 60^{0} (m = 0) and 120^{0} leave the lattice invariant; there is invariance under translation d along line AB,

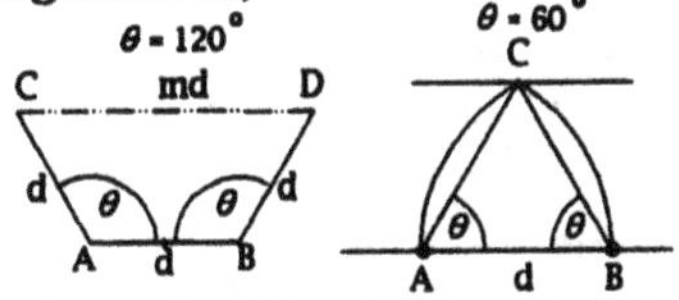

Figure I.5.b-1: EXAMPLES OF ALLOWED ROTATIONS.

Problem I.5.b-1: Since m is arbitrary, why does this give the constraint on θ?

Problem I.5.b-2: Was the requirement that θ be an integral divisor of 2π used?

Problem I.5.b-3: Redraw these diagrams carefully,

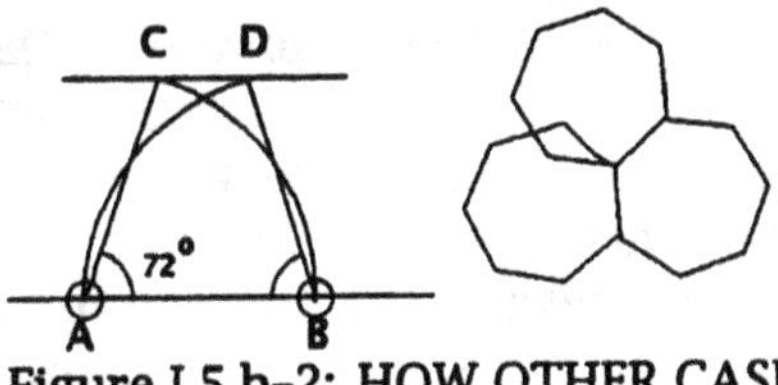

Figure I.5.b-2: HOW OTHER CASES FAIL,

showing that angle $2\pi/5$ does not work, and that an attempt to tile the plane with septagons is also impossible. Draw figures of eight and nine sides (and so on) to see what goes wrong. Explain. Why is an attempt to tile the plane with regular figures (that is cover it with a single repeated object with no holes and no overlap) relevant?

I.5.c Complex numbers, rotations and translations

Since the points of a plane can also be labeled by complex numbers we might expect another proof using them [Jaswon and Rose (1983), p. 69].

Problem I.5.c-1: Denote by $\underline{z}$ a lattice vector starting from the origin, at which there is an n-fold symmetry axis perpendicular to the plane. So by definition there are n lattice vectors of length $|\underline{z}|$ at this point,

$$\underline{V}_l = \underline{z},\ \underline{z}\exp(\frac{2\pi i}{n}),\ \underline{z}\exp(\frac{4\pi i}{n}),\ \ldots, \qquad (I.5.c-1)$$

(these being $|\underline{z}|$ times the n n'th roots of 1 — why?). Since a two-dimensional lattice is given by two lattice vectors each of these can be written as the sum of any two, thus

$$\underline{z}\exp(\frac{4\pi i}{n}) = m_1\underline{z} + m_2\underline{z}\exp(\frac{2\pi i}{n}), \qquad (I.5.c-2)$$

with the m's integers. Show that this gives

$$m_1 = (\cos\frac{4\pi}{n} - m_2\cos\frac{2\pi}{n}) + i(\sin\frac{4\pi}{n} - m_2\sin\frac{2\pi}{n}), \qquad (I.5.c-3)$$

so

$$\sin\frac{4\pi}{n} - m_2\sin\frac{2\pi}{n} = 0,\ \ m_2 = 2\cos\frac{2\pi}{n}, \qquad (I.5.c-4)$$

and as a check

$$m_1 = \cos\frac{4\pi}{n} - 2\cos^2\frac{2\pi}{n} = -1, \qquad (I.5.c-5)$$

so the only values giving integral m_2 are

$$n = 1, 2, 3, 4, 6, \qquad (I.5.c-6)$$

as before (fortunately). Check that the same result obtains using any of the lattice vectors on either side of the equation. How was the discreteness of the lattice used? Draw diagrams illustrating this. This gives an interesting relationship between complex numbers and Euclidean geometry — which has real coordinates [Mirman (1995b), sec. 2.2, p. 32].

Problem I.5.c-2: Find other ways (sec. I.6.b.ii, p. 36) of proving this [Bravais (1969), p. 52; Hamermesh (1962), p. 45; Sternberg (1994), p. 31]. Can these proofs be obtained from any of the others? Which generalizes most easily to other dimensions? Must an equivalent result hold for every dimension [Senechal (1989), p. 14; Senechal (1995), p. 51]? Why? How about other geometries? What postulates of Euclidean geometry give these results? How? Can one postulate be changed to give a different set of angles? To allow any angle?

I.5.d Why are rotational symmetries of lattices limited?

In summary, there is a restriction on symmetry angles because a lattice contains a finite number of points in any volume (the total number is, of course, infinite) and translational invariance requires that it be invariant not under every translation, but only discrete ones — the distance moved must be finite, not infinitesimal. Thus all displacements leaving it invariant are integer multiples of some minimum one. The result therefore does not hold for, say, a continuous fluid, or for molecules.

Now a rotation takes one point to another and since points are finite distances apart, rotations leaving the lattice invariant can be only through discrete angles. But there is a minimum value (and a maximum, 2π) for this angle since there is a minimum distance between points, giving a maximum value for the order of rotation. Why 6? This gives $\theta = \frac{\pi}{3}$, so the triangle formed by points A, B and C in fig. I.5.b-1, p. 25, is equilateral. A smaller angle would mean that AC is smaller than AB, which was assumed the minimum. If it were smaller we would repeat the argument using AC instead of AB. The essential point is that there is a minimum distance, and this requires that the triangle be equilateral.

Why not $m = 5$? Rotations around either A, B or C must give invariance, so all lines are equal, requiring an equilateral triangle or a square. Thus the angle must be $\frac{\pi}{3}$, $\frac{\pi}{2}$, or an integral multiple of them. All lawful rotations really reduce to these two.

Problem I.5.d-1: Draw a triangle (what kind?), copy it and show that it tiles the plane (no holes, no overlap). Try it with pentagons, septagons, octagons, nonagons, decagons, ..., and notice that they cannot be fit together to fill the plane. Why?

Problem I.5.d-2: How do these conditions give the requirement that the cells fill the plane? Is the converse true?

I.6 POINT GROUPS

Having defined the territory to which our explorations are limited, we now start to develop the tools necessary, the groups of symmetry of these objects [Fedorov (1971), p. 24]. First we study only rotation and reflection symmetries, so only point groups — ones that leave a point fixed. Such symmetries of lattices are (quite?) few, thus their symmetry groups are. Point groups of lattices, the crystallographic point groups, do not include translations, but are restricted by the invariance under translations — only a subset of point groups is compatible with this invariance. The set of molecular point groups is larger — they are not so restricted.

What are the point groups, why can rotations and reflections be combined to give only them, and which are crystallographic? First, what are the transformations on crystals and molecules, and what groups do they give [Borchardt-Ott (1993), p. 48; Cornwell (1984), p. 323; Heine (1993), p. 128; Ladd (1989), p. 57; Sands (1993), p. 13]? We start with these; the groups of the full set of symmetries of lattices, and more generally crystals, the space groups — found from the point groups plus the (discrete) translations — are considered later (chap. III, p. 132).

I.6.a The symmetry operations on crystals and molecules

Given an object, say a crystal or molecule, what transformations are possible on it? This set is quite large. We might for example interchange two atoms, leaving all others fixed, or rotate the spin of some single electron, while doing nothing else. Some of these operations might be interesting, and lead to useful insight and results. Here however we study only transformations that act on the object as a whole, rotating it as a rigid body, reflecting it in a plane, or inverting it through a point, and these are required to leave it invariant (there is no way of telling whether the object has been transformed). In all these cases the *relative* positions of atoms (and subatomic particles) are fixed; these are the, and all, point operations that leave a lattice invariant. What operations are they? (Reflection and inversion cannot physically be carried out in the same way that a rotation can; what we mean is that we consider an object obtained by replacing each particle by an identical one with coordinates given by the operation — for symmetry, these two objects must be indistinguishable.)

I.6.a.i *Naming the point symmetry transformations*

To give transformations we need labels for them [Bradley and Cracknell (1972), p. 27; Hamermesh (1962), p. 61; Sands (1993), p. 54]. There are two systems, the Schoenflies and the International.

A lattice has rotational symmetry about a finite number of axes, and the symmetry can be different for different axes. The one with the largest symmetry — with the smallest angle of rotation giving a symmetry (the rotation of highest order) — is the principal axis (usually named z); for several such, any one is chosen. This is drawn vertically (from top to bottom of the page), thus the plane in which the axis lies is called the vertical plane; that perpendicular to it is the horizontal plane. An axis bisecting the angle between the principal axis and one of its neighbors is a diagonal axis, and its plane a diagonal plane — the plane through the diagonal axis, and perpendicular to the plane formed by the principal and diagonal axes. A diagonal mirror (plane) — of symmetry — containing the principle axis is a plane bisecting the angle between adjacent axes of two-fold symmetry (perpendicular to the principle axis), not one along a diagonal axis.

Problem I.6.a.i-1: Should the vertical plane be defined as "a" plane containing the principal axis, not "the" plane?

I.6.a.ii *The point symmetry transformations*

We come then to the point symmetry operations [Bishop (1993), p. 27; Burns (1977), p. 2; Burns and Glazer (1990), p. 2; Chen (1989), p. 380; Cotton (1990), p. 18; Jaswon and Rose (1983), p. 15; Joshua (1991), p. 5; Koster (1957), p. 180; Tinkham (1964), p. 51; Tsukerblatt (1994), p. 1; Wood (1977), p. 1; Wooster (1973), p. 3]; seeing their effect on a comma may be particularly useful [Megaw (1973), p. 120]. These are

1. The identity E (of course),

2. Rotations through $\frac{2\pi}{n}$, where n is an integer (for a crystal only one of five). For each n there are n distinct rotations. These are labeled, in the font we use, $\mathbf{C_n}$ in the Schoenflies notation, $\boldsymbol{n}$ in the International notation. In the next figure is a hand, and at the top right, it is shown rotated, of course not leaving it invariant, by $\frac{\pi}{4}$ around its lower left corner (marked by O), about an axis perpendicular to the drawing. In the lower center the rotation is of $\frac{\pi}{2}$ about a vertical axis (along the mirror), so its back is seen.

3. Reflections. The transformed object is the original seen in a mirror. There are symbols for general reflections, and ones in horizontal, vertical and diagonal mirrors. These are, in the Schoenflies notation, σ, specifically $\sigma_{\mathbf{h}}$, $\sigma_{\mathbf{v}}$, $\sigma_{\mathbf{d}}$, $\sigma_{\mathbf{xy}} = \sigma_{\mathbf{h}}$ and $\sigma_{\mathbf{xz}}$ (= $\sigma_{\mathbf{v}}$). In the International notation a reflection in a vertical or diagonal plane is denoted by $\boldsymbol{m}$, in a horizontal plane by $/\boldsymbol{m}$. The reflection (by $\sigma_{\mathbf{v}}$) of the hand is at the top left.

4. Inversions. The object is reflected through the origin; the transformed object has points with coordinates $(-x_j, -y_j, -z_j)$, where those for the original object are (x_j, y_j, z_j), for each point j. This operation is denoted by $\mathcal{I}$. The inversion shown is through the origin, at the lower left of the hand, marked O (with the hand displaced for clarity).

5. Improper rotations (rotary-reflections) are products of a rotation and a reflection. Thus improper rotation $\boldsymbol{S_n}$ is the product of a $\mathbf{C_n}$ rotation and a reflection in the plane normal to the $\mathbf{C_n}$ axis. The hand is reflected in the mirror (in the plane), and is therefore now a left hand, and is then rotated by $\frac{\pi}{4}$ about an axis perpendicular to the page through O, giving $\boldsymbol{S_4}$. In the International notation $\bar{\boldsymbol{n}}$ denotes a rotary-inversion, a product of a $\mathbf{C_n}$ rotation and an inversion. These symbols thus refer to

different operations, and the set of operations (for which symbols are defined) in the two systems of notation are therefore different.

These transformations are illustrated acting on a stylized hand:

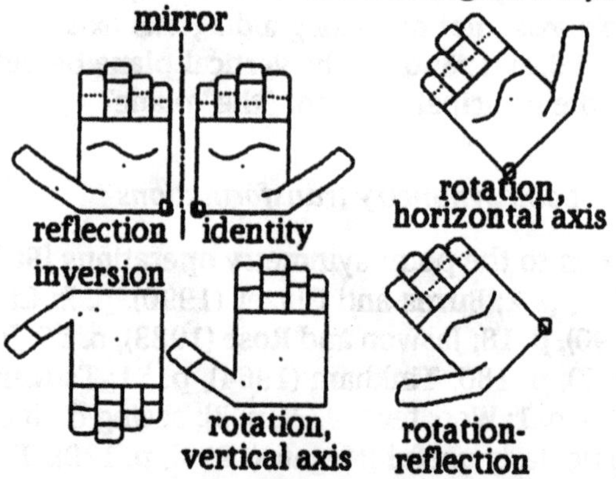

Figure I.6.a.ii-1: TRANSFORMATIONS OF A HAND.

Elements of this limited set of point-group transformations give several groups, so lattices. The choice of elements to form a group is not arbitrary — products are in the group and some products are not allowed. Ways of combining these transformations to obtain allowed ones and the groups that various choices give are needed next [Cotton (1990), p. 33; Tsukerblatt (1994), p. 13]. Groups containing only rotations are the finite subgroups of the rotation group, and these we know (sec. I.4.b, p. 8). Thus we must still find how to combine them with reflections, in all (allowed) ways giving distinct groups.

Problem I.6.a.ii-1: Why must the point through which the (symmetry) inversion (of a lattice) takes place lie on the (a?) rotation axis?

Problem I.6.a.ii-2: Explain why a reflection (of a lattice) in any plane besides a vertical, horizontal or diagonal, cannot give a symmetry. Notice that the horizontal mirror is perpendicular to the principle axis, but the other two contain it, thus the distinction between them is somewhat artificial. The vertical mirror is taken to be a plane containing another coordinate axis, the diagonal does not. Of course with the definition of these planes the choice of axes is not fully arbitrary.

Problem I.6.a.ii-3: Not all the transformations have been illustrated (in particular the $\bar{4}$). Draw the others. There is a minimal set: all transformations can be written as products of members of it. Give it. Is it unique? Write all other transformations as such products.

Problem I.6.a.ii-4: A cube (fig. I.4.d.i-1, p. 18) provides examples of these symmetry operations. Find all the vertical, horizontal and diagonal rotation axes. Also find the mirrors. Are there diagonal ones? Why? How do the rotary-reflections and rotary-inversions compare?

Problem I.6.a.ii-5: Notice that these transformations leave at least one point fixed. Check that if the fixed points were different for different operations of a group, it would include translations (so would be a space, not a point, group). Is it possible that no points are fixed? Which transformations give only one? What do the others leave fixed?

Problem I.6.a.ii-6: These coordinate axes are inverted through the origin, reflected in the yz plane, rotated by $\pi/6$ around z, and then the rotated figure is reflected in the yz plane, giving the effect of operation S_6,

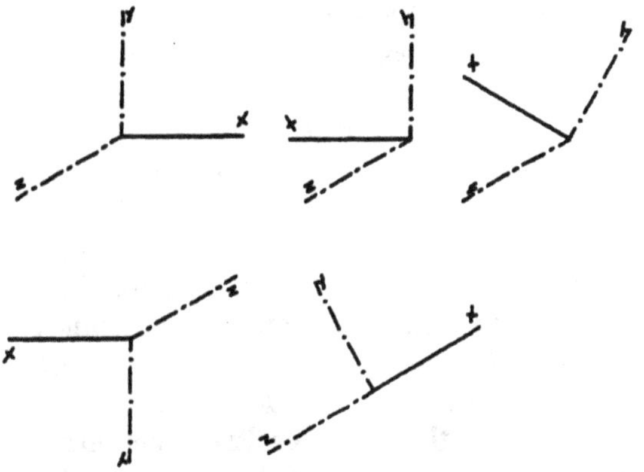

Figure I.6.a.ii-2: TRANSFORMATIONS OF COORDINATE AXES.

Which operation gives which set of axes? Verify these.

Problem I.6.a.ii-7: Draw a benzene molecule (C_6H_6), which is a ring (pb. X.5.c-7, p. 545), and put in the symmetry axes and planes [Burns (1977), p. 2; Burns and Glazer (1990), p. 2].

Problem I.6.a.ii-8: Show that all point-group symmetry operations are included [Armstrong (1988), p. 136; Burn (1991), p. 23; Lockwood and Macmillan (1983), p. 119; Martin (1987), p. 182]. Pick the fixed point as the origin and note that all symmetry rotations around any axis through it are listed. Only reflections in a plane through the origin, and an inversion in it are left. Do we have all?

Problem I.6.a.ii-9: These operations are not all independent; some are products of others (improper rotations are examples). But there is a stronger result: show that any rotation can be written as a product of reflections [Burn (1991), p. 28, 36; Martin (1987), p. 35, 185]. For example, prove that the product of reflections in two vertical mirrors (ones containing the vertical axis) with angle ϕ between them is equivalent to a rotation of 2ϕ about the principal axis [Chen (1989), p. 383; Hamermesh (1962), p. 37]. Why should the number of such mirrors and the angle ϕ be limited? This (clearly?) means that the existence of two such mirrors requires that there be an n-fold axis, with

$$n = \frac{\pi}{\phi}. \qquad \text{(I.6.a.ii-1)}$$

And if the group has such an axis and one vertical mirror then there are n vertical mirrors with angle

$$\phi = \frac{\pi}{n} \qquad \text{(I.6.a.ii-2)}$$

between adjacent ones. These geometrical results impose restrictions on groups — or group theory restricts geometry. Can geometries be found that evade these restrictions, or do they follow from the definition of geometry (whatever that is)? Why?

Problem I.6.a.ii-10: That a set of points forms a lattice means that it is highly restricted. This one word — lattice — leads by itself to several symmetry requirements. Thus every lattice point is a center of inversion [Burns and Glazer (1990), p. 53], which should be both obvious and proven (or at least explained), but more interestingly, there is another center of inversion halfway between every two lattice points. Why? What is a center of inversion? In studying unit cells (chap. IV, p. 185), first in two dimensions, then in three, list all centers of inversion. Moreover, and perhaps not unrelated, if there is a two-fold rotation axis at every lattice point (is there always?) then halfway between two lattice points is another such axis. The same is true for mirror planes. Also if a group has two-fold axes with angle ϕ between them, it has an n-fold axis (where?), with

$$n = \frac{\pi}{\phi} \qquad \text{(I.6.a.ii-3)}$$

[Chen (1989), p. 381]. What does this say about the allowed angles between two-fold axes? Conversely if it has an n-fold and one two-fold axis there are n two-fold axes (where?) with angle

$$\phi = \frac{\pi}{n}, \qquad \text{(I.6.a.ii-4)}$$

between any two adjacent ones. Prove these with a drawing. Can they be proved analytically? Again list all two-fold axes and mirrors in the unit cells as they are met, assuming various possibilities for these at the lattice points (how many different cases are there?). Why are the equations of this problem the same as those of the previous one?

Problem I.6.a.ii-11: In what way do these results follow from the postulates (which?) of (Euclidean) geometry? Can any be avoided?

I.6.a.iii *Why only some transformations are considered*

Why are (only) these the operations, not all of which are rigid body transformations, that we consider? They are the symmetry transformations on a lattice, a set of points. Thus we do not study other operations that might be relevant to a crystal, such as changing the type of atom or spin direction at certain positions — which if, say, the spins at adjacent points were opposite would give invariance under the combined operation of spin flip and translation (chap. VIII, p. 393). All operations considered here can be performed physically, without tearing the lattice, either by rotations or by using a mirror for reflections. Thus they define, and determine properties of, lattices which more general transformations do not. And these are the most general ones that do.

Problem I.6.a.iii-1: Justify the last two sentences. Find additional operations on points in a crystal (not considering them as atoms, say), that also satisfy (chap. III, p. 132).

I.6.b Naming and describing the point groups

Having given the point symmetry transformations, we now find the groups formed from them. In three dimensions there are a mere (?) 32 point groups. To be able to deal with groups we must name them, but to do so we must describe them [Bishop (1993), p. 35; Blichfeldt (1917), p. 64; Borchardt-Ott (1993), p. 111; Bradley and Cracknell (1972), p. 24; Burns (1977), p. 10; Burns and Glazer (1990), p. 55, 72; Chen (1989), p. 380; Cotton (1990), p. 39; Elliott and Dawber (1987), p. 183; Evarestov and Smirnov (1993), p. 39; Falicov (1966), p. 82; Fässler and Stiefel (1992), p. 214; Hamermesh (1962), p. 32; Hargittai and Hargittai (1987), p. 375; Heine (1993), p. 128; Hilton (1963), p. 52; Hochstrasser (1966), p. 51; Inui, Tanabe, and Onodera (1967), p. 169; Jansen and Boon (1967), p. 334; Janssen (1973), p. 70; Jaswon and Rose (1983), p. 26; Jones (1975), p. 60; Joshi (1982), p. 230; Koster (1957), p. 181; Ladd (1989), p. 57; Lax (1974), p. 34; Leech and Newman (1969), p. 13; Lockwood and Macmillan (1983), p. 123; Lomont (1961), p. 132; Ludwig and Falter (1988), p. 18; Lyndon (1989), p. 61; Lyubarskii (1960), p. 18; Martin (1987), p. 211; Megaw (1973), p. 118; Miller (1972), p. 23; Murnaghan (1963), p. 328; Sands (1993), p. 26; Senechal (1990), p. 31, 127; Shubnikov and Belov (1964), p. 31; Streitwolf (1971), p. 60; Tinkham (1964), p. 54; Tsukerblatt (1994), p. 26; Weyl (1989), p. 149; Wood (1977), p. 53]. Here we give the point groups of crystals, listing their elements, with examples of objects of each symmetry. Actual crystals going with various point groups provide another way of supplying concrete meaning [Hargittai and Hargittai (1987), p. 376; Ladd (1989), p. 70; Sternberg

(1994), p. 42]. The group symbols (which can be determined with a flow chart [Burns and Glazer (1990), p. 72]), are listed in the two systems the Schoenflies and the International (in parentheses). The International symbol is often abbreviated and the full symbol is also given. From the description it should be clear that these are crystallographic point groups, and all. In appendix B, p. 631, we list the groups, giving both notations for all and the crystal systems (sec. I.6.d, p. 42) for each. As molecules have no restriction on rotations, they have an infinite number of point groups. (This leaves open the question whether among the infinite number of states of the known, or possible, molecules, there is at least one giving a structure invariant under each of these groups, or whether some are not, perhaps cannot be, allowed?)

We first describe the groups containing only rotations, which we know. Then we combine these with reflections and the inversion. For lattices, the point groups are obtained using all allowed combinations of symmetry operations giving nonisomorphic groups: rotations of order 1,2,3,4 and 6, reflections in horizontal, vertical and diagonal mirrors, and the inversion. The following are the point groups.

Problem I.6.b-1: The generators of a group are those of its elements (preferably a minimal set) from which all other elements are generated — all are products of the generating elements. These are important, for one reason as restrictions on, and information about, properties of those physical objects for which the group is a symmetry group can often be obtained using only this smaller set. Show that the generators (with the identity always understood) of the point groups are those given in tbl. B-1, p. 633 [Burns (1977), p. 378; Lomont (1961), p. 140]. Here C_{2y} indicates a two-fold rotation around a horizontal axis, taken as along y, and $C_3[111]$ a three-fold rotation around the line with direction cosines [111], which is the diagonal of the cube. That these are the generators can be shown by taking sufficient products of them for each group and verifying that they give the group, either by identifying the products with the group elements, or by finding the group multiplication table and showing that it is the same as that of the group. Perhaps better is a geometrical approach; study each figure (even construct a model for each), find the effect of the generators, and then show that their products give all symmetries of the object. Finally ways must be found to break the symmetry, reducing it to a subgroup, and repeating the process for that.

Problem I.6.b-2: To what extent can the generators of the crystallographic point groups be used to show that they are not isomorphic (and to find the ones that are)? To what extent can they be used to show that all these groups have been found?

I.6.b.i *Point groups with only rotations*

There are eleven crystallographic groups whose elements are rotations, only. Both a rotation through $\frac{2\pi}{n}$, and the groups that have only (powers of) this operation, are denoted by C_n in the Schoenflies, and n in the International notation. To distinguish them we write the operation as $\mathcal{C}_n$ (*n*), the group as C_n (n). There are five groups with only rotations about a single axis, of orders 1,2,3,4 and 6, C_1 (1), C_2 (2), C_3 (3), C_4 (4), C_6 (6), respectively. Four more contain also a two-fold rotation about an axis perpendicular to the principal one, in addition to the $\mathcal{C}_n$ operations, these are dihedral groups D_2 (222), D_3 (32), D_4 (422), D_6 (622). In the International notation, which is more explicit than that of Schoenflies, the first number gives n, the order of the principle axis, the others the classes of two-fold axes perpendicular to it. So these have respectively three classes of two-fold rotations, a three-fold axis and one class of two-fold axes, a four-fold and two such classes of two-fold ones, and a six-fold plus two such classes of two-fold rotations [Hamermesh (1962), p. 43; Heine (1993), p. 134].

The other groups [Jaswon (1965), p. 20; Jaswon and Rose (1983), p. 37] containing only rotations (subgroups of SO(3)) are those of the regular polyhedra (sec. I.4.c, p. 10), and are named by the solids whose (rotational) symmetry they express. Tetrahedral group T (23) gives the symmetry of a regular tetrahedron [Mirman (1995a), sec. III.2.f, p. 76]. That of the octahedron (and cube), the octahedral group [Harter (1993), p. 227; Janssen (1973), p. 79; Kettle (1995), p. 139; Mirman (1995a), sec. III.2.g, p. 78], is denoted by O (432). For a regular icosahedron the rotational symmetry group is I or Y [Elliott and Dawber (1987), p. 190]. This has a five-fold axis so is not a crystallographic group. It is, however, favored by many viruses [Burns and Glazer (1990), frontpiece and p. 230; Hargittai and Hargittai (1987), p. 407].

Problem I.6.b.i-1: For the groups of the hexagonal system, tbl. B-1, p. 633, lists the generators as $\mathcal{C}_2$ and $\mathcal{C}_3$, rather than $\mathcal{C}_6$. Why?

Problem I.6.b.i-2: Find and explain the numbers of two-fold axes and the number of their classes, and the total number of classes, for each group [Elliott and Dawber (1987), p. 192; Inui, Tanabe, and Onodera (1967), p. 173; Jaswon and Rose (1983), p. 171]. Why does D_2 have three two-fold axes, and all in different classes [Hamermesh (1962), p. 43]? Of course D_1 is meaningless.

Problem I.6.b.i-3: Check that tetrahedral group T(23) consists of operations E, $3\mathcal{C}_2$, $8\mathcal{C}_3$, and has four classes [Elliott and Dawber (1987), p. 188; Hamermesh (1962), p. 50; Inui, Tanabe, and Onodera (1967), p. 172; Jaswon and Rose (1983), p. 38]. Though listed here as three sets, there are twelve different operations. What are they? About which axes are the rotations? What are the classes? The International labels of T and D_3 are quite similar (with the three-fold axis listed first for the latter, probably for consistency with the other dihedral groups, so for distinctness, second for the former), but note that the number of operations and classes is larger for T. Are the International symbols for the tetrahedral and octahedral groups correct?

Problem I.6.b.i-4: Check that all crystallographic subgroups of SO(3) are on this list.

Problem I.6.b.i-5: For plane rotations $R(\theta)$ through θ [Lyndon (1989), p. 64], show that

$$R(\frac{2\pi}{2})R(\frac{2\pi}{3})^{-1} = R(\frac{2\pi}{6}), \quad R(\frac{2\pi}{3})R(\frac{2\pi}{4})^{-1} = R(\frac{2\pi}{12}), \qquad (I.6.b.i-1)$$

so if a group contains rotations of order 2 and of order 3 it must contain one of order 6. Why? But it cannot contain rotations of order 3 and of order 4. Thus every group consists of rotations that are powers of some elementary one, where the angle of that is (not surprisingly)

$$\theta = \frac{2\pi}{6}, \frac{2\pi}{4}, \frac{2\pi}{3}, \frac{2\pi}{2}, \frac{2\pi}{1}, \qquad (I.6.b.i-2)$$

(the last gives the trivial group having only the identity). Verify.

Problem I.6.b.i-6: It is interesting to speculate if the properties, and virulence, of viruses are related to group theory.

I.6.b.ii *Restrictions on orders of point groups from their characters*

As usual characters, and their implications, provide helpful information about the groups, giving specifically the allowed orders of the point groups.

Lattice vectors are integer multiples of the primitive lattice vectors (sec. I.3, p. 5; sec. II.2.a, p. 68). The matrices representing the point group operators change these integer vectors into other integer vectors, thus have integer entries (sec. I.5.a, p. 24; sec. III.7, p. 169). So [Janssen (1973), p. 72] every point group (representation) is equivalent to an integral matrix group (a group defined by its realization as a set of matrices with only integral entries), thus the characters of point group representations are integers. A point group represented by matrices with integral entries is known as an arithmetic point group. For the (three-dimensional reducible) representation acting on the coordinates, of those groups containing only rotations (no reflections), we use the matrix for a rotation around z [Mirman (1995a), eq. X.4.a.ii-6, p. 280] to get the characters — these are class functions and the class is independent of the axis. The characters then are given by

$$\zeta(\theta) = 1 + 2\cos\theta, \qquad (I.6.b.ii-1)$$

and the only integral ones are,

$$\zeta = 3, 2, 1, 0, -1, \tag{I.6.b.ii-2}$$

for the angles of rotation, respectively,

$$\theta = 0^o, 60^o, 90^o, 120^o, 180^o; \tag{I.6.b.ii-3}$$

the only ones of crystallographic point groups [Heine (1993), p. 131].

The order, g, of a finite group must be a divisor of $\sum_\theta P(\zeta(\theta))$, summed over all group elements; $P(z)$ is any polynomial with positive-definite integral coefficients, since there are g terms, and each character is an integer (also characters of finite groups are algebraic integers) [Burrow (1993), p. 89; James and Liebeck (1993), p. 245; Ledermann (1987), p. 141; Lomont (1961), p. 134; Streitwolf (1971), p. 61]. To prove this we need only consider $\sum_\theta (\zeta(\theta))^n$, for all integral n. For $n = 0$, this equals g. Consider $n = 1$. Then taking the representation as completely reduced (as we always can), its characters, in terms of the irreducible representations in its decomposition, are, with p denoting the irreducible representations,

$$\sum_\theta \zeta(\theta) = \sum_{n_p} \sum_\theta \zeta_p(\theta). \tag{I.6.b.ii-4}$$

Since all characters $\zeta_1(\theta)$ of the identity representation are 1, we can write this as

$$\sum_\theta \zeta(\theta) = \sum_{n_p} \sum_\theta \zeta_1(\theta)\zeta_p(\theta) = n_1 g, \tag{I.6.b.ii-5}$$

by the character orthogonality condition [Mirman (1995a), sec. VII.5.c, p. 195], where n_1 is the number of times the identity representation appears in the decomposition. This give the important result that, except for the identity representation, the sum of the characters over all elements is zero for an irreducible representation (over the complex numbers [Mirman (1995a), pb. VII.7.a.ii-2, p. 210]). For $n = 2$, the sum of the squares of the characters equals g [Mirman (1995a), sec. VII.6.b, p. 200]. For larger values of n, we consider the representation matrices of a product of a representation with itself, which reduces to a sum of irreducible representations. The character of representation matrix $M(\theta)^n$ equals that of $M(\theta^n)$, which is the character of $M(\theta')$, for some other group element. This is true for each element, so the sum over characters is equivalent to that for $n = 1$, and all terms in the polynomial are either zero, or proportional to g, thus any such polynomial is divisible by the order of the group.

Now to find the allowed orders consider some representation of the group, with ζ_1 the character of the identity, and define

$$P(z) = (z - \zeta_2)(z - \zeta_3)\dots(z - \zeta_t), \qquad \text{(I.6.b.ii-6)}$$

where the ζ's are the remaining distinct characters, which are integral, so $P(z)$ has integral coefficients. Now substituting each character for z and summing over all gives

$$\sum_{\theta} P(\zeta(\theta)) = P(\zeta_1) = (\zeta_1 - \zeta_2)(\zeta_1 - \zeta_3)\dots(\zeta_1 - \zeta_t), \qquad \text{(I.6.b.ii-7)}$$

so g is a divisor of $P(\zeta_1)$. We now use the allowed values of the characters, and the requirement that there be a three-dimensional representation to act on the coordinates, to get

$$P(\zeta_1) = (3-0)(3-1)(3+1)(3-2) = 24. \qquad \text{(I.6.b.ii-8)}$$

The possible orders of the point groups containing only rotations therefore must be divisors of 24, and these are

$$g = 1, 2, 3, 4, 6, 8, 12, 24. \qquad \text{(I.6.b.ii-9)}$$

This greatly limits the point groups, in agreement with the previous derivations (sec. I.4.b, p. 8; sec. I.6.b.i, p. 35).

While only these orders are allowed, this does not mean that every group with one of them can be a point group (although there is, at least, one point group for each of these orders). For groups that contain only rotations, representation matrices have determinant 1, and their characters must be ones listed. Also each point group must have a representation that performs a rotation on the coordinates, thus is three-dimensional, real and faithful [Mirman (1995a), sec. V.4.a, p. 158]. These conditions determine which of the finite groups with one of these orders can be a point group — it is not acceptable unless it has a representation with these properties. From the set of finite groups [Ford (1972), p. 129] we pick out those that are point groups; their representations are then known as we know those of cyclic, dihedral and symmetric groups.

The remaining point groups are obtained by adding the reflections, and for each, there is a factor group [Mirman (1995a), sec. IV.8.c, p. 138] of index 2 which is one of the point groups that contain only rotations. Using this, and the representations for the latter groups, the representations of all point groups can be determined. This is considered explicitly below (sec. V.2, p. 247).

Problem I.6.b.ii-1: In proving the divisibility of an arbitrary polynomial, how was the fact that these are operations for a group used?

Problem I.6.b.ii-2: Why does the required representation have to be faithful?

Problem I.6.b.ii-3: For each allowed order, find a representation that acts on coordinates and determine its group, showing that there are point groups for every such order.

I.6.c Point groups with reflections

The other point groups contain improper transformations — ones that reverse the sign of (an odd number of) coordinates, the inversion and reflections — but we cannot simply add (adjoin is a better word) these to each of the pure-rotation groups for that gives duplications. Also the mirror usually contains, or is perpendicular to, the principal axis (this also being reflected) else there will be too many mirrors and axes, giving a group that is not a rotation subgroup. So we now add reflections and inversions to the point groups we have found (there are no others) obtaining the complete set of point groups, the finite subgroups now of O(3) [Neumann, Stoy and Thompson (1994), p. 183; Sternberg (1994), p. 33].

Problem I.6.c-1: Why is adjoin a better word?

Problem I.6.c-2: Can a mirror not contain, or be perpendicular to, the principal axis?

I.6.c.i *Cyclic groups with reflections*

We start with cyclic groups.

Problem I.6.c.i-1: If a cyclic group has rotational subgroup C_n, and also reflection σ_h in the horizontal plane (perpendicular to the principal symmetry axis), the group is C_{nh} ($\frac{n}{m}$), the m standing for mirror (so care is needed to distinguish the order of the operation, here n, and the symbol naming an operation, here m). These groups are C_s ($= C_{1h}$) ($\frac{1}{m}$), C_{2h} ($\frac{2}{m}$), C_{3h} ($\frac{3}{m}$), C_{4h} ($\frac{4}{m}$), C_{6h} ($\frac{6}{m}$). Draw an object, say a set of axes or a hand, and apply the transformations of each group to it. It should be clear that these groups are not isomorphic to each other or to cyclic groups. To show that none are isomorphic to a dihedral group, take your hand (with the principal axis along a finger) and see how the set of rotations of π around each axis perpendicular to your hand differs from the single reflection in a horizontal mirror. Draw axes and notice what happens to them under each of these groups.

Problem I.6.c.i-2: There are groups with reflections in the vertical plane, σ_v (the principal axis lies in the mirror); these are named C_{nv} (nm). For $n = 1$ there is no principal axis so no distinction between horizontal and vertical, thus the only new groups are C_{2v} (mm2), C_{3v} (3m), C_{4v} (4mm), C_{6v} (6mm), with two, three, four and six mirrors respectively. For C_{2v} discuss how the x and y axes appear after a rotation, a reflection, and after the product of these two. What is the difference between the two mirrors? By analyzing the effects of their transformations on, say, axes (or a hand) show that that there are no isomorphic groups among the ones discussed. Two pairs of hands, one with the hands related by a π rotation, the other, with the two pressed together (as if in prayer [Hargittai and Hargittai (1987), p. 56]), illustrate that the π rotation, and a reflection in a vertical mirror, do not give the same result — although the abstract groups are isomorphic. Sets of such pairs can be used for larger groups. Also pairs of hands in the horizontal plane can be used, one obtained by reflection (so from one person), the other by rotation (requiring two people, preferably identical twins). The pair of praying hands, rotated, now gives an object invariant under reflection in the horizontal mirror.

I.6.c.ii *Dihedral groups with reflections*

Next are the groups obtained by adjoining reflections to dihedral groups [Hamermesh (1962), p. 55]. The product of σ_h with one of the π-rotations perpendicular to the principle axis gives σ_v, so a k-fold principal axis plus σ_h leads to k vertical mirrors (with reflections σ_v). These give the groups D_{nh} $(\frac{n22}{mmm})$, and as σ_h commutes with all elements of D_n, they are direct products [Mirman (1995a), pb. IV.8.b-3, p. 138],

$$D_{nh} = D_n \otimes C_{1h}; \tag{I.6.c.ii-1}$$

if n is even $(= 2p)$, D_{nh} contains the inversion so

$$D_{nh} = D_{2p} \otimes S_2, \tag{I.6.c.ii-2}$$

where S_2 is the group containing the identity E and inversion I. This gives four additional groups, D_{2h}, D_{3h}, D_{4h}, D_{6h},

$$D_{2h}(\frac{222}{mmm}),\ D_{4h}(\frac{422}{mmm}),\ D_{3h}(\bar{6}m2),\ D_{6h}(\frac{622}{mmm}), \tag{I.6.c.ii-3}$$

where the inclusion of an S_n transformation is indicated by $\bar{n}$.

Problem I.6.c.ii-1: For D_2, take a plane containing two two-fold axes, and a horizontal mirror. By either drawing, or using coordinates — preferably both — subject a point in the first quadrant to the reflection and rotations, and show that there results a point identical to one given by a reflection in a vertical mirror. Thus the presence of the horizontal mirror requires a vertical one also. There is an equivalent plane, so there are two (mutually perpendicular) vertical mirrors. Repeat this for the other dihedral groups.

Problem I.6.c.ii-2: Consider the effect of these groups on, say, axes or a hand and show that they are nonisomorphic and none are isomorphic to a previously-given group. Also check that the direct products are correct and that the inversion appears where stated.

Problem I.6.c.ii-3: For D_n, a vertical mirror can be added that bisects the angle between neighboring two-fold axes. Verify that the rotations (which?) then require n such. This gives the groups D_{nd} (d for diagonal). Show that for $n > 3$, there is a rotation of order greater than 6, which is noncrystallographic. Thus there are two new groups D_{2d} ($\bar{4}2m$), and D_{3d} ($\bar{3}\frac{2}{m}$). Check that these are different from any previous ones. What happens if a mirror containing a two-fold axis is added to D_n? Why is there no D_{4d} or D_{6d}?

Problem I.6.c.ii-4: Show that for n even, D_{nh} also contains the inversion through the point of intersection of the axis and the mirror (the center of symmetry). Can other vertical mirrors be added?

Problem I.6.c.ii-5: Transformation S_n is a product of a rotation of $\frac{\pi}{n}$ with a horizontal reflection,

$$S_n = C_n\sigma_h = \sigma_h C_n. \qquad \text{(I.6.c.ii-4)}$$

The groups containing only S_n transformations are denoted by S_n, giving S_2 ($\bar{1}$), S_4 ($\bar{4}$) and S_6 ($\bar{3}$). Adding S_1 to any of the above groups (obviously?) gives another on the list. Also S_n, for $n \geq 4$, gives a rotation of order greater than 6. Thus we can adjoin transformations S_2 and S_3. Also S_6, found by adjoining S_3 to C_3 (which does contain an inversion), is isomorphic to no previous group. Show that adjoining the rotation-reflections S_n to dihedral groups gives ones on the list. What would adjoining S_2 to C_3 or S_3 to C_2 give? Check that these elements cannot be adjoined to other cyclic groups (to give crystallographic groups). Thus this gives only two new groups, S_4 and S_6. Why is S_6 denoted by $\bar{3}$?

Problem I.6.c.ii-6: There is one further element, inversion S_2 (why this symbol?). Note that C_i (= S_2) appears above, and adding the inversion to any of these groups gives one already listed.

I.6.c.iii *Adding reflections to the groups of regular polyhedra*

Finally is the adjunction ("joining to") of reflections to the groups of the regular polyhedra [Hamermesh (1962), p. 58]. These must be such as to produce no new rotations.

Problem I.6.c.iii-1: For the tetrahedron, show that there is a group T_d with transformation S_4 and a (diagonal) mirror that bisects the angle between two horizontal two-fold axes, plus a group T_h with transformation S_6 and a (horizontal) mirror that bisects the angle between three-fold axes. Also

$$T_h = T \otimes C_i. \tag{I.6.c.iii-1}$$

Check that these are both new, and no other new groups are obtained by adding reflections or an inversion. Adding a reflection gives T_d ($\bar{4}3m$), adding an inversion to T gives T_h ($\frac{2}{m}\bar{3}$).

Problem I.6.c.iii-2: Octahedral group O allows only one new group, O_h, with

$$O_h = O \otimes C_i. \tag{I.6.c.iii-2}$$

Where is the mirror? Adding an inversion to O gives O_h ($\frac{4}{m}\bar{3}\frac{2}{m}$).

Problem I.6.c.iii-3: Icosahedral group Y (which is noncrystallographic) also allows only one new group, Y_h (or I_h), and

$$Y_h = Y \otimes C_i, \tag{I.6.c.iii-3}$$

(with an inversion). Where is the mirror?

Problem I.6.c.iii-4: Verify that there are five point groups with no unique principal axis, but with four axes each of order 3. Show that the largest of these has order 48. Which is it?

I.6.d There are thus 32 point groups

This completes the construction of the crystallographic point groups — symmetry groups of lattices (holohedry groups) or their subgroups — thus showing that they are all. These groups form sets, for reasons we must see, crystal systems, of which there are seven. The largest in each is the holohedry group; the others are subgroups of the seven holohedry groups (sec. II.2.d, p. 69). Some holohedry groups are subgroups of others [Borchardt-Ott (1993), p. 114; Bradley and Cracknell (1972), p. 187; Chen (1989), p. 389; Yale (1988), p. 106, 111]. The symmetry groups of lattices, but not of all crystals, are holohedry groups. In three-dimensions there are fourteen distinct lattices, the Bravais lattices, but only seven crystal systems to which these belong, so more than one lattice can have the same holohedry group (chap. II, p. 66).

Problem I.6.d-1: What is the definition of a point group? This gives the thirty-two point groups. Go over the list and check that elements cannot be adjoined to any without getting one already given; these are all the (crystallographic) point groups, they are all nonisomorphic, and all do satisfy the definition. Verify that tbl. B-2, p. 634, is correct, including the number, and assignment of elements, to classes. The last (largest) group listed for each system is the holohedry group. Note that all other groups of the system are subgroups of it. Also check that the symbols in the International, and where possible the Schoenflies, notation are correct — they do properly specify the transformations and symmetries.

Problem I.6.d-2: It is interesting — and relevant — that the point groups can also be obtained using the requirement that each have a representation that acts on a three-dimensional, real, vector — that it be a subgroup of SO(3) [Lomont (1961), p. 132]. How is this derivation related to the present discussion?

Problem I.6.d-3: How many point groups are there in two dimensions (pb. II.3.a-3, p. 72); what are they [Hahn (1989), p. 11; Lyndon (1989), p. 74]? Are they related to those in three dimensions?

Problem I.6.d-4: One way of obtaining the holohedry groups is to consider the set of permutations of the vertices leaving invariant the unit cells whose corners are the points of the (seven) primitive lattices (chap. II, p. 66), which can easily be found from the relative dimensions and angles given for each system in tbl. B-2, p. 634. These are symmetric groups, or subgroups, and they can readily be determined.

Problem I.6.d-5: Here we gave one way of showing that these are, and are all, the point groups. It should be compared with others, including ones finding all holohedry groups of the lattices, and all their subgroups (why is this relevant?) [Burns and Glazer (1990), p. 69]. Do they differ in assumptions, or only terminology?

Problem I.6.d-6: An analytic method of finding these groups would be useful.

Problem I.6.d-7: These thirty-two groups have been shown (by the reader) to be nonisomorphic by considering objects and checking that the sets of objects given by the transformations of the different groups are different. An analytical proof would find the effect of the transformations on the coordinates of an object.

Problem I.6.d-8: The most group-theoretical proof that these are nonisomorphic is one in which the group tables are found (which is useful in any case) and checked to see that they are different (carefully verifying that one cannot be transformed into another by relabeling); some were previously listed [Mirman (1995a), chap. II, p. 29, chap. III, p. 67]. A computer program to do this might be interesting. Would the results of the previous problem help, or conversely? Can this be used to prove that the list of these groups is complete?

Problem I.6.d-9: There are thirty-two point groups, but many are isomorphic as abstract groups — the elements are the same but have different physical meaning (as rotations through π or inversions, for example). Thus C_{2v} and C_{2h} are isomorphic, and isomorphic to D_2. So these three point groups actually give only one abstract group. Go through tbl. B-2, p. 634, and list all isomorphic ones [Lomont (1961), p. 138]. How many nonisomorphic abstract groups are there in the set of point groups? This emphasizes that whether groups are identical is (partly) a matter of definition. Here we have a few of the very many examples of groups identical abstractly, but with different meanings (realizations [Mirman (1995a), sec. V.3.c, p. 157]).

Problem I.6.d-10: Find the number of point groups in n-dimensional space. Give them. Can a computer program do this? Perhaps the methods of the previous problems can be generalized to arbitrary dimensions (or geometries?).

Problem I.6.d-11: It should be clear that the point group of a lattice must be one of the eleven with a center of symmetry — a lattice is invariant under inversion [Burns and Glazer (1990), p. 74]. This emphasizes that crystals, some of which are invariant only under smaller point groups, are different than — are more than — lattices. These point groups with a center of symmetry are called centrosymmetric, or Laue groups [Ladd (1989), p. 85].

Problem I.6.d-12: Each group is a subgroup of a symmetric group. Write their transformations as permutations [Chen (1989), p. 386], and verify that they are groups — that the transformations form closed sets (and if not obvious, that they satisfy the other group axioms [Mirman (1995a), sec. I.4.a, p. 14]).

Problem I.6.d-13: Of course, all (or most?) of these results are dependent on Euclidean geometry. Or are they? Are there ways of avoiding them, say using different postulates? What types of geometries would such revisions give? What would the analogous restrictions be on symmetry groups (of presumably objects analogous to lattices)? Is any, or all, parts of the last sentence(s) relevant?

I.6.e Why are these all the point groups?

There are other arguments that perhaps show more clearly how geometry limits these groups. That all purely-rotational crystallographic point groups have been given follows from the list of finite subgroups of SO(3) (sec. I.4.b, p. 8). An inversion, and a rotation-reflection, can be written as a product of a rotation and a reflection, so the only question is whether all mirrors have been included. When a mirror is inserted a symmetry axis is also reflected, but since we have all these the only mirrors are those adding no axes. Thus they must either contain or be perpendicular to an axis, or bisect the angle between two of the same order. Also mirrors must not give groups already found.

For cyclic groups the only mirrors are vertical and horizontal; these have all been listed. This is also true for dihedral groups, but now horizontal directions are distinguished, those of the two-fold axes, so where are the mirrors with respect to them? They are either along each of these axes (vertical mirrors), or bisect the angles between every pair (diagonal mirrors). However a two-fold axis plus a horizontal mirror containing it has the same effect as a vertical mirror (pb. I.6.c.ii-1, p. 40), and these were tabulated first. Thus only diagonal mirrors give new groups, and all have been listed.

For a cube the only (sets of) mirrors either contain a three-fold axis, going through diagonally opposite vertices, or are parallel to opposite faces, midway between them, containing one each of the other two types of axes. If a cube is divided into tetrahedra, their symmetry is that of tetrahedral group T, and as the cube contains a mirror through the vertices, it also contains a second equivalent one, perpendicular to it, through one side of the tetrahedron. With this mirror symmetry the group is T_d, with d for diagonal, since this mirror, as with the D_d groups, bisects the angle between two two-fold axes. The mirror parallel to the sides gives group T_h. Since these mirrors bisect angles between three-fold axes, they become S_6 axes. And as the group of the cube contains these, it also contains inversion I.

Problem I.6.e-1: Check these using the figures in appendix C, p. 637, (or others), verifying that they do have these mirrors, and can have no others.

Problem I.6.e-2: Note that for octahedral group O if one mirror is present, the other is also. This gives O_h. For it, the C_3 axes become S_6 axes, and it contains the inversion.

Problem I.6.e-3: Show that, as with D_{2d}, in T_d the two-fold rotation axes become four-fold rotation-reflection axes [Hamermesh (1962), p. 59]. Also the mirrors pass through the three-fold axes, so these are two-sided.

Problem I.6.e-4: Check that the inversion can be added to the icosahedral group. Does it then contain mirrors? Where? How about S_n axes?

Problem I.6.e-5: Which point groups are simple [Mirman (1995a), sec. IV.9, p. 143]? Does this have geometrical significance?

Problem I.6.e-6: If a group is not simple (or Abelian) it is a product of other groups (sec. I.7, p. 50). Relate dihedral group D_n to cyclic group C_n. Why, geometrically, are there invariant subgroups, when there are?

Problem I.6.e-7: Find the classes (sec. I.7.c, p. 57) for each point group [Hamermesh (1962), p. 23; Lyubarskii (1960), p. 20; Tsukerblatt

(1994), p. 21]. Explain their geometrical significance. Write all point groups that can be, as products (sec. I.7, p. 50). What are the invariant subgroups? In particular give the groups containing reflections as direct products [Mirman (1995a), pb. IV.8.b-3, p. 138], with one term a pure rotation group. Explain the form of these groups.

Problem I.6.e–8: Every finite group is a subgroup of a symmetric group (Cayley's theorem [Mirman (1995a), sec. IV.3.d.ii, p. 113]). For each point group give the smallest one of which it is a subgroup. For each of these symmetric groups find all other subgroups (if any) and explain why these are not (crystallographic) point groups. Are the other operations of these symmetric groups relevant to crystals? Why? What are the effects of the other subgroups? The possible mirrors have been found geometrically (using pictures). This might be used to check them group-theoretically (using algebra). That the set of regular polyhedra is limited implies that larger symmetric groups cannot be embedded in the rotation group. Is this rigorously true? Can it be shown directly, and explained (pictorially)? Can it be extended to find all finite subgroups of SO(n), thus the regular polyhedra in any (real) space? Or is three dimensions unique?

Problem I.6.e–9: The elements, and generators, of the icosahedral group and its subgroups, are not included in the tables of appendix B, p. 631 — it is not crystallographic. However it would be useful to determine these, and answer the above questions for them. This might be done geometrically, by studying the symmetries of the icosahedron (or dodecahedron), perhaps by constructing an actual model, or group-theoretically by using the group table and identifying the elements with the symmetry operations, or perhaps with a computer(-aided design) program.

I.6.f The molecular point groups

Molecules are free of the crystallographic restriction, allowing rotations beyond those of order 1,2,3,4 and 6, so have further possible point-group symmetries [Borchardt-Ott (1993), p. 134; Ladd (1989), p. 92; Tsukerblatt (1994), p. 26]. Flow charts and rules can be used to establish the group of a molecule [Bishop (1993), p. 46; Burns (1977), p. 13; Burns and Glazer (1990), p. 73; Cotton (1990), p. 54; Hargittai and Hargittai (1987), p. 83; Kettle (1995), p. 174; Sands (1993), p. 43; Tsukerblatt (1994), p. 48]. Diagrams of molecules with these symmetries illustrate both groups and molecules [Borchardt-Ott (1993), p. 147]. We mention here these other groups, but as they, except for the icosahedral group, are all cyclic or dihedral, not much is added (at least in terms of making lists).

All cyclic groups are now possible: C_n, C_{nh}, C_{nv}, for all integral n, including $C_{\infty v}$ [Hochstrasser (1966), p. 70; Kettle (1995), p. 404]. Likewise all dihedral groups, including $D_{\infty h}$, are allowed (which does not mean they all occur — since there are not an infinite number of molecules, not all do, although for any group there might, or perhaps might not, be some excited state with that symmetry). Also allowed are the symmetry groups of the regular polyhedra, including those with reflections.

Problem I.6.f-1: Why are all other molecular point groups either cyclic or dihedral?

Problem I.6.f-2: What type of molecules have infinite-dimensional point groups? How do ones invariant under cyclic, and dihedral, groups differ?

Problem I.6.f-3: What are the point groups of H_2O, CO_2, CO, NH_3, CH_4 (chap. X, p. 512)?

Problem I.6.f-4: Given a molecule we can determine its symmetry group. But often we want to go the other way. From experimental data we find the group, and then the molecule having the symmetry. For each crystallographic point group construct a molecule with that symmetry. There are numerous examples of molecules with various symmetry groups [Hargittai and Hargittai (1987)]. Do these all exist? Repeat this for (some of) the noncrystallographic groups.

I.6.g Objects invariant under the point groups

Group operations are defined abstractly, and can be given more concretely in terms of, say, coordinates. Why discuss other realizations also? Objects invariant under the crystallographic point groups provide another realization of — another way of making real — the groups and their transformations [Mirman (1995a), sec. V.3.c, p. 157]. Formally they add nothing new; pictorially, and in terms of developing intuition, they can add much. They can — if the reader is able to extract the information, and learn from them — make concrete the meaning of the transformations, and how these govern the nature of objects invariant under them.

To show the meanings of the point group operations, their differences, and of the point groups, there are in appendix C, p. 637, diagrams of objects invariant under these groups. There are other sets of such objects in the literature [Bishop (1993), p. 41; Burns and Glazer (1990), p. 298; Hargittai and Hargittai (1987); Harter (1993), p. 112, 152; Kettle (1995)], and it is useful to compare the various sets, in particular for the discussion and problems below. Since these are two-dimensional projections of three-dimensional objects, it is important to notice which lines are out of the plane.

What do we learn from these diagrams?

Notice the difference between S_2 and C_{2h} (fig. C-23, p. 648). For the former, the object is invariant under the combined operation of a π rotation around an axis and a reflection in the plane perpendicular to it, for the latter it is invariant under both separately. This emphasizes, and pictures, S_2 as a product, and shows the difference in the type of objects invariant under both operations, and ones invariant only under their product. Group S_2 differs from C_{1h} (fig. I.6.a.ii-1, p. 30) in having an inversion, while C_{1h} has a symmetry plane.

Compare also D_2 and D_{2h} (fig. C-24, p. 649); in particular commas differ for the two cases. For D_4 (fig. C-13, p. 644; fig. C-14, p. 644), the object is invariant under a $\frac{\pi}{4}$ rotation about the z axis, as well as a π rotation about any of four mutually perpendicular axes perpendicular to it, but not under a reflection in the xy plane; in the diagram the commas break the horizontal mirror symmetry. The object invariant under D_{2d} should be compared with that invariant under S_4 (fig. C-15, p. 645).

Look at the cubic system. The tetrahedron, which is an object invariant under tetrahedral group T, is drawn with solid lines (fig. C-6, p. 640); there is a three-fold axis (dashed), and a two-fold one. A tetrahedron is invariant under T_d. The symmetry is then broken to T (fig. C-9, p. 642), and these objects can be compared. For T_h (fig. C-8, p. 641), the tetrahedron is inverted, and the two are superimposed. The cube is an object invariant under O_h. Here a tetrahedron is inserted, the transformations are applied, giving two tetrahedra inscribed in a cube.

The pictures should clarify such aspects as the difference between Abelian and non-Abelian groups, and the difference in the requirements of horizontal, vertical and diagonal mirrors on objects. Also they picture that while groups may be isomorphic abstractly (for example, the two-element groups), they can be very different geometrically.

Problem I.6.g-1: For each of the groups, note which other operations are generated by the stated transformations; this gives the full set of symmetry elements. Also find a minimum set of operations from which all others follow — the generators of the group [Mirman (1995a), sec. III.3, p. 82]. Thus check tbl. B-1, p. 633. Every element listed in tbl. B-2, p. 634 of each group should be a product of its generators.

Problem I.6.g-2: One way of checking that the generators are correct is to build models of these objects, or others invariant under the groups, perhaps computer models, first checking that a model is invariant under the transformations of the generators, then applying various products of these — applying them in succession — to the model, verifying that all group transformations are obtained, and (of course) all leave the object invariant.

Problem I.6.g-3: The combined sets of generators of S_2 and C_2 is the set of generators of C_{2h} (fig. C-23, p. 648). Objects invariant under the latter look different than ones invariant under the first two — an object invariant under a group differs from ones invariant under subgroups, and not (obviously) obtainable from them. Can you guess what the object invariant under C_{2h} looks like from knowing the objects invariant under these subgroups? Would it help to know the full set of transformations of the groups, not merely the generators? Try this with C_{4h} (fig. C-12, p. 643) and its subgroup C_4 and subgroup S_2 (fig. C-23, p. 648). Another case is S_6 and subgroups C_3 and S_2. Try other examples. Also C_{3h} (fig. C-18, p. 646) and subgroups C_3 and C_{1h} (fig. I.6.a.ii-1, p. 30) give an example with a reflection rather than inversion. Does this matter? Take a cube and reduce its symmetry to that of C_{1h}.

Problem I.6.g-4: Some of these groups are direct products (sec. I.7, p. 50) [Mirman (1995a), pb. IV.8.b-3, p. 138]. Does this affect objects invariant under them? Could you tell, by looking at such an object, at all such objects, that the group is a direct product, and find what groups it is a product of? Do the answers change for semi-direct products (sec. I.7.a, p. 50; sec. III.2.e, p. 136)?

Problem I.6.g-5: Dihedral groups are obtained by adding a π rotation to the cyclic ones. This is much like a reflection, but not quite. Look at these objects and state and explain the differences in the effect of these operations. Also state and explain how the addition of mirrors to cyclic and dihedral groups result in different symmetry for the objects. In what way do these pictures show why the dihedral groups (except for ?) are non-Abelian?

Problem I.6.g-6: Groups, as abstract objects, have various entities associated with them, multiplication tables, classes, cosets, subgroups, and so on. The symbols can be realized (perhaps interpreted is a better word here) in various ways, say as reflections or rotations or inversions. These objects are concrete realizations of the transformations and groups, giving different interpretations of the operators, but all geometrical. From them, is it possible to obtain these associated entities? Is it useful? Do they have meaning in this context? How do the different interpretations aid this analysis?

Problem I.6.g-7: Other objects that realize these same groups have been, and will be, discussed, some, like molecules and crystals, of intrinsic interest. Thus it is useful to answer these various questions for them; the comparison of the different objects as a source of information about the groups might be particularly helpful. The information obtained from crystals and molecules, and the different forms of these objects with the same symmetry, should be compared, keeping in mind how such pictures, and comparisons, clarify the types, and symmetry, of crystals.

Problem I.6.g-8: Expressing the transformations as permutations, explain, for each group, in terms of objects invariant under it, why the permutations that appear do so, and why the others of the smallest symmetric group of which it is a subgroup, do not.

Problem I.6.g-9: How do these objects help answer the questions raised in this chapter? Have these questions been answered, either using such objects, or in other ways?

I.7 STRUCTURE OF POINT GROUPS

Having listed the point groups we now turn to their structure, in particular whether, and how, they can be expressed as products, and then their classes.

I.7.a Point groups as semi-direct products

An important aspect of the structure of a group is its relationship to its subgroups. Many point groups are related in a simple way (pb. III.2.e-1, p. 137), they are direct, or semi-direct products of subgroups [Altmann (1977), p. 267; Bradley and Cracknell (1972), p. 186] — one of the subgroups in the product is invariant under all group operators (for a semi-direct product only one of the two, if both are then the group is a direct product).

Space groups (as well as the inhomogeneous rotation, and Poincaré, groups) are also semi-direct products, so the illustrations given here help make concrete concepts useful elsewhere.

The symbol $\wedge$ denotes a semi-direct product, with the group before the symbol invariant. Thus

$$G = A \wedge B, \qquad \text{(I.7.a-1)}$$

states that G is the semi-direct product of invariant subgroup A and subgroup B. For direct products $\otimes$ is used.

Problem I.7.a-1: Check that cyclic group C_6 can be written as the direct product

$$C_6 = C_3 \otimes C_2. \qquad \text{(I.7.a-2)}$$

Also $C_n, n < 4$, cannot be written as a product. Are there cyclic groups that can be written as semi-direct products? Which groups cannot be written as direct products?

Problem I.7.a-2: Give the four-group [Mirman (1995a), sec. III.2.c, p. 71] as a direct product (pb. V.4.c.i-2, p. 268). Compare with C_6.

Problem I.7.a-3: Write the dihedral groups as products. In particular show that

$$D_2 = C_2 \otimes C_2', \qquad \text{(I.7.a-3)}$$

$$D_3 = C_3 \wedge C_2', \qquad \text{(I.7.a-4)}$$

$$D_4 = C_4 \wedge C_2' = D_2 \wedge C_2'', \qquad \text{(I.7.a-5)}$$

$$D_6 = C_6 \wedge C_2' = D_3 \wedge C_2''. \qquad \text{(I.7.a-6)}$$

The subgroups are not arbitrary, but are sets of specific operators, indicated by the primes. What are these subgroups? Can D_5 be written as a product? Why? Give the explanation for any dihedral group.

Problem I.7.a-4: Show that tetrahedral group T (sec. I.6.b.i, p. 35) is (isomorphic to?)

$$T = D_2 \wedge C_3' = C_2 \otimes C_2'' \wedge C_3', \qquad \text{(I.7.a-7)}$$

(using what cyclic subgroups?). Check this using the transformations of a tetrahedron. For the last equation, put in parentheses.

Problem I.7.a-5: Octahedral group O has the form

$$O = T \wedge C_2'' = D_2 \wedge C_3' \wedge C_2''. \qquad \text{(I.7.a-8)}$$

In the last expression how are the two cyclic subgroups related? In a semi-direct product, one subgroup is invariant. But in the last equation there are two such products. Which subgroups are invariant, and under what groups?

Problem I.7.a-6: Symmetric group S_n is a semi-direct product of (the invariant) alternating group A_n and C_2 which is the identity plus a transposition [Mirman (1995a), pb.IV.3.d.i-2, p. 112],

$$S_n = A_n \wedge C_2. \qquad \text{(I.7.a-9)}$$

Groups T and O are A_4 and S_4. Compare this result to those of the previous problems. Relate the operations.

Problem I.7.a-7: One check of these relations is that the orders of the groups on both sides of each equation are the same.

Problem I.7.a-8: This discussion is limited to the crystallographic point groups. It is a useful exercise to extend it to, at least a few, others. Since the number of finite groups is infinite, this cannot be done for all. Or can it?

I.7.a.i *Dihedral groups*

The exploration of the structure of dihedral group D_n (pb. I.4.b-1, p. 9) is next [Altmann (1977), p. 269; Hamermesh (1962), p. 43]. Its generators are a π/n rotation C_n, around the principle axis, taken as the z axis, plus a π rotation, C_2', around an axis in the xy plane (pb. I.6.b.i-2, p. 35). The dihedral groups are defined by rotations of $2\pi j/n$ (around z), and a two-fold rotation perpendicular to the z axis, however there must be n of these latter, taken into one another by the $2\pi j/n$ rotations.

Why are these groups semi-direct products? Any transformation of a dihedral group can be written as a product of rotations about perpendicular axes, thus the group is a product of subgroups about these axes. It is clear that a two-fold rotation leaves the set of n rotations perpendicular to it invariant — the C_n subgroup is invariant. However a C_2 rotation goes into a C_2 rotation under a C_n one, but a different such rotation. Thus the C_2 subgroup is not invariant, and the product is semi-direct.

Problem I.7.a.i-1: For the alternate form of D_n, n even, with a $D_{n/2}$ invariant subgroup (pb. I.7.a-3, p. 51), what is the C_2'' group? How many C_2'' groups are there? Show that the operators of all but one can be written as products of the other operators of the group, so that only one C_2'' subgroup gives a generator.

I.7.a.ii *Cubic groups*

Tetrahedral group T can be expressed as the product of D_2 and C_3 (eq. I.7.a-7, p. 51), with C_3 the group of $\frac{2\pi}{3}$ rotations around the diagonal axes of the cube, which bisect the angles of the three orthogonal axes of the D_2 rotations. A tetrahedron, with solid lines,

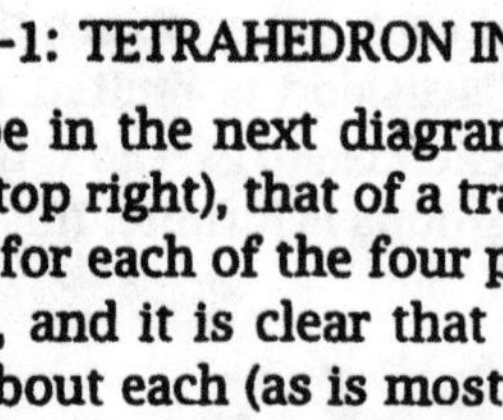

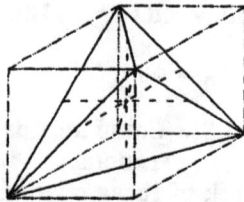

Figure I.7.a.ii-1: TETRAHEDRON INSCRIBED IN A CUBE,

is inscribed in a cube in the next diagram. A dashed three-fold axis (from bottom left to top right), that of a transformation of group C_3, is shown. There is one for each of the four pairs of corners. Also drawn, dashed, are C_2 axes, and it is clear that the tetrahedron is invariant under a π rotation about each (as is most noticeable from the vertical C_2 axis). There are three of these, one through each pair of opposite sides, so mutually perpendicular. The rotations about them are the transformations of D_2. These are taken into each other by the C_3 rotations. Thus D_2 is invariant. However the C_3 rotations are taken into other sets of C_3 rotations by the D_2 transformations. Thus C_3 is not invariant, and T is a semi-direct product.

For octahedral group O, the C_2' cyclic group (eq. I.7.a-8, p. 51) contains a π rotation around a line bisecting diagonally opposite sides. It is a symmetry of the cube, but not of the tetrahedron. However while these do not leave the tetrahedron invariant, they do leave the set of transformations of T unchanged, thus T is an invariant subgroup, and O is a semi-direct product.

Problem I.7.a.ii-1: Construct a cube and check this.

I.7.a.iii *The structure of improper point groups*

What are the structures of groups that, besides rotations, also contain improper transformations (reflections and the inversion)? We start with ones containing the inversion; operators of such groups are of two kinds, proper rotations, R, and improper ones, S, and each S can be written

$$S = \mathcal{I}R, \tag{I.7.a.iii-1}$$

where $\mathcal{I}$ is the inversion. Clearly the set of proper rotations forms an invariant subgroup, and the product of two S's is an R. Thus the group G can be given as as a direct sum,

$$G = R \oplus \mathcal{I}R, \tag{I.7.a.iii-2}$$

where R is the proper subgroup, and the coset representative [Mirman (1995a), sec. IV.6, p. 128] is inversion $\mathcal{I}$. A point group with the inversion is the direct product of pure rotation group R with C_i.

We write the group containing just the identity and $\mathcal{I}$ as C_i (also S_2), it being C_2 with this particular realization. There is another realization of C_2, the identity plus the horizontal reflection, which is group C_{1h} (also C_s).

There is a difference in the improper groups. Consider C_{4h} and S_4. The former is the sum of two sets of operators, those of C_4, and those obtained by multiplying the C_4 operators by $\mathcal{I}$—noting that a C_2 operation times $\mathcal{I}$ is the same as a horizontal reflection and that

$$S_4 = C_4\sigma_h, \quad S_4^{\,3} = \mathcal{I}C_4. \tag{I.7.a.iii-3}$$

The S_4 group is obtained by replacing the C_4 rotations of the cyclic group by S_4 rotation-reflections. It does not contain either an inversion or a reflection, and is a direct sum of cyclic rotation group C_2 with the set obtained by multiplying this by operation S_4. Groups containing $\mathcal{I}$ describe bodies with a center of symmetry, those without it are for objects that have no such point.

Group C_{3h} also does not contain $\mathcal{I}$. It is a direct product of C_3 and the C_2 group of the identity and a horizontal reflection. For C_{3v}, also with no $\mathcal{I}$, there is in addition to the operators of C_3 three vertical reflections. This is a semi-direct product of C_3 and the C_2 group of the identity plus one reflection. The other two reflections are generated by the C_3 rotations. Under reflections the rotations are taken into each other, so this is a semi-direct product. Notice that the horizontal reflection and also the S_3 transformation leave the set of C_3 rotations invariant, so C_{3h} is a direct product.

Problem I.7.a.iii-1: Check these statements using, for example, the diagrams in appendix C, p. 637.

Problem I.7.a.iii-2: The structure of the improper groups can now be given. There are two types, with and without the inversion. For the former show that [Altmann (1977), p. 268],

$$C_{2h} = C_2 \otimes C_i; \quad C_{3i} = C_3 \otimes C_i; \qquad \text{(I.7.a.iii-4)}$$

$$C_{4h} = C_4 \otimes C_i; \quad C_{6h} = C_6 \otimes C_i, \qquad \text{(I.7.a.iii-5)}$$

$$D_{2h} = D_2 \otimes C_i; \quad D_{3d} = D_3 \otimes C_i; \qquad \text{(I.7.a.iii-6)}$$

$$D_{4h} = D_4 \otimes C_i; \quad D_{6h} = D_6 \otimes C_i, \qquad \text{(I.7.a.iii-7)}$$

$$T_h = T \otimes C_i; \quad O_h = O \otimes C_i. \qquad \text{(I.7.a.iii-8)}$$

For the groups without inversion, with the definitions,

$$C_s{}^h = E \oplus \mathcal{I}C_2 = E \oplus \sigma_h, \qquad \text{(I.7.a.iii-9)}$$

$$C_s{}^v = E \oplus \mathcal{I}C_2{}' = E \oplus \sigma_v, \qquad \text{(I.7.a.iii-10)}$$

$$C_s{}^d = E \oplus \mathcal{I}C_2{}'' = E \oplus \sigma_d, \qquad \text{(I.7.a.iii-11)}$$

where $\oplus$ is the direct sum, the set containing all elements of both sets which mutually commute, we should get,

$$C_s = C_1 \otimes (E + \mathcal{I}C_1) = C_1 \otimes (E + \sigma_h) = C_1 \otimes C_s{}^h; \qquad \text{(I.7.a.iii-12)}$$

$$S_4 = C_2 \oplus S_4 C_2; \quad C_{3h} = C_3 \otimes C_s{}^h; \qquad \text{(I.7.a.iii-13)}$$

$$C_{2v} = C_2 \otimes C_s{}^v; \quad C_{3v} = C_3 \wedge C_s{}^v; \qquad \text{(I.7.a.iii-14)}$$

$$C_{4v} = C_4 \wedge C_s{}^v; \qquad \text{(I.7.a.iii-15)}$$

$$D_{2d} = D_2 \otimes C_s{}^d; \quad C_{6v} = C_6 \wedge C_s{}^v; \qquad \text{(I.7.a.iii-16)}$$

$$D_{3h} = D_3 \wedge C_s{}^d; \qquad \text{(I.7.a.iii-17)}$$

$$T_d = T \wedge C_s{}^d. \qquad \text{(I.7.a.iii-18)}$$

I.7.b Classes of the point groups

The other structural aspect that we need is that of the classes of the point groups. We can find some information about this from the abstract groups to which they are isomorphic [Lomont (1961), p. 139].

As cyclic groups are Abelian, each element is in a class by itself. Groups C_{1h}, C_{2h}, C_{3h}, C_{4h}, C_{6h}, are likewise Abelian, giving for C_{nh}, $2n$ elements, in $2n$ classes. Also S_4 and S_6 are Abelian. These have four, and six, elements, in four and six classes.

Groups C_{2v}, C_{3v}, C_{4v}, C_{6v}, which are isomorphic to abstract dihedral groups, are somewhat more complicated [Hamermesh (1962), p. 52], having vertical mirrors, and for C_{nv}, there are n of these. Thus it contains $2n$ elements. For n odd, all mirrors are equivalent, and all reflections are in the same class (which immediately indicates that these are not Abelian). Rotations about the two-sided principle axis fall into $p+1$ classes (where $n = 2p+1$), E, $C_{2p+1,k}$, $C_{2p+1,-k}$, $k = 1, 2\ldots, p$. This gives a total of

$$p + 2 = \frac{n+3}{2} \qquad \text{(I.7.b-1)}$$

classes. For $n(=2p)$ even, the reflections form two classes of p elements each. The rotations belong to classes E, C_2, $C_{2p,k}$, $C_{2p,-k}$, $k = 1, 2\ldots, (p-1)$. The total number of classes is

$$p + 3 = \frac{n+6}{2}. \qquad \text{(I.7.b-2)}$$

The proper dihedral groups are defined by rotations of $\frac{2\pi j}{n}$ (around z), and a two-fold rotation perpendicular to the z axis. There must be n of these latter, taken into one another by the $\frac{2\pi j}{n}$ rotations. Generators of D_n are a rotation C_n, around the principle axis, z, plus a π rotation, C_2' around an axis in the xy plane. The C_n rotations form a class, as do each set of $C_{n/p}$ rotations about the principle axis for each integral n/p. Also for n even, each two-fold rotation about in an axis in the xy plane plus its inverse forms a class. These have $\frac{n}{2}$ operations (so n must be even). Dihedral group D_n, n odd, has n $\frac{2\pi j}{n}$ rotations C_j, $j = 1, \ldots, n$, around the principle axis, belonging to a class plus the inverses to these forming another class, and one set of n rotations, C_2' around axes perpendicular to z. This set forms one class (pb. I.6.b.i-2, p. 35).

Next we turn to the improper dihedral ones. Finding the classes for the direct product groups is simple. For a direct product with C_2, the number of elements and the number of classes is doubled. Each class goes into two, one containing the proper transformations, the second these multiplied by $\mathcal{I}$. This gives the classes of groups D_{2h}, D_{3h}, D_{4h}, D_{6h}.

Finally for the dihedral groups there is left D_{2d} and D_{3d} (isomorphic to D_4 and D_6). The two-fold axes are equivalent, since each axis of a neighboring pair is transformed into the other by a reflection in the mirror midway between them. Likewise, using two-fold rotations, all reflection planes are seen to be equivalent. Also $S_{2n}{}^{2k+1}$ and $S_{2n}{}^{-(2k+1)}$ are conjugates for

$$\sigma_d S_{2n}^{2k+1} \sigma_d{}^{-1} = \sigma_d \sigma_h C_{2n}{}^{2k+1} \sigma_d$$

$$= \sigma_h \sigma_d C_{2n}{}^{2k+1} \sigma_d = \sigma_h C_{2n}{}^{-(2k+1)} = S_{2n}{}^{-(2k+1)}. \qquad (I.7.b\text{-}3)$$

Thus for even $n(= 2p)$, $D_{2p,d}$ has

$$n + 3 = 2p + 3 \qquad (I.7.b\text{-}4)$$

classes: E, the $C_{2p}{}^{p} = C_2$ rotation around the principle axis; $p - 1$ classes of pairs $C_{2p}{}^{k}$, $C_{2p}{}^{-k}$, $k = 1, 2, \ldots, p - 1$; the class of $2p$ two-fold rotations around horizontal axes; the class of $2p$ reflections σ_d; and p classes of pairs of rotation-reflections $S_{2n}{}^{2k+1}$ and $S_{2n}{}^{-(2k+1)}$, $k = 0, 1, \ldots, p - 1$.

If n is odd, $n = 2p + 1$, D_{nd} contains inversion $\mathcal{I}$, and it can be written as the direct product

$$D_{2p+1,d} = D_{2p+1} \otimes C_i. \qquad (I.7.b\text{-}5)$$

It has $2p + 4$ classes, double the number of D_{2p+1}, as discussed above.

The remaining crystallographic groups are the cubic ones, the tetrahedral and octahedral groups. Now

$$T_h = T \otimes C_i, \qquad (I.7.b\text{-}6)$$

so again the number of classes is twice that of T. Likewise

$$O_h = O \otimes C_i. \qquad (I.7.b\text{-}7)$$

There is left only one group, with horizontal reflections. Its classes are

$$T_d : E, 8C_3, 6\sigma_d, 6S_d, 3C_3, \qquad (I.7.b\text{-}8)$$

or, in different notation,

$$T_d : E; 4C_3, 4C_3{}^2; 6\sigma_d; 3S_4, 3S_4{}^2; 3C_3 = 3S_3{}^2. \qquad (I.7.b\text{-}9)$$

As in group D_{2d}, the two-fold axes become four-fold rotation-reflection axes. The symmetry planes pass through the three-fold axes, so these are two-sided. All reflection planes are equivalent, and all S_4 axes are equivalent. This gives the five classes for the 24 elements of T_d.

Problem I.7.b-1: A class of a group is a set of elements that go into themselves under any similarity transformation by a group element. For each object, for example in appendix C, p. 637, apply transformation $t_i^{-1}t_jt_i$, for all i, j, and check that the classes given are correct. Of course, it is not necessary to use all group elements to verify this. Why?

I.8 DOUBLE GROUPS

The point groups consist of rotations and reflections. They act not merely on the crystal lattice, a set of points, but also on the objects at these points, atoms, molecules, or variations of these. Such objects have structure, thus their behavior under the transformations affects the symmetry (and properties) of the crystal.

Orbital angular momentum undergoes the same rotational transformations as the crystal itself. However, in addition, the particles can have half (odd) integral spin. Transformations on these are induced by rotations, but do not belong to the rotation group, but to its covering group (sec. V.5.b.iii, p. 277), SU(2) [Mirman (1995a), sec. X.2, p. 270; (1995b), sec. 7.1.2, p. 124]. So we have to also study point groups consisting of reflections and the inversion, plus a set of discrete SU(2) transformations — but only the finite subgroups that can transform crystals. These are closely related to the point groups already considered, and are called, for obvious reasons, double groups [Bradley and Cracknell (1972), p. 418; Chen (1989), p. 416; Cornwell (1969), p. 191; Cornwell (1984), p. 142, 342; Elliott and Dawber (1987), p. 199, 269; Falicov (1966), p. 103; Hamermesh (1962), p. 357; Heine (1993), p. 137; Inui, Tanabe and Onodera (1990), p. 176; Jansen and Boon (1967), p. 261; Janssen (1973), p. 155; Jones (1975), p. 256; Joshi (1982), p. 254; Joshua (1991), p. 67, 184; Koster (1957), p. 238; Lax (1974), p. 62; Lomont (1961), p. 161; Ludwig and Falter (1988), p. 29; Streitwolf (1971), p. 164; Tinkham (1964), p. 75; Tsukerblatt (1994), p. 246].

The double point groups are related to the ordinary point groups in the same way as the SU(2) group is related to SO(3): they are homomorphic to it, and they have twice the number of representations.

Problem I.8-1: Prove these statements [Cornwell (1984), p. 142].

I.8.a Definition of double groups

For a spin-$\frac{1}{2}$ object (more generally, one of half-odd-integral spin, say a nucleus having an odd number of nucleons), the product of the rotation representation-matrices for angles θ_1, θ_2, with

$$\theta_1 + \theta_2 = 2\pi, \tag{I.8.a-1}$$

is not the unit matrix, as for integral spin, but the negative of the unit. Thus, in general, we have for representation matrices

$$R_1 R_2 = \omega R_3, \tag{I.8.a-2}$$

where

$$\omega = 1, \text{ integral spin}; \quad \omega = \pm 1, \text{ odd half-integer spin.} \tag{I.8.a-3}$$

These matrices form not an ordinary (vector) representation of the rotation group, but a projective one (sec. V.4, p. 259). For a finite group, and so for this, a projective representation is an ordinary representation of a larger group, which here we would expect to have twice the number of elements (since ω can take only two values). This is the double group.

The double group is related to the point group (the single group) by a two-to-one homomorphism, following from that of SU(2) and SO(3). Its representation basis states are two-component spinors, or functions obtained by the decomposition of the product of these with the basis states of the point group. However it is convenient to define the double group so as to avoid the explicit relationship between SU(2) and SO(3), and it is, in fact, a rather fictitious group (although a group nevertheless). For half-integer spin a rotation of 2π is equivalent not to the identity E, but to its negative $-E$. Therefore we double the number of operators, considering $-E$ as a group element F different from E, rather than its negative, and similarly for all other elements of the group (so element R becomes two, R and FR). The result then is a group of twice the number of elements. We label a double group by the use of a star attached to the group symbol; thus $D_4(422)$ becomes $D_4{}^*(422^*)$.

I.8.b Construction of double groups

The construction of double groups illustrates their meaning and properties [Hamermesh (1962), p. 357].

The simplest group is $C_2(2)$, with two elements, the identity and the rotation through π, with matrices

$$E = \begin{pmatrix} 1 & 0 \\ 0 & 1 \end{pmatrix}, \quad C_2 = \begin{pmatrix} -1 & 0 \\ 0 & -1 \end{pmatrix}, \tag{I.8.b-1}$$

this obeying

$$C_2{}^2 = E. \tag{I.8.b-2}$$

There are other ways of realizing this group, for example

$$E = \begin{pmatrix} 1 & 0 \\ 0 & 1 \end{pmatrix}, \; C_2 = \begin{pmatrix} 1 & 0 \\ 0 & -1 \end{pmatrix}. \tag{I.8.b-3}$$

The first realization can be interpreted as a rotation of π around z, changing the sign of the x and y axes — we can also write these as 3×3 matrices with the elements of the last row and column, labeled by z, all 0 except that the $(3, 3)$ element is 1. The second realization corresponds to a reflection of the y axis (in the xz plane), changing the coordinate system from right to left-handed. Here it is the first interpretation, and realization, that is relevant.

For double group $C_2{}^*(2^*)$, a rotation through 2π can change the sign of the statefunction so now there are two matrices for E, and correspondingly for C_2,

$$E = \begin{pmatrix} 1 & 0 \\ 0 & 1 \end{pmatrix}, \; C_2 = \begin{pmatrix} i & 0 \\ 0 & -i \end{pmatrix}, \tag{I.8.b-4}$$

and

$$F = \begin{pmatrix} -1 & 0 \\ 0 & -1 \end{pmatrix}, \; C_2' = \begin{pmatrix} -i & 0 \\ 0 & i \end{pmatrix}. \tag{I.8.b-5}$$

If we choose the first set we get

$$C_2 E = C_2, \tag{I.8.b-6}$$

while

$$C_2 C_2 = -E. \tag{I.8.b-7}$$

For the second,

$$C_2' F = -C_2', \tag{I.8.b-8}$$

and

$$C_2 C_2' = E. \tag{I.8.b-9}$$

Thus these operators, which are needed to account for the sign change under 2π rotations, do not satisfy the multiplication rules for C_2. The group describing a system that has C_2 symmetry, but with spin-$\frac{1}{2}$ particles, is not C_2, and we need to describe the correct group.

Problem I.8.b-1: Why is the π-rotation interpretation the one that is relevant?

Problem I.8.b-2: Show that the transformations

$$E=\begin{pmatrix}1&0\\0&1\end{pmatrix},\ C_2=\begin{pmatrix}i&0\\0&-i\end{pmatrix},\ R=\begin{pmatrix}-1&0\\0&-1\end{pmatrix},\ RC_2=\begin{pmatrix}-i&0\\0&i\end{pmatrix},\tag{I.8.b-10}$$

form a group and construct its multiplication table. Which group is it [Mirman (1995a), chap. II, p. 29]? Verify that it is Abelian. There are representations in which distinct operators have the same representation matrix, for example the scalar. It turns out, fortunately, that there are representations of this double group in which R has the same characters, so matrices, as E (it is helpful that the character of C_2 is zero). These representations are the same as those of C_2 — they are the single-valued representations of C_2. Those for which the character of R is the negative of that of E are the double-valued representations of C_2, but of course the single-valued representations of double group $C_2{}^*$.

Problem I.8.b-3: A more interesting (and instructive) example is double group $D_2{}^*(222^*)$ of the (Abelian) group $D_2(222)$, with transformations

$$E=\begin{pmatrix}1&0\\0&1\end{pmatrix},C_z=\begin{pmatrix}i&0\\0&-i\end{pmatrix},C_y=\begin{pmatrix}0&1\\-1&0\end{pmatrix},C_x=\begin{pmatrix}0&i\\i&0\end{pmatrix},$$

$$R=\begin{pmatrix}-1&0\\0&-1\end{pmatrix},C_zR=\begin{pmatrix}-i&0\\0&i\end{pmatrix},C_yR=\begin{pmatrix}0&-1\\1&0\end{pmatrix},C_xR=\begin{pmatrix}0&-i\\-i&0\end{pmatrix}.\tag{I.8.b-11}$$

Check that

$$C_xC_y=C_zR=-C_z,\tag{I.8.b-12}$$

but

$$C_yC_x=C_z,\tag{I.8.b-13}$$

so $D_2{}^*(222^*)$, unlike its single group $D_2(222)$, is not Abelian.

I.8.c Double group classes

Next we have to find the classes of the double groups [Heine (1993), p. 139; Koster (1957), p. 239; Tsukerblatt (1994), p. 246]. The set of elements obtained from a class of the single group form either a single class, or two. The question is whether the two sets, obtained from a single class, form a single class of the double group, or two different ones? Since the double group contains more elements, it contains more classes. How many more?

The rotation matrices [Mirman (1995a), eq.X.5.a-3, p. 285], defining z as the axis of rotation, gives the character of rotation ϕ, in representation l [Hamermesh (1962), p. 336; Joshua (1991), p. 67] as a geometric series,

$$\eta^j(\phi) = \sum_{-j}^{j} exp(im\phi) = \sum_{0}^{j} cos(im\phi) = \frac{sin[(j+\frac{1}{2})\phi]}{sin[\frac{1}{2}\phi]}, \quad \text{(I.8.c-1)}$$

and for rotation $\phi + 2\pi$,

$$\eta^j(\phi + 2\pi) = \frac{sin[(j+\frac{1}{2})(\phi + 2\pi)]}{sin[\frac{1}{2}(\phi + 2\pi)]} = (-1)^{2j}\eta^j(\phi), \quad \text{(I.8.c-2)}$$

so for half-integer j,

$$\eta^j(0) = 2j + 1, \quad \eta^j(2\pi) = -(2j+1). \quad \text{(I.8.c-3)}$$

Since ϕ and $\phi + 2\pi$ do not have the same character in general, they cannot be in the same class; in particular E and $-E$ (denoted by F) are in different classes. However

$$\eta^j(\pi) = 0, \quad \text{(I.8.c-4)}$$

so rotations of π and 3π can be in the same class. We write the characters as

$$\eta(\boldsymbol{R}) = -\eta(\boldsymbol{R}F) = 2cos\frac{\theta}{2}cos\frac{(\omega + \phi)}{2}. \quad \text{(I.8.c-5)}$$

If the characters are different, the transformations belong to different classes. Thus $\boldsymbol{R}$ and $\boldsymbol{R}E$ are in different classes unless their characters are zero; for any class of the single group, with character η, there are two classes of the double group to which the members can be mapped, with characters η and $-\eta$, unless $\eta = 0$. This holds for $\theta = \pi$, allowing, but not requiring, the class of the single group to be mapped into but one of the double group. If there is another rotation in the group of angle π about an axis perpendicular to the first, but only in this case, does one class map into a single one of the double group.

Problem I.8.c-1: If they are in the same class then there is a group element $\boldsymbol{S}$, such that

$$\boldsymbol{R} = \boldsymbol{SRFS}^{-1}. \quad \text{(I.8.c-6)}$$

Picking the axis of rotation of $\boldsymbol{R}$ as the z axis, show that $\boldsymbol{S}$ is a rotation of π about an axis perpendicular to the z axis. Hence the elements are in different classes unless they are rotations through π, and there is another group element giving a rotation of π about an axis perpendicular to that of the rotation. Thus the π rotation $\boldsymbol{C}_2$, and the 3π rotation

FC_2, are in the same class if, but only if, there is another 2-fold axis perpendicular to the axis of C_2. So the number of classes of a double group is greater than, but generally not twice as great as, the number in the group [Jones (1975), p. 261].

I.8.d The single group is a factor group

The ordinary (or single) group G is not a subgroup, but a factor group of the double group G^* [Mirman (1995a), sec. IV.8.c, p. 138] with (E,F) forming the invariant subgroup. Every element of a class of G^* is mapped into the same class of G under this homomorphism. But as we see more than one class of G^* can be mapped into one of G. For example, E and F are in different classes, but are mapped into the class of E. In fact two classes, but no more, can be mapped into a single class of G, the exception is if the class of G consists of π rotations, and in addition there is in G a π rotation axis perpendicular to one of the class. Then the mapping is of a single class of G^* into G [Streitwolf (1971), p. 169].

To show this we denote that element of G^* mapped into element $\boldsymbol{R}$ of G by $M(R)$. Then for two elements $\boldsymbol{R}$, $\boldsymbol{S}$ in the same class

$$\boldsymbol{R} = \boldsymbol{T}\boldsymbol{S}\boldsymbol{T}^{-1}, \tag{I.8.d-1}$$

so

$$M(\boldsymbol{R}) = \pm M(\boldsymbol{T})M(\boldsymbol{S})M(\boldsymbol{T})^{-1}; \tag{I.8.d-2}$$

since there are but two signs there can be but two classes. This means that either $M(\boldsymbol{R})$ and $M(\boldsymbol{S})$ are in the same class, or $M(\boldsymbol{R})$ and $-M(\boldsymbol{S})$ are. Now all rotations of a class have the same angle, but different axes [Mirman (1995a), sec. IV.5.d, p. 125]. For SU(2) the only case for which

$$M(\boldsymbol{R}) = -M(\boldsymbol{T})M(\boldsymbol{S})M(\boldsymbol{T})^{-1} \tag{I.8.d-3}$$

is that with $\boldsymbol{T}$ a rotation of π. Thus only for this is there a possibility of mapping a single class of G^* into one of G. The SU(2) matrix relating $M(\boldsymbol{R})$ and $-M(\boldsymbol{R})$ corresponds to a rotation of π about an axis perpendicular to that of $\boldsymbol{R}$ — in this case, and in this alone, one class is mapped to one class.

So D_4(422) (sec. V.2.c, p. 249) has eight operations and five classes: $(E), (C_4, C_4{}^3), (C_2), (2C_2'), (2C_2'')$; its double group $D_4{}^*(422^*)$ has sixteen elements, and seven classes: (E), (F), $(C_4, FC_4{}^3)$, $(C_4{}^3, FC_4)$, $(2C_2, 2FC_2)$, $(2C_2', 2FC_2')$, $(2C_2'', 2FC_2'')$. The elements C_2 and FC_2 are in the same class for each two-fold rotation, while C_4 and FC_4 are not.

Another example is given by O(432) and O*(432*), the latter having $2 \times 24 = 48$ elements, and eight classes. Group O has representations of dimension 1,1,2,3,3, so

$$1^2 + 1^2 + 2^2 + 3^2 + 3^2 = 24, \tag{I.8.d-4}$$

thus O*(432*) has representations of dimension 1,1,2,3,3,2,2,4 so, correctly,

$$1^2 + 1^2 + 2^2 + 3^2 + 3^2 + 2^2 + 2^2 + 4^2 = 48. \tag{I.8.d-5}$$

The double group has the representation of the ordinary group, plus a set of extra ones.

Problem I.8.d-1: Relate the arguments of this section to those of the previous one.

Problem I.8.d-2: In point groups there are n-fold rotations, and often also two-fold rotations. For D_n a two-fold axis is perpendicular to an n-fold one. Then

$$C_2 C_n{}^k C_2{}^{-1} = C_2 C_n{}^k C_2 = C_n{}^{n-k}. \tag{I.8.d-6}$$

Thus $C_n{}^{n-k}$ and $C_n{}^k$ are in the same class. Verify this. Check that this can also be shown by using a reflection plane through the n-fold axis (if there is one) and explain the relationship between these demonstrations. Does this agree with what we have already found?

Problem I.8.d-3: Now what about the double group? Here the inverse of C_2 is, not C_2, but C_2F; a rotation of 2π changes the sign of the basis vector, so the inverse of a rotation through π is 3π, given by C_2F. Thus check that this similarity transformation is

$$C_2 C_n{}^k C_2{}^{-1} = C_n{}^{n-k} F, \tag{I.8.d-7}$$

and it and $C_n{}^k$ are in the same class. This implies that $C_n{}^k$ and $C_n{}^{n-k}F$ are in one class, while $C_n{}^k F$ and $C_n{}^{n-k}$ are in another, which means the double group has twice as many classes as the single group.

Problem I.8.d-4: Verify this explicitly for D_2.

Problem I.8.d-5: Of course this is only an implication, and there is an exception. If n is even, then $C_n{}^{n/2}$, a rotation through π, is in a class by itself, and $C_n{}^{n/2}$, $C_n{}^{n/2}F$ form a single class in double group D_n*. Check this by direct computation. This occurs because $C_n{}^{n/2}$ is a rotation through π, while $C_n{}^{n/2}F$ is a rotation through π in the opposite direction, thus, even for a spinor, the inverse of the first rotation. This can only happen for rotations through π. So if there is a two-fold axis perpendicular to an n-fold one, with n even, the number of classes is less than doubled. Because the number of classes is not doubled, the number of representations is not doubled either. Check these and compare with the arguments above. Which is clearer and more general?

Problem I.8.d-6: Verify that since for two-valued representations

$$M(SF) = -M(S), \tag{I.8.d-8}$$

the characters have opposite signs, so if these two elements are in the same class, their characters are 0. Show that this occurs for any rotation through π about a two-fold axis. Check that it is the case for D_2.

Problem I.8.d-7: Since we know the finite subgroups of the rotation group we know all double point groups (how, since these are subgroups of SU(2)?), and all the combinations of axes and rotations around them, and reflections. Check that this is a complete list, and having given all classes for each one on the list, we now have all classes for all double point groups.

Problem I.8.d-8: The multiplication tables of the double groups can be found either from the results of the previous problem, or from those of the ordinary groups [Mirman (1995a), chap. II, p. 29]. Try it both ways for $C_2{}^*, C_{3v}{}^*, C_{4v}{}^*, O^*, O_h{}^*, T_d{}^*$ [Joshua (1991), p. 68, 184, 249].

I.9 SIMPLICITY AND SYMMETRY IN A COMPLEX ENVIRONMENT

While solids are complex, atoms often see symmetry, they can find at least part of the rotational and translational symmetry of space. This was one of the main points of our discussion.

Problem I.9-1: Explain the reason(s) that symmetry appears even in a complex environment. Why are the properties of crystals determined, in part, by the symmetry of space? Under what conditions does at least a remnant of symmetry remain? Although at one level the answers are simple, there are deeper problems. Why do objects, here atoms or molecules, or collections, arrange themselves so as to be symmetrical, to allow symmetry to be retained, at least in part? What properties of these objects, not merely allow, but require it? Are there objects for which this is impossible? Why? Explain also the properties of these objects that allow (require?) them to combine in symmetrical arrangements, but which restricts these arrangements. Can the objects be generalized so as to permit further types of symmetry? Why? And how does all this depend on geometry? Would it matter if space were spherical, say, or of other such shapes? How do these answers depend on the dimension of space? Are the answers to these questions physical or geometrical? Why? Which? Why do physics and geometry lead to these answers?

Problem I.9-2: It is interesting that these many sets of relations, like those involving eqs. I.8.d-4, I.8.d-5, p. 65, which require the simultaneous satisfaction of several equations, and with the solutions integers, are actually satisfied, and in many different situations, in fact all [Mirman (1995a), sec. IX.2.c, p. 253]. Otherwise mathematics would be inconsistent, and the universe impossible. That may be the reason they are satisfied. But it seems far more likely that they could not, certainly not all sets, be satisfied simultaneously, and with the proper solutions. Fortunately it turns out that they are. Why?

Chapter II

Crystal Structures and Bravais Lattices

II.1 CRYSTALS AND LATTICES

A crystal is a lattice, with points occupied by physical objects, atoms, ions, or sets of these, the basis (sec. I.3.a, p. 6). As a lattice it has groups under which it is invariant — this limits the types of lattices, so (by definition) crystals. Knowing the point groups, we can now classify lattices — these are given by the point groups (the accepted classification is related to the point groups) — and find the symmetry groups for each. How then are lattices related to their groups, and why are they so related? What other conditions are there, what is the rationale for their determination and classification of lattices and crystals? While *crystal* is a physical concept, *lattice* is a mathematical one, independent of this particular use [Borchardt-Ott (1993), p. 20; Brown, *et al*, (1978), p. 8; Fässler and Stiefel (1992), p. 209; Johnston and Richman (1997), p. 153; Megaw (1973), p. 58], although it is often easier to picture the basis as sets of atoms or balls, than as sets of points [Rosen (1977), p. 49]. Lattices have symmetry, but crystals which are built on them may have less — but richer — symmetries. How, and why, do symmetries of crystals differ from those of lattices, how are they restricted by and related to them and what is the significance of these? And what are the limitations on symmetry, what combinations of symmetries are possible and distinct [Ladd (1989), p. 75]? Geometry limits physics, and is itself limited [Mirman (1995b,c)], but at present it is not clear that geometry fully explains what the universe is.

Here we start to consider such questions, illustrated by the determination, in different ways, of the points, lines, planes and operations going with the set of symmetries of an object. Then to provide examples, we do this explicitly for the two lattices of highest symmetry, the cubic, and the hexagonal (and it is useful to note that there are lattices of highest symmetry, two distinct ones, as well as of lowest, by convention only one, for reasons we should understand).

Problem II.1-1: A lattice can be defined as a regular arrangement of points, of infinite extent, with each in the same vector environment as all other points [Ladd (1989), p. 109]. Is the definition redundant? Is the word "vector" needed? Why? What does "environment" mean? Is it necessary that there be a finite distance between points?

Problem II.1-2: Are lattices determined by their symmetries, or defined by them?

II.2 LATTICES AND CRYSTAL SYSTEMS

It is part of the folklore of crystallography that there are (in three dimensions) fourteen Bravais lattices [Altmann (1994), p. 51; Bhagavantam and Venkatarayudu (1951), p. 119; Borchardt-Ott (1993), p. 63; Burns and Glazer (1990), p. 34; Chen (1989), p. 465; Cornwell (1984), p. 198; Gasson (1989), p. 254; Inui, Tanabe and Onodera (1990), p. 239; Koster (1957), p. 194; Ladd (1989), p. 115; Lines (1965), p. 189; Lomont (1961), p. 198; Sands (1993), p. 55; Sternberg (1994), p. 311]. Why are there more than one of these — and restrictions on them, thus fewer than infinity? And why fourteen? There are seven crystal systems — seven classes of lattices — specified by their unit cells. These, the primitive unit cells — containing a single lattice point each — give the primitive lattices. In addition there are lattices with more than one point per unit cell, seven (in three dimensions), giving altogether fourteen Bravais lattices. These have been found and shown to be all by Bravais well over a century ago [Bravais (1969)] — before it was clear that there are atoms, that crystals consist of atoms, a view which underlies our present understanding of them. Thus this tells us something fundamental about geometry, and perhaps about the way we wish to understand it. Here we list and describe lattices, indicating why their number is limited, why they are these, and these are all, and why we divide lattices into systems, and then allow more than one in a system. There are subtilties here, which emphasize that part, but only part, of this (and any) classification is a matter of definition (sec. IV.4, p. 212).

First we have to define a lattice, and relate it to its symmetry group, and it is the symmetry of the lattice that defines it — in part; the choice of what lattices are, and how they are specified, is shared by us and by nature — by the geometry nature does, or must, choose [Mirman (1995b)]. The symmetry can be stated by the transformations that leave the lattice invariant, or by the relationships between sides and angles. Mathematically, these are equivalent. Experimentally it is easier to find (believe in?) invariance than equalities of lines or angles, which can be known only within experimental precision. Thus definitions based on invariance have advantages. Here the discussion is of mathematics, so we do both — this is useful pedagogically. But someone considering experiments should be aware of this distinction.

II.2.a Definition of a lattice

A lattice is defined by the repetition of lattice vectors, not only a specific lattice, but the concept: a lattice is a set of (an infinite number of) vectors (or the points at their ends) which are, in n dimensions, integer sums of n (linearly-independent) vectors — these then are the lattice vectors (sec. I.3, p. 5). (In two dimensions a lattice is often called a net; however we prefer one term for every dimension, so use lattice in all cases.)

Problem II.2.a-1: It is a property of a lattice that all lattices points are identical. Show that it follows from this definition that they are. The only way of distinguishing two points is by drawing vectors from them to all closest points. If this is done for any two lattice points, the sets of vectors are the same (in what sense?).

Problem II.2.a-2: In the definition, integer implies all integers. Why? Restate the definition making this explicit.

Problem II.2.a-3: A lattice is not merely limited by its symmetry group, it is a group [Janssen (1973), p. 72; Lomont (1961), p. 201; Yale (1988), p. 99, 192]. Check that, with the law of combination vector-addition, a lattice obeys all group axioms. What type of group is it (pb. II.7.f-17, p. 125)?

II.2.b Unit cells are parallelepipeds

A lattice is (given by) the repetition, in all directions, of a unit cell (sec. I.3.b, p. 7), and there are only a few cells that can be repeated to give a lattice — cells that fill space (without distortion, without overlap, and without holes; we do not consider whether removing such requirements leads to anything interesting). The Bravais lattices are determined by their unit cells; these are restricted, thus so are the lattices, for reasons which we (must?) impose.

All unit cells, used here, are parallelograms, in three dimensions, parallelepipeds (three-dimensional figures with opposite sides parallel).

These need not be taken as the unit cells, but they define, and exhaust, the lattices; once we have determined, and differentiated, the lattices we can choose unit cells in other ways, which need not give parallelepipeds (sec. IV.2.c, p. 189). Since this term is used in different ways, care is needed. Lattices are sets of points; distorting the unit cell does not move the points, nor allow different arrangements — but may change the (apparent) symmetry.

Though we might require the (identical) unit cells to have (an average of) one point per cell, this is too restrictive. There are lattices with more than one; these are different from lattices with a single point. Beyond seeing why, and why there are restrictions on the number of points, we have to give the rules for picking the lattices; these limit and determine them, so limit crystals.

Problem II.2.b-1: In what way does the definition of a lattice require that there be unit cells that are parallelepipeds? What does "in all directions" mean here?

Problem II.2.b-2: Why are the only unit cells of lattices those that can fill space? Do all such cells give lattices? Why do all space-filling parallelepipeds give lattices? It is implied that the orientations of all unit cells are the same. Is this a hidden assumption, or does it follow from ... ?

II.2.c What is a Bravais lattice?

A Bravais lattice is defined as a lattice that partitions the space into identical parallelogram-shaped regions (in three-dimensions parallelepipeds), with these regions the smallest such ones (called Bravais unit cells) that do so. The lattice is not a sublattice of any lattice that performs such a partition (here a sublattice has unit cells larger than those of the lattice — so the sublattice has fewer unit cells per volume than the lattice). There is an (extremely) subtle difference between a lattice and a Bravais lattice. A lattice is defined as an infinite set of points invariant under a set of (integral) translations between them. The Bravais lattices are the set of all lattices in a given space, here (two or) three-dimensional Euclidean space. We ignore the question whether "Euclidean" is necessary.

II.2.d The holohedry groups and their subgroups

The lattice vectors of a lattice give three lengths and three angles. It is convenient (or at least standard) to organize lattices according to how many lengths, and how many angles, are equal. This gives the set of crystal systems. Each is also given by its (largest) symmetry group; the reasons for this and how these classifications turn out to be identical are studied below. The symmetry group is that of the lattice; the group of the unit cell may be different. However that must be such as to give a lattice with the required symmetry. It is important in checking the symmetry, and the symmetry elements, to be aware of this distinction.

There are in three dimensions thirty-two point groups but only seven crystal systems. Thus each system has several symmetry groups. The largest symmetry group is called the holohedry group of the system (sec. I.6.d, p. 42). The stem "-hedral" apparently comes from the Greek word for seat (as in cathedral — the seat of a bishop), and since chairs were probably quite simple then it has come to imply surface, say having a specified number of surfaces (as in dihedral). So holohedral implies having a whole number of surfaces, the whole number needed for the most complete symmetry.

Problem II.2.d-1: Show that the thirty-two point groups form seven sets with the members of each subgroups of some largest one [Lax (1974), p. 176; Yale (1988), p. 112] — which is what, and describing which system(s?)? This illustrates that all the thirty-two point groups are crystallographic groups, in that each describes the symmetry of a (crystal) lattice. We discuss the lattices given by each of the holohedry groups, thus showing that there are lattices for each. It would also be interesting to take each of these systems and reduce its symmetry to that given by each subgroup, showing that the thirty-two point groups do describe thirty-two different crystals (why this word and not "lattices"?). Also seeing how this is done might raise the question of whether the word *holohedry* (or the above explanation of it) is completely appropriate.

Problem II.2.d-2: A crystal might have less symmetry than that given by the holohedry group of its lattice, for example if instead of all atoms being identical, there were two types, or not all lattice sites were occupied, or, as we shall see, if for each lattice point there are several atoms of the crystal. Then its symmetry would be that of a point group which is a subgroup of its holohedry group. If there are only half the number of faces interchanged by the group (half the number of group elements) the group (of the crystal system) is hemihedric, and for a fourth it is tetartohedric [Yale (1988), p. 112]. Why are these names used? Why are these the only possibilities? Using the tables in appendix B, p. 631, for each crystal system, classify the groups in this manner. What would happen if we took a crystal, say cubic, and placed the (different types of) atoms such that the symmetry were less than that given by the tetartohedric group? Try drawing, or constructing, this. Do all systems have a hemihedric or a tetartohedric subgroup? Why? When we discuss breaking of symmetry by putting at lattice points objects of lower symmetry we have to distinguish between a mathematical and a physical object. Thus we can consider an object with one symmetry placed on a lattice with a different one. But physically this changes the potential energy of the objects, thus distorting the lattice [Burns and Glazer (1990), p. 83]. The (physical) lattice on which we placed the objects is not the same as the one holding them, so care is needed in specifying what we mean.

Problem II.2.d-3: For a cube take the spins of the atoms (inserted at the vertices) to alternate, up and down (chap. VIII, p. 393), to the extent possible. What is the symmetry group of the crystal? How is it related to that of the cube with all atoms spinless? What would happen if all atoms had spins, but in the same direction? What is the symmetry if only half the lattice points are occupied, if the half are chosen in a regular manner; in a nonregular manner? Suppose only a quarter were occupied, would there be a difference? Could there be fewer sites occupied and still have a cube (with symmetry)? Check that all symmetry groups of a cube are in tbl. B-1, p. 633 and tbl. B-2, p. 634 under the "cubic" heading. Do any give another crystal system? Find arrangements of atoms (and of types of atoms, or ones with spins) placed at the vertices of a cube that give a crystal invariant (only) under each of the groups listed under "cubic".

II.3 LATTICES IN TWO DIMENSIONS

To classify lattices we start with two dimensions [Bhagavantam and Venkatarayudu (1951), p. 19; Cotton (1990), p. 350; Field and Golubitsky (1992), p. 207; Joshua (1991), p. 41; Ladd (1989), p. 111; Lovett (1990), p. 5; Megaw (1973), p. 139], figures in a plane. Vectors $\underline{a}$ and $\underline{b}$ (with lengths $|\underline{a}|, |\underline{b}|$) from any vertex (taken at a lattice point) to its nearest neighbors in the same cell, either those on neighboring vertices or in the center, are chosen so [Armstrong (1988), p. 149]

$$|\underline{a} - \underline{b}| \leq |\underline{a} + \underline{b}|. \quad \text{(II.3-1)}$$

These (and in three dimensions $\underline{c}$) are taken throughout to be the primitive lattice vectors (the ones of shortest length).

II.3.a The two-dimensional lattices

The lattices are then

a) Oblique:

$$|\underline{a}| < |\underline{b}| < |\underline{a} - \underline{b}| < |\underline{a} + \underline{b}|, \quad \text{(II.3.a-1)}$$

b) Rectangular:

$$|\underline{a}| < |\underline{b}| < |\underline{a} - \underline{b}| = |\underline{a} + \underline{b}|, \qquad (II.3.a-2)$$

c) Centered Rectangular:

$$|\underline{a}| < |\underline{b}| = |\underline{a} - \underline{b}| < |\underline{a} + \underline{b}|, \qquad (II.3.a-3)$$

d) Square:

$$|\underline{a}| = |\underline{b}|| < |\underline{a} - \underline{b}| = |\underline{a} + \underline{b}|, \qquad (II.3.a-4)$$

e) Hexagonal:

$$|\underline{a}| = |\underline{b}| = |\underline{a} - \underline{b}| < ||\underline{a} + \underline{b}|. \qquad (II.3.a-5)$$

Problem II.3.a-1: Check that

$$|\underline{a}| = |\underline{b}| < |\underline{a} - \underline{b}| < |\underline{a} + \underline{b}| \qquad (II.3.a-6)$$

does not give a different lattice. The parallelogram is a rhombus (an equilateral parallelogram); its diagonals bisect each other at right angles giving again a (not the) centered rectangular structure, with the vectors now $\underline{a} - \underline{b}$ and $\underline{a} + \underline{b}$.

Problem II.3.a-2: Draw figures for all these six cases. Check that

$$|\underline{a} - \underline{b}| \leq |\underline{a} + \underline{b}| \qquad (II.3.a-7)$$

can always be imposed, that all figures are different, and that the list includes all lattices that have rotational invariance under a subgroup of the rotation group compatible with the crystallographic restriction (sec. I.5, p. 23) — the order of rotation is 1,2,3,4 or 6 — with the center of rotation at a lattice point. Would there be a change if any of these conditions were dropped? While the rectangle can be centered the square cannot be (and still give a new lattice). A centered square is still a square, though smaller and rotated through $\frac{\pi}{4}$. Check this. A point placed elsewhere in these figures requires additional points to maintain the symmetry giving the same lattice, though with an (irrelevant, by definition) change of scale. Try putting in points and verify this.

Problem II.3.a-3: From these it should be possible to find all point groups in two dimensions (pb. I.6.d-3, p. 43) [Hahn (1989), p. 11; Lovett (1990), p. 7; Lyndon (1989), p. 74].

Problem II.3.a-4: Show analytically that these have the stated shapes, and the requirements are necessary to get these shapes.

Problem II.3.a-5: Construct these lattices, using balls and sticks, or cut-out pieces of paper, or by drawing them (perhaps with a computer), and check that placed together the unit cells fill — without gaps —

space, here of two dimensions — a necessary condition for a lattice. Why? Are there other (regular) figures that do? Draw some and see if any can; explain what goes wrong with other ones.

Problem II.3.a-6: Take the lattice with the (unit cell of) least symmetry, the oblique parallelogram having symmetry only under rotations of π and 2π, and (thus?) inversion [Mirman (1995c), sec. 4.2.6, p. 60]. Show that a figure with less symmetry — it is not a parallelogram — does not give a lattice. Why? What is the difference between this symmetry and that of the rectangular lattice?

Problem II.3.a-7: Each figure is a special case of the oblique lattice, and some are special cases of other special cases, obtained by setting a "<" to "=", giving a lattice with greater symmetry. Does it always? Which lattices go into which under this operation? Give the symmetry groups of these lattices, and verify that each more general lattice has a symmetry group that is a subgroup of the symmetry group of a special case. Start with the largest groups possible (how many?), find the lattices, then the subgroups and their lattices, and check that these are the correct generalizations.

Problem II.3.a-8: To obtain a crystal from a lattice, we put on each of its points a material object, a set of atoms, say. Is it possible to put the objects elsewhere than on (the) lattice points? What does "put on a point" mean for a set? For the unit cells of each lattice put atoms on (at?, near?) all four corners, such that adjacent atoms are different (the only case of interest; why?). Find the symmetry groups of the resultant crystals. How do these differ from those of the lattices? Are they related to those of other lattices? Why? If instead of a physical object, a set of points were assigned to each lattice point, what would the object be?

Problem II.3.a-9: Show that a straight line that goes through two lattice points, goes through infinitely many [Armstrong (1988), p. 153]. Would you expect this? Explain. Would this be true if space were not Euclidean?

Problem II.3.a-10: What lattices are given by the vectors [Armstrong (1988), p. 153]

$$\underline{a} : (1, \sqrt{3}), \quad \underline{b} : (-1, \sqrt{3}), \qquad \text{(II.3.a-8)}$$

$$\underline{a} : (1, 0), \quad \underline{b} : (2, -4), \qquad \text{(II.3.a-9)}$$

$$\underline{a} : (-2, 0), \quad \underline{b} : (-1, 3). \qquad \text{(II.3.a-10)}$$

Draw them.

Problem II.3.a-11: What are the one-dimensional lattices (sec. III.8.a, p. 173) [Ladd (1989), p. 111]?

II.3.b Why we consider all these lattices, and all distinct

The oblique case includes all lattices, and we would get all crystals if we limited ourselves to just it — but would not get much information about them. So we consider lattices, of course still members of the oblique class, but with greater symmetry. To obtain as much information as possible about a lattice from its symmetry, we must classify them by the largest groups under which they are invariant. We also have to decide whether, and to what extent, we should distinguish lattices of the same symmetry, but which are in some ways distinguishable, and to see why we wish that classification, but nothing more (here).

The rotational symmetries have orders 1,2,3,4,6, for rotations of $\frac{2\pi}{6}$ and of $\frac{\pi}{4}$ — a figure cannot have both at once (pb. I.6.b.i-5, p. 36) — and their multiples; to study the lattices we need figures that have each. That with order 2 is a rectangle, with order 4 a square, so we regard it as a distinct class, while the other two orders give a hexagonal lattice, and it is also assigned to its own class. In what relevant ways do the oblique and rectangular lattices differ? While they both are invariant under a rotation of π, the rectangular is also unchanged by reflection in a mirror placed along either lattice vector, the oblique is not, thus they are (taken) different. At the top left is an oblique lattice and to its right the lattice reflected in a mirror along the middle horizontal lattice vector (labeled m), and these are different,

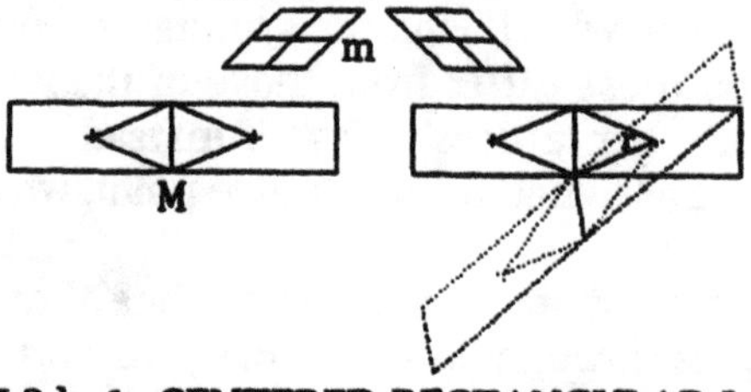

Figure II.3.b-1: CENTERED RECTANGULAR LATTICE.

The centered rectangular lattice on the bottom left has symmetry under reflection in a mirror placed on one, but not both, lattice vectors, differing from the rectangular lattice. It, here with only two cells drawn, each with a lattice point at the center, shows the reflection in the vertical mirror along the lattice vector in the middle (labeled M), and then this is overlaid on the right by the lattice (shown dotted) obtained by reflection in the mirror along the solid lattice vector to the center (labeled r). It is invariant under the first reflection, but not the second. The rectangle, square and hexagon then are all the figures, with one point per unit cell, that have all combinations of rotational and reflection symmetries — given an allowed symmetry there is on this list a (single) figure that displays it, and there are figures, here the square and the rectangle, that distinguish between symmetries (to the extent possible; if a figure is invariant under rotations, it is invariant under their products).

For the hexagonal lattice the (defining) unit cell (which is always taken as a parallelogram) shows, but does not have, the symmetry of the lattice — it is not invariant under (any) $\frac{2\pi}{6}$ rotation, but only under one of π. This lattice also has two rotational symmetries, of $\frac{2\pi}{6}$ and $\frac{2\pi}{3}$. In the next diagram are three unit cells, CabcC (with solid lines), CcdeC, CefaC, with dashed edges obtained from the solid one by rotation; these have four sides so are not hexagons, but fitted together they form a hexagon. Sixfold symmetry is shown by the six lines from C, three being dotted. The unit cells fill space, as can be seen, giving a hexagonal lattice with sixfold symmetry around point C,

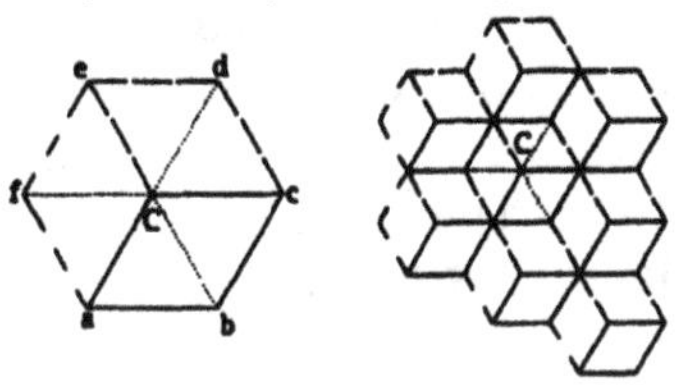

Figure II.3.b-2: SYMMETRY OF THE HEXAGON.

II.3.c Why are points added to unit cells?

These lattices have unit cells with only a single point, except the centered rectangle, which has two. Why add a point, and in what sense is this a different lattice? And why not add more?

Problem II.3.c-1: For the centered rectangle, the unit cell can be taken to have only one point; it then becomes an oblique parallelogram. Draw it and check. However the lattice has more symmetry than this shows, and that symmetry carries physical (and mathematical) information, and has physical consequences, which we want to know. Thus this oblique unit cell throws away important information — that is why we use the centered one, even though it has more than one point. The centered rectangle gives a rectangular lattice, not an oblique one; it has the same symmetry group as the former, not the latter. What information is thrown away going to an oblique cell? But it also differs from the rectangular lattice; check that no rectangular lattice can be constructed using this set of points. Show that the square lattice, tilted at $\frac{\pi}{4}$, gives a (all?) centered rectangular lattice(s) [Armstrong (1988), p. 154]; the latter can also be used to give the hexagonal lattice. In what sense does it carry information not available in these (thus, say, in what ways are crystals based on the lattices different)? A (not the) centered rectangle, tilted at $\frac{\pi}{4}$, is a rhombus, so the two-point cell can be replaced by a rhombic cell with a single point. How does it differ from the oblique lattice? Draw this and check. This lattice has the same group as the rectangle; however they are labeled as different lattices. Why? Notice in fig. II.3.b-1, p. 74, that although the centered rectangle can be replaced by a primitive unit cell (with one point), the rhombus, this does not show reflection symmetry in the mirrors along the horizontal lattice vectors, indicating why the nonprimitive cell (one with more than a single point) is preferred. All these are lattices: any point can be reached from any other by a sum of the lattice vectors, with *integral* coefficients. In particular this should be checked for the centered rectangle, to show that every vertex of a rectangle can be reached from every center by such a translation, and conversely.

Problem II.3.c-2: Show that the unit cell for the hexagonal lattice consists of two adjacent equilateral triangles. For rectangles, points can be put on the inside only at the center, else the symmetry would change. Prove that a hexagonal lattice can be centered (only) by dropping a perpendicular from each vertex to the opposite side, and putting points at $\frac{1}{3}$ and $\frac{2}{3}$ of its length, as we see,

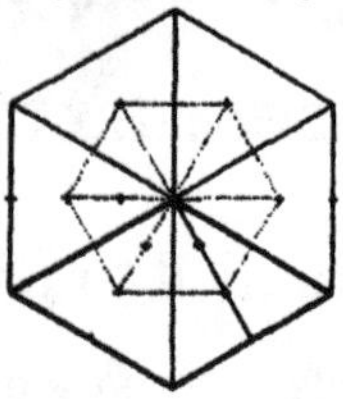

Figure II.3.c-1: CENTERED HEXAGON.

What is wrong with doing that?

II.3.d Where can points be put simultaneously in two dimensions?

Nonprimitive lattices are obtained by the addition of points to unit cells, but this is possible only for a few emplacements. That there are limitations on inserting points can be seen by considering two-dimensional lattices; this forms the foundation of the general analysis as these lattices restrict those in three dimensions explaining the allowed ones. Here then we study the reasons why points can, and cannot, be put at various loci.

Where can additional points be placed in the unit cell? What are the restrictions? Adding points to opposite sides of a two-dimensional rectangular lattice gives the top lattice (points labeled by boxes are the original ones, the others are those added). It can be seen that this is a lattice (indeed without the lines, and the different ways of drawing them, we would not be able to distinguish points), though it is still rectangular, so not new,

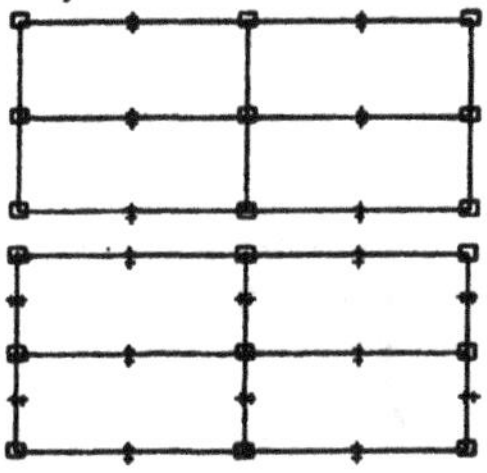

Figure II.3.d-1: WHERE POINTS CAN, AND CANNOT, BE PLACED.

Suppose we add points to both pairs of sides. The result is the bottom lattice (?). Here points with vertical lines through them are at unit distance from their nearest vertical neighbors, the other points are at a distance of 2 from their vertical neighbors. The two sets of points are distinct, this is not a lattice — a translation of one unit in the vertical direction does not leave it invariant — so points cannot be added to both pairs of sides simultaneously. This provides an example of a regular arrangement of points that is not a lattice, so cannot give a crystal. It is conceivable that atoms can be placed on these points, giving a physical object. Yet it is interesting that despite this possibility nature does (seem to) prefer lattices as the mathematical figures on which to build physical objects. Or is that just a illusion?

So we are restricted in placing additional points. For some the resultant figure is a lattice not different from one already listed, for others it is not a lattice. But it is possible to add points as the centered rectangle shows.

Problem II.3.d-1: That the object on the bottom in the figure is not a lattice means that if atoms are placed at the points the resultant object is not a crystal. Are there physical objects with atoms arranged like this, or reasons why there are not? Are crystals preferable to such objects? Or might such objects be redefinable to make them crystals?

Problem II.3.d-2: Discuss the physical reasons, to the extent that they can be determined, why nature prefers lattices as the mathematical figures on which to build physical objects. Or is it just that human beings, or physicists, emphasize such objects to the extent that others are overlooked, or forced into our (overly restrictive) categories?

Problem II.3.d-3: Why can we not (or should it be do we not?) add points to one pair of sides, and also the center, simultaneously? Can we add to both pairs and the center? Is it useful? Why?

Problem II.3.d-4: We can now summarize why there are five — only — lattices in two dimensions (which is important since lattices in higher dimensions must obey the two-dimensional conditions for all plane sections containing points — why?). Symmetry rotations (of what?) can only be about a lattice point, or the center of the unit cell; why? The unit cells must obey the crystallographic conditions (why, since these are supposed to apply to lattices, not unit cells?), and have opposite angles equal (why?). If all four are equal, the angles are $\frac{\pi}{2}$, if there are two pairs the angles are $\frac{\pi}{3}$ and $\frac{2\pi}{3}$. Thus angles are limited, and a symmetry rotation of $\frac{\pi}{2}$ gives equal sides. The crystallographic condition allows different lattices, but few. There is only one further operation, reflection. Mirrors along a lattice vector or two give only two different lattices (why?), for a total of five. What about inversion? There are limits on the number of operations, on the angles, and on the points and lines left unchanged. Another way of putting this is that points can only be added at special positions (to give new lattices?), the centers — of faces of the unit cell — and there are restrictions on simultaneous insertion of points. Discuss the relationship between these two ways of limiting lattices, symmetry and insertion of points. How does one set give the other? Explain why there can be only one distinct lattice with more than one point, the centered rectangular.

II.3.e The symmetry groups of two-dimensional lattices

Having the lattices we can now consider the operations leaving them invariant, and the groups given by these operations.

Problem II.3.e-1: The unit cell of the first is an oblique parallelogram, with no conditions on sides or on angles (except that they not give a higher symmetry). It has no symmetry except the identity (a rotation through 0°) and inversion — this is the same as a rotation of π around one of its points, which can easily be proven (and which has an interesting consequence of giving what is regarded as a fundamental law of nature, the TCP theorem [Mirman (1995c), sec. 4.2.6, p. 60]). Check that the unit cell does not have symmetry under rotations around any point except its center, though the lattice does. Why is there inversion symmetry? About which point? Why? Can there be lattices without such? Where are the symmetry axes for the lattice? What is the symmetry group?

Problem II.3.e-2: The next lattice has a unit cell invariant under rotations through $\frac{2\pi}{2}$ (around what points?), and also (necessarily?) under reflections in both lines that bisect two opposite sides (and so are perpendicular to them), and inversion through its center — why? The lattice is also invariant under reflections through the lattice vectors. Which of these operations are independent? This unit cell is a rectangle. Draw a rectangle and verify this. What is its point group; give a name used for finite groups [Mirman (1995a), chap. II, p. 29]? State all elements of the group. There are other names for lattice symmetry groups (as in appendix B, p. 631), related to the properties of the lattices, which we will consider. It is clear what the relevant ones are. Repeat this for the centered rectangle. How do the groups differ? Notice that the lattice (but not the unit cell) is invariant under reflections through lattice vectors (as expected; why?), thus through the line from a corner to the center of a cell.

Problem II.3.e-3: If the symmetry angle is $\frac{2\pi}{4}$ the figure is a square. Are the other symmetries (if any) the same? What is its point group? How is it related to that of a rectangle? What are its transformations?

Problem II.3.e-4: Finally (why?) the rotation can be through $\frac{2\pi}{6}$, giving a hexagon. A computer program to draw a hexagon and confirm this provides a useful foundation for programs that will be helpful later. What are the other symmetries? What is its point group? From this we can get equilateral triangles. Is there a difference in the lattices they give? Why? Can both tile the plane? What is the unit cell for the hexagonal lattice?

Problem II.3.e-5: Explain why the definition of a lattice and these limitations on its symmetries allow no other unit cells (perhaps better, every set of points meeting the definition of a lattice allows one of these unit cells). Why can we, although we do not, limit the cells to have only a single point? Could there be other two-dimensional objects, with some symmetry, that do not fit into this classification? If so would they have unit cells? How would they differ from a lattice? Are there limitations on such objects (if the ones given do not exhaust the possible limitations)?

II.3.f Finding the two-dimensional Bravais lattices analytically

We have found the Bravais lattices by geometrical arguments, but they can also be found analytically.

Problem II.3.f-1: Write the primitive lattice vectors as

$$\underline{V}_1 : (a,0),\ \underline{V}_2 : (b\cos\theta, b\sin\theta) \qquad \text{(II.3.f-1)}$$

and subject them to all point-group symmetries. If a group gives a lattice, the resultant vectors are sums of these two, with integral coefficients. Find all values of a/b, and all of θ, that satisfy, and find the lattices they give. Check that these agree with the ones found. How is the centered rectangle obtained?

II.3.g What makes two lattices distinct?

Lattices of the same form and same symmetry group can be considered different, while lattices that differ (such as in size or angles) may be regarded as the same (sec. IV.4, p. 212). What are the rules for distinguishing lattices, and what is their rationale? Lattices (here restricted to two or three dimensions) are taken as distinct if they have different symmetry groups, or if they cannot be taken into each other (without changing the symmetry) by a change of ratios of lengths of lattice vectors, or of angles between them. Thus all oblique lattices (the only two-dimensional case for which the angle can be changed) are the same, no matter what the angle (except $\frac{\pi}{2}$) or the ratio of the magnitudes of their lattice vectors (except when they give a hexagonal unit cell). These all have the same symmetry group, and all can be obtained from any one by variation of angles (so of ratios of sides). All squares are the same, sizes being irrelevant.

The rectangular and centered rectangular lattices are topologically different: they cannot be distorted into each other. The square and rectangle can be, but they have different symmetry groups. Also for the rectangle, the angle between the two primitive lattice vectors is $\frac{\pi}{2}$, for the centered rectangle it never is. And the sets of symmetry elements are not identical. For the rectangular lattice there is a mirror along every lattice vector, for the centered case the vector to the center is not a line of mirror symmetry (although it is for the square, indicating why it does not give a new lattice). Also the rectangular lattice is invariant under rotations about, and inversions through, lattice points and reflections in mirrors containing them, but the unit cell is not (it is invariant under operations through its center, but these are derived from the generators defining the group which leave lines through lattice points invariant). For the centered case the unit cell is also invariant under rotations about, and inversions through, the lattice point at the center and reflections through the mirror at the center, thus it has the same symmetry group as the lattice as a whole. So while the symmetry groups are the same there are different relationships between group elements, lattice vectors, and unit cells.

The distinction then is geometrical — the symmetry groups and the topology are properties of the geometry — but also conventional. We regard lattices as different, because it is useful mathematically and physically, if they differ in ways other than their groups, but regard them as the same even if they differ (physically the classification is useful because objects of the same lattice share enough important properties to make it worthwhile to keep them together, and differ in important ways from objects with different lattices; the, perhaps interesting, physical reasons for this are not considered here, but will be suggested by the studies of applications in later chapters). The only other way of getting different lattice vectors is to vary their relative magnitudes. We can always change the scale of one axis with respect to the other so do not regard these as different (which is not to say that in a particular application this need be irrelevant), as long as the symmetry is not changed. We might use a finer classification, thus for a rectangle if there was a meaningful difference between the x and y axes (as there might be if the figure were part of a three-dimensional one) we could regard lattices as different depending on whether the magnitude ratio of their vectors were greater or less than 1. While such may at times be useful, this is not part of the convention used for distinguishing lattices.

Problem II.3.g-1: There is another group-theoretical aspect. The points of a unit cell (properly their coordinate vectors) form a group representation basis; sets are mixed by the group operations. However the center of the cell is a special point, one invariant under the group, so its coordinate vector is a basis vector of the identity representation. One of the lattice vectors of the centered rectangle connects a point of a nonscalar representation, a vertex, with the center, which belongs to a scalar representation. This is not true for the rectangular lattice and in this way these two lattices are different. Is this distinction an accident here, or does it also hold for (all) three-dimensional lattices? It is worth checking, and discussing, as these lattices are studied (sec. VII.2.e, p. 342).

Problem II.3.g-2: While it should be clear that all two-dimensional lattices, distinguished by these rules, have been given, and that they are all different, it is worth seeing if others can be constructed, and what goes wrong. Check that there are no other cases for which the unit cell has a single point. Then put points in for each, first putting in a single point, then more than one. To get a new lattice the figure must give a lattice, it must have the same symmetry group as the lattice from which it is derived, or else it would be in a different system, and it must not give a lattice that is the same, except of a different size or orientation. It should then be clear that these are all the lattices in two dimensions — putting in more points does not give new ones. These are the Bravais lattices.

Problem II.3.g-3: Thus we have found the Bravais lattices by taking the two primitive vectors, and their sum and difference, and all the figures obtained by different combinations of equal signs, eliminating those that are only special cases of, or are topologically equivalent to, others. Verify that all lattices obtained by doing this are the ones that we have considered.

Problem II.3.g-4: There is an additional way of studying lattices using a point group symmetry, that of reflection. All these unit cells (must) have symmetry under inversion, but we wish cells that can be obtained only by using reflection. These are all but the oblique. Let the primitive translation vectors be, with $\hat{\underline{i}}$ a unit vector,

$$\underline{a} = a_x\hat{\underline{i}} + a_y\hat{\underline{j}}, \quad \underline{b} = b_x\hat{\underline{i}} + b_y\hat{\underline{j}}, \tag{II.3.g-1}$$

which under a reflection in the x axis become

$$\underline{a}' = a_x\hat{\underline{i}} - a_y\hat{\underline{j}}, \quad \underline{b}' = b_x\hat{\underline{i}} - b_y\hat{\underline{j}}, \tag{II.3.g-2}$$

which are also lattice vectors if the lattice is invariant under this mirror symmetry. One possibility is that

$$\underline{a}' = n_{aa}\underline{a} + n_{ab}\underline{b}, \quad \underline{b}' = n_{ba}\underline{a} + n_{bb}\underline{b}, \tag{II.3.g-3}$$

with the n's integers. Does this give new lattices if all n's are positive? Another possibility is

$$\underline{b}' = \underline{a} - \underline{b}. \tag{II.3.g-4}$$

This is a lattice vector (one taking a point to an equivalent point, so a sum of $\underline{a}$ and $\underline{b}$ with integer coefficients). Presumably only this form gives a lattice vector — others, with different coefficients for the vectors, are sums of these with integer coefficients. Then

$$b_x{}' = a_x - b_x = b_x, \quad b_y{}' = a_y - b_y = -b_y, \tag{II.3.g-5}$$

giving

$$a_y = 0, \quad b_x = \frac{1}{2}a_x, \tag{II.3.g-6}$$

so a possible set of translation vectors,

$$\underline{a}' = a\hat{\underline{i}}, \quad \underline{b}' = \frac{1}{2}a\hat{\underline{i}} + b_y\hat{\underline{j}}. \tag{II.3.g-7}$$

That is for this to give a lattice invariant under reflection in the x axis, the lattice vectors must be so related. The lattice for which these are the primitive basis vectors is the centered rectangle (the oblique lattice for which we can also choose these vectors does not have reflection symmetry). Draw these vectors and verify that they do give this figure, and that it has this symmetry (but does not have symmetry under a reflection in the other lattice vector). This is the reason for choosing a lattice with more than one point in a primitive cell: the unit cell (not merely the lattice) has reflection symmetry in a mirror through a lattice point, which no rectangular lattice with a single point in a unit cell does. Show that a reflection in the plane through the y axis gives no other lattices. Can there be reflections in other lines? Also show that imposing reflection symmetry on the square does not give a new lattice. Why would we think of

$$\underline{b}' = \underline{a} - \underline{b}? \qquad \text{(II.3.g-8)}$$

A lattice is given by the set of points generated by translations from any one using the primitive lattice vectors any number of times. Thus its lattice vectors are sums of the primitive ones, with positive integer coefficients. This suggests using negative coefficients, which is possible since the difference of two vectors can be smaller than either so that $\underline{a} - \underline{b}$ cannot be obtained by repeating the translations given by the two primitive vectors. But why not take this smaller vector as the primitive vector (this cannot be repeated indefinitely — there is a smallest vector by atomicity, or mathematically by definition of a lattice)? The reason that there is an additional two-dimensional Bravais lattice, and one with more than one point, is that this unit cell, only, has reflection symmetry. Check that inversion (a product of a rotation and reflection) does not give a new lattice. Why not consider

$$\underline{b}' = 2\underline{a} - \underline{b}, \qquad \text{(II.3.g-9)}$$

say? Other coefficients, besides 1, should give nothing (new). This is an analytic proof that these are all the lattices. Requiring reflection symmetry results in a rectangular lattice, of which there are two. So this is the only case in which there is a smallest unit cell with more than one point.

Problem II.3.g-5: Repeat these arguments considering rotations instead of reflections. Verify that the only lattices possible are oblique, rectangular (and square) and hexagonal.

Problem II.3.g-6: Can these arguments be generalized to three (and higher) dimensions? Would a computer program help?

Problem II.3.g-7: We do not consider whether it is possible to have lattices in non-Euclidean spaces. In particular, our space has signature 3+1 (eq. IV.3.h-1, p. 211), so it is (locally) flat but not Euclidean. This raises the questions whether there can be (analogs of) lattices in such a space with nondefinite signature, and if so, what are they like?

Problem II.3.g-8: Several proofs, or explanations, have been given for the Bravais lattices. How are they related? Which are the most rigorous? Why? Which can be (most easily) generalized to higher dimensions, like three dimensions? Which show (most clearly) what restrictions can be removed to generalize lattices to other reasonable figures? Are there any such?

II.4 THE SEVEN THREE-DIMENSIONAL CRYSTAL SYSTEMS

There are in three dimensions only seven primitive lattices (giving seven crystal systems), and another seven nonprimitive ones, fourteen altogether, which we describe and (hopefully) justify. A crystal consisting of (sets of) atoms or molecules, perhaps with internal degrees of freedom, belongs (by definition) to one of these systems but may have lower symmetry than the lattice of points on which it is built. The systems are given by the sets of angles and of sides that are equal; further equalities are (generally) not allowed for then the lattice would belong to a system of higher symmetry. Objects with symmetry groups that are subgroups of the holohedry group are regarded as being in the same system if the conditions on sides and angles remain the same.

The seven crystal systems (in order of increasing symmetry starting with that of least symmetry) are the triclinic, monoclinic, orthorhombic, tetragonal, cubic, and trigonal and hexagonal [Bhagavantam and Venkatarayudu (1951), p. 119; Borchardt-Ott (1993), p. 108; Bradley and Cracknell (1972), p. 37, 81; Burns (1977), p. 255, 371; Burns and Glazer (1990) p. 22, 290; Chen (1989), p. 465; Cornwell (1984), p. 198; Cotton (1990), p. 368; Evarestov and Smirnov (1993), p. 43; Fässler and Stiefel (1992), p. 221; Hahn (1989), p. 10; Hamermesh (1962), p. 60; Hilton (1963), p. 128; Inui, Tanabe and Onodera (1990), p. 239; Janssen (1973), p. 110; Jaswon (1965), p. 36; Jaswon and Rose (1983), p. 73; Joshua (1991), p. 49; Kettle (1995), p. 245; Ladd (1989), p. 69; Lax (1974), p. 169; Lines (1965), p. 189; Lockwood and Macmillan (1983), p. 146; Lomont (1961), p. 143; Lovett (1990), p. 9; Ludwig and Falter (1988), p. 33; Lyubarskii (1960), p. 348; Megaw (1973), p. 142; Sands (1993), p. 50; Senechal (1990), p. 39; Shubnikov and Koptsik (1974), p. 203; Sternberg (1994), p. 309; Streitwolf (1971), p. 52; Wherrett (1986), p. 223; Yale (1988), p. 108]. They are defined by their holohedry symmetry groups, these leading to conditions on the sides and angles which can also be used to give the systems.

The stem "clinic" apparently comes from the Greek word for couch or bed, which clinics have a lot of (and related to the Greek for lean); here it means a line or axis (couches were apparently simple in those days), specifically one oblique to a fixed line (as in "incline", which, like many a couch, is oblique to the horizontal). The unit cell of a *tri*clinic lattice has three such axes, meeting obliquely, a *mono*clinic has one. A rhombus is an equilateral parallelogram (apparently from the Greek "to whirl", which perhaps what was done with the first rhombuses; the Indo-European root is the word for turn or bend), so an *ortho*rhombus has orthogonal sides (the base is square). The prefix "ortho-" means straight, vertical, perpendicular, correct (as in orthodoxy, although some may disagree that orthodoxy is always, or even ever, correct). The stem "gonal" means cornered or angled (and appears in "diagonal" — between corners — and orthogonal — right cornered). There are four for a *tetragonal* lattice, three for the *trigonal* and six for the *hexagonal* ones. What these numbers count, we have to see.

There are different ways of obtaining the systems and lattices, and we consider more than one. These provide different insights, and indicate different approaches to the removal of conditions (including that the dimension of space is three) so generalizing the concepts, in perhaps useful ways.

II.4.a Labeling axes and faces

Except for the cubic case, not all axes of the lattice (so not all faces of a unit cell) are the same. Some are longer than others; the angles between pairs are different. Thus we need labels for the axes [Borchardt-Ott (1993), p. 10; Megaw (1973), p. 102]. The z axis is that of greatest symmetry (or any of these). The x and y axes are perpendicular to z, thus not necessarily along, or perpendicular to, a face. A set of (three) lines is given by their direction cosines with respect to this coordinate system, which has unit vectors $\underline{\hat{i}}$, $\underline{\hat{j}}$, $\underline{\hat{k}}$. The labels however are chosen as integers (for lattice vectors). Thus, with brackets used for the labels, [abc] is the line of vector $\underline{t}$ from the origin to point (a,b,c), with

$$\underline{t} \bullet \underline{\hat{i}} = a, \quad \underline{t} \bullet \underline{\hat{j}} = b, \quad \underline{t} \bullet \underline{\hat{j}} = c, \qquad \text{(II.4.a-1)}$$

and these are taken to be the smallest such integers (they are all divided by their greatest common divisor).

The labels for faces, Miller indices [Altmann (1994), p. 71; Borchardt-Ott (1993), p. 248; Burns and Glazer (1990), p. 313; Hilton (1963), p. 8; Jaswon (1965), p. 32, 41; Jaswon and Rose (1983), p. 64; Jones (1975), p. 33; Joshua (1991), p. 60; Ladd (1989), p. 33; Lax (1974), p. 190; Lovett (1990), p. 11; Megaw (1973), p. 103; Sands (1993), p. 66; Wood (1977), p. 30], are given in parentheses, (hkl). Here h,k,l are the inverses of the intercepts of the plane and the coordinate axes, and these are again divided by their greatest common divisor (since we are only interested in orientation). If a plane is parallel to an axis it does not intercept it (that is it intercepts it at infinity), so the corresponding value is 0. Any plane parallel to the xy plane is thus labeled (0,0,1), for ones parallel to the xz plane the label is (0,1,0), the plane parallel to the z axis at $\frac{\pi}{4}$ to the x axis is (1,1,0).

Problem II.4.a-1: While in general not all these indices for a plane can be taken as integers, for a crystal, by periodicity all, for planes through lattice points, can be taken as rational numbers (ratios), thus by multiplication can be replaced by integers. Prove this [Sands (1993), p. 66].

II.4.b Classification of lattices by their sides and angles

Lattices are distinguished from each other by how their sides and angles are related. For the sides there are three possibilities: all equal, two equal or all different. But restrictions on angles are more complicated, for not only is the number equal important, but so are their values (specifically whether they are, or are not, multiples of $\frac{\pi}{2}$ or $\frac{\pi}{3}$). It is here that symmetry enters, for (as in two dimensions) we distinguish lattices by the conditions their symmetry groups place on the angles and relative lengths of the sides.

All lattices have an inversion, and the combination of this with rotations gives the reflections and rotation-reflections, so these add nothing new. We need consider then only the requirements imposed on sides and angles by rotations. These give seven systems, seven primitive lattices (whose unit cells contain only a single point).

These relationships, for each lattice, are given in tbl. B-2, p. 634.

Problem II.4.b-1: There are more than three lines for the unit cell of a lattice. Why are there only three conditions?

II.4.c Description of the seven systems

The lattice with the least symmetry is the triclinic, which has three (different, oblique) axes so no rotation about any one leaves it invariant. Its only symmetry is an inversion through a lattice point (taken as the orgin). For it not to have a higher symmetry, no two of the three angles of the unit cell can be equal, although if no more than one equaled $\frac{\pi}{2}$, if that were possible, the symmetry would not change if the other two were equal; it can be see that there can be some equalities without changing the symmetry — the restrictions on the other systems are quite strong. The holohedry group consists of just the identity plus the inversion, $S_2(\bar{1})$, which has subgroup $C_1(1)$.

Problem II.4.c-1: Draw both the unit cell and lattice, and check. Show that inversion (through which point?, points?) is a symmetry. The triclinic lattice is defined to have the least symmetry, yet it does have this one. Why? What type of object would not have even inversion symmetry? Check that the holohedry group of the triclinic system is correct. Note that in the special case of angles equal (no more than one is $\frac{\pi}{2}$), the holohedry group is unchanged.

Problem II.4.c-2: For three unequal sides, the greatest (and least) symmetry possible is a two-fold rotation. Check that a cell has symmetry under this if, and only if, one side is perpendicular to the other two (forming the base), so two angles equal $\frac{\pi}{2}$. The rotation axis of this lattice, the monoclinic, is through the center (and through each lattice point; why?) and perpendicular to the base; it is the heavy dashed line on the left object,

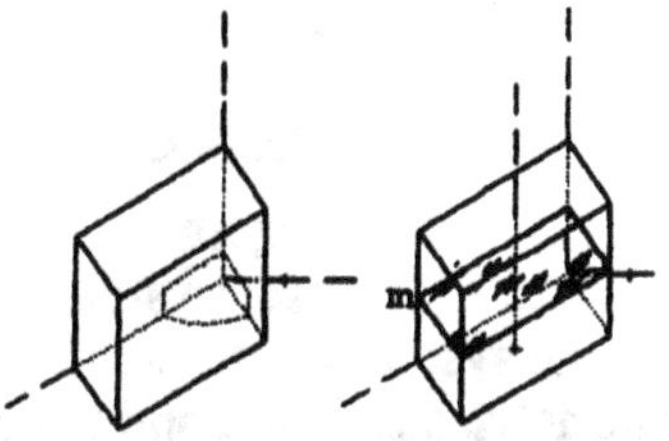

Figure II.4.c-1: MONOCLINIC UNIT CELL.

This has more symmetry than the triclinic, so is different. Why can there not be two perpendicular twofold axes? Why is the lattice also invariant under reflection in a plane, labeled m, enclosed by the heavy wavy lines, perpendicular to the axis, shown on the right object? The three sides have different lengths, and two of the angles (those between the symmetry axis and the other two) are right angles, the third (between the second and third axes) is not. Why? The holohedry group of the monoclinic system is $C_{2h}(\frac{2}{m})$, which has subgroups $C_{1h}(m)$ and $C_2(2)$. Verify that the conditions on the sides and angles gives the holohedry group, and conversely, also that the figure (or one the reader might draw) is invariant under (only) this group.

Problem II.4.c-3: If there are two $\frac{\pi}{2}$-rotational symmetry axes (and no sides equal) than all three angles equal $\frac{\pi}{2}$ (why?), the system is orthorhombic, and again having more symmetry than the previous two it is taken as different. Why are two $\frac{\pi}{2}$-symmetry axes perpendicular? Note that all cases with $\frac{\pi}{2}$-symmetry have been given. The only other rotational symmetry is of $\frac{\pi}{3}$, for the hexagon, but this has sides equal, so we now have all lattices with three unequal sides. The orthorhombic lattice has two or more twofold axes (which is the same as two mirrors — planes in which reflection is a symmetry; why?). Show that if there are two such axes there is a third; this lattice has three twofold axes (through the midpoints of the faces), as drawn, for the unit cell, on the right, of the diagram. These three, through each lattice point, and the center of the unit cell, have to be perpendicular. Why? For this system the three sides have different lengths. The holohedry group of the orthorhombic system is $D_{2h}(\frac{222}{mmm})$, which has subgroups $C_{2v}(mm2)$ and $D_2(222)$. Check that these conditions, and the holohedry group, are equivalent, and that the figure is invariant under the group,

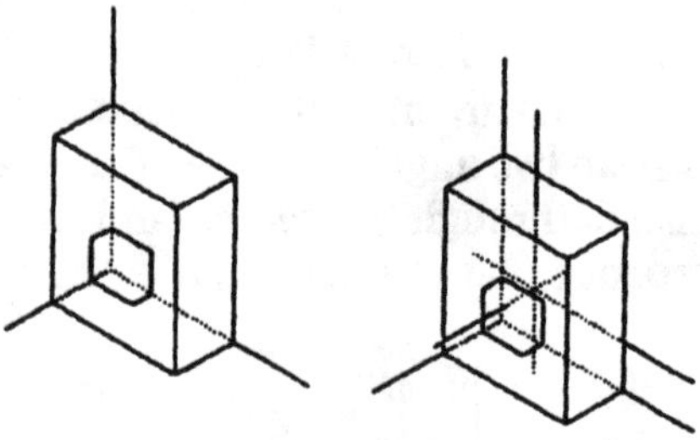

Figure II.4.c-2: ORTHORHOMBIC LATTICE UNIT CELL.

Problem II.4.c-4: The rotational symmetries specified determine the first two of these systems. Even if some sides are equal, the lattices are still assigned to the same systems as those with unequal sides. Explain. So we need other symmetries to give conditions to specify the remaining systems. A fourfold rotation gives a square base, and the third side perpendicular to it (why?), so all angles are $\frac{\pi}{2}$, and two sides are equal, the third, parallel to the symmetry axis, is different. The lattice is tetragonal, which also has further symmetries, including a two-fold rotation so adding that would not result in a different lattice; thus it has a larger holohedry group (than what?). Check these. The tetragonal system, with four, thus eight (?), identical corners (vertices) has unit cells invariant under fourfold rotations, so, like the lattice, two equal sides, and all right angles, as is also true for the symmetry axes. The holohedry group of the tetragonal system is $D_{4h}(\frac{422}{mmm})$, with subgroups $D_{2d}(\bar{4}2m)$, $C_{4v}(4mm)$, $D_4(422)$, $C_{4h}(\frac{4}{m})$, $S_4(\bar{4})$, $C_4(4)$. Justify these: show that the symmetries given by these symbols are those of the lattice (and the unit cell?). What is the significance of the subgroups? In particular, show that the relations among the sides and angles are required by the symmetry group, and conversely, and that a subgroup that gives fewer conditions would allow dropping more than one, giving the orthorhombic lattice — there is no (intermediate) lattice with fewer restrictions and less symmetry than the tetragonal, but more than the orthorhombic. Draw a unit cell, and a lattice, and check that they (both?) are invariant under the group.

Problem II.4.c-5: Note that two fourfold rotations (necessarily?) perpendicular to each other require that all sides be equal, giving the cube, and all angles equal (and equal to $\frac{\pi}{2}$) so this is the lattice with the greatest (?) symmetry (the largest holohedry group). There is thus no way to get other lattices using (only?) two-fold or four-fold rotations. The cubic lattice is defined as having four threefold-rotation axes, these corresponding to the four body-diagonals (fig. I.4.d.i-1, p. 18). These diagonals go through the center of the unit cell; this point bisects each of the three mutually perpendicular lines (each is perpendicular to a pair of opposite edges). Show that this set of axes does give a cube [Burns and Glazer (1990) p. 27, 67]. Notice that all groups of the cubic system have this threefold operation, but not all have fourfold rotations (the holohedry group does). Why? Thus these rotations, not the more familiar $\frac{\pi}{4}$ ones, define the system. A rotation of $\frac{2\pi}{3}$ around one diagonal interchanges axes $x \Rightarrow y$, $y \Rightarrow z$, $z \Rightarrow x$. What do the others do? Thus all three sides are equal. Some of these rotations put minus signs in for the permuted axes (which means what?). The unit cell can be unchanged only if all axes are mutually orthogonal, so it is a cube. What is the action of the other symmetry group generators? The holohedry group of the cubic system is $O_h(\frac{4}{m}\bar{3}\frac{2}{m})$, with subgroups $T_d(\bar{4}3m)$, $O(432)$, $T_h(\frac{2}{m}\bar{3})$, $T(23)$. Check all this.

Problem II.4.c-6: The trigonal system has a single threefold symmetry axis, and no higher (pure) rotational axis. Here we consider only the primitive unit cell (with one point), though it can be given by a nonprimitive cell (sec. II.6.d, p. 108). This primitive cell is called the rhombohedral unit cell. It has all sides and all angles equal, but these are not right angles, as seen from the two views of the cell on the top in the next diagram. Why are these angles less than $\frac{2\pi}{3}$? How does this differ from the triclinic? On the bottom of the diagram the unit cell is shown with the threefold axis (going through opposite vertices) — it makes equal angles to the lattice vectors, these being the sides of the unit cell starting at the point from which the symmetry axis emerges. Next to it are unit cells that have been rotated through $\frac{\pi}{3}$ and $\frac{2\pi}{3}$ (notice the position of the lines), and can be seen to be the same (except for the lines that were drawn differently to show the effect of rotations),

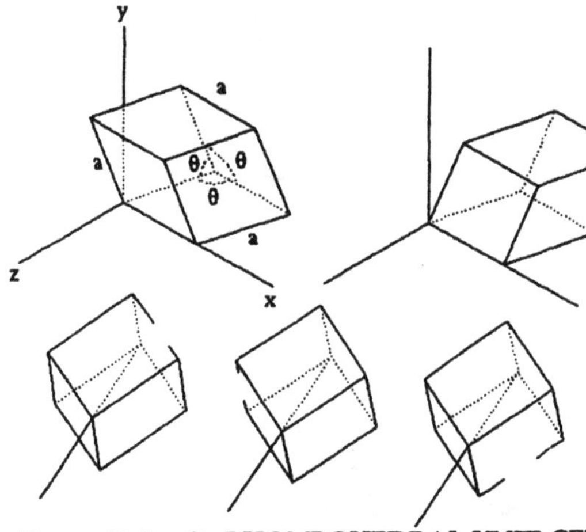

Figure II.4.c-3: RHOMBOHEDRAL UNIT CELL.

Also it is clear that this cell does fill all space, as a drawing or model (constructed by the reader) shows. Check that while it has a threefold symmetry axis so belongs to the trigonal system, it does not posses a sixfold axis, so is not of the hexagonal system — a point to which we shall return (pb. II.6.d-1, p. 109). Verify that the holohedry group of this system is $D_{3d}(\bar{3}\frac{2}{m})$, which has subgroups $C_{3v}(3m)$, $D_3(32)$, $S_6(\bar{3})$, and $C_3(3)$. Check that the conditions and the group give the same results, and that the figures are properly invariant. A three-fold rotation leads to an equilateral triangle, thus two equal sides, and the third side perpendicular, giving two angles equal, and equal to $\frac{\pi}{2}$ — another way of specifying the lattice. How are these two ways related? Draw this and check. This gives further symmetries (what?), but no other conditions (why?), and thus the trigonal lattice. The cube also has three-fold rotations, though more. How are these two systems related?

Problem II.4.c-7: Finally (?) we can add a six-fold rotation, which includes a three-fold one thus the trigonal conditions, but now the third angle equals $\frac{2\pi}{3}$, giving a hexagonal lattice. For the next diagram the unit cell on the top left has, of course, a base that is not a hexagon. By fitting unit cells together, as seen on the top right, a base which is a hexagon is obtained, and it is clear that this tiles space, as shown on the bottom, so gives a lattice. The lattice has rotational symmetry of $\frac{\pi}{3}$ about the center of the hexagon, which is at a corner of the unit cell — this is invariant only under rotations of π, but its sides and angles result in a hexagonal lattice when these cells are fitted together. The hexagonal unit cell is defined as that giving (but not possessing) a single sixfold axis of rotation, the largest possible, so in this sense it also has (leads to?) the greatest symmetry, as we see in

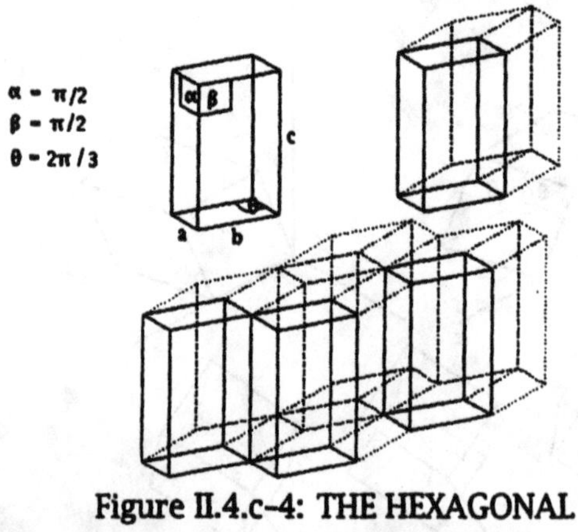

Figure II.4.c-4: THE HEXAGONAL LATTICE.

This is (plainly?) the same as a single threefold axis of rotation (where?) plus a reflection in the plane parallel to it — why? As this rotation interchanges the other two axes, they must be identical, and by the standard argument (?) both perpendicular to the symmetry axis. Since a threefold rotation interchanges them, the angle between them is $\frac{2\pi}{3}$. If two sides are equal, but the third not, two angles are $\frac{\pi}{2}$, the third is $\frac{2\pi}{3}$. With all angles determined, and no freedom to relate sides, there can be no lattices different from these. Verify the statements, and that the holohedry group of the hexagonal system is $D_{6h}(\frac{622}{mmm})$, which has subgroups $D_{3h}(\bar{6}m2)$, $C_{6v}(6mm)$, $D_6(622)$, $C_{6h}(\frac{6}{m})$, $C_{3h}(\bar{6})$ and $C_6(6)$. What is the significance of the subgroups? Why is there more than one lattice of greatest symmetry (the cubic and hexagonal)? Are they related?

Problem II.4.c-8: Review each group to check that they have the subgroups given, and no other (point?) group.

Problem II.4.c-9: In summary, check that these systems all have different symmetry groups, and these groups are nonisomorphic; these are the correct holohedry groups of the crystal systems. However there is still a slight problem, some of these holohedry groups are subgroups of holohedry groups of other systems, but not all subgroups give different systems, though some do. Why, and why not?

Problem II.4.c-10: What are the holohedry groups of the two-dimensional lattices (sec. III.8, p. 172) [Field and Golubitsky (1992), p. 208]?

II.4.d Why the systems have the names they do

While the names for the crystal systems are suggestive they may not seem completely explicit. How can we tell a system from its name? The tri*clinic* has three lines in*clin*ed so there can be no conditions on it. The *mono*clinic has *one* such line thus the other two must be orthogonal, so two angles are $\frac{\pi}{2}$, one not. This is sufficient to differentiate it from the tri*clinic* so no conditions are placed on the edges. The *ortho*rhombic is a rhombus (an equilateral parallelogram) with all lines *orthogonal*, thus all angles $\frac{\pi}{2}$, and again there need be, and are, no conditions on the sides. The tetra*gonal* is perhaps somewhat incompletely named, having *four* (actually all) angles equal (to $\frac{\pi}{2}$), like the ortho*rhombic*, but actually having four (really eight) sides equal, thus bounded by two squares, so one condition on the three (defining) edges. The *tri*gonal system has *three* angles (and sides) equal, but not $\frac{\pi}{2}$ and less than $\frac{2\pi}{3}$; these would give more symmetry. The *hex*agonal has an angle of $\frac{2\pi}{3}$, and two of $\frac{\pi}{2}$, and two, but not (necessarily?) three sides equal, for it must give a *six*-sided figure upon rotation. The cube (the hexahedron; why?) is obvious.

II.4.e The restrictions on projections of lattices from the point groups

The elements of the limited set of point-group transformations give several groups, so lattices. First we note conditions placed on the two-dimensional projections of the lattices by the allowed symmetry groups. There are three cases allowed by the crystallographic condition that the order of a symmetry rotation $n = 1,2,3,4,6$. If $n = 2$ the rotations in the group are the identity and that through π — every two-dimensional lattice has this symmetry. If also there is an $n = 4$ rotation, the lattice is invariant under rotations through $\frac{\pi}{2}$, only, and these leave distances invariant so the lattice consists of (repetitions of) squares. If $n = 3$ we take the axis on a lattice point and the unit cell is obtained from equilaterial triangles. We get the same result for the axis at the center of the unit cell, but it can be at no other point. If $n = 6$ then the group has an element of order 3 giving again equilaterial triangles. These then are the two-dimensional projections of the three-dimensional lattices.

Problem II.4.e-1: Draw these figures and check that they have the symmetry groups stated and that there are no other figures with these groups (and of course with any other such groups).

Problem II.4.e-2: Why cannot the group simultaneously contain rotations of orders 3 and 4?

Problem II.4.e-3: Justify the statements about the points at which axes can be placed.

Problem II.4.e-4: Why, for $n = 6$, are triangles obtained as for $n = 3$? Is there a difference?

II.4.f What determines these systems?

There are seven crystal systems — seven primitive lattices. Why? What criteria have been used for calling them distinct? Why can we not, or have we not, picked a finer classification? Why are there not more systems? Or fewer?

Problem II.4.f-1: The crystal systems are defined by the number of sides that are equal (or not) and by the number of angles that are equal (or not), and also by which are $\frac{\pi}{2}$ (tbl. B-2, p. 634). Why is there this last condition? Could further systems be obtained by removing it, while leaving the number of equalities the same?

Problem II.4.f-2: There is another way of obtaining the seven crystal systems. If a lattice has a set of axes of some order then it must have other axes also [Fedorov (1971), p. 28], but the axes are limited [Bravais (1969), p. 58]. Also the symmetry planes are then determined [Bravais (1969), p. 55; Fedorov (1971), p. 38]. Thus if there is a threefold axis, and perpendicular to it a twofold axis, there must be three twofold axes. Explain this, and also find the other conditions. This leads to the classification of crystal systems according to their sets of axes of largest order [Bravais (1969), p. 79; Burns (1977), p. 372; Lines (1965), p. 195]. Show that these are ones with

1) three fourfold axes;
2) a sixfold axis;
3) a single fourfold axis;
4) a threefold axis;
5) three twofold axes;
6) a single twofold axis; and
7) no axis of symmetry.

Which go with which of the lattices? For each show that the given set of axes necessitates the other axes and symmetry elements of the lattice (the other elements of the holohedry group). Try also to find lattices that do not fit into this system — say which have more symmetry axes. What goes wrong? Why is it not necessary to include reflections and inversions?

Problem II.4.f-3: The reason that there are only seven crystal systems while there are thirty-two point groups is that a crystal system is defined by its holohedry group of which there are seven. Check that all point groups appear (so each is physically realizable, if the crystal system is) and that all point groups are subgroups of these seven holohedry groups. The holohedry groups are also related (sec. I.6.d, p. 42). Show that most point groups are subgroups of the holohedry group of the cubic system, $O_h(\frac{4}{m}\bar{3}\frac{2}{m})$. What are the exceptions? Might these have been expected? What are these subgroups of? Find the subgroup relations of the holohedry groups.

Problem II.4.f-4: If some holohedry groups are subgroups of others, while other subgroups are not holohedry groups, why do we distinguish them? Why not take the largest as the holohedry group, and regard the rest as its subgroups? Show that all holohedry groups are symmetry groups of lattices, but the other point groups are not. In what way? Would we then expect them to be relevant to anything?

Problem II.4.f-5: These systems have been defined by their numbers of sides of equal lengths, of equal angles, and of angles equal to $\frac{\pi}{2}$. They can be found from the cubic and hexagonal (not surprisingly; they have the largest holohedry groups), in fact from the most symmetric one, the cubic, by dropping different restrictions. Changing one edge of the cube (so it is no longer the same as the other two) gives the tetragonal cell, changing a second gives the orthorhombic cell. If one of the three angles does not equal $\frac{\pi}{2}$ the cell is monoclinic, if a second is not $\frac{\pi}{2}$ it is triclinic, and as this has no symmetry besides (the necessary) inversion it does not matter what the third angle is. This can be seen going from left to right in

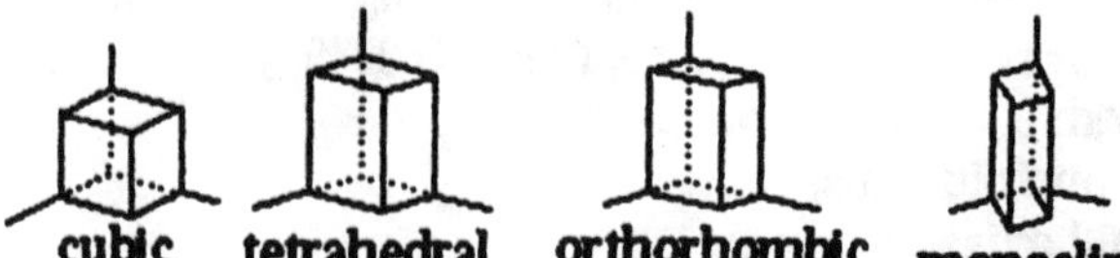

Figure II.4.f-1: DROPPING RESTRICTIONS CHANGES LATTICES.

How are the other systems obtained from the cubic? If instead of changing the length of one edge, the cube, on the left, is extended along a diagonal (how?) the result is the rhombohedron shown in two views obtained by extension along two different diagonals. Of course these two are identical, as can be seen by close examination, but the difference in the way they look in two-dimensional projections can clarify the effect of this transformation,

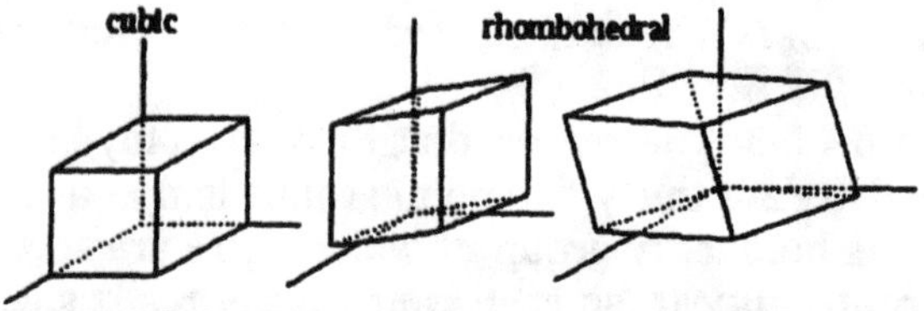

Figure II.4.f-2: CUBE GOES TO A RHOMBOHEDRON.

It would be useful to redraw this, but for a cube contracted along a diagonal, rather than stretched. Do the rhombohedrons differ, beyond their sizes? In all these cases removal of a restriction, even if the change is infinitesimal, gives a different system. There is one other possibility; if the angle for the tetrahedron (or the cube, it does not matter) is changed from $\frac{\pi}{2}$ to $\frac{2\pi}{3}$, which is not an infinitesimal change, the resultant cell is the unit cell of the hexagonal lattice. Draw this and check. It would be interesting (using extensible rods, or a computer program) to try this experimentally. Note that the condition on the sides for the hexagonal lattice is listed as an inequality, but it really means here not "does not", but "need not" — for which there is no symbol. Does it make a difference if all sides are equal? Check that all (relevant) possible ways of dropping restrictions have been listed. Notice that when a restriction is changed a visually distinguishable figure is obtained (although if the change is small enough, a microscope might be needed), and that all such figures have been given (up to relative scaling of sides, or changing of angles — which may be physically important, but are not used in this classification). Because they are related by infinitesimal changes the different systems can be taken to have the same topology as the cube (why?); but can the hexagonal?

Problem II.4.f-6: There is another way of seeing that these are different crystal systems, and are all of them. In two dimensions the different primitive lattices have different symmetry groups. For three dimensions we regard some groups as giving different lattices, but others as subgroups of the (holohedry) groups that label (and determine) the lattices. Why are some groups taken as holohedry groups, while others as merely their subgroups? There are two largest crystal point groups, and being distinct they give different systems. Take their subgroups. Draw a diagram to show how they are related [Yale (1988), p. 111]. Some belong to the same system, but others do not. How can we distinguish these cases? All holohedry groups contain the inversion. Why? However not all groups containing the inversion are holohedry groups. There are three holohedry groups that are simultaneous subgroups of two distinct holohedry groups. Thus these (can reasonably be taken to) give crystal systems different from the systems of the two groups of which they are subgroups. The holohedry groups of the tetragonal system, a subgroup of that of the cubic, and of the triclinic, a subgroup of the monoclinic, are not so distinguished. Why do they give different systems? Check that in both cases, in going to the subgroup the largest rotation (or rotation-reflection) symmetry is lost. Thus it makes sense to regard these as giving different systems, does it not? Explain why this loss of symmetry agrees with the dropping of one of the restrictions on sides and angles, and conversely. There are now the cases in which a subgroup contains the inversion, but the system is not different. Observe that in all these cases the subgroup does require the same relations on the sides and angles as the holohedry group, thus does not give a different system, and also that in these cases the rotation (or rotation-reflection) of largest order is the same in the holohedry group, and in the subgroup. What is the significance of this? Compare these arguments with those of pb. II.4.f-4, p. 93, as well as the other arguments for these systems. Using this classification then these seven (primitive) systems are all (sec. II.7, p. 111). State what "this classification" means; what are the rules giving it? Might they be changed to give different ways of organizing crystals? Would these make sense? Do they give classifications of crystals or of lattices?

Problem II.4.f-7: A lattice is defined as a set of (identical) points; is it possible to have a lattice (not a crystal) whose symmetry group is not the holohedry group? Is this related to there being seven primitive lattices?

Problem II.4.f-8: There is an important assumption (well) hidden here, that the holohedry group of a lattice is the same at all points. Prove it [Yale (1988), p. 109].

II.4.g Where can axes be placed?

A lattice can have several rotational axes of symmetry. Need they be related, can they be placed anywhere with respect to each other, and with any values of angles? It is useful to understand these relations — they determine the allowed lattices (so limit the physical objects wishing to be, or be determined by, lattices); an understanding of them thus increases that of the objects — and geometry.

The points and planes of symmetry are placed on the lattice points, however these need not be all — there can be intermediate ones. But for each latter point or plane there is an equivalent one of the former, so putting them in does not change the lattice, or increase the number of different lattices [Bravais (1969), p. 53, 84]. Thus intermediate axes and planes can be, and are, ignored.

A rotation around the axis of greatest symmetry (the principle axis) labeled z of $\frac{2\pi}{n}$, $n = 1,2,3,4,6$, leaves the lattice unchanged, so every other axis must be a member of some set, the elements of which go into each other under these rotations. Thus if there are other symmetry axes besides z, they come in sets of 2,3,4 or 6, with the angle between these axes π, $\frac{2\pi}{3}$, $\frac{\pi}{2}$, $\frac{\pi}{3}$, respectively.

A two-fold axis perpendicular to z is possible; it just inverts the z axis giving the same set of rotations about it. This is the reason that, besides the cyclic (rotation) groups, the dihedral groups are finite subgroups of the rotation group. Can there be an axis that is not along z of order higher than two?

Problem II.4.g-1: With the principle axis of order k, any other axis would be in a set of k members. This is possible if each of the set is of order 2. Higher order would mean that the principle axis would itself be in a set, each axis with order k, thus increasing the number of members of the other set (or increasing the number of such sets). Axes of different order belong to different classes of the symmetry group (why?), as would axes of the same order if there were no rotations taking them by a similarity transformation into each other (why?). It would be an interesting exercise to go through the list of all finite groups (of low order), starting with ones whose group tables are easily available [Mirman (1995a), chap. II, p. 29], looking at the class structure, to find those that could serve as symmetry groups of regular polyhedra, using this analysis of the relationship between the axes. Presumably this would give the same set of point groups and also limit where axes could be placed, and their orders, and thus the shapes of the regular polyhedra (sec. I.4.c, p. 10). Compare with the results of Bravais (1969).

II.4.h What symmetry elements must a lattice contain?

If a lattice has a set of symmetry elements then it is required to have further ones (so it is usually defined by a minimal set, giving the group generators, the others then being found from it). Rotations about one axis, say, produce from a perpendicular axis further perpendicular ones. It is useful to understand how symmetry elements are required by the presence of other elements.

Problem II.4.h-1: If the two-dimensional lattice in the plane perpendicular to a two-fold axis is rectangular, show that the (three-dimensional) lattice has three mutually perpendicular two-fold symmetry axes [Bravais (1969), p. 61]. Also if there are two such axes, there is a third mutually perpendicular to them.

Problem II.4.h-2: For a lattice with a threefold symmetry axis, the lattice on the plane perpendicular to it consists of equilaterial triangles [Bravais (1969), p. 67]. Show that each of their sides is a twofold symmetry axis.

Problem II.4.h-3: If a lattice has two threefold axes that are not parallel, it has four (these forming the diagonals of a cube), and only four [Bravais (1969), p. 71].

Problem II.4.h-4: Check that a lattice that has four threefold symmetry axes also has three fourfold axes [Bravais (1969), p. 76]. Draw them, showing how they are related. Also if there are two fourfold axes, there are three, only, and at right angles. If a lattice has both a threefold and a fourfold axis, then it has three fourfold and four threefold axes [Bravais (1969), p. 77]. Further it also has six twofold axes which bisect the right angles formed by pairs of fourfold axes. How many crystal systems does this apply to? Which? Where are the symmetry planes for such a lattice?

Problem II.4.h-5: A lattice with a sixfold axis can have no other symmetry axes except perpendicular twofold ones [Bravais (1969), p. 79]. Must it have the latter?

Problem II.4.h-6: It is difficult from these results alone to develop an understanding of the geometry. It would therefore be interesting to (try to) repeat this discussion for higher-dimensional Euclidean spaces, and for non-Euclidean ones [Brown *et al*, (1978); Fejes Toth (1964); Schwarzenberger (1980), p. 110; Senechal (1990), p. 105]. What, if anything, is learned this way? What kind of "crystals" could there be in these spaces? Could there be analogs in, or projections into, ordinary space, perhaps non-crystalline, or quasi-crystalline [Jaric (1989); Lifshitz (1997); Senechal (1995)]? What axioms of Euclidean geometry give these results? Which axioms give which results? Which, if any, of the set of axioms of Euclidean geometry (pb. I.4.c.iv-5, p. 17) are unnecessary for obtaining these lattices, and only them? Can these concepts be generalized to other geometries? What would they give?

II.5 PRIMITIVE AND NON-PRIMITIVE BRAVAIS LATTICES

For each crystal system there are a set of lattices, the primitive ones — with a single point per unit cell — in three dimensions seven, and, as in two dimensions, others with more points, giving another seven (centered) nonprimitive lattices. To determine these we consider where extra points can be placed, where they cannot, why they can, and need, be added, why placements are restricted, and then for each system, find the lattices these give [Brown, *et al*, (1978), p. 19; Burns and Glazer (1990), p. 34; Jaswon (1965), p. 45; Jaswon and Rose (1983), p. 80]. But, what is centered?

If we can add points can there be restrictions on the number and types of lattices; can we not just add more and more and get more and more lattices? How do seven types of unit cells give fourteen lattices, and how do we know that these fourteen are the only ones — and in what sense? Requirements are that the figure with additional points still be a lattice, have, and show, the symmetry of the crystal system from which it is constructed (or else it would belong to a different system) and that the unit cell cannot be replaced by one with fewer points without losing symmetry information — and it must be different as discussed above (sec. II.3.g, p. 80) from listed lattices (it cannot be transformed into another without changing the symmetry, and it has different topology). What do these lead to?

II.5.a The conditions on added points in three dimensions

For three dimensions there are various ways of finding the conditions on where, and how many, points can be added to a primitive lattice to give a different one. All provide understanding and, fortunately, all give the same set of nonprimitive lattices.

II.5.a.i *Conditions given by two-dimensional lattices*

Conditions on three-dimensional lattices come from those on the two-dimensional ones obtained from each plane section (containing lattice points); the points in each plane must form a two-dimensional lattice, and these are restricted. The procedure is to add points at every position allowed by these sections and then check whether the resultant sets of points form a lattice, and a new one. A lattice identical to a primitive one, though of different size and orientation, is of no interest (here). A lattice is new only if topologically different — it cannot be distorted into a known one.

Problem II.5.a.i-1: Why must the points in every plane form a lattice?

Problem II.5.a.i-2: The additional Bravais lattices are constructed from the primitive ones by adding points either to pairs of opposite faces or the inside (where?) of the primitive-lattice unit cell. Why are both not done simultaneously?

II.5.a.ii *The points allowed by the two-dimensional sublattices*

Except for a few loci, if a point is added to a figure then a set is needed to preserve the symmetry; the resultant figure, though smaller, can be distorted into the first — they are not topologically distinct — so it is not new. This holds for a plane along the face of a unit cell perpendicular to a symmetry axis; for a parallel face, symmetry requires that a point lie on the line through the center of the face parallel to the axis, and that there be equivalent points on all four faces leading to just a smaller cell. So only one point can be added to the face, in the middle.

Points on the interior of planes through diagonally opposite pairs of corners, except at the center of the cell, must come in sets, giving a cell, which if part of a lattice, would not be new. Thus only one point can be added to the interior of a unit cell, at its center. The only *new* three-dimensional lattices obtained by adding points are face-centered and body-centered ones.

Problem II.5.a.ii-1: Check this for each crystal system. It is also worthwhile checking that other planes do not give restrictions (other than those mentioned here). Can it be rigorously proven in general that only points at the centers of faces or the center of the unit cell can be added?

II.5.a.iii *Adding points simultaneously*

This gives the positions at which points might be added. Now at which positions can points be added, simultaneously, and do the resultant unit cells form lattices, and new ones? There are restrictions in two-dimensions (sec. II.3.d, p. 76); we should expect (related) ones in three-dimensions.

The only positions allowing added points are the center of the unit cell, and the centers of opposite faces. All unit cells are parallelepipeds, thus have six faces. In three dimensions one, or all three, pairs of faces can be centered, but not two [Sands (1993), p. 57, 158]. To show that two cannot be, we have an arbitrary parallelepiped with two pairs of faces centered (indicated by dotted lines, which are not part of the figure, but are added to emphasize the faces). Each corner has a lattice point. The (dashed) lattice vector from the center of one face to a lattice point at a vertex is drawn. For a lattice this vector should bring any other lattice point to a lattice point. But as can be seen in

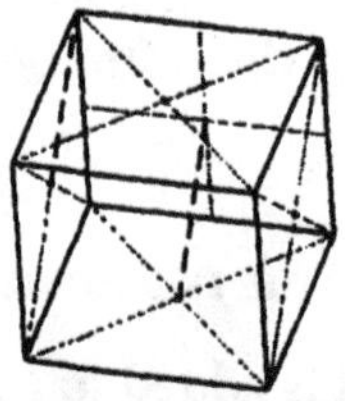

Figure II.5.a.iii–1: TWO-FACE CENTERING IS IMPOSSIBLE,

this is not true for a point on the bottom centered face, for which the (identical dashed) vector is shown, and does not go to anything. It would be true however if all faces were centered for it goes to the center of the third face (as shown by the crossed dotted lines) which is where another point should be placed. For one-face centering this second point would not be inserted. So one or three pairs of faces can be centered, but not two.

Body-centering and face-centering are mutally exclusive. Consider a diagonal plane through opposite edges, going through the center of the unit cell; if we put a point at the center (of the unit cell) and the centers of that pair of faces whose diagonals also lie on this plane, and so are perpendicular to it, the result is not a lattice: a lattice vector taking a point at the center of a face to a lattice point at the vertex does not take the point at the center to another lattice point. Hence it is impossible to put points at the center of the unit cell, and on the faces.

Points cannot be put at the centers of the edges. Taking a plane through a face and putting points at the centers of both pairs of diagonally opposite edges does not give a lattice; the argument is the same as for two-face centering. Placing points at the center of diagonally opposite edges and an additional point at the center of the unit cell, gives a unit cell of the same or lower symmetry, so nothing new.

Points that are placed on faces parallel to a symmetry axis can only be put at the center, by symmetry, or else must be in sets. For a plane perpendicular to an axis, if it is the only one, points can be placed elsewhere, but in sets to keep the symmetry. But in these cases the sides of the unit cell could be taken through the new points, giving an equivalent cell, but smaller, so nothing new.

If points are placed inside the unit cell then they must be in sets and on the line perpendicular to the symmetry axes and faces, and going through the centers of these faces. Then again we can take the sides of the unit cell through these points, giving an equivalent, but smaller, unit cell (a change of symmetry, say from a square to a rectangle, would give a lattice of a different, but listed, system).

The only ways to add points then is to put one at the center of the unit cell (body-centered), or points at the centers of one pair (base-centered), or of all three pairs, of opposite faces (face-centered) [Lines (1965), p. 190]. Thus the number of nonprimitive lattices is limited. For each crystal system we have to see which of these actually gives lattices, of the same symmetry, that are not equivalent to ones of smaller unit cells.

Problem II.5.a.iii-1: Draw diagrams to illustrate and check these discussions about where points cannot be placed, and make the arguments precise. If any cases have been left out, add them, and give proofs.

Problem II.5.a.iii-2: Check, especially with diagrams, perhaps computer generated, that if points are placed at the centers of one pair of opposite faces, or of all three pairs, or at the center of the parallelepiped, the resultant set of points forms a lattice (which does not mean that it has the symmetry of the unit cell, or that it gives a lattice distinct from others). Is this true for all crystal systems, or just most?

Problem II.5.a.iii-3: New lattice points can only be placed on symmetry axes; at the points where the axes pierce a face, or at the center. Is this an accident?

Problem II.5.a.iii-4: A unit cell of a crystal has symmetry under a subgroup of the rotation group; show that we can always take the cell such that its center is the center of rotation. Thus if we put an atom at the center the symmetry group is unchanged, but the lattice is. Why must the group also be unchanged by reflections (through what?)? In what way is the lattice (regarded as a geometrical object) different?

Problem II.5.a.iii-5: The lines from the center to each point go into each other under the group transformations. We might consider generalizing by putting extra points on these lines. Show that this, if it gives a lattice, gives smaller unit cells, but no new ones. Also if we put points on the lines connecting vertices of the same faces, halfway between them, the symmetry group is not changed, but the lattice might be. However it could be that the symmetry group is unchanged (or that there is still symmetry, but less) if we do not consider all operations taking a point into any other, but combine the points into sets, and consider only operations taking sets into each other. The individual points may not form a lattice (why?), but the sets (assigning one point to each set) might. Need the assigned point be in the set?

Problem II.5.a.iii-6: Show that any further points can be placed only in the center of the cell, or in the centers of (some) faces. Points placed elsewhere must come in sets (which is not necessarily required for these special positions) in order to retain the symmetry of the crystal system. But then there would be unit cells, with a single point, giving the same lattice; nothing new would be obtained — the unit cell would merely have a smaller size.

Problem II.5.a.iii-7: Explain why in some cases setting angles to special values, $\frac{\pi}{2}$ or $\frac{\pi}{3}$ say, gives a new lattice, in other cases it does not, and what determines this for each lattice.

Problem II.5.a.iii-8: For an n-dimensional parallelepiped, how many pairs of faces can be centered? What are the general rules for adding points?

II.5.b Thus there are limits on lattices

These results mean that we cannot add points indefinitely. There are only a small number of possibilities. However while we can add points to these positions only, we may not be able to place them at all for every system (and still get a new lattice with the same symmetry). This is why the number of lattices is restricted.

In three dimensions there are seven crystal systems, seven primitive unit cells — those with a single point. Allowing more than one point per cell gives another seven unit cells. Stacking unit cells on each other — so that there is translational symmetry, with discrete translations, in all directions — gives thus fourteen Bravais lattices. In considering each lattice, as described, we must check that it is actually a lattice, that it is different (in relevant ways) from others, and also that the set obtained by this procedure gives all such lattices for each system. To get the list of lattices, we look at each system individually.

Problem II.5.b-1: This statement of how to get the lattices is somewhat different than those above. How are they related?

II.5.c Finding the Bravais lattices analytically

The Bravais lattices are those that express the point group symmetries. We have found them by geometrical arguments, but there are also analytic ones.

Problem II.5.c-1: Write the primitive lattice vectors as

$$\underline{V}_1 : (A, 0, 0), \;\; \underline{V}_2 : (B\cos\alpha, B\sin\alpha, 0),$$
$$\underline{V}_3 : (C\cos\beta\cos\gamma, C\cos\beta\sin\gamma, C\sin\beta) \qquad \text{(II.5.c-1)}$$

and apply all point group symmetries to them [Burns and Glazer (1990), p. 281]. For these to give a lattice the resultant vectors must be sums of these three, with integral coefficients. Find all ratios of A, B, and C and all values of α, β, γ that satisfy, obtaining thus the primitive lattice vectors for all the lattices, and find the lattices they give [Chen (1989), p. 466; Cornwell (1984), p. 200; Lomont (1961), p. 199]. They should agree with the ones found. Do these give nonprimitive lattices?

Problem II.5.c-2: Using the conditions on the sides and angles (tbl. B-2, p. 634), find the lattice vectors for all fourteen lattices. Compare with the results of the previous problem. Using these lattice vectors find the symmetry groups. Also find the angles between the lattice vectors for each lattice. To what extent do these show that the lattices are different?

II.6 DESCRIPTIONS OF THE FOURTEEN LATTICES

Having determined (all?) the rules, we are now able to describe the Bravais lattices and show for each system that they are lattices, and all allowed do exist for that system. It is useful to look for holes in these arguments and compare them (the arguments, but also the holes) with more rigorous discussions [Bravais (1969)].

II.6.a Triclinic, monoclinic, orthorhombic and tetragonal systems

The triclinic has a unit cell invariant only under the identity and inversion. There is just a single unit cell, the primitive — adding points gives another triclinic unit cell (it cannot have less symmetry and still give a lattice). This cell is

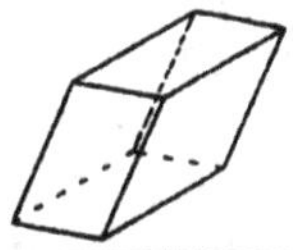

Figure II.6.a-1: TRICLINIC UNIT CELL.

Problem II.6.a-1: The monoclinic system has two lattices, the primitive, shown in two different views on the top, with the coordinate axes in solid lines outside the cell, and one with a unit cell of three points, with two in the centers of opposite faces, at the intersection of the dotted lines (drawn only for guidance) at the bottom left of the diagram. These two faces are parallel to the symmetry axis (chosen); either pair, but of course not both, can be used (and since the faces are different, the two choices give lattices that are not regarded as different — why?). Draw a diagram, or use this,

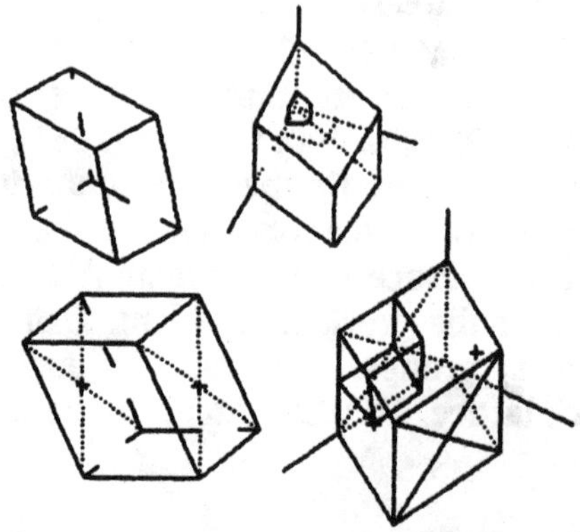

Figure II.6.a-2: MONOCLINIC UNIT CELLS,

to check that this gives a lattice (for both choices of centered faces), but if points are placed in the center of both pairs of these faces the result is not a lattice — the points are distinguished, so there are two sets of points whose environments are different. Also points in the other pair of faces would, as in the analogous (which?) two-dimensional case, give nothing new; would it give just a smaller primitive cell, or would the result not be a lattice? At the bottom right of the diagram, points have been added to the centers of all three faces. As can be seen this gives a smaller monoclinic primitive unit cell. Is this also true for a point in the center? Try other possibilities to check that they, when possible, also give nothing new. So there is but one way of getting a nonprimitive unit cell, thus there are two monoclinic lattices, the primitive and the face-centered. Note however, that there are two pairs of faces that can be centered, so two different unit cells can be obtained from the primitive one, but the resultant lattices are regarded as the same (which does not imply that crystals constructed by putting the same atoms on different pairs of faces need have the same properties). These differently-centered lattices are however distinguished as being of different settings [Burns and Glazer (1990), p. 146], leading to different space groups (sec. III.9.a.i, p. 181). Find the translations leaving the lattices invariant.

Problem II.6.a-2: For the orthorhombic system there are four lattices. The primitive is on the left in the diagram, the body-centered lattice with a point in the middle of the cell (at the intersection of the dashed lines, which are lattice vectors) is next to it; the base-centered lattice has two added points, at the centers of any two opposite faces. For the face-centered lattice (far right) there are six extra points, each at the center of one of the six faces — all faces are centered. Prove that all these are lattices, and new,

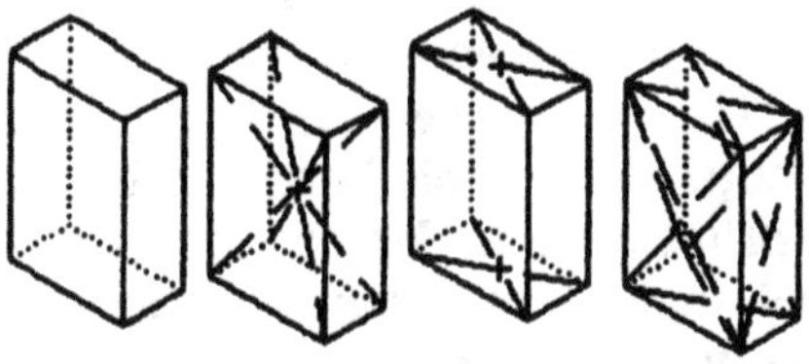

Figure II.6.a-3: UNIT CELLS OF THE ORTHORHOMBIC SYSTEM.

Check that adding four points to the centers of four of the six faces does not produce a lattice. What goes wrong if in the final lattice another point is added at the center? Explain why there are no other ways of adding points. What translations leave each lattice invariant?

Problem II.6.a-3: Next is the tetragonal system. Besides the primitive cell on the left, there is one other, body-centered, on the right, with a point in the center of the primitive cell,

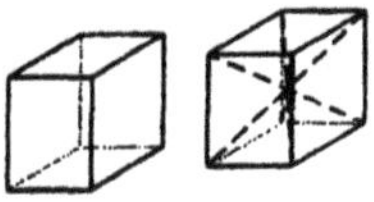

Figure II.6.a-4: UNIT CELLS OF THE TETRAGONAL SYSTEM.

Prove that this gives a (new) lattice, and that there are no other new ones. Why can points not be added to the faces? Find the translations that leave each of these invariant.

II.6.b The cubic system

The cubic crystal system has three lattices [Yale (1988), p. 101],

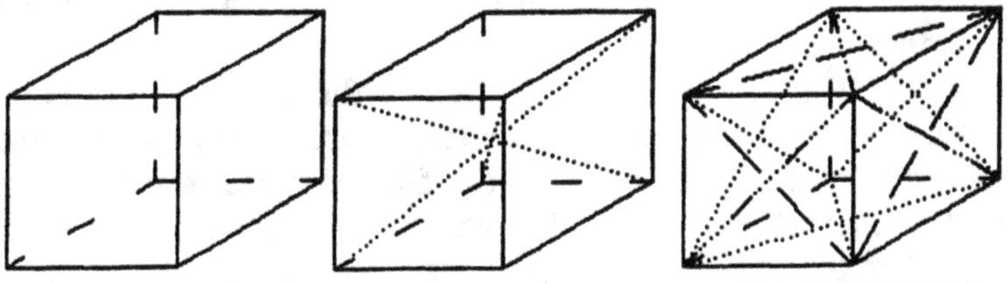

Figure II.6.b-1: UNIT CELLS OF THE CUBIC SYSTEM,

(with points at the intersections of the lines), the primitive, the body-centered (bcc — body-centered cubic) and the face-centered (fcc — face-centered cubic). The body-centered cubic is obtained by adding a point at the center of the unit cell. For the face-centered cubic lattice, one point is inserted at the center of each face, a total of six additional points.

Problem II.6.b-1: Check that these give (new) lattices. Find the translations that leave each invariant. What would go wrong if both of the two centered lattices were combined, that is points were added to the center of each face, and to the center of the unit cell?

Problem II.6.b-2: To see why we want lattices with more than one point per unit cell note that a cubic lattice with a point at the center is different from a primitive one. We can take a unit cell with only a single point, but that shows less symmetry than the cube. Thus, as in two dimensions, the unit cell does not show all properties (that it is supposed to) of the lattice — we lose information. Draw a cube with a point at the center, and a unit cell containing only a single point, and check that the latter shows less symmetry than the former (though the resultant lattice has the same symmetry).

Problem II.6.b-3: One way of specifying a lattice is by using lattice vectors. Draw the lattices given by these vectors to check that they are [Streitwolf (1971), p. 57]
primitive:

$$\underline{a}:\lambda(1,0,0),\ \ \underline{b}:\lambda(0,1,0),\ \ \underline{c}:\lambda(0,0,1); \qquad \text{(II.6.b-1)}$$

fcc:

$$\underline{a}:\lambda(0,1,1),\ \ \underline{b}:\lambda(1,0,1),\ \ \underline{c}:\lambda(1,1,0); \qquad \text{(II.6.b-2)}$$

bcc:

$$\underline{a}:\lambda(-1,1,1),\ \ \underline{b}:\lambda(1,-1,1),\ \ \underline{c}:\lambda(1,1,-1); \qquad \text{(II.6.b-3)}$$

λ is the lattice constant. Explain specifically what λ means in these three cases. Show that each lattice point has six, twelve and eight nearest neighbors, respectively. Nearest neighbors are lattice points that are reachable with a single lattice vector, others are connected to the point by a sum of smaller lattice vectors. What are the distances to them? Calculate the angles between the vectors for the three cases, and note that they are fixed — and different. So we consider these lattices distinct.

Problem II.6.b-4: The holohedry group of the cubic system is (isomorphic to?) symmetric group S_4 plus the improper transformations. If alternating vertices of the cube have different types of objects, giving two interpenetrating lattices, the symmetry is reduced. Check that each of the lattices still has the symmetry given by the eight C_3 rotations, and three C_2 rotations. Which C_2 rotations are these? This is the (which?) tetrahedral group (isomorphic to alternating group) A_4 which (with the improper transformations) is the hemihedry group of the cubic system. Is the alternating-group name relevant here? Which permutations are no longer symmetries? Why? Also each of the two lattices, again separately, still belongs to the cubic system, but the unit cell is now face-centered cubic. Note that it still has more symmetry than any other lattice that might be constructed by breaking the cubic symmetry:

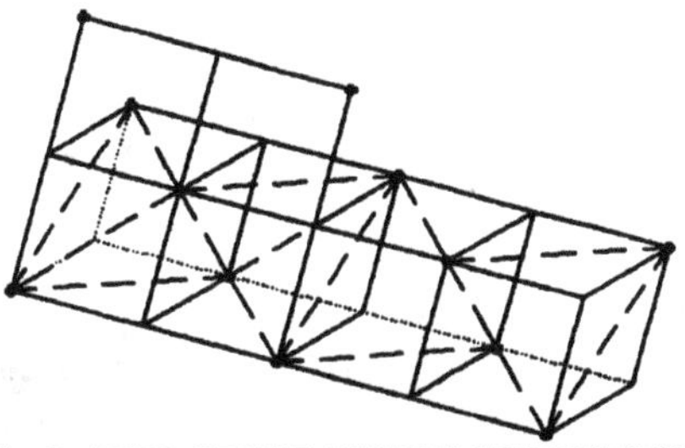

Figure II.6.b-2: THE CUBIC SYSTEM WITH DISTINCT VERTICES.

Can the symmetry be broken in other ways? Are there ways of breaking it to give systems with other symmetry groups? Would these still be of the cubic system? Why?

II.6.c The trigonal and hexagonal systems

To consider the relationship between the trigonal and hexagonal systems [Burns and Glazer (1990), p. 43; Jaswon and Rose (1983), p. 84; Sands (1993), p. 59] we first study the hexagonal unit cell (fig. II.4.c-4, p. 90); the two sides of the base are equal, but the height need not be (notice that two angles are $\frac{\pi}{2}$, the third $\frac{2\pi}{3}$). On the right this cell is shown twice rotated about an axis through the $\frac{2\pi}{3}$ angle (note the three vertical dotted lines) and the three cells are fitted together — the resultant figure is invariant under rotations of $\frac{2\pi}{6}$ about this axis, which the unit cell is not; the lattice would not be invariant if the angle were other than $\frac{2\pi}{3}$, or if the base were not equilateral. Also these cells are shown placed together to fill the plane, and this again requires an angle of $\frac{2\pi}{3}$ and an equilateral base. These indicate why the unit cell is given by, and determines, though does not explicitly show, the symmetry.

Problem II.6.c-1: Draw a diagram to exhibit that the unit cell rotated six times about the axis (which?) gives a lattice invariant under this $\frac{2\pi}{6}$ rotation. The hexagonal lattice has symmetry under both $\frac{2\pi}{3}$ and $\frac{2\pi}{6}$ rotations, though about different axes. Which are these? Axes may appear different, but the lattice points on them are identical, as required for such points — if two edges of one cell are at an angle of $\frac{2\pi}{6}$, the two edges meeting at that point of the neighboring cell are at angle $\frac{2\pi}{3}$ (if two angles of the base are $\frac{2\pi}{3}$, the other two have to be $\frac{2\pi}{6}$).

Problem II.6.c-2: How can points be added? The lines from the center of the base to its lattice points at the vertices are perpendicular, so a point placed at the center does not give a hexagonal lattice (what is it — why?). Draw a diagram and prove this, and also that this rules out centering both pairs of side faces, which requires centering the base in addition to obtain a lattice. Centering only one pair is not possible because rotations of $\frac{2\pi}{6}$ take one face into another. For body-centering, retaining the symmetry requires that in the horizontal plane through the center of the cell, lattice points be placed exactly as those in the base, but shifted by the vector going from a vertex to the center of the base. But then the lattice vector from a point on the base to one on this plane does not make an angle of $\frac{\pi}{2}$ with the base, so again this lattice is not hexagonal. Try with a diagram to violate these conditions and explain why they cannot be (and still get the required results — which are what?).

Problem II.6.c-3: Points cannot be added to both pairs of faces parallel to this axis without adding ones to the third pair, giving a smaller, but similar, (hexagonal) cell. This indicates why there is only one hexagonal lattice. Add points elsewhere [Burns and Glazer (1990) p. 43], say the middle of edges, and check that no new lattices are obtained.

II.6.d The rhombohedral unit cell

The rhombohedral unit cell (pb. II.4.c-6, p. 89), can be labeled another way: use a hexagonal unit cell, but now nonprimitive, with three lattice points, and refer the lattice points of the primitive rhombohedral cell to the axes given by the hexagonal lattice vectors. So

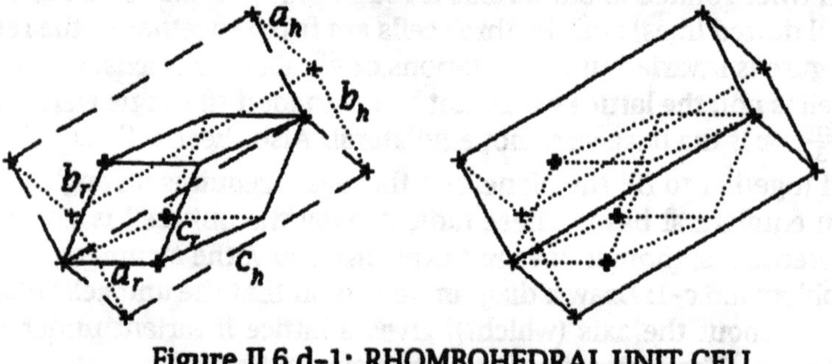

Figure II.6.d-1: RHOMBOHEDRAL UNIT CELL IS A NONPRIMITIVE HEXAGONAL UNIT CELL.

On the left the (nonprimitive) hexagonal unit cell, which contains internally three lattice points, indicated by crosses, and connected by lattice vectors, is drawn (with dashed lines); the (primitive) rhombohedral unit cell is depicted with solid lines (and on the right with dotted lines which may make it easier to see how these are related). The lines that would be unseen if the object were solid are drawn dotted. Thus centering the hexagonal lattice is possible, but it gives another system, the trigonal system, though differently labeled — the nonprimitive hexagonal unit cell can be replaced by the primitive rhombohedral one — thus is not new; this is perhaps not surprising since "centering" is not quite appropriate, the points not being placed there. The (single) trigonal system is thus (conventionally) labeled in two ways, and also two names, trigonal and rhombohedral, are used, so discussions of it may seem to imply (incorrectly) two systems, or that it is part of the hexagonal system. That both of these cells fit together to tile space is (merely) indicated in

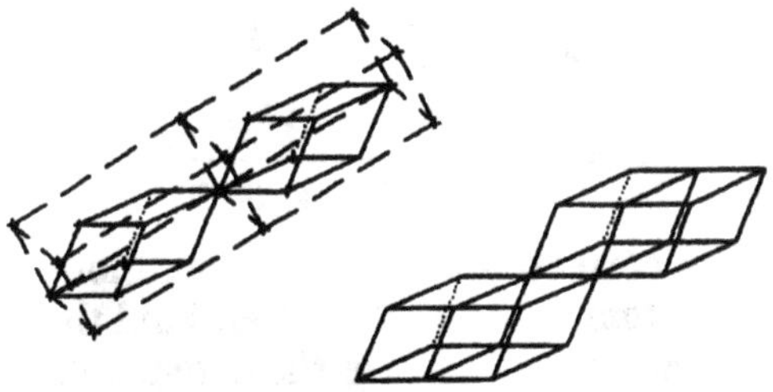

Figure II.6.d-2: RHOMBOHEDRAL UNIT CELLS FILL SPACE.

To consider this analytically, let the rhombohedral lattice vectors be $\underline{a}_r$, $\underline{b}_r$, $\underline{c}_r$, with equal magnitude, and identical angle α between members of each pair; the hexagonal lattice vectors are $\underline{a}_h$, $\underline{b}_h$, $\underline{c}_h$, with $\underline{c}_h$ perpendicular to $\underline{a}_h$ and $\underline{b}_h$, these with angle $\frac{2\pi}{3}$, and

$$|\underline{a}_h| = |\underline{b}_h|, \tag{II.6.d-1}$$

$$\frac{\pi}{2} \neq \alpha = \beta = \gamma < \frac{2\pi}{3}, \tag{II.6.d-2}$$

with the restrictions on the angles needed else the cell would belong to a more symmetric system.

Problem II.6.d-1: Verify the previous paragraph. For this the threefold symmetry axis makes equal angles with all three coordinate axes, as can be measured with a protractor. Draw this and measure. This unit cell is primitive and is said to be specified by rhombohedral axes of reference. It is the (primitive) unit cell of the trigonal system. A unit cell can also have three points; the threefold axis is the principle axis (perpendicular to the base) of length $|\underline{c}_r|$. For it (with the base in the plane perpendicular to the principle axis)

$$|\underline{a}_r| = |\underline{b}_r|, \ \frac{\pi}{2} = \alpha = \beta, \ \gamma = \frac{2\pi}{3}. \tag{II.6.d-3}$$

This cell is thus specified by hexagonal axes. Then [Burns and Glazer (1990) p. 45; Megaw (1973), p. 190, 207]

$$\underline{a}_r = \frac{1}{3}(2\underline{a}_h + \underline{b}_h + \underline{c}_h), \ \underline{b}_r = \frac{1}{3}(-\underline{a}_h + \underline{b}_h + \underline{c}_h), \ \underline{c}_r = \frac{1}{3}(-\underline{a}_h - 2\underline{b}_h + \underline{c}_h); \tag{II.6.d-4}$$

$$\underline{a}_h = \underline{a}_r - \underline{b}_r, \quad \underline{b}_h = \underline{b}_r - \underline{c}_r, \quad \underline{c}_h = \underline{a}_r + \underline{b}_r + \underline{c}_r; \qquad \text{(II.6.d-5)}$$

$$d = \frac{|\underline{c}_h|}{|\underline{a}_h|} = \frac{[9 - 12sin^2\frac{\alpha}{2}]^{\frac{1}{2}}}{2sin\frac{\alpha}{2}}, \qquad \text{(II.6.d-6)}$$

$$\underline{a}_h = 2\underline{a}_r sin\frac{\alpha}{2}; \qquad \text{(II.6.d-7)}$$

$$\underline{a}_r = \frac{1}{3}\underline{a}_h[3 + d^2]^{\frac{1}{2}}, \qquad \text{(II.6.d-8)}$$

$$\alpha = 2arcsin\frac{3}{2[3 + d^2]^{\frac{1}{2}}}. \qquad \text{(II.6.d-9)}$$

Check that these are correct, and that the conditions on each set of lattice vectors follow from those on the other. Find the matrices transforming one set to the other (how are these two sets of matrices related?). Each set contains one parameter (really two, but one is just the, here irrelevant, scale), the ratio of the sides, for one, the angle for the other. These equations relate them — the angle of the rhombohedron is determined by how much the hexagonal unit cell is stretched relative to its base. Verify that these points do form a lattice, and that it has threefold symmetry (its symmetry group is that of the trigonal system), the sixfold symmetry reduced by the insertion of points at alternating levels, at $\frac{1}{3}$ and $\frac{2}{3}$ of the distance between the bases, in addition to those at the corners — at $(a/3, 2a/3, 2c/3)$ and $(2a/3, a/3, c/3)$, which is the same as $\pm(a/3, 2a/3, 2c/3)$. Draw this and check. Why can points be put at, only, these positions? Check that these unit cells form lattices and do belong to the systems stated.

Problem II.6.d-2: The hexagonal and trigonal systems are thus related. This raises the question of whether there might also be such relationships between other systems. As we have seen, centering others does not reduce the symmetry, giving the same system. But might we put points at arbitrary positions, giving another system? Certainly not arbitrary, for the resultant set of points must form a lattice. Go through each of the systems, make a list of positions where points can be placed, and see whether there are analogous relationships for other systems, or whether the hexagonal-trigonal pair is unique. Explain the reasons.

Problem II.6.d-3: It would be interesting to do this in four dimensions, find the (analogs of the) crystal systems, make a list for each of the positions at which points can be placed, thus getting all lattices, and determining whether there are pairs that are related. Repeat it for n dimensions. An algorithm, computer-implemented, would be especially useful.

Problem II.6.d-4: Draw the lattice given by the vectors

$$\underline{a} : (k, 0, l), \ \underline{b} : (\sqrt{\frac{3}{2}}k, -\frac{k}{2}, l), \ \underline{c} : (-\sqrt{\frac{3}{2}}k, -\frac{k}{2}, l), \qquad \text{(II.6.d-10)}$$

and check that it belongs to the rhombohedral (or trigonal) system [Burns (1977), p. 374; Cornwell (1984), p. 200]. Show that these are equivalent to

$$3\underline{a} : (2k, k, l), \ 3\underline{b} : (-k, k, l), \ 3\underline{c} : (-k, -2k, l). \qquad \text{(II.6.d-11)}$$

In what way do they differ? Find the angles between the vectors. Compare these with the stated sides and angles (tbl. B-2, p. 634).

Problem II.6.d-5: Both the hexagonal and trigonal systems have a single lattice each, the primitive ones. This gives fourteen altogether. Check that these are (and have been shown to be) lattices. Find the translations that leave each invariant [Streitwolf (1971), p. 59]. Summarize why further points can not be added.

II.7 THE SEVEN CRYSTAL SYSTEMS AND THEIR SYMMETRY GROUPS

Lattices are classified into seven systems (called crystal systems, although they are really systems of lattices). These are distinguished by their symmetry groups, so here we summarize which of the point groups determine lattices, and why others do not. The symmetry group of a lattice, but not necessarily of a crystal, must contain the inversion. However not every point group containing the inversion (tbl. B-2, p. 634) can be such a symmetry group (pb. II.4.f-6, p. 95). The inversion-containing groups C_{4h}, C_{6h}, S_6, and T_h do not give crystal systems. These are subgroups of the holohedry groups of the systems. Point groups not containing the inversion are also holohedry subgroups. The seven allowed lattice symmetry groups, each giving a crystal system, are then

$$S_2, C_{2h}, D_{2h}, D_{4h}, D_{3d}, D_{6h}, O_h. \qquad \text{(II.7-1)}$$

For example in fig. C-12, p. 643, is an object invariant under C_{4h}, but not D_{4h}. To reduce the symmetry added structure is needed, in this case arrows. But with the structure, the object is no longer a lattice. The points (at the vertices) are not identical, some being closer to the tips of arrows than others. Groups C_{4h}, C_{6h}, S_6, and T_h, as we will see, are not symmetry groups of lattices because they are not rich enough — a lattice, being simple, has a large symmetry group; these do not contain all required transformations. This is illustrated by this example of a lattice to which has been added structure, so that it is no longer a lattice, giving it symmetry group C_{4h}.

II.7.a Systems, groups and lattices

There are seven crystal systems, thirty-two point groups, eighteen non-isomorphic abstract groups, and fourteen Bravais lattices. In tbl. B-2, p. 634, notice that C_2 appears three times, realized (and named) differently. These three groups are isomorphic as abstract groups, but have different geometrical meaning. Why do two belong to the same system, but the third to a different one? The transformations of these three, acting on the coordinates, are (with the rotation around the z axis)

$$I=\begin{pmatrix}-1&0&0\\0&-1&0\\0&0&-1\end{pmatrix},\ C_2=\begin{pmatrix}-1&0&0\\0&-1&0\\0&0&1\end{pmatrix},\ \sigma_h=\begin{pmatrix}1&0&0\\0&1&0\\0&0&-1\end{pmatrix}. \tag{II.7.a-1}$$

The trace of the first, of the triclinic system, is -3, of the other two, of the monoclinic, are -1 and 1. There is no similarity transformation connecting these, illustrating why they are not (realizations of) the same group.

A refinement of this classification [Senechal (1990), p. 92] has matrices

$$\sigma_x=\begin{pmatrix}1&0&0\\0&-1&0\\0&0&1\end{pmatrix},\ \sigma_{xy}=\begin{pmatrix}0&1&0\\1&0&0\\0&0&1\end{pmatrix}. \tag{II.7.a-2}$$

giving reflections in the xz plane and in the line $x = y$. These have the same characters — reflections are equivalent — so there are similarity transformations taking one to the other, but with real matrices — the entries in the matrix giving the similarity transformation are real numbers. However there is no matrix giving such a conjugation with integer entries. This classification, distinguishing between operations that can be conjugated with real but not integer matrices, is important, and we say that these reflections are in the same geometrical equivalence class, but different arithmetical ones (sec. III.7, p. 169) [Janssen (1973), p. 119; Sternberg (1994), p. 314].

Problem II.7.a-1: Find the transformation between σ_x and σ_{xy}. A general similarity transformation cannot have integer entries [Sternberg (1994), p. 313]. Why?

Problem II.7.a-2: The point groups listed in tbl. B-2, p. 634, should belong to eighteen nonisomorphic classes [Janssen (1973), p. 121]: C_1, C_2, C_3, C_4, C_6, D_2, D_3, D_4, D_6, $C_4 \otimes C_2$, $C_6 \otimes C_2$, $D_2 \otimes C_2$, $D_4 \otimes$

C_2, $D_6 \otimes C_2$, T, O, $T \otimes C_2$, $O \otimes C_2$. This emphasizes that groups can be isomorphic as abstract groups, but have different geometrical, and physical, meaning, so be regarded as distinct.

II.7.b Derivation of the Bravais lattices from their symmetry

The Bravais lattices can be obtained rigorously using their symmetry [Sternberg (1994), p. 314]. Lattices with different symmetry groups are different, but as we see there is generally more than one lattice for each holohedry group, and it is interesting to see why in this approach.

The analysis uses a plane lattice, the lattice in the xy-plane that is the projection of the three-dimensional lattice. And this projection must be a lattice itself. The primitive lattice vectors in the xy-plane, that perpendicular to the principle axis of rotation along z, are $\underline{a}$ and $\underline{b}$. There is one further lattice vector, not in the plane, $\underline{c}$, taken as the one with the smallest z component. Its component in the plane is c_p (and vector $\underline{c}$).

For the first lattice all primitive lattice vectors, and angles between them, are unequal. This is the triclinic, with symmetry group S_2, the smallest possible holohedry group for a lattice.

II.7.b.i *The monoclinic lattices*

Next we add a π rotation (around z), denoted by $R(\pi)$. The effect of a symmetry rotation on a lattice vector is to produce another lattice vector, which can also be obtained by adding a lattice vector to the first, so $\underline{c} - R(\pi)\underline{c}$ is a lattice vector. And since the z component is invariant,

$$\underline{c} - R(\pi)\underline{c} = 2\underline{c}_p, \qquad \text{(II.7.b.i-1)}$$

as the rotation simply reverses the direction of the component.

Now if $\underline{c}_p$ is a lattice vector it is an integral sum of $\underline{a}$ and $\underline{b}$, which can be subtracted to give zero, so $\underline{c}$ lies along z. If not, $2\underline{c}_p$, linking two lattice points, is a lattice vector so that

$$\underline{c}_p = \frac{1}{2}\underline{a},\ \frac{1}{2}\underline{b},\ \text{or}\ \frac{1}{2}(\underline{a}+\underline{b}), \qquad \text{(II.7.b.i-2)}$$

since these are the smallest possibilities. We can specify the lattice vectors such that $\underline{c}_p = \frac{1}{2}(\underline{a}+\underline{b})$. The most general case, with a π rotation, is for the lattice vectors to have unequal lengths (taking $|\underline{a}| < |\underline{b}|$), with the angle between $\underline{a}$ and $\underline{b}$ neither $\frac{\pi}{2}$ nor $\frac{\pi}{4}$. Then a symmetry transformation can only take lattice vectors into themselves, or reverse their directions. A transformation that sent $\underline{c} \Rightarrow -\underline{c}$ would be a rotation about a line in the xy plane, but this can only be a symmetry if the angle between the lattice vectors in the plane is $\frac{\pi}{2}$ or $\frac{\pi}{4}$, which we excluded (for now). Thus there are no further symmetry rotations of the specified lattice, although there is a (horizontal) mirror, so the symmetry group of the lattice is C_{2h}. The lattices can only be further distinguished (up to change of scale) by the angle between $\underline{a}$ and $\underline{b}$. They are therefore called monoclinic lattices (having but *one* free incl*ina*tion). And as we see, this argument shows that there are two. The one with $\underline{c}$ along the axis of rotation z, is the primitive monoclinic lattice, that with $\underline{c}$ at an angle to z is the face-centered monoclinic.

Why is the nonprimitive lattice face-centered? The only other possibility is body-centered. Putting a lattice point in the center of the cell would give a point with an environment different from those in the plane, so the set would not be a lattice. We now have all three lattice vectors specified, thus there cannot be points on the other faces; only one pair of opposite faces can be centered, as we know (sec. II.6.a, p. 103).

This illustrates why there are two lattices for this system. The system is defined by its largest (holohedry) group, here C_{2h}. There are two types of lattices with this symmetry group, taking lattices with different angle between $\underline{a}$ and $\underline{b}$, or with different lengths of the primitive lattice vectors, as the same, except for values that give an increase in symmetry. This does not mean that these differences are unimportant, certainly not for physics. But they are given by continuous variables, so must be ignored for a discrete classification. The two monoclinic lattices are distinct, so are different enough to distinguish. There are two because with this pair of plane lattice vectors there are two distinct, discrete possibilities for the third lattice vector. In particular, there are no transformations of space, rotations, reflections, inversions, or stretchings or contractions, that take one into the other — and they are not given by special cases of a continuous variable, like an angle.

Problem II.7.b.i-1: Repeat this discussion with diagrams for the two lattices and check that it is correct. Explain how we can tell which lattice is primitive, which centered.

II.7.b.ii *Tetragonal lattices*

We next add to the symmetry group a rotation around z of $\frac{\pi}{2}$, $R(\frac{\pi}{2})$. Then $\underline{a}$ and $\underline{b}$ are of equal length and orthogonal. Of the four possibilities consider first

$$\underline{c}_p = \frac{1}{2}\underline{a} \text{ or similarly } \underline{c}_p = \frac{1}{2}\underline{b}. \qquad \text{(II.7.b.ii-1)}$$

Then

$$\underline{c} - R(\frac{\pi}{2})\underline{c} = \frac{1}{2}(\underline{a} - \underline{b}), \qquad \text{(II.7.b.ii-2)}$$

since this rotation moves $\underline{a}$ to $\underline{b}$, giving a vector smaller than these which are taken as the smallest, these two cases are ruled out. Thus

$$\underline{c}_p = 0, \quad \text{or} \quad \underline{c}_p = \frac{1}{2}(\underline{a} + \underline{b}). \qquad \text{(II.7.b.ii-3)}$$

For the length of $\underline{c}$ different from those of $\underline{a}$ and $\underline{b}$ there are no other symmetry rotations, except for ones of π around x and y. This is clearly true for $\underline{c}$ along z. In the other case this rotation around x leaves $\underline{a}$ unchanged and reverses the direction of $\underline{b}$, as well as $\underline{c}$. But these three are equivalent lattice vectors, so the lattice is invariant. The lattices are also invariant under reflection in the xy plane, and in that bisecting $\underline{c}$ (diagonal, horizontal and vertical planes). Their symmetry group is thus D_{4h}. Having four equal angles, they are the tetragonal lattices.

Point group C_{4h} is not a symmetry group of a lattice (although it contains the inversion). Comparing the tetragonal and monoclinic lattices indicates why. A π rotation around x, or y, reverses the direction of the other lattice vector in the xy plane, so cannot be a symmetry unless the resultant vector is also a lattice vector. This is not true unless the lattice vectors in this plane are orthogonal, so the rotation does not give a symmetry for the monoclinic lattice. But their being orthogonal is a strong enough restriction so that the π rotation is a symmetry. Hence it leads to the invariance group being dihedral; C_{4h} is too small to include every symmetry transformation.

The two lattices then are the primitive tetragonal (the P-lattice) and the body-centered tetragonal (the I-lattice). Why body-centered? For it

$$\underline{c}_p = \frac{1}{2}(\underline{a} + \underline{b}). \qquad \text{(II.7.b.ii-4)}$$

The projection in the xy plane of $\underline{c}$ goes to the middle of the unit cell in this plane, which is the projection of the center of the three-dimensional unit cell, as well as the opposite face. Suppose that the lattice point were on the face. There would then also be one in the xy plane, and one lattice vector would lie along x, the other would go from the origin to the lattice point in the center, so these two would not be orthogonal, as we assumed — the lattice then must be body-centered.

Although there are two tetragonal lattices we can actually get four (sec. IV.5.b.iii, p. 223), but these are equivalent to the primitive and body-centered ones [Burns (1977), p. 259; Megaw (1973), p. 195, 204]. Besides these two, the base of the tetragonal cell (which is square) can be centered (although not the sides). However this is equivalent to the primitive, but rotated by $\frac{\pi}{4}$. Likewise all faces can be centered, but this gives merely a differently-oriented body-centered lattice. The tetragonal lattice can be obtained from the cubic by stretching or shrinking one side. The cubic has three distinct lattices, but the change of length of one side makes two lattices equivalent. Also by symmetry there cannot be only one pair of faces of the cube centered, but as this system has less symmetry, such centering is permitted.

Problem II.7.b.ii–1: Draw diagrams to check these arguments.

Problem II.7.b.ii–2: The C-lattice is obtained if the pair of bases are centered (sec. IV.5.b, p. 221; sec. IV.6.b, p. 243). Show that it is equivalent to the P-lattice but with primitive lattice vectors rotated by $\frac{\pi}{4}$ to those of the P-lattice. The primitive cell is larger for the C than for the P-lattice (and usually we take the smallest unit cell, so consider the P, not the C-lattice). What is the ratio of their volumes? Check that if a pair of side faces are centered, the result is not a lattice. If all faces are centered the resultant lattice (the F-lattice) is equivalent to the I-lattice, but differently oriented. However the I-lattice should have a smaller unit cell.

II.7.b.iii *The orthorhombic lattices*

The monoclinic lattice can be generalized another way. A second C_2 rotation, with axis perpendicular to that of the first (of course) can be added. Thus the symmetry group of this lattice includes D_2, which does not contain the inversion, so cannot be the full symmetry group, which implies that D_{2h} is, as we have to show.

A plane lattice invariant under reflection in a line lying in it is either rectangular or shaped like a diamond — the primitive lattice vectors can be chosen to give a unit cell that is either a rectangle or a rhombus.

Problem II.7.b.iii–1: This can easily be proved by drawing. Take the mirror on the x axis and denote the reflection by r. Then for lattice vector $\underline{f}$ in the plane check that we get that $\underline{f} + r\underline{f}$ lies on x, $\underline{f} - r\underline{f}$ on y. These are orthogonal, thus linearly independent — there are lattice vectors on the x and y axes. These may not be primitive. The shortest of these axes are denoted by $\underline{u}$ and $\underline{v}$. Then any lattice vector $\underline{l}$ in the plane can be written

$$\underline{l} + r\underline{l} = m\underline{u}, \quad \underline{l} - r\underline{l} = n\underline{v}; \qquad \text{(II.7.b.iii–1)}$$

m and n are integers. So

$$2\underline{l} = m\underline{u} + n\underline{v}. \qquad \text{(II.7.b.iii–2)}$$

If $\underline{u}$ and $\underline{v}$ are primitive, all m's and n's are even. The lattice is then rectangular. Otherwise (why?) we can take $m = 1$ (or $n = 1$), so $\frac{1}{2}\underline{u} + \frac{n}{2}\underline{v}$ is a lattice vector. As $\frac{1}{2}\underline{u}$ is not a lattice vector, n cannot be even. Taking the smallest value, show that we find

$$\underline{a} = \frac{1}{2}(\underline{u} + \underline{v}) \text{ and } \underline{b} = r\underline{a} = \frac{1}{2}(\underline{u} - \underline{v}) \qquad \text{(II.7.b.iii-3)}$$

to be primitive lattice vectors. These give a diamond-shaped lattice, with the unit cell a rhombus. Why is the unit cell a rhombus? Check this with a drawing.

Problem II.7.b.iii-2: We use this to find the three-dimensional lattice. The most general case is for different lengths of $\underline{a}$, $\underline{b}$, and $\underline{c}$. First take the plane lattice rectangular. Then as for the tetragonal case (why?),

$$\underline{c}_p = 0, \ \frac{1}{2}\underline{a}, \text{ or } \frac{1}{2}(\underline{a} + \underline{b}), \qquad \text{(II.7.b.iii-4)}$$

($\frac{1}{2}\underline{b}$ gives nothing additional, being just the result of interchanging the x and y axes). For the diamond shaped lattice, $\underline{c}_p = \frac{1}{2}\underline{a}$ cannot occur. Why? The π symmetry rotation about an axis in the xy plane, $R_p(\pi)$, must interchange $\underline{a}$ and $\underline{b}$, as it cannot lie along one (as can be seen by drawing, or simply from the fact that these are not distinguished, so a symmetry axis along one requires such an axis along the other, which gives too much symmetry — why?). That rotation would give

$$\underline{c} + R_p(\pi)\underline{c} = \frac{1}{2}(\underline{a} + \underline{b}), \qquad \text{(II.7.b.iii-5)}$$

since this sum is invariant under the rotation, but this is not a lattice vector. Why? There are thus five cases left (but as two are equivalent, there are four lattices), three for the rectangular lattice, and two for the diamond-shaped.

Problem II.7.b.iii-3: Now for

$$\underline{c}_p = 0, \qquad \text{(II.7.b.iii-6)}$$

$\underline{c}$ is perpendicular to $\underline{a}$ and $\underline{b}$, which by D_2 symmetry requires them to be orthogonal, and the lattice is rectangular. Why? This is the primitive orthorhombic lattice, being based on a rhombus whose sides are orthogonal. If the plane lattice is rectangular,

$$\underline{c}_p = \frac{1}{2}\underline{a} \qquad \text{(II.7.b.iii-7)}$$

gives the face-centered orthorhombic lattice (points in the centers of all faces); as $\underline{c}$ goes to a lattice point out of the plane, and its projection on the plane lies along one of the lattice vectors, it can only go to the center of a face, and since $\underline{a}$ can be taken as any lattice vector in the plane, there must be a point at the center of every face. The case

$$\underline{c}_p = \frac{1}{2}(\underline{a}+\underline{b}) \qquad \text{(II.7.b.iii-8)}$$

gives the body-centered lattice — this is the projection of the center of the unit cell. The diamond plane lattice, which also has

$$\underline{c}_p = \frac{1}{2}(\underline{a}+\underline{b}), \qquad \text{(II.7.b.iii-9)}$$

but with different vectors, gives the base-centered orthorhombic lattice (points added to only one pair of opposite faces). Why? For this the faces are distinguished, as one lattice vector lies on the edge of one, the other not. Thus $\frac{1}{2}(\underline{a}+\underline{b})$ is the projection on the (unit cell of the) plane of a point on the opposite face of the unit cell. Therefore only these two faces are centered. Draw these and check.

Problem II.7.b.iii-4: Show that these (all) have D_2 symmetry, plus the inversion, and no other symmetry generators, so their invariance group is D_{2h}.

Problem II.7.b.iii-5: Verify that these are all lattices.

II.7.b.iv *The trigonal and hexagonal lattices*

These two lattices are related (sec. II.6.c, p. 107), so we consider them together. There is one rotation not yet discussed, that of $\frac{2\pi}{3}$, denoted by $R(\frac{2\pi}{3})$, and here it is the largest one, so its axis is the principle axis, z. Lattice vectors $\underline{a}$, $R(\frac{2\pi}{3})\underline{a}$, $R(\frac{2\pi}{3})^2\underline{a}$, in the xy plane, form an equilateral triangle (pb. II.4.c-6, p. 89). Any two of these are the sides of a parallelogram that is the unit cell of the plane lattice — it contains no other plane lattice vector, for since these two are also sides of an equilateral triangle any such vector would be shorter than $\underline{a}$. Thus the primitive lattice vectors in the plane are $\underline{a}$ and $R(\frac{2\pi}{3})\underline{a}$.

Again we consider the third lattice vector (the one not in the plane) $\underline{c}$ and its plane projection $\underline{c}_p$. As before $|\underline{c}_p| \le |\underline{a}|$. Also $R(\frac{2\pi}{3})\underline{c} - \underline{c}$ is a lattice vector in the plane. And

$$|R(\frac{2\pi}{3})\underline{c}-\underline{c}| = |R(\frac{2\pi}{3})\underline{c}_p - \underline{c}_p| = \sqrt{3}|\underline{c}_p| \le \sqrt{3}|\underline{a}|. \qquad \text{(II.7.b.iv-1)}$$

What lattice vectors are there in the plane with length $\sqrt{3}|\underline{a}\,|$? These are

$$\underline{0},\ \underline{a},\ \underline{a}+R(\frac{2\pi}{3})\underline{a},\ R(\frac{2\pi}{3})\underline{a},\ -\underline{a},\ -\underline{a}-R(\frac{2\pi}{3})\underline{a},\ \underline{a}-R(\frac{2\pi}{3})\underline{a}, \qquad \text{(II.7.b.iv-2)}$$

with all, except $\underline{0}$, having the same length. So

$$\underline{c}_p = 0, \tag{II.7.b.iv-3}$$

or is one of the other six. Since it is irrelevant which is taken, we get

$$R(\frac{2\pi}{3})\underline{c}_p - \underline{c}_p = -\underline{a}, \text{ giving } \underline{c}_p = \frac{2}{3}\underline{a} + \frac{1}{3}R(\frac{2\pi}{3})\underline{a}. \tag{II.7.b.iv-4}$$

The first gives the hexagonal lattice, with symmetry group D_{6h}, the second the trigonal, with symmetry group D_{3d}. There is no further freedom, so these two lattices are fully determined, thus there is only one in each system, the primitive ones.

We notice that S_6 and C_{3h}, and as we see next, T_h, though they contain the inversion, are not rich enough to be lattice symmetry groups.

Problem II.7.b.iv-1: Why?

Problem II.7.b.iv-2: Check that these are all lattices, and the correctness of the symmetry groups, using this construction.

II.7.b.v *The cubic system*

All crystallographic point groups which can be lattice symmetry groups have now been studied, except for the tetragonal, T, and octahdral, O, sets. Thus we now take a system whose symmetry includes T. This can be considered as the orthorhombic system with more symmetry, as D_2 is a subgroup of T, so we start with the results for that system.

The lattice vectors are $\underline{a}$, $\underline{b}$ and $\underline{c}$. Among the elements of T are rotations $R(\frac{2\pi}{3})$, so

$$R(\frac{2\pi}{3})\underline{a} \neq \pm\underline{a} \tag{II.7.b.v-1}$$

therefore

$$R(\frac{2\pi}{3})\underline{a} = \pm\underline{b}, \text{ and } R(\frac{2\pi}{3})^2\underline{a} = \pm\underline{c}. \tag{II.7.b.v-2}$$

Thus these are all of the same length, and this is similar to the orthorhombic system, but now the rectangular parallelepiped (unit cell) is a cube. The base-centered lattice is not invariant under the rotations $R(\frac{2\pi}{3})$ (which mix faces, so that a single pair cannot be centered). We are left then with the primitive cubic, the body-centered cubic, and the face-centered cubic, in which all pairs of faces are centered. The symmetry group is O_h.

Problem II.7.b.v-1: Show that the symmetry group is correct.

Problem II.7.b.v-2: Explain why T is not a symmetry group (with and without the inversion).

II.7.c Seven lattice symmetry groups and fourteen lattices

There are, in three dimensions, seven crystallographic point groups capable of being holohedry groups, so there are seven crystal systems. And there are fourteen distinct lattices (with distinct being determined by the conventions listed). These then are the set of lattices that we and the laws of geometry working together have decided exist in three-dimensional space. We leave open the questions of what laws, what geometry, and why we, as crystallographers, physicists, mathematicians, geometers, group theorists, have so decided — and whether, if we were to categorize ourselves differently, we would categorize lattices differently.

II.7.d Decreasing the symmetry

A lattice with cubic symmetry is a set of points. On these we can place physical objects, usually atoms or molecules although buildings in a city are another possibility, giving (for some objects) a crystal. And this can reduce the symmetry. What symmetries result from various different arrangements, and types of atoms?

Problem II.7.d-1: Is it reasonable to consider placing buildings in a city on a lattice? Are three-dimensional lattices relevant?

Problem II.7.d-2: Can (spinless) atoms be put on (all) vertices of a cube such that the symmetry is of octahedral group O(432), only, with no reflection or inversion symmetry? Suppose they have spin, what types of symmetry breaking are possible?

Problem II.7.d-3: Break the symmetry by using two types of atoms, as in pb. II.2.d-3, p. 71, and state the reflections and rotation-reflections that remain, and their interpretation.

Problem II.7.d-4: For a crystal use not atoms, but molecules. If one atom of each molecule occupies a lattice point, what can be said about the points occupied by the other atoms? Can some other point of the molecule, at which an atom is not placed, be at a lattice point? Now suppose the molecule does not have spherical symmetry. Take it with all molecular point groups (sec. I.6.f, p. 46) in turn (or a reasonable number) and find the symmetry group of the resultant crystal. Could (must) the symmetry be less if the molecular symmetry group were also cubic? Try this for all three cubic lattices.

Problem II.7.d-5: How does the symmetry of a tetragonal lattice change if a small sphere is placed at the center of the base of every unit cell? Of one of the rectangular faces [Lines (1965), p. 200, 279]? Similarly place regular tetrahedra (how?) in a cubic lattice.

Problem II.7.d-6: Draw a cube and show that four points can be chosen that form the vertices of a tetrahedron, so the cube is two tetrahedra placed together — an obvious generalization (what?) of the case for a square. Relate this to the subgroup structure of the holohedry group of the cube.

Problem II.7.d-7: Draw a cube showing the full symmetry, then color it (sec. VIII.4, p. 411) to break the symmetry to each of the subgroups of the holohedry group [Lax (1974), p. 61].

II.7.e Dilatations relating lattices

There are transformations (dilatations or dilations) that change lengths. Several Bravais lattices are related to others by a change in a length, which for some axes, could give distortion [Kettle (1995), p. 187].

Problem II.7.e-1: Make a list of these relations.

Problem II.7.e-2: State the rules relating the lattices. Do the transformations interchanging Bravais lattices this way form a group? If so, what is its multiplication table (or perhaps, more properly, the rule for combining operations); which group is it, if it is a group; what are its representations; do these have geometrical or physical significance? Consider the set of these transformations plus the transformations of all point groups, and answer the same questions for it. (Continuous groups called conformal groups [Barut and Raczka (1986), p. 412] have dilatations and rotations as subgroups and it might be interesting to answer the questions in the context of these groups.) There are also transformations that change angles and the discussion of these groups can be enlarged to include such transformations (although there is a slight problem that the unit cells of some Bravais lattices include more points than those of others). Is there any way of finding a group whose transformations give all Bravais lattices starting from one? Does it have significance? It might also be interesting to extend these questions to other dimensions, and to crystals, which can have less symmetry than a lattice.

II.7.f Crystals and lattices

A lattice is a set of points, a crystal is a set of sets of atoms (or molecules or ions, for example, or more generally sets of identical points) with each set placed at (or assigned to) a lattice point. Hence the concept of a crystal (or at least the collection of crystals) is richer than that of a lattice. The Bravais lattices are the configurations in which objects must arrange their atoms to give crystals. Of course we cannot prevent one from arranging its atoms in some other way if it wished. However we would not then regard it as a crystal — and would not be able to apply (many of) these results. As a crystal can have less symmetry than the lattice on which it is based, there are (many) more types of crystals than of lattices. It should not be surprising if nature makes good use of this diversity. However the possibilities are even greater than have been considered.

Problem II.7.f–1: Design arrangements (if there are such) that violate one condition, only, on the Bravais lattices, for each condition. (It would probably help to first state explicitly the conditions.) What are the resultant symmetries? Do the objects obtained seem like they might occur in nature? Why? Would they be interesting, or even economically valuable?

Problem II.7.f–2: It is useful to verify that objects, many have been considered here, illustrating the point group symmetries do actually have the symmetry of the crystallographic systems stated (tbl. B–2, p. 634). For each object draw a unit cell (sec. I.3.b, p. 7) containing it and check that the cell, including the object, gives a lattice invariant under the proper group (and no larger one), and belongs to the given system. Notice the difference between objects invariant under the largest group of the system (the holohedry group), and those invariant only under a subgroup. Explain, for each case, how the symmetry of the holohedry group is broken by the object. Draw other such objects (three-dimensional models, and computer programs to draw models of such, would be interesting). A computer-aided design program, say, that carries out the operations, can be helpful.

Problem II.7.f–3: The objects illustrated here are artificial, used only to show aspects of the theory. Thus studying more realistic ones might shed further light into crystals, lattices, geometry, and physics. The reader will have to decide whether this light illuminates any useful properties. Construct unit cells, with atoms, invariant under these groups. Compare with the pictured objects. Check that they belong to the proper systems (tbl. B–2, p. 634). Discuss, again, the difference between objects invariant under a holohedry group, and ones invariant only under a subgroup.

Problem II.7.f–4: Is it possible, by just looking at these objects, to tell which crystal systems they belong to, and why we distinguish systems rather than differentiating objects only by their symmetry groups? Can you tell why objects are grouped together into systems, why it makes sense and is useful to do so, and also why the groupings are reasonable and useful (mathematically, physically, biologically, economically)?

Problem II.7.f–5: Above (sec. II.7.d, p. 120; sec. II.7.e, p. 121) the relationships between the different systems was discussed, indicating how to get each from the cubic. Draw diagrams showing how to do this, for each of the three cubic lattices. Which of the resultant figures gives a lattice, and which are new? Are all fourteen lattices obtained this way? Why?

Problem II.7.f-6: As in the problem for the cubic system (sec. II.6.b, p. 105), find the lattice vectors for the other eleven Bravais lattices [Chen (1989), p. 466].

Problem II.7.f-7: The lattice constants of these problems are real. We do not consider what would be obtained if they were not (if they were complex, for example, if possible).

Problem II.7.f-8: As a check note that we cannot have only certain pairs of faces centered without destroying a symmetry for certain systems. Thus verify that we have all possible cubic lattices. For the monoclinic systems only one pair of faces receives a point. Which pair it is does not matter; it would simply change the labeling to use another pair (although this is not unimportant, especially if the faces were distinguished in some interesting ways). Explain. It should thus be clear that the system has all lattices allowed for it, as does the cubic. For the tetragonal system face-centering is not possible because it would require two faces be centered to prevent loss of symmetry, and this does not give a lattice. Check this. These lattices are both obtained from the cubic by dilations. How does this prevent some nonprimitive cubic lattices from going into nonprimitive ones of these systems?

Problem II.7.f-9: Can the (different kinds of) atoms of a crystal be placed on a lattice, so the crystal (really object) has no (point group) symmetry? Can there be crystals made up of (say, two) kinds of atoms, each of which belongs to a crystal system, with the crystal systems different for the two types? Are there constraints? What could be said of the entire crystal in such a case? Are there actual crystals (or, more generally, objects) for which these questions are relevant?

Problem II.7.f-10: That there are five lattices in two dimensions, and fourteen in three, is a property of our (Euclidean, flat) geometry. Explain how it is related to the axioms of geometry. Construct geometries in which the numbers of lattices are different, or prove that such is impossible. Explain the results.

Problem II.7.f-11: The invariance transformations of crystals are associative (fortunately), allowing these to form groups [Mirman (1995a), sec. I.4.a, p. 14]. What properties of lattices and crystals (and geometry?) require (allow?) this? These transformations can be regarded in two ways, either transforming the crystal leaving the observer fixed, or transforming the observer, leaving the crystal fixed. Here they are taken as the same (except the representation matrices are inverses). The operation transforming the crystal can be interpreted as relating two identical ones, relatively transformed, or it can be regarded as a physical transformation on the object. Is this last view relevant to associativity? (In reality, an actual physical transformation can change the crystal, so we must be careful about what we are doing.)

Problem II.7.f-12: Give the group table for each point group (preferably using the symbols for physical transformations on crystals, perhaps from actual pictures, or better three-dimensional physical objects), and check that each does obey the group axioms. For many, if not all, these are listed [Mirman (1995a)], but maybe with different symbols. Compare the results with those.

Problem II.7.f-13: A rotation is performed by a series of infinitesimal rotations. Thus a motion picture can be made showing it. Doing this for a reflection, or an inversion, requires more imagination. One possibility is to replace the coordinates of the points in an object by complex numbers (which is not a trick, but an idea which can be useful [Mirman (1995b), sec. 7.2, p. 124; Streater and Wightman (1964)]); the imaginative reader can perhaps think of other ways. After doing so, write a computer program that shows the effect of each of the point group transformations, especially inversions and reflections, on objects, first lattices then more interesting configurations of which there are many such possibilities [Armstrong (1988); Burn (1991); Hargittai and Hargittai (1987); Lockwood and Macmillan (1983); Shubnikov and Koptsik (1974)]. Apply all transformations of every group to each object, thus showing visually which are invariant under each group, and which not, and why. Also use this to show that these transformations form groups. This can be extended to groups that are not (crystallographic) point groups to show why they are not. For example, a transformation can not only change the position or orientation of an object, but its color, particularly the colors of its parts (sec. VIII.4, p. 411); varying these continuously would be attractive. Thus a picture can simultaneously show the object before and after transforming, clarifying whether it is invariant under the transformation. Likewise two transformations, and their product, can be shown using a set of colors, thus proving (not rigorously, of course) that the product has the same effect as the transformations in succession. How is associativity shown?

Problem II.7.f-14: Computer programs can also be used to show, algebraically, invariance of objects under transformations and groups, and to find the ones under which each object is invariant, say by assigning coordinates to lattice points, or points in the object [Janssen (1973), p. 114]. They might easily be generalizable to other dimensions (and geometries?), presumably helping to find the lattices in such spaces.

Problem II.7.f-15: The symmetric and alternating groups have automorphism groups [Mirman (1995a), sec. IX.6, p. 267]; what are they? How are they related to the geometrical symmetries (which are, or are subgroups of, a few of these)? That of S_6 is relatively larger than those for the other symmetric groups. Does this have geometrical relevance?

Problem II.7.f-16: Several proofs, or explanations, have been given for the Bravais lattices. How are they related? Which are the most rigorous? Why? Which can be (most easily) generalized to other dimensions or spaces with other signatures (eq. IV.3.h-1, p. 211) [Janner and Ascher (1969a,b, 1971); Mackay and Pawley (1963)]? Which show (most clearly) what restrictions can be removed to generalize lattices to other reasonable figures? Are there any such? Which best show how the properties of geometry determine those of lattices and crystals (or could it be that these are only due to definitions)?

Problem II.7.f-17: A lattice is a group (pb. II.2.a-3, p. 68), specifically an infinite discrete, Abelian group. Are all infinite discrete, Abelian groups lattices? Here we have found all lattices in (Euclidean) spaces of dimension one, two and three. Thus we have found all, or a specific subset, of all such groups with one, two and three generators. For each of these, give its presentation [Mirman (1995a), sec.III.3, p. 82], and its group table (which presumably can be given in some compact form, since a lattice is a repetition of a unit cell). Is it possible to tell from these whether a lattice is primitive or nonprimitive? Do they show which crystal system a lattice belongs to? Can these be used to show that the lattices we have found are all, and different? Can they be used to determine if the set of lattices exhaust the set of infinite, discrete Abelian groups with a given number of generators? Is it possible to find general rules on presentations and group tables that would allow the determination of the set of lattices in any dimension, specifically 1,2 and 3? If so, it would be useful to find all lattices in dimensions 4,5,6, ..., and even obtain a rule for all in a space of arbitrary dimension. Either show that this gives all these groups with the specified number of generators, or determine how ones not given differ from these, leading to rules for their determination.

Problem II.7.f-18: Here we have imposed totally arbitrary conditions and obtained a set of lattices into which we shall fit the crystals that occur in nature. Thus, for example, by requiring that a physical object be a lattice we have ruled out two-face centering. Yet most (all?) objects with symmetry (seem to?) fall into our classification scheme. Nature (apparently?) does agree with our esthetic taste. It is interesting to wonder why.

II.8 THE SYMMETRIES OF CUBIC AND HEXAGONAL LATTICES

Given an object how do we determine its symmetries: the operations, and the points, lines and planes they leave unchanged? To illustrate, we start with the cube, some of whose symmetries are obvious, others less so. Stating explicitly the reasons for the obvious symmetries helps develop rules, and intuition, which can be used in less straightforward cases. As a single object a cube is not restricted by the lattice condition that limits the angles of rotation. However cubes fill space — they can be joined together to form a lattice — showing that they do obey this condition.

II.8.a The rotational symmetries of the cube

What are the symmetries of the cube [Armstrong (1988), p. 37; Joshi (1982), p. 235] (fig. I.4.d.i-1, I.4.d.i-2, p. 19)? Rotations about lines through the center of the cube (which is always fixed) to any point but a vertex, or the center of a face or edge, do not leave the cube invariant, moving vertices, but not to other ones. A rotation taking a face into itself about an axis through the center of the face, with vertices and edges going into adjacent ones, has order four. There are three pairs of (identical) faces, so three such axes (any one, labeled z, is taken as the principle axis). Also there are rotations for which every face is taken into a different one. There are axes along the diagonals of the cube (four), $\frac{\pi}{4}$ from the z axis, and also axes in the planes bisecting the faces, through the centers of the edges. A rotation about an axis through the midpoint of an edge moves the diagonals so interchanges pairs of them. Twice the rotation must therefore be (equivalent to) the identity (that is 2π), so its angle is π. The final set are rotations about axes along the diagonals (through vertices), these through $\frac{2\pi}{3}$. They interchange two faces, but only for this angle, and integral multiplies of it, are faces moved into one another, so these, but only these, are symmetries. Why $\frac{2\pi}{3}$? Diagonal axes fix two of the eight vertices. Consider a corner on the same face as one of these two. A rotation must keep it on the same face (say the top), take it to the corner on the opposite end of the diagonal of a perpendicular face, which is impossible because the axis goes through this corner so is fixed, or take it to the corner opposite on the diagonal through the center, and perpendicular to the rotation axis. This cannot occur for all three pairs of vertices without turning the cube inside out (which is not a rotation). Thus the three corners on one face are mixed, so three such rotations gives the identity, each therefore is of $\frac{2\pi}{3}$. That the cube has threefold symmetry can be seen by looking directly at a vertex [Martin (1987), p. 199]. The three edges (emphasized) going into each other under a $\frac{2\pi}{3}$ rotation are shown in fig. I.4.d.i-1, p. 18. Also shown are the four threefold axes at $\frac{2\pi}{3}$ from each other, each triplet being mixed by the threefold rotation about the fourth axis. Each of the sets of axes is mixed among itself by the rotations of the other sets.

Why are the axes through the vertices and the centers of the edges $\frac{\pi}{4}$ from the z axis? There is rotational symmetry in the xy plane, so these axes must each be in sets of four, $\frac{\pi}{4}$ apart, mixed by these rotations. And a rotation around a diagonal axis takes one of the other symmetry axes into a symmetry axis. A rotation around an axis through the center of an edge cannot move a line through the center of one face into a line through the center of a perpendicular face. Thus these rotations move one axis by $\frac{\pi}{2}$, so are the only twofold symmetries.

This establishes that the only axes are through the centers of the faces, through the middle of edges and along the diagonals, these are the only rotations around them, and that they leave a cube invariant, thus they are all its rotational symmetries.

II.8.b The symmetry group as a subgroup of symmetric groups

The symmetry rotations permute vertices of the cube, so form a subgroup of symmetric group S_8; of course, all finite groups are symmetric group subgroups — Cayley's theorem [Mirman (1995a), sec. IV.3.d.ii, p. 113]. However they are more restricted since they can only permute the four diagonals, so the four pairs of diagonally opposite vertices, thus giving a subgroup which has 24 elements, as does S_4, suggesting that the subgroup is isomorphic to S_4. The rotational symmetry group of the cube is in fact (isomorphic to) S_4, and is called here octahedral group O(432).

Rotations of $\frac{\pi}{2}$, thus of cycle length four corresponding to the permutations of single cycles like (1234), are about axes through the centers of the three pairs of opposite faces, and rotations can be positive or negative, so there are six. The rotations of $\frac{2\pi}{3}$, corresponding to permutations like (123)(4), are about axes through opposite vertices, there being four of these, so eight such rotations. Rotations through π, of cycle length two, form two classes, one corresponding to the transpositions, the other to the product of commuting transpositions, like (12)(34), and there are six and three of these respectively. The axes are through the centers of opposite sides. There are six pairs of opposite sides, so three of these rotations correspond to others already listed.

Problem II.8.b-1: Need a group with the same number of elements as some subgroup of a symmetric group be isomorphic to it?

Problem II.8.b-2: Label the diagonals, and relate each S_4 transformation to a rotation. Check these statements. What is the effect of the other S_8 elements [Mirman (1995a), sec. IV.3.d.iii, p. 114]?

Problem II.8.b-3: Repeat this discussion for the symmetry groups of the other lattices considered as subgroups of symmetric groups.

II.8.c The representation matrices of the symmetry group

The representation matrices for the twofold rotations, labeled with the direction cosines of their axes (sec. II.4.a, p. 85) are [Burns and Glazer (1990) p. 282]

$$[100]:\begin{pmatrix}1&0&0\\0&-1&0\\0&0&-1\end{pmatrix},[010]:\begin{pmatrix}-1&0&0\\0&1&0\\0&0&-1\end{pmatrix},[001]:\begin{pmatrix}-1&0&0\\0&-1&0\\0&0&1\end{pmatrix}$$

$$[110]:\begin{pmatrix}0&1&0\\1&0&0\\0&0&-1\end{pmatrix},[101]:\begin{pmatrix}0&0&1\\0&-1&0\\1&0&0\end{pmatrix},[011]:\begin{pmatrix}-1&0&0\\0&0&1\\0&1&0\end{pmatrix}. \tag{II.8.c-1}$$

Applying these to the vector $\begin{pmatrix}1\\1\\1\end{pmatrix}$, going to the vertex of the cube (taken to have center at the origin and length of each side 2), gives

$$[100]\Rightarrow\begin{pmatrix}1\\-1\\-1\end{pmatrix},[010]\Rightarrow\begin{pmatrix}-1\\1\\-1\end{pmatrix},[001]\Rightarrow\begin{pmatrix}-1\\-1\\1\end{pmatrix},$$

$$[110]\Rightarrow\begin{pmatrix}1\\1\\-1\end{pmatrix},[101]\Rightarrow\begin{pmatrix}1\\-1\\1\end{pmatrix},[011]\Rightarrow\begin{pmatrix}-1\\1\\1\end{pmatrix}. \tag{II.8.c-2}$$

Problem II.8.c-1: Show that we can take the rotations and transpositions to correspond according to

$$[100]\Rightarrow(12),\ [010]\Rightarrow(34),\ [110]\Rightarrow(23). \tag{II.8.c-3}$$

Then

$$[001]\Rightarrow(12)(34). \tag{II.8.c-4}$$

Find the correspondence for the other two. Thus the rotation around [001] is the same rotation as the product of the first two. Axis [110] is through the centers of opposite sides. Interpret the other ones of these rotations, and verify that the product rules, found from the symmetric group, agree with those found geometrically, and also using the matrices.

II.8.d Adjunction of the inversion

In addition to rotations a cube also is invariant under inversion through its center, and reflections in each of the three planes through the center parallel to the (six) sides, plus reflections in the six (diagonal) planes through opposite pairs of sides. Also there are the rotation-reflections, eight sixth-order ones, and six fourth-order ones.

An example of a sixth-order one, about the axis [111] is

$$S_6 = \begin{pmatrix} 0 & -1 & 0 \\ 0 & 0 & -1 \\ -1 & 0 & 0 \end{pmatrix}. \qquad \text{(II.8.d-1)}$$

As there are four of these axes, there are eight such operators.

Problem II.8.d-1: Draw the cube and check that this produces a rotation of $\frac{2\pi}{3}$ about a diagonal, followed by a reflection in the perpendicular plane.

Problem II.8.d-2: The fourth-order rotation-reflection around [100] is given by

$$S_4 = \begin{pmatrix} -1 & 0 & 0 \\ 0 & 0 & 1 \\ 0 & -1 & 0 \end{pmatrix}. \qquad \text{(II.8.d-2)}$$

Check that it gives a rotation around a line through the centers of opposite faces, followed by a reflection (which?).

II.8.e Generalizations of the cube

The generalization of the cube to n-dimensional space is obvious. How do the results here generalize?

Problem II.8.e-1: What are the symmetry elements, axes, angles and mirrors, and symmetry groups, as functions of n? Explain. Show that the group in n-space is a subgroup of the n-dimensional rotation group. Also, being a finite group, it is a subgroup of a symmetric group. Relate the transformations of the smallest symmetric group to the geometrical transformations. Are the groups isomorphic? Why? Is the symmetric group a subgroup of a rotation group (of m-space, $n \leq m$)? By taking slices, figures of lower dimensions can be obtained, in particular a three-dimensional cube. Can the symmetry elements and group of the latter be found this way? Might other slices give interesting results? Going from a (reasonable large n) down to three dimensions might be possible using different chains of slices. Hopefully the final results are the same no matter how obtained. Are there (interesting, qualitative) differences in these results as a function of n, or between arbitrary n and the three-dimensional case? The orthogonal group in four dimensions is not simple, but only semisimple [Mirman (1995a), pb. X.7.e-4, p. 300]. Would this be relevant? An algorithm, or better a computer program, to answer these questions, as a function of n, would be helpful.

II.8.f The symmetries of the hexagonal lattice

The other lattice of highest symmetry is the hexagonal. It is also interesting as the symmetry of the lattice is not that of the unit cell, this always being a parallelepiped that when copied to fill space gives the lattice (fig. II.3.b-2, p. 75). The lattice, but not the unit cell, has sixfold symmetry around an axis perpendicular to the base through any lattice point; it consists of nonprimitive cells obtained by adjoining three unit cells relatively rotated by $\frac{2\pi}{3}$ around an axis through any vertex, these being lattice points. It also has twofold symmetry about a parallel axis through the center of the base of the unit cell (which does not contain a lattice point). Or we can take it to have threefold symmetry about the lattice point, plus this twofold symmetry, their product giving the sixfold symmetry. Since the threefold and sixfold rotations can be in either direction, the rotational symmetries are $2C_6$, $2C_3$, C_2. Instead of these we can take the rotation axes to go through a lattice point. Then there are twofold rotational symmetries around axes coinciding with the edges of the unit cell — this gives each edge as an axis.

There are in addition twofold rotational symmetries about axes parallel to the base and through the center of the unit cell. These are perpendicular to pairs of opposite sides, there being two, and to opposite edges, two more. The products of these give two additional sets of rotations, $3C_2'$ and $3C_2''$. And there is inversion symmetry through the center of the unit cell, and reflection symmetry in the three planes through the center and parallel to the sides. Further there are the rotation-reflection symmetries given by products of these.

Problem II.8.f-1: Draw a diagram using the lattice vectors [Cornwell

(1984), p. 200; Streitwolf (1971), p. 59]

$$\underline{a} : \alpha(1,0,0),\ \ \underline{b} : \alpha(-\frac{1}{2},\sqrt{\frac{3}{2}},0),\ \ \underline{c} : \beta(0,0,1), \qquad \text{(II.8.f-1)}$$

and verify that these give a hexagonal lattice. Find the angles between the vectors and compare with the conditions on the sides and angles given in tbl. B-2, p. 634.

Problem II.8.f-2: Discuss how to break the symmetry (physically) to give the subgroups of the holohedry group.

Problem II.8.f-3: Can the hexagonal lattice be generalized to n dimensions (space must be filled for the system to be called a lattice)? Can this analysis be extended to arbitrary dimensions? A computer program to do so, or general algorithms, would be useful.

Problem II.8.f-4: What is the smallest symmetric group for which the hexagonal holohedry group is a subgroup? Relate the permutations to the geometric operations. If not all are included, explain why the others do not give symmetries. Repeat for n dimensions (if possible).

Problem II.8.f-5: Repeat the discussion of this section for the other lattices.

Chapter III

Space Groups

III.1 GROUPS WITH DISCRETE TRANSLATIONS

Lattices are translationally invariant but — more fundamentally thus — are invariant under rotations, inversions and translations together. This results in both strong restrictions and, especially when sets of atoms are associated with lattice points, a rich array of crystal types. These symmetries are described by space groups. The subject is interesting not only because of the understanding it offers of lattices and crystals (and art, and other objects), but because it requires new concepts and illustrates significant aspects of group theory and relationships among its entities.

Molecules are richer than crystals in one sense since their set of point groups is larger, but crystals are richer in having an additional symmetry, translation, and so more, and more complicated, types of groups describing them. There are two sets, finite point groups and infinite — though discrete — translation groups, and their combinations, space groups. These we start to study here, describing them and considering why they are what they are and their relationship to the point groups.

Translation, point, and space groups are subgroups of the inhomogeneous rotation group of three-dimensional space, Euclidean group E(3) [Armstrong (1988), p. 136; Chen (1989), p. 461; Evarestov and Smirnov (1993), p. 31; Janssen (1973), p. 105; Johnston and Richman (1997), p. 142; Mirman (1995a), sec. XIII.4.b, p.382; Neumann, Stoy and Thompson (1994), p. 159; Nikulin and Shafarevich (1987), p. 66] — a *subgroup of* the inhomogeneous Lorentz group (of 3+1-dimensional space), the Poincaré group [Mirman (1995c)] — implying that we would find all space groups if we just listed the crystallographic subgroups of

E(3). But some space groups contain operations, the nonprimitive operations, glide reflections and screw rotations, which are not (apparently) in the inhomogeneous rotation group. How can a subgroup (apparently) contain operations not in the group? Why do these appear? How do they affect the classification of space groups, and the properties of the objects they describe? Perhaps better, what properties of these objects require them to be described by groups with such operations? And how can objects for which these operations are necessary exist in a space whose invariance group does not (seem to) contain them? Most disturbing, if subgroups can (presumably) contain elements not in the group, can we ever find all subgroups?

III.2 SPACE GROUPS: DEFINITIONS AND NOTATIONS

Space groups are the sets of symmetry transformations that are possible on crystals, discrete translations, rotations, the inversion, and reflections. They are inhomogeneous groups [Mirman (1995a), sec. II.3.h, p. 46], with the translations an invariant subgroup, and a point group (which is homogeneous), not necessarily a subgroup, required (by definition) to be crystallographic. The restriction to crystallographic groups is needed for the study of crystals (and lattices), but it does not rule out the possibility that other space groups are interesting and useful.

The theory and applications of space groups are quite extensive so unfortunately too much to consider now in breadth or depth [Altmann (1994), p. 53; Armstrong (1988), p. 145; Borchardt-Ott (1993), p. 183; Bradley and Cracknell (1972), p. 37, 81; Burn (1991) p. 205; Burns (1977), p. 251; Burns and Glazer (1990), p. 76; Chen (1989), p. 460; Cornwell (1984), p. 222; Falicov (1966), p. 150; Fässler and Stiefel (1992), p. 232; Fedorov (1971); Hargittai and Hargittai (1987), p. 327; Hilton (1963), p. 157; Jaswon (1965), p. 62; Jaswon and Rose (1983), p. 98; Janssen (1973), p. 108; Ladd (1989), p. 143; Lax (1974), p. 231; Lockwood and Macmillan (1983); Martin (1987), p. 78; Megaw (1973), p. 148; Nikulin and Shafarevich (1987), p. 149; Sands (1993), p. 69; Senechal (1990), p. 59; Shubnikov and Koptsik (1974); Streitwolf (1971), p. 51; Wherrett (1986), p. 223; Wood (1977), p. 150; Yale (1988), p. 113; Zak, *et al* (1969)]. Thus all we do is initiate a study of them to form a foundation for fuller consideration, especially introducing the two kinds of space groups, and analyzing how they are related to their point groups [Borchardt-Ott (1993), p. 209], and to objects they describe.

Of course, space groups (as considered here) are defined up to equivalence [Kocinski (1983), p. 74].

The space groups and the point groups are fully discussed, with their properties explicitly given, in the International Tables [Hahn (1987)]. There is a teaching edition [Hahn (1989)], extracted from the Tables, listing the nomenclature and symbols, making these easily available. Discussions of the tables, and explanations of how to use them are also available [Borchardt-Ott (1993), p. 198; Burns and Glazer (1990), p. 133, 174; Senechal (1990), p. 97]. We only state the symbols and terms that we use. For a complete list and discussion, these sources should be consulted.

III.2.a Denoting space group operations

Space groups have two different types of operations, homogeneous (point group) transformations (rotations, reflections, inversions, and products), and translations, so the notation for their operators should show both. An operation is labeled by a Seitz operator [Altmann (1977), p. 107; Altmann (1994), p. 59; Bradley and Cracknell (1972), p. 45; Burns and Glazer (1990) p. 77, 134; Johnston and Richman (1997), p. 145], $\{R|\underline{t}\}$, which is defined by its action on the coordinates,

$$\{R|\underline{t}\}r = R\underline{r} + \underline{t}, \tag{III.2.a-1}$$

giving homogeneous transformation R and translation $\underline{t}$, which together form an inhomogeneous transformation.

Problem III.2.a-1: Check that this symbol properly states the group laws:

a) identity: $\{1|0\}$ and $\{E|0\}$ in the International and Schoenflies notation, respectively,

b) multiplication:

$$\{R|\underline{t}\}\{P|\underline{u}\}\underline{r} = \{R|\underline{t}\}(P\underline{r} + \underline{u}) = \{RP|R\underline{u} + \underline{t}\}\underline{r} = RP\underline{r} + R\underline{u} + \underline{t}, \tag{III.2.a-2}$$

c) existence of an inverse:

$$\{R|\underline{t}\}\{R^{-1}| - R^{-1}\underline{t}\} = \{1|0\}. \tag{III.2.a-3}$$

Explain (geometrically) why the multiplication rule and inverse have these forms. Are there options in these choices? How is associativity stated?

III.2.b The affine group

Given a space, there are various transformations that can be performed on it (tearing is not considered), such as distortions, scale changes (stretchings or contractions), rotations, reflections, translations. Distortions are given by nonlinear transformations, and these form (presumably) nonlinear groups. They leave the topology unchanged, though they change distances and angles. They are not very familiar, and (perhaps) not relevant here; rotations (orthogonal transformations) and translations are. There is thus a series starting with the most general groups, perhaps those leaving a topology invariant, to the special ones we consider. There is one more general than these, but still linear (pb. III.8-2, p. 173). It is the affine group [Burns (1977), p. 250; Lyndon (1989), p. 34; Neumann, Stoy and Thompson (1994), p. 127; Streitwolf (1971), p. 50; Yale (1988), p. 158]. This is the set of transformations that leave parallel lines parallel (thus angles invariant), but not necessarily distances. The (here discrete) orthogonal and translation groups that we consider (and their combinations) are subgroups. The relevant affine group is the real affine group, since the coordinates are real numbers. It is the group of general linear transformations on a real three-dimensional space, GL(3,R), but here only a subgroup of it is relevant since we consider only discrete rotations and translations.

Problem III.2.b-1: Write a transformation, not a rotation, reflection or translation that leaves angles invariant.

Problem III.2.b-2: Is it possible to have a group under which distances, but not angles, are unchanged?

III.2.c Types of space groups

A crystal is described (in part) by its space group, and also by its point group, these related to subgroups of space groups. There are two types, symmorphic and nonsymmorphic space groups. They differ in the types of operations they contain, in how they are related to their associated point groups, and the kinds of objects they describe — both can give symmetries of crystals, but only symmorphic ones can give symmetries of lattices. What are these groups, how do they differ, why are there two types (there is only one type, in fact only one, inhomogeneous rotation group), what are these extra transformations, and why?

III.2.d Translations are invariant

Perhaps the most fundamental aspect of space groups (following from that of the Poincaré group, the group of transformations — and interestingly also of invariance — of our geometry [Mirman (1995a), sec. II.3.h, p. 45; (1995b,c)]) is that there is an Abelian subgroup, the translations, and that it is an invariant subgroup [Mirman (1995a), sec. IV.7, p. 132] (or factor group [Mirman (1995a), sec. IV.8.c, p. 138]); any space group operation acting on a translation gives a translation. It is on this that much of the analysis is based.

Any translation can be obtained from any other by a similarity transformation with a rotation, so rotations (and reflections) plus translations generate the space group (every element of the group can be written as a product of a translation with a point-group generator — which does not imply that the individual terms in the products form subgroups, as we shall see). For two-dimensions the rotations can be reduced to one, R_1, so every crystallographic group in the plane not containing an inversion is generated by two elements, a translation and R_1. If there is an inversion then every transformation that inverts the object (or coordinate system) is a product of the inversion with a translation or rotation; this group is generated (in a way we must see) by these three elements [Armstrong (1988), p. 140].

The group formed by the $\{R|\underline{t}\}$ is infinite dimensional (unless we consider translations going from one unit cell to another as identical, that is we take the elements modulo the lattice translations between unit cells — a convention whose feasibility, consistency and implications we do not analyze). These $\{R|\underline{t}\}$ do not form a continuous set, but rather an infinite discrete one, indexed by the point group symmetries, and the translations.

Problem III.2.d-1: Prove, using the Seitz notation, that the translations are invariant (if that is what this proves) [Burns and Glazer (1990) p. 170].

Problem III.2.d-2: For the inhomogeneous rotation group, with operator $R_z(\theta)$ giving a rotation of θ around z, $T_x(a)$ a translation of a along x, and $f(x,y)$ any basis state, show that

$$R_z(\theta)T_x(a)R_z(-\theta)f(x,y) = f(x + acos\theta, y - asin\theta), \quad \text{(III.2.d-1)}$$

giving a translation of a at an angle of θ to the x axis. Do this using the group expressions for these operators, and those of the Lie algebra [Mirman (1995a), sec. X.7.b, p. 295]. Thus the effect of a rotation on a translation is another translation, so these form an invariant subgroup. Try this for an arbitrary rotation, around any axis. Also prove it for any dimension, in a way independent of the dimension. Here both the rotations and the translations form subgroups (but only one is invariant; why?).

III.2.e Semi-direct products

A direct product group consists of two subgroups such that all elements of one commute with all of the other. The two sets of transformations are independent. A group G that is a semi-direct product [Altmann

(1977), p. 257; Mirman (1995a), sec. III.5.c, p. 102] consists of two subgroups R and T, one of which is invariant under the other. If these have generators r and t, with t belonging to the invariant part,

$$r_i t_j = t_k, \tag{III.2.e-1}$$

$$t_i t_j = t_k, \tag{III.2.e-2}$$

$$r_i r_j = r_k. \tag{III.2.e-3}$$

For the translations, the t's form an Abelian group.

Problem III.2.e-1: Prove that every (factorizable) point group is a semi-direct product [Altmann (1977), p. 267]; a semi-direct product might be direct. One way of doing this is by going through the list of point groups and checking that each satisfies (sec. I.7.a, p. 50). This may not be the best way. Find better ways. Why, geometrically, is one subgroup invariant, but not both? Of course it is obvious that every point group containing an improper element (reflection or inversion) is a semi-direct product (in addition one subgroup may itself be). Why is it obvious?

III.3 SYMMORPHIC SPACE GROUPS

The simpler kind of space group is the symmorphic one. A symmorphic group is a semi-direct product of a point group and a translation group (the invariant subgroup), and both are symmetry groups of the crystal. The point group is a subgroup of the symmorphic space group, for a suitable choice of the origin in the lattice. A space group is symmorphic if there is a single origin that allows every element to be written as a product of a homogeneous transformation (a rotation, reflection, or inversion) and an inhomogeneous transformation, a translation, and all with the same origin. A nonsymmorphic group cannot be so written, no matter where the origin is.

We refer to the object whose symmetry we study as (generically) a crystal. For a nonsymmorphic group it cannot be a lattice, for a symmorphic group it may, but we have to consider whether every symmorphic group, or only a subset, can be a symmetry group of a lattice. Thus at this stage we must be general so use the term crystal.

Problem III.3-1: Prove, from this definition requiring a single origin, that every symmorphic space group is (isomorphic to) a semi-direct product group [Cornwell (1984), p. 226; Jansen and Boon (1967), p. 241; Shubnikov and Koptsik (1974), p. 255], in particular of a point group and the group of (lattice) translations, and using the fact that the space group is a symmetry group, show that both subgroups are also.

III.3.a A symmorphic group in two-dimensions

An example of a symmorphic group is that of a crystal with a two-dimensional square lattice (sec. VII.6, p. 372), covering the xy plane, [Falicov (1966), p. 158] which is used for many examples [Jansen and Boon (1967), p. 259; Joshi (1982), p. 277]. The points are taken to hold objects with spin (or some other vector) in the z direction so that we can ignore a mirror, and all 2-fold axes in the xy plane. In other words we really do regard this as a two-dimensional object, not as a plane projection of a three-dimensional one. It can be visualized as

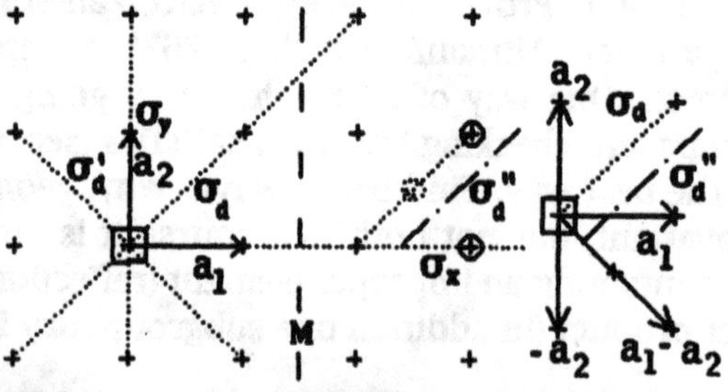

Figure III.3.a-1: TWO-DIMENSIONAL SQUARE LATTICE.

What transformations leave it invariant [Mirman (1995a), sec. II.2.f, p. 37]? First is a lattice translation, $\{E|\underline{t}\}$, where

$$\underline{t} = n_1\underline{a}_1 + n_2\underline{a}_2; \qquad \text{(III.3.a-1)}$$

the $\underline{a}$'s are vectors giving the minimum displacements along the x and y axes under which the crystal is unchanged — the displacement between neighboring lattice points on these axes (for a square lattice,

$$a = a_1 = a_2), \qquad \text{(III.3.a-2)}$$

and the n's are integers. Pick an origin at one of these points. Then the homogeneous symmetry operations are (with the subscripts on the σ's giving the line along which the mirror lies),

$$\{E|0\}, \{C_4|0\}, \{C_4{}^{-1}|0\}, \{C_4{}^2|0\}, \{\sigma_x|0\}, \{\sigma_y|0\}, \{\sigma_d|0\}, \{\sigma_d'|0\}. \qquad \text{(III.3.a-3)}$$

The mirror planes are shown as dotted lines, and four-fold rotations about the z axis perpendicular to the plane are indicated as usual by a (four-sided) symbol, a square, for clarity here an open one.

With R a point group operation, space group operations are of the form $\{R|\underline{t}\}$; we have to verify that R is one of the point group operations listed, if the group is symmorphic. There are mirrors bisecting the lines between neighboring points parallel to both x and y axes; the lattice is invariant under reflections in them. Are these of the form {R| $\underline{t}$}, with one of the listed R's, and with all translations (all $\underline{t}$'s) the same? Take the one midway between lattice points parallel to the y axis (shown by dashed line M); that reflection is the same as the reflection in the y axis (the mirror σy), but displaced by $\frac{1}{2}\underline{a}_1$. These mirrors are related by a similarity transformation giving the translation,

$$\{E|\frac{1}{2}\underline{a}_1\}\{\sigma y|0\}\{E|-\frac{1}{2}\underline{a}_1\} = \{E|\frac{1}{2}\underline{a}_1\}\{\sigma y|-\frac{1}{2}\underline{a}_1\} = \{\sigma y|\underline{a}_1\}, \quad \text{(III.3.a-4)}$$

since the mirror reverses the direction of the translation, and this shows that it has the required product form using the same origin for all transformations.

Problem III.3.a-1: Verify that the operations listed are symmetry elements of the lattice (and all of them?), and that they are not symmetries for a different origin (a different fixed point — in the unit cell). What point group do they form? Note that it is a subgroup of the space group. Explain why this is obvious. Also all translations for these operations are the same, $\underline{t} = 0$. Check that if the set of transformations included a lattice translation then the operations would still be symmetries. To be sure that this is correct implement these operations (on a diagram) and it should turn out that $\{\sigma y|\frac{1}{2}\underline{a}_1\}$ is the same as $\{\sigma y|0\}$ followed, or preceded, by $\{E|\underline{a}_1\}$. The effect of a reflection in mirror M is the same as a product of that in σy and a translation of $\underline{a}$ along x, so is an operation of the space group.

Problem III.3.a-2: Obviously (?) these homogeneous transformations are symmetries if taken about, not a lattice point, but one in the center of a square (displaced by $\frac{1}{2}(\underline{a}_1 + \underline{a}_2)$ from the origin). Check that all operations are of the form {R|$\underline{t}$}, where

$$\underline{t} = 0, \ \underline{a}_1 \ \text{or} \ \underline{a}_2, \quad \text{(III.3.a-5)}$$

and R is one of the listed (point group) transformations. This means the group is symmorphic. Why? Are there other points around which these transformations can be taken to give symmetries? Explain. Observe that this is true for the diagram.

Problem III.3.a-3: Write the group table (though it is infinite), and verify that it satisfies the group axioms. Can you tell by looking at it that the group is symmorphic? Does the table show the crystal class (here a square)? From it find the point group. How are the tables of the point and space groups related? Can the given figure be obtained from the group table — either or both?

Problem III.3.a-4: By applying the group transformations to a formula for the coordinates of the lattice points, show that the lattice is invariant under the group. Can you also show that it is symmorphic?

Why? From such a formula is it possible to find the transformations of the invariance group? Find the points that are the centers of rotation, and the lines that are mirrors.

III.3.b Glide planes

There is one further operation, whose significance will become more interesting when nonsymmorphic groups are discussed. This is a glide reflection. Consider a mirror parallel to σ_d, but displaced by $\frac{1}{4}(\underline{a}_1 - \underline{a}_2)$ from it (shown by the dotted and dashed line σ_d''). A reflection in this plane, followed by translation $\frac{1}{2}(\underline{a}_1 + \underline{a}_2)$ along it, leaves the lattice invariant (as shown by the three circled points; that on the x axis is reflected, bringing it to the center of the dotted circle, which is not a lattice point, and then translated, which does bring it to a lattice point, that circled). Neither transformation alone is a symmetry, only the product. This plane, σ_d'', is a glide plane. We then have, with $\sigma_d\underline{a}$ the translation given by reflecting $\underline{a}$ in the mirror,

$$\{E|\frac{\underline{a}_1 - \underline{a}_2}{4}\}\{\sigma_d|\frac{\underline{a}_1 + \underline{a}_2}{2}\}\{E|\frac{-\underline{a}_1 + \underline{a}_2}{4}\}$$

$$= \{E|\frac{\underline{a}_1 - \underline{a}_2}{4}\}\{\sigma_d|\sigma_d(-\frac{\underline{a}_1 + \underline{a}_2}{4}) + \frac{\underline{a}_1 + \underline{a}_2}{2}\}$$

$$= \{E|\frac{\underline{a}_1 - \underline{a}_2}{4}\}\{\sigma_d|\frac{-\underline{a}_2 + \underline{a}_1}{4} + \frac{\underline{a}_1 + \underline{a}_2}{2}\} = \{\sigma_d|\underline{a}_1\}, \qquad \text{(III.3.b-1)}$$

a lattice translation, so a symmetry, and equivalent to one of the $\{R|\underline{t}\}$, with the same origin. With this shift of the origin, the glide becomes equivalent to a symmetry translation, and this same shift transforms all glides to such form. The group is thus symmorphic. For a nonsymmorphic group not all glide reflections (generally non-primitive elements) can be reduced to this form.

Problem III.3.b-1: Find all glides for this lattice, and show that they can be transformed in this way, by this single shift of the origin, and that every other symmetry transformation is, and is left as, a product of a homogeneous transformation with a translation.

Problem III.3.b-2: Redo this entire analysis for a three-dimensional square lattice. First take the points as structureless so that the lattice can be regarded as a plane in three-dimensional space, and then consider an actual (although thin) solid (which is not a three-dimensional lattice). How are the groups of these various problems related? Which are subgroups of which? Are all symmorphic? Try this then for a cube (and a cubic lattice).

Problem III.3.b-3: A space group without glides or screws is (given by) a semi-direct (but not a direct) product of a point group, and the (discrete) translations, these forming an ideal, an invariant subgroup [Mirman (1995a), sec. XIII.2.i, p. 375], and the translations are the basis vectors of a representation of the point group. The inhomogeneous rotation and the Poincaré (inhomogeneous Lorentz) groups are semi-direct product groups; that such space groups are also should not be surprising since they are subgroups of these. Check that this makes sense and that it describes the action of such a space group; in particular there are an infinite number of (discrete) translations, but all point group representations are finite-dimensional. In what sense do the translations form a point-group representation basis space? Check also that the Seitz notation actually does express this property — which has what significance? As space groups are discussed, and in particular as individual ones and their representations are considered, it should be noted whether they actually conform to this requirement of the correctness of the Seitz notation placed on them. What places the requirement on them?

III.4 NONSYMMORPHIC SPACE GROUPS

Nonsymmorphic groups have besides the primitive operations — those that appear in symmorphic groups — additional elements, the non-primitive ones. Each nonsymmorphic group includes at least one such nonprimitive transformation. The point group operations are not symmetries of the object whose symmetries are described by the nonsymmorphic space group, and the point groups formed by their transformations are not subgroups, but rather factor groups. Thus a nonsymmorphic group cannot be written as a semi-direct product of groups. The systems to which they are relevant are, in a sense, richer than those of the symmorphic groups.

There are two symmetry elements that are neither homogeneous transformations nor pure translations: screw rotations (about screw axes) [Hargittai and Hargittai (1987), p. 336; Jaswon (1965), p. 71; Jaswon and Rose (1983), p. 104; Ladd (1989), p. 158] and glide reflections (with respect to glide planes, mirrors through which reflections, but not symmetry reflections, take place) [Hargittai and Hargittai (1987), p. 328; Jaswon (1965), p. 85; Jaswon and Rose (1983), p. 119; Ladd (1989), p. 160; Martin (1987), p. 62]. Transformations that are translations, or homogeneous, or products, are primitive, these two are nonprimitive [Altmann (1977), p. 105; Altmann (1994), p. 55; Borchardt-Ott (1993), p. 177; Bradley and Cracknell (1972), p. 44; Burns and Glazer (1990), p. 84; Hilton (1963), p. 146; Jones (1975), p. 132; Lax (1974), p. 231; Megaw (1973), p. 149; Sands (1993), p. 70; Senechal (1990), p. 25; Shubnikov and Koptsik (1974), p. 82, 103; Wooster (1973), p. 201]. Though these are not unfamiliar — certainly screws are quite familiar — the reasons they cannot be written as products may not be obvious.

With these nonprimitive operations, (at least) two (sets of) constants are needed to specify the crystal. One set, as always, gives the lattice size, the other the size of the translation going with this operation. For a glide plane there are two points (related by the glide) displaced from the mirror plane. However the value of the displacement (the ratio of the displacement to the lattice constant) is arbitrary, so requiring another constant; for a screw it is the thickness that must be specified. But if a group contains several of these operations the constants are not independent, they cannot be arbitrarily given for each. Because of symmetry, because the transformations take points into each other, they are related. (Actually even for a symmorphic group there generally need be more than one constant, as angles and ratios of sides have to be specified — but here this additional type of constant appears so it must be noticed.)

The word symmorphic implies a symmetrical shape (invariant under point group transformations), thus the shape, or form, of a crystal, while not unsymmetrical under nonsymmorphic group transformations, has a more complicated, and interesting, invariance. Why, and how, this arises, we have to see. To do so, we first consider the new types of operations.

III.4.a Screw axes

Screw rotations in space groups are not rotations of screws as that term is commonly used, for rotations of screws are continuous, while the operations in space groups are discrete. Although the term screw rotation is what is used, so that is correct, a better analogy to a screw axis is an infinite (circular) spiral staircase [Rosen (1977), p. 60] — a staircase of steps around a circular cylinder. Moving from a step to one directly above or below is a symmetry operation. This can be done by moving in a curved path, requiring a rotation. However rotations are not symmetries of the staircase, nor is a translation along the axis for that alone would not go from one step to its neighbor. An object placed on one step and then subject (only) to a rotation, or such a translation, would be unsupported (and fall under gravity). The symmetry transformation from one step to its neighbor is neither a rotation nor a translation, but a product of these. The symmetry translation between points above or *below* each other is a product of symmetry transformations going from one step to the next, as illustrated in

Figure III.4.a-1: STAIRCASE OF STEPS FORMING A SCREW.

The axis of the staircase is a screw axis; symmetry groups of (objects containing) screw axes are nonsymmorphic.

A screw axis is labeled n_m; the rotation is $\frac{2\pi}{n}$, the translation $\frac{m}{n}$ of h — the distance between two points connected by a single translation (in the staircase, the distance between a step and one directly above it). Thus for a 6_3 axis a rotation of $\frac{2\pi}{6}$ (= 60°) gives a translation of $\frac{3}{6}$(= $\frac{1}{2}$) of h along the screw axis. A 6_5 axis, gives, for this rotation, a translation of $\frac{5}{6}$ of h, that is a rotation $\frac{2\pi}{5}$ moves from one step to an adjacent one. So for a 3_2 axis, the first step is at 0°, at the origin. The next is at $\frac{2\pi}{3}$ (= 120°), at height $\frac{2h}{3}$, the following at $\frac{4\pi}{3}$ (= 240°), with height $\frac{4h}{3}$. The third rotation brings it back to 0°, but now at a height $\frac{6h}{3}$ = $2h$, and this is equivalent to the identity. Notice that there are steps at $(0^o, 0)$, $(120^o, \frac{2h}{3})$, $(240^o, \frac{4h}{3})$ and also at $(0^o, h)$, $(120^o, \frac{5h}{3})$, $(240^o, \frac{6h}{3})$. A primitive transformation from one point to an equivalent one (say the corresponding one in the next cell) leaves the crystal unchanged, so is equivalent to the identity. This is not true for a nonprimitive operation, which requires several turns.

This is illustrated with a 2_0 (symmorphic) screw on the left:

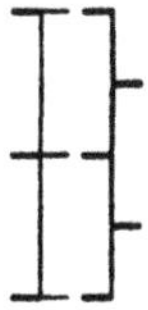

Figure III.4.a-2: SYMMORPHIC AND 2_1 SCREWS.

Corresponding steps (drawn heavily) are connected by a translation. To go from the first to the second step requires a 180° rotation, with no translation. On the right is a 2_1 (nonsymmorphic) screw. Two neighboring steps are taken into each other by 180° rotation plus a translation of half the distance between corresponding steps. A translation without a rotation bypasses the intermediate step.

Problem III.4.a-1: Draw a diagram for the 6_3 screw. Notice that though there is a step above the initial one, the screw bypasses it, so takes two complete turns to reach a step in that line — two complete turns is equivalent to the identity. The first turn of the screw goes to the step at $(60^o, \frac{1}{2}h)$. To get to the first step above the origin requires an additional turn of $(300^o, \frac{1}{2}h)$, which gives a symmetry, but leaves out the intermediate operations. Thus the emplacement of the points in the crystal (the steps) results in them being related by a 6_3 screw, and the space group operations connecting them form a screw that requires two turns for an operation equivalent to the identity. Demonstrate this with the diagram. Explain it using the staircase.

Problem III.4.a-2: Show that a right-handed screw n_m is the same as the left-handed n_{n-m} one [Borchardt-Ott (1993), p. 182]. This means that the only screws are the right- and left-handed versions of n_0 (clearly symmorphic), n_1, n_2, and n_3. Why? Draw the stairs for all these screws, plus the corresponding ones of the other handedness. This should clarify the reason that, not one, but several turns may be equivalent to the identity. How?

III.4.a.i *The space group with a screw and the point group*

What is the space group of a system with a screw axis, is there a point group, how are they related? The symmetry group contains translations. The staircase, for example, is invariant under a translation from one step to another (not the neighboring one). Also there are space group transformations that are products of a translation and a rotation, although neither alone is a member of the group; these screw operations (which are symmetries) move from one point of the crystal to another, in the staircase from a step to a neighbor. The set of these translations and screws is the space group describing the spiral stairs; the space group of a crystal contains these. There are rotations involved, even though the space group has no subgroup of rotations — it is a group whose elements are translations, and products of translations and rotations. In general for a nonsymmorphic group, there is a point group, here rotation group SO(2), that is, not a subgroup, but a factor group, of the space group.

Consider the translations T_1 from a step to any directly above or below it. This set is an invariant subgroup of the space group. It is a subgroup because these transformations leave the staircase invariant; the point group of rotations is not a subgroup because a rotation does not give invariance. It is a normal subgroup because it commutes with rotations (and translations). Take the corresponding set of translations for the neighboring step, T_2. Its members are related to those of the first set by a rotation,

$$T_2 = R(\frac{2\pi}{n})T_1, \qquad \text{(III.4.a.i-1)}$$

where n is an integer; and similarly for each step in one full turn. Thus we have a set of cosets [Mirman (1995a), sec. IV.6, p. 128] T_1, $R(\frac{2\pi}{n})T_1$, and so on. These cosets form a group, the factor group of the space group. And this group is isomorphic to a finite subgroup of, here, SO(2) — but SO(2) is not a subgroup of the space group.

This space group consists of translations, but not of all translations, rather of a finite number of sets of these, each containing an infinite number of discrete translations (for the staircase these translations are labeled by the steps in one complete turn). These sets are related to each other by a set of discrete rotations. There are no rotations in the space group, but the elements of the group define a set of rotations which is the factor group, and is a crystallographic point group since the restrictions on the rotations in it are the same as those for the point group. The conditions being the same, the groups are isomorphic.

For every space group (symmorphic and nonsymmorphic) there is a point group that is a factor group, with the translations the invariant subgroup. For a symmorphic group that factor group is a subgroup, and the point group of symmetries of the lattice, but it is not for a nonsymmorphic group.

Problem III.4.a.i-1: What would happen if n in the equation for T_2 were not an integer?

Problem III.4.a.i-2: Prove that if space groups are isomorphic, their point groups are also [Armstrong (1988), p. 153].

III.4.a.ii *Limitations on crystallographic screws*

A screw rotates through $\frac{2\pi}{n}$, and for a crystallographic space group, $n = 1, 2, 3, 4$ or 6 [Fedorov (1971), p. 57]. The reason that screw rotations have the same limitations as crystallographic point-group rotations is that they are essentially the same [Burns and Glazer (1990) p. 85]. That a screw rotation is a product of a rotation with a translation does not affect the argument giving the limitation (a screw rotation moves an object out of the plane, but the condition holds on the projection down to the plane). At point A in the next diagram there is a screw axis perpendicular to the page (shown dotted), of angle $\theta = \frac{2\pi}{n}$. This takes point Q to P at height τ above the plane, and picking lattice points at equal distances from A, it takes B to C. Now angle BAC $= \phi = \frac{2\pi}{p}$ is in a plane of lattice points so there is a restriction,

$$p = 1, 2, 3, 4 \text{ or } 6. \qquad \text{(III.4.a.ii-1)}$$

Angle θ is between the line to Q and the projected point, that on the plane under P. Hence screw angle θ is the same as ϕ, which is the symmetry angle of the crystal, giving

$$n = 1, 2, 3, 4 \text{ or } 6 \qquad \text{(III.4.a.ii-2)}$$

(essentially the translations off the plane do not affect the restrictions on it), as we see in

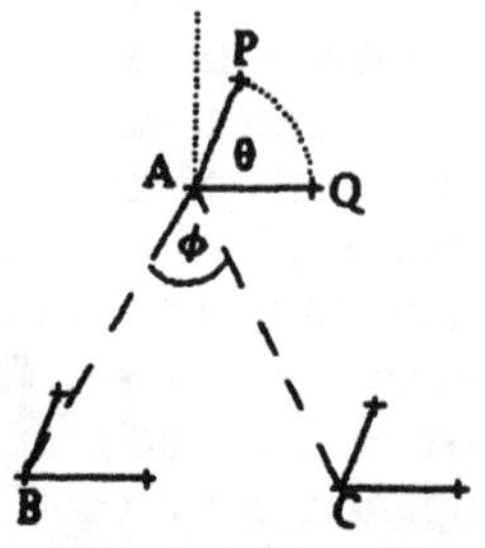

Figure III.4.a.ii-1: CRYSTALLOGRAPHIC SCREW ANGLES ARE LIMITED.

Problem III.4.a.ii-1: Is there really a screw with $n = 1$?

Problem III.4.a.ii-2: There are limitations (fortunately) on the screws, and the set of screws, and their positions, that can appear in a space group [Fedorov (1971), p. 57]. It is a useful exercise to determine them.

Problem III.4.a.ii-3: Previously (sec. I.5, p. 23) other proofs were given that the rotational symmetries of a lattice are limited to these orders. Check that these hold here also — or modify them so that they do.

III.4.a.iii *Pairs of enantiomorphic screws*

Screw rotations introduce new complexity into space groups. One aspect is that a screw rotation can be either right- or left-handed. Thus space groups that contain screw rotations are paired if not all screws in the group are paired, and the corresponding screws are of opposite chirality [Hargittai and Hargittai (1987), p. 54]. These screws are enantiomorphic pairs. Groups with unpaired screws are chiral space groups, and form enantiomorphic pairs. Crystals having them as their symmetry groups thus have a built-in handedness; they could show optical activity, say.

III.4.b The nonprimitive glide reflection

The other nonprimitive operation is a glide reflection [Martin (1987), p. 62], a reflection in a plane (in two dimensions, a line) followed (or preceded) by a translation (a glide) parallel to the plane, the glide plane. An example of an object invariant under a glide reflection is a sine wave [Hargittai and Hargittai (1987), p. 330]. To illustrate a "crystal" with a glide plane [Armstrong (1988), p. 140], we have a rectangular lattice that is clearly unchanged if it is reflected in one of the mirrors (labeled m) then translated along m by one box (or translated, then reflected),

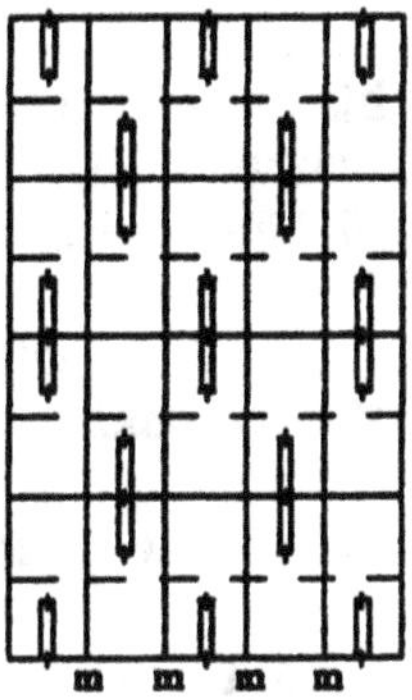

Figure III.4.b-1: GLIDE PLANE.

Neither the reflection nor the translation leaves it invariant, only their product (though translations by two boxes do). This is a glide reflection, and line m is a glide plane. Notice that there is no mirror for which these objects are reflections of each other. They are related only by a glide. And the cells of course fill space.

The invariance group of this figure includes the glide reflections in the labeled vertical planes and the dashed horizontal ones; here there are both horizontal and vertical glides. Both form invariant subgroups (their glides are of different lengths). The point group is the factor group of the space group using the invariant subgroup of glide reflections. Translations by one box are not symmetries, but the translations by two boxes form an invariant symmetry subgroup. In addition here there are other elements, diagonal translations. The unit cell of the lattice is a single box, for the "crystal", the lattice with the rectangular object, it is 2×2 boxes.

Examples of glide planes and screw axes within actual crystals are given below (sec. III.5, p. 159, sec. XI.2.j, p. 590).

Problem III.4.b-1: Show, pictorially and analytically, that a sine wave has a glide reflection as a symmetry. Do other trigonometric functions have such symmetries? Why?

Problem III.4.b-2: Are there other glides in the figure? How about other symmetry operations? There are mirrors. Where? What is the complete invariance group? What are its factor groups? We have given two sets of cosets, formed from glide planes and glide translations by two boxes. How are they, and the factor groups they give, related? What are the generators of the invariance group?

Problem III.4.b-3: How would these results change if the lattice were square, instead of rectangular?

Problem III.4.b-4: Take a walk along a straight line, at a constant pace, in sand. Check that your footprints are related to each other by a glide reflection [Martin (1987), p. 64]. Would this be true if you were to walk in a circle? Would there then be any (group of) transformations relating the footprints? This can be extended: suppose the pace were not constant, say it varied sinusoidally, what would be the symmetry transformations and group? Must these form a group? Is this likely to be relevant to real physical materials? Why?

Problem III.4.b–5: Show that these glide reflections and screw rotations, for the cases considered (and for nonsymmorphic groups, in general) cannot be written in the form $\{R|\underline{t}\}$, where R is a point group symmetry of the system, and $\underline{t}$ is the *same* translation for *all* elements R.

Problem III.4.b–6: Prove that there can be no other symmetry operations (of lattices?), beyond those of the point groups and the translations, except for screw rotations and glide reflections [Martin (1987), p. 65].

Problem III.4.b–7: Using the Seitz notation show that the translations form an invariant subgroup of a space group [Burns and Glazer (1990) p. 170]. (What does this actually prove?) Write the elements of the factor group in this notation.

III.4.c Why are these the only non-primitive operations?

Screw rotations and glide reflections are nonprimitive, the other operations, which belong also to symmorphic groups, are primitive. Nonprimitive ones are of the form $\{R|\underline{t}\}$, where translation $\underline{t}$ depends on homogeneous transformation R. Thus, while for primitive operations, we can pick an origin so that *all* operations (except translations) are of the form $\{R|0\}$, this is not true for nonprimitive ones. There are only two such operations. Why these, and why only these?

Here we mean something slightly different (and more specific) than usual by translation, rotation and reflection. Usually we talk about a rotation of say $\frac{\pi}{6}$ or a translation $\underline{t}$ along some direction. But here the rotation is not merely of $\frac{\pi}{6}$, but $\frac{\pi}{6}$ between $\frac{\pi}{4}$ and $\frac{10\pi}{24}$. Likewise the translation is of a specific distance along a specific line, between given initial and final points. Thus for the spiral staircase the height of the initial and final points, so the specific translation, depends on the initial and final steps, thus on the specific rotation. It is in this sense that the translation depends on the rotation, and similarly for glides.

Why are these the only two nonprimitive operations? There are four operations, rotations, reflections, inversions and translations. Inversions are products of the first two so we need not consider them.

If a translation depends on the homogeneous transformation there are only two possible classes of nonprimitive elements, for rotations and for reflections. However we have specified that the translation is along the axis of rotation or the plane of reflection. Could it be perpendicular, or at some angle? Consider such a translation for the rotational case. That would give not a circular spiral, but an ever widening (or narrowing) one, in violation of the required translational invariance, so we eliminate it from consideration (which does not necessarily mean that nature has). We might try to save translational invariance in a sense by taking the spiral to widen and then narrow. But again we do not consider this.

Problem III.4.c-1: Is there any reason real materials should not have these properties? What would they be like? How would they differ from ordinary crystals? Are there any that are like this? Would they be described by (symmetry) groups? Repeat this analysis for glide planes. Analyze the possibility of the existence of real materials differing from our specifications. This emphasizes that the symmetry elements we study are limited by our definitions, but not necessarily by nature. Or might there be some reason why nature likes our definitions?

Problem III.4.c-2: Would it make any sense, or difference, to take the translation not along the axis of rotation (or glide plane), but displaced from, though parallel to, it.

Problem III.4.c-3: There is a basic reason why nonprimitive transformations cannot be constructed from inversions.

Problem III.4.c-4: In elementary physics we learn also of other types of quantities, for example one given by a vector attached to a specific line or point, such as a current or torque. Yet this does not seem to lead (here) to any new types of transformation. Why?

III.4.c.i *These are the — only — nonprimitive operations*

To show that glide reflections and screw rotations are the only nonprimitive elements we consider a transformation on $\underline{r}$,

$$\underline{r}' = R\underline{r} + \underline{t}, \tag{III.4.c.i-1}$$

given by point-group transformation R and translation $\underline{t}$ [Lax (1974), p. 231; Martin (1987), p. 194]. A homogeneous transformation can be rewritten without a translation by moving the origin by $\underline{b}$, defining

$$\underline{s} = \underline{r} + \underline{b}, \quad \underline{s}' = \underline{r}' + \underline{b}, \tag{III.4.c.i-2}$$

so that

$$\underline{s}' = R\underline{s}, \tag{III.4.c.i-3}$$

else the transformation does not belong to a point group (it is inhomogeneous). Thus

$$\underline{s}' = R\underline{s} + \underline{t} + \underline{b} - R\underline{b}. \tag{III.4.c.i-4}$$

For the transformation to be homogeneous the last three terms must sum to zero, giving

$$\underline{t} = (R - 1)\underline{b}. \qquad \text{(III.4.c.i-5)}$$

If R is a proper rotation there are three eigenvalues, at least one of which equals 1 (every rotation has an axis) [Goldstein (1953), p. 118]. Denoting the three eigenvectors of R by $\underline{e}_i$,

$$R\underline{e}_i = \lambda_i \underline{e}_i, \qquad \text{(III.4.c.i-6)}$$

and expanding $\underline{t}$,

$$\underline{t} = \sum u_i \underline{e}_i, \qquad \text{(III.4.c.i-7)}$$

and similarly $\underline{b}$, we get the condition for homogeneity as

$$\underline{b} = \sum (\lambda_i - 1)^{-1} u_i \underline{e}_i. \qquad \text{(III.4.c.i-8)}$$

Since at least one eigenvalue equals 1, this equation cannot be satisfied unless each

$$\lambda_i = 1 \Rightarrow u_i = 0. \qquad \text{(III.4.c.i-9)}$$

Thus if $\underline{t}$ is perpendicular to the rotation axis it can be eliminated by a shift of origin (as can any perpendicular component), but if it is parallel it cannot be — the transformation must be inhomogeneous. So for rotations the general transformation is either a pure rotation, or a screw rotation (it is interesting to compare this with Chasle's theorem [Goldstein (1953), p. 124] that the most general displacement of a rigid body is a translation plus a rotation).

What have we done by shifting the origin? Suppose we take a point and rotate the vector from the origin to it — move the point in an arc on a plane — then translate the point perpendicular to the rotation axis. This result says that there is an origin, that is a rotation axis parallel to the original one, for which the transformation of the point is a pure rotation. This is not surprising as the point moves on a plane, and it can be moved from one position on the plane to another by a pure rotation. However if it moves off the plane then there is no pure rotation, with axis *parallel* to the original one, that takes it from the first position to the final one.

Problem III.4.c.i-1: Draw a point undergoing the transformation leaving it on the plane and find, geometrically and analytically, the origin for which the transformation is a pure rotation, and the path (including a diagram) and the angle with respect to the shifted origin. Show that rotations about, and inversions in, a point other than the origin give a translation, which explains that one in the equation (which equation?). For the screw rotation why is there a requirement that the axes be parallel?

Problem III.4.c.i-2: An improper rotation (one including a reflection) has determinant -1, and (clearly) eigenvalues (1,1,-1) or (-1,-1,-1) [Goldstein (1953), p. 122]. The first gives a mirror plane with eigenvalue 1 for any vector in the plane, the second an inversion. Again by shifting the origin (perpendicular to the mirror) we can eliminate any (component of a) translation perpendicular to the mirror, but not parallel to it (for an inversion any translation can always be eliminated — of course an inversion does not introduce a direction). Thus there are inhomogeneous reflections with translations in the plane of the mirror that cannot be eliminated. Explain this in terms of the motion of a point. Again check it with a drawing. Explain also why this completes the proof that the only nonprimitive elements are screws and glides.

Problem III.4.c.i-3: Why does Chasle's theorem (seem to) ignore the possibility of glide reflections?

Problem III.4.c.i-4: A slightly different derivation uses the multiplication rule

$$\{R|\underline{t}\}\{S|\underline{v}\} = \{RS|\underline{t} + R\underline{v}\}, \qquad \text{(III.4.c.i-10)}$$

for a similarity transformation to get the new operation,

$$\{S|\underline{v}\}^{-1}\{R|\underline{t}\}\{S|\underline{v}\} = \{S^{-1}RS|S^{-1}(\underline{t} + R\underline{v} - \underline{v})\}. \qquad \text{(III.4.c.i-11)}$$

Explain why this leads to the result.

Problem III.4.c.i-5: Why must the unit cell of a crystal whose space group is nonsymmorphic have more than one atom (the space group of a crystal, not of a lattice, can be nonsymmorphic) [Lax (1974), p. 253]? Give rigorous proofs (geometrical and analytic). Why is the space group of a lattice always symmorphic?

III.4.c.ii *Nonprimitive operations are determined by primitive ones*

There being no other (primitive) point-group operations, there are no other nonprimitive operations.

In summary, by definition, a lattice is invariant under a set of translations, so these form a symmetry group of it, and translations form an invariant subgroup of the inhomogeneous rotation group. Thus the space group elements take a subgroup of translations into another set of translations. But these sets can only be related by point group operations, rotations and reflections (inversions are not relevant). Hence for all space groups the factor group with respect to translations must be a point group, with the elements of the latter giving the nonprimitive space-group elements, thus these are screw rotations or glide reflections (only).

As isometry ("equal measure") is a transformation that leaves distances and angles invariant [Johnston and Richman (1997), p. 148]. The fundamental outcome is that the isometries of (our) geometry, transformations leaving bodies invariant, are (only) translations, rotations, reflections, inversions, screws and glides, and products (including rotary-reflections) [Armstrong (1988), p. 140; Farmer (1996), p. 15; Lockwood and Macmillan (1983), p. 97; Lyndon (1989), p. 23; Schattschneider (1990), p. 34; Yale (1988), p. 62].

Problem III.4.c.ii–1: Do these results give a condition on physical objects (that can exist in our geometry), on the way we describe them, or a restriction on the type of materials to which these mathematical objects apply, or on geometry?

Problem III.4.c.ii–2: Are there more isometries in spaces of higher-dimension [Brown, *et al*, (1978); Senechal (1990), p. 115; Shubnikov and Koptsik (1974), p. 305]? Would it matter if space were not flat? Suppose the metric were not positive-definite (eq. IV.3.h–1, p. 211)? Can the axioms of geometry [Neumann, Stoy and Thompson (1994), p. 116, 159; Nikulin and Shafarevich (1987), p. 45] be modified in other ways (pb. II.4.h-6, p. 97; pb. II.7.f-10, p. 123; pb. IV.3.h-8, p. 211) to give different numbers and types of isometries? What properties of lattices and crystals would this lead to? Why should (our) geometry have these isometries (only)?

III.4.d The factor group of a space group gives a point group

How do we know that the factor group formed using the (an?) invariant translation subgroup is isomorphic to a point group [Burns and Glazer (1990) p. 171]? Why should it not be? Given an ideal T of group G, we multiply it by a minimal set of elements Q_i (here rotations and reflections) outside T to get group G. Since we started with an ideal, the set Q_i forms a group [Mirman (1995a), sec. IV.8, p. 137]; but is it a point group, might the minimal set be something different? The elements of invariant subgroup T are $\{E|\underline{t}_n\}$, where n labels a (discrete) set of lattice translations. The only operations available are translations and point group operations, but for nonprimitive elements the homogeneous transformations depend on the translations; can this prevent the factor group from being a point group? To get other cosets from the translation ideal its elements are multiplied by homogeneous transformations. Write space group G as a sum of cosets of invariant translation subgroup T, giving factor group G/T,

$$G = \{E|\underline{\tau}_1 = 0\}T + \{R_2|\underline{\tau}_2\}T + \ldots + \{R_h|\underline{\tau}_h\}T, \qquad \text{(III.4.d–1)}$$

displaying the explicit dependence of the elements of the factor group $\{R_k|\underline{\tau}_k\}T$ on translations τ. Now the elements of a coset are

$$\{R_k|\underline{\tau}_k\}T = \{R_k|\underline{\tau}_k\}\{E|\underline{t}_n\}T = \{R_k|R_k\underline{t}_n + \underline{\tau}_k\}T, \qquad \text{(III.4.d-2)}$$

so that all in a coset have the same homogeneous operator. Also all elements multiplied by the same translation are in the same coset,

$$\{R_i|\underline{\tau}\}\{E|\underline{t}_n\} = \{R_i|R_i\underline{t}_n + \underline{\tau}\}. \qquad \text{(III.4.d-3)}$$

It does not matter if $\{R_i|\underline{t}\}$ is nonprimitive.

Each coset representative is a transformation consisting of a point group operation, and a translation, this determined by the point group transformation. The difference between symmorphic and nonsymmorphic groups is that for the former all translations of the coset representatives can be set to zero, for the latter there is at least one that cannot be — not all can be set to zero *simultaneously*. Thus for a symmorphic group, taking all translations of the cosets zero gives a point group, and this is a subgroup of the space group. For a nonsymmorphic group, the point group obtained by setting these translations to zero is not a subgroup, the set obtained by setting as many translations to zero as possible contains a translation so cannot be a point group — in fact not a group as can be most easily seen if only one of these translations cannot be eliminated, for then the product of two members of the set contains a translation, but there is no further element with a translation, so this product is not in the set.

Problem III.4.d-1: This gives an isomorphism between the elements of factor group G/T and the set of homogeneous transformations and these form a point group (why?), with elements $\{R_k|0\}$, where k runs over the same range as for G. Thus the factor group also must be a point group.

Problem III.4.d-2: Why must the point group so obtained be limited to a crystallographic point group (the order of rotations is 1,2,3,4 or 6)?

III.4.d.i *Cosets and coset representatives*

The cosets of a group obtained with the invariant subgroup form a group, the factor group, but the set of coset representatives, the elements chosen one from each coset, do not form a group [Mirman (1995a), pb. IV.6.a-1, p. 129]. This can be illustrated using the spiral staircase. The first coset, the invariant subgroup of translations, contains all translations from a step to those above and below it; these are on a single line. Thus these steps lie on an infinite line. This is true for all cosets; they differ in that the lines are displaced by $\frac{2\pi k}{n}$,

$k = 1, 2, \ldots, n-1$. The coset representatives are the set of translations, chosen arbitrarily, one from each of these lines. Thus the first one may go from step 0 to step 1, the second from step 33 to step 34, these rotated with respect to step 0, the third from step 9 to step 10, and so on. These do not form a group since the identity, taking step 0 to itself, is not in any of the sets of translations, except the first.

The cosets (for the staircase) are the sets of sets of translations, that is the sets of infinite lines (each considered to be made up of a set of lines of finite length). The product of the set of translations at angles $\frac{2\pi}{n}$ and $3\frac{2\pi}{n}$ is the set of translations at $4\frac{2\pi}{n}$. These sets (the cosets) form a group; their products are in the set.

III.4.d.ii *The difference in factor groups of symmorphic and nonsymmorphic groups*

The difference between a symmorphic and a nonsymmorphic space group is that for the former the operations multiplying the invariant subgroup form a point group which is isomorphic to the factor group. For the latter these operations do not form a point group, although there is a group that is isomorphic to a point group, the factor group. Thus they are products of rotations and translations, but point groups do not contain the latter. In a symmorphic group all elements are (expressible as) either point group operations or pure translations.

We realize the translations here not as a set of vectors (vectors are given by their components, but are not attached to a point) but rather as a set of objects given by both vectors and a point to which they are attached (compare force and torque). This allows the introduction of a new kind of transformation, one that simultaneously transforms a vector into another, and changes the point to which it is attached. It is these which appear in nonsymmorphic groups.

III.4.e How nonprimitive elements are restricted

Screws and glides give simultaneous rotations, or reflections, and translations. How are these translations restricted by the conditions on the homogeneous transformations [Lax (1974), p. 234]? Since, for some n,

$$R^n = E, \tag{III.4.e-1}$$

the identity, we can write for an element of the space group,

$$\{R|\underline{v}\}^n = \{R^n|\underline{v} + R\underline{v} + R^2\underline{v} + \ldots + R^{n-1}\underline{v}\} = \{E|\underline{t}\}, \tag{III.4.e-2}$$

where $\underline{t}$ is a lattice translation, using the multiplication rule (pb. III.2.a-1, p. 134), as the crystal is invariant under the group.

A component of $\underline{v}$ perpendicular to the rotation axis or mirror can be taken as zero (sec. III.4.c.i, p. 149), but for the parallel component, $\underline{v}_p$, the n'th power of $\{R|\underline{v}_p\}$ is a lattice translation $\underline{t}$. Thus the translation $\underline{v}_p$ associated with homogeneous transformation R must be an integral multiple of a lattice translation divided by n,

$$\underline{v}_p = \frac{k}{n}\underline{t}, \quad 0 \leq k \leq n, \qquad \text{(III.4.e-3)}$$

giving the relationship between the order of the rotation and the magnitude of the translation.

Problem III.4.e-1: For a glide reflection this can be achieved in several ways. Show that as the square of a reflection is the identity, with lattice translations denoted by $\underline{a}$, $\underline{b}$, $\underline{c}$, the possible glide translations [Burns and Glazer (1990), p. 89] are (sec. III.4.b, p. 146)

1. axial glides (along lattice axes),

$$\underline{a}/2, \underline{b}/2, \underline{c}/2, \qquad \text{(III.4.e-4)}$$

2. diagonal glides (along diagonals),

$$(\underline{a}+\underline{b})/2, (\underline{a}+\underline{c})/2, (\underline{b}+\underline{c})/2, (\underline{a}+\underline{b}+\underline{c})/2, \qquad \text{(III.4.e-5)}$$

3. diamond glides (so called for obvious reasons) [Jaswon (1965), p. 100; Jaswon and Rose (1983), p. 133; Lockwood and Macmillan (1983), p. 160],

$$(\underline{a}+\underline{b}/4), (\underline{a}+\underline{c})/4, (\underline{b}+\underline{c})/4, (\underline{a}+\underline{b}+\underline{c})/4. \qquad \text{(III.4.e-6)}$$

Draw diagrams illustrating, and explaining, these (pb. III.5-2, p. 161). Modify fig. III.4.b-1, p. 147, if necessary, so that it contains diagonal glides.

Problem III.4.e-2: The isometries of (our) geometry as we have (hopefully) shown, are (only) translations, rotations, reflections, inversions, screws and glides, and products (sec. III.4.c.ii, p. 151). As we see this statement is incomplete; "glide" is not fully defined, but presumably with the previous problem all isometries of space are specified unambiguously and completely. Is this true? Why does (our) geometry allow these glides, and these only? Does this hold for every dimension? For every signature of space? How about other types of geometries?

Problem III.4.e-3: Explain why, if the lattice is invariant under nonprimitive operation $\{R|\underline{t}\}$, then $\{R|\underline{t}\}^n$ is (equivalent to) the identity, for some n, which means it equals a lattice translation. Need n be the smallest power for which

$$R^j = E? \qquad \text{(III.4.e-7)}$$

III.4.f Example of a two-dimensional nonsymmorphic group

What type of object has a nonsymmorphic group? This

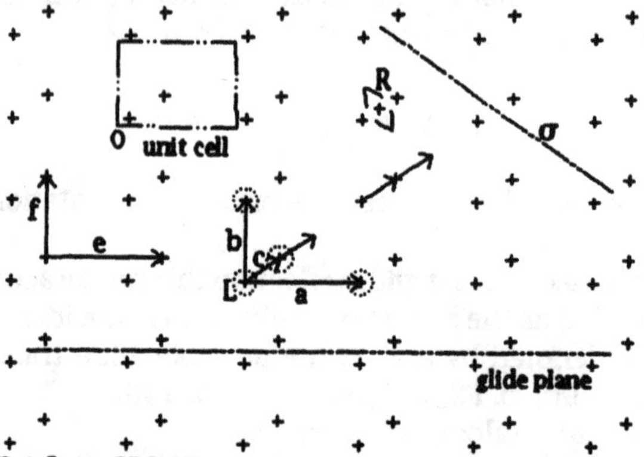

Figure III.4.f-1: CRYSTAL WITH A NONSYMMORPHIC GROUP,

is an example of a two-dimensional crystal with a nonsymmorphic space group, which also illustrates the difference between a crystal and a lattice. This crystal has a rectangular lattice [Falicov (1966), p. 168].

A square version is

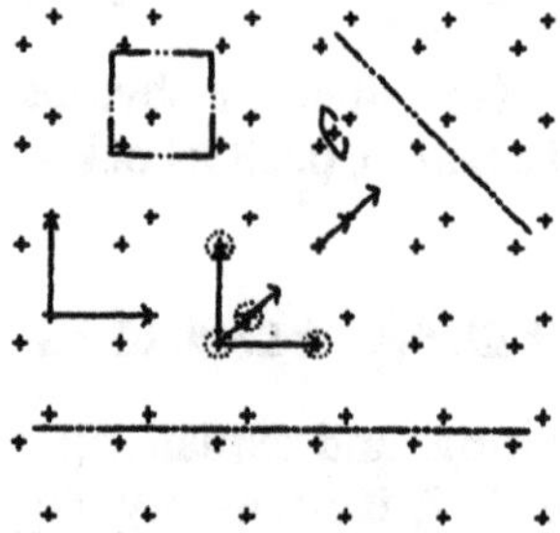

Figure III.4.f-2: A SQUARE CRYSTAL WITH A NONSYMMORPHIC GROUP.

The points can be taken as the atoms of the crystal — however these points do not form a lattice. Consider the four points that are circled (for identification). Vectors $\underline{a}$ and $\underline{b}$ are lattice vectors; integral multiples of them go from one point to an identical one. But $\underline{c}$ is not a lattice vector; $2\underline{c}$ does not go to another point — there is nothing at the end of it (as shown by its repetition). Yet this is a crystal, and it is built on a lattice. The point at the tail of these three vectors, L, is a lattice point; the points found using all sums of integral multiples of $\underline{a}$ and $\underline{b}$ form a lattice. A unit cell, which has two points, is indicated, and it is clear that repetitions of it fill, with no gaps, the entire plane. The crystal then is constructed by assigning pairs of atoms, displaced from each other, to the lattice points — these pairs are the basis. We take the crystal again (sec. III.3.a, p. 138) as a two-dimensional set of point particles (there are no operations reversing the sign of z). So $\underline{a}$ and $\underline{b}$ define a Bravais lattice with rectangular symmetry, and are orthogonal. The unit cell of this lattice contains two points; it can also be taken as the rectangle whose sides are these vectors.

III.4.f.i *How the crystal differs from the lattice*

The crystal, but not the lattice, consists of two overlapping rectangular sets of points, displaced from each other. Integral multiples of lattice vectors $\underline{e}$ and $\underline{f}$ do take points of the second lattice to other points of the same lattice; only $\underline{c}$, which connects points of one lattice with those of the other is not a lattice vector.

We take these two sets of points identical (we place the same type of atoms on the two lattices). Because the objects at the lattice points, the sets of atoms, now have structure, there are symmetries of the lattice that are not symmetries of the crystal; the symmetry of the crystal is, in that sense, less than that of the lattice. However there is now an additional transformation, taking the points of the two lattices into each other, which is a symmetry (provided they are identical). Also there are products of this, with other transformations which are not necessarily, but which can be, symmetries. Thus the group of the crystal is different from that of the lattice, and here it contains a different type of symmetry transformation, a glide reflection.

The crystal is invariant under reflections in the mirrors, such as σ, bisecting the lines between neighboring points of the two lattices. It is also invariant under two-fold rotations (indicated by a figure with two sides) about point R, with the z axis perpendicular to the plane, and any equivalent point. Both of these transformations take the points of one lattice into those of the other. Thus the crystal is not invariant under (all?) symmetry transformations of either lattice (alone), but rather under point group transformations that interchange points of the two lattices. There is an additional symmetry, also of this type, a glide reflection in glide plane $\sigma_{\mathbf{x}}$ (and equivalent ones).

III.4.f.ii *The space group operations*

One space group operation is the identity, $\{E|0\}$. Taking the origin at O (or R, the point with a two-fold axis), we get space group operations $\{C_{2z}|0\}$ and mirror $\{\sigma|0\}$. There is also the glide reflection in mirror plane $\sigma_{\mathbf{x}}$, which can be written as $\{\sigma_{\mathbf{x}}|\underline{c}_x\}$ since it is given by a reflection after a translation by the x component of $\underline{c}$. These are not of the form $\{R|\ \underline{t}\}$, for a single $\underline{t}$; two $\underline{t}$'s are required. This indicates a nonsymmorphic group. There is no point group that is a subgroup of this space group because setting translations to zero does not give a point group. Some of the products of the point group operations are not members of the set, so it is not a group;

$$\{R|\underline{t}\}\{R'|\underline{t}'\} = \{RR'|\underline{t} + \underline{t}'\} \qquad (III.4.f.ii-1)$$

is not a symmetry of the crystal.

III.4.f.iii *Shifting the origin cannot make the group symmorphic*

However there is one possible hole in this argument. Might it be possible to shift the origin so that all operations would then form a point group with respect to the new origin? The origin is shifted by vector $\underline{\tau}$, with components $\underline{\tau}_x$ and $\underline{\tau}_y$, and the subscripts on the mirrors indicating the lines along which they lie. Then

$$\{E| - \tau\}\{E|0\}\{E|\tau\} = \{E|0\}, \qquad (III.4.f.iii-1)$$

$$\{E| - \tau\}\{\sigma_{\mathbf{x}}|\frac{1}{2}\underline{a}_1\}\{E|\tau\} = \{\sigma_{\mathbf{x}}|\frac{1}{2}\underline{a}_1 - \underline{\tau}_y\}, \qquad (III.4.f.iii-2)$$

$$\{E| - \tau\}\{\sigma_{\mathbf{y}}|\frac{1}{2}\underline{a}_1\}\{E|\tau\} = \{\sigma_{\mathbf{y}}|\frac{1}{2}\underline{a}_1 - \underline{\tau}_x\}, \qquad (III.4.f.iii-3)$$

$$\{E| - \tau\}\{C_{2z}|0\}\{E|\tau\} = \{C_{2z}| - 2\tau\}. \qquad (III.4.f.iii-4)$$

There is no $\underline{\tau}$ for which all of these are of the form $\{R|\ \underline{t}\}$, where

$$\underline{t} = n_1\underline{a} + n_2\underline{b}. \qquad (III.4.f.iii-5)$$

Hence the group is nonsymmorphic. Note that $\{\sigma_{\mathbf{x}}|\frac{1}{2}\underline{a}_1 - \underline{\tau}_x - \underline{\tau}_y\}$ depends on both an x and a y component. Thus although the other operators can individually be put in the form $\{R|\ 0\}$ or $\{R|\ \underline{t}\}$ by proper choice of $\underline{\tau}$, it cannot. This is the characterization of a glide translation of a nonsymmorphic group.

Problem III.4.f.iii-1: Explain why both x and y components appear here. How is that related to the fact that g is a glide translation?

Problem III.4.f.iii-2: Find the factor group with respect to the translations, and show that it does form a group and is a point group. Which point group?

Problem III.4.f.iii-3: What is the symbol for the space group of this rectangle (sec. III.8.b, p. 175)?

Problem III.4.f.iii-4: Can we tell by looking at the figure that the group is nonsymmorphic?

Problem III.4.f.iii-5: Here the space group is nonsymmorphic because the lattice has at its points objects with structure (there are two interpenetrating lattices). Need this always be true; can a lattice with (structureless) points have a nonsymmorphic group? Explain. Are there requirements on this structure for the group to be nonsymmorphic? How about for it to be a group? Does this hold for every dimension [Brown, *et al*, (1978)]? Thus it is the space group of the crystal, not the lattice, that is nonsymmorphic. The lattice is just a rectangular one, with (necessarily) a symmorphic space group. But because every lattice point has assigned to it an (identical, of course) object with structure, the set of symmetry elements is different — there is another transformation (with what limitations?) between the lattice point and the atom displaced from it (that is between the two interpenetrating lattices). The totality of these form a nonsymmorphic space group.

Problem III.4.f.iii-6: Determine the (infinite) group table and verify that it satisfies the group axioms. By looking at the table can you tell that the group is nonsymmorphic? Is it possible to see that it has a glide, and find it? Can the crystal class be found from the table? Does it show that the crystal consists of two lattices displaced from each other? Can it be used to find the point group, and its group table? How are the group tables related? Is there a way of obtaining the diagram from the group table — either or both?

Problem III.4.f.iii-7: Find a formula for the coordinates of the lattice points, and by applying the group transformations, show that the lattice is invariant under the group. Does this show that it is nonsymmorphic? Why? Does such a formula give the transformations of the invariance group? Determine the points that are the centers of rotation, and the lines that are mirrors, and the glides.

Problem III.4.f.iii-8: Find, and analyze, a three-dimensional generalization of this. Repeat for n dimensions.

Problem III.4.f.iii-9: Are there nonsymmorphic space groups in all dimensions [Brown, *et al*, (1978)]? Do they all have the same nonprimitive elements?

III.5 THE DIAMOND STRUCTURE

A quite illustrative, and important, three-dimensional structure with a nonsymmorphic space group, Fd3m (O_h^7), no. 227 [Hahn (1987); Lockwood and Macmillan (1983), p. 189] is that of diamond [Burns and Glazer (1990) p. 155; Hargittai and Hargittai (1987), p. 392; Herring (1942), p. 538; Inui, Tanabe and Onodera (1990), p. 262; Janssen (1973), p. 124; Jaswon and Rose (1983), p. 82; Jones (1975), p. 155; Lax (1974), p. 256; Megaw (1973), p. 80; Rosen (1977), p. 89; Sands (1993), p. 137; Shubnikov and Koptsik (1974), p. 216; Streitwolf (1971), p. 71;

Wherrett (1986), p. 237]. Other crystals have this diamond structure, including ones of silicon and germanium not surprisingly, though diamond is the most famous. The lattice itself is face-centered cubic (sec. II.6.b, p. 105). The representations of this group are discussed below (sec. VII.7, p. 384).

The length of the sides of the unit cell is taken as 1, with the origin at a vertex. For each lattice point there is a pair of carbon atoms. We refer to the two atoms going with each lattice point as "companions". The atoms are at (n_x, n_y, n_z) and $(n_x + \frac{1}{4}, n_y + \frac{1}{4}, n_z + \frac{1}{4})$, for the vertices, and $(n_x + \frac{1}{4}, n_y - \frac{1}{4}, n_z + \frac{1}{4})$ and $(n_x + \frac{1}{2}, n_y, n_z + \frac{1}{2})$ for the centers of the faces, for all integer n (for the unit cell the n's are 0 or 1). For example there is a pair at $(0,0,0)$, $(\frac{1}{4}, \frac{1}{4}, \frac{1}{4})$, and another pair at $(\frac{1}{4}, \frac{1}{4}, -\frac{1}{4})$, $(\frac{1}{2}, \frac{1}{2}, 0)$.

Each unit cell contains eight atoms. For the face-centered cubic lattice there are eight corners, each shared by eight cells — the corners thus contribute one point — and six faces, each going with two cells, so the faces contribute three points, giving four lattice points. In diamond each point holds a pair, thus there are eight atoms per unit cell.

The unit cell of diamond with points at its corners and centers of its faces is obtained from the face-centered lattice drawn with heavy solid lines, with atoms at the lattice points (ones of each pair of companions), as shown in

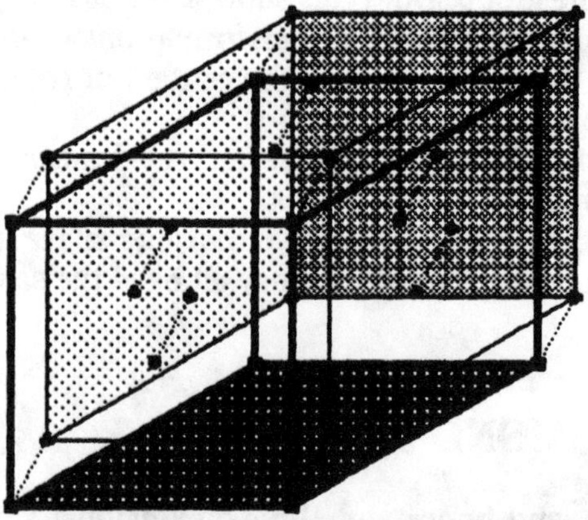

Figure III.5-1: UNIT CELL OF DIAMOND.

A second face-centered lattice, formed by the others of each pair, is shown with lighter lines. Pairs of companions are connected by dotted lines, perhaps most clearly seen at the top left (front) and bottom right (rear). The two lattices are different without being distinguishable; here they are drawn distinct (and with shading to emphasize the planes). For clarity the points of a pair are drawn close together, but they are actually further apart,

The unit cell of a lattice is repeated to form the lattice, but the unit cell of diamond consists of two lattice unit cells, displaced from each other. Thus the unit cells overlap, that drawn with light lines penetrates the next cell drawn with heavy lines, and conversely. Unit cells of lattices are like solid blocks (say of wood) placed together. Here they have to be made of wire.

Problem III.5-1: The diamond has screw axes. One, a 4_1 axis — the rotation is $\frac{2\pi}{4}$, the translation $\frac{1}{4}$ of a unit cell (sec. III.4.a, p. 142) — is parallel to z (taken as the line from front to back) and through $(\frac{1}{2}, 0, \frac{1}{4})$ [Burns and Glazer (1990) p. 155]. Show that a rotation around it by $\frac{\pi}{2}$ followed (or preceded) by a translation of $\frac{1}{4}$ of a unit cell along z takes point $(\frac{1}{4}, \frac{1}{4}, \frac{1}{4})$ to $(\frac{1}{2}, \frac{1}{2}, 0)$, that is the "companion" of the corner point at $(0, 0, 0)$ to the (?) face-centered point. Where does it take $(0, 0, 0)$? Check that a point is there. Also find its effect on the point at the center of a face, and its companion, and verify that it takes both to points at which there are atoms — it is a symmetry of the crystal. This indicates why there is this symmetry — there are two different (but indistinguishable) types of points, those at lattice points and their companions, and these are mixed. It is also relevant that there are points at both the vertices and centers of faces, as can be seen. Explain why it is necessary to consider just one lattice point to show that the screw rotation gives a symmetry. Apply this transformation two and three times. Where do they take the points? A screw can be modeled by a spiral staircase. Trace out such a set of stairs using points in the diagram to represent their edges.

Problem III.5-2: The diamond crystal also has glide planes. There is one perpendicular to the 4_1 axis (so parallel to the xy plane) at z $= \frac{1}{8}$, marked by a dot in front (left bottom) of the cell. We see the relationship of the glide plane (glazed) and the screw axis (dashed) next. Show that a reflection across the plane plus a translation of $(\frac{1}{4}, \frac{1}{4}, 0)$ takes $(\frac{1}{4}, \frac{1}{4}, \frac{1}{4})$ to $(\frac{1}{2}, 0, \frac{1}{2})$, again to a point in the center of a face — and note that this translation is not along one of the lattice vectors. What effect does it have on the latter point? Where does (0,0,0) go? Does that seem reasonable? Why is the plane at $z = \frac{1}{8}$? This is a diamond glide (pb. III.4.e-1, p. 155). Thus a diamond glide reflection is a symmetry of the diamond, as we see

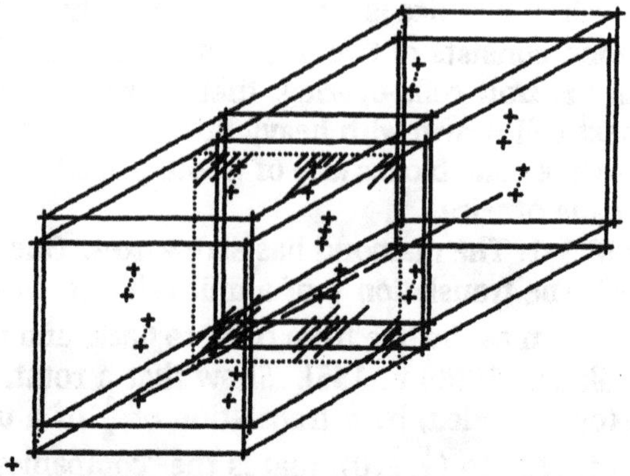

Figure III.5-2: DIAMOND GLIDE PLANE AND SCREW AXIS.

What effect does the glide applied twice have?

Problem III.5-3: Find all the elements of the space symmetry group of diamond [Inui, Tanabe and Onodera (1990), p. 262].

Problem III.5-4: These (screw plus glide) symmetry operations are of the form $\{R|\underline{t}\}$ and $\{\sigma|\underline{t}'\}$, where R is a rotation, σ a reflection, and the $\underline{t}$'s translations. But here, $\underline{t}$ and $\underline{t}'$ differ — the group is nonsymmorphic. Take the product of these two transformations, say acting on point $(\frac{1}{4},\frac{1}{4},\frac{1}{4})$, and show that it is of the form $\{S|\underline{t}''\}$, where S is, as expected, a rotation-reflection (which?), and $\underline{t}''$ a third translation. How is it related to $\underline{t}$ and $\underline{t}'$? Check that there is no way to shift the origin to make these translations the same. Conversely, set the translations equal and show that one of the resultant operations is not a symmetry of the crystal.

III.5.a The point group of the diamond space group

For a symmorphic space group the set of point group *symmetry* transformations of the unit cell forms a group, and this group is a subgroup of the space group. Here, however, we see that the product of elements, screw rotations or glide reflections, gives elements that take points in the unit cell to points outside it. Moreover the elements of a point group are of finite order. But these operations are not; no power of a screw rotation or glide reflection equals the identity. What is the point group of the space group and what is its relevance?

The holohedry group of the cubic lattice is octahedral group O_h; however this symmetry is broken because each point has two atoms (there are two interpenetrating lattices) so diamond is invariant only under the subgroup, tetrahedral group T_d — that the crystal be invariant only under a subgroup of the holohedry point group is necessary for a space group to be nonsymmorphic. What are the point symmetry transformations, and why are these less than those of the holohedry group?

For O_h, the group elements are $E, 8C_3, 3C_2, 6C_2, 6C_4$, and $I, 8S_6$, $3\sigma_h$, $6\sigma_d$, $6S_4$, while for T_d they are E, $8C_3$, $3C_2$, $6\sigma_d$, $6S_4$. Thus the $6C_2$, $6C_4$ rotations, the inversion, $3\sigma_h$ reflections, and $8S_6$ rotation-reflections are lost. What happens to them?

Problem III.5.a-1: At each point there is a pair of atoms and this destroys inversion invariance (which is not surprising as a screw is not invariant under inversion — is this relevant?). Are the screws paired? This also ruins invariance under reflections in mirrors perpendicular to the faces. Draw the diamond and check this, and also that reflections in diagonal mirrors are still symmetries. (Where are these mirrors?) Make sure that in addition this eliminates the S_6 symmetries, but leaves the S_4 ones. Why is there this difference? Does the presence of glide planes also prevent inversion symmetry? Why? Give a simple geometrical explanation of the reasons the eliminated symmetries, and the retained ones, are these. The set of remaining symmetries, T_d, must be a subgroup of O_h. How does the explanation show this? For each atom, (check that) its four nearest neighbors lie on the vertices of a tetrahedron. Is this relevant? Where is the atom with respect to the tetrahedron?

Problem III.5.a-2: We see that although the crystal is not invariant under any point group transformation, there are symmetries given by the product of these with translations. Prove that the set of such symmetries do form a group, and find the factor group of the space group of symmetries. Further show that it is a subgroup of the holohedry group of the lattice, and that it is proper (smaller). Also verify that it belongs to the same crystal system as that of the lattice. Check all this explicitly for the diamond structure, geometrically and analytically (using the coordinates of the atoms, perhaps by computer). The point group of the space group, not its subgroup but a factor group, is a subgroup of the holohedry group of the cube. However while there is a one-to-one correspondence between the transformations of this point group and the subgroup of the holohedry group (here the tetrahedral group) they are not the same. The positions of mirrors and axes of rotations, which in one case, but not the other, give symmetries, are different, as is clear, and should be checked, for diamond.

Problem III.5.a-3: Find the site-symmetry group for each atom — the group of transformations leaving the atom fixed, and leaving the environment it sees unchanged. What is its relationship to the point symmetry group? The holohedry group?

Problem III.5.a-4: Why for a space group to be nonsymmorphic is it necessary that the crystal be invariant only under a subgroup of the holohedry point group?

III.5.b Where are the nonprimitive elements placed?

Given the symmetry elements for a crystal (say, diamond), we can see that they are correct. But if, conversely, we take an object, how do we know what the symmetry elements are, where to place them, and what the space group is? Diamond illustrates how to answer these.

III.5.b.i *Placing glides*

For glides, the reflection plus translation must take an atom to another plane with an atom at the final point. There are two possibilities, either the glide plane is (halfway) between the points (say, atoms) of a pair, as illustrated with diamond, or it is between pairs. The cubic system has mirrors parallel to each pair of faces, and perpendicular to the remaining four, plus diagonal mirrors (sec. II.8, p. 126), and the lattice has mirrors along the faces of the unit cell. A reflection in a mirror through the center of the unit cell reverses the direction of the vector linking companion points (clearly so for diamond), and no translation parallel to this plane brings an atom at a resultant position to a place where an atom was before. Hence glide planes must be between the points of the pairs. If the two companions merge, these planes become the mirrors on the faces of the unit cell.

Problem III.5.b.i-1: Draw this and check that these, and only these, are the glide planes, and their reasons. Is this true in all cases, or just diamond? Explain how the positions of the atoms result in this glide leaving the structure invariant.

III.5.b.ii *Placing screws*

A screw takes a point to a line on which there is another point. Here again it is derived from a symmetry element of the cube, a translation, but this merely requires that it be perpendicular to a plane, and does not specify where.

Problem III.5.b.ii-1: Explain, in terms of the positions of the atoms, where the screws are. Does this agree with what was stated above?

Problem III.5.b.ii-2: What is the transformation given by the product of a screw with a glide reflection? For diamond take several points and find the effect of this transformation on them. Does this give any further restrictions?

Problem III.5.b.ii-3: A glide reflection requires two planes on each of which are corresponding sets of points, relatively displaced, with the plane of reflection halfway between these. If the translational displacement becomes zero, the glide plane becomes a mirror, and conversely each mirror can be changed to a glide plane by translating one image with respect to the other — although since there are other atoms besides this pair in the crystal, this does not mean that every mirror in the lattice can be replaced by a glide, does it? Likewise a screw can be reduced to a translation, and conversely. Explain how. However (presumably) not all mirrors and translations can necessarily be so replaced, and still give a symmetry, and arbitrary numbers and positions of glides and screws cannot appear, for glides, and screws, and glides and screws, are related. For the groups considered here, and for others, starting from the symmorphic groups of each system obtain the nonsymmorphic ones (sec. III.9, p. 180), and find for each all ways of replacing mirrors by glides, and translations by screws. Are there general rules, or algorithms? Is this true for arbitrary spaces?

III.5.c Why is the space group of diamond so rich?

The diamond structure appears rather simple, just a cube, with points at the centers of all the faces, and a pair of identical atoms at each point, yet its space group is quite rich having many elements, including screw rotations and glide reflections. Why?

Two aspects are essential, all atoms are identical thus operations taking them into each other can be symmetries, and the lattice is cubic, having the most symmetry operations of lattices, and moreover face-centered cubic, with a large number of atoms, so allowing a large number of transformations mixing them. The carbon atoms can be (and are) regarded as spherically symmetric, so the space group transformations leave not only their positions, but the atoms themselves, thus the crystal, invariant. But there is one other aspect: each lattice point holds a pair, and the second atom has coordinates of the form $\frac{n}{4}$, while the one at the center of a face has coordinates like $\frac{n}{2}$. And $\frac{1}{4}$ and $\frac{1}{2}$ are simply related — an essential point as can be seen from the transformations above, especially for the presence of diamond glides.

Thus it is because the arrangement is so simple, all atoms identical, in a nice symmetrical system, and with the coordinates simply related, that the space group is so complicated. As in many other areas, it is from the simplicity that the complexity arises.

Problem III.5.c-1: Give the (infinite) group table (in what way?) and verify that it satisfies the group axioms. Can you tell from the table that the group is nonsymmorphic? How? Is it possible to tell what the nonprimitive elements are (their type, their positions and orientations, the angles of rotations, and lengths of the glides)? Does the table indicate that there are two interpenetrating lattices? Does it show the lattice class (fcc)? Use this table to find that of the point group (and its name), and explain the relationship between the tables. To what extent does this relationship explain the structure of the space group, its richness, and that it is nonsymmorphic? Do either of these tables, or their relationship, show that both glides and screws are present? Do they tell anything further about them, including their number and placement? Can either, or both, be used to help draw the crystal?

Problem III.5.c-2: Apply the group transformations to a formula for the coordinates of the atoms, and show that the crystal is invariant under the group. Does this reveal that it is nonsymmorphic? Why? Is it possible to find the transformations of the invariance group (so the group) from the formula? Find (completely) the points that are the centers of rotation, the planes that are mirrors, the screws, the glides.

Problem III.5.c-3: Write a computer program to produce a motion picture showing the action of each group transformation (a product of infinitesimal transformations) on the diamond structure. This should demonstrate its invariance. Also given the crystal, develop a program that finds, algebraically and pictorially, the invariance group.

Problem III.5.c-4: What is the (simplest) generalization of diamond to four dimensions? To n dimensions? Repeat the analysis of this section for it, and for a general (most general) geometry.

III.5.d Spinel

Another mineral with same space group as diamond, Fd3m ($O_h{}^7$), is spinel (Al_2MgO_4 [Megaw (1973), p. 221]). Each carbon atom in diamond is replaced by seven atoms of the spinel motif (molecule is not quite the right term); there thus are eight of these motifs in a unit cell. (A motif, an object or design that is repeated, is related to the word motive since the object is moved to achieve repetition; here we consider only motifs that fill, or tile, the plane, thus the proper mathematical term for a motif is a tile [Schattschneider (1990), p. 95].) However this replacement alone is insufficient to retain the space group. The set of seven atoms is not spherically symmetric so their arrangement is altered by the group operations. Thus for the crystal to be invariant under the group, the object at each point, here the set of seven atoms, must also be.

The eight magnesium atoms in a unit cell of spinel are at the same positions as the eight carbons in diamond (pure magnesium [Megaw (1973), p. 75] has a primitive hexagonal structure with space group $P6_3/mmc$, no. 194 of the International tables [Hahn (1987)]). There are sixteen aluminum atoms, four attached to each pair of magnesium atoms [Burns and Glazer (1990) p. 157].

Problem III.5.d-1: Where does the glide reflection that takes the Mg atom at $(\frac{1}{4},\frac{1}{4},\frac{1}{4})$ to the face center $(\frac{1}{2},\frac{1}{2},0)$, take the attached Al atom? Is this position correct? Check that the other atoms in the set at the (0,0,0) vertex also go into atoms attached to the one at the center of the face. Thus the space group interchanges the magnesium atoms, and simultaneously moves the other atoms into the proper relationships with the displaced magnesium atoms. The crystal, not merely the set of Mg atoms, is invariant under it, because of the arrangement of the other atoms. Using this find the group table, and show that it is the same as for diamond. That the space group is the same as that of diamond comes then from two factors, the identical Mg atoms are at the same positions as the C atoms in diamond, and the other atoms are properly numerous and are in the proper positions so they are interchanged in the same way as the Mg atoms to which they are associated. We leave it to the reader to consider why the positions of these atoms is fortunately *exactly* (to infinite precision) what is required to retain the space group symmetry of the Mg atoms (in this material) alone, and what the relationship of this is to the fact that the Mg atoms do arrange themselves, for spinel but not for pure magnesium, so as to be invariant under this space group. Of course these same questions can be raised about carbon in diamond, and we can ask what properties of these two systems result in them having the same group? Can general rules be found to tell what elements, and what combinations of elements, have this structure? Can they be found for any space group? Conversely, given an element, or a set, can rules be found to determine the space group of the crystal they (presumably) form?

Problem III.5.d-2: Modify the computer programs that produce motion pictures for diamond, so that they apply to spinel. Combine them, with the various types of atoms in different colors.

III.6 INHOMOGENEOUS ROTATION GROUPS AND NONSYMMORPHIC GROUPS

A space group, symmorphic or nonsymmorphic, is a subgroup of the inhomogeneous rotation group. Yet this (which can be called symmorphic) is a semi-direct product of the invariant translation subgroup and the rotation group. Does a nonsymmorphic group also have such a structure? How can it be a subgroup, why does a group allow a subgroup which seems so different, how are these groups related?

Problem III.6-1: For a space group, symmorphic or nonsymmorphic, prove that the set of symmetry operations $\{R|\underline{t}\}$, for all — which means what for a nonsymmorphic group? — point group operations R, form a group. Note, however, that for a nonsymmorphic group the translations depend on R. Show that the group formed by the $\{R|\underline{t}\}$ acts on the translations, but leaves the translation subgroup invariant.

III.6.a Are nonsymmorphic space groups semi-direct products?

The group of transformations of (three) space, the inhomogeneous rotation group, is a semi-direct product. Nonsymmorphic groups also are transformations of (objects in) the (same) space. Does this imply that they are also semi-direct products? A symmorphic group is such a product — of the point and translation symmetry groups of the crystal. A semi-direct-product group is one with two disjoint sets of elements, both groups, with one invariant under all group operations. A nonsymmorphic group is a semi-direct product of the factor group, a point group, and the translations, but the difference is that the factor group is not a subgroup and not a symmetry group of the crystal. How can a subgroup of a semi-product group not be one?

III.6.b Why are there nonsymmorphic groups?

Why are there nonprimitive operations and groups that contain them? One answer is that crystals have properties that require these groups to describe them (and fortunately there are groups able to do so). But what is it about these groups that lead them to have such properties? How can a subgroup of the inhomogeneous rotation group (which, extending the terminology, is symmorphic) be nonsymmorphic? All operations, primitive and nonprimitive, of a space group are in the inhomogeneous rotation group, obviously, but not all operations of this full group are in the space subgroup. The inhomogeneous rotation group has operations $\{R|\underline{t}\}$, but also has operations R and $\underline{t}$, so any $\{R|\underline{t}\}$ can be written as a product of a homogeneous transformation and a translation, where both are in the group. In going to a subgroup, we may include an $\{R|\underline{t}\}$ without also including R or $\underline{t}$, thus $\{R|\underline{t}\}$ cannot be written as a product of elements in the subgroup. For a nonsymmorphic group we pick a particular subset not all of whose elements are homogeneous transformations, some are products of these and translations. And we leave out other elements of the inhomogeneous rotation group. This subset is a group, but by the (peculiar?) choice of members is nonsymmorphic. The reason that the "symmorphic" inhomogeneous rotation group can have a nonsymmorphic subgroup is this exclusion of elements.

We are considering two entities, space described by the inhomogeneous rotation group, and the crystal, which picks the subgroup that it wants to describe it.

Problem III.6.b-1: For some screws and glides of the diamond structure, for example, find transformations that, if included, would give a symmorphic group.

Problem III.6.b-2: What axioms of geometry (pb. III.4.c.ii-2, p. 152) require that space be described by a symmorphic group (which has subgroups that are nonsymmorphic)? The (discrete) points of a crystal can be regarded as forming a space. How are the axioms of Euclidean geometry changed to give its geometry? How do they allow nonsymmorphic groups? Is it possible to have a geometry for a space whose points form a continuum that is nonsymmorphic?

III.7 GEOMETRIC AND ARITHMETIC EQUIVALENCE OF CRYSTAL CLASSES

The objects that we consider, abstract groups, realizations of these, representations, entities on which they act such as basis vectors, lattices and crystals, are associated. Their relationships may depend on how we regard the systems, thus abstract groups may be identical (isomorphic), but not their realizations, differing, say, for reflections and π rotations. One of our aims is organization; by organizing the entities we can deal with many at once, finding rules that apply to all. Thus we study the relationships of these objects, and implications for their properties. Here we introduce another classification, that of geometrical and of arithmetical equivalence, so of geometrical equivalence classes and of arithmetical equivalence classes [Brown, *et al*, (1978), p. 11; Janssen (1973), p. 73, 119; Kocinski (1983), p. 74; Lifshitz (1997), p. 1188; Senechal (1990), p. 92; Sternberg (1994), p. 309; Schwarzenberger (1980), p. 34].

Point groups, though they may be isomorphic as abstract groups, differ if realized differently (pb. I.6.d-9, p. 43). They are the abstract groups that can be realized as operations on coordinate vectors, thus specifically with three-dimensional representations (n-dimensional in n-dimensional space), and are regarded, not abstractly, but as these realizations. Such (realization) defining representations are equivalent to ones with orthogonal matrices, for they are subgroups of orthogonal group O(3). The matrices, because they act on a lattice, have integral entries (sec. I.6.b.ii, p. 36). All point groups have three-dimensional representations equivalent to ones with such matrices; every point group is equivalent to an integral matrix group (one whose matrices have integral entries) — it can be realized as an arithmetic point group (that is an integral matrix group).

Two point groups are geometrically equivalent if they are equivalent as representations of the same abstract point group. This means that they are equivalent under a (nonsingular, of course) transformation with real entries, since they have real entries as they act on coordinates (although not all matrices of orthogonal groups need have real entries [Mirman (1995b), sec. 7.2, p. 124]). Being subgroups of O(3) their matrices are orthogonal, thus a transformation is by an orthogonal matrix [Mirman (1995a), sec. X.2.b, p. 272]. Geometrically-equivalent point groups are conjugate subgroups [Mirman (1995a), sec. IV.3.a, p. 109] of O(3), being transformed into each other by an element of O(3). This leads to the concept of geometrical crystal class; each such class contains the groups that are geometrically equivalent. The groups of different geometrical classes can be isomorphic (as abstract groups). Of course, all members of a class are isomorphic. Arithmetic point groups are arithmetically equivalent if they are conjugate — so they have to be conjugate under a similarity transformation by a matrix with all integer entries (we might consider rational entries, but these can all be made integral, with the matrix unimodular). The equivalence classes defined by this are the arithmetic crystal classes. That is we can choose, say, different axes, giving different point groups, but these both are integral matrix groups, so related by a similarity transformation with integer entries, thus belong to the same arithmetic crystal class. Arithmetic equivalence implies geometric equivalence, but not conversely. There are thirty-two point groups (geometric classes of point groups); eighteen (abstractly) nonisomorphic.

Crystal systems are geometrically equivalent if they can be transformed into each other with non-singular matrices — the primitive lattice vectors of one can be taken into those of another by such a matrix. They are arithmetically equivalent if the transforming matrix is a unimodular matrix whose elements are integers. Two Bravais lattices, which are given by sums of the primitive lattice vectors with integer coefficients, are equivalent if they are arithmetically equivalent — they can be obtained from each other with a unimodular matrix whose elements are integers. The Bravais lattices are the arithmetic classes of the crystal systems.

Different geometrical classes can have point groups that are isomorphic. In tbl. B-2, p. 634, the two-element cyclic group is listed three times. These belong to different geometrical crystal classes (sec. II.7.a, p. 112).

The arithmetic classes of the space groups are determined by the symmorphic ones; there is a one-to-one correspondence between the symmorphic space groups and the arithmetic classes [Janssen (1973), p. 120]. There often is more than one space group in each arithmetic class; the symmorphic one defines the class, the others are nonsymmorphic. Although the space groups of a class are not equivalent, their (factor) point groups are arithmetically equivalent.

Problem III.7-1: It was stated that the similarity transformation between two matrices with integral entries also has integral entries. Presumably this is true.

Problem III.7-2: Show that the necessary and sufficient condition for two lattices (how might they differ?) to be equivalent is that they be related by a unimodular matrix with integer elements. For example write the lattice vectors as column vectors, and consider the transformations leaving the lattice invariant. Why is the matrix unimodular? Can this method be used to find the Bravais lattices? All? Only? Does it give the same results as above (chap. II, p. 66)?

Problem III.7-3: Show that point groups that are the factor groups (or subgroups) of isomorphic space groups are arithmetically equivalent [Janssen (1973), p. 119].

Problem III.7-4: Explain the meaning of "equivalent as representations of the same abstract point group". Might "realization" [Mirman (1995a), sec. V.3.c, p. 157] be a better word than "representation"? Show that there is an orthogonal transformation linking two geometrically equivalent point groups. Why?

Problem III.7-5: Check that the crystal systems given in tbl. B-2, p. 634, belong to different geometrical crystal classes — the primitive lattices cannot be obtained from another one by a similarity transformation (with real entries?), and the groups listed for each do belong to the same ones.

Problem III.7-6: Prove that there are 32 geometric crystal classes, thus (why?) 32 point groups [Janssen (1973), p. 73], but there are fourteen arithmetic classes. Relate the classes to the Bravais lattices. Also the assignment of the point groups to crystal classes should be a consequence of these results, and (hopefully) is the same as that given in appendix B, p. 631. This provides one explanation of why there are fourteen Bravais lattices (in three dimensions), giving a rigorous (?) proof. How is it related to all preceding arguments? Why do they follow from this, and conversely (presuming that they do)?

Problem III.7-7: Find the geometrical and arithmetical classes, and the lattices in two and n-dimensional spaces. Are the results consistent?

III.8 THE SPACE GROUPS IN ONE AND TWO DIMENSIONS

Space groups describe objects with discrete symmetries in space. Insight into these is increased by understanding how they vary with the nature of the space, here only its dimension. These objects also are categorized by the dimension of the space in which they exist, and are different for different dimensions, at least those that we are familiar with, dimensions 1, 2 and 3. So we consider these three spaces, describing their groups, and the objects whose symmetry they give. The interest is not only physical and mathematical, but, especially for linear and planar objects, aesthetic [Audsley (1968); D'Avennes (1978); Farmer (1996), p. 89; Gillon (1969); Grafton (1992); Hargittai and Hargittai (1987); Hessemer (1990); Hornung (1975); Jones (1987); Martin (1987), p. 87; Schättschneider (1990); Simakoff (1993); Shubnikov and Koptsik (1974), p. 350].

The ornamental groups of the plane [Martin (1987), p. 88] (as they might be called) are of three types, plus combinations [Shubnikov and Koptsik (1974)], so this is merely a brief introduction. Pictures and diagrams of interesting examples from many fields, including art and architecture, are widely available [Fejes Toth (1964), p. 40; Field and Golubitsky (1992); Hargittai and Hargittai (1987), p. 327; Lockwood and Macmillan (1983); Shubnikov and Koptsik (1974)]. One type (often called rosette groups) has no translations [Shubnikov and Koptsik (1974), p. 11] — these (rose-like) groups are point groups, the cyclic and dihedral groups (sec. I.4.a, p. 8). The groups with one translation are those of patterns (with various symmetries) that are endlessly repeated along a line — the symmetry groups of one-dimensional space (in a sense). They are the frieze groups [Armstrong (1988), p. 164; Burn (1991), p. 208; Farmer (1996), p. 39; Fejes Toth (1964), p. 16; Hargittai and Hargittai (1987), p. 332; Johnston and Richman (1997), p. 149; Lockwood and Macmillan (1983), p. 13, 76, 108; Lyndon (1989), p. 40; Martin (1987), p. 78; Menten (1975); Shubnikov and Koptsik (1974), p. 79], with the name coming from that type of decoration. The third type, the groups that have two translations, are known as wallpaper groups [Armstrong (1988), p. 155; Burn (1991), p. 213; Farmer (1996), p. 43; Field and Golubitsky (1992), p. 55; Grossman and Magnus (1992), p. 160; Ladd (1989), p. 144; Martin (1987), p. 88; Schattschneider (1978); Schattschneider (1990); Shubnikov and Koptsik (1974), p. 134] — it is these to which "two-dimensional space group" usually refers.

Problem III.8-1: Show that there are seven frieze groups, and describe them. There are in addition 24 symmetry groups of two-sided bands, those with a second-order screw axis (second-order, of course)

[Hargittai and Hargittai (1987), p. 335; Shubnikov and Koptsik (1974), p. 95]. Does this differ from a glide?

Problem III.8-2: All these groups are isometries (equal measures) — they leave distances (so sizes) invariant — but are not the only transformations of space. There are also ones that change distances, so include dilations (sec. II.7.e, p. 121), these being similarities (they transform figures into ones that are similar, though not congruent) [Burn (1991) p. 31 Martin (1987), p. 136, 194; Rosen (1977), p. 62; Yale (1988), p. 68]. Check that frieze groups are isometries. Beyond these we might consider, but not here, transformations that introduce distortions, but leave various relationships unchanged, leading to affine transformations (sec. III.2.b, p. 134), the affine group [Burn (1991) p. 24, 185; Burns (1977), p. 250; Lyndon (1989); Martin (1987), p. 167; Shubnikov and Koptsik (1974), p. 236] and topology. It might seem that dilations are not relevant to physical objects. However we can consider (among other possibilities) a "crystal" made up of layers each identical (or perhaps transformed into each other by space group transformations) except for a change of scale. Or can we? Should we? Are there other physical cases to which they might be relevant? How about more general transformations, affine and others? Certainly there are physical properties dependent on size, and we can relate ones identical except for size, provided ....

III.8.a The symmetry groups of linear objects (frieze groups)

An object (figure, motif (sec. III.5.d, p. 166)) is repeated indefinitely along a line. What are the possible symmetries of the pattern? By definition of this, a frieze, discrete translations are one. In addition there can be rotations about a point in the object, but clearly only of $\frac{\pi}{2}$, reflections through that point, reflections in the mirror on the line of translation, and glide reflections. The combinations of these gives the frieze groups. The plane in which the frieze lies is not transformed, only the object in it. If it were, if we consider flipping it to the other side, we would be studying two-sided bands, a generalization which we must leave to the reader.

There are two one-dimensional patterns (invariant under discrete translations),

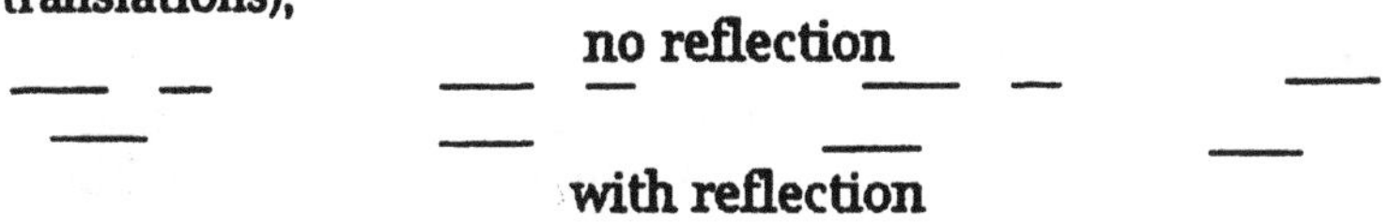

Figure III.8.a-1: ONE-DIMENSIONAL "CRYSTALS",

without, and with, symmetry under inversion (or reflection, there being no difference here) in each of the lattice points (of the one-dimensional lattice), where there are, for one pattern, distinct points of inversion (perhaps not surprisingly).

We write the symbol for the group (which is not the only convention) for a one-dimensional space group starting with F (p, for primitive, which is what all unit cells are, is also used, but since it is used elsewhere we prefer F). The first number after it is 2 if there is symmetry under rotations of π about an axis perpendicular to the plane, 1 if not. In the second position is m if there is a transverse (cutting the line) mirror, 1, if not. An m in the third position indicates the presence of a longitudinal mirror — along the axis. A g there indicates a glide, and if neither symmetry is present, then there is no symbol.

The friezes are two-dimensional objects invariant under a single translation. The first of the frieze groups, F_1, has only translational symmetry; the second, F_2, has also a point of symmetry about which a rotation of π leaves the frieze invariant. The group F_{1m} has transverse mirrors, while F_{11m} has a longitudinal mirror, but neither has rotational symmetry. Group F_{2mm} has a point for which the pattern is invariant under both a reflection and a rotation and also additional lines of symmetry (transverse and longitudinal mirrors). For F_{2mg} there is both a point of rotational symmetry and a line of reflection symmetry (a mirror perpendicular to the translation line), and these are displaced (by half of a translation), plus a glide. The seventh group is F_{11g}, which has a glide reflection. These are

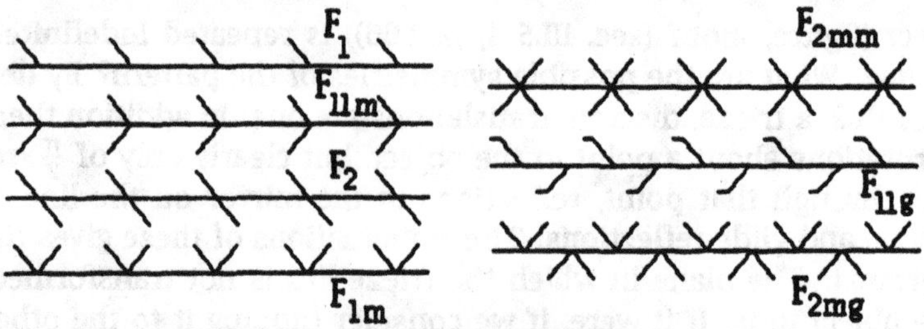

Figure III.8.a-2: THE FRIEZE GROUPS.

Problem III.8.a-1: Prove that all listed operations are symmetries (and the only ones?). Explain why the presence of some symmetries require others, so some figures have more than one. List these. Show that these are groups, that there are no other frieze groups, and that they are nonisomorphic. Check all this using diagrams, including the group symbols. For each draw a figure with a motif that has a higher symmetry than the figure on which it is placed (the "crystal").

Problem III.8.a-2: There is a point group (a factor group, though not always a subgroup) for each frieze group (named here by the abstract group to which it is isomorphic). Show that these are [Burn (1991) p. 209, 212]:

point group:	C_1	C_2	D_1	D_1	D_2	D_1	D_2
frieze group:	C_∞	D_∞	D_∞	C_∞	D_∞	$C_\infty \otimes C_2$	$D_\infty \otimes C_2$.

Explain the notation for the space groups, correlate them with the diagrams, justify writing the groups that include translations as infinite cyclic or dihedral groups, and explain why some patterns have a cyclic group, others a dihedral one. Can you tell by looking at a picture which type the group is? Which of these factor groups are subgroups? Why?

Problem III.8.a-3: Would there be a difference if instead of an infinite line, the pattern formed a circular one? How about elliptical?

Problem III.8.a-4: For each of these groups give the group table and verify that it satisfies the group axioms. Can you tell by looking at them which have glides? Find the table of the point group for each, and explain how the tables are related. Do the relationships show which groups have glides? Also show how each figure, or other such ones, follows from the corresponding table(s) — to the extent that any do.

Problem III.8.a-5: Find a formula for the coordinates of the lattice points, and apply the group transformations to it showing that each pattern is invariant under the stated group. From such a formula, is it possible to find the transformations of the invariance group? Find the centers of rotation, the mirrors, the glides.

Problem III.8.a-6: Check all this, algebraically and pictorially, using computer programs. The diagrams given here are very simple. However there can be quite complicated art work — there are many examples — that are invariant under these groups. Write computer programs to generate such patterns.

III.8.b Space groups with two translations

In two dimensions there are only 17 space groups (having two translations); beyond their intrinsic interest (in physics, mathematics, art, architecture, for example) they illustrate methods for obtaining such groups. (Here we consider only groups in a plane; there are more groups for planes in three-dimensional space [Lockwood and Macmillan (1983), p. 137].) A two-dimensional crystal system has a holohedry group (sec. II.3.e, p. 78), and Bravais lattices — one for each system except the rectangular which has two. Each holohedry group, as well as each subgroup, gives a space group. These are the two-dimensional symmorphic space groups. And there is a nonsymmorphic operation, a glide (a screw is inherently three-dimensional); so there are two-dimensional nonsymmorphic space groups.

The symbol (used below) for a two-dimensional space group describes the symmetry; first is either p or c for a primitive or centered unit cell, then the integer giving the highest order of rotation, followed by m (a mirror), g (a glide-reflection, if it results in a nonsymmorphic group), and 1 if a symmetry element is absent (suppressed for the short symbol). These group labels are given in parentheses at the bottom of each diagram, right above is the shortened symbol (used if it does not cause ambiguity).

III.8.b.i *Description of the two-dimensional space groups*

What are these groups, and how are they obtained? Here we merely describe them, thus sketching how they are found. Essentially, for each of the five Bravais lattices (sec. II.3, p. 71) we have a holohedry group and its subgroups. Each gives a space group. In addition, in some cases, we can add glides. Thus the question is for which lattices and groups can glides be added, how many, and where? Using all two-dimensional point groups, and placing all allowed glides, we obtain all two-dimensional space groups.

There is only one oblique Bravais lattice, so it is primitive, the holohedry point group is (in international notation short form) 2, with subgroup 1, and there can be no glide planes (in two dimensions actually lines). Thus the two space groups, both symmorphic, for this system are $p2$ and $p1$, with p indicating primitive (unnecessary here, but included for consistency). Why are there no glide planes? For the oblique lattice there is no reflection symmetry thus no glides.

The rectangular lattice has holohedry point group $2mm$, with subgroup m, and two lattices, primitive and centered. This gives symmorphic space groups $p2mm$ and $c2mm$, with c indicating centered (clearly needed), and also pm and cm. There are in addition nonsymmorphic groups; a glide reflection is possible. These groups are pg, $p2gg$ and $p2mg$, with g indicating a glide. How are they obtained, and why are these the combinations? Here either one or two mirrors are replaced by a glide plane (removing either then gives subgroups), but only for the primitive lattice. Why is it not possible to replace a mirror by a glide for the centered lattice? The centered rectangular lattice, with two points per unit cell, can be replaced by a primitive one so with a single point, but this is oblique, thus losing symmetry information (sec. II.3, p. 71). However the oblique lattice does not allow a reflection. Thus if we have a lattice with a glide plane it must be rectangular —

we need consider only the primitive unit cell; the centered lattice adds nothing additional.

The square has holohedry point group $4mm$ with subgroup 4 so its space groups are (with full and short symbols) $p4gm$ ($p4g$), $p4mm$ ($p4m$), going with the holohedry group, and $p4$ ($p4$), for the subgroup.

Finally there is the hexagonal lattice, with holohedry point group $6mm$, and subgroups 6 and $3m$, this having subgroup 3. The space groups are $p6mm$ (with short symbol $p6m$), for the holohedry group, $p6$ for the first subgroup, $p31m$ and $p3m1$ for the second (two since there is an arbitrariness in labeling axes), and $p3$ for the last. For this lattice there are no glide planes.

Problem III.8.b.i-1: For the square, why can both mirrors not be replaced by glide planes (as is possible for the rectangular case)?

Problem III.8.b.i-2: If the hexagonal lattice allowed glide planes — if it had a nonsymmorphic group — there would have to be unit cells with more than one atom. Show that such a unit cell can always be replaced by a (smaller) primitive unit cell that is still hexagonal, so does not give a different lattice. Why does the square have a glide plane?

Problem III.8.b.i-3: It should thus be clear that these are all the (nonisomorphic) two-dimensional space groups. Explain why it is clear. Check also that these really are groups, and all nonisomorphic.

Problem III.8.b.i-4: There is another way of obtaining these. The symmorphic ones are semi-direct products of the translation ideal and a point group. For each lattice the translations form an Abelian group. Write the group tables for the translation groups. Then use the elements of each point group, acting on each of these translation groups, to form the cosets, thus the product groups. Those point groups that leave invariant the same lattice as the translation group give space groups. How do glides enter? However if the point group and the translation group do not belong to the same lattice, the product (group?) should not be correct. What goes wrong? Explain. What are the factor groups of these space groups? Which are subgroups? Does that seem correct? Why?

Problem III.8.b.i-5: There are various derivations of these groups [Armstrong (1988), p.155; Burn (1991), p. 213; Martin (1987), p. 88]. How are these related? Which is most rigorous? Why? How can each be obtained from the others? Which are most easily generalized to other geometries? Can they be?

Problem III.8.b.i-6: Count the number of groups; check that there are seventeen, thirteen symmorphic and four nonsymmorphic. Prove that they are nonisomorphic [Armstrong (1988), p. 161], which is not completely trivial as there are different space groups (which?) with isomorphic point groups.

III.8.b.ii *Notation and illustrations for the wallpaper groups*

There are seventeen plane space groups with two translations, the wallpaper groups (although that application is not our primary concern) [Hargittai and Hargittai (1987), p. 345; Johnston and Richman (1997), p. 166; Ladd (1989), p. 144; Lockwood and Macmillan (1983), p. 16, 114, 137; Lyndon (1989), p. 74; Schattschneider (1978); Shubnikov and Koptsik (1974), p. 134]; the group symbol for a pattern can be found using a flowchart [Farmer (1996), p. 58]. Unfortunately only a small fraction of the information about these groups [Schattschneider (1978)] can be given here.

Diagrams illustrating objects with symmetry under these groups are given in appendix D, p. 650.

Problem III.8.b.ii-1: The difference between groups with glides, and those without, is that for the latter glides are unnecessary; all can be replaced by other operations. So where a glide is shown, but the group is symmorphic, it is unnecessary, but for nonsymmorphic groups there are glides (not necessarily all) that cannot be replaced. Check that this is really true for these patterns.

Problem III.8.b.ii-2: Note the crystal system and type of lattice for each group.

Problem III.8.b.ii-3: For which groups must the full symbol be used — the shortened form is ambiguous? Why?

Problem III.8.b.ii-4: Check that each lattice has all symmetries indicated, and no others (except perhaps products), thus is properly named.

Problem III.8.b.ii-5: Using the same motif, are there combinations of copies of it differently rotated that have the same symmetries?

Problem III.8.b.ii-6: Notice that there are two dimensions, no diamond glides (sec. III.4.e, p. 154); why? Can there be diagonal glides [Burns (1977), p. 323]?

Problem III.8.b.ii-7: Find the point groups that the lattices are invariant under. How are these related to the factor groups of the space groups of symmetries? How are the space groups generated from these point groups?

Problem III.8.b.ii-8: For group $p4$, a rotation of $\frac{\pi}{2}$ about a point labeled as having symmetry under a π rotation leaves the lattice invariant, but translated. Why do we not then say that there is another symmetry, a rotation followed by a translation, in analogy to glide-reflections? Are there other such cases?

Problem III.8.b.ii-9: Use these diagrams to explain why the space groups are all nonisomorphic [Armstrong (1988), p. 161].

Problem III.8.b.ii-10: Write the group table for each group and verify that it satisfies the group axioms. Can you tell by looking at them which groups are, and are not, symmorphic? From these tables find those of the corresponding point groups. How are they related? Is it possible to tell from the tables, and from the relationships, which groups are nonsymmorphic? Explain.

Problem III.8.b.ii-11: Using formulas for the coordinates of the lattice points [Hahn (1989), p. 3] — the lattice points of these patterns have to be specified — apply the group transformations, and show that each pattern is (hopefully) invariant under the given group (and no larger one). Can you also tell which are symmorphic? Why? From such formulas is it possible to find the transformations of the invariance group? Determine the rotation axes, mirrors, and glides. For each find the invariance point group (the site-symmetry group, which varies with position), and check that they are the same as those obtained from the space group tables (how?). Use these tables to draw (other) figures invariant under the groups.

Problem III.8.b.ii-12: Write a computer program that shows the action of the various transformations on figures (like these) to determine which groups leave each invariant. Check that the results are the same as previously given. Conversely, write a program, that given the group, draws a figure left invariant. Check that it is correct using the first program. Use these to show that the figures are not invariant under other groups (except subgroups). There is much art invariant under space groups (two-dimensional, but perhaps it is possible to develop sculptures invariant under three-dimensional space (?) groups). Write computer programs to generate such patterns. One commercial use is to design wallpaper and floor tilings. Are there others? Are they economically valuable?

Problem III.8.b.ii-13: Art work can be scanned into a computer program that determines its invariance groups, if such a program can be written. One that includes the colors would be especially interesting, as would a program that finds and compares the groups when colors are distinguished, and when they are not (although there has to be at least two, unless the scanner is not working properly).

Problem III.8.b.ii-14: The work of M. C. Escher provides many examples of two-dimensional space groups. It is a useful exercise to find the groups for the various drawings and use the results to check the accepted list [Schattschneider (1990), p. 330]. These have colors, and sometimes more than one motif.

Problem III.8.b.ii-15: There are in the references, and a vast number of other places (few of which appear to have anything to do with mathematics, or with formal presentations), objects with symmetry under space groups, particularly for one and two dimensions. Be aware of them as they are encountered, and find the point and space groups leaving each invariant.

III.9 THE THREE-DIMENSIONAL SPACE GROUPS

There are 230 (nonequivalent) space groups in three dimensions [Borchardt-Ott (1993), p. 183; Burns and Glazer (1990), p. 76; Hilton (1963), p. 168; Janssen (1973), p. 121; Megaw (1973), p. 148; Sands (1993), p. 145; Senechal (1990), p. 59; Zak, *et al* (1969), p. 39]. They and their properties are given in the International Tables [Hahn (1987)], with some in the teaching version [Hahn (1989)]. Individual groups, all [Lockwood and Macmillan (1983), p. 153], or some [Ladd (1989), p. 153; Megaw (1973), p. 153], are discussed in depth, and there are computer programs for them [Davies and Dirl (1994a,b)]. It is unnecessary to list them since tables are widely available, often with their names, and the point group and crystal system each belongs to [Burns and Glazer (1990), p. 303; Hahn (1987); Lockwood and Macmillan (1983), p. 190; Zak, *et al* (1969)], and their original derivation can be consulted [Fedorov (1971)].

A space group is denoted (in the International Notation) by the symbol for its point group preceded by a symbol giving its lattice type, *P* (for primitive), *I* (for interior) denoting a body-centered lattice, and *F* (for all faces centered); for single face-centering the letter is *A*, *B* or *C* for the centered faces perpendicular to the x,y,z axes respectively [Bradley and Cracknell (1972), p. 46; Lax (1974), p. 235; Megaw (1973), p. 151; Zak, *et al* (1969), p. 9].

What determines — what limits, what necessitates — space groups?

III.9.a The symmorphic space groups

We start by counting the symmorphic groups [Jaswon and Rose (1983), p. 100]. The Bravais lattices are different so are invariant under different space groups. And for each there are several point groups, the holohedry group and its subgroups. These give distinct space groups. Thus the cubic system has three lattices, the primitive, the body centered and the face centered. There are five point groups (tbl. B-2, p. 634), thus the cubic system has fifteen symmorphic space groups.

Problem III.9.a-1: It is a simple matter of counting to find 66 groups this way.

III.9.a.i *Ambiguities of setting give more space groups*

These are not all the symmorphic groups. There is a matter of ambiguity of setting (labeling) [Janssen (1973), p. 119; Jaswon and Rose (1983), p. 100; Lax (1974), p. 238]. For the end-centered orthorhombic cell, say, which has a 2-fold symmetry axis — labeled by [001] (sec. II.4.a, p. 85) — we can take the centered face as (001), or either (100) or (010). The latter two are distinct but not different, while they do differ from the first which is perpendicular to the symmetry axis [001]; the other two are parallel to it. Thus there are two different space groups for this cell, Cmm2 (the C meaning centered, but now centered along the C face), the other called either Amm2 or Bmm2 (these centered along the A and B faces).

The word setting refers to the choice of which face we center, and there is an ambiguity in the statement that the cell is face centered leading to more than one space group.

Problem III.9.a.i-1: It should be easy to check that there are seven such cases so that the total number of symmorphic space groups is 73 [Bradley and Cracknell (1972), p. 45].

Problem III.9.a.i-2: It is a useful exercise to derive all these. That is for each crystal system, and point group, give the names of the space groups, and their generators and group tables. The names are (almost) trivial. In the International Notation simply attach to the front of the point group symbol given in tbl. B-2, p. 634, the letters P, and where appropriate I and F, and C (and in a few cases where there is more than one setting A or B).

III.9.a.ii *Why are these the symmorphic space groups?*

A space group is given by the set of point group transformations plus the set of translations that leave the object invariant. The translations are determined by the lattice, and of course are different for each lattice. Thus the set of groups is obtained by combining each point group of a system with a lattice of the system (the only possibility since no other combinations would result in a lattice) to get a set of nonisomorphic space groups. However in some cases the names of the lattices do not give unambiguous ones — there are different settings. These give different groups.

There are no further ambiguities or choices, so no other symmorphic groups. For additional groups, nonsymmorphic ones, it is necessary to add further operations. The only available are the nonsymmorphic glides and screws, thus the groups containing them cannot lengthen our list of symmorphic groups.

Problem III.9.a.ii–1: Although we do not distinguish lattices of the same symmetry with different ratios of the lengths of their lattice vectors (sec. IV.4, p. 212), when centering there are cases in which this matters, leading to different labels for the groups. Check that all resultant space groups, due to different settings, are distinct.

III.9.b Enantiomorphic space groups

Some space groups contain screws, and screws have handedness — they are either right- or left-handed (sec. III.4.a.iii, p. 146). There are groups that have screws without partners of opposite chirality — these are chiral space groups. For each there is another, identical, except that the screws are replaced by ones of opposite handedness, giving enantiomorphic pairs of space groups. There are 11 such, of course all nonsymmorphic [Burns and Glazer (1990), p. 303; Lax (1974), p. 236]. As these are mirror images of each other they do not contain the inversion.

There is a slight semantic problem here. The members of a pair are really the same, each is but the mirror image of the other. Thus they can be regarded as identical. However it is conventional to take them different. In our world, for example in the biology of creatures on Earth, there is a difference between right-handed and left-handed molecules, not fundamentally, but certainly in their abundance. And that, at least, differentiates in a practical way crystals of opposite handedness. This is one reason why it is useful to distinguish crystals and molecules with different chirality, and the groups describing them.

III.9.c Deriving the space groups by enumeration

What does it mean to derive the space groups? For each we want its name, that is its symbol in the International Notation (this being linked closely to its operations and other properties). For symmorphic groups this is simple. We put in front of the symbol for each point group of a system the letters P, I, F, A, B, C, as appropriate. To get nonsymmorphic groups we must find which of these symmorphic groups allow glides and screws, how many, where, and of what type.

Then we wish the full set of operations of a group, with the generators specified. Again this is not too difficult for the symmorphic groups.

Next a unit cell is drawn, with a set of points arranged to show the symmetries. These are in, and above and below, the plane, and must be distinguished, and their coordinates specified.

The interesting problem is putting in the glides and screws to get the nonsymmorphic groups. There are limitations on screws (sec. III.4.a.ii, p. 145). Also the only way of obtaining a glide plane is to replace, in the symmorphic group from which the nonsymmorphic one is found, a mirror with a glide. Likewise a screw can only be obtained by replacing an axis of rotation by a screw. Now the elements are related by symmetry. Thus if there is a screw perpendicular to an axis of n-fold symmetry, there must be n screws taken into each other by this rotation, and similarly for glides. This is also true for inversion and reflection symmetries, although if these are present they rule out certain nonprimitive operations; screws are incompatible with an inversion. So each mirror in turn can be replaced by a glide, with all other mirrors related by symmetry also replaced. Then two mirrors, if not related by symmetry, can be replaced in the same way, and so on, until all mirrors (that can be) are exchanged for glides. The process is repeated for screws, and for screws and glides. As this is done it is necessary to check that all members of each set of elements related by all (remaining) symmetries are replaced. Also the operators must close to form a group, in particular a space group. Since mirrors or rotations are replaced, the set remaining forms a point group with fewer elements than without these nonprimitive operations, so less than the holohedry group of the system, and holohedry groups cannot be point groups for nonsymmorphic space groups.

Problem III.9.c-1: Since there are 73 symmorphic space groups, the number of nonsymmorphic ones must be 157. Finding them is more difficult. Ambitious readers can derive these and show there are a total of 230 space groups (in three dimensions), say by working out the required properties of each group individually [Lockwood and Macmillan (1983), p. 153]. In doing so it should be proved, both analytically, and geometrically (from diagrams) that they are groups, and distinct. Also for each, find a crystal whose group it is.

Problem III.9.c-2: For every space group, find the two-, one-, and (sec. III.8, p. 172) zero-dimensional (rosette) groups, that are its subgroups.

Problem III.9.c-3: Write group tables for several groups, and verify that they satisfy the group axioms. Can you tell by looking at them which groups are symmorphic, which have screws, which have glides, which have both? From these find the tables of the point groups. How are the tables related? Is it possible to tell from the tables, and from the relationships, which groups are nonsymmorphic, and why? Explain. Can tables for the point groups be used to construct those for the space groups, in particular can they show where screws and glides can (must?) be placed? Is it possible to write a single group table for all space groups such that individual ones are obtained, say, by removing specific rows and columns?

Problem III.9.c-4: Using formulas for the coordinates of the lattice points, by applying the group transformations, show that each lattice is invariant under its group. Is it possible to determine whether the group is symmorphic? Why? From such formulas can the transformations of the invariance groups be found? List the centers of rotation, the mirrors, glide planes and screws (positions and angles).

Problem III.9.c-5: Use a computer program that shows the action of the various transformations on the figures to determine which groups leave each invariant. Check that the results are the same as previously. Conversely write a program that given the group, draws a figure left invariant. Check that it is correct using the first program. Use these to show that the figures are not invariant under other groups (except subgroups).

Problem III.9.c-6: Repeat these problems for a space of four dimensions. Is it possible to discuss them for a space of arbitrary dimension (and signature)? How about arbitrary geometry?

Problem III.9.c-7: Translation groups of lattices are infinite discrete groups (sec. IV.5.g, p. 238). Space groups are also infinite discrete groups. Presumably it is not possible to write space groups as (pure) translation groups of lattices in some dimension. However are all infinite discrete groups space groups in the proper dimension, or might there be others that cannot be so classified?

Chapter IV

Representations of Translation Groups

IV.1 REPRESENTATIONS AND THE ROLE THEY PLAY

While lattices and crystals themselves are characterized by their symmetry groups, physical entities associated with them are described by statefunctions — representation basis vectors. Thus we come to the study of representations. These are interesting here not merely because of their physical applicability, but also because they require new concepts in group theory, and generalization of the meaning of representation. They therefore increase our understanding of mathematics, and the reasons that it is designed to have the properties we consider — which, as happens so frequently, raises the question why it can be so designed? Why is mathematics so capable, often seemingly with so little effort, of being applied to physics?

First we explore representations of translation groups, important in themselves, and as a foundation for the study, here particularly, of the representations of space groups, considered below (chap. VII, p. 337). And these illuminate the meanings, definitions, assumptions, limitations, of the concepts and entities that we have introduced. Lattices are spaces, so with geometries, and these representations help highlight the difference between representations defined over different geometries.

Point and translation groups are discrete groups [Yale (1988), p. 86], the former finite, but translation groups, though they give invariance only under discrete operations, are of infinite order, so provide examples of infinite discrete groups [Mirman (1995a), sec. II.3.j, p. 48]. These two types of groups describing symmetries of lattices and crystals are thus linked, and understanding these relationships and how they affect representations beyond being necessary is instructive. Why this is so instructive is itself interesting.

IV.2 REPRESENTATIONS OF TRANSLATIONS

The translation group [Brown, *et al*, (1978), p. 415] is Abelian, so its representations are quite simple, with basis vectors of the form $exp(ikx)$ [Mirman (1995a), sec. V.5.a, p. 162]. This solves the problem of finding their representations in free space. But the spaces that we are interested in, lattices (spaces of discrete sets of points), are not invariant under all translations, but only those that are integer multiples of a set of primitive ones, thus have richer and more complicated geometries than that of free space, so while the corresponding representations are similar to those, they are also richer and more complicated, and there are more types. The translation groups of lattices, being subgroups of the translation group of Euclidean space, are themselves groups, so they also are lattices (pb. II.2.a-3, p. 68; pb. II.7.f-17, p. 125).

Eigenstates of Abelian operators are one-dimensional and are given by $exp(i\underline{k} \bullet \underline{x})$, for all real $\underline{k}$'s, where the $\underline{k}$'s determine the representation — they are the translation-representation labels. These, up to equivalence, are the only representations — in Euclidean (free) space. Not all are representations of the crystal translations — a subgroup of the group of free-space translations. What are the representations of the translations for a crystal?

To find them we need not only the crystal lattice, but an additional, related, one.

IV.2.a The reciprocal space

Translation basis vectors are (unnormalized)

$$|\underline{k}) = exp(i\underline{k} \bullet \underline{x}); \qquad \text{(IV.2.a-1)}$$

the (components) of the representation labels, the $\underline{k}$'s, run over all real numbers, for a free particle — over a three-dimensional Euclidean space [Mirman (1995a), sec. XIII.4.b, p. 382; (1995b), sec. 2.1.5, p. 31]. Thus we have, besides the space in which we (and crystals) exist, a reciprocal one identical to it, momentum space, the space of the $\underline{k}$'s [Megaw (1973), p. 110]. For a lattice however, the set of vectors to the points of the space (the lattice points) is discrete. What then is the reciprocal space, the set of points giving the representation labels? Are the spaces still identical; why?

A crystal is invariant under any sum of integral multiples of three non-coplanar translations by lattice vectors $\underline{a}_1$, $\underline{a}_2$ and $\underline{a}_3$. Eigenfunctions of crystal translations thus must obey

$$exp(i\underline{k} \bullet \underline{x}) = exp(i\underline{k} \bullet (\underline{x} + \underline{a})) \qquad \text{(IV.2.a-2)}$$

or

$$exp(i\underline{k} \bullet \underline{a}) = 1, \qquad \text{(IV.2.a-3)}$$

which means

$$k_j = \frac{2\pi n_j}{a_j}, \qquad \text{(IV.2.a-4)}$$

for each integer n_j. These give the representations of the crystal translations, one for every set of three integers, so are discrete. These translation eigenfunctions are defined only at the lattice points, thus we still must give the basis functions everywhere (sec. IV.2.e, p. 193).

Such basis functions have absolute value 1. For free space, they can be any complex numbers with this absolute value. However for a crystal, translations — a subset of those of free space — are determined modulo the lattice vectors. Only ones between points in a unit cell are relevant here; those from a point to an identical one in a different cell provide no information or restrictions (for our present considerations). Thus, for crystal translations, we have to determine representation basis vectors that reflect this property — ones that are defined over a unit cell and are periodic functions of the lattice vectors, so identical in all unit cells.

This specifies a new space, that of the $\underline{k}$'s, and a new set of vectors, giving the discrete points of this space. Starting with the lattice vectors $\underline{a}_i$, we define a reciprocal lattice whose primitive (reciprocal) lattice vectors $\underline{b}_j$ are given by

$$\underline{a}_i \bullet \underline{b}_j = 2\pi\delta_{ij}, \quad i,j = 1,2,3. \qquad \text{(IV.2.a-5)}$$

The three $\underline{b}$'s have lengths

$$b_i = \frac{2\pi}{a_i}. \qquad \text{(IV.2.a-6)}$$

In this sense they are the reciprocals of the $\underline{a}$'s. They lie along the same lines as the three $\underline{a}$'s if, but only if, the lattice vectors are orthogonal.

Why this definition? It is chosen so that translations by a reciprocal vector leave the basis function unchanged,

$$exp(i(\underline{k} + \underline{b}) \bullet \underline{a}) = exp(i\underline{k} \bullet \underline{a})exp(i\underline{b} \bullet \underline{a}) = exp(i\underline{k} \bullet \underline{a}). \qquad \text{(IV.2.a-7)}$$

Thus

$$[exp(i\underline{b} \bullet \underline{a})]^m = 1, \qquad \text{(IV.2.a-8)}$$

for the infinite set of integers m. This means that the eigenfunctions of these translations are roots of unity, the particular set of roots depending on, and determining, the reciprocal lattice, so (in part?) the lattice itself.

Problem IV.2.a-1: Show that

$$\underline{b}_1 = \frac{2\pi(\underline{a}_2 \times \underline{a}_3)}{(\underline{a}_2 \times \underline{a}_3) \bullet \underline{a}_1}, \qquad \text{(IV.2.a-9)}$$

and similarly for the other two $\underline{b}$'s. The denominator is the volume defined by the three $\underline{a}$'s, that of the unit cell (what unit cell?), so is the same for all three arrangements of reciprocal lattice vectors in it. Give a geometrical interpretation of this formula.

Problem IV.2.a-2: How do we know that the reciprocal lattice is actually a lattice? Is it?

Problem IV.2.a-3: Prove that the symmetry groups (point groups — why?) of a lattice, and its reciprocal space, are the same. They belong to the same crystal system.

Problem IV.2.a-4: For which lattices are the reciprocal lattice vectors parallel to the lattice vectors? Check (including examples) that for these vectors to be parallel the lattice vectors must be orthogonal. For the other cases, how are the two lattices related [Bradley and Cracknell (1972), p. 86]?

IV.2.b Lattice points and vectors

A crystal is (partially) specified by the vectors $\underline{a}$ giving the translations under which it is invariant or by the coordinates of its points (a word whose ambiguity we ignore here). These should be related of course. However to identify a crystal, we do not need — fortunately — coordinates of all points (an advantage of crystals over other objects).

Lattice points are located using all sums of the primitive lattice vectors, with integral coefficients, n_i, which we write as a column vector, n. Then a (lattice) vector from the origin to a lattice point is

$$t = An, \qquad \text{(IV.2.b-1)}$$

where A is a matrix that specifies (in this way) the Bravais lattice [Jones (1975), p. 29; Zak, *et al* (1969)]. The reciprocal lattice, with lattice vector r, is likewise given by matrix B and row vector m, so

$$r = mB, \qquad \text{(IV.2.b-2)}$$

and since these are reciprocal, with I the 3×3 unit matrix,

$$BA = 2\pi I. \tag{IV.2.b-3}$$

This determines B, thus the reciprocal lattice.

IV.2.c Brillouin zones

The $\underline{k}$ vectors labeling the translation group representation define a lattice (in reciprocal space), one invariant under the translations,

$$\underline{K}_m = m_1\underline{k}_1 + m_2\underline{k}_2 + m_3\underline{k}_3, \tag{IV.2.c-1}$$

where the m's are (all) integers; these $\underline{k}$ vectors are the primitive reciprocal lattice vectors. The unit cell of this lattice, centered on a lattice point, is called its (first) Brillouin zone [Birman (1974), p. 42; Bradley and Cracknell (1972), p. 86; Burns (1977), p. 287; Cornwell (1969), p. 78; Cornwell (1984), p. 209; Cracknell (1975), p. 16; Elliott and Dawber (1987), p. 293; Falicov (1966), p. 147; Fässler and Stiefel (1992), p. 212; Heine (1993), p. 265; Jaswon (1965), p. 56; Jones (1975), p. 29; Joshi (1982), p. 270; Joshua (1991), p. 51; Lax (1974), p. 185, 443; Lyubarskii (1960), p. 90; Tinkham (1964), p. 270]. A reciprocal lattice then consists of a series of larger and larger Brillouin zones. If we know these — only the first is necessary — we know the reciprocal lattice vectors and thus the lattice vectors, so the lattice; a lattice is determined by its first Brillouin zone. Note that unit cells of lattices refer to parallelepipeds (sec. II.2.b, p. 68), but unit cells that are Brillouin zones need not be.

A Brillouin zone consists of parts, these related by the symmetry point group (sec. VII.2.d, p. 342). Thus we can often consider not the entire zone, but the smallest piece from which the others are obtained by the action of the point group — the representation domain [Altmann (1977), p. 203; Bradley and Cracknell (1972), p. 147].

Since unit cells can have various shapes, the Brillouin zone is defined as the cell that is most symmetric about the origin (in $\underline{k}$ space — the point labeled by zero momentum). The (most common type of the) first Brillouin zone (often called the Wigner-Seitz cell) is the set of points in reciprocal space, exhibiting the symmetry of the lattice, that lie closer to its origin than to any other reciprocal lattice point (sec. IV.3, p. 196). This unit cell, centered around this (zero-momentum) lattice point, is bounded by planes perpendicular to, and bisecting, the lines connecting this origin with all nearest lattice points (thus each stops at the first lattice point it reaches); the cell therefore encloses only a single lattice point. It differs from the cells considered above for Bravais lattices (sec. II.2.b, p. 68) in that it is not generally a parallelepiped (but some parallelepipeds, the nonprimitive ones, have more than one lattice point). This emphasizes that unit cells can have different shapes, the choice depending on what is most useful in a given situation.

Problem IV.2.c-1: This first Brillouin zone is bounded by planes (in two-dimensions, lines) that bisect, and are orthogonal to, the vectors from the origin to all neighboring reciprocal lattice points. Need these be nearest neighbors (pb. II.6.b-3, p. 106)? If not, how distant can they be? This cell exhibits (?) the point-group symmetry of the reciprocal, thus of the direct, lattice. Why? These cells, one for each lattice point, fill space (as a lattice does). Why?

Problem IV.2.c-2: Draw the first Brillouin zone for each of the fourteen Bravais lattices [Burns and Glazer (1990), p. 290].

IV.2.c.i *Brillouin zones are of the lattice*

The Brillouin zone is for the reciprocal of the lattice, not for the crystal, from its definition. It is the same for all crystals formed from a single lattice, thus its symmetry point group is the holohedry group of the system (sec. II.2.d, p. 69) [Elliott and Dawber (1987), p. 293]. Whether the lattice has a basis (sec. I.3.a, p. 6) is irrelevant; the reciprocal space of a crystal, but not a lattice, contains less information than the direct space, the latter having structure not appearing in its reciprocal.

The symmetry group of the Brillouin zone is the symmetry group of the lattice — the holohedry group of the crystal system — thus is always symmorphic (sec. III.3, p. 137). However representations of space groups are determined by the point group, which need not be the holohedry group. Thus the relevant symmetry operations are not those of the Brillouin zone for these may include transformations under which the crystal is not invariant. It is important to know the effect of an operation — while it may seem to give a symmetry of the reciprocal lattice that could tell nothing about the crystal [Altmann (1994), p. 116].

Problem IV.2.c.i-1: Check that centered lattices do give different Brillouin zones than their primitive ones.

IV.2.c.ii *Symmetry of points in the Brillouin zone*

Each point of a Brillouin zone (or equivalently, the vector from the origin to the point) is moved to another indistinguishable point by each element of the (holohedry) point group. However there are points taken into themselves by some group transformations. The center of the zone is invariant under the rotations and inversion. Thus these transformations generate, for each vector to a point in the zone, a set of vectors, in general equal in number to the order of this group — but there are special cases.

A general point is one for which the set containing the vector to that point has one distinct member for each operator of the point group. Special points are those for which some point-group operations take the vector into itself; a set containing one of these has fewer members than the point group. The set of vectors related by the group is pivotal (sec. VII.2.e, p. 342) especially in finding the representations of the space groups, and in applications.

IV.2.c.iii *Boundaries of Brillouin zones*

Representations of lattice translations are specified by $\underline{k}$, but adding to $\underline{k}$ a reciprocal lattice vector gives a representation which is equivalent (eq. IV.2.a-7, p. 187) — justifying the introduction of this reciprocal space. Thus $\underline{k}$ is defined only up to the addition of reciprocal lattice vectors. The addition of such a vector gives an equivalent, but longer, one. A $\underline{k}$ that is shorter than any vector equivalent to it lies in the first Brillouin zone. Suppose that it lies on the boundary of a zone (in particular the first). It may then be possible to add to it a reciprocal lattice vector that leaves it on the boundary. There is no vector equivalent to one inside the first Brillouin zone that also lies inside this zone (but, of course, there are equivalent vectors lying in other Brillouin zones), however there are equivalent vectors of the first zone with equal length, these lying on the boundary.

As we will see many characteristics of space groups, and so crystals, come from a property of the boundaries of the first zone. If we consider an angle its range is, say,

$$0 \leq \theta < 2\pi, \qquad \text{(IV.2.c.iii-1)}$$

including one end point but not the other, since they are identical. A Brillouin zone being that of a lattice is invariant under reflections and inversions, so the entire boundary must be in the zone. However, because points on opposite edges are related by reciprocal lattice vectors, they must be identified. Thus for the representations we only consider half the boundary in the zone.

This means that a lattice translation from one end point to the other (in higher dimensions from one boundary point to an equivalent one) may go from a point in the zone to one that can also be obtained from the first by a point-group transformation — there are point group transformations of some boundary points that take them to ones to which they are equivalent. This is not true for interior points, and applies only to special points on the boundary, but such special points exist, and have an important effect on the group and the crystal (sec. VII.5.e, p. 370). And this differs greatly from free space.

In particular, as we will see, space-group representations can contain states of different momenta.

IV.2.c.iv *Examples of two-dimensional Brillouin zones*

To illustrate how these zones are constructed we start with a few two-dimensional ones. That of the first lattice, the rectangular, is

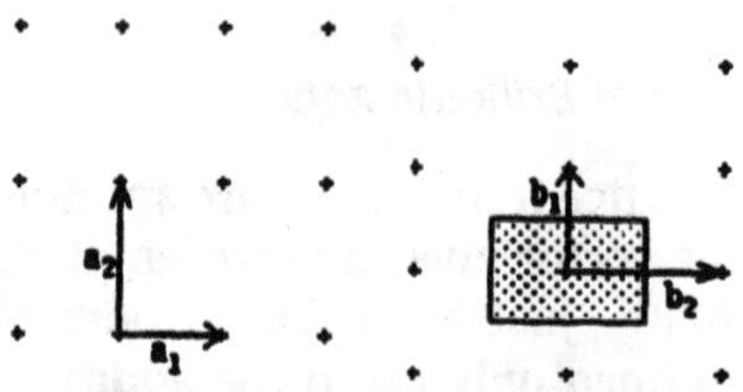

Figure IV.2.c.iv-1: RECTANGULAR-LATTICE BRILLOUIN ZONE.

The reciprocal lattice is on the right; its first Brillouin zone is dotted (either can be taken as the direct lattice, the other then contains the Brillouin zone). The boundaries are perpendicular to the reciprocal lattice vectors. Also, in this case the two lattices are oriented at $\frac{\pi}{2}$ to each other, with magnitudes of the vectors related by

$$|\underline{b}_i| = \frac{2\pi}{|\underline{a}_i|}. \quad \text{(IV.2.c.iv-1)}$$

The hexagonal lattice, with the reciprocal one on the right, is

Figure IV.2.c.iv-2: HEXAGONAL-LATTICE BRILLOUIN ZONE,

and again the first Brillouin zone is shaded. Clearly the reciprocal lattice vectors are perpendicular to, and bisect, the boundaries of the first zone. Magnitudes of the vectors are now related by

$$|\underline{b}_i| = \frac{4\pi}{\sqrt{3}|\underline{a}_i|}. \quad \text{(IV.2.c.iv-2)}$$

The centered rectangle is on the left,

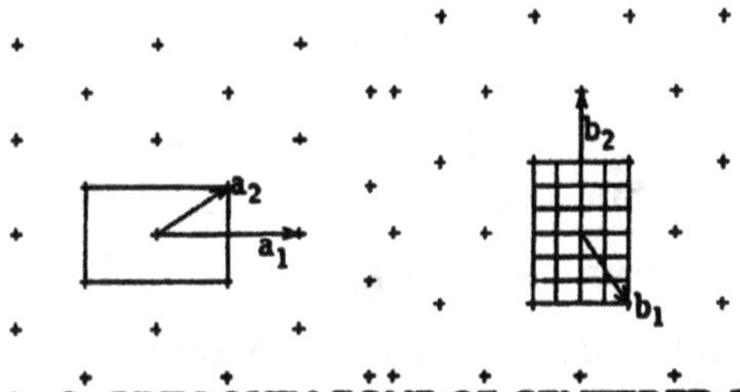

Figure IV.2.c.iv-3: BRILLOUIN ZONE OF CENTERED RECTANGLE.

Its reciprocal lattice, on the right, is identical, but rotated. These vectors have magnitudes whose relationship is

$$|\underline{b}_i| = \frac{2\sqrt{2}\pi}{|\underline{a}_i|}. \qquad \text{(IV.2.c.iv-3)}$$

Problem IV.2.c.iv-1: One vector is not perpendicular to the boundary of the Brillouin zone. Why? We must be careful about the vectors that we are discussing.

IV.2.d The representation basis states of the translations

Since free space is invariant under translations so must probabilities be, thus the translation eigenfunctions have to be of the form

$$\{E|\underline{t}\} = exp(-i\underline{k} \bullet \underline{t}); \qquad \text{(IV.2.d-1)}$$

$\underline{k}$ and $\underline{t}$ can have any real values. These, and only these, are invariant, except for phase changes, thus the magnitude of a statefunction is spatially invariant (time is irrelevant, of course, since all objects are static). For a lattice, there is invariance but only under lattice transformations: a translation smaller than a lattice vector relates points which are distinct — they have different relationships to the lattice points. More picturesquely, for a crystal the displacements to the atoms are different; an object, say an electron, sees a potential that varies with position, but is invariant under lattice translations. Translation eigenfunctions cannot (except for trivial cases) be invariant under translations within a unit cell, so can not be the same as free-space eigenfunctions. What then are the eigenfunctions for a lattice?

IV.2.e Bloch's theorem

The translation basis vectors of a lattice are

$$|\underline{k}\rangle = u(x)exp(i\underline{k} \bullet \underline{x}), \qquad \text{(IV.2.e-1)}$$

with

$$exp(i\underline{k} \bullet \underline{a}) = 1, \qquad \text{(IV.2.e-2)}$$

for a lattice translation. (The vector sign on arguments to functions are suppressed, unless needed.) A lattice translation acting on this gives the same state — but an arbitrary translation does not. Thus

$$|\underline{k}) = u(x+a)exp(i\underline{k} \bullet (\underline{x}+\underline{a})) = u(x+a)exp(i\underline{k} \bullet \underline{x}). \qquad \text{(IV.2.e-3)}$$

Invariance under lattice translations therefore requires

$$u(x+a) = u(x), \qquad \text{(IV.2.e-4)}$$

so the Bloch function, $u(x)$, has the periodicity of the lattice — any lattice translation leaves it invariant. The probability depends only on $u(x)$, and varies within a cell, but is invariant under translations taking a point to an equivalent one in another cell. Statefunctions thus are products of exponentials that are functions of discrete representation labels, with functions — Bloch functions — defined over the first Brillouin zone and these are then extended to ones defined over the entire crystal by requiring that they have the periodicity of the lattice. That the representation basis vectors of the translation, so the space, groups are of this form — or, as it is often stated, that the crystal statefunctions are of this form — is Bloch's theorem [Altmann (1977), p. 118; Bradley and Cracknell (1972), p. 120; Burns (1977), p. 292; Cornwell (1984), p. 207; Heine (1993), p. 266; Inui, Tanabe and Onodera (1990), p. 247; Jansen and Boon (1967), p. 249; Joshi (1982), p. 267; Lax (1974), p. 182; Lomont (1961), p. 208; Ludwig and Falter (1988), p. 231; Streitwolf (1971), p. 135; Tinkham (1964), p. 38].

The Bloch functions are defined over the unit cell of the lattice, not of the reciprocal lattice, so the point group governing them, and so the space group, need not be holohedry groups. The irreducible representations of translation subgroup T of space group G then are given by the vectors $\underline{k}$ of the first Brillouin zone of the Bravais lattice Γ. The representation matrices $M^{\underline{k}}$ are

$$M^{\underline{k}}(\{E|\underline{t}\}) = exp(-i\underline{k} \bullet \underline{t}), \qquad \text{(IV.2.e-5)}$$

with $\underline{t}$ one of the translations [Bradley and Cracknell (1972), p. 146].

The theorem assumes, of course, that space is translationally invariant, which it would not appear to be if the crystal were in a field [Bradley and Cracknell (1972), p. 120], or if it were a domain of a larger crystal.

IV.2.f The translation irreducible representations

The statefunction of an object, an arbitrary sum of eigenfunctions, depends on space; its Fourier transform is a sum of functions of momenta, the $\underline{k}$'s. Translation eigenfunctions are more complicated than those of free space, being exponentials times functions that are periodic in the Brillouin zones but are not determined by the translation group (symmetry) within each. Thus they, through their Bloch functions, are specific for each (electromagnetic) potential function, so depend on the crystal, not the lattice — different types of atoms create different potentials. These eigenfunctions give the translation group representation within a zone; each (for one dimension) is a function of position within a zone modulated by an exponential of wavenumber $k \sim 1/\lambda$, with λ the wavelength. The range of λ starts from infinity (with $\underline{k} = 0$), and decreases to the length of the zone (length of the lattice vector). Smaller λ's are absorbed into the Bloch function $u(x)$; we can redefine $u(x)$ so that it is a function of position within a zone times an exponential with a wavelength larger than the lattice vector — wavelengths vary from infinite to a minimum, with smaller ones in the Bloch function. Thus the range of wavelengths is divided into two, one for the exponential, the other for the Bloch function. Hence k varies from 0 to a maximum given by $k_m \sim 1/\lambda_m$, where λ_m is the zone length; k_m is the length of the first Brillouin zone (the length of the reciprocal-space lattice). For three dimensions, the eigenfunction is a product of three of these.

Functions $u(x)$ depend on the crystal, so on its space group (whose representations we thus have to find) and have symmetry determined by that. Because they are dependent on the potential they differ, not merely for different crystals, but for different sets of atoms — different atoms in the same crystal structure see different values of the potential. Different crystals with the same space group have different atoms, so different potentials in a zone. Thus an essential task is to find the restrictions on the u's for each space group for the first Brillouin zone, actually only for part of it, the representation domain — the u's for the other parts are related to this by the symmetry transformations (sec. IV.2.c, p. 189).

It is the first Brillouin zone that determines the representations of the translation group, for these basis functions must be invariant under translations from one zone to the equivalent point in any other — but not to points within a zone. We therefore need these functions in only this one zone.

This emphasizes a reason for introducing an additional space, the reciprocal space. This is the space of representation labels, and because of the symmetry of the lattice, these must have symmetry which is shown by that of the reciprocal space. Because of the periodicity, the range of these labels is limited, thus the representations are, meaning that only a portion of the reciprocal space, the first Brillouin zone, is needed and possible, for their range. Each component of $\underline{k}$ varies from 0 to its value on the boundary of the Brillouin zone.

IV.3 THE WIGNER-SEITZ CELLS

One advantage of a crystal is that its properties are determined by a single part — knowing the properties of, and in, the first Brillouin zone, even the representation domain, is all that is necessary, those of the infinite crystal are known from that. But what is this smallest part; what criteria are used to chose it? Representations of groups are specified by these geometrical sections. What properties of geometry determine group representations, and conversely? How good a foundation for extension to other geometries, to other groups, do these objects provide? What do they tell us about lattices (not only in 3-space) and about what we have necessarily (?) chosen as lattices?

The usual forms of the first Brillouin zones are (called in physics) Wigner-Seitz cells. These, at least in some configurations, are also known as Dirichlet domains, parallelohedra, proximity cells, Voronoï cells [Bradley and Cracknell (1972), p. 90; Burns and Glazer (1990), p. 50; Gasson (1989), p. 226; Holden (1991), p. 174; Koster (1957), p. 194; Lax (1974), p. 177; Lines (1965), p. 146; Senechal (1990), p. 43; Senechal (1995), p. 41]; some authors may use other names. Although there are fourteen Bravais lattices there are only five categories of Wigner-Seitz cells [Fejes Toth (1964), p. 114; Gasson (1989), p. 249; Senechal (1990), p. 13, 45; Shubnikov and Koptsik (1974), p. 224], these differing for the different lattices by their relative dimensions, producing twenty-four cells [Burns and Glazer (1990), p. 290; Senechal (1990), p. 51].

These and their assignment to lattices are given in many places, often with explicit diagrams. However there is no indication of how they are obtained, or why the lattices have the cells shown. Apparently this is knowledge that all solid-state physicists and crystallographers are born with.

Care is required in these discussions. These have been called parallelohedra because when placed face-to-face and parallel they fill space (obviously necessary for them to be unit cells of lattices). However they are not all parallelepipeds (sec. II.2.b, p. 68), as not all have pairs of opposite sides, and even when sides can be identified as opposite they need not be parallel, nor are all parallelepipeds parallelohedra. Thus the meaning of unit cell is different here than previously.

So we consider why there are only five categories (but twenty-one, or twenty-four, cells depending on convention or fashion), why they are these, and how they are related to their lattices. One way of stating our aim is that we wish to find all polyhedra whose repetitions fill space (these polyhedra tessellate space; this is an extension of the word often used for a plane to any dimension, and with a restriction to a regular tessellation — using only a single object). Of course, this is a special case; we can fill space using sets consisting of different types of polyhedra. With unit cells of lattices, for obvious reasons, we are limited to one. Except for the cube, the Platonic solids (sec. I.4.c, p. 10) do not fill space. Solid here means an object with no holes, and which does not allow overlapping.

The Wigner-Seitz cell of a lattice is the smallest closed object that is invariant under the same symmetry group as the lattice, centered on a lattice point, and including only that single point. These cells, one for each lattice point, fill space. They are determined by their construction: their faces are on the planes that are perpendicular to, and bisect, the lines from lattice points to their nearest (or next nearest) neighbors; they are cells of smallest volume whose faces lie on these planes.

A Wigner-Seitz cell must be convex by construction — it contains the complete line connecting any two points. A star (polyhedra) is not convex, since there are such lines outside it.

Objects that tessellate space [McMullen (1980)] are important in understanding lattices, space groups, and the properties of crystals, such as the behavior of electrons traversing through them. However it may also be possible to generalize these concepts, say to higher dimensions or quasicrystals [Burns and Glazer (1990), p. 239; Lifshitz (1997)], perhaps with interesting consequences. Thus it is useful to study them in depth, for that may clarify how to generalize, or find corresponding ones where these might be interesting, or ones that do not correspond, but for interesting reasons. And they emphasize that there are fourteen Bravais lattices partially by convention. Physics may not always like our conventions so it is important to understand what these are, and what freedom we have.

IV.3.a The centered rectangular lattice — what determines its cell?

To illustrate why this definition of the Wigner-Seitz cell is used we start in two dimensions considering the one nonprimitive lattice, the centered rectangular (sec. II.3.a, p. 71). We would expect that most properties of a crystal, and of the objects in it like electrons, can be determined by studying only half of the rectangle given by the diagonal and two sides — most properties, but perhaps not all; this triangle does not have the symmetry of the cell. But the properties in the other half should be related. This, at least, hints that the entire rectangle is not needed. There is a smaller cell that can be used; what is it?

Consider a centered rectangle of solid lines, with sides 6 and 2, thus area 12, on the left, with the smaller cell reproduced on the right to show that it fills space,

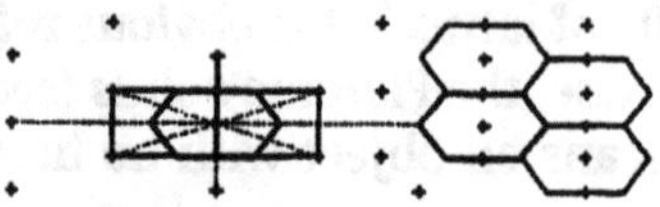

Figure IV.3.a-1: WIGNER-SEITZ CELL OF CENTERED RECTANGLE. The primitive lattice vectors from the lattice point at the center lie on the dotted lines, the bisectors perpendicular to these vectors, forming the Wigner-Seitz cell, are drawn heavier. The area of this cell, 6, is smaller than that of the centered rectangle. The cell displays the symmetry of the rectangle — it is invariant under the same group. It is also different from both the primitive (uncentered) rectangle, and its Wigner-Seitz cell, which is itself a rectangle, thus demonstrating that this lattice is not primitive. The cell has six sides, but is not a hexagon (there are two sets of lengths). Why six? The rectangle can be divided into two triangles related by a symmetry rotation — either can be taken as the representation domain (sec. IV.2.c, p. 189). Thus the triangle determines, at least almost, the crystal and its properties. So the Wigner-Seitz cell must also be so divisible, giving six sides.

As can be seen, the cell (one copy heavier to show its shape) fills space, is centered on a lattice point, and is the smallest cell with the symmetry of the group of the rectangle filling space that includes lattice points. Hence the properties of this cell determine those of the lattice, and the statefunction of an object in it gives the statefunction over the entire crystal. Not containing any unneeded, so redundant, regions, it is the proper cell to use. Because the crystal is invariant under a space group, it is not possible to arbitrarily specify a statefunction over the rectangular lattice (needed for the description of all relevant physical situations), for its values in different parts are related. But it is possible over this cell, and knowing that we know it everywhere. A rhomb like that in fig. II.3.b-1, p. 74, does not show that the lattice is centered rectangular, thus we cannot assume that a statefunction, for example, can be given in it and known everywhere. We may have to add conditions on the statefunction to make it the correct one for the crystal.

This is the rationale for the Wigner-Seitz cell — the smallest tiling cell centered on a lattice point with, and showing, the symmetry of the lattice. Thus it is the one that allows the properties to be determined completely and uniquely.

We shall see that all two-dimensional Wigner-Seitz cells have either four or six sides, because of the condition on the angles between lattice vectors, a fundamental property in determining the three-dimensional cells.

Problem IV.3.a-1: To show that the ratio of the sides is irrelevant (except when it is 1), redraw the diagram stretched horizontally.

IV.3.b The space-filling parallelohedra

In restricting the parallelohedra to number only five, we regard distinct ones to be in the same category [Fejes Toth (1964), p. 119]. What determines these categories, why have we decided to so associate different ones, and why do we take categories as different? The five parallelohedra fill space. Clearly stretching or contracting sides or changing angles leaves them still touching, and does not cause overlapping or holes (regions not in a cell). Thus the resultant objects also fill space, one of the requirements for Wigner-Seitz cells. A category is defined by the numbers of each type of face (which can only be only hexagonal or rectangular), the numbers of sides (of each face), four or six, the types of vertices (determined by the number of edges meeting at a vertex), and the numbers of each type; these values are not all independent, of course. Names of categories of parallelohedra refer to their most symmetrical members (if these have special names), the others are obtained by their deformations, changes of angles or of lengths (stretchings, dilations).

These cells filling space are congruent (identical) and homothetic [Fejes Toth (1964), p. 115]. That is they are similar and similarly oriented — "homo" meaning like, and "thetic" coming from the word to place, put, set (a thesis puts the results of research into a treatise; it puts forth a proposition). They are placed the same way. While a Platonic solid has but one type of surface, parallelohedra are not so restricted, but only two types appear, rectangular and hexagonal. Vertices can be the intersection of three edges or four. These numbers are the same for all of a category, and the sets of these numbers differ between categories. The faces of these cells, unlike those of the Platonic solids, are not all identical — except for the cube there is more than one type. But there are restrictions on them.

The parallelohedra are one regular (Platonic) solid, a semi-regular (Archimedean) solid (sec. I.4.c.iv, p. 16) [Ghyka (1977), p. 51; Holden (1991), p. 46], a dual to an Archimedean solid, the hexagonal prism, and one other solid tessellating space. The five tessellating polyhedra are the (dilations of the) cube (a Platonic solid), an Archimedean solid (the truncated octahedron — also known as the tetrakaidecahedron of

Kelvin or Kelvin's polyhedron [Ghyka (1977), p. 54, 84]), the rhombic dodecahedron [Schattschneider (1990), p. 246], which is the dual to the Archimedean solid, the cubooctahedron, (or cuboctahedron), the elongated dodecahedron [Coxeter (1973), p. 257], and the hexagonal prism. A prism is a solid object with identical, parallel, ends (bases), and with sides that are parallelograms. An object whose (projected) sides are, not parallelograms, but triangles, is an antiprism; these are obviously not capable of filling space (although pairs may be). Many familiar prisms of optics are thus really antiprisms, with one base a point. The hexagonal prism has hexagonal bases, and six rectangular sides, and can clearly fill space [Ghyka (1977), p. 84]. What is referred to as a cube in these discussions is actually a rectangular prism; it has rectangular bases, and thus four rectangular sides. The cube is a special case. Two parallelohedra are not regular or semiregular, the rhombic dodecahedron and the elongated dodecahedron, although one is dual to a semiregular object.

We can start with the most symmetric parallelohedron, based on the cube. The elongated dodecahedron is essentially a rectangular prism for which the rectangular ends have each been replaced by four rectangles. This cannot be done with the hexagonal prism as a hexagon cannot be split into either hexagons or rectangles. So the number of categories can be regarded as four, rather than five, with the elongated dodecahedron giving a category, of which the rectangular prism is special case (although with different numbers and kinds of faces), and the cube a special case of that.

The elongated dodecahedron is actually misnamed. It has rectangular and hexagonal faces and cannot be obtained from the dodecahedron which has pentagonal ones; also the number of vertices differ. The truncated octahedron has fourteen faces, eight hexagons and six squares [Ghyka (1977), p. 57]. It has thirty-six edges, and twenty-four vertices, at each of which three edges meet.

The parallelohedra, with V the number of vertices, F the number of faces and E the number of edges, are [Coxeter (1973), p. 29]

	V	E	F	Rectangular faces	Hexagonal faces
cube	8	12	6	6	0
hexagonal prism	12	18	8	6	2
rhombic dodecahedron	14	24	12	12	0
elongated dodecahedron	18	28	12	8	4
truncated octahedron	24	36	14	6	8

Table IV.3.b–1: PARAMETERS FOR THE PARALLELOHEDRA,

as we have to show.

IV.3.c Dual figures

Polyhedra are related [Gasson (1989), p. 249], and this can be studied using Euler's topological equation (pb. I.4.c.ii-1, p. 13),

$$V - E + F = 2. \quad \text{(IV.3.c-1)}$$

For parallelohedra and their duals,

	V	E	F
cube	8	12	6
octahedron	6	12	8
rhombic dodecahedron	14	24	12
cuboctahedron	12	24	14

Table IV.3.c-1: PARALLELOHEDRA AND THEIR DUALS.

Thus the cube and the octahedron, both Platonic solids, are dual (as we know) by interchange of vertices and faces, with edges the same (sec. I.4.c, p. 10). The rhombic dodecahedron and the (Archimedean) cuboctahedron are likewise dual [Holden (1991), p. 50].

IV.3.d Inscription in, and circumscription about, spheres

Semi-regular parallelohedra can be inscribed in spheres [Holden (1991), p. 41] — a sphere with center at the lattice point touches each vertex of the object. The semi-regular (three-dimensional) objects that can be inscribed in spheres are the prisms, infinite in number, the five Platonic solids, and the thirteen Archimedean solids.

The Wigner-Seitz cells of this set have all faces either rectangles or hexagons. These include the hexagonal and rectangular prisms, the latter having as a special case the cube (the only Platonic solid that fills space — because it is a special case of a rectangular prism).

A cell can also circumscribe a sphere — each line through the center of every face is tangent to the sphere; the sphere touches every face at its center. The Platonic solids can be both inscribed in, and circumscribed about, spheres. This is not true for Archimedean solids, which can all be inscribed. However each has a dual which can circumscribe [Holden (1991), p. 51]. One Wigner-Seitz cell is dual to an Archimedean solid.

Problem IV.3.d.-1: If we also consider star-shaped objects, dropping the limitation to ones that are convex (sec. IV.3, p. 196), then it may be impossible to find all that fill space [Senechal (1990), p. 12]. However it would be useful to find such ones, and see if general results can be obtained.

IV.3.e Illustrations of parallelohedra

Three of the parallelohedra are less familiar than the other two, so it is useful before further discussion to see what they are like.

For the rhombic dodecahedron,

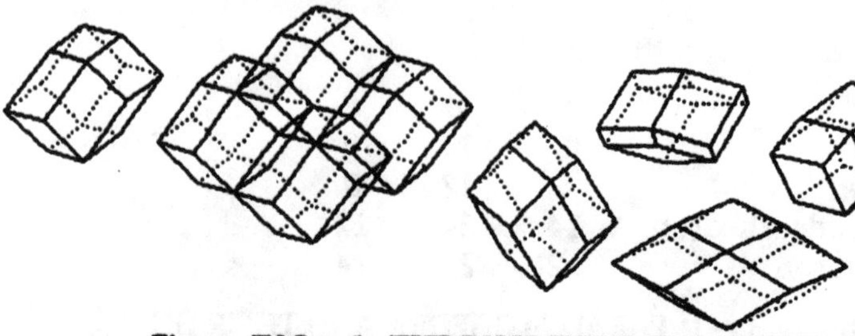

Figure IV.3.e-1: THE RHOMBIC DODECAHEDRON,

several versions are shown; copies of the one at the left have been placed together on its right — these do tessellate space. To the right of these are other such cells, all of the same category, but looking quite dissimilar, indicating that physical objects with these cells, although not differentiated by our classification, could have quite different properties.

While the truncated octahedron is semiregular — three faces meet at each corner, the rhombic dodecahedron is not, there are vertices with three faces, others with four. This is also true of the elongated dodecahedron, so it also is not semiregular. The truncated octahedron is semiregular, but not regular, since all faces are not the same.

The elongated dodecahedron can be seen from the diagram to fill space. Here are two views, on the left an oblique one, on the right the same object in isometric view:

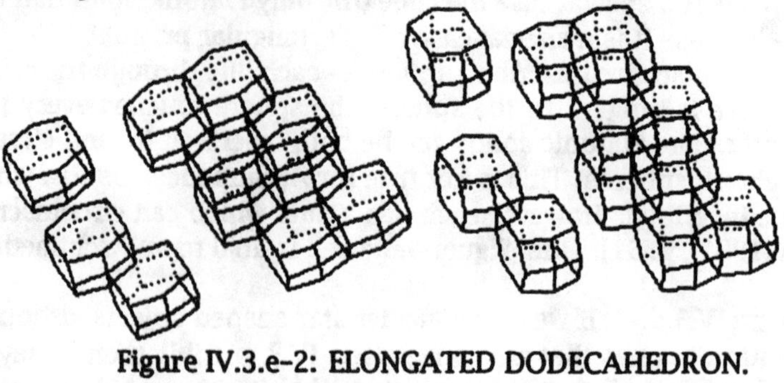

Figure IV.3.e-2: ELONGATED DODECAHEDRON.

The set of copies of the truncated octahedron clearly shows that it fills space [Holden (1991), p. 155, 164],

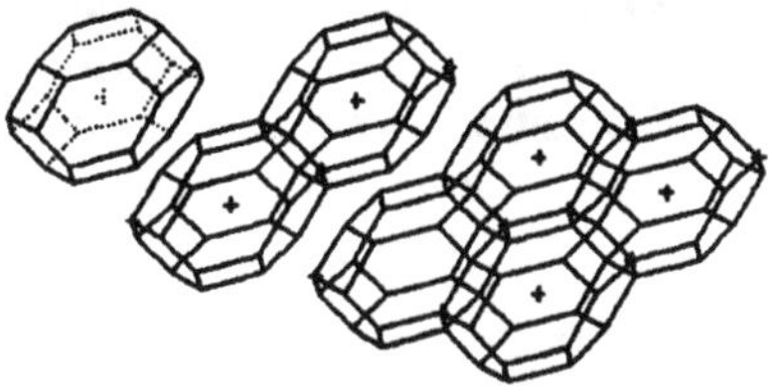

Figure IV.3.e-3: TRUNCATED OCTAHEDRON.

Problem IV.3.e-1: A rhombic dodecahedron can be constructed starting with a face the ratio of whose diagonals is $\frac{1}{\sqrt{2}}$ [Coxeter (1973), p. 26]. Build one and compare with the diagrams. Also use it to check the various properties of these objects considered here.

IV.3.f Parallelohedra give lattices

The number of categories of parallelohedra are limited. One reason is that all give lattices, and the number of lattices is limited. Another reason, as usual, is convention, raising the questions of what conventions are necessary, why necessitates them, what these are, and why we have decided as we have? Here we study how their relationships to lattices limits them.

Wigner-Seitz cells are defined so that a vector from a lattice point to a nearest neighbor (pb. II.6.b-3, p. 106) is perpendicular to a face and goes through the center of the face (for next-nearest neighbors perhaps through a vertex). This face shares edges with other faces and each face has a similar vector from the origin going through its center. By regularity, a rotation about every one of these vectors, taking the bounding faces into each other, leaves invariant the set of nearest neighbors, and thus the entire set of lattice points, since each can be taken as the origin. A set of points invariant under all such rotations forms a lattice — the vectors from any point to its nearest neighbors are taken into each other by these rotations, so are primitive lattice vectors of a lattice.

Each lattice then defines a parallelohedron, and each parallelohedron defines a lattice. There are fourteen lattices, so we would want no more than fourteen parallelohedra. However we combine them into categories, whose members are related by variations of lengths and angles, so there are fewer. Each lattice has a Wigner-Seitz cell by construction. Further each smallest space-filling cell with the symmetry of a lattice defines a lattice, with one point at its center.

Problem IV.3.f-1: Is the requirement "smallest cell" necessary?

IV.3.g Determining parallelohedra from their properties

Why are these five categories the possible Wigner-Seitz cells, and only these? Because they are required to have symmetry of lattices — and lattices are limited — and because they must fill space. We now need to make explicit how these restrictions determine the categories of cells.

IV.3.g.i *The two-dimensional projections of the cells*

A three-dimensional parallelohedron, projected into a plane of lattice points through its center, gives a Wigner-Seitz cell of a two-dimensional lattice — its edges are the lines at which the plane cuts the faces of the parallelohedron, the vertices are the lattice points where the plane intersects a cell edge. Two-dimensional cells are fewer, and simpler, than for three dimensions, so the restrictions on them are used to get restrictions on three-dimensional cells, and thus these cells.

A hexagonal prism, and its projection onto (for example) the plane on which its base lies defines a lattice whose points are its vertices, plus that at its center. The lines connecting them form the Wigner-Seitz cell for the two-dimensional lattice. This can be compared with fig. IV.2.c.iv-2, p. 192, and it is seen that the projection does give a Wigner-Seitz cell,

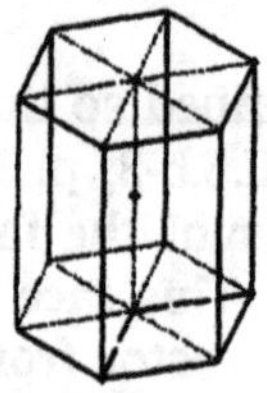

Figure IV.3.g.i-1: PROJECTION OF A HEXAGONAL PRISM.

The only Wigner-Seitz cells in two dimensions are rectangles and hexagons (the only polygons giving lattices), so three-dimensional cells can be only those that have these projections. This is one reason there are few. These two-dimensional cells, thus three-dimensional ones, need not be regular, their sides need not all be equal — as with the centered rectangle. Two dimensions then gives a fundamental restriction: faces of three-dimensional Wigner-Seitz cells have either four or six sides.

Problem IV.3.g.i-1: Prove that the projections, which are shown only for one cell, do give a lattice.

IV.3.g.ii *The cells are centrosymmetric*

Lattices always have one symmetry: inversion through a lattice point (and also the midpoint of a lattice vector). Thus a unit cell, either centered about a lattice point, or with corners at lattice points, is inversion invariant — it is symmetric about its center. And each face is also centrosymmetric [McMullen (1980), p. 114]: it is unchanged by an inversion through its center (and similarly for each edge). Two-dimensional cells are centrosymmetric, three-dimensional ones must also be. A Wigner-Seitz cell likewise, and each face of one, is centrosymmetric — each surface is defined as perpendicular to a vector joining lattice points, so the vector goes through the center of the cell. Every face is centrosymmetric because inversion of the cell through its center takes the face into another, on the other side, which, by inversion invariance of the lattice, is identical to it; a face of a Wigner-Seitz cell and that obtained by inversion must have identical projections onto the plane parallel to them through the center. This can only be if the face is centrosymmetric — invariant under inversions through its own center.

Problem IV.3.g.ii-1: A body with all faces centrosymmetric gives a lattice. Take the points at the ends of those vectors from the center of the body that are perpendicular to, and bisected by, the faces. Show that objects identical to the body can be placed with their centers at each of these points. Continue in the same way, filling space and check that these points form a lattice — they have identical environments, (and necessarily?) inversion invariance, and fill space.

Problem IV.3.g.ii-2: Prove that inversion symmetry through a lattice point requires it through a point midway between two lattice points (and, as is obvious, conversely), and this requires the face to be centrosymmetric, using both two-, and three-dimensional lattices.

Problem IV.3.g.ii-3: In two dimensions the cell must therefore have an even number of sides, so can be only a rectangle or hexagon. How, and why, is this relevant to the crystallographic restriction on the angles of a lattice (sec. I.5, p. 23)?

Problem IV.3.g.ii-4: These statements should be checked with every diagram of a parallelohedron.

IV.3.g.iii *The faces form bands*

From centrosymmetry comes a fundamental property. The faces form bands [Fejes Toth (1964), p. 115; Senechal (1990), p. 46]. The edge between adjacent faces is parallel to the opposite edges of both — the opposite edges are related by inversion so must be parallel. Thus from any edge there is a path of parallel edges, with finally the last edge, so the last face, the same as first. All faces in a band need not be identical — their numbers of sides can differ.

By centrosymmetry of faces opposite edges are parallel, thus all edges of a band are. By centrosymmetry of cells, for each face there is an identical one obtained by inversion through the lattice point at the cell's center. Faces are of finite size, the cell has only a finite number, so traversing through the band leads back to the first member — the band of faces is closed.

A face has either four or six sides, and two of its edges belong to each band it is in, so it belongs to either two or three bands [Fejes Toth (1964), p. 115].

Both parallelepiped unit cells (sec. II.2.b, p. 68) and Wigner-Seitz cells consist of a set of closed bands of centrosymmetric faces. The centers of the faces of a single band lie in a plane, which contains the center of the cell, and is perpendicular to each edge of the band, which it, of course, bisects. The relationship between the plane and an edge follows immediately from symmetry. For two adjacent faces, their centers and the midpoint of their mutual edge form a plane; by symmetry it contains the center of the cell, and moreover coincides with the corresponding plane obtained from one of these faces and its other neighbor. Thus all centers of a band lie in a single plane, which contains the center of the cell. There are at the center of each cell a set of these (intersecting) planes. A discrete set of planes parallel to these contains the centers of every cell of the lattice (these centers determine a set of planes, each parallel to one of the set at the center of any cell).

The projection of a cell onto the plane perpendicular to the faces of a band through its center is a unit cell of a two-dimensional lattice (the lines formed by the intersection of the faces with the plane bound this cell), thus has either four or six sides — the center of the cell is a lattice point, the vector from the center to the center of a face goes to a lattice point, and the angles between lattice vectors are limited giving these as the allowed numbers. So a band has four or six faces lying on it [McMullen (1980), p. 114] — their sides give the edges of the two-dimensional Wigner-Seitz cell.

Each band intercepts all other bands (each pair of bands has two faces in common) since the planes through them go through the center of the cell, and the projections onto planes through their centers again are two-dimensional cells. As these planes go through the center of the cell, each intercepts every other one (in a line going through lattice points), and each plane goes through lattice points, thus every one contains a two-dimensional lattice. A band contributes two edges to this cell, from the two opposite faces. As the cell has only faces with four or six sides, each can lie on only two or three bands.

An object whose faces do not form bands is a tetrahedron,

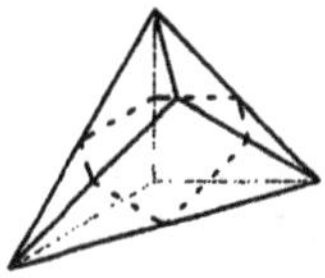

Figure IV.3.g.iii-1: FACES OF A TETRAHEDRON DO NOT FORM BANDS;

its faces (four each with three edges, and no centers) are not centrosymmetric. Edges of adjacent faces are not parallel and a path bisecting the edges (shown dashed) does not lie on a plane.

Compare this with the elongated dodecahedron (drawn three times for clarity), showing three of the paths, each forming a plane, which includes the center of the cell, and marking the set of faces in a band, each containing four or six faces. The rightmost figure is drawn rotated to show one path more clearly. This does not connect opposite edges that are parallel, so is not an acceptable path, and the center of the cell does not lie on it. Each face lies in two or three bands,

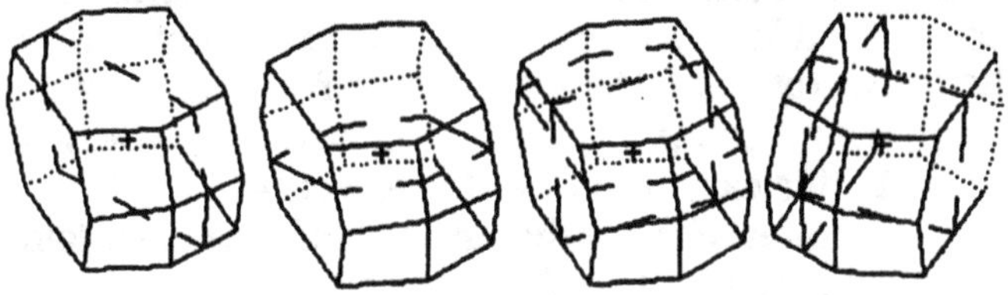

Figure IV.3.g.iii-2: BANDS FOR THE ELONGATED DODECAHEDRON.

Problem IV.3.g.iii-1: These statements should be checked for these diagrams, drawing the remaining paths, and then with every diagram of a parallelohedron.

IV.3.g.iv *Sharing by adjacent cells*

Adjacent cells share whole faces (by symmetry under translations and inversions through the centers of the cells), so the centers of these faces coincide. The cells thus form parallel layers. These cells tile space because this sharing precludes holes and overlaps. Tessellate is probably a better word because in common use tile refers to two dimensions (we would not want to tile an entire room), but both terms are used, and are here restricted to regular tilings, those with a single object.

IV.3.g.v *Counting nearest neighbors*

Wigner-Seitz cells are constructed by drawing lattice vectors from any point to each of its nearest neighbors — the lattice points that can be reached with primitive lattice vectors, and the ones obtained from these with any symmetry-group operation (which leave lengths unchanged) — leading to the cell showing the symmetry of the lattice. In some cases the next-nearest neighbors are also used. Lattice points that need multiples of these vectors are not included.

For each vector we draw the surface that is its perpendicular bisector; these surfaces, for all vectors going to these neighbors, form an enclosed space, the Wigner-Seitz cell. To find the number of surfaces of this cell, we count the number of such neighbors, as considered here for the Wigner-Seitz cells going with the cubic, and trivially, the hexagonal, lattices.

A point of a hexagonal lattice is surrounded (in a plane) by six others, and it has points above and below (on the axis perpendicular to the plane), thus eight altogether so its Wigner-Seitz cell has eight faces.

Taking the lattice point of the primitive cubic lattice at a corner, there are three lattice vectors going to three of the other corners. But a corner belongs to four cubes, each pair of which share an edge, thus it has six nearest neighbors (on both sides of the three axes) and its Wigner-Seitz cell has six faces. The perpendicular bisectors of the primitive lattice vectors, and the set obtained from them by the operations of the symmetry group, O_h, form a closed box, which has the same symmetry (in this case the same shape, a cube).

Using the center of the body-centered cube as the origin, the vectors go the eight corners and to the centers of the neighboring cubes, six altogether, so there are 14 faces for its cell. Each corner is surrounded by four others, the face it defines therefore shares edges with theirs, plus the edges of the surfaces given by the vectors to the two nearest centers — it has six edges and is hexagonal. The vector to a center is surrounded by four corners, so its perpendicular bisector shares edges with the ones they give, and that face is rectangular, actually square. Of the 14 faces then six are rectangular and eight hexagonal, as (with a set of primitive lattice vectors drawn heavily) we see in

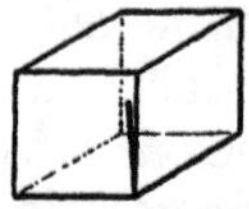

Figure IV.3.g.v-1: NEAREST NEIGHBORS FOR BODY-CENTERED CUBE.

For the face-centered cube take the point at a corner. The three primitive lattice vectors go to the centers of the faces meeting at that point, so those to the nearest neighbors are these, their negatives, and the set obtained by rotation of $\frac{\pi}{2}$ around any of the axes (edges of the cube), a total of 12 — these primitive vectors plus those given by O_h. Each of these points is surrounded by four others, so all surfaces are rectangular,

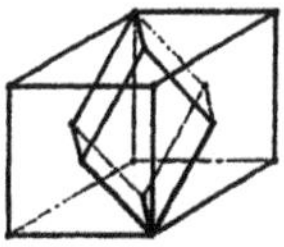

Figure IV.3.g.v–2: THE FACE-CENTERED CUBIC LATTICE.

The body-centered cube has a cell obtained using six next-nearest neighbors, while the cell for the face-centered one uses only nearest neighbors. Twice the vectors for these go to corners, which are thus not included, but half the corners are not reached this way. Why are they not used? Each of them can be obtained from one of the other corners by a reflection in a plane through the center of the cube, thus are equivalent to points that are twice the distance as nearest neighbors, so not included. This is not true for the six next-nearest neighbors of the body-centered cube, which are not equivalent to any excluded points — vectors to them must be used.

Why does this matter? The vectors to the corners, so their perpendicular bisectors, are orthogonal. Thus using only them would give a Wigner-Seitz cell that is a cube (though rotated from the lattice), losing information about the lattice, that it is centered. Is all (possible) information obtained by including the next-nearest neighbors? Why stop there? Why is this not necessary for the face-centered lattice? It is not possible to use other points since bisectors of the vectors to them lie on the vertices or outside of the closed cells given by the nearest and next-nearest neighbors, thus they could not be used to construct the smallest cell. This follows from their being equivalent under reflections through the sides or centers of the cube to points reached by multiples of primitive lattice vectors, or their transforms by operations leaving the lattice invariant.

The plane bisecting the vector between the centers excludes points further from the origin than it is, but which are included in the volume defined by the bisecting planes perpendicular to the vectors to the corners. Thus it gives a smaller cell than if it were omitted, which is what is required.

These are the reasons for including, and excluding, next-nearest neighbors, and excluding all those further away.

The primitive tetragonal lattice is obtained by stretching or compressing a cube changing one length, so has the same number of nearest neighbors, and a Wigner-Seitz cell of the same category, but less symmetrical.

Problem IV.3.g.v–1: *Draw diagrams of the cubic and hexagonal lattices and verify these statements.*

IV.3.h The number of sides of parallelohedra

One reason that there are so few categories of parallelohedra is the limitation on the number of their faces. The minimum is six — there are three primitive lattice vectors plus their negatives, each giving a side. The maximum number, as we see from their construction is fourteen. The number of sides must of course be even. There are Wigner-Seitz cells with six, eight, twelve (twice) and fourteen sides. Ten is missing.

Why these numbers? The faces form bands, each with either four or six cells, and each face lies on two or three bands (two or three paths marking the bands cross at its center); for a cube, only, all faces lie on only two bands. Each pair of bands shares faces, two. A band crosses all others, sharing two faces with each, and it can have either four or six faces.

All faces are contained in either two or three bands, so we need consider only these cases. To show this take two opposite surfaces belonging to three bands (if none, the object is a rectangular prism), called the top and bottom. If there are faces on the sides that are not on any of these three, then none of their edges can be horizontal for the bands would be vertical so contain the top and bottom faces. The plane through the midpoints of these edges, and the center of the cell, is inclined to the vertical and does not pass through the edges of the top and bottom faces. Thus it must extend beyond them, which is impossible, so there can be no surfaces not lying on one of the three bands to which the top and bottom belong, and these bands contain every face of the Wigner-Seitz cell.

Suppose that there are two bands, each with four faces. This gives eight faces, two shared, so the cell has six surfaces (the rectangular prism). If one band has four, the other six faces, two shared, the cell has eight faces (the hexagonal prism). Two bands of six faces each is not possible. A band would share two faces with the other, leaving four unshared, which cannot be. This explains the nonoccurrence of ten faces.

The other cases are for objects with three bands. All three cannot have four faces; if a band contains four faces these are perpendicular, two such bands give six perpendicular surfaces, but there cannot be another band containing faces not included and which are perpendicular to these. There can only be objects with two bands of four faces, and one of six, or a band of four faces, and two bands of six, or finally three bands of six. So we get the other three parallelohedra, these are all possible (since we have constructed them), and only these are possible.

Problem IV.3.h–1: There is a slight hole in this argument that there are only three more parallelohedra. It is possible that given the number of bands, and the number of faces in each, there is more than one way to arrange them, so more objects. Explain why this does not occur (in three dimensions?).

Problem IV.3.h–2: The Wigner-Seitz cells are defined as those space-filling solids centered on a lattice point, showing the symmetry of the lattice, and having the smallest volume of any such objects. Prove that their volumes are indeed the smallest. Is it possible to center them on other points, though always including a single lattice point? Could this reduce volumes?

Problem IV.3.h–3: For four dimensions find the possible number of bands (assuming that the faces still form bands), the allowed set of numbers of members for them, and of members, the sets of bands that can go together, thus the number of faces of the Wigner-Seitz cells, presuming that (objects corresponding to) these exist. Then find the cells, giving in particular the number of faces, edges of each, vertices, and number of lines meeting at them. Draw each cell.

Problem IV.3.h–4: Repeat for n dimensions [Coxeter (1973), p. 257; Fejes Toth (1964), p. 124; Ghyka (1977), p. 68], finding the cells that tessellate space [Lyndon (1989), p. 78, 92; McMullen (1980)]. Can general rules be found (or computer programs written)?

Problem IV.3.h–5: The six-generator orthogonal algebra in four dimensions (the equivalent of the three-generator rotation algebra, the angular momentum operators, of three dimensions), the algebra of SO(4), splits into two sets of three each — the algebra is, uniquely, semisimple, not simple [Mirman (1995a), pb. X.7.e-4, p. 300; (1995b), sec. 7.1.2, p. 124]. Does this affect the tiling, or Wigner-Seitz cells, if any? How?

Problem IV.3.h–6: The signature of a space is given by the set of signs in the formula for distance,

$$d^2 = \sum x_i^2 - \sum x_j^2. \qquad \text{(IV.3.h–1)}$$

The signature of Euclidean space is positive-definite, all signs are positive. Is it possible to tile a space with signature (1,1)? What are, and what are the properties of, the (objects corresponding to) Wigner-Seitz cells? Repeat for (2,1), (2,2), ... (m,n). Our space has signature (3,1), and a study of it might be especially interesting.

Problem IV.3.h–7: These spaces are all real, the coordinates are real numbers. Repeat these problems for complex spaces. Can the complex plane be tiled? What are the cells? How about a two-dimensional complex space? Can general rules be found for n dimensions? Are there differences between an n dimensional complex space, and a $2n$-dimensional real one [Mirman (1995b), sec. 2.2.1, p. 33]?

Problem IV.3.h–8: The axioms of geometry that give the set of Bravais lattices (pb. II.4.h-6, p. 97; pb. II.7.f-10, p. 123; pb. III.4.c.ii-2, p. 152) undoubtedly also give tessellations and existence of Wigner-Seitz cells, the limitations on them, their number, and the particular set of cells. However might other axioms be needed? Might some be dropped? Is this true in other dimensions, and other types of spaces [Fejes Toth (1964), p. 84; Lyndon (1989); McMullen (1980)]?

Problem IV.3.h-9: What kind of cells tile a sphere, or toroids [Ceulemans and Fowler (1995)]? Are there Wigner-Seitz cells?

Problem IV.3.h-10: In n dimensions the space corresponding to a sphere (a hypersphere) would be tiled by solids. Is this possible? Describe them, and their Wigner-Seitz cells, if any.

Problem IV.3.h-11: Is it possible to tile the projective plane [Lyndon (1989), p. 121]? Does it have Wigner-Seitz cells?

Problem IV.3.h-12: That the rectangular prism is a limit (which?) of the elongated dodecahedron, though regarded as a separate category, emphasizes that our classification is (partly) conventional. The cube which is used to denote the category is a special case of the rectangular prism. Is it possible that there are other cases in which one category is a special case, or limit (if their is a difference), of another? Why? One possibility is that two cases are different limits of an object that is not a Wigner-Seitz cell (in what way?) except for these limits. Also it may be that the cells are three-dimensional projections of special cases of a category of Wigner-Seitz cells in higher dimensions, or different three-dimensional projections of the same cell. The Wigner-Seitz cells might be different depending on which dimension is chosen.

IV.4 LATTICES CAN HAVE SEVERAL WIGNER-SEITZ CELLS

Lattices are determined by geometry, and also by convention. A lattice has a Wigner-Seitz cell, but because we have decided to regard two lattices as the same, though their geometries are not identical, have we therefore decided that their Wigner-Seitz cells are the same? Can we? If the same lattice (type) can have several Wigner-Seitz cells, why do we ignore the distinction and consider all members of the set as one?

IV.4.a Emphasizing the importance of understanding what is geometry, what is convention

One of the interesting aspects of Wigner-Seitz cells is that it shows how classification can depend not only on mathematics, but also on convention, on prejudice, on thoughtlessness. Distinctions, even important physical ones, can be overlooked because the same word is applied to different objects, perhaps only because of the lack of thought, lack of care. That there are fourteen Bravais lattices is part of the folklore of physics and mathematics. But are there?

It might be useful to keep in mind any lessons learned here in studying other fields, not only in physics and mathematics. Our thinking is shaped by the language we use. Often, quite often, we use the same word for very different entities, and thus fail to notice that they are different, and how different they are, so treat them as if they were really similar. And other times we apply different words to concepts and objects that are quite similar, and so fail to note how related they are. Language is essential, and also misleading and dangerous.

IV.4.b The centered tetragonal lattice

The centered tetragonal lattice introduces a new aspect, which we shall explore in depth. First we state it to clarify and motivate this exploration, its implications and its lessons.

If the ratio of height of the cell to the length of an edge of the base is large enough, there are vectors from the center of the cell to its eight corners plus those (parallel to the base) to the centers of the four neighboring cells, thus 12 vectors, giving a Wigner-Seitz cell of 12 faces. From the centered cube the four vectors parallel to the base remain when the cell is stretched, as expected since this plane is not affected by the length change giving the tetrahedron, but of the ten vectors of that cube only eight remain.

Consider a very long cell (relative to the edge of the base). The vector to a corner is almost parallel to the z axis (the principle axis), and the perpendicular bisector of this vector is almost parallel to the top face of the tetrahedron. The four surfaces that are perpendicular bisectors of the vectors to the upper corners meet well below the center of the vector to the middle of the next cell along z. Thus the perpendicular bisector of this vector linking the centers of these cells lies outside the closed volume whose upper boundary is formed by the four surfaces — the vector along the z axis between the centers does not give a face of the Wigner-Seitz cell. For a very short cell, the perpendicular bisectors of the vectors to the vertices are almost parallel to the sides of the tetrahedron, so the midpoint of the vector along z is inside the volume they give, thus the surface perpendicular to that vector is a face of the Wigner-Seitz cell. This cell is of the same category as that of the centered cube. At the border between these cells the planes perpendicular to the vectors to the corners intersect in a line perpendicular to, and bisecting, that between the centers, the line of a bounding plane. For a longer cell the plane is above this line, thus not included, for a shorter it is below providing a boundary. The Wigner-Seitz cells of the short and long lattices belong to different categories.

So a lattice may have more than one (type of) Wigner-Seitz cell depending on the ratio of the sides. The body-centered tetragonal has two. Also as the ratio of the sides is changed the lattice passes through a value for which the symmetry is higher, becoming cubic. We consider the body-centered tetragonal lattice to be a single lattice, but clearly it is not. A fundamental property, the form of its Wigner-Seitz cell, is different for different ranges of the ratio of sides. However our convention is to ignore this, and then with this definition, there is but one of these lattices. Notice that in the derivation of the body-centered tetragonal lattice (sec. II.7.b.ii, p. 114) one component of a lattice vector is determined, but not the other. We have decided in our convention to ignore this (except when it leads to higher symmetry). Thus we put these different lattices into one set. This emphasizes that the number, and types, of lattices (and of course, other structures) is determined both by convention and geometry. Once we have picked the convention, geometry decides. But the possible conventions are also determined by geometry. Given, say, a tetragonal system, there are only a finite number of discrete ways of defining subsystems.

While we may regard lattices with different Wigner-Seitz cells as identical, physical properties, say the behavior of electrons in crystals with these lattices, can be different, and in important ways. Thus in any situation it is necessary to investigate whether a particular convention is appropriate, or whether, say, a finer, or coarser, classification is better — or indeed whether the classification is at all relevant.

IV.4.c Implications of multiple Wigner-Seitz cells

A lattice then can have several Wigner-Seitz cells, depending on the ratios of the sides. What is the significance of this? It emphasizes that the number of lattices is — partially — conventional. We may have choices as to whether we wish to regard lattices as different or not, and that choice has consequences for our views of physical objects.

Close to the boundary value of a ratio, surfaces exist but can have infinitesimal size. We started by regarding lattices identical, even though their Wigner-Seitz cells differ. But we can also regard them as different if their Wigner-Seitz cells differ. Physically the size of a surface of a (say) Wigner-Seitz cell might matter. The quantities determining the properties of crystals, electron wavelengths, vibrational amplitudes of crystal objects, say, determine physically relevant sizes, lengths for example. A crystal with an infinitesimal surface, specifically with a length smaller than one of these physical quantities, may have very different properties than a crystal with a surface of larger size. Thus we might wish to classify lattices differently depending on the ratio of, say, the length of an edge of the Wigner-Seitz cell and some physical quantity, like amplitude or wavelength. This might still give a discrete classification. Such a classification would then be partially geometrical, partially conventional, and now partially physical.

The amplitude of oscillation, for example, depends on the masses, potential and temperature. Thus these oscillations could be different if that amplitude were smaller than a length of the Wigner-Seitz cell than if it were larger. So the properties of the crystal could change at the temperature at which these were (about) equal, and as there are several lengths, there could be several such temperatures. Crystals that differed very slightly in the (relevant) ratio of sides could have different properties in certain temperature ranges, causing, perhaps, relatively rapid variation of properties with temperature.

Pressure can change the dimensions of the unit cell, the ratio of the sides. We would expect that properties of a crystal, including temperature dependence, vary simply (at least roughly) with pressure. However it is possible that around values for which the symmetry of the unit cell changes, or the Wigner-Seitz cell changes, the variation with pressure could be quite complicated. Certainly simplicity should not be assumed without examination.

This emphasizes that the classification of (not only) lattices depends to some extent on convention and is determined, in part, by physics. A classification useful in one context may have irrelevant detail in another, but not enough in a different one, so hiding physically important aspects. Thought about the best choice is always useful, something not necessarily clear just from lists, which can distract from the possibility of choice.

Problem IV.4.c–1: It might be argued that geometry allows only fourteen Bravais lattices (in three dimensions) and it is only when we go to a reciprocal space (which is where we use Wigner-Seitz cell cells), that is define functions over the direct space, specifically basis functions of the representations of the space groups of the lattices, is it possible, even necessary (especially for physics), to have a finer classification. Is this argument correct?

IV.5 WIGNER-SEITZ CELLS FOR THE LATTICES

Having considered the general theory of these cells, we describe them for each lattice, and (hopefully) show that they do have the required properties, and that there are multiple ones. Since the sets of lattices and of reciprocal lattices are the same, these can be taken as the unit cells for the lattices stated, whether direct or reciprocal. We give the form of the Wigner-Seitz cell, which for the reciprocal lattice is the first Brillouin zone, and the lattice vectors, and study the properties relevant to group-theoretical analyses and physical applications, including special points, lines and planes (sec. IV.2.c.ii, p. 190). First we investigate geometry.

IV.5.a The cubic system

To study the Wigner-Seitz cells we can start either from the least symmetric lattice, the trigonal, or the most, the cubic. We consider the cubic first [Koster (1957), p. 196], since it illustrates most, but not all, of the concepts involved (sec. II.6.b, p. 105, sec. XI.2.h, p. 579).

IV.5.a.i *The simple cube*

For a cube the three primitive vectors go from any corner (the origin) along the three sides meeting at that corner (taken as the axes) so are

$$\underline{a}_1 : a(1,0,0), \underline{a}_2 : a(0,1,0), \underline{a}_3 : a(0,0,1), \qquad \text{(IV.5.a.i-1)}$$

giving the reciprocal vectors (eq. IV.2.a-5, p. 187),

$$\underline{b}_1 : \frac{2\pi}{a}(1,0,0), \underline{b}_2 : \frac{2\pi}{a}(0,1,0), \underline{b}_3 : \frac{2\pi}{a}(0,0,1), \qquad \text{(IV.5.a.i-2)}$$

and the first Brillouin zone of the reciprocal lattice is also a cube.

Problem IV.5.a.i-1: Check that the sides of this cell are orthogonal to, and bisect, the vectors from lattice points to their nearest neighbors.

IV.5.a.ii *The face-centered cubic lattice*

More interesting than the simple cubic is the face-centered cubic lattice [Gasson (1989), p. 264; Koster (1957), p. 197], with primitive lattice vectors

$$\underline{a}_1 : \frac{a}{\sqrt{2}}(1,1,0), \underline{a}_2 : \frac{a}{\sqrt{2}}(0,1,1), \underline{a}_3 : \frac{a}{\sqrt{2}}(1,0,1). \qquad \text{(IV.5.a.ii-1)}$$

These vectors are not along symmetry axes: rotating about a vector from a corner to the center of a face leaves a cube a cube, but moves it, so does not give a symmetry of the lattice. This is illustrated in fig. IV.3.g.v-2, p. 209. Clearly rotations around these axes move the cube. Inside the face-centered cube is a unit cell (whose edges are the primitive lattice vectors), this with only a single point — remembering that each point counts as $\frac{1}{n}$ if a cell shares it with n others. It is not cubic (as expected since the primitive lattice vectors do not lie on symmetry axes), a reason the larger (cubic) unit cell is used.

The Wigner-Seitz cell is a rhombic dodecahedron [Coxeter (1973), p. 25; Senechal (1990), p. 51] which has twelve faces, each a rhomb (rhombus) — a quadrilateral that has four equal sides with opposite pairs parallel (sec. II.3.c, p. 75). Each cell has three primitive lattice vectors, with the corner (the origin) shared by eight cells, but each lattice vector is shared by two of these cells, giving twelve lattice vectors, so twelve perpendicular bisectors, twelve faces for the Wigner-Seitz cell. Rotations of $\frac{\pi}{4}$ take the sides of the cube, so the primitive lattice vectors, into each other. Thus they move an edge of a face of the Wigner-Seitz cell to another, so each face has four edges. This (with the dashed cube displaced for clarity) is

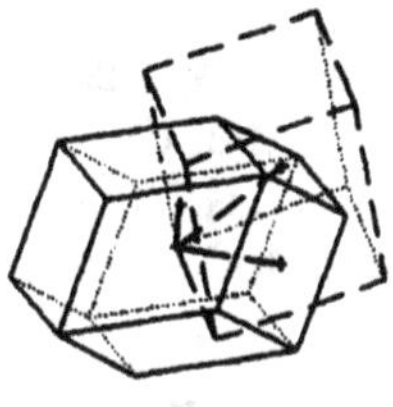

Figure IV.5.a.ii-1: THE WIGNER-SEITZ CELL OF THE FCC LATTICE.

Two versions of the object are

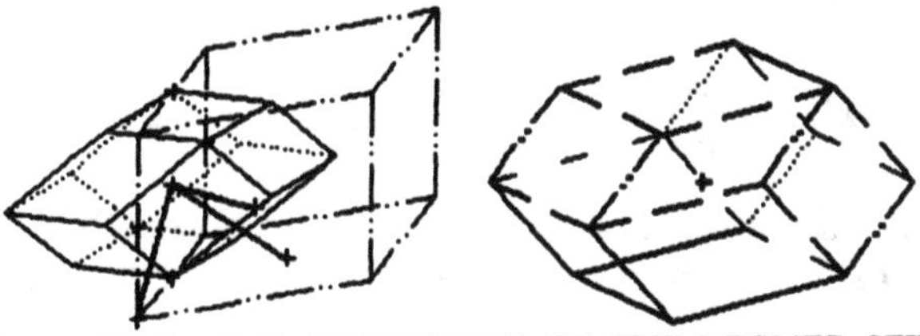

Figure IV.5.a.ii-2: TWO VIEWS OF THE WIGNER-SEITZ CELL.

On the left is a Wigner-Seitz cell attached to a cube with the origin at a lattice point at the center of a face, with three primitive lattice vectors, drawn heavily, going to a corner, and to the centers of two other faces. The cube is tilted slightly, for clarity, so in two dimensions does not look quite cubic. On the right is another rhombic dodecahedron, to emphasize certain aspects: note that a band of faces is shown with the parallel edges drawn with dots and dashes, and its "hidden" lines (that would not be seen if it were a solid) dotted. This band has six faces, each a quadrilateral. Their other edges are dashed.

Problem IV.5.a.ii-1: To check the primitive lattice vectors consider a (two-dimensional) square with the origin at its lower left corner. The vectors to the point at the center of the bottom line, and to the center of the square are $\frac{a}{2}(1,0)$ and $\frac{a}{2}(1,1)$. Verify this and use these to obtain the given vectors for three dimensions.

Problem IV.5.a.ii-2: It is not surprising that the Wigner-Seitz cell has twelve faces; show that each lattice point has twelve nearest neighbors. Nearest neighbors are lattice points that are reachable with a single lattice vector, others are connected to the point by a sum of lattice vectors. The Wigner-Seitz cell is defined by the requirement that its faces be the perpendicular bisectors of these lattice vectors, and as there are twelve, there are twelve faces. It should be clear from the primitive lattice vectors that these twelve lattice vectors are

$$\pm\underline{a}_1,\ \pm\underline{a}_2,\ \pm\underline{a}_3,\ \pm(\underline{a}_1-\underline{a}_2):\pm\frac{a}{\sqrt{2}}(1,0,-1),$$

$$\pm(\underline{a}_1-\underline{a}_3):\pm\frac{a}{\sqrt{2}}(0,1,-1),\ \pm(\underline{a}_2-\underline{a}_3):\pm\frac{a}{\sqrt{2}}(-1,1,0). \quad \text{(IV.5.a.ii-2)}$$

These vectors all have the same form; relabeling the axes interchanges them. We might therefore expect that the faces of the Wigner-Seitz cell are all of the same form, rectangles. The Wigner-Seitz cell of twelve rectangular faces is the rhombic dodecahedron (sec. IV.3.b, p. 199).

IV.5.a.iii *The body-centered cubic lattice*

Next is the body-centered cubic lattice [Gasson (1989), p. 262; Koster (1957), p. 198]. The primitive lattice vectors are

$$\underline{a}_1:\frac{a}{\sqrt{3}}(1,1,1),\underline{a}_2:\frac{a}{\sqrt{3}}(1,1,-1),\underline{a}_3:\frac{a}{\sqrt{3}}(1,-1,-1). \quad \text{(IV.5.a.iii-1)}$$

The body-centered square lattice, with the sides of length a, illustrates this. From the origin, at the center, to the two right corners the vectors are

$$\underline{a}_1^2:\frac{a}{\sqrt{2}}(1,1),\underline{a}_2^2:\frac{a}{\sqrt{2}}(1,-1). \quad \text{(IV.5.a.iii-2)}$$

So

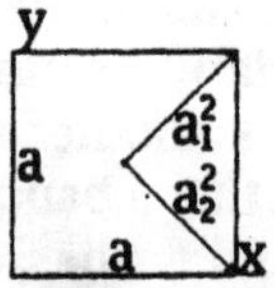

Figure IV.5.a.iii-1: BODY-CENTERED SQUARE LATTICE.

The generalization to three-dimensions, on the left is clear,

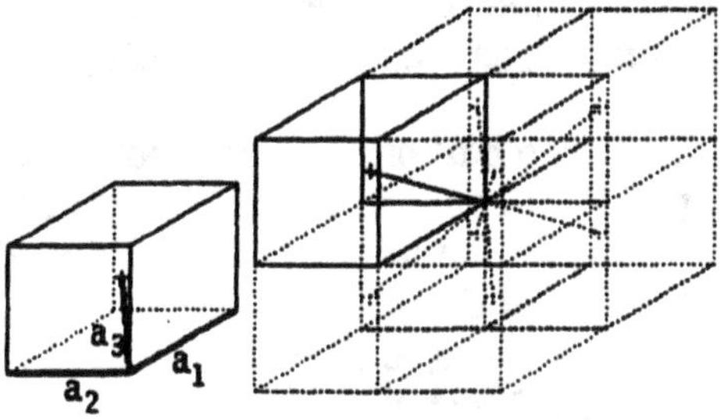

Figure IV.5.a.iii-2: BODY-CENTERED CUBIC LATTICE AND VECTORS,

while on the right there is a unit cell, a cube (with solid lines) with a lattice point at the center. The origin, from which the lines are drawn, is shown surrounded by eight body-centered cubes, these, except for the first, drawn dotted.

A body-centered cube, and a Wigner-Seitz cell placed at the center of the cube (in a lattice any lattice point can be taken as the center of the cube), looks like

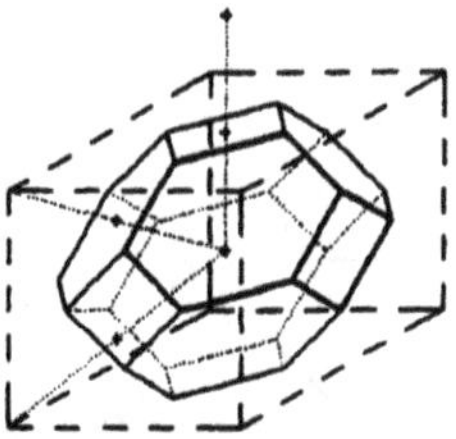

Figure IV.5.a.iii-3: BODY-CENTERED CUBE AND WIGNER-SEITZ CELL.

The cell has two types of faces, square and hexagonal (one of each drawn heavily). Lines (dotted) are drawn from the center of the cube, and of the unit cell, to the nearest lattice point, through one of each type, and it can be seen that the faces bisect these lines, and are perpendicular to them.

As we see the Wigner-Seitz cell of this lattice is a volume bounded by six squares, and eight hexagons. Each edge of a cube (and each line between centers) is an axis of four-fold symmetry, and each of the three axes is bounded by two sides, accounting for the six squares. Hexagons occur because axes through the center are of three-fold symmetry; again there are four each with two ends, passing through any corner of the cube (the origin). A face perpendicular to one of these axes shares edges with squares. Edges of squares are perpendicular which would not be possible if the face were a triangle; it is a hexagon. It cannot be larger as there are not enough squares with which to share edges. This cell is a truncated octahedron [Coxeter (1973), p. 30]. For it, six rectangular faces are perpendicular to the vectors to the centers of the neighboring cells (a cube has six sides), the eight hexagonal ones are perpendicular to the vectors from the center of the cell to its eight vertices. Since all sides of a cube are identical, as are all vertices, this gives the reason for six and eight.

We give here a mere sketch of one way, but not the only way [Burns and Glazer (1990), p. 50], of constructing this cell. At the top is a diagram showing the vectors to the lattice points of the boundary of the cube (which is half the size of the lattice, since the unit cell is bounded by planes bisecting these vectors, and perpendicular to them), and the lattice vector to the center. Drawn is the hexagon that is perpendicular to it, thus bisecting the lattice vector. We take two such vectors, say that shown, the other in the negative y direction, and find their intercepts, to determine the corner of this hexagon. To the right of this is a diagram showing the two hexagons perpendicular to these two vectors. The construction is continued with the rightmost object. This is then rotated about a coordinate axis and superimposed to get the bottom left object. Continuing, we obtain the right bottom figure, a closed unit cell, showing the band of hexagons circling the unit cell, and between them the squares,

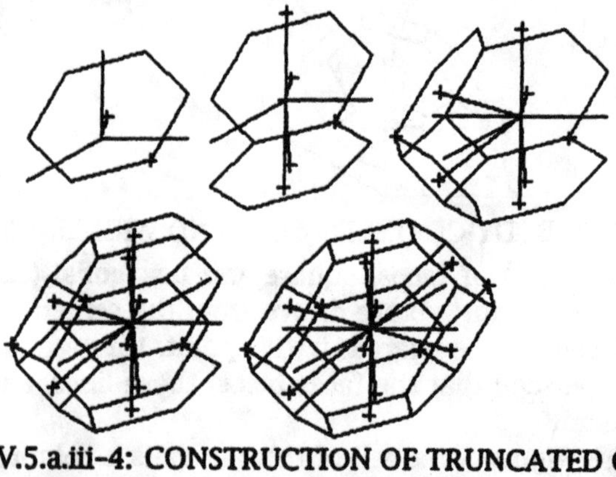

Figure IV.5.a.iii-4: CONSTRUCTION OF TRUNCATED OCTAHEDRON.

Problem IV.5.a.iii-1: Check that the vectors are correct, that they do go to the center of the cube, and that every lattice point can be reached by a sum of these three, with integral coefficients. Also this cell, though it is not a cube, does show the full cubic symmetry (how?). The faces are drawn as hexagons, which clearly works, and has maximum symmetry. Note however that there are two sets of edges, one bounding two hexagons, the other a square and a hexagon (thus the cell is not regular, but only semi-regular). Because of three-fold symmetry all edge lengths are equal. Is it possible to get a (reasonable) cell if they have different lengths?

Problem IV.5.a.iii-2: Verify that a lattice point of this system has fourteen nearest and next-nearest neighbors, giving fourteen faces of the Wigner-Seitz cell. However these faces are of two kinds, eight hexagonal, and six rectangular. Eight of the shortest lattice vectors are

$$\pm \underline{a}_1, \ \pm\underline{a}_2, \ \pm\underline{a}_3, \ \pm(\underline{a}_1 - \underline{a}_2 + \underline{a}_3) : \pm\frac{a}{\sqrt{3}}(1,-1,1), \qquad \text{(IV.5.a.iii-3)}$$

with the other six

$$\pm(\underline{a}_1 + \underline{a}_3) : \frac{2a}{\sqrt{3}}(1,0,0), \ \pm(\underline{a}_2 - \underline{a}_3) : \frac{2a}{\sqrt{3}}(0,1,0),$$

$$\pm(\underline{a}_1 - \underline{a}_2) : \frac{2a}{\sqrt{3}}(0,0,1), \qquad \text{(IV.5.a.iii-4)}$$

these lying along the coordinate axes. Note that these last are mutually perpendicular, but this is not true of other pairs of the fourteen vectors. Perhaps this is not unrelated to their faces, their perpendicular bisectors, being rectangular. The first eight vectors can be interchanged by relabeling axes, as can the other six, but not members of the two sets. They have different lengths (why?); is there any other way of seeing this?

IV.5.b The tetragonal system

The next most symmetric system is tetragonal [Koster (1957), p. 199], and it is interesting to see how it is related to the cubic, as well as to see how the Wigner-Seitz cells are related. That of the primitive lattice is also a primitive tetragonal lattice. For the cubic lattice the Wigner-Seitz cell of the body-centered lattice is face-centered, and conversely. However there are only two tetragonal lattices. But the all-face-centered lattice is equivalent to the body-centered one (sec. II.7.b.ii, p. 114). The same relationship between the centered cubic lattices holds here, except that they become equivalent so now, while the lattice and its reciprocal are still dual in the same way, they are also the same, but differently oriented.

IV.5.b.i *The primitive tetragonal lattice*

For the primitive tetragonal lattice the vectors are

$$\underline{a}_1 : c(1,0,0), \qquad \text{(IV.5.b.i-1)}$$

$$\underline{a}_2 : a(0,1,0), \qquad \text{(IV.5.b.i-2)}$$

$$\underline{a}_3 : a(0,0,1). \qquad \text{(IV.5.b.i-3)}$$

The reciprocal vectors (eq. IV.2.a-5, p. 187) are thus

$$\underline{b}_1 : \frac{2\pi}{c}(1,0,0), \quad \text{(IV.5.b.i-4)}$$

$$\underline{b}_2 : \frac{2\pi}{a}(0,1,0), \quad \text{(IV.5.b.i-5)}$$

$$\underline{b}_3 : \frac{2\pi}{a}(0,0,1), \quad \text{(IV.5.b.i-6)}$$

and the first Brillouin zone is also tetragonal.

A unit cell of the lattice, and the Wigner-Seitz cell attached, is

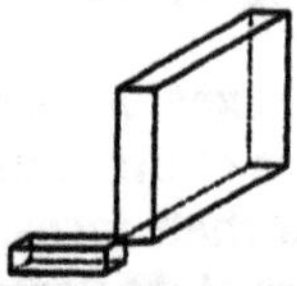

Figure IV.5.b.i-1: WIGNER-SEITZ CELL OF TETRAGONAL LATTICE;

both are tetragonal, and are obtained by stretching (or compressing) the primitive cubic lattice, with unit cell, along one axis.

A (base) face-centered tetragonal lattice (with square base, although that may not be clear from the diagram) is drawn with solid lines, the heavy dashed lines show the lattice vectors. These can be seen to form a primitive lattice, rotated by $\frac{\pi}{4}$, with a cell smaller than the face-centered one, so conventionally that is the one used,

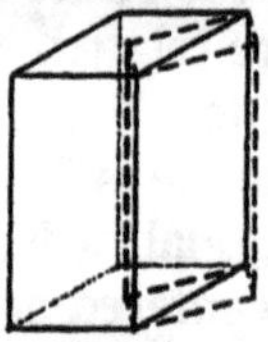

Figure IV.5.b.i-2: THE BASE-CENTERED TETRAGONAL IS PRIMITIVE.

IV.5.b.ii *The tetragonal Wigner-Seitz cell is not a stretched cubic Wigner-Seitz cell*

The tetragonal lattice can be obtained from the cubic by a change of length of one side. Does this change the Wigner-Seitz cell of the cubic into that for the tetragonal? At the upper left is a unit cell of a body-centered cubic lattice with its Wigner-Seitz cell, an elongated dodecahedron, and lattice vectors shown by lines of dots and dashes. These are, of course, perpendicular to the surfaces of the Wigner-Seitz cell. The other two lattices are obtained by compressing, and by stretching, the unit cell, and with it the Wigner-Seitz cell of the cube. These do not give perpendicular bisectors of the lattice vectors — the sides are no longer perpendicular to the vectors — so changing the length of one side of a cubic lattice gives a tetragonal one, but the cell so obtained from a Wigner-Seitz cell is not a Wigner-Seitz cell,

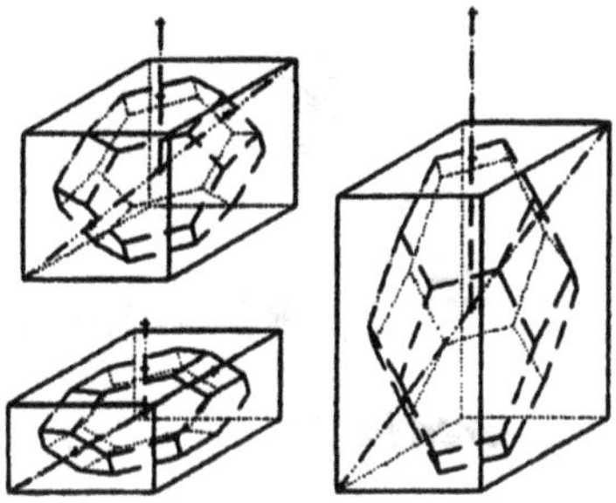

Figure IV.5.b.ii-1: STRETCHED WIGNER-SEITZ CELL IS NOT ONE.

Problem IV.5.b.ii-1: Explain why applying to the Wigner-Seitz cell the transformation that changes the length of a side, so for example, a cubic to a tetragonal lattice, does not give a Wigner-Seitz cell.

IV.5.b.iii *The body-centered tetragonal lattice*

The other tetragonal lattice is the body-centered one (tetragonal-I). The primitive lattice vectors, in terms of their components with respect to axes along the sides, and the origin at the center, are

$$\underline{a}_1 : (a, a, c), \underline{a}_2 : (a, a, -c), \underline{a}_3 : (a, -a, -c). \qquad \text{(IV.5.b.iii-1)}$$

Reciprocal vectors

$$\underline{b}_1 : \frac{2\pi}{2}(\frac{1}{a}, 0, \frac{1}{c}), \underline{b}_2 : \frac{2\pi}{2}(0, \frac{1}{a}, \frac{-1}{c}), \underline{b}_3 : \frac{2\pi}{2}(\frac{1}{a}, \frac{-1}{a}, 0), \qquad \text{(IV.5.b.iii-2)}$$

give again a body-centered tetragonal lattice — the lattice must belong to the same system, it can not be primitive, there is no other of this system. And, of course, the vectors clearly give this lattice. Interestingly for $c = a$, the reciprocal lattice is the (rotated) face-centered cubic (pb. II.6.b-3, p. 106; pb. II.7.b.ii-2, p. 116); the direct one is, fortunately, the body-centered cubic.

An elongated dodecahedron has eight rectangular faces, these perpendicular to the vectors to the eight vertices, and four hexagonal ones, perpendicular to the vectors parallel to the base to the centers of the neighboring cells. All vertices of the tetragonal unit cell are identical, and four, but not six, sides are. Thus we get the number of sides of the Wigner-Seitz cell as eight and four.

Next to show the lattice vectors, we draw these from the center of the cell, dashed and heavy, taking $a = 1$ and $c = 2$. Three unit cells of the lattice with solid lines are shown, with the square base facing front. Centers of cells are connected by lines with dots and dashes which can also be taken as lattice vectors,

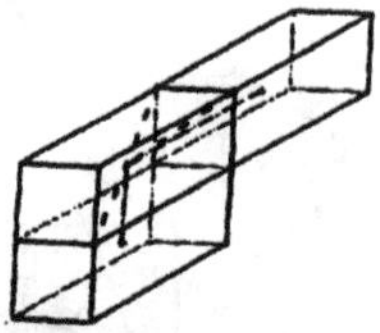

Figure IV.5.b.iii-1: BODY-CENTERED TETRAGONAL LATTICE.

IV.5.b.iv *The all-face-centered lattice*

There is one further lattice, with all faces centered,

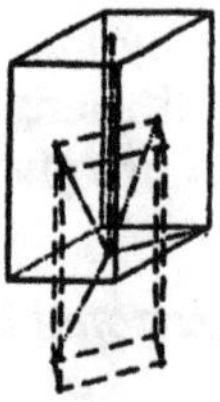

Figure IV.5.b.iv-1: REPLACING THE FACE-CENTERED LATTICE.

A cell, in solid lines is shown and has attached a body-centered parallelepiped, dashed. There are lattice vectors from its center to the corners of the smaller cell which, like the other lattice vectors, are drawn with lines of dots and dashes. This cell is smaller, so again is the one — conventionally — chosen. As we see in fig. IV.5.b.i-2, p. 222, a base-centered cell can be replaced by a smaller primitive one. The four tetragonal lattices reduce to two, the primitive and the body-centered.

IV.5.b.v *Why the tetragonal has fewer lattices than the cubic*

To see the effect of changing the ratio of sides, so the system, we draw on the left of the next diagram a base-centered cell with a primitive one attached, both of the tetragonal system. This cell is shrunk, giving the all-face-centered cell on the right, a cube. However the dashed cell, which is body-centered, is not cubic. So for the cubic case the face-centered and body-centered lattices are distinct, unlike the tetragonal lattice which having less symmetry, allows the identification of cells with different ratios of sides. A body-centered tetragonal unit cell can be replaced by another (using the points of the same lattice), which is face-centered tetragonal. The lattice can be regarded as either body- or face-centered, but with the former smaller, it is (usually) chosen,

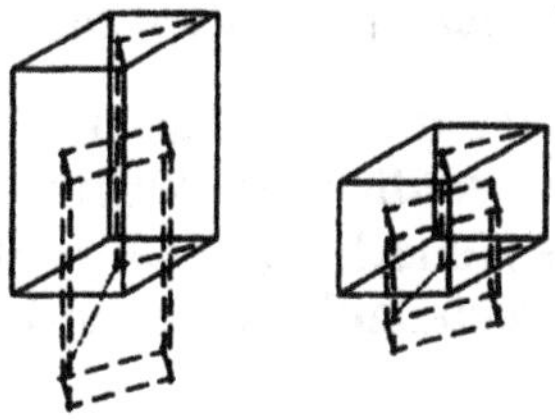

Figure IV.5.b.v-1: SHRINKING A FACE-CENTERED CUBE.

If this is done for the cube the smaller cell is not of the same system, so is unacceptable. The cubic, being more restricted, having a strict criterion on ratios of lengths, so allowing less freedom of assignment, has more lattices than the tetragonal system. This emphasizes that classifying the tetragonal ones (for example) as belonging to one set is (partly) conventional.

The elongated dodecahedron is shown in fig. IV.3.e-2, p. 202, the truncated octahedron in fig. IV.3.e-3, p. 203. An elongated dodecahedron, attached to the unit cell for which $\frac{c}{a} = 2$ (slightly tilted so the top and bottom look a little different), is an example of how these unit and Wigner-Seitz cells are related,

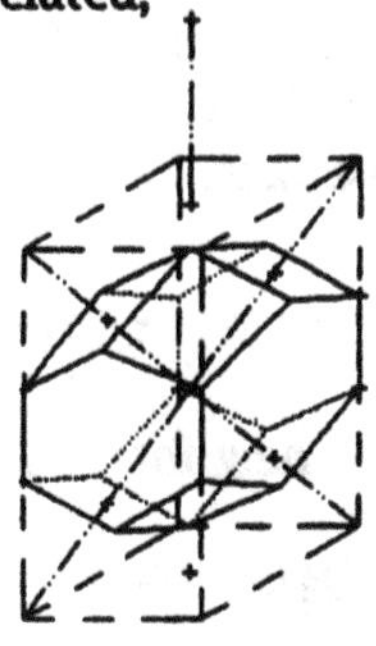

Figure IV.5.b.v-2: ELONGATED DODECAHEDRON FOR THE LATTICE.

The top view of the next cell is shown with the boundary surfaces; the unit cell is dotted to make the faces of the Wigner-Seitz cell easier to see,

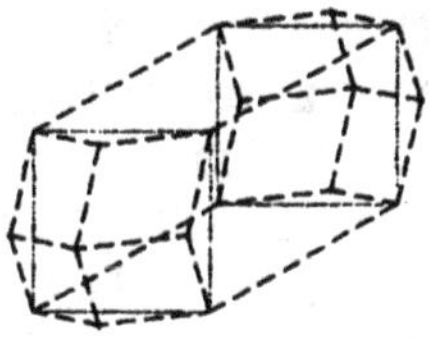

Figure IV.5.b.v-3: TOP VIEW OF LATTICE AND WIGNER-SEITZ CELL.

IV.5.b.vi *The differing Wigner-Seitz cells of the lattice*

The tetragonal lattice, whose unit cell has height c and square base with side of length a, has two different Wigner-Seitz cells depending on the relative dimensions [Senechal (1990), p. 50]. It is an elongated dodecahedron (with eight rectangular faces and four hexagonal ones) if

$$c > \sqrt{2}a, \tag{IV.5.b.vi-1}$$

and a truncated octahedron (with six rectangular and eight hexagonal faces) for

$$c < \sqrt{2}a \tag{IV.5.b.vi-2}$$

(the same as the body-centered cubic lattice — which we might expect since this is also a body-centered lattice). They are distinct objects with different numbers of faces, and differing numbers of sides. A unit cell with

$$c = \sqrt{2}a \tag{IV.5.b.vi-3}$$

belongs to a different system, the cubic.

To find this ratio, we take the x and y axes along the sides of the square base, perpendicular to the z axis, that parallel to the side of length c, the origin at the center of the unit cell, and use lattice vectors rotated by $\frac{\pi}{4}$ about z from these axes,

$$\underline{b}_1 : \frac{1}{2}(a, a, c), \underline{b}_2 : \frac{1}{2}(a, a, -c), \underline{b}_3 : \frac{1}{2}(a, -a, -c). \tag{IV.5.b.vi-4}$$

The lattice vectors to the nearest neighbors are $\pm\underline{b}_1, \pm\underline{b}_2, \pm\underline{b}_3$, and

$$\pm(\underline{b}_1 - \underline{b}_2) : \pm(0, 0, c), \pm(\underline{b}_1 + \underline{b}_3) : \pm(a, 0, 0), \pm(\underline{b}_2 - \underline{b}_3) : \pm(0, a, 0), \tag{IV.5.b.vi-5}$$

twelve altogether. Their perpendicular bisectors are the faces of the Wigner-Seitz cell. The last pairs are perpendicular to the sides of the unit cell, the others go to its corners. The diagonal of the base equals $\sqrt{2}a$, so for $c = \sqrt{2}a$ the projection of the unit cell of the lattice onto the plane through the diagonal and parallel to z is square. This is the relevant plane since the primitive lattice vectors from the center of the unit cell go to the corners thus lie in this plane, and their perpendicular bisectors meet at the center of the top side of the cell if this projection is square. So this is the value at which the face of the Wigner-Seitz cell becomes zero, giving one type cell on one side of the value, another type on the other.

The cell with $c = \sqrt{2}a$ is

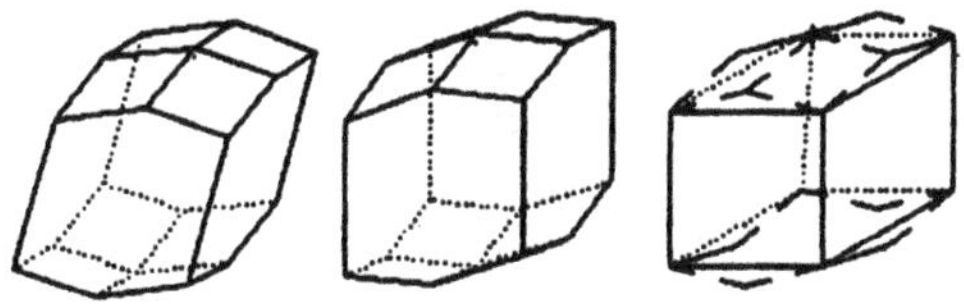

Figure IV.5.b.vi-1: THE TETRAGONAL LATTICE THAT IS CUBIC.

Problem IV.5.b.vi-1: Draw a diagram to show that these are the vectors to the nearest neighbors.

Problem IV.5.b.vi-2: In deriving this lattice (sec. II.7.b.ii, p. 114), what freedom is there that allows these different types of lattices (so these different Wigner-Seitz cells), and only these?

IV.5.b.vii *How cells change*

How does one object suddenly turn into another, with different numbers and types of faces, as the ratio of lengths of the cell that determines it changes, and in a continuous manner (sec. IV.4, p. 212)? How do faces and edges appear and disappear? For the tetragonal lattice, at the boundary value, the angle between the top edges of the Wigner-Seitz cell along the sides becomes zero. Hence two lines become one, the number of edges of the side face goes from six to four, and the hexagonal face becomes rectangular.

From the origin at the center of the unit cell four lattice vectors go to the lattice points at the centers of four of the neighboring cells, along x and y. Their perpendicular bisectors are the planes forming these faces of the unit cell, so are four of the faces of the Wigner-Seitz cell. For $c > \sqrt{2}a$ the lattice vector from the origin along z to the center of the unit cell above passes lattice vectors at the corners of the cell, so its perpendicular bisector is not a face of the Wigner-Seitz cell. The four lattice vectors go from the origin to the four corners of the top face (and likewise the bottom). This gives four perpendicular bisectors, so four faces on the top and four on the bottom, twelve altogether. By symmetry the four top faces are identical and rectangular. However they are at angles to side faces so these faces of the Wigner-Seitz cell are bounded by two lines (at an angle to each other) at the top (and also the bottom), plus two on the sides, thus have six edges, giving a cell with eight rectangular, and four hexagonal, faces.

To study this we take the plane along z parallel to one pair of sides, and bisecting the other pair, through the center of the unit, and Wigner-Seitz, cells,

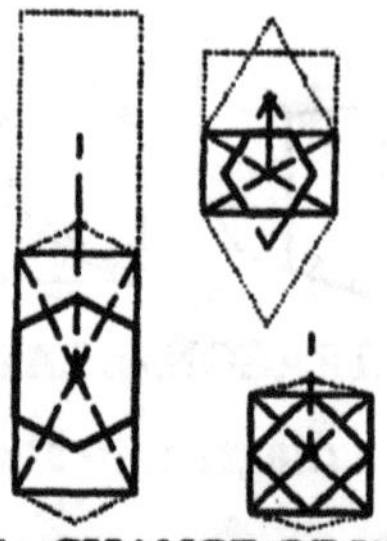

Figure IV.5.b.vii–1: CHANGE OF WIGNER-SEITZ CELL TYPE.

A cell with $c > \sqrt{2}a$, with the projection onto this plane of its Wigner-Seitz cell drawn heavily, is on the left. Lattice vectors are dashed; their perpendiculars, as well as the unit cell above, are dotted. For this cell the perpendicular bisectors of the vectors to the vertices meet below the bisector of the vector between the centers — the top face of the unit cell — so it is not part of the Wigner-Seitz cell. This face has four sides. The projection is that of an elongated dodecahedron. For the cell on the top right $c < \sqrt{2}a$; the perpendicular bisectors of the vectors to its vertices meet above that of the vector between the centers, so part of that is included in the Wigner-Seitz cell. The resultant face, of six sides, shown is one of those of the truncated octahedron. At the boundary value of c/a the perpendicular to the vector between the centers passes through the point of intersection of the perpendiculars to the two vectors to the vertices. This is shown on the bottom right, for which the side of the unit cell and the face of the Wigner-Seitz cell, are both squares, as the edges of the Wigner-Seitz cell meet just at the midpoint of the vector to the cell above. The lengths of the top and bottom sides are now zero, so the hexagonal face becomes rectangular.

From the diagrams [Burns and Glazer (1990), p. 292] we can understand why there is more than one Wigner-Seitz cell. For small c/a the perpendicular bisector of the vector from the center of a lattice cell to that of the one above lies below the intersection of the perpendiculars to the corners, so forms a surface of the Wigner-Seitz cell, as do the four perpendicular bisectors from the center to those of the neighboring ones at the same level. By symmetry, since each is bounded by four bisectors of the vectors to the corners, these six surfaces are rectangular. However the two surfaces in the bases, identical to each other, and the four on the sides, likewise mutually identical, are not the same, so the Wigner-Seitz cell consists of two sets of rectangles, with the areas for each set different. The latter eight surfaces are each bounded by six others — on each of its two sides there are edges shared with two identical surfaces, plus two rectangular ones that bisect the vectors through the base, and through one side. Thus this figure has six rectangular surfaces, and eight hexagonal ones. It is a truncated octahedron.

When $c = \sqrt{2}a$, the areas of the rectangular faces in the bases becomes zero, as the bisectors of the vectors to the corners meet at the base. Above this value each long edge of the unit cell intersects two of the planes bisecting the vectors to the vertices of the unit cell. At $c = \sqrt{2}a$ these two lines meet at the edge so the length of the line between them becomes zero, and the hexagonal surface becomes rectangular. For this value, the corners of this rectangular surface are at the edges of the unit cell, but as c/a decreases they move toward the center of the face of the cell. This explains how surfaces appear and disappear as the ratio of sides changes, and how a Wigner-Seitz cell of one category becomes a cell of another.

To illustrate this we have on the left a cell with $c = 2a$, an elongated dodecahedron (with eight rectangular faces and four hexagonal ones), on the right $c = a$, a truncated octahedron (with six rectangular and eight hexagonal faces), in the middle $c = 1.414a = \sqrt{2}a$, a rhombic dodecahedron, with twelve rectangular faces (only parts of some lattice vectors are drawn),

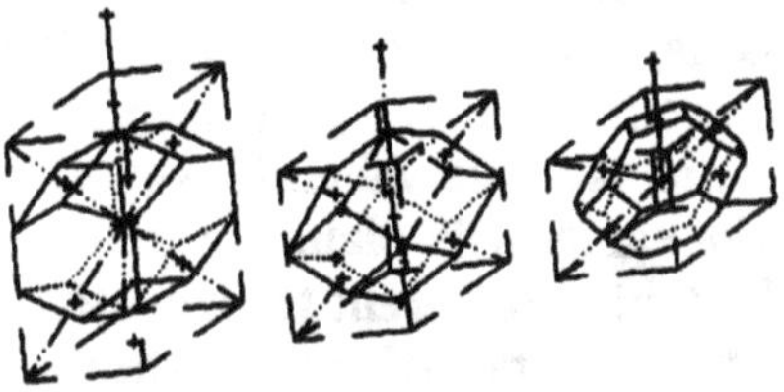

Figure IV.5.b.vii-2: WIGNER-SEITZ CELL CHANGING TYPE.

Problem IV.5.b.vii-1: Draw diagrams and check these statements.

IV.5.b.viii *Surfaces appear when edges disappear*

The value of c/a for which two new surfaces appear is the same as that for which four edges disappear, resulting in eight surfaces changing from rectangular to hexagonal, and four from hexagonal to rectangular. Why must both occur at the same value? If the edges disappeared first it would give an object with twelve rectangular faces, a rhombic dodecahedron, which is possible. But if the top and bottom faces vanished first the object would have ten faces, eight of which would have five sides, which is not possible. It is fortunate that both occur at the same value.

To see why this occurs, consider a two-dimensional section of the unit cell through two of its edges so along a diagonal of a base. The c/a boundary value is that at which the perpendicular bisectors of the vectors from the center of the two-dimensional figure to its corners intersect at the midpoint of the top base (and the two others at the bottom). For a larger side they intersect below, for a smaller above. Thus this is the value at which new surfaces appear. But for the perpendiculars to so meet, the two-dimensional figure must be a square — the lengths of the height and the diagonal of the base (which is $\sqrt{2}$ times the length of a side of the base) must be equal. This explains why the ratio is $\sqrt{2}$. If the figure is a square, then the bisectors meet at the top, and also just meet at the sides — the surfaces appear and the lengths become zero simultaneously, as they must. This also shows why the boundary has to be a square in this two-dimensional projection of the Wigner-Seitz cell, that bisecting these faces, giving a projection of them onto the plane bisecting them, and going through the points of intersection at the bases.

IV.5.b.ix *Cells as limits of others*

So we have an example of a crystal system with a Wigner-Seitz cell that changes into different one at a (single) value of a parameter. The Wigner-Seitz cell is different on the two sides of the transition, but the crystal system is the same, by convention, although it becomes different at an isolated point (becoming cubic), perhaps not surprisingly, since the cube exists only for an isolated value. However, though the crystal system is (taken as) the same, it is plausible that physical properties of some objects can be (quite) different for the different categories of Wigner-Seitz cells on the two sides of the transition.

While we treat the Wigner-Seitz cells as distinct, and they are quite different, here we see that the rhombic dodecahedron is the limit of the elongated dodecahedron and also the truncated octahedron (however since these are categories of objects it is — apparently — correct to say that a rhombic dodecahedron is a limit of an elongated dodecahedron and a truncated octahedron).

Problem IV.5.b.ix-1: A computer-generated motion picture showing how one lattice turns into another, and how one Wigner-Seitz cell becomes a different one, would clarify this behavior.

IV.5.c The orthorhombic system

This system [Koster (1957), p. 202] is obtained from the tetragonal by allowing the lengths of the sides of the base to be different, although all angles remain at $\frac{\pi}{2}$. We would expect therefore that it has the same Wigner-Seitz cells as the tetragonal, but the lesser symmetry permits more. In fact this has the most of any system, with ones in all five categories. It also has, besides the primitive cell, all three types of centering, body, two-face and all-face. The orthorhombic system has enough symmetry to allow these, but not enough to lead to cells being equivalent.

There is an additional aspect that did not appear for the cubic and tetragonal systems. For the latter a unit cell could have more than one Wigner-Seitz cell, actually two, depending on the ratio of lengths of sides. This occurs for this system also, but the converse also occurs [Senechal (1990), p. 51]. A Wigner-Seitz cell (category) can go with more than one unit cell — there are differently-centered lattices of this system with the same Wigner-Seitz cell (category).

IV.5.c.i *The simple orthorhombic lattice*

Taking the primitive vectors of the orthorhombic-P lattice to define the axes, these are

$$\underline{a}_1 : d(1,0,0), \ \ \underline{a}_2 : b(0,1,0), \ \ \underline{a}_3 : a(0,0,1). \qquad \text{(IV.5.c.i–1)}$$

The faces of the Wigner-Seitz cell are perpendicular to these axes and form a rectangular parallelepiped, whose most symmetric form is the cube, which names the cell category for this lattice. The Wigner-Seitz cell is a rectangular prism.

Problem IV.5.c.i-1: With a drawing (or three-dimensional model), show the relationship between the lattice and its Wigner-Seitz cell.

IV.5.c.ii *The two-face-centered orthorhombic lattice*

For the two-face-centered lattice (sometimes known as the one-face centered lattice, but it is a pair that is centered), the orthorhombic-C lattice, the Wigner-Seitz cell is of the hexagonal category. The most symmetric is the hexagonal prism.

With the origin at the center of a face, the primitive lattice vectors go to the two corners of the face and to the center of the face of the adjacent cell, this vector defining the z axis. Since all lengths are unequal, these three primitive vectors have arbitrary lengths, and the two in the face have arbitrary angle thus are

$$\underline{a}_1 : (d,b,0), \ \ \underline{a}_2 : (d,-b,0), \ \ \underline{a}_3 : (0,0,a). \qquad \text{(IV.5.c.ii–1)}$$

Clearly then, the Wigner-Seitz cell has two surfaces parallel to the face, these perpendicular to the z axis, and six perpendicular to the xy plane, four perpendicular to the lattice vectors going to the four corners of the face, and two perpendicular to the vectors going from its center to the centers of the faces of the four bordering unit cells in the xy plane. So there are eight vectors to the eight nearest neighbors.

Problem IV.5.c.ii–1: Draw this and check. Also check analytically. In particular verify that these are the correct primitive lattice vectors, and that these faces are their perpendicular bisectors.

IV.5.c.iii *The all-face-centered orthorhombic lattice*

The unit cell of the orthorhombic-F lattice is a rectangular parallelepiped, with sides of unequal lengths; a corner is used for the origin and the axes are along the sides. The primitive lattice vectors go to the centers of the three faces meeting at that corner. These then are

$$\underline{a}_1 : (d, b, 0), \;\; \underline{a}_2 : (d, 0, a), \;\; \underline{a}_3 : (0, b, a). \qquad \text{(IV.5.c.iii–1)}$$

There are fourteen nearest neighbors. The Wigner-Seitz cell is (of the category of) a truncated octahedron, the same as the body-centered cube (sec.IV.5.a.iii, p. 218). This relationship can be seen by setting values in these vectors equal, giving the vectors for cubic lattice.

Problem IV.5.c.iii–1: How do we know that this is the correct set of nearest neighbors? Find the vectors to these. Check that the lengths of these vectors form sets of equal values, that they are the smallest possible, vectors to other points are longer, and a vector of one set cannot be written as a sum of vectors from others. Explain why this gives the result. Verify that these vectors do give the truncated octahedron. Show that no other set of primitive lattice vectors gives a cell with a smaller volume. Draw a diagram, or construct a model.

IV.5.c.iv *The body-centered orthorhombic lattice*

The orthorhombic-I lattice [Koster (1957), p. 204] has three Wigner-Seitz cells, sharing one with the all-face-centered lattice — the lattice does not uniquely specify the Wigner-Seitz cell, nor does that cell uniquely specify the lattice. The origin is at the center of the unit cell, and again the edges of its faces define the axes. The primitive lattice vectors go to three of the eight corners of the cell so are

$$\underline{a}_1 : (a, b, d), \;\; \underline{a}_2 : (a, b, -d), \;\; \underline{a}_3 : (a, -b, -d). \qquad \text{(IV.5.c.iv–1)}$$

For the tetragonal lattice there were two lengths and the Wigner-Seitz cell depended on their ratio. Here there are three, so more ratios thus more Wigner-Seitz cells.

Problem IV.5.c.iv–1: First take

$$a^2 > b^2 + d^2, \;\; a > b > d; \qquad \text{(IV.5.c.iv–2)}$$

clearly it does not matter which side is labeled a. Why? Find the vectors giving the eight front and back faces (parallel to the z axis), the other four determining the top and bottom faces, and the two perpendicular to y. Show that this cell is a rhombic dodecahedron, the same category as the face-centered cubic lattice.

Problem IV.5.c.iv-2: For the next case take

$$a^2 < b^2 + d^2, \quad a > b > d, \tag{IV.5.c.iv-3}$$

and show that besides the twelve of the previous case there are two more faces, front and back, finding the vectors giving them. Check that with fourteen faces the cell is a truncated octahedron, the same (form) as for the body-centered cubic lattice.

Problem IV.5.c.iv-3: The third Wigner-Seitz cell of this lattice is the elongated dodecahedron, with twelve sides, like that of the body-centered tetragonal lattice. This differs from the (twelve-sided) rhombic dodecahedron, which has all rectangular sides, in having hexagonal ones. Thus the lattice shares (categories of) Wigner-Seitz cells with two other body-centered lattices, but also with a face-centered one. Draw these and show how they are (or are not) related.

Problem IV.5.c.iv-4: Again we have to verify that these are the correct set of nearest neighbors. Check, for each case, that the lengths of the vectors form sets of equal values, that they are the smallest possible, vectors to other points are longer, a vector of one set cannot be written as a sum of vectors from others, and that the cell has the smallest volume. Also explain how we know that the faces have the shapes given. While we have counted the numbers of faces for these cells, that does not prove they are the (correct) objects stated. Prove it. This can be related to the other lattices with these Wigner-Seitz cells.

Problem IV.5.c.iv-5: The tetragonal-I system (sec. IV.5.b.iii, p. 223), allows the value of c/a at which the Wigner-Seitz cell changed type to be found by taking a two-dimensional section along a diagonal of the bases. For the various orthorhombic cases can this be done? Would this show why for one lattice there are three categories of Wigner-Seitz cells, and why one category appears for two different lattices?

Problem IV.5.c.iv-6: While there is more than one lattice of this system going with a Wigner-Seitz cell category, and more than one type of centering giving a cell in each category, are the cells of a category really the same? Are there rules that specify the cell (not merely the category) belonging to a lattice, and uniquely so? What would (qualitatively) distinguish the cells? That is can we introduce a finer classification of the Wigner-Seitz cells so that the category going with a lattice is unique (and conversely?)? Of course this has a slight problem that a crystal system (plus centering) does not uniquely specify a lattice, but rather a category.

Problem IV.5.c.iv-7: In the derivation (sec. II.7.b.iii, p. 116) of this lattice, what freedom is there that allows these different types of lattices (so these different Wigner-Seitz cells), and only these?

IV.5.d The other systems

These three systems show the basic properties of these cells and their relationships. Since the listing of the cells is available in many places [Burns and Glazer (1990), p. 292; Koster (1957); Senechal (1990), p. 51], here we just briefly describe them, providing the reader with the useful opportunity to verify these well-known (and hopefully correct) lists.

Problem IV.5.d-1: There is but one lattice of the hexagonal system (hexagonal-P), and its Wigner-Seitz cell is hexagonal. Draw a diagram showing their relationship. Is there a difference between this cell and the hexagonal ones of other lattices — if you had the cell would you know that it belonged to the hexagonal lattice? How much freedom is there in specifying the hexagonal lattice? What freedom does this give for its Wigner-Seitz cell?

Problem IV.5.d-2: The Wigner-Seitz cell of the primitive monoclinic lattice (monoclinic-P) is hexagonal. Why? How does it differ from that of the hexagonal lattice (qualitatively?)? Draw it attached to the lattice. How does it vary as the parameters describing the lattice (which are what?) vary?

Problem IV.5.d-3: There is only one other monoclinic lattice (monoclinic-B), the two-face centered one, but it has five Wigner-Seitz cells, the rhombic dodecahedron which appears once, the truncated octahedron and elongated dodecahedron, each appearing twice. For the latter two there are disjoint ranges of the ratios of the sides giving the same cell category. These are examples of versions of a single lattice different enough that they not only go with different Wigner-Seitz cells, but ones going with a single cell (category) are distinguishable — a finite classification gives more than one lattice to a cell. However the lattice is regarded as only a single one. Find the Wigner-Seitz cells for each range of parameters giving the lattice, and verify that there are five. How do the two lattices of each category that has two differ? Do the cells (of the same category) that go with them differ qualitatively? How? Why? Does it make sense to regard two lattices that have the same Wigner-Seitz cell as different? Why? Are there parameter ranges for the lattice such that a cell changes category as the parameters are varied, and then changes back? Find the primitive lattice vectors for all these cases, and draw the cells attached to the lattices (or construct three-dimensional models), and illustrate this behavior.

Problem IV.5.d-4: A computer program that produces a movie of this would be enlightening. What has to be known (about the — properties of the — lattices) to write such a program?

Problem IV.5.d-5: The rhombohedral system has two lattices, the primitive, with Wigner-Seitz cell a rhombic dodecahedron (perhaps not surprisingly), and the nonprimitive, whose Wigner-Seitz cell is a truncated octahedron. Check this, both by drawing (or writing a computer program) and analytically.

Problem IV.5.d-6: Repeat this for the trigonal system. How are the results related (sec. II.6.c, p. 107)?

Problem IV.5.d-7: The (single) triclinic system has only a primitive lattice, but three Wigner-Seitz cells, a rhombic dodecahedron, a truncated octahedron and an elongated dodecahedron. Find the parameter ranges giving them. Use the primitive lattice vectors to draw the cells and attach them to the lattices. This system may not appear interesting, but there is more structure hidden away then might be guessed from a single lattice (with determined ratios of sides) of this system. Why?

Problem IV.5.d-8: All these results are well-known and widely accepted, and may even be true. However it would be useful to prove everything (assuming that everything can be proven).

Problem IV.5.d-9: Given a Wigner-Seitz cell, is it possible qualitatively (or semi-quantitatively) to tell which lattice it goes with? Why?

Problem IV.5.d-10: In the derivation of these lattices (sec. II.7.b, p. 113) what arbitrariness allowed several cells for the lattices that have them? Does the derivation show why other lattices do not have arbitrariness?

Problem IV.5.d-11: Write a computer program to find (and draw) all Wigner-Seitz cells, attaching them to their lattices. Can this be done for spaces of arbitrary dimension (for which the drawings would be more challenging)? How about arbitrary signature, and arbitrary geometry?

Problem IV.5.d-12: Find all Wigner-Seitz cells that are limits of other Wigner-Seitz cells, in particular, give the (ranges of) parameters for which such limits occur. Are there interesting aspects to these values? Are there fundamental differences in spaces of other dimensions?

IV.5.e Why are there five categories of Wigner-Seitz cells?

One way to show that we have all the cells is to consider how they are related by starting with the most symmetrical lattices, the cubic and the hexagonal. All others can be obtained from these two by decreasing the symmetry. As we do that Wigner-Seitz cells change into others. This raises two questions: What other cells are so generated, and could the Wigner-Seitz cell found not be the smallest, so the correct one is not a cell so generated?

All Wigner-Seitz cells are actually obtained by this procedure. Any lattice, no matter what its ratio of sides or values of angles, can be obtained from the two most symmetrical by decrease in symmetry. The order in which symmetry is changed, which ratios and which angles are changed first, determines the cell reached. That is why there may be several cells for a lattice, depending on these values. But are all uncovered this way?

Which ones are so found? All parallelohedra have surfaces with either four or six sides (sec. IV.3.g.i, p. 204). Starting from the cubic, it is possible to break a side into two, at an angle to each other, changing a rectangular face to a hexagonal one. By the regularity needed for a lattice, this must be done (of course) for opposite sides, and also for faces sharing an edge not changed, that is for all four sides of the rectangular parallelepiped, giving the elongated dodecahedron. But this cannot be carried further, for it would result in surfaces with too many sides. For the same reason it cannot be done for the hexagonal prism, clearly not for the hexagonal sides. If a rectangular side were made hexagonal, all six of these must be, since neighboring ones are required to. But this would introduce a new edge connecting the new vertices, reducing the sides to rectangular again. But then the two hexagonal surfaces would no longer be the same so the rectangular surface would have sides of different lengths. This would give a pair of hexagonal objects with different size bases, with their larger bases attached to each other, so an unacceptable type, not a hexagonal prism — it could not fill space. Such a figure (whose quadrilaterals are not rectangles), is on the left, and then a set of several overlaid is shown on the right,

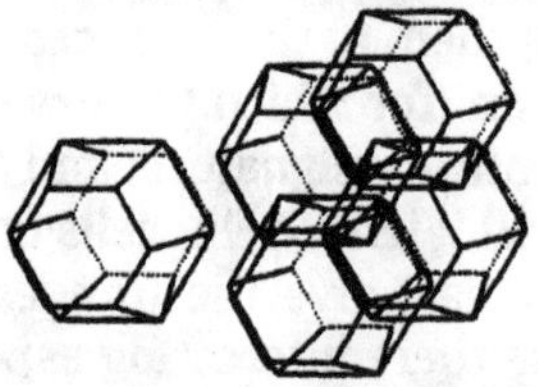

Figure IV.5.e-1: THE HEXAGONAL PRISM CANNOT BE DISTORTED.

It can be seen that they interpenetrate, so do not tile space. Changing the hexagonal faces of the prism to rectangular ones gives the cubic form, so no new cell.

A rectangular face (and obviously a hexagonal face) of the truncated octahedron cannot have the number of its sides increased, for that would give faces with more than six edges. If sides of a rectangle were broken in two there would be a new edge thus changing the rectangular face to two rectangular ones. This would also divide hexagonal faces into two, each with five edges, which is not a possible Wigner-Seitz cell.

Can some hexagonal faces of the truncated octahedron be changed to rectangular? Doing so for one face would make neighboring ones nonregular with sides of different lengths, even if it were also done to the opposite vertex of the neighboring side. Thus it is not possible. If two edges of a hexagonal surface are replaced by one, the neighboring rectangle would become five sided, so the object would no longer be a Wigner-Seitz cell.

And can some faces of the rhombic dodecahedron be made hexagonal? Breaking lines to increase the number of edges leads to new vertices and edges connecting them. Instead of making the surface hexagonal, this increases the number of surfaces, resulting in two rectangular ones; these have sides of different lengths, so the figure is not semiregular. Such a figure cannot tessellate space since a new edge would have to coincide with an old one, but they are of different sizes.

There are no other categories of space-filling cells (in three dimensions).

Problem IV.5.e–1: Give constructions for obtaining all (which means what?) Wigner-Seitz cells from the two prisms.

Problem IV.5.e–2: Generalize these arguments to other dimensions (starting with two), and spaces.

IV.5.f Obtaining three-dimensional cells from their cross-sections

There is another argument for these cells [Fejes Toth (1964), p. 115; Senechal (1990), p. 47], which we mention so the reader can work out the details. We wish to combine the allowed two-dimensional cells to give three-dimensional ones. These lie in parallel layers (sec. IV.3.g.iv, p. 207); how can sets of rectangles and hexagons be overlayed to give the required cells?

For two rectangles, one above the other, there are three possibilities:

R-I. The edges of one can be directly above those of the other.

R-II. The edges can be displaced by half the width of the rectangle so that the edge of one lies above the line through the center of the other parallel to that edge.

R-III. The vertex of one can lie above the center of the other, so displaced by both half the width and half the height.

The reason is that each lattice point, which the center of the cell is, must be in a neighborhood identical to every other lattice point. Except for these three cases, the vertices and center of a cell would be closer to one side of the underlying cell, than to the other side, which is not allowed for a lattice.

For the hexagon there are four cases:

H-I. For one, again, the edges of one hexagon overlay the corresponding ones of the other.

H-II. In the second, the vertex of one is over the center of the other, and their edges, when projected onto the same plane, make an angle of $\frac{2\pi}{6}$.

H-III. The third has the vertex of one over the center of an edge of the other, again with the projection of edges at an angle of $\frac{2\pi}{6}$.

H-IV. The last case has the vertex of one on the inside of the other.

While it looks like there are seven cases, there are actually only five as R-II and R-III are the same as H-I and H-III respectively.

Problem IV.5.f-1: Show that the categories of cells so obtained are the same as those found by the previous arguments. Which categories do each of these give?

Problem IV.5.f-2: How are these various arguments related?

Problem IV.5.f-3: Generalize all these arguments to other dimensions (starting with two).

IV.5.g Lattices as spaces and representations of their translation groups

The Wigner-Seitz cells determine the representations of the symmetry translation groups of the lattices, which are subgroups of the translation group of Euclidean space (showing that translation groups can have much structure). Thus we are determining a set of subgroups. There is another view of this. Our considerations are based on the (unconscious?) bias that space is Euclidean (so continuous). But lattices form spaces, though their points are discrete. And discrete spaces, as we see, have more structure (in a sense?), thus there are more of them (in each dimension). So here we are finding the representations of the translation groups of these various spaces, the lattices. Such groups are infinite discrete groups, therefore our work is broader than crystallography, being a topic in pure group theory — obtaining the representations of infinite discrete groups. The Wigner-Seitz cells give the representations of a class of these groups, and are interesting for that reason also (even more?). We leave open the question whether this method can be generalized to other groups (like these?)?

Problem IV.5.g-1: Do the Wigner-Seitz cells give a set of subgroups of the Euclidean translation group, or do they give all (if such is possible)?

Problem IV.5.g-2: Does this method give the representations of the translation groups of these discrete spaces, or the groups themselves, or is there no difference?

Problem IV.5.g-3: Can the method be generalized, or might it be that every infinite discrete group [Mirman (1995a), sec. II.3.j, p. 48] is a lattice in the proper dimension. Why? If not, can those groups that are lattices be specified?

Problem IV.5.g-4: While we have been treating lattices as geometrical objects they can be considered in other ways. The integers form a one-dimensional lattice, and complex numbers with integer real and imaginary parts a two-dimensional one. Why? This can easily be generalized to other dimensions. Thus all our results follow not from geometry, but from arithmetic. Or could it be that there is no difference? Also lattices are themselves groups (pb. II.2.a-3, p. 68; pb. II.7.f-17, p. 125). Thus our group-theoretical results follow not from geometry, or arithmetic, but from group theory. How are all these related? Why?

Problem IV.5.g-5: Questions have been raised above, and will be below, about how various results follow from the postulates of Euclidean geometry. Lattices are also spaces, and we can consider their geometries. Thus we can ask whether the properties and objects found (or introduced) for Euclidean space have analogs in such spaces? Are there symmetry point groups? Space groups? Is it possible to define a lattice in a space that is itself a lattice? What freedom do we have in these choices, analogous to that in Euclidean space for which there are several lattices, and different symmetry groups? Are there relationships between a lattice and symmetry groups defined over it (if such are possible), and the groups of the lattice regarded as an object in Euclidean space (which we have been considering)?

Problem IV.5.g-6: Find (an algorithm for obtaining) the Wigner-Seitz cells for all infinite discrete groups (that can be written as lattices).

Problem IV.5.g-7: A computer program that finds (draws?) all these cells for any dimension should be fruitful. In particular, it might be a good tool for studying group theory.

IV.6 BRILLOUIN ZONES FOR THE LATTICES

Physical applications of these groups use (generally the first) Brillouin zones, the smallest volume of reciprocal — momentum — space from which the representations are generated. These Wigner-Seitz cells, have been described, giving the representations of the translation groups. But lattices also have point groups. Symmetry of a lattice relates the statefunctions its points label. Symmetry, so the point group that the statefunctions are basis vectors of, depends on the point — the value of the momentum. For momentum zero, at the origin of the reciprocal space, the statefunction is a representation basis vector of the holohedry group of the lattice. But for other positions — other momentum values — point-group transformations change the momentum, thus are not symmetries; the statefunction belongs to the group consisting only of the identity. However there are positions — points, lines and planes — labeling basis vectors that are invariant under a subgroup of the holohedry group. These play a central role in the development of space-group representations, and so in physical applications.

Positions of the Brillouin zone specifying basis vectors that are not invariant under any (nontrivial) point group operation are general positions, special points (or lines or planes), generically special positions (sec. VII.2.e, p. 342), are those for which statefunctions (space-group basis vectors) are basis vectors of a (nontrivial) subgroup of the holohedry point group.

To (merely) illustrate this we describe the (first) Brillouin zones of some Bravais lattices [Bradley and Cracknell (1972), p. 95; Burns (1977), p. 313; Cornwell (1969), p. 82; Cornwell (1984), p. 211; Cracknell (1975), p. 80; Heine (1993), p. 268; Inui, Tanabe and Onodera (1990), p. 243; Jones (1975), p. 109; Joshi (1982), p. 289; Joshua (1991), p. 53; Koster (1957), p. 196; Lax (1974), p. 443; Wooster (1973), p. 226; Zak, *et al* (1969)]. We now know the geometrical objects that are the cells, and here we consider the general and special positions within each, and the groups, the holohedry group of the lattice, and the subgroups, for each position in the zone, that produce a set of distinct vectors from each basis vector.

Problem IV.6-1: Why for momentum zero, is the holohedry group the relevant one?

IV.6.a The cube and its relatives

The Brillouin zones of the objects with greatest symmetry are those for the cubic system, and we start with these (sec. II.6.b, p. 105, sec. IV.5.a, p. 216, sec. XI.2.h, p. 579).

IV.6.a.i *The Brillouin zone of the simple cube*

To illustrate the first Brillouin zone for this lattice we take the origin at the center, (always) labeled Γ, and the $\underline{k}$ axes parallel to the sides of the cube. The special points and lines — for which the point group, of transformations around that point or line, is larger than for a general point of the cube, are labeled. These points have $\underline{k}$ coordinates

$$\Gamma : \frac{\pi}{a}(0,0,0), X : \frac{\pi}{a}(0,1,0), M : \frac{\pi}{a}(1,1,0), R : \frac{\pi}{a}(1,1,1), \quad \text{(IV.6.a.i-1)}$$

and are

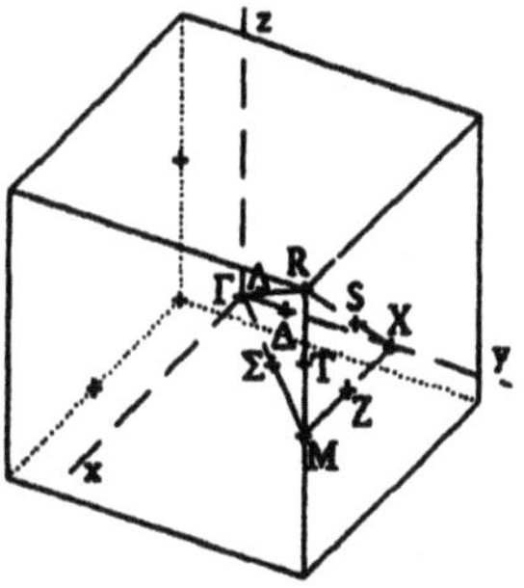

Figure IV.6.a.i-1: BRILLOUIN ZONE OF THE SIMPLE CUBIC LATTICE.

Problem IV.6.a.i-1: Check that the positions shown are special, that there no others, and if there are, put them in.

Problem IV.6.a.i-2: For each, and for the special lines, Λ, *S*, *T*, *Z*, find the point group that leaves basis vectors invariant, and the largest group giving a distinct set of vectors (how are these related?), and compare them to those for a general point and to the holohedry group [Cornwell (1984), p. 232; Jones (1975), p. 109]. Check that these are the special positions, and all. Some transformations give vectors of the set that are equivalent to others of the same set under the addition of a reciprocal lattice vector (eq. IV.2.a-7, p. 187). Are there such reciprocal lattice vectors that connect points on the boundary of this Brillouin zone?

IV.6.a.ii *The Brillouin zone of the body-centered cubic lattice*

The next lattice is the body-centered cubic (sec. IV.5.a.iii, p. 218), whose Brillouin zone is the Wigner-Seitz cell of a face-centered lattice, as in

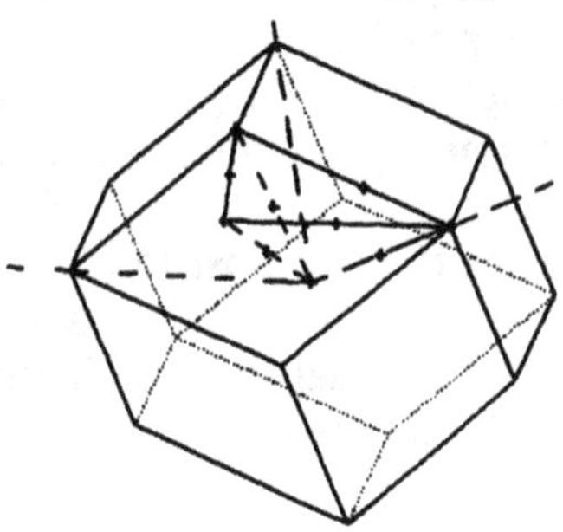

Figure IV.6.a.ii-1: BRILLOUIN ZONE OF BCC LATTICE,

and,

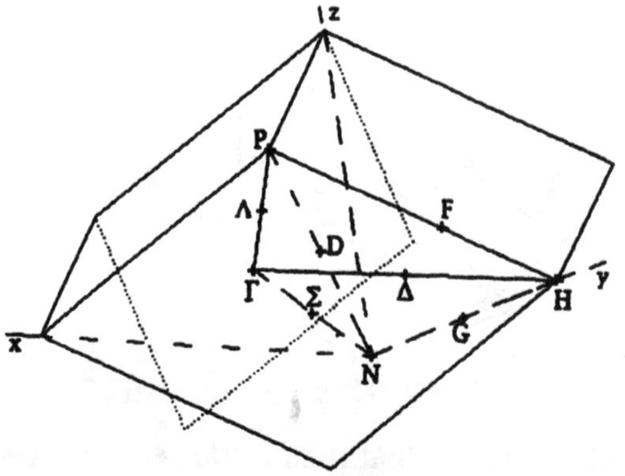

Figure IV.6.a.ii-2: BCC BRILLOUIN ZONE WITH LABELS,

where, for clarity only part of the Brillouin zone (the rear and top) is shown with the special and general points marked. These two diagrams can be related by matching the axes.

Problem IV.6.a.ii-1: The vectors for the first Brillouin zone of the reciprocal lattice are

$$\underline{b}_1 : \frac{2\pi}{a}(1,0,1), \underline{b}_2 : \frac{2\pi}{a}(0,1,-1), \underline{b}_3 : \frac{2\pi}{a}(1,-1,0), \qquad \text{(IV.6.a.ii-1)}$$

and a diagram will show that this is face-centered cubic. Check that these do reach the centers of the faces, and are a required complete set (all lattice points can be reached by integral sums of these).

Problem IV.6.a.ii-2: Check that the special points are (only?)

$$\Gamma : \frac{\pi}{a}(0,0,0), H : \frac{\pi}{a}(0,2,0), N : \frac{\pi}{a}(1,1,0), P : \frac{\pi}{a}(1,1,1). \quad \text{(IV.6.a.ii-2)}$$

What are the special lines and planes, if any? Draw a cube putting these, and the special lines, in. For each special position, find the symmetry point group, and compare it to that for a general point. Are there reciprocal lattice vectors that connect points on the boundary of the Brillouin zone?

Problem IV.6.a.ii-3: Check that the positions shown are special, and that there an no others, and if there are, put them in.

Problem IV.6.a.ii-4: Build a model and verify that the construction of the zone is correct. Check that the correct numbers of hexagons and squares are obtained.

IV.6.a.iii *The Brillouin zone of the face-centered cubic lattice*

Finally for this system is the face-centered cubic (sec. IV.5.a.ii, p. 216). The first Brillouin zone is a body-centered cubic lattice, so the reciprocal lattice vectors are

$$b_1 = \frac{2\pi}{a}(1,1,-1), b_2 = \frac{2\pi}{a}(-1,1,1), b_3 = \frac{2\pi}{a}(1,-1,1). \quad \text{(IV.6.a.iii-1)}$$

An example of this system is the diamond structure (sec. III.5, p. 159), which is that of many crystals, including of course, diamond. The Wigner-Seitz cell of the lattice [Birman (1974), p. 345; Ludwig and Falter (1988), p. 204], is a rhombic dodecahedron. The general and special points of the Brillouin zone, a truncated octahedron (for the reciprocal lattice), are

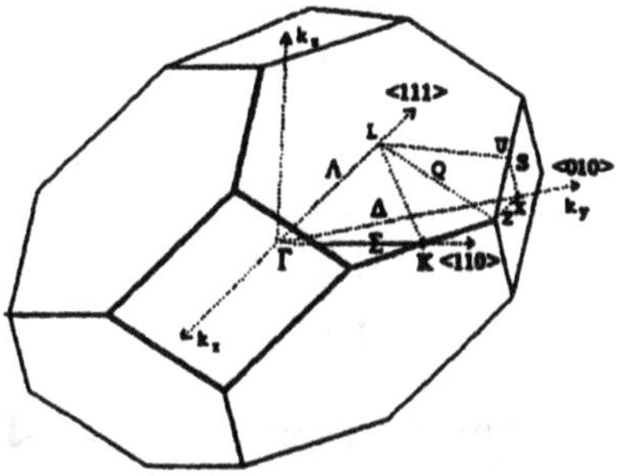

Figure IV.6.a.iii-1: BRILLOUIN ZONE OF THE FCC LATTICE.

Problem IV.6.a.iii-1: Check that these are the reciprocal lattice vectors.

Problem IV.6.a.iii-2: Draw the Wigner-Seitz cell putting in all special positions [Koster (1957), p. 197], checking the diagram (sec. VII.7, p. 384). Are there reciprocal lattice vectors that connect points on the boundary of the Brillouin zone?

IV.6.b The Brillouin zones of the tetragonal lattices

Tetragonal lattices are next [Inui, Tanabe and Onodera (1990), p. 245; Koster (1957), p. 199; Lax (1974), p. 449; Zak, *et al* (1969), p. 115]. The cell for the primitive lattice is

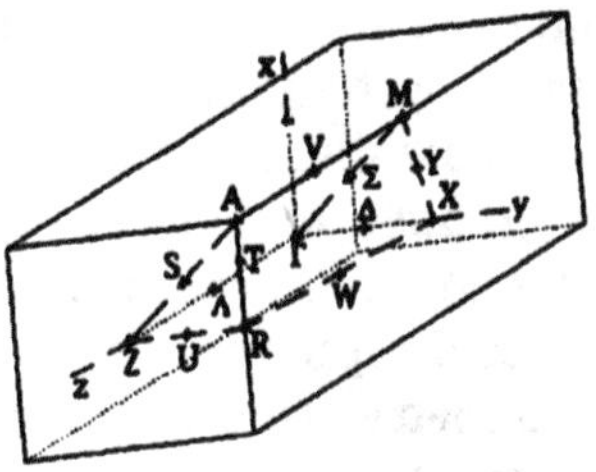

Figure IV.6.b-1: PRIMITIVE-TETRAGONAL BRILLOUIN ZONE.

For the centered lattice there are two Wigner-Seitz cells, these going with different ratios of the sides. The sets of special points of the two are not the same, emphasizing again that these lattices, classified as a single one, are different in essential ways.

Problem IV.6.b-1: In the next diagram there is a cell obtained by changing the length of the cube along the x axis, with the labels for the special points and lines left in place,

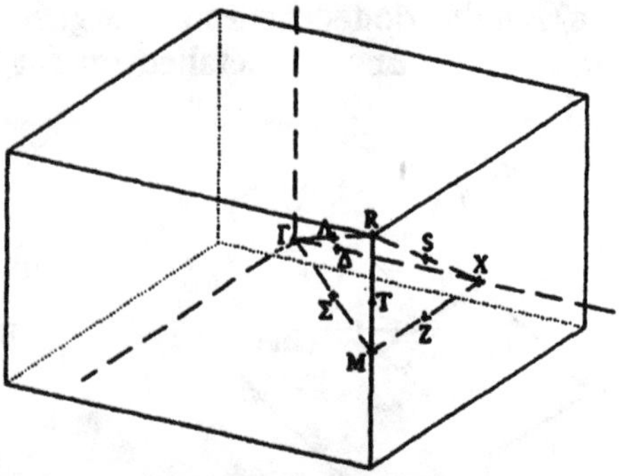

Figure IV.6.b-2: STRETCHING DOES NOT GIVE A PROPER CELL.

Which, if any, of these are still special? Again we see that the Brillouin zone for the tetragonal lattice is not simply a stretched version of that for the cubic (sec. IV.5.b.ii, p. 222). Explain why (any?) positions that correspond do so, why those that are no longer special lose that status, and why others gain it.

Problem IV.6.b-2: Give the groups of a general point and of each special point and line, finding all these. How do they compare to the ones for a cube? Are they related? Are there reciprocal lattice vectors that connect points on the boundary of the Brillouin zone? How do any such compare with the corresponding ones of the cube?

Problem IV.6.b-3: For the tetragonal-I lattice there are two categories of Wigner-Seitz cells, depending on the ratio $\frac{c}{a}$ (sec. IV.5.b.vi, p. 226). The special positions for the two cases do not completely correspond, differing for example in numbers, so these two lattices, although both classified as being in the same system, do differ, which is important (certainly) physically. Explain how these positions change at the value for which the Wigner-Seitz cell does, and either verify that there is no discontinuity, or explain why any that occurs is not unphysical. How do special positions appear or disappear? The symmetry groups for special positions (the site-symmetry groups) differ from that of a general point. Why should this group change if a general point becomes special, or a special becomes general? Do all pairs of corresponding special points and lines of the two cells have the same symmetry group, or could the symmetry change even though the point or line remains special?

Problem IV.6.b–4: One cell is

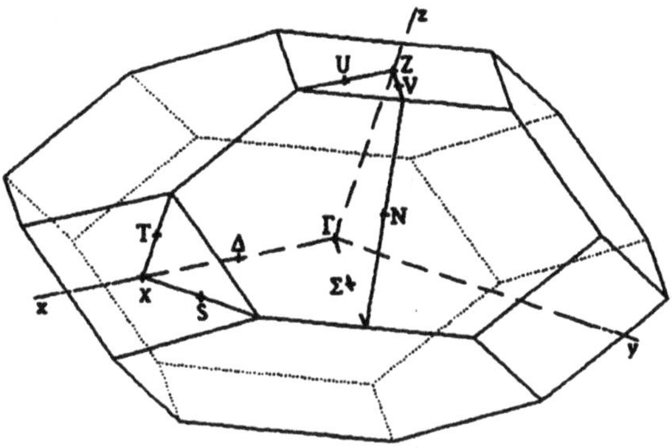

Figure IV.6.b–3: FIRST BRILLOUIN ZONE FOR THE TETRAGONAL-I,

which is that for $c < \sqrt{2}a$. The other, for $c > \sqrt{2}a$, looks like

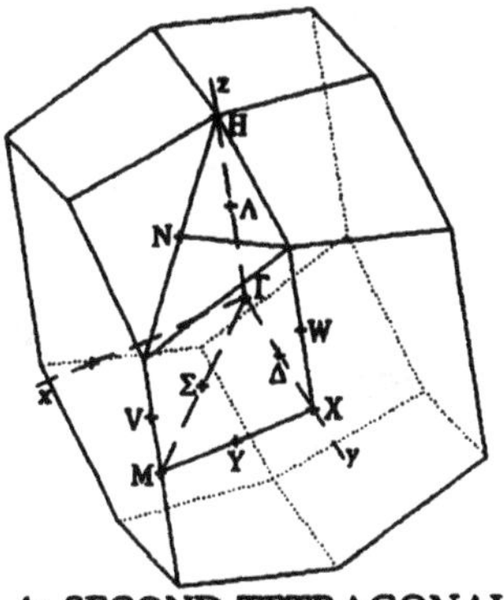

Figure IV.6.b–4: SECOND TETRAGONAL-I BRILLOUIN ZONE.

Check that the positions shown are special, that there an no others, and if there are, put them in. These diagrams are schematic and it is useful to redraw them so that they actually form correct Wigner-Seitz cells, putting in the special positions at the proper places. A question, which might be relevant physically, is what are the ranges of the ratios of the sides of these cells? Presumably they cannot be arbitrarily distorted and still give Wigner-Seitz cells (of this system). Or can they?

IV.6.c The other systems

These illustrate the essential ideas and since there are diagrams of (many of) the zones widely available [Bradley and Cracknell (1972),

p. 95; Koster (1957), p. 196; Lax (1974), p. 443; Zak, *et al* (1969), p. 39] it is unnecessary to give them here (the Brillouin zone for the hexagonal lattice is shown in sec. XI.2.j, p. 590). They provide useful exercises, in checking that the diagrams are correct, but also in finding the various ratios of sides (and values of angles) allowed for the cells of each system.

Brillouin zones have points, lines and planes that are taken into themselves by some operations of the point group, called special. Can nonprimitive operations change these, make general points special, or special general? The symmetry group of the Brillouin zone is the holohedry group of the lattice, thus is always symmorphic (sec. IV.2.c.i, p. 190). The nonprimitive elements of the symmetry group of the direct space (that of the crystal) can therefore have no effect on the points of the reciprocal space, thus of the Brillouin zone. It is determined by the lattice, not the crystal.

Problem IV.6.c-1: It is possible that not all zones (with their special positions) are available. Fill in the gaps.

Problem IV.6.c-2: There are cases (sec. IV.5.c.iv, p. 232) for which a category of a Wigner-Seitz cell of a system occurs more than once. In these cases (all?, none?) are the special positions the same (or at least do they correspond) for the different occurrences? Why?

Problem IV.6.c-3: While everyone knows (or believes they do) that these (given here or in the literature) are the special positions (perhaps because everyone copies from the same source), a rigorous proof could be useful. Also explain why these special positions are the correct ones.

Problem IV.6.c-4: As lattices are distorted in various ways, changing their Brillouin zones, or symmetry, the special positions change. For the different cases, explain how these distortions cause such positions to arise or disappear (suddenly?).

Problem IV.6.c.-5: Likewise, as lattices are changed, their sets of screws and glide planes also are. For the different cases, explain how these distortions cause screws and glide planes to appear or disappear (suddenly?). What happens to such an element if it vanishes? A computer-generated motion picture showing these would be interesting.

Chapter V

Representations: Point Groups and Projective

V.1 FORMULATING REPRESENTATIONS

Representations are usually the part of group theory relevant to applications, and to the development of representations. Those of space groups need representations of the translation groups, which we have just studied, and of the point groups. But there are also new aspects introduced by space-group representations, that are needed to find them, and are interesting — and valuable — in themselves. Here we collect some useful material, about the representations of point groups, their double groups, and about representations. These new concepts, projective representations and extensions, while needed for some representations of space groups, also are relevant to point and translation groups, to Lie groups and probably far beyond. Unfortunately we can only introduce these subjects, but this introduction is not merely a preliminary to the study of space groups representations, but to a topic that keeps recurring in many places.

V.2 REPRESENTATIONS OF THE CRYSTALLOGRAPHIC POINT GROUPS

Having knowledge of the structure of the point groups, we are able next to consider their representations [Bradley and Cracknell (1972), p. 56; Lomont (1961), p. 132; Lyubarskii (1960), p. 85]. We start with their characters, how they are found, how they are used.

V.2.a The characters of point groups

Perhaps the most useful constituent of the representations of point groups in applications are the character sets [Bhagavantam and Venkatarayudu (1951), p. 268; Burns (1977), p. 50, 379; Cornwell (1984), p. 323; Elliott and Dawber (1987), p. 265; Fässler and Stiefel (1992), p. 214; Heine (1993), p. 448; Koster (1957), p. 181; Wooster (1973), p. 145, 176]. Here we outline a way of finding them [Tsukerblatt (1994), p. 185; Wilson, Decius and Cross (1980), p. 312]; in appendix E, p. 655, these are listed for all crystallographic point groups.

The method for computing the characters is based on the representation matrices, but characters, being realization independent, do not depend on how they are found. The point-group transformations are defined on coordinates (those of atoms in a crystal or molecule, say), but representations give transformations of basis vectors. For each representation there are sets of functions of coordinates that transform as basis vectors of that representation (tbl. E.2-1, p. 657). Hence we can calculate the characters using these functions and the known action of transformations on coordinates, and then they are determined no matter how basis vectors are interpreted — and we use different interpretations in applications.

Some groups are simple enough so their characters can be found immediately; we start with these.

V.2.b Representations and characters of cylic groups

Cyclic groups, being Abelian, have only one-dimensional representations, their representation matrix elements and characters are thus identical. If element T is of order n, so

$$T^n = E, \quad \text{(V.2.b-1)}$$

the character to the n'th power equals 1, so is an n'th root of unity. These are complex numbers giving the representation matrix elements in the form

$$T = exp(\frac{2\pi i}{n}). \quad \text{(V.2.b-2)}$$

The element T^k (thus its character) is then

$$T^k = exp(\frac{2\pi ki}{n}). \quad \text{(V.2.b-3)}$$

As k runs from 1 to n this gives the n n'th roots and the representations for all elements in the cycle. This character, and matrix element, are complex (pb. VI.4.a-1, p. 297), necessary for the irreducible representation to be one-dimensional [Mirman (1995a), pb. V.5.a-4, p. 163].

V.2.c Dihedral group representations and characters

Dihedral group D_n (sec. I.7.a, p. 50) consists of a cyclic subgroup C_n of order n, and a subgroup of order 2 [Mirman (1995a), sec. II.4.c, p. 50; Wilson, Decius and Cross (1980), p. 316].

Problem V.2.c-1: Show that each element T_n^k of C_n and its inverse T_n^{n-k} form a class (sec. I.7.c, p. 57). Explain this geometrically using the interpretation of D_n as the group of n rotations of angle $\frac{2\pi}{n}$ plus n 2-fold rotations about axes in the plane perpendicular to the principle axis. Also the latter set are all in the same class for n odd, or in two equinumerous classes for n even. What is the geometrical significance of the two classes in the latter case? For a cyclic group all elements are in classes by themselves, but for a cyclic subgroup this is not true. Why? Count the number of classes (pb. I.4.b-1, p. 9); it should equal [Hamermesh (1962), p. 43]

$$N = \frac{n+3}{2} \text{ and } \frac{n+6}{2}, \tag{V.2.c-1}$$

for the odd and even cases, respectively. The order of the group is 2n. Since the cyclic elements all belong to the same class as their inverses (the classes are ambivalent), the characters equal their complex conjugates so are real, thus can be only 1 and -1. For odd n these characters are all 1, for even n some can be -1.

Problem V.2.c-2: The number of one-dimensional representations is limited here to a mere 2 when n is odd, one being the completely symmetric representation, for the other the character of one member of the n-th order cyclic subgroup equals -1. Why is there such a limitation? Does it matter which element is taken to have a negative character? Having chosen this element what are the characters of the others of this subgroup? If n is even there are four one-dimensional representations. What are the characters of the elements of the n-th order cyclic subgroup for these four? What are the dimensions of the other representations? It turns out that these are also limited; all others have dimension 2. Why?

Problem V.2.c-3: Since the number of (equivalence classes of) representations equals the number of classes [Mirman (1995a), sec. VI.4.a, p. 176], if there are l_d representations of dimension d presumably

$$\sum l_d = \frac{n+3}{2}, \quad n \text{ odd}, \tag{V.2.c-2}$$

$$\sum l_d = \frac{n+6}{2}, \quad n \text{ even}. \tag{V.2.c-3}$$

Also the sum of the squares of the dimension equals the order [Mirman (1995a), pb. VI.2.b-1, p. 175],

$$\sum d_i^2 = 2n = g, \tag{V.2.c-4}$$

the order of the group (with what ranges for the sums?). From this show that

$$l_d = 0, \quad d > 2. \tag{V.2.c-5}$$

Also

$$l_2 = \frac{n-1}{2}, \quad \text{odd}, \tag{V.2.c-6}$$

$$l_2 = \frac{n-2}{2}, \quad \text{even}. \tag{V.2.c-7}$$

The transformations of the dihedral group are interpreted here as rotations, and thus the matrix elements can be taken real, the representation matrices are then not only unitary, but orthogonal. These are thus (?) all of the form

$$U = \begin{pmatrix} cos\phi & -sin\phi \\ \pm sin\phi & \pm cos\phi \end{pmatrix} \tag{V.2.c-8}$$

so the characters of the two-dimensional representations are (why?)

$$\eta_2 = 0 \text{ or } 2cos\phi. \tag{V.2.c-9}$$

Problem V.2.c-4: For the two-dimensional representations, the characters of the two-fold rotations are zero and those for the principle cyclic subgroup are

$$\eta_n^k = exp(\frac{2\pi ik}{n}). \tag{V.2.c-10}$$

The characters that give distinct representation should be

$$\eta_e = 1, 2, \ldots, \frac{n}{2} - 1, \quad \text{for n even}, \tag{V.2.c-11}$$

and

$$\eta_o = 1, 2, \ldots, \frac{n-1}{2}, \quad \text{for n odd}. \tag{V.2.c-12}$$

Problem V.2.c-5: Work out the representation matrices, and characters, for these two-dimensional representations.

V.2.d Computation of characters

Finding the characters of the other point groups is more complicated. So we consider general methods, based on the action of transformations on coordinates. The character of representation matrix T is the sum of its diagonal elements. What does a transformation of coordinates do? It takes the coordinate to itself, its negative, or mixes it with other coordinates. For the latter there are two possibilities: a rotation takes a coordinate, x^α, into a sum of itself plus others of the same particle (more generally point) or an interchange of these replaces x^α with a coordinate of a different point, x^β. The diagonal elements are

$$x^\alpha \Rightarrow x^\alpha, \quad T_{\alpha\alpha} = 1, \qquad \text{(V.2.d-1)}$$

$$x^\alpha \Rightarrow -x^\alpha, \quad T_{\alpha\alpha} = -1, \qquad \text{(V.2.d-2)}$$

$$x^\alpha \Rightarrow x^\alpha cos\theta + y^\alpha sin\theta, \quad T_{\alpha\alpha} = cos\theta, \qquad \text{(V.2.d-3)}$$

$$x^\alpha \Rightarrow x^\beta, \quad T_{\alpha\alpha} = 0; \qquad \text{(V.2.d-4)}$$

so for a particle left fixed the diagonal element is ± 1 or $cos\theta$, while ones interchanged contribute 0 to the character.

An inversion changes the sign of the three coordinates adding -3 to the character; a reflection changes the sign of one coordinate, the other two are unchanged, so adds 1. A rotation through θ leaves one coordinate fixed, and multiplies the other two by $cos\theta$; it gives a sum of diagonal elements,

$$Sum = 1 + 2cos\theta. \qquad \text{(V.2.d-5)}$$

Thus the contribution to the character of $\mathbf{C_n}^k$ for each rotated particle is

$$\eta_k = 1 + 2cos\frac{2\pi k}{n}. \qquad \text{(V.2.d-6)}$$

A rotary-reflection, $\mathbf{S_n}^k$, a rotation followed or preceded by a reflection, adds

$$\eta_k' = -1 + 2cos\frac{2\pi k}{n}. \qquad \text{(V.2.d-7)}$$

The character of the identity transformation is the number of (linear combinations of) coordinates forming basis vectors. The expression for the rotations immediately gives the characters for all cyclic groups.

These contributions (in three dimensions) to character η_R of transformation R, for each atom not interchanged with another, are then [Wilson, Decius and Cross (1980), p. 105]

Proper Elements	η_R	Improper Elements	η_R
$E = C_1{}^k$	3	$\sigma = S_1{}^1$	1
$C_n{}^k$	$1 + 2cos(\frac{2\pi k}{n})$	$S_n{}^k$	$-1 + 2cos(\frac{2\pi k}{n})$
$C_2{}^1$	-1	$I = S_2{}^1$	-3
$C_3{}^1, C_3{}^2$	0	$S_3{}^1, S_3{}^5$	-2
$C_4{}^1, C_4{}^3$	1	$S_4{}^1, S_4{}^3$	-1
$C_6{}^1, C_6{}^5$	2	$S_6{}^1, S_6{}^5$	0

Table V.2.d-1: CONTRIBUTIONS TO CHARACTERS.

Problem V.2.d-1: For dihedral groups D_n [Ledermann (1987), p. 65, 209; Wilson, Decius and Cross (1980), p. 316], check from this table that the two-fold axes are all in the same class, for n odd, for n even they are in two classes each with $\frac{n}{2}$ members (and each rotation about the principal axis is in the same class as its inverse), say by applying the transformations to a physical model. Why are there two classes? For one-dimensional representations the characters of C_n follow from the table, and since each is in the same class as its inverse, all are real, so are ± 1 for n even but only 1 for n odd. Explain why this shows that there are only two one-dimensional representations for n odd, and four for n even. What are their characters? All other representations are two-dimensional. Also check that for the set of twofold rotations about the axes perpendicular to the principle axis, the characters are 0 as seen using the character orthogonality relations [Mirman (1995a), sec. VII.5.c, p. 195]. This gives the characters for the dihedral groups, and it should agree with the results of the previous section.

Problem V.2.d-2: Find the relationship between the characters of a group that is a direct product of groups, and those of its subgroups [Wilson, Decius and Cross (1980), p. 320]. Check that this is correct for those direct-product groups whose characters we have found.

Problem V.2.d-3: It should now be straightforward to calculate the character tables for all groups formed from the cyclic and dihedral groups with the adjunction of the inversion and reflections (sec. I.6.c, p. 39).

Problem V.2.d-4: It is interesting that for some of these groups there are two representations of the same dimension with identical character sets, except (essentially) shuffled. For one, C_n is realized (in part — why?) as

$$C_n \sim exp(\frac{2\pi i}{n}), \qquad \text{(V.2.d-8)}$$

while $C_n{}^2$ is realized as

$$C_n{}^2 \sim exp(\frac{4\pi i}{n}); \qquad (V.2.d-9)$$

for the other, these two are interchanged. Why are there two — only, why do these two, only, give inequivalent representations — and how does the representation decide which to pick? Check that (for these groups) C_n and $C_n{}^2$ can have either realization, are in different classes, and that further powers of the transformations do not give additional classes, so no further arbitrariness. Then show that the character sets both satisfy all requirements (including orthogonality) on them, in particular they are mutually orthogonal, so go with nonequivalent representations. Either representation can be given subscript 1, the other then gets 2. Is there a reason why one choice is more natural?

Problem V.2.d-5: Repeat this discussion for the tetrahedral and octahedral groups (sec. I.6.c.iii, p. 41).

Problem V.2.d-6: Octahedral group O_h (pb. I.6.c.iii-2, p. 42), of order 48, has four one-dimensional, two two-dimensional, and four three-dimensional representations. This can easily be checked (or derived) from these formulas for the characters.

Problem V.2.d-7: Next collect the elements into classes and use class multiplication [Mirman (1995a), sec. IV.5.f, p. 126]. The product of all elements of one class with all of another is a set of elements; these belong to one, or several classes, such that if an element appears, so do all members of its class (why?). Thus with C_l the set of elements of class l,

$$C_k C_l = \sum a_{klm} C_m; \qquad (V.2.d-10)$$

the a's are the class multiplication coefficients. Then [Lomont (1961), p. 61; Wooster (1973), p. 180] using the properties of the characters, η_l, that we now know, and defining x_l by

$$x_l \eta_E = h_l \eta_l, \qquad (V.2.d-11)$$

where η_E is the character of the identity, and h_l the number of elements in class l, verify that [Hamermesh (1962), p. 109]

$$x_k x_l = \sum a_{klm} x_m, \qquad (V.2.d-12)$$

or

$$\sum_m (a_{klm} - x_l \delta_{km}) x_m = 0. \qquad (V.2.d-13)$$

Do these have to be modified if the characters are complex? This set of equations can be solved to find the x's, and from them the

η's. Use of orthogonality relations and knowledge of the characters of the smaller representations simplifies the procedure. Moreover for the three-dimensional representation, the basis vectors are the coordinates of (say) vertices. The effect of the transformations on these can be simply found (only one element of each class is needed) giving the matrix, so the character. It is seen that many of these characters are zero (or have other simple values) and the reasons for these should be clear (and explained). Understanding these reasons gives the characters immediately. Also they serve as a check on the correctness of the procedure and results. These results can (and should) additionally be checked using the orthogonality relations, and the results of all other derivations of them. How are these derivations related? Which are preferable? Which provide more insight? Are clearer? Are easier? Are better for checking?

Problem V.2.d–8: The characters for the subgroups of O_h follow from those of it, though some transformations do not appear in the subgroups, and representations that are irreducible for the full group need not be for a subgroup.

Problem V.2.d–9: Repeat the derivation of the characters for the icosahedral group, and its subgroups (sec. I.6.b.i, p. 35).

V.2.e The table of characters

In tbl. E.2–1, p. 657, are listed for each crystallographic point group all representations, using both notations for them, and the characters for each generator. On the right are functions of the coordinates, and functions of axial vectors (which can be taken to be angular momenta, say realized as products of coordinates and derivatives with respect to them), that are basis functions for each representation. It is these that are used to compute the characters.

In the character tables two-dimensional representations are listed as two one-dimensional ones marked by braces (so the character of the identity is given as 1 twice rather than 2 once). This emphasizes that representations irreducible over real numbers can be reducible over complex ones (note that both characters and basis vectors are complex). The proper linear combinations giving the real forms can be found from the basis vectors.

Problem V.2.e–1: Try this for a few cases.

Problem V.2.e–2: The tables were copied from the sources given, and these did not state where they copied them from. Hence there is always a possibility that errors have crept in, and it would be wise to check these values. The number of representations, and their dimensions, can be obtained (almost) uniquely from the decomposition of the regular representation — the set of group transformations [Mirman (1995a), chap. VI, p. 170]. Thus it can be verified that all representations have been given, and since the character sets are different (which must be checked — why?), each only once. And it should be easy to check that these are (likely) representations, by direct multiplication (stating first how the character of a product is related to the characters of the terms in the product). Also the values of the characters should be verifiable by the methods above, as should the orthogonality relations. Taking models of objects with a symmetry group, and understanding the effect of the group transformations on these, allows reasonable guesses about transformation matrices and characters. This provides another useful check. What is the most rigorous way of checking? The simplest? Why? Do the characters of the (irreducible) representations have geometrical significance (in terms of these models)?

V.2.f Representations of the tetrahedral and octahedral groups

The other two crystallographic point groups containing only rotations are the tetrahedral and octahedral groups (sec. I.6.b.i, p. 35). The latter is symmetric group, S_4, the former its subgroup, A_4. The representations of the symmetric groups have been given (in principle) [Mirman (1995a), chap. VIII, p. 214], so those of these groups are known. The largest representations of the former group are three dimensional, and it remains to relate these representations to the transformations on crystals.

Problem V.2.f-1: The representations of alternating groups are those of symmetric groups, except that they can be reducible. The one-dimensional symmetric representation must remain that for A_4, the antisymmetric representation necessarily (why?) breaks up into two equivalent realizations of the symmetric representation. Explain why they are equivalent, but different realizations. We can see from tbl. E.2-1, p. 657, that it has three one-dimensional representations, the completely symmetric one, and the two obtained from the antisymmetric representation of the symmetric group, which are complex conjugates of each other. Why complex conjugate? How do they differ? Can there be representations over the reals? There is also a three-dimensional representation, the one (that has an interpretation of) acting on coordinates. It is obtained from the two three-dimensional representations of the symmetric group S_4 [Mirman (1995a), pb. VIII.4.c.iii-4, p. 231], which become equivalent (when the odd permutations are dropped), as can be seen from their representation matrices. Why? Are there other realizations (that are, in some way (?), equivalent)?

V.2.g Adjunction of reflections and the inversion

The other point groups are found from these by adjunction of reflections and the inversion. These are all direct products, with the pure rotation group a factor group of index 2 [Mirman (1995a), sec. IV.6.b, p. 131], as is the two-parameter group of the identity and the inversion (or a reflection). The representations thus are products of those of the pure rotation group and of the improper transformation, these latter all being one dimensional and with matrix elements (and characters) 1 and -1. The number of representations is doubled, the matrices of the proper transformations (the rotations) being the same in each member of each pair, but with the inversion represented by a diagonal matrix with entries 1, for one member of the pair, and -1 for the other member. The matrices of the other improper transformations are given by those of the relevant products.

Problem V.2.g-1: Find the representation matrices and characters for all improper cyclic and dihedral groups. Are these irreducible?

Problem V.2.g-2: What is the effect of adjoining the inversion to the tetrahedral group? Give the representation matrices and characters. There are two extended groups T_d and T_h. How do their representations differ?

Problem V.2.g-3: Repeat for the octahedral group.

Problem V.2.g-4: Find the representations of both the proper and the improper icosahedral groups (sec. I.6.b.i, p. 35; pb. I.6.c.iii-3, p. 42). How many of the latter are there?

V.3 REPRESENTATIONS OF DOUBLE GROUPS

The representations of double groups (sec. I.8, p. 59) include those of the single group, plus additional ones [Bradley and Cracknell (1972), p. 418; Hamermesh (1962), p. 357; Heine (1993), p. 137; Jansen and Boon (1967), p. 273; Koster (1957), p. 238], as we might guess from the representations of SU(2), which are the representations of O(3), plus another set, equal in number [Mirman (1995a), sec. XI.5, p. 329], and also from the classes of the double groups being more numerous than of the single groups. Here we consider the representations of double point groups.

Character tables for double groups can be constructed in a way similar to those for single groups [Burns (1977), p. 180, 417].

The double group has more representations than the single group. For these matrix F is representated by the negative of the unit matrix. For improper double groups, we use the representations of the proper double group to find those of the improper group by taking direct products, as usual.

Problem V.3-1: Of course, character tables in the references should be checked carefully.

Problem V.3-2: Construct character tables for the double groups, say by finding general algorithms for cyclic and dihedral groups, and then the other point groups. Innovative methods of finding characters would be educational, and perhaps useful in other ways.

V.3.a The representations of the double cyclic groups

The double cyclic groups $C_n{}^*$ provide the simplest examples of how to find representations of double groups, and their relationship to those of the (single) group. For $C_n{}^*$ the elements are

$$C_n, C_n{}^2, \ldots, C_n{}^n = F, C_n{}^2F, \ldots, C_n{}^{2n} = F^2 = E. \qquad \text{(V.3.a-1)}$$

Thus a generator of cycle length n in a cyclic group, has cycle length $2n$ in the corresponding double group. After n applications it returns the system to its original configuration and the statefunction to the negative of the original value. The group is still Abelian and now has $2n$ (one-dimensional) representations. The basis vectors are

$$1, e^{\frac{1}{2}\frac{2\pi i}{n}}, e^{\frac{2\pi i}{n}}, e^{\frac{3}{2}\frac{2\pi i}{n}}, \ldots, e^{i(n-\frac{1}{2})\frac{2\pi i}{n}}. \qquad \text{(V.3.a-2)}$$

Those with integer coefficients of $\frac{2\pi i}{n}$ are representations of C_n, while those with half-integer coefficients are the new representations that appear when we go to the double group:

$$C_n{}^n e^{im\frac{2\pi i}{n}} = \pm e^{im\frac{2\pi i}{n}}; \qquad \text{(V.3.a-3)}$$

the plus sign is for m an integer, the minus for m a half (odd) integer.

Problem V.3.a-1: Clearly the representations of $C_2{}^*$ are

E	F	C_2	C_2F
1	1	1	1
1	1	−1	−1
1	−1	i	$-i$
1	−1	$-i$	i

Table V.3.a-1: REPRESENTATIONS OF $C_2{}^*$,

where the first two rows are the single-valued representations (the faithful representations of C_2), the last two are the double-valued representations (the unfaithful representations of C_2, but the faithful ones of $C_2{}^*$). For the single-valued representations there is no distinction between corresponding operators (E, and F, and similarly for the others), while for the double-valued representations these have opposite signs, thus the basis vectors and the eigenvalues are doubled. The number of representations, like the number of operators, is doubled. Note that for all four representations, the orthogonality relations are satisfied (what orthogonality relations?). The last two are not orthogonal, why?

Problem V.3.a-2: Verify that the characters of FS and S (for all S) have the same sign for integer representations, opposite signs for half-integer representations. Thus the characters, treated as vectors, of the two-valued representations are orthogonal to those of the one-valued representations. Should we expect this?

V.3.b Representations of $D_2{}^*$

If a double group is non-Abelian how are representations affected? This serves as an example [Hamermesh (1962), p. 359].

Problem V.3.b-1: Show that the eight elements of $D_2{}^*$ are in five classes, E; F; C_x, C_xF; C_y, C_yF; C_z, C_zF (pb. V.2.d-1, p. 252), so there are five representations; four are one-dimensional, coming from the Abelian D_2 group. The sum of the squares of the dimensions equals the order of the group [Mirman (1995a), pb. VI.2.b-1, p. 175],

$$\sum d_i^2 = 8, \qquad \text{(V.3.b-1)}$$

so the remaining representation has dimension 2. Verify that the characters are

E;	F;	$C_x,$	C_xF;	$C_y,$	C_yF;	$C_z,$	C_zF
1	1	1	1	1	1	1	1
1	1	−1	−1	−1	−1	1	1
1	1	1	1	−1	−1	−1	−1
1	1	−1	−1	1	1	−1	−1
2	−2	0	0	0	0	0	0

Table V.3.b-1: CHARACTERS OF $D_2{}^*$,

with the first four also the representation matrices of the single-valued representations (those of D_2), and the last new. The explicit matrices for the last representation are given, as the realization (pb. I.8.b-3,

p. 62), and the basis functions are $exp(\pm\frac{i\phi}{2})$. What is ϕ? That the double group is non-Abelian thus introduces a higher-dimensional representation. Its physical meaning will have to be considered, and we shall have to see why there is a two-valued representation when the underlying group has only single-valued ones.

Problem V.3.b-2: Verify that for each S, the characters of FS and S have the same sign for integer representations, opposite signs for half-integer ones. Again the characters, treated as vectors, of the two-valued representations are orthogonal to those of the one-valued representations. Should we expect this? Is it true in general?

Problem V.3.b-3: Is there a generic way of getting these characters, or is it just a matter of trying to guess the representation matrices?

V.3.c Representations of other double groups

With these results we can now find the representations of other double groups [Hamermesh (1962), p. 357]. There are not many.

Problem V.3.c-1: Show that $D_3{}^*$ has 12 elements in six classes, and that the dimensions of the double-valued representations are 1,1 and 2. What is the physical significance of double-valued one-dimensional representations (for which E and F are different)?

Problem V.3.c-2: Find the classes and representations for $D_4{}^*$ [Bradley and Cracknell (1972), p. 421; Hamermesh (1962), p. 361].

Problem V.3.c-3: Find the classes and representations for $D_6{}^*$.

Problem V.3.c-4: Repeat for T*.

Problem V.3.c-5: Repeat for O*.

Problem V.3.c-6: These groups lack an inversion. How does that affect the representations? Adjoin the inversion to all these point groups, and find the representations.

Problem V.3.c-7: Repeat this for the improper double point groups that do not have an inversion (sec. I.7.b, p. 55).

V.4 PROJECTIVE REPRESENTATIONS

A representation consists of a set of matrices (and the vectors on which they act) that obey the same product rules as the group operators — by definition. This requirement is useful, often necessary, for studying the group and the physics it describes. But it is imposed by us, not necessarily by groups or by physics. Might the condition be too restrictive? Could there be sets of matrices with product rules that are "almost" the same as those of the group operators, and which are also useful, perhaps necessary, in studying groups and obtaining information about systems to which they are applicable?

The physical objects of quantum mechanics, probabilities for example, are given by absolute values or by matrix elements of operators between statefunctions and conjugates of other statefunctions suggesting that if representation-matrix products differed by phases from those of the group, they might still be physically relevant. There is thus some freedom in the concepts of representations and the entities of which it is composed. This is fortunate, for the freedom is needed — or, perhaps, the latitude exists because it is needed.

We thus turn to projective (collineatory, multiplier, ray, weighted) representations [Altmann (1977), p. 71; Barut and Raczka (1986), p. 400; Bradley and Cracknell (1972), p. 156; Chen (1989), p. 418; Cornwell (1984), p. 155; Hamermesh, p. 458; Inui, Tanabe and Onodera (1990), p. 88; Janssen (1973), p. 37, 257; Lax (1974), p. 245; Lomont (1961), p. 227; Ludwig and Falter (1988), p. 208; Lyubarskii (1960), p. 95; Rotman (1995), p. 201; Sternberg (1994), p. 157; Weyl (1931), p. 180].

V.4.a Definition of projective representation

A projective representation of a group with the product

$$A_iA_j = A_k \tag{V.4.a-1}$$

is one whose matrices obey

$$R(A_i)R(A_j) = \omega(i,j;k)R(A_k), \tag{V.4.a-2}$$

where the ω's, functions of i and j, are constants (that is they commute with all R's). The set of ω's is a factor system. If all ω's are 1, we have an ordinary (vector) representation. If not, these matrices belong to a projective (ray) representation. For some groups — but not all — it is possible to redefine the representation matrices to make every projective representation into a vector one. Here we emphasize those groups, and their representations, for which this redefinition is not possible.

Problem V.4.a-1: Iterating the product gives

$$R(A_i)R(A_j)R(A_k) = \omega(j,k;m)R(A_i)R(A_m)$$

$$= \omega(j,k;m)\omega(i,m;n)R(A_n). \tag{V.4.a-3}$$

The ω's are not completely arbitrary. Verify that, because group multiplication is associative,

$$R(A_i)(R(A_j)R(A_k)) = (R(A_i)R(A_j))R(A_k), \tag{V.4.a-4}$$

the factors obey

$$\omega(j,k;m)\omega(i,m;n) = \omega(i,j;l)\omega(l,k;n), \qquad \text{(V.4.a-5)}$$

where l and m depend on, and are determined by, i, j, k. Notice that there are more equations than ω's (how many?). Thus not all product laws need allow solutions, that is projective representations. Fortunately (of course?) there is always one solution, that for which all ω's equal 1. Clearly if a representation is projective, the ω's obey this equation. Show the converse: if there is a set of ω's satisfying it, then the ω's give a projective representation [Altmann (1977), p. 72; Hamermesh (1962), p. 460]. This assumes (does it not?) that we have a vector representation and assign the ω's to make it a ray representation. How does this enter the proof?

V.4.b Factor systems

A set of ω's for a projective representation is a factor system of that representation. We need not consider every possible factor system [Kurosh (1960b), p. 121]. Using projective representation R, we define a new one with factor system υ, with the matrices related by

$$R'(i) = C_i R(i), \qquad \text{(V.4.b-1)}$$

(where obviously none of the C's can be zero, luckily) so

$$\upsilon(i,j;k) = \frac{C_i C_j}{C_k}\omega(i,j;k). \qquad \text{(V.4.b-2)}$$

Factor systems, ω and υ, so related, are said to belong to the same class. It is only necessary to study one system from each class. This transformation is sometimes known as a gauge transformation [Altmann (1977), p. 74; Birman (1974), p. 70], but it is important to remember that it is completely distinct from gauge transformations of electromagnetism and the equivalent transformations for gravitation which are (necessary) properties of massless representations (only) of the Poincaré group [Kupersztych (1976); Mirman (1995c), sec.3.4, p. 43].

Problem V.4.b-1: Verify that the $\upsilon(i,j;k)$ are solutions to the associativity condition if the $\omega(i,j;k)$ are, with the C_i any (nonzero) constants. Thus we can multiply the matrices of a projective representation by arbitrary constants and the resultant matrices still form a projective representation, something very different than for vector representations. The latter are quite restricted, but by allowing the factor in the product rule, we have introduced much leeway in the choice of the representation matrices.

Problem V.4.b-2: Every projective representation is equivalent to a unitary one, with factor system in the same class [Bradley and Cracknell (1972), p. 158].

V.4.b.i *The multiplicator*

The number, m, of classes of factor systems of a finite group G is finite, and the classes can be put in one-to-one correspondence with the elements of an Abelian group M of order m, called the multiplicator (or multiplier) of the finite group. The set, eq. V.4.a-5, p. 261, has an infinite number of solutions, despite there being more equations than variables. However the number of factor systems is finite, thus there are transformations of the factors leaving each system invariant, implying a group. Also the factor systems can be multiplied, the multiplication is commutative, so they form an Abelian group.

Problem V.4.b.i-1: We multiply factors from two classes, denoted by α, β, to get the set γ,

$$\omega(i,j;k)^{\gamma} = \omega(i,j;k)^{\alpha}\omega(i,j;k)^{\beta}; \qquad \text{(V.4.b.i-1)}$$

the factors go with the same group elements. Show that this set of numbers also satisfies the associativity condition, so is a factor system, and that the class to which ω^{γ} belongs is only determined by the classes to which ω^{α} and ω^{β} belong, not by the choices of factor systems in the classes (or group elements). Thus if we label the classes by K_i (with K_o containing the factor system of the vector representation,

$$\omega = 1), \qquad \text{(V.4.b.i-2)}$$

the product of the K's is defined. And they form an Abelian group. Why? Check that it is actually a group (with elements the classes of factor systems), that its order is the number of classes, and give (as much information as possible about) its multiplication table. This is the multiplicator group of group G. We might expect that this multiplicator group is closely related to the structure of G, and knowing it essentially determines the factor systems (especially since they are limited to roots of unity — why?).

Problem V.4.b.i-2: Check that for a group of order g, in any representation there are, of course, g^2 factors, but the associativity condition gives g^3 equations.

V.4.b.ii *The Abelian group formed by the factor classes*

The Abelian group does more than label the classes, it gives the product rules of the classes in the following sense: if the elements of the Abelian group M obey

$$A_r = A_p A_q, \tag{V.4.b.ii-1}$$

and we assign to elements p and q of M the factors ω_p and ω_q, then there is a factor ω_r assigned to r, such that for all i and j,

$$\omega_r(i,j;k) = \omega_p(i,j;k)\omega_q(i,j;k). \tag{V.4.b.ii-2}$$

The order of every element of M is a factor of the order of G. If M_p has order α then a factor system can be chosen from the class corresponding to M_p whose values are all α_pth roots of unity, this called a normalized factor system [Altmann (1977), p. 78]. The factor system is always taken to be that one for which the product of the identity with itself equals the identity, without any multiplicative factor (a standard factor system [Altmann (1977), p. 78]) — we do not consider whether other choices are meaningful or useful.

Since an equivalence transformation can be found to make the projective representation unitary, every projective representation is equivalent to a unitary one with a factor system in the same class. The identity element of M corresponds to the class containing a factor system whose values are all unity. Thus the ordinary representations of G are the irreducible unitary projective representations of the class corresponding to the identity of M.

Problem V.4.b.ii-1: Show that the number of these classes is not greater than g^n, where [Hamermesh (1962), p. 462]

$$n = g^2. \tag{V.4.b.ii-3}$$

Write this bound for all g's from 1 to 10.

V.4.b.iii *All factors give projective representations*

Factors ω obey the associativity condition (eq. V.4.a-5, p. 261). Suppose we have a set of constants that obey these equations. Do they form an acceptable system — is there a projective representation with these as a factor system [Hamermesh (1962), p. 460]? In fact, for every such system, there is a projective representation; the converse is true by definition.

To construct such a representation for factor system ω of group G of order g, define matrix X with (group-operator) entries

$$X_{ij} = \omega(k,i;j)A_k, \tag{V.4.b.iii-1}$$

where

$$A_i = A_j A_k, \quad A_k = A_j^{-1}A_i, \tag{V.4.b.iii-2}$$

for all elements of G, these labeling the rows and columns of X. Then

$$X = \sum D(A_t)A_t, \tag{V.4.b.iii-3}$$

summing over all group elements, where $D(A_i)$ is obtained using the independence of the group elements, and defined by

$$X_{ij} = \sum D(A_t)_{ij}A_t = \omega(k,i;j)A_k, \tag{V.4.b.iii-4}$$

so

$$D(A_k)_{ij} = \omega(k,i;j), \tag{V.4.b.iii-5}$$

giving the matrix going with, not — at least at this point — representing, operator A_k. It contains only one nonzero element in each row and each column thus, happily, is nonsingular. To show that matrices $D(A_i)$ form a representation, of course (allowed to be) reducible, with factor system ω, we need their products,

$$D(A_j)_{km}D(A_i)_{mn} = \omega(j,k;m)\omega(i,m;n) = \omega(i,j;l)\omega(l,k;n)$$
$$= \omega(i,j;l)D(A_l)_{kn}, \tag{V.4.b.iii-6}$$

using the associativity condition, and the product

$$A_jA_i = A_l. \tag{V.4.b.iii-7}$$

So, as these matrices are nonsingular, and

$$D(A_j)D(A_i) = \omega(i,j;l)D(A_l), \tag{V.4.b.iii-8}$$

they form a projective representation of the group, with factor system $\omega(i,j;l)$.

Thus the necessary and sufficient condition for the ω's to be a factor system is that the g^2 ω's obey the g^3 equations given by the associativity condition. However these equations do not uniquely determine the ω's. Sets of solutions are related, but there are also different sets not related to each other. There are an infinite number of solutions because we can obtain a new set by multiplying each element by a constant (eq. V.4.b-2, p. 261). This really does not change the group, in fact being little more than a change of notation, but sometimes one with physical significance. Two factor systems related by this are called associated, or equivalent, factor systems. If they cannot be so related they are inequivalent, or of different type. Each type then is said to be in a single class. The number of classes is finite.

Problem V.4.b.iii-1: How are the matrices belonging to equivalent factor systems related?

V.4.b.iv *The factors are roots of unity*

There is much freedom in the choice of factors. But actually they are almost completely determined [Hamermesh (1962), p. 462].

Problem V.4.b.iv-1: Take the determinant of the product equation (eq. V.4.b.iii-8, p. 264), with

$$det|D(R)| = d_R, \qquad \text{(V.4.b.iv-1)}$$

to establish that

$$(\omega_{R,S})^g = \frac{d_R d_S}{d_{RS}}. \qquad \text{(V.4.b.iv-2)}$$

Using the constants

$$c_R = d_R^{-1/g}, \qquad \text{(V.4.b.iv-3)}$$

to give the equivalent factor system ω', show that

$$(\omega_{R,S}')^g = 1. \qquad \text{(V.4.b.iv-4)}$$

This means that the members of set ω' are the g roots of unity. Each such factor system contains an equivalent set for which this is true. Since the number of such different sets is finite, so is the number of classes of factors (pb. V.4.b.ii-1, p. 263). Thus the different classes of factor systems are determined by the subset of these roots of unity that they contain. Different subsets give different classes.

Problem V.4.b.iv-2: Show, without using the results of the last problem, that any factor system can be written (in slightly different notation)

$$\omega(i,j;k) = exp(2\pi i\frac{a(i,j)}{g}), \qquad \text{(V.4.b.iv-5)}$$

where g, the order of the group and $a(i,j)$ are both integers, and

$$0 \le a(i,j) \le g-1, \qquad \text{(V.4.b.iv-6)}$$

$$a(E,j) = a(i,E) = 0. \qquad \text{(V.4.b.iv-7)}$$

The factors then can always be taken as phases that are roots of unity; thus the number of classes of equivalent factor systems is finite, for any (finite) group. Further from associativity [Altmann (1977), p. 90],

$$a(j,k;m) + a(i,m;n) = a(i,j;l) + a(l,k;n), \mod g. \qquad \text{(V.4.b.iv-8)}$$

V.4.b.v *Properties of factor systems and projective representations*

In many ways projective representations are like vector ones, but there are also significant differences. Here we mention some properties, and consider the similarities and differences.

Problem V.4.b.v–1: The concepts of reducibility and irreducibility, and reduction of representations, are identical for projective and vector representations, as well as the definition of character and the orthogonality relations — since these follow from Schur's lemmas, which are unaffected (why?) by the presence of a factor system [Mirman (1995a), pb. VII.3.c-2, p. 190]. Check this. State the orthogonality relations for the matrices and characters of projective representations [Altmann (1977), p. 81; Bradley and Cracknell (1972), p. 157].

Problem V.4.b.v–2: While for vector representations, the number of equivalence classes of representations equals the number of similarity classes [Mirman (1995a), sec.VI.4.a, p. 176], this is not true for projective ones. Neither the total number of representations, vector plus projective, nor the number of the latter, equals the number of similarity classes. The total can not since the number of vector representations does. And the number of inequivalent projective representations does not equal the number of vector ones. At which step does the proof [Mirman (1995a), sec.VII.6.c, p. 202] break down? How are the number of (properly-defined) classes and the number of irreducible ray representations related [Inui, Tanabe and Onodera (1990), p. 92]?

Problem V.4.b.v–3: Unlike vector representations, for projective representations the characters of the members of a class need not all be the same, although they can differ only in phase [Altmann (1977), p. 84; Inui, Tanabe and Onodera (1990), p. 92]. (Note that we are, unfortunately, using the word class in several different ways.)

Problem V.4.b.v–4: The basis vectors of the regular representation, which are the group operators, break into sets inequivalent under similarity transformations, and these sets can either be interpreted as similarity classes of operators, or as basis vectors of irreducible representations, so the number of these are equal. However for projective representations the representation matrices are not submatrices of those of the regular representation. Why? Thus the basis vectors of the projective representations are not sums of group operators, since such sums do not transform under these representations; they obviously cannot. How are they related to the group operators [Inui, Tanabe and Onodera (1990), p. 93]? Actually they are related to group operators, being sums of operators from a larger group, the one we find (sec. V.5, p. 270) by group extension.

V.4.c Some groups with projective representations

To illuminate these concepts (or at least dispel some of the darkness) we consider ray representations of a few elementary groups. One aspect that we wish to make concrete is that of class, why systems fall into classes, and why there are different ones. In particular, there is one class that gives vector representations (all groups — at least those that we are familiar with — have vector representations, otherwise we would take them to have no representations at all, a possibility that we do not consider). Clearly there is only one such class of vector representations.

Problem V.4.c-1: Why? How much freedom is there in choosing the factor system within this class.

V.4.c.i *Projective representation of cyclic groups and products*

The first example is the cyclic group of order n, so

$$A^n = E. \tag{V.4.c.i-1}$$

Its representation matrices satisfy

$$D(A)^n = \omega 1, \tag{V.4.c.i-2}$$

and can be redefined

$$D' = \omega^{-1/n} D, \tag{V.4.c.i-3}$$

to satisfy

$$D'^n = 1. \tag{V.4.c.i-4}$$

All representations of cyclic groups are equivalent to vector representations [Altmann (1977), p. 76; Hamermesh (1962), p. 463]. Here we see that while there are various factor systems they all belong to one class, and that contains a system with all factors 1, thus gives ordinary representations.

Next is the group of two operators, and their powers,

$$A^n = E, \text{ and } B^m = E, \tag{V.4.c.i-5}$$

with all elements commuting. We apply the same argument to both A and B, absorbing the constants but still have

$$D(A)D(B) = \omega D(B)D(A). \tag{V.4.c.i-6}$$

The matrices can commute, but up to a constant. The D's are already determined, so can not be redefined to eliminate ω. This group, then, has a projective representation. However ω is not arbitrary.

Problem V.4.c.i-1: Show that ω satisfies

$$\omega^n = 1 \text{ and } \omega^m = 1. \qquad \text{(V.4.c.i-7)}$$

Hence projective representations appear only if n and m are not relatively prime. If they are, all representations are equivalent to vector representations and are, as we know, one-dimensional. Check this for a few values of n and m, taking especially some relatively prime values and show specifically that they have only vector representations.

Problem V.4.c.i-2: The simplest.case for which there is a projective representation is for $n = m = 2$ (the Abelian four-group [Mirman (1995a), sec.III.2.c, p. 71]; one realization is the product of space inversion and time reversal). For it [Hamermesh (1962), p. 464]

$$A^2 = E, \; B^2 = E, \; C = AB = BA. \qquad \text{(V.4.c.i-8)}$$

Then

$$\omega = \pm 1. \qquad \text{(V.4.c.i-9)}$$

For

$$\omega = 1 \qquad \text{(V.4.c.i-10)}$$

the representation is the usual vector one. Check that these two factors belong to different classes. This illustrates that there can be more than one class, a requisite for projective representations. Why? For

$$\omega = -1, \qquad \text{(V.4.c.i-11)}$$

let

$$D(C) = iD(A)D(B). \qquad \text{(V.4.c.i-12)}$$

Show that the squares of all matrices equal 1, and that they anticommute (despite the fact that the group elements commute),

$$D(A)D(B) = -D(B)D(A). \qquad \text{(V.4.c.i-13)}$$

They can therefore be realized by the unit matrix plus the Pauli σ's [Mirman (1995a), pb. II.4.g-1, p. 56], giving a two-dimensional representation — although Abelian groups have vector representations that are one-dimensional only (over the complex numbers [Mirman (1995a), pb. V.5.a-4, p. 163], but there is a complex number here). Further show that this is irreducible and the only irreducible (ray) representation. There is a difference between vector and projective representations (for example in dimensionality) and results found for the former need not hold for the latter.

V.4.c.ii *Dihedral groups*

Other examples are given by the dihedral groups (sec. I.7.a.i, p. 52; sec. V.2.c, p. 249) [Mirman (1995a), sec.III.2.h, p. 79]. The two cyclic subgroups of D_n are generated by A and S [Hamermesh (1962), p. 465], with

$$A^n = E, \quad S^2 = E. \tag{V.4.c.ii-1}$$

We denote their representation matrices by the same symbols. Again we absorb constants in the definition of the matrices for each cyclic subgroup, leaving only

$$SAS = cA^{-1} \tag{V.4.c.ii-2}$$

with a constant. Again taking powers gives

$$c^n = 1, \tag{V.4.c.ii-3}$$

so c is an n'th root of unity, which we write as

$$c = exp(2\pi i\frac{m}{n}), \tag{V.4.c.ii-4}$$

with m of course an integer between 0 and n-1, and we have different choices of c depending on m. Multiplying A by $exp(2\pi im'/n)$ we get

$$SAS = exp(2\pi i\frac{m-2m'}{n})A^{-1}. \tag{V.4.c.ii-5}$$

Problem V.4.c.ii-1: Check that for n odd we can choose

$$m - 2m' = 0, \quad \text{or} \quad n, \tag{V.4.c.ii-6}$$

so that all representations are (equivalent to) vector ones. In this case there is only one class; the value of m is irrelevant. For n even, and m even this is also true, while for m odd all cases reduce to

$$c = exp(\frac{2\pi i}{n}). \tag{V.4.c.ii-7}$$

Hence for even n there are two classes, labeled by the parity of m, one having the vector representations while the other has

$$SAS = exp(\frac{2\pi i}{n})A^{-1}, \tag{V.4.c.ii-8}$$

again independent of m, and this is the only projective representation for dihedral groups. What properties does this have? What are its representations? Why, in these cases, do different values of m not give different classes?

Problem V.4.c.ii-2: Observe that for $n = 2$ this (nicely) reduces to the four-group, and gives the proper representation.

Problem V.4.c.ii-3: Show that the dihedral-group projective representations are two-dimensional, and can be put in the form, with the c just defined (and with what values of r?),

$$A = \begin{pmatrix} c^r & 0 \\ 0 & c^{1-r} \end{pmatrix}, \quad S = \begin{pmatrix} 0 & 1 \\ 1 & 0 \end{pmatrix}. \qquad \text{(V.4.c.ii-9)}$$

Is there anything else that can be said about them? Would we want to?

V.5 CENTRAL EXTENSIONS

Are projective representations really a new type of object, or are they actually ordinary representations, but with that disguised? To study this we have to relate group G to other groups.

Groups are related in various ways. Group extension is one [Altmann (1977), p. 34; Birman (1974), p. 19; Florek (1994); Florek (1998); Kurosh (1960a), p. 76; Kurosh (1960b), p. 121; Robinson (1993), p. 301; Rotman (1995), p. 154; Shubnikov and Koptsik (1974), p. 244]. The groups we consider here, point and space groups, colored and uncolored groups (sec. VIII.4, p. 411), are so related. However this concept is general. The one type of extension that is presently relevant, that providing the projective representations needed for the space groups, is special — this is probably fortunate since it allows us to obtain useful results that may not hold in general. Whether other types of extensions furnish useful and relevant information about physics, crystals or space (or other types of) groups, is therefore left open.

Every group G with an invariant subgroup has associated another group, its factor group, which is helpful in studying G and in finding its representations. We would expect that we can go in the other direction: from group G find group G_M whose factor group G is, and it, and the relationship, are also suitable for the study of both G and G_M. But then we need an invariant subgroup. The center of a group is the invariant subgroup that commutes with all elements of the group; it is the set of elements each of which is a class by itself [Mirman (1995a), sec.IV.4, p. 116].

Problem V.5-1: Why are these two definitions of center the same?

Problem V.5-2: From G, a finite group of order $|G|$, with multiplicator M of order m (sec. V.4.b.i, p. 262), we construct G_M, of order $m|G|$, with M (isomorphic to) a subgroup of the center of G_M. Then G_M is a central extension of G (as G is extended to another group by use of its center) with kernel M [Barut and Raczka (1986), p. 619; Hall (1959),

p. 218; Kurosh (1960b), p. 145; Rotman (1995), p. 201]. Explain why G_M/M is isomorphic to G. So this process is in a sense the inverse to finding a factor group. The kernel of a mapping is the set of elements that map to the identity. Check that M is the kernel of this mapping (since it maps to the identity of G). Note that G is a factor group of G_M, but it is not (need not be?) a subgroup: Why? While central extensions seem to be the most important (or most common for applications), they are not the only ones [Altmann (1977), p. 37; Kurosh (1960b), p. 148; Shubnikov and Koptsik (1974), p. 244].

Problem V.5-3: This gives an important theorem of Schur on projective representations [Altmann (1977), p. 88; Bradley and Cracknell (1972), p. 158]. Prove that by restricting the irreducible representations of G_M to G (that is by considering the representation matrices for only the operators of G_M corresponding to those of G) we obtain all unitary irreducible projective representations of G. Thus to obtain these we can use the representations of G_M, giving a method (not yet completely specified) for finding ray representations.

V.5.a Nonuniqueness of central extensions

The product of two numbers is unique. A central extension is, in a sense, a product of two groups. Is it also unique? Groups are more complicated than numbers and the answer is no [Kurosh (1960a), p. 76]. Given two groups, they can give more than one central extension. The cyclic group of order 4, with

$$A^4 = E, \qquad \text{(V.5.a-1)}$$

has invariant subgroup $\{E, A^2\}$, the cyclic group of order 2. The factor group has elements that are the sets

$$\{E, A^2\}, A\{E, A^2\} = \{A, A^3\}, \qquad \text{(V.5.a-2)}$$

and is another realization of the cyclic group of order 2. The cyclic group of order 4 is then the central extension of the cyclic group of order 2 by the cyclic group of order 2.

Problem V.5.a-1: There is another Abelian group of order four, the four-group (pb. V.4.c.i-2, p. 268). It also has invariant subgroups, all are cyclic groups of order 2. Thus it too is a central extension of this group by itself. Write the expressions for the elements of the group in terms of those of its factor groups.

V.5.a.i *Symmetric group S_3 as a central extension*

Another example provides an additional interesting point.

Problem V.5.a.i–1: Consider the cyclic group of order 6; it has a unique cyclic subgroup of order 3, which is invariant. It is a central extension of this group by the (a?) cyclic group of order 2, the factor group. But there is another product of these two groups. This is symmetric group S_3, which has a normal subgroup, alternating group A_3, which is (isomorphic to) the (a?) cyclic group of order 3. The factor group is of order 2. Write the expressions for the elements of the two central extensions in terms of those of the factor groups. How do they differ?

Problem V.5.a.i–2: Can other symmetric groups also be written as central extensions [Hamermesh (1962), p. 466]? Can their factor groups give additional extensions?

V.5.a.ii *Why central extensions are not unique*

Why, if we can multiply numbers uniquely, can we not multiply groups uniquely? The answer is that numbers are fully defined, group operators are not. So taking two cyclic groups of order two — E, A and E, B — we get their product E, A, B, AB. If we regard A and B as distinct this gives the four-group. But suppose we set, as we can,

$$B = A^2. \qquad \text{(V.5.a.ii–1)}$$

This gives the cyclic group of order 4. To get this latter group from the former we impose an extra condition, relating B and A. This freedom to impose such conditions, which do not affect the factor groups, leads to nonuniqueness.

Problem V.5.a.ii–1: It is useful to look at more examples of central extensions to see whether this is always true, and also whether there are other reasons for nonuniqueness.

V.5.a.iii *Central extension of the four-group*

The four-group (pb. V.4.c.i-2, p. 268), labeled G, which is itself a central extension, has vector representations all of dimension one — it is Abelian. But it has a projective representation of dimension two (sec. V.4.c.ii, p. 269). Why [Altmann (1977), p. 96; Weyl (1931), p. 182]? To explore this we consider a central extension of it, F_2, so

$$G = F_2/H. \qquad \text{(V.5.a.iii–1)}$$

A nontrivial invariant subgroup, H, has at least two elements and we start with just two, e and t, and of course

$$t^2 = e. \qquad \text{(V.5.a.iii–2)}$$

Elements of F_2 can then be written as pairs,

$$(E,e),(E,t),(A,e),(A,t),(B,e),(B,t),(C,e),(C,t), \qquad \text{(V.5.a.iii-3)}$$

and all commute. We take first

$$(C,e) = (A,e)(B,e). \qquad \text{(V.5.a.iii-4)}$$

However

$$(A,t)(B,e) = (C,t), \text{ and } (A,t)(B,t) = (C,e). \qquad \text{(V.5.a.iii-5)}$$

This group is then just the direct product of the two invariant subgroups and contains nothing interesting.

It is clear that this can be done for H of any order, and will yield nothing really new. What other possibilities are there? First H might have more structure, it may be more complicated than just a cyclic group. Second, the element of H in the product can depend on the operators of G appearing;

$$(A_i,t_\alpha)(A_j,t_\beta) = (A_iA_j, f(t_\alpha t_\beta, i, j)), \qquad \text{(V.5.a.iii-6)}$$

where f is some function giving one of the t's which is determined not only by α and β, but also by i and j. The simplest group we can take for H is the four-group itself, having elements e, a, b, c. So we get

$$(A_i,e)(A_j,e) = (A_iA_j,e), \qquad \text{(V.5.a.iii-7)}$$

and the A's again form a subgroup. Now

$$(A,a)(B,b) = (C,c), \text{ and } (A,a)(B,c) = (C,b). \qquad \text{(V.5.a.iii-8)}$$

Again this is the direct product and not interesting. Giving H more structure will not change this. In particular this does not lead to a group with a two-dimensional (vector) representation (which becomes a projective representation of G).

To find something new we need a different approach. Since the four-group has a projective representation of dimension 2, there is a noncommutative group of which it is a factor group, and this representation is a vector representation of it. The smallest noncommutative group is symmetric group S_3, of order 6. If the four-group is a factor group of it the other factor has the impossible order 1.5. How about a group of order 8, so the other factor has order 2? We can realize the group using the Pauli σ 's, giving two invariant subgroups,

$$H = \{E, \sigma_x\} \qquad \text{(V.5.a.iii-9)}$$

and

$$F_2 = \{E, i\sigma_z, (i\sigma_z)^2, (i\sigma_z)^3, (i\sigma_z)^4 = E\}. \qquad \text{(V.5.a.iii-10)}$$

Then, for example,

$$\sigma_x(i\sigma_z) = -(i\sigma_z)\sigma_x, \qquad \text{(V.5.a.iii-11)}$$

so the product group is not Abelian. This is a central extension of the four-group G [Hamermesh (1962), p. 464].

Problem V.5.a.iii-1: What is the center? Prove that it is a central extension.

Problem V.5.a.iii-2: Why is this group not a direct product? Which of the above steps, leading to direct products, does not apply here? How do the factor systems of the four-group give this group, and conversely?

Problem V.5.a.iii-3: This realization by the σ's gives one representation, which being two-dimensional appears twice in the regular representation. That has order eight so there are four other representations, all one-dimensional. What are they? Are they representations of the four-group? How would we know that a vector representation of this group is a projective representation of the factor group? Why are their representations related? Which group is this [Mirman (1995a), chap. III, p. 67]? Is it listed in appendix B, p. 631?

Problem V.5.a.iii-4: Find the extensions of every cyclic group [Hall (1959), p. 225; Hamermesh (1962), p. 463; Weyl (1931), p. 182]. Extend it to the improper groups.

Problem V.5.a.iii-5: What are the extensions of an arbitrary Abelian group [Kurosh (1960b), p. 126, 146]?

Problem V.5.a.iii-6: Determine the central extensions of the dihedral groups (sec. V.4.c.ii, p. 269) [Hamermesh (1962), p. 465]. What are the central extensions of the improper groups?

Problem V.5.a.iii-7: Find these extensions for the other crystallographic groups.

Problem V.5.a.iii-8: Determine the extensions for the infinite cyclic and dihedral groups, including the improper ones (sec. I.6.f, p. 46).

V.5.b Central extensions and factor systems

A projective representation of a group is a vector representation of a larger group (thus in this sense is not a new type of object). So we have a way of obtaining from a group larger ones, by finding groups whose representations give projective representations of the original group — the extensions of it. Conversely, if we know the group extension we can find the ray representations (however that a group has an extension need not imply that it has ray representations). Thus we use the properties of the representations to find group extensions, which give the representations.

Projective representations have factor systems. How are these, the factor systems, so the projective representations, related to the vector representations of the group extension? For every factor system there is a central extension, for every central extension a factor system [Kurosh (1960b), p. 121]. This implies, since there are several (classes of) factor systems, that a group has several extensions — central extensions are not unique. We thus have to relate factor systems and extensions.

Group G has normal subgroup H, so factor group F (since we are only considering central extensions); G is a central extension of H by F. Denoting the elements of these three groups by g_i, h_j, f_k, the cosets are g_iH, with $g_o = e$, the unit, and from each we chose a representative, an element r_i of G that is in coset $f_i = g_i$H. The product of two representatives r_i and r_j is r_{ij} — the representative of coset g_{ij}H. So

$$r_i h_i r_j h_j = r_{ij} h_{i,j}, \qquad \text{(V.5.b-1)}$$

where $h_{i,j}$ is an element of the normal subgroup, by definition of representative, and this defines $h_{i,j}$. That is, the product of two representatives is given by the product of the coset multipliers times the product of the subgroup elements, and the latter is a subgroup element. Likewise

$$r_k h_k r_{ij} h_{ij} = r_{ijk} h_{i,j;k}. \qquad \text{(V.5.b-2)}$$

Taking the unit element of F (= H) relates the unit elements

$$g_o = h_e. \qquad \text{(V.5.b-3)}$$

Problem V.5.b-1: Using associativity for G (eq. V.4.a-4, p. 260), show that

$$h_{m,jk} h_{n,im} = h_{l,ij} h_{n,lk}. \qquad \text{(V.5.b-4)}$$

Thus these obey the equations for factor systems; the elements of the normal subgroup serve as factors. Any set that obeys these equations is a factor system. So every extension gives a factor system. Does every factor system give an extension? The set G — which is hopefully a group — consists of the sets $f_i h_a$, with h_a running over the elements of H, and i over the cosets (so G has the right order to be an extension of H by F). Check that the product of two such elements is such an element.

Problem V.5.b-2: Prove that G is a group [Kurosh (1960b), p. 124].

Problem V.5.b-3: Show that there is a normal subgroup of G that is isomorphic to H and that factor group G/H is isomorphic to F. Thus G is an extension of H by F. So for each factor system there is an extension — however there can be more than one, as we have seen, for there is freedom in choosing the coset representatives.

Problem V.5.b-4: Relate the Abelian group associated with the factor systems (sec. V.4.b.i, p. 262) to this procedure.

Problem V.5.b-5: Prove that every projective representation is a vector representation of every central extension of the group.

V.5.b.i *The multiplicator and central extensions*

Why do we introduce multiplicator group M (sec. V.4.b.i, p. 262); how is it related to the central extensions of G [Bradley and Cracknell (1972), p. 158]? The order of every element of M is a factor of the order of G_M since M is isomorphic to (though need not be) a subgroup of G_M, and this is true for every element of any subgroup of a group [Mirman (1995a), sec. IV.6.b.i, p. 131]. Also if element M_p of M has order α_p, it is possible to pick a representative factor system from the class corresponding to M_p whose values are all α_p roots of unity (sec. V.4.b, p. 261). This means that every projective representation is equivalent to a unitary one with a factor system in the same class. The identity, M_1, of M corresponds to that class containing a factor system with all values 1, so the representations of G are the unitary irreducible projective representations of the class corresponding to M_1, these being the vector representations.

Now, with the order of G denoted by g, and that of M by m, there is a group G_M, of order gm, so that M is isomorphic to a subgroup of the center of G_M. If we take M to be the subgroup, then G_M/M is isomorphic to G, that is G is a factor group of G_M. Group G_M is a central extension of G (because M is in its center), with kernel M (since the elements of M map to the identity).

This gives the basic result that every unitary projective irreducible representation of G is found by taking all irreducible representations of G_M, and considering only the elements of G_M that belong to G. This relates the representations of a group G, the projective as well as, of course, the vector ones, to the vector representations of another group, its extension, of which G is a factor group. The extension is found by taking the cosets with respect to the group given by the classes of the factor systems (that is obtained from the multiplicator group). The projective representations are vector representations of its extension.

V.5.b.ii *The meaning of projective representation*

Thus we can see the meaning of projective representations. They are not really representations of the group, but of a larger group, the extension. Since the groups are related their representations are, and that they are related in this way follows from the form of the relationship of the groups, the one whose representation is projective is a factor group of the other for which the representation is a vector representation. What we have to consider is why groups so related are interesting, how knowing the projective representations of a group (that is its extensions and their representations) helps us to learn about it, and in particular how it helps in finding the representations of these two groups.

V.5.b.iii *The covering group*

The projective (as well as vector) representations of a group are all vector representations of another group, this called its covering group [Altmann (1977), p. 85; Birman (1974), p. 68; Robinson (1993), p. 306; Rotman (1995), p. 208], or representation group. A well-know example is the (continuous) group SU(2), which is the cover of the rotation group SO(3) (sec. I.8, p. 59) [Mirman (1995a), sec.X.2, p. 270].

The group found by central extension is the covering group.

Problem V.5.b.iii-1: The mapping is not completely defined, since there can be several different elements of the covering group that can map to an element of the group. Why? Thus the covering group is defined up to an isomorphism (automorphism?). Check this for the homomorphism between SO(3) and SU(2). How about the finite groups that we have considered?

V.5.b.iv *Why only one factor system is considered*

This discussion lays the basis for finding the central extensions of a group, and its representations, thus all projective representations of the group itself [Altmann (1977), p. 88]. These, the explicit matrices and states, remain to be found. We do not try to do this in general, considering only those special cases needed for finding the representations of the space groups [Bradley and Cracknell (1972), p. 159]. Space group representations are labeled by the (star of) a momentum vector (sec. VII.2.d, p. 342), thus by a single factor system. We therefore need not consider all extensions, but only the one given by this particular system. It is taken as the normalized factor system of its equivalence class.

V.5.c Forming a central extension using a specific factor system

If we have a group with projective representations, we want to find an extension for which these are vector representations. How do we do so? For group G with operators A, the relevant factor system is labeled υ, and the factors can be written (pb. V.4.b.iv-2, p. 265),

$$\upsilon(A_i, A_j) = exp(\frac{2\pi i a(i,j))}{g}). \tag{V.5.c-1}$$

The product for the set, denoted by Z_g, of the cyclic group of integers $0, 1, 2, \ldots, (g-1)$, is addition modulo g [Mirman (1995a), pb.II.f.4-1,m, p. 55]. Define group G^* of the $g|G|$ elements that consist of the pairs (A_j, α),α being one of these integers, with multiplication given by [Altmann (1977), p. 91; Bradley and Cracknell (1972), p. 159]

$$(A_j, \alpha)(A_k, \beta) = (A_jA_k, \alpha + \beta + a(A_j, A_k)). \tag{V.5.c-2}$$

Problem V.5.c-1: Confirm that G^* is a group. What is the identity? What is the inverse of an element? Verify also that the elements (E, α) form a subgroup (isomorphic to Z_g), but the $(A_j, 0)$ do not. Thus G is not a subgroup of G^*. How are they related? Check that

$$(E, \gamma)(A_j, \alpha) = (A_j, \alpha)(E, \gamma) = (A_j, \alpha + \gamma). \tag{V.5.c-3}$$

This means that (E, α) is a subgroup of g elements in the center of G^*, so with Γ the matrices of a representation of G^*, matrix $\Gamma(E, \alpha)$ is a scalar multiple of the unit matrix, by Schur's lemma. It should be clear that G^* is a central extension of G. What is the kernel? It is interesting that this product involves a product of elements, a sum of different ones and to that is added a number given by a product from a factor group — emphasizing the generality of multiplication.

Problem V.5.c-2: Let Γ be irreducible unitary representation matrices of G^*. Show that it is always possible to find a representation of G^* for which [Altmann (1977), p. 92]

$$\Gamma(E, \alpha) = exp(\frac{2\pi i\alpha}{g})I, \tag{V.5.c-4}$$

where I is the unit matrix (what unit matrix?). Is this true for all representations? It is on this that the rest of the discussion is based, so we do not obtain all projective representations; we will see that this is sufficient to give those necessary for the space groups. Then

$$\Gamma(A_k, \beta) = \Gamma(A_k, 0)\Gamma(E, \beta) = \Gamma(A_k, 0)exp(\frac{2\pi i\beta}{g}), \tag{V.5.c-5}$$

from the definition of group product. Writing

$$\Gamma(A_k, 0) = \Omega(A_k), \tag{V.5.c-6}$$

we find

$$\Omega(A_i)\Omega(A_j) = \Omega(A_iA_j)exp(\frac{2\pi i a(i,j)}{g}), \qquad \text{(V.5.c-7)}$$

which means that Ω is a projective representation of G. As Γ is unitary, so is Ω. It is irreducible; otherwise (E, α) would be represented by a scalar matrix for all α, so all matrices Γ would be in the same reduced block, contrary to the hypothesis that Γ is irreducible. Check that if Ω is a unitary irreducible projective representation of G, then

$$\Gamma(A_k, \alpha) = \Omega(A_k)exp(\frac{2\pi i\alpha}{g}) \qquad \text{(V.5.c-8)}$$

is a representation of G^*. The matrices of the projective representations have now been given explicitly (using those of G). This means that all irreducible projective representations of G can be found from the vector representations of G^* [Altmann (1977), p. 95], which makes G^* useful for the study of G. (Also, this shows that all irreducible projective representations of G can be lifted into G^* — lift being defined by this procedure of finding the vector representations of the larger group from the projective ones of the smaller.)

Problem V.5.c-3: We have considered only special cases here. It would be interesting to determine a general algorithm for finding all central extensions of a group — that is given a group G, find all central extensions, their representations (labels, matrices and states), then use these to find all projective representations of G. It might also be valuable to extend the concept, and calculations, to extensions that are not central [Altmann (1977), p. 37; Kurosh (1960b), p. 148]. Are these of use or interest?

Chapter VI

Induced Representations

VI.1 INDUCING AND SUBDUCING TO FIND REPRESENTATIONS

Having become acquainted with the space groups, we continue to prepare to study their representations. These are interesting because of their applications in the real world, but also because they require new concepts in group theory, enhancing our capabilities for they are relevant, often necessary, elsewhere. We start with these new aspects of the theory, in part to develop ideas and methods needed for the representations, which we then consider.

First we need a procedure for finding (these) representations. Here our emphasis is general, not only as an introduction to the subject, but to provide material that is widely useful. In the next chapter we consider specifically space groups, and the application of the method to them — which provides examples that could clarify this material so referring back and rereading the relevant sections may be useful.

Space groups have factor groups (sec. III.4, p. 141), translation and point groups, which are not necessarily subgroups, with known representations. Thus the construction of the representations of the space groups would be simplified if they could be built upon — induced from — the representations of these smaller, and simpler, groups. There is a detailed theory of the procedure for finding representations of a group from those of its factor groups. Thus studying induced representations has greater value than merely (?) for its present application to space groups.

Inducing a representation is a way of finding representations of a group from those of its subgroups. The reverse process is finding the representations of a subgroup from one of a group when its elements are limited to those in the subgroup: the representations of the subgroup are subduced from those of the group. The former procedure requires the latter — inducing depends on the representations that are subduced, so that is what we begin with.

VI.2 SUBDUCED REPRESENTATIONS

A representation of subgroup H of G is subduced from a representation of G by considering only those matrices representing elements of H [Altmann (1977), p. 137; Birman (1974), p. 49; Bradley and Cracknell (1972), p. 176; Evarestov and Smirnov (1993), p. 20; Jansen and Boon (1967), p. 133; Janssen (1973), p. 138; Lomont (1961), p. 219; Streitwolf (1971), p. 39]. In that sense it may seem trivial. What (less than obvious) aspects prevent it from being so? While the representation of G may be irreducible the representation of H subduced from it need not be; clearly if the representation of G is reducible, so is the subduced representation. Thus we want the irreducible representations in the decomposition of the subduced representation, and we want to know how they are related, to the representation of G, and to each other. The subduced representations are limited and are also closely related, which is both interesting and helpful — having general knowledge reduces the effort needed in specific cases.

Problem VI.2-1: Suppose we subduce a representation of subgroup H from a representation of G, and then from that subduce a representation of M, a subgroup of H. Prove that subduction is transitive, that is the representation obtained by this process is the same as that obtained by subducing directly [Altmann (1977), p. 137; Bradley and Cracknell (1972), p. 182]. What does it mean to say that subduction is distributive [Mirman (1995a), pb. XIII.2.c-1, p. 371]? Prove that it is.

VI.2.a Conjugate representations

If H is a normal subgroup, then $g^{-1}hg$ belongs to H, for all h of H, and g of G (so not all are in H); this is not true for a noninvariant subgroup, an important distinction. This allows the definition that two subduced representations of normal subgroup H, $D(h)$ and $D_g(h)$, are conjugate [Altmann (1977), p. 48; Birman (1974), p. 51; Jansen and Boon (1967), p. 142; Lomont (1961), p. 221; Streitwolf (1971), p. 39] if they are transformed into each other by an element of G,

$$D_g(h) = D(g^{-1}hg). \qquad \text{(VI.2.a-1)}$$

That is g gives an automorphism of H [Mirman (1995a), sec. III.4.e, p. 91], outer if g is not in H, else inner, and the corresponding representation of the transformed group, D_g, is conjugate to D. Of course, since H is normal both h and $g^{-1}hg$ are elements of H. Thus D and D_g have the same representation matrices, but assigned differently to subgroup elements. This definition of conjugate representations does not apply for nonnormal subgroups.

Notice the difference between conjugate and equivalent representations [Mirman (1995a), sec. V.4.b, p. 159]. Two representations are equivalent (here under a transformation by an element of G) if their representation matrices for each element are related by a similarity transformation

$$D(h)' = D(g)^{-1}D(h)D(g), \qquad \text{(VI.2.a-2)}$$

where $D(g)$ is a matrix (say a representation matrix of element g); the group element h is the same, but the matrix is different. But they are conjugate if the group elements are transformed, and so the matrices are the same but elements to which they are assigned are different. If g is in H then the conjugate representations are also equivalent (under g).

Conjugate representations need not be equivalent since $D(g)$ may be nonexistent. The expression for conjugation then could not be factorized into the one for equivalence. That factorization requires that $D(g)$ be a representation matrix of H. For g not in H, the matrix of an irreducible representation, $D(h)$ of H, need not even have the same dimension as any representation $D(g)'$ of G, and then the conjugate representations clearly could not be equivalent — equivalence may not be definable (for this element). Also, if H is not an invariant subgroup, g may take elements of one class of H into different classes of the transformed group.

If a representation is equivalent to itself under all elements of G, it is self-conjugate.

Though corresponding elements of conjugate subgroups are different, they are related. Elements h and $g^{-1}hg$ are in the same class in H, for both elements are in H — but for all g and h only if H is normal — and classes are invariant under similarity transformations. Of course they are always in the same class in G. So these elements have the same character, if H is normal. For a noninvariant subgroup, they are not, in general, in the same subgroup, the same class or have the same character. The loss of such restrictions makes analysis of the general case harder.

An example is tetrahedral group, A_4, an invariant subgroup of octahedral group, S_4 (sec. I.4.b, p. 8). This has three one-dimensional representations, and one three-dimensional one (tbl. E.2-1, p. 657). The octahedral group has representations of one, two and three dimensions, all of which decompose into representations of the subgroup, and obviously two-dimensional representations (at least) could not provide similarity transformations under which either the three-dimensional or one-dimensional subgroup representations are equivalent.

Problem VI.2.a-1: Check that a conjugate representation to a representation is actually a representation [Jansen and Boon (1967), p. 142].

Problem VI.2.a-2: The dimensions of conjugate representations are the same, and they are either all reducible or all irreducible.

Problem VI.2.a-3: There are point groups with representations that are one-dimensional with complex numbers assigned to some group elements, the complex conjugates assigned to others (tbl. E.2-1, p. 657). Other representations also have the same numbers, but assigned to elements differently. For them find invariant subgroups, if any, and show that there are pairs of conjugate representations of the subgroup, but because of the assignment of the numbers to elements, these representations are not equivalent. Are there other cases of nonequivalent representations (there)?

VI.2.b Orbit of a representation

For each representation of a normal subgroup subduced from one of the group, there are a set of conjugate representations. These form the orbit of the representation [Altmann (1977), p. 166; Jansen and Boon (1967), p. 142; Kocinski (1983), p. 29; Lomont (1961), p. 221; Streitwolf (1971), p. 40] — in the decomposition of the representation of G, the set of conjugate (inequivalent, irreducible) representations of H is the orbit (of every one). The order of an orbit is the number of conjugate (inequivalent, irreducible) representations of H it contains. The order of the orbit of a self-conjugate representation is one.

The decomposition of a representation then breaks into sets — orbits — each of mutually conjugate representations. Each of these sets occurs several times in the decomposition; the number of occurrences is the multiplicity, m, of the orbit in the irreducible representation of G [Kocinski (1983), p. 30]. So each irreducible representation appears the same number of times m, that is the orbit occurs m times. The number of sets of mutually equivalent representations equals the number of nonequivalent representations in the orbit.

This meaning of orbit is somewhat different from that of a set of vectors generated from one by group operators giving the star of that vector (sec. VII.2.c, p. 341) and of the orbit of a pole (sec. I.4.d.i, p. 17).

Problem VI.2.b-1: Why must all conjugate representations be in the decomposition of a single representation of G? That is the decomposition contains only complete orbits.

Problem VI.2.b-2: All representations of an orbit have the same dimension. The order of an orbit is no greater than the number of representations with that dimension. Why? What does "number of representations" mean here? Can it be less?

VI.2.c Little groups

The methods considered here depend on a particular subgroup of G, the little group [Altmann (1977), p. 164; Jansen and Boon (1967), p. 143; Janssen (1973), p. 139; Kocinski (1983), p. 30; Lomont (1961), p. 230; Streitwolf (1971), p. 43] (also called the stability group of the element, isotropy group [Sternberg (1994), p. 13], or various other things).

For the inhomogeneous rotation group, or the Poincaré group, there are transformations that change a momentum vector, and others that leave it fixed. Every momentum vector is invariant under all rotations with it as an axis. These rotations form a group, a subgroup of the rotation group. Such a subgroup leaving a particular element of invariant subgroup T unchanged is called a little group. For these inhomogeneous groups, T is the subgroup of translations, or momenta. The little group depends on the element of T. Little groups play a central role in the induction process (sec. VI.3.a, p. 289) because they are the subgroups upon which the procedure is built, allowing a systematic method for finding the required representations [Mirman (1995c), sec. 2.2, p. 12].

There is another definition (or perhaps meaning) of little group. Consider two conjugate subduced representations $D(h)$ and $D_g(h)$. The set of elements g of group G under which these two, and so all conjugate to them, are not only conjugate but equivalent is the little group of the representation — perhaps better, of the orbit. These conjugate representations are equivalent under the little group, but not necessarily under G. The little group, a subgroup of G, contains H, but may be larger than it.

Problem VI.2.c-1: Are the two meanings of little group related? How?

Problem VI.2.c-2: The little group consists of H plus the set of other elements of G that commute with at least one element of H (except the identity?) — if so, why?

Problem VI.2.c-3: Another way of defining a little group of a representation is that it consists of (the elements of) normal subgroup H, and the left cosets of H in G that generate (all, but only) equivalent representations of it (that is the representations are equivalent under the representatives of the cosets). Explain why various definitions are the same (or should it be equivalent?).

Problem VI.2.c-4: Why should a subduced representation have a little group, one larger than H but smaller than G? In the decomposition of an irreducible representation of G a particular irreducible representation of H can occur several times. Find examples. The elements of G in H are block-diagonal(ized); not all others can be. Why? Hence these other elements mix blocks corresponding to a representation, giving a set of conjugate representations. The reducible representation of H consisting of the entire set of conjugate representations is thus taken into an equivalent reducible representation; these reducible representations are equivalent under G, but the individual irreducible members of the set need not be. Why? Now it may happen that there are elements of G, not in H, also block-diagonalized, with the same blocks, when H is (they are block-diagonal on the same basis states that H is). Then these elements, plus those of H of course, form the little group of H. Why?

Problem VI.2.c-5: How do we know that little groups are actually groups [Jansen and Boon (1967), p. 143; Janssen (1973), p. 139]? Need a little group of a representation of an invariant subgroup be itself normal in the group? Why?

Problem VI.2.c-6: If a representation of H is self-conjugate, the little group of that representation is G. What is the little group if there are no representations conjugate to it?

Problem VI.2.c-7: If no group elements leave a vector invariant, its little group is the identity. This is not true for a representation, but we would expect that in similar cases the little group is an identity of something. What?

Problem VI.2.c-8: Take the little group L of a representation of a normal subgroup of G and form the left cosets of it. (Need these give a factor group?) Show that each coset generates (in what sense?) a different member of the orbit of that representation [Jansen and Boon (1967), p. 144]. Thus the number of members of the orbit, the number of subgroups conjugate to any one, equals the index [Mirman (1995a), sec. IV.6.b, p. 131] of the little group in G, the number of elements of G divided by that of L. Why does this turn out to be an integer?

Problem VI.2.c-9: If a subgroup of G is a little group for some representation of H, need it be a little group for all (non-conjugate) representations — that is all orbits? If not are the little groups related? How?

VI.2.d The multiplicity of an orbit is representation independent

In the decomposition of irreducible representation R of G, there are several sets of mutually equivalent subduced, irreducible representations of normal subgroup H, each of which gives an orbit. The orbit thus appears several times, the number being its multiplicity, m. This multiplicity is independent of the representation of the orbit that defines it — it is the same for each (conjugate) representation of the orbit. Also, as is necessary for the previous statements to have meaning, all conjugate representations appear the same number of times — that is the decomposition consists of complete orbits, only. Seeing why provides information about this decomposition.

The reducible representation that is the orbit is labeled R^s, and the irreducible components of this subduced representation as R_i^s, with their multiplicity m_i (the number of times each appears in the decomposition of R). So the decomposition is

$$\mathrm{R} = \sum m_i R_i^s. \tag{VI.2.d-1}$$

Consider representation R^{gs} conjugate to R^s,

$$R^{gs}(h) = \mathrm{R}^{-1}(g)R^s(h)\mathrm{R}(g) = \sum m_i^g R_i^{gs}(h), \tag{VI.2.d-2}$$

and for the irreducible components

$$R_i^{gs}(h) = R_i^s(g^{-1}hg), \tag{VI.2.d-3}$$

where h belongs to H (only because H is normal), and g is any element of G. Notice the difference in the expressions for conjugation of the reducible and irreducible subduced representations of H — since the reducible representation is a (subset of the matrices forming a) representation of G, while the irreducible representation need not be (its representation matrices are blocks in the representation matrices of G). Now R^s, being reducible (in general) is self-conjugate under G, and equivalent to all its conjugate representations R^{gs}. Thus their decompositions into irreducible parts are the same (up to possible equivalence and rearrangements),

$$\sum m_i R_i^s \sim \sum m_i^g R_i^{gs}. \tag{VI.2.d-4}$$

This means that if a representation appears in the decomposition, all those of its orbit do, so the sum over the irreducible representations contains only complete orbits. Thus all representations in an orbit appear the same number of times

$$m_i^g = m_i, \tag{VI.2.d-5}$$

and the m_i's are equal.

VI.2.e There is but one orbit in the decomposition

Only one orbit occurs in the decomposition of R (perhaps more than once). Since R is an irreducible representation of G, conjugation transformations of the irreducible members of each orbit merely rearrange its constituents, and do not mix representations belonging to different orbits; they do not mix inequivalent representations. If there were more than one orbit, the representation matrices would be block-diagonal, with the blocks consisting of representations from each. Then R would be reducible, contrary to stipulation. However since some conjugate representations can also be equivalent, each can occur several times — and these are taken into each other by conjugation transformations. So the orbit can appear several times, as the decomposition may contain more than one appearance of its members.

An example is given by the A_3 subgroup of S_3 (sec. VI.4.a, p. 294) whose two-dimensional representation decomposes into the two representations of A_3 (with complex characters), which are inequivalent as their characters are different (tbl. E.2-1, p. 657). These two form the orbit, being conjugate under the three transpositions of S_3, any of which can be taken as the coset representative. The orbit does not depend on this choice. The decomposition of the two-dimensional representation must include both representations of the orbit since the representation of S_3 is taken into itself by every one of its operators (of course) and these are the elements under which the representations of the orbit are conjugates.

Problem VI.2.e-1: Why are conjugate, reducible, representations equivalent?

Problem VI.2.e-2: The number of times a representation (here an irreducible one of a normal subgroup) appears in the decomposition of a reducible representation (of that subgroup, but irreducible in G) is given in terms of the characters of the two representations [Mirman (1995a), sec. VII.6.a, p. 199]. From this, show that all conjugate representations appear the same number of times [Jansen and Boon (1967), p. 144]. How does H being normal enter?

Problem VI.2.e-3: Check that the little group of a subduced representation is independent of the element of H. All little groups of an orbit are isomorphic (but consist of different sets of elements of G).

Problem VI.2.e-4: For each of the examples of induced representations given below (sec. VI.4, p. 293), check the correctness of these statements.

VI.2.f The set of subduced representations of a normal subgroup — Clifford's theorem

The results for the set of related representations can now be summarized (Clifford's theorem [Jansen and Boon (1967), p. 146]).

Consider irreducible representation R of G and the representation R^s of normal subgroup H subduced from it. Then R^s is either irreducible, or decomposes into irreducible components all of the same dimension. Take R_1^s, one of the irreducible components of R^s; all other irreducible components in the decomposition of R^s are inequivalent conjugates under G of R_1^s. Every such conjugate occurs the same number of times (up to equivalence) in this decomposition.

This can be stated in terms of orbits. Every such R^s is a multiple of the direct sum of all irreducible representations of some fixed orbit of H. Other representations, that are not conjugate to it, belong to different orbits, and occur only in the decomposition of different irreducible representations of G.

With each orbit — with each maximal set of inequivalent, conjugate irreducible representations of H — we can associate a maximal set of conjugate little groups.

Problem VI.2.f–1: Check that all these statements follow from the previous discussions.

Problem VI.2.f–2: Define "conjugate little groups". Relate the little group for one member of the orbit to the set (what set?) of left-coset representatives of that little group. Show that if the little group is normal in G, all members of the orbit have the same little group. Of course, this holds if the little group is either H or G. However if the little group is not invariant in G, then the intersection of the little groups of an orbit (the set of elements they have in common) is invariant.

VI.3 INDUCED REPRESENTATIONS

The representations of space groups are found from those of their factor groups, the procedure being that of induction of representations. This technique is useful for these groups, but is more general, applicable and often (almost) necessary for others, including continuous groups. The analysis is the same for finite, discrete-infinite, and continuous groups, since for all the representations are finite-dimensional, which is required (at least by the notation). We study this here, both for its present importance and as an introduction to the subject [Altmann (1977), p. 137; Barut and Raczka (1986), p. 205; Bradley and Cracknell (1972), p. 171; Burrow (1993), p. 77; Chen (1989), p. 38; Coleman

(1968); Cornwell (1984), p. 118; Evarestov and Smirnov (1993), p. 22; Fulton and Harris (1991), p. 32; Harter (1993), p. 255; Inui, Tanabe and Onodera (1990), p. 82; Jansen and Boon (1967), p. 132; Janssen (1973), p. 139; Kocinski (1983), p. 31; Lomont (1961), p. 223; Ludwig and Falter (1988), p. 113; Streitwolf (1971), p. 41].

The method of induction would not be sufficiently useful unless only a reasonable amount of work was needed to obtain the representations, and these were irreducible and all. Thus it is not a method relevant and helpful for every group. Is it ever useful? There is (at least) one type of group — fortunately an important one — for which it does give irreducible representations, and all. Groups for which it works include semi-direct products (sec. III.2.e, p. 136), especially with the invariant subgroup Abelian. Space groups are such.

VI.3.a Representation finding using induction

How might we find the representations of the inhomogeneous rotation group, or the inhomogeneous Lorentz (Poincaré) group [Mirman (1995b,c)]? A reasonable way is to start with the known representations of simpler (or smaller) groups (subgroups or factor groups) and then find the actions of the remaining operators on their states. For the inhomogeneous rotation group we fix one vector, one translation or momentum (depending on the realization), say that along z, and find the representations of all rotations about this vector. The group of these rotations, in this case SO(2), is the little group of the vector — the group leaving the vector unchanged.

For the Poincaré group, we likewise fix a momentum and consider the group of rotations around it; for a massive object, an electron or nucleon for example, we fix the three-momentum — at zero, taking the object at rest, and consider all rotations of its spin, giving the little group which is SO(3) — and similarly for the other types of representations. Having these known representations, we find those of the full group by applying to the ones of the little-group the set of transformations that change the three-momentum, for a massive object the boosts that give it nonzero velocity. To do this we need the vectors generated from the fixed one by these transformations, the orbit of that vector (sec. VII.2.c, p. 341), for this case rather obvious, but not so for the space groups. (Rationales for this method, and for the choices that go into it, are mentioned elsewhere [Mirman (1995c), sec. 2.2.3, p. 14].) Thus we start with known representations, and have to find only the action of a subset of group operators on their states.

For space groups we fix a point of the reciprocal lattice (sec. IV.2.a, p. 186), thus a momentum vector, and use the point group of rotations and inversions about it that leaves the lattice invariant, the little group for the point (the stability group — keeping a point stable, that is fixed, and likewise the lattice invariant, or stabilizer, or isotropy group — giving the transformations under which the lattice is isotropic, the same in all these finite set of directions, in crystallography, the site-symmetry group — the group giving the symmetry at a particular site, or point [Evarestov and Smirnov (1993); Senechal (1990), p. 29]).

This outlines the approach we use for the space groups, based on little groups [Altmann (1977), p. 158; Bradley and Cracknell (1972), p. 149; Chen (1989), p. 475; Cornwell (1984), p. 120; Inui, Tanabe and Onodera (1990), p. 85; Janssen (1973), p. 142; Kocinski (1983), p. 30; Lomont (1961), p. 230; Ludwig and Falter (1988), p. 204]. (For nonsymmorphic groups, especially, the meaning of the little group needs more care (sec. VII.2.h, p. 344)).

To induce representations of G requires that there be a (rich enough) subgroup whose representations are known, or easily found — and such that the representations of G can be simply constructed from those of it (the identity is a subgroup with representations that are easily found, but they do not help). Fortunately this is often the case. In particular the representations of both point groups and space groups are important, so having found those of the former (perhaps by induction) we use them to find those of the latter — they are rich enough.

Essentially what this does it to start with a subgroup (or factor group) with known representations, and find the matrices of the group elements that are not in it, from those that are. An alternate view is that the action of these unknown matrices on the subgroup basis vectors is determined, giving the unknown representation matrices. Finding this action, of only a subset of representation matrices, on the known basis vectors is easier than finding all matrices of the group representation. For the method to be useful there has to be a large enough factor group with known representations, with the action of the remaining elements on their basis vector computable relatively simply.

So we need a coset decomposition of each (space) group, and the set of vectors obtained from those of the (corresponding) little group using the operations of the relevant (point) group, and from these we find the (space-group) representations. For the inhomogeneous rotation group (and somewhat less so for the Poincaré group) this is straightforward, for symmorphic space groups almost obvious, but for nonsymmorphic space groups it becomes more stimulating. The first step in the procedure then is to find reasonable little groups, and their representations, which needs several (sub)steps.

This procedure is the method of induced representations; the representations of the (space) group are induced (lead into) from those of its factor groups. This requires a general definition.

Problem VI.3.a-1: What is the site-symmetry group of those points of a cube that have one [Senechal (1990), p. 30]?

VI.3.b Definition of induced representation

An induced representation of group G which has order (number of elements) o(G), is one obtained from a representation of a factor group H, with order o(H), according to a procedure which we now describe and (hopefully) justify.

Using H we form cosets and take a coset representative from each, these labeled $r_1, r_2, \ldots, r_{o(G)/o(H)}$; the subscripts give the cosets, thus

$$G = r_1 H + r_2 H + \ldots + r_{o(G)/o(H)} H, \qquad \text{(VI.3.b-1)}$$

with $r_1 = E$. Using matrices M, a d-dimensional representation of H, a representation of G is given by the $d(o(G)/o(H))$ by $d(o(G)/o(H))$ matrices, for all g of G,

$$\begin{aligned} \Gamma(g)_{kt,jv} &= M(r_k g r_j^{-1})_{tv}, \quad \text{if } r_k g r_j^{-1} \varepsilon \text{ H}, \\ &= 0, \qquad\qquad\qquad \text{otherwise.} \end{aligned} \qquad \text{(VI.3.b-2)}$$

The matrix element is nonzero only if $r_k g r_j^{-1}$ is an element of H, with k, j running from 1 to o(G)/o(H), and t, v from 1 to d. Thus the representation matrices of G are partitioned into blocks labeled by k, j (with elements labeled by t, v), each a representation matrix of the subgroup, with 0 for blocks having no corresponding subgroup matrix (for those values of j, k that do not label a subgroup matrix), since $M(r_k g r_j^{-1})$ is not otherwise defined for them; the k, j (block) matrix element of Γ is $M(r_k g r_j^{-1})$, or 0. Each row, and each column, of blocks, that is each k, and each j, has but one nonzero block. This whole analysis assumes that for every matrix, for each k and each j, there is in fact a block that is nonzero. Otherwise the representation matrix would be singular, which is not acceptable — every group representation matrix has an inverse.

The representation of G induced from an irreducible representation of H is generally reducible. Also it is determined up to equivalence since the choice of the coset representative for each coset is arbitrary — any element of each coset can be chosen.

Problem VI.3.b-1: What representation of G is induced from a representation of H, if H is G [Jansen and Boon (1967), p. 135]? Suppose H is the identity?

Problem VI.3.b-2: Need the requirement be imposed that there be a nonzero block?

Problem VI.3.b-3: Why is there only one nonzero block for each row and each column?

Problem VI.3.b-4: State explicitly the relationship between the representation matrices of G given by different coset representatives. Show that they belong to equivalent representations.

Problem VI.3.b-5: For fixed g, show that

$$M(r_k g r_j^{-1})M(r_v g r_s^{-1}) = M(r_k g r_j^{-1} r_v g r_s^{-1}); \qquad \text{(VI.3.b-3)}$$

that is if the two group operators on the left ε H, so does that on the right, and if one does not, neither does that on the right so these M's do form a representation of H (it would be uncomfortable if one side of the equation were zero, the other not). Also the matrices belonging to the subgroup are block-diagonal, while the other matrices are not. Does this seem reasonable?

Problem VI.3.b-6: Can the procedure be limited so that the representation induced from an irreducible representation is itself irreducible?

VI.3.c The matrices give a representation of the group

First we must check that these matrices do provide a representation of G [Jansen and Boon (1967), p. 134; Streitwolf (1971), p. 41]. The double index notation of the matrix elements is reminiscent of that for direct products. So, with g an arbitrary element of G, h of the subgroup, r a coset representative, and

$$g'' = gg', \qquad \text{(VI.3.c-1)}$$

with

$$o = o(G)/o(H), \qquad \text{(VI.3.c-2)}$$

we have

$$\Gamma(g'')_{kt,jv} = \{\Gamma(g)\Gamma(g')\}_{kt,jv} = \sum_{l=1}^{o}\sum_{u=1}^{d} \Gamma(g)_{kt,lu}\Gamma(g')_{lu,jv}$$

$$= \sum_{l=1}^{o}\sum_{u=1}^{d} M(r_k g r_l^{-1})_{tu} M(r_l g' r_j^{-1})_{uv}$$

$$= \sum_{l=1}^{o} M(r_k g r_l^{-1} r_l g' r_j^{-1})_{tv} = M(r_k g g' r_j^{-1})_{tv}$$

$$= M(r_k g'' r_j^{-1})_{tv}, \text{ or } 0, \qquad \text{(VI.3.c-3)}$$

where the sum over l includes only those values for which both $r_k g r_l^{-1}$ and $r_l g' r_j^{-1}$ are in H. We have to prove the last equality, that is show that the sum has only one term, that there is only one l for which both terms are in H. Now $r_k g r_l^{-1} \varepsilon$ H means $r_k g \varepsilon Hr_l$, that is $r_k g$ belongs to coset Hr_l. But an element can belong to only one coset [Mirman (1995a), pb. IV.6.a-4,5, p. 129], so the sum over l contains no more than one term. Thus there is only one nonzero block for each row and each column, and these matrices do form a representation of G.

Why does this provide a representation? If g and g' are members of the subgroup their product is correct because M is a representation. Suppose one or both are not. These representation matrices must have the same product rules as the group. This requires that the terms in the product, and the product itself, either be in the subgroup, or, when one is a subgroup element, the product and at least one term not be in the subgroup. The method works because the definition insures this. If one term is not in the subgroup, then clearly this product cannot be either. If both terms are, then the product must also be, since the two terms are formed by insertion of a subgroup element and its inverse.

It follows from the unitarity of subgroup representation matrices M, which we can always require [Mirman (1995a), sec. VII.2.a, p. 182], that the Γ's, the group representation matrices, are unitary,

$$= M(r_k g r_j^{-1})_{tv} = M(r_j^{-1} g^{-1} r_k)^*_{vt} = \Gamma(g^{-1})^*_{jv,kt}$$

$$\text{if } r_j g^{-1} r_k \,\varepsilon\, \mathrm{H}, \ = 0, \ \text{ otherwise,}$$

$$\Gamma\,(g^{-1})^*_{jv,kt} = \Gamma(g)_{kt,jv}, \qquad \text{(VI.3.c-4)}$$

as required.

VI.4 EXAMPLES OF INDUCED REPRESENTATIONS

Before proceeding to a discussion of the properties of induced representations it is useful to consider some examples [Bradley and Cracknell (1972), p. 188]. Beyond clarifying the concepts, these also reveal aspects that may not be apparent from general definitions. The method is quite useful, but it also has limitations, and examples help us search for these.

Problem VI.4-1: If the subgroup is the identity, the representation induced is the regular representation [Mirman (1995a), chap.VI, p. 170].

VI.4.a Representations of S_3 as examples of induction

Symmetric group S_3 (which is C_{3v}, under which name it is listed in the tables) is a semi-direct product of alternating group A_3 which is (isomorphic to) cyclic group C_3, and S_2, which is C_2 (pb. I.7.a-6, p. 51), with A_3 invariant. We start the study of examples using this to find the representations of S_3 and compare them with the known ones [Mirman (1995a), sec. VIII.6, p. 238]. These subgroups are both Abelian (but not S_3) so their representation matrices are their characters (tbl. E.2-1, p. 657).

We write the operators of A_3 as

$$E,\ P = C_3 = (123),\ P^2 = C_3{}^2 = (132),\ P^3 = E, \qquad \text{(VI.4.a-1)}$$

and those of C_2 as E and $T = (12)$. Also $T' = (23)$ and $T'' = (13)$. The coset representations are E and T. We have, for the set of elements $r_k V r_j^{-1}$, with V denoting the elements of the group, for all these examples,

V	E	P	P^2	T	T'	T''
EVE	E	P	P^2	T	T'	T''
EVT	T	T''	T'	E	P^2	P
TVE	T	T'	T''	E	P	P^2
TVT	E	P^2	P	T	T''	T'.

(VI.4.a-2)

The only members of the invariant subgroup are

$$r_k V r_j^{-1} = E,\ P,\ P^2. \qquad \text{(VI.4.a-3)}$$

So for the completely symmetric representation of A_3,

$$E = \begin{pmatrix} 1 & 0 \\ 0 & 1 \end{pmatrix},\ T = T' = T'' = \begin{pmatrix} 0 & 1 \\ 1 & 0 \end{pmatrix},\ P = P^2 = \begin{pmatrix} 1 & 0 \\ 0 & 1 \end{pmatrix}. \qquad \text{(VI.4.a-4)}$$

These can be diagonalized to give the (of course irreducible) symmetric and antisymmetric representations of S_3; both subduce the same representation of A_3, the symmetric one. It is, obviously, self-conjugate. The orbit then has order 1, and the little group is S_3.

More interesting are the other representations of A_3. The representation matrices that the first gives are

$$E = \begin{pmatrix} 1 & 0 \\ 0 & 1 \end{pmatrix},\ T = \begin{pmatrix} 0 & 1 \\ 1 & 0 \end{pmatrix},\ T' = \begin{pmatrix} 0 & \varepsilon^* \\ \varepsilon & 0 \end{pmatrix},\ T'' = \begin{pmatrix} 0 & \varepsilon \\ \varepsilon^* & 0 \end{pmatrix},$$

$$P = \begin{pmatrix} \varepsilon & 0 \\ 0 & \varepsilon^* \end{pmatrix},\ P^2 = \begin{pmatrix} \varepsilon^* & 0 \\ 0 & \varepsilon \end{pmatrix}; \qquad \text{(VI.4.a-5)}$$

for the second ε and ε^* are interchanged. These matrices cannot be reduced so the representation is two-dimensional. That they do give a representation can be seen by checking their products, which are the same as those of S_3. This representation subduces the two complex representations of A_3 (not surprisingly); these are conjugate but not equivalent, and the orbit has order 2, containing these two.

These are not one of the usual forms which are real, so there is still something undone. First we diagonalize T, which we have identified as transposition (12). Then using

$$\varepsilon + \varepsilon^* = 2cos(\frac{2\pi}{3}) = -1, \quad \varepsilon - \varepsilon^* = 2isin(\frac{2\pi}{3}) = i\sqrt{3}, \qquad \text{(VI.4.a-6)}$$

the matrices become

$$E = \begin{pmatrix} 1 & 0 \\ 0 & 1 \end{pmatrix}, \; T = \begin{pmatrix} 1 & 0 \\ 0 & -1 \end{pmatrix}, \; T' = \frac{1}{2}\begin{pmatrix} -1 & -i\sqrt{3} \\ i\sqrt{3} & 1 \end{pmatrix},$$

$$T'' = \frac{1}{2}\begin{pmatrix} -1 & i\sqrt{3} \\ -i\sqrt{3} & 1 \end{pmatrix}, P = \frac{1}{2}\begin{pmatrix} -1 & -i\sqrt{3} \\ -i\sqrt{3} & -1 \end{pmatrix}, P^2 = \frac{1}{2}\begin{pmatrix} -1 & i\sqrt{3} \\ i\sqrt{3} & -1 \end{pmatrix}. \qquad \text{(VI.4.a-7)}$$

The characters are 2, 0, -1, as they should be. However this cannot give the orthogonal form of the matrices [Mirman (1995a), sec. VIII.6, p. 238]. The complex entries cannot be removed by a similarity transformation, so these are not suitable as representation matrices for symmetric or orthogonal groups (as usually interpreted; we do not consider whether they are relevant elsewhere).

Let us then start with the two-dimensional form of the representation of A_3, using the diagonal matrices for P and P^2 (eq. VI.4.a-5). We undiagonalize these — find matrices that when diagonalized give the diagonal matrices, which are

$$E = \begin{pmatrix} 1 & 0 \\ 0 & 1 \end{pmatrix}, \; P = \frac{1}{2}\begin{pmatrix} -1 & \sqrt{3} \\ -\sqrt{3} & -1 \end{pmatrix}, \; P^2 = \frac{1}{2}\begin{pmatrix} -1 & -\sqrt{3} \\ \sqrt{3} & -1 \end{pmatrix}. \qquad \text{(VI.4.a-8)}$$

Abelian matrices can always be diagonalized to give one-dimensional representations — over the complex numbers [Mirman (1995a), pb. V.5. a-4, p. 163]. If we use only real numbers, they can have higher-dimensional representations. Then, inducing from this two-dimensional representation, we get

$$E = \begin{pmatrix} 1 & 0 & 0 & 0 \\ 0 & 1 & 0 & 0 \\ 0 & 0 & 1 & 0 \\ 0 & 0 & 0 & 1 \end{pmatrix}, \; T = \begin{pmatrix} 0 & 0 & 1 & 0 \\ 0 & 0 & 0 & 1 \\ 1 & 0 & 0 & 0 \\ 0 & 1 & 0 & 0 \end{pmatrix}, \; T' = \frac{1}{2}\begin{pmatrix} 0 & 0 & -1 & -\sqrt{3} \\ 0 & 0 & \sqrt{3} & -1 \\ -1 & \sqrt{3} & 0 & 0 \\ -\sqrt{3} & -1 & 0 & 1 \end{pmatrix},$$

$$P = \frac{1}{2}\begin{pmatrix} -1 & \sqrt{3} & 0 & 0 \\ -\sqrt{3} & -1 & 0 & 0 \\ 0 & 0 & -1 & -\sqrt{3} \\ 0 & 0 & \sqrt{3} & -1 \end{pmatrix}, \quad \text{(VI.4.a-9)}$$

and so on. Diagonalizing these gives two copies of the two-dimensional representation, as can be seen from the characters, and in orthogonal form. So we first had to bring the A_3 matrices to two-dimensional orthogonal form, since those lead to a four-dimensional representation which is then reduced. The one-dimensional A_3 representations are complex, and give a complex two-dimensional S_3 representation, which cannot be changed to a two-dimensional real one. In this four-dimensional form the representation subduces both complex representations of A_3, each twice. The orbit, consisting of these two conjugate and inequivalent representations, occurs twice.

The two representations of alternating group A_3 have matrix elements (characters) $1, \varepsilon, \varepsilon^*$, and $1, \varepsilon^*, \varepsilon$, or in real form 1,1,-1, and 1,-1,1 (tbl. E.2-1, p. 657), which are inequivalent for the characters of the classes are not the same. But they are conjugate under the group of the identity and transposition (12), which interchanges elements P and P^2. However for the two-dimensional representation form, these two classes are the same, and the characters both become -2. From the requirement that the sum of the squares of the representation dimensions equal the order of the group [Mirman (1995a), pb. VI.b-1, p. 175], here 3, we find that there are three one-dimensional representations, and that there is no two-dimensional representation. Care is needed.

This illustrates the method, and its limitations, and perhaps advantages. Representation matrices are obtained, but they are not real, and not in the usual (desired) form. Further work has to be done to bring them into such a form. However it does give matrices that we might not have obtained in other ways, and perhaps not have known existed. It is possible that in some cases this could be useful. But this example provides a cautionary note — it should be remembered for there may be similar issues when the method is used for applications.

There is another difficulty with this method: it prefers an invariant subgroup. For S_n, that is subgroup A_n, but A_n is simple, $n > 4$, thus has no invariant subgroups. So half the matrices of S_n, those belonging to A_n, might have to be obtained by other methods (say by multiplication of neighboring transpositions [Mirman (1995a), sec. VIII.2.c, p. 217]). It is questionable whether there is any advantage to then using a different method to find the other half of the set.

The little group, L, of the orbit (of representations) of subgroup S, is the (not necessarily proper) subgroup of group G, under which conjugate representations of S are equivalent. It can be larger than the subgroup induced from; thus

$$G \sqsupset L \sqsupset S, \qquad \text{(VI.4.a-10)}$$

and these three may be distinct ($\sqsupset$ states that S is a subgroup of L). Here conjugation by elements of C_2 interchanges elements P and P^2 of A_3. Its representation matrices perform a similarity transformation on the representation matrices of A_3 giving another representation. But there is only one two-dimensional representation, and one symmetric one-dimensional representation; also P and P^2, which are mixed by this, are in the same class (in S_3, but not in A_3). Thus the conjugate representations in the decomposition of the two-dimensional representation are equivalent in S_3, but not in A_3. The little group is A_3. The reducible representation of S_3 contains two nonconjugate representations, so two orbits, each having one representation. For the decomposition of the four-dimensional representation, the two two-dimensional ones are equivalent, and these are mixed by the transposition. The little group consists of all operators under which these representations are equivalent in the subgroup, here only those of A_3, it thus being the little group for all representations subduced from S_3. For the four-dimensional form, the transpositions mix the two representations in the decomposition, which are equivalent, so the little group is S_3, but this representation is reducible.

Problem VI.4.a-1: Instead of the first of these representations labeled by E, we can use the second, its complex conjugate. Show that this gives the same representation, merely interchanging T' and T'', and so P and P^2, giving an automorphism of S_3 [Mirman (1995a), sec. IX.6, p. 267]. Show that the representations stated to be irreducible, are actually so. Since each element of C_3 is in a class by itself, and the characters are not the same, these two — conjugate — representations are not equivalent, giving an example showing that conjugate representations need not be equivalent, which seems to disagree with the last paragraph. Whether representations are equivalent may depend on whether they are taken over the complex numbers, or the reals [Mirman (1995a), pb. V.5.a-4, p. 163], which also is relevant to their dimension, and whether they are irreducible.

Problem VI.4.a.-2: That A_3 has a two-dimensional representation which according to the restriction on the sum of the squares of the representation dimensions is nonexistent, implies that it is a projective representation, so a vector representation of a (central) extension. Is this true? Why? What else can be said about it? Does this, plus the previous problem, suggest any general rules? What is the covering group?

Problem VI.4.a-3: Check that the matrices obtained from the E representations do give representations (of S_3).

Problem VI.4.a-4: Is it possible to write the two-dimensional complex matrices as four-dimensional real ones, and then reduce these to give the orthogonal matrices?

Problem VI.4.a-5: The representations found are not in the form of symmetric group representations that we are familiar with. However the symmetric group is but one realization of the abstract group. Another is the group of transformations of a triangle, with the cyclic subgroup that of rotations, while the transpositions interchange vertices. But we have found representations with complex entries, which we might not expect for such transformations. Consider a statefunction defined over a circle, but perhaps not the entire circle, just three points. Construct a model of a physical system for which these representations, in the form given, are relevant. Is it surprising that there are different types of representations?

Problem VI.4.a-6: For an Abelian group, what is the largest dimension of the representation matrices over the reals?

VI.4.b Induced representations of D_4

Next is dihedral group D_4 [Bradley and Cracknell (1972), p. 188; Cornwell (1984), p. 124], a semi-direct product (denoted by $\wedge$) of two cyclic groups, C_4 and C_2, with C_4 invariant (sec. I.7.a, p. 50),

$$D_4 = C_4 \wedge C_2' = D_2 \wedge C_2''. \qquad \text{(VI.4.b-1)}$$

The operators of C_4 (tbl. B-2, p. 634) are

$$C_4 : \{E, C_{4z}, C_{2z}, C_{4z}^{-1}\}, \qquad \text{(VI.4.b-2)}$$

with

$$C_{2z} = C_{4z}^{2}, \text{ and } C_{4z}^{-1} = C_{4z}^{3}. \qquad \text{(VI.4.b-3)}$$

For C_2,

$$C_2 : \{E, C_{2x}\}. \qquad \text{(VI.4.b-4)}$$

Also

$$C_{2y} = C_{2z}C_{2x}, \; C_{2c} = C_{4z}C_{2x}, \; C_{2d} = C_{4z}^{-1}C_{2x}. \qquad \text{(VI.4.b-5)}$$

We need $r_k V r_j^{-1}$ for all eight V of D_4, and all four pairs r_k, r_j of coset representatives. The only nonzero blocks are those for which $r_k V r_j$ are in C_4. Only one element of each of the five classes has to be considered. For the 1,1 element,

$$r_1 = E, \; r_k V r_j^{-1} = V; \text{ all } V \text{ of } D_4. \qquad \text{(VI.4.b-6)}$$

For the 1,2, 2,1, and 2,2 elements ($C_{2x}VC_{2x}$),

V	E	C_{2x}	C_{2y}	C_{2z}	C_{2c}	C_{4z}	C_{2d}	C_{4z}^{-1}
EVC_{2x}	C_{2x}	E	C_{2z}	C_{2y}	C_{4z}	C_{2c}	C_{4z}^{-1}	C_{2d}
$C_{2x}VE$	C_{2x}	E	C_{2z}	C_{2y}	C_{4z}^{-1}	C_{2d}	C_{4z}	C_{2c}
$C_{2x}VC_{2x}$	E	C_{2x}	C_{2y}	C_{2z}	C_{2d}	C_{4z}^{-1}	C_{2c}	C_{4z}.

(VI.4.b-7)

An induced representation of G is given by the 2 by 2 matrices

$$\Gamma(V)_{kt,jv} = M(r_k V r_j^{-1})_{tv}, \text{ if } r_k V r_j^{-1} \varepsilon C_4,$$

$$= 0, \text{ otherwise,} \qquad \text{(VI.4.b-8)}$$

for all V of G, with coset labels k, j running from 1 to 2, and t, r from 1 to 1. Both subgroups are Abelian so all representations are one-dimensional; they therefore have four and two representations respectively. We have the characters, the representation matrices (tbl. E.2-1, p. 657). The induced representations are all two-dimensional, since those of the subgroup are all one-dimensional, although most are reducible. Since the sum of the squares of the dimensions of the irreducible representations equals the order of the group, here 8, there are four one-dimensional irreducible representations, and one two-dimensional one.

The matrices for the identity are all 1. The one-dimensional forms of the representations of C_4 listed last in the table are denoted here by E' and E''. For one-dimensional representations,

$$k = j = 1, \; r_1 = E; \; k = j = 2, \; r_2 = C_{2x}. \qquad \text{(VI.4.b-9)}$$

Thus we have the matrices from C_4 representation A,

$$E = \begin{pmatrix} 1 & 0 \\ 0 & 1 \end{pmatrix}, \; C_{4z} = \begin{pmatrix} 1 & 0 \\ 0 & 1 \end{pmatrix}, \; C_{4z}^{-1} = \begin{pmatrix} 1 & 0 \\ 0 & 1 \end{pmatrix},$$

$$C_{2z} = \begin{pmatrix} 1 & 0 \\ 0 & 1 \end{pmatrix}, \; C_{2x} = \begin{pmatrix} 0 & 1 \\ 1 & 0 \end{pmatrix} = C_{2y} = C_{2c} = C_{2d}, \qquad \text{(VI.4.b-10)}$$

and from representation B,

$$E = \begin{pmatrix} 1 & 0 \\ 0 & 1 \end{pmatrix}, \; C_{2z} = \begin{pmatrix} 1 & 0 \\ 0 & 1 \end{pmatrix}, \; C_{4z} = \begin{pmatrix} -1 & 0 \\ 0 & -1 \end{pmatrix}, \; C_{4z}^{-1} = \begin{pmatrix} -1 & 0 \\ 0 & -1 \end{pmatrix},$$

$$C_{2x} = \begin{pmatrix} 0 & 1 \\ 1 & 0 \end{pmatrix} = C_{2y}, \; C_{2c} = \begin{pmatrix} 0 & -1 \\ -1 & 0 \end{pmatrix} = C_{2d}, \qquad \text{(VI.4.b-11)}$$

The matrices of both of these representations can be immediately diagonalized. The matrices that are diagonal stay so, the others become proportional to

$$\begin{pmatrix} 1 & 0 \\ 0 & -1 \end{pmatrix}.$$

Thus for the first line coming from the first set, all matrices have value 1, so go with representation A_1 of D_4. The second line of this set has all matrices 1, except -1 for C_{2x}, C_{2y}, C_{2c}, C_{2d}. This belongs to A_2. For the second set, the first line gives all matrices 1, except for -1 for C_{2c}, C_{2d}, C_{4z}, ${C_{4z}}^{-1}$. This goes with B_1. The second line again has all 1's except -1 for C_{2x}, C_{2y}, C_{4z}, ${C_{4z}}^{-1}$. This is representation B_2.

From representation E' we obtain

$$E = \begin{pmatrix} 1 & 0 \\ 0 & 1 \end{pmatrix}, C_{4z} = \begin{pmatrix} i & 0 \\ 0 & -i \end{pmatrix}, {C_{4z}}^{-1} = \begin{pmatrix} -i & 0 \\ 0 & i \end{pmatrix},$$

$$C_{2z} = \begin{pmatrix} -1 & 0 \\ 0 & -1 \end{pmatrix}, C_{2x} = \begin{pmatrix} 0 & 1 \\ 1 & 0 \end{pmatrix}, C_{2y} = \begin{pmatrix} 0 & -1 \\ -1 & 0 \end{pmatrix},$$

$$C_{2c} = \begin{pmatrix} 0 & i \\ -i & 0 \end{pmatrix}, C_{2d} = \begin{pmatrix} 0 & -i \\ i & 0 \end{pmatrix}. \quad \text{(VI.4.b-12)}$$

Since the diagonal matrices are not proportional to the unit matrix, as for the other representations, this cannot be reduced to one-dimensional representations, so is two-dimensional. The E'' representation also gives a two-dimensional representation of D_4, but as there is only one, the two are equivalent. This is not surprising since it differs only by the interchange of C_{4z} and ${C_{4z}}^{-1}$, but these are equivalent.

The positions of the zero matrix entries are the same for all representations, it is only in the values of the nonzero entries that the representations differ.

For the first representation listed we get, as we should, the A representation of C_4 subduced twice, for the second the B subduced twice. From the last, the two E representations are subduced once each. These two are conjugate, but not equivalent, and the orbit, appearing once, consists of them. The little group is just C_4.

Problem VI.4.b-1: Find the representation matrices using E'', and show that the representations are equivalent.

Problem VI.4.b-2: Verify that all these are representations, using the group multiplication table, and also by interpreting the operators as transformations, finding basis vectors that have geometrical meaning, and checking that the matrices act properly on them.

Problem VI.4.b-3: Use this method to obtain the representations of D_3.

Problem VI.4.b-4: Using the representations of D_3, find those of D_6.

VI.4.c The representations of tetrahedral group T (A_4)

While the representations of A_n cannot generally be found by induction from a normal subgroup, there is one more case for which the method works (pb. I.7.a-4, p. 51),

$$T = D_2 \wedge C_3' = C_2 \otimes C_2' \wedge C_3'. \qquad \text{(VI.4.c-1)}$$

For these we use the notation for the permutations realizing the groups related to that of the transformations listed in appendix B, p. 631 by

$$E,\ C_3 = (123),\ \ C_3^{\,2} = (132), \qquad \text{(VI.4.c-2)}$$

for C_3, and for D_2,

$$E,\ C_2 = (12)(34),\ C_{2y} = (13)(24),\ C_{2x} = (14)(23). \qquad \text{(VI.4.c-3)}$$

The coset representatives are E, (123) and (132). Since there are three of these, the induced representation is three-dimensional. Is it irreducible? To study this we do not need all matrices, and thus give an incomplete list:

V	E	(12)(34)	(13)(24)	(14)(23)
EVE	E	(12)(34)	(13)(24)	(14)(23)
(123)VE	(123)	(134)	(243)	(142)
EV(123)	(123)	(243)...	...	
(123)V(123)	(132)	(124)...	(143)	
(123)V(132)	E	(14)(23)	(12)(34)	(13)(24)
(132)VE	(132)	(234)...	...	
EV(132)	(132)	(143)...	...	
(132)V(123)	E	(13)(24)	(14)(23)	(12)(34)
(132)V(132)	(123)	(142)...	...	

V	(123)	(132)	(124)	(142)
EVE	(123)	(132)	(124)	(142)
(123)VE	(132)	E	(13)(24)	(143)
EV(123)	(132)	E	(14)(23)	(234)
(123)V(123)	E	(123)	(142)	(12)(34)
(123)V(132)	(123)	(132)	(234)	...
(132)VE	E	(123)	(243)	(14)(23)
EV(132)	E	(123)	(134)	(13)(24)
(132)V(123)	(123)	(132)	(143)	...
(132)V(132)	(132)	E	(12)(34)	... .

(VI.4.c-4)

Representation A of D_2 gives again the symmetric and antisymmetric representations. We consider then a B representation which gives

$$E = \begin{pmatrix} 1 & 0 & 0 \\ 0 & 1 & 0 \\ 0 & 0 & 1 \end{pmatrix}, \ (12)(34) = \begin{pmatrix} -1 & 0 & 0 \\ 0 & -1 & 0 \\ 0 & 0 & 1 \end{pmatrix}, \ (13)(24) = \begin{pmatrix} 1 & 0 & 0 \\ 0 & -1 & 0 \\ 0 & 0 & -1 \end{pmatrix},$$

$$(14)(23) = \begin{pmatrix} -1 & 0 & 0 \\ 0 & 1 & 0 \\ 0 & 0 & -1 \end{pmatrix}, \ (123) = \begin{pmatrix} 0 & 1 & 0 \\ 0 & 0 & 1 \\ 1 & 0 & 0 \end{pmatrix}, \ (132) = \begin{pmatrix} 0 & 0 & 1 \\ 1 & 0 & 0 \\ 0 & 1 & 0 \end{pmatrix},$$

$$(124) = \begin{pmatrix} 0 & 0 & -1 \\ 1 & 0 & 0 \\ 0 & -1 & 0 \end{pmatrix}, \ (142) = \begin{pmatrix} 0 & 1 & 0 \\ 0 & 0 & -1 \\ -1 & 0 & 0 \end{pmatrix}. \qquad \text{(VI.4.c-5)}$$

This is irreducible as seen from the characters and dimension; it is the τ representation of T.

Problem VI.4.c-1: Check that the statement about representation A is correct.

Problem VI.4.c-2: Finish this, find all matrices, verifying that this is a representation, and that it is really irreducible. It is a representation of two (or more?) different, although isomorphic, groups (that is different realizations of the same abstract group). Show that it is a representation of both, using the proper realizations for them. Perform a similarity transformation to bring the matrices into orthogonal form [Mirman (1995a), sec. VIII.6, p. 238]. Does the transformation matrix have only real entries? This (probably) requires finding the neighboring transpositions for S_4, and then from them computing the matrices for A_4, or at least a few. It is however possible to check that these matrices are at least plausible. Which B representation was used to get these matrices? How do the induced representations given by the other B representations differ? Are the representations related by an automorphism? Which?

Problem VI.4.c-3: Write D_2 as the product of two cyclic groups. Express these in terms of permutations.

Problem VI.4.c-4: In this form the matrices are difficult to interpret as permutations. Do they have any interpretation, perhaps geometrical or physical, maybe in terms of the tetrahedral group?

Problem VI.4.c-5: For all these representations, find the conjugate subduced representations of the subgroup, giving their orbits. Which are equivalent? What are their little groups?

VI.4.d Induced representations of octahedral group O

The previous examples have a limitation: the representations induced from are one-dimensional. To study a case in which they have larger dimension (here 3) we consider octahedral group O [Bradley and Cracknell (1972), p. 199], which is symmetric group S_4, a semi-direct product

$$O = T \wedge C_2' \qquad \text{(VI.4.d-1)}$$

of the invariant tetrahedral group T and cyclic group C_2 (pb. I.7.a-5, p. 51), whose representations are discussed above (sec. V.2.f, p. 255). For the tetrahedral group (alternating group A_4), there are three one-dimensional representations, the completely symmetric one, plus the two complex conjugates of each other, obtained from the antisymmetric representation of S_4, and a three-dimensional one. One interesting question is whether the induced representations are of a form given previously [Mirman (1995a), sec. VIII.6, p. 238], and whether we wish them to be?

The tetrahedral group has elements that can be realized as permutations (123), (12)(34), and so on. The coset representatives we take are E and C_2, which is transposition (12).

We expect that the A and E representations of T induce the O representations of one and two dimensions, as before. We thus consider only representation τ, whose matrices we have just found. Since A_4 consists of permutations with an even number of transpositions, we need list only some in the table; the other terms give zero and the particular permutation does not matter. (It is useful here to write permutations as (12)(3)(4), and so on). Of the listed permutations, those that belong to T are (12)(34), (13)(24), (14)(23), (123), (132), (124) and (142).

Thus

V	E	(12)	(12)(34)	(13)(24)
EVE	E	(12)	(12)(34)	(13)(24)
$(12)VE$	(12)	E	(34)	(1324)
$EV(12)$	(12)	E	...	
$(12)V(12)$	E	(12)	(12)(34)	(14)(23)

V	(14)(23)	(123)	(132)	(124)	(1234)
EVE	(14)(23)	(123)	(132)	(124)	(1234)
$(12)VE$		(23)	(13)	(24)	
$EV(12)$	...				
$(12)V(12)$	(13)(24)	(132)	(123)	(142)	(1342).

(VI.4.d-2)

The matrices are 6×6, so not irreducible. We expect that they reduce to two 3 × 3 representations. So we obtain

$$E=\begin{pmatrix}1&0&0&0&0&0\\0&1&0&0&0&0\\0&0&1&0&0&0\\0&0&0&1&0&0\\0&0&0&0&1&0\\0&0&0&0&0&1\end{pmatrix},\ (12)=\begin{pmatrix}0&0&0&1&0&0\\0&0&0&0&1&0\\0&0&0&0&0&1\\1&0&0&0&0&0\\0&1&0&0&0&0\\0&0&1&0&0&0\end{pmatrix},$$

$$(12)(34)=\begin{pmatrix}-1&0&0&0&0&0\\0&-1&0&0&0&0\\0&0&1&0&0&0\\0&0&0&-1&0&0\\0&0&0&0&-1&0\\0&0&0&0&0&1\end{pmatrix},\ (13)(24)=\begin{pmatrix}1&0&0&0&0&0\\0&-1&0&0&0&0\\0&0&-1&0&0&0\\0&0&0&-1&0&0\\0&0&0&0&1&0\\0&0&0&0&0&-1\end{pmatrix},$$

$$(14)(23)=\begin{pmatrix}-1&0&0&0&0&0\\0&1&0&0&0&0\\0&0&-1&0&0&0\\0&0&0&-1&0&0\\0&0&0&0&1&0\\0&0&0&0&0&-1\end{pmatrix},\ (123)=\begin{pmatrix}0&1&0&0&0&0\\0&0&1&0&0&0\\1&0&0&0&0&0\\0&0&0&0&0&1\\0&0&0&1&0&0\\0&0&0&0&1&0\end{pmatrix},$$

$$(132)=\begin{pmatrix}0&0&1&0&0&0\\1&0&0&0&0&0\\0&1&0&0&0&0\\0&0&0&0&1&0\\0&0&0&0&0&1\\0&0&0&1&0&0\end{pmatrix}. \qquad \text{(VI.4.d-3)}$$

Problem VI.4.d-1: These now have to be reduced into two. How do these two differ? Are they related by an automorphism? Which? Verify that the two are irreducible. Find the remaining matrices.

Problem VI.4.d-2: Check that our expectations of what the A and B representations give are correct.

Problem VI.4.d-3: Prove that all these are actually representations, and of both the symmetric and octahedral groups, using the proper realizations for them.

Problem VI.4.d-4: For each irreducible representation, find the subduced irreducible representations, and their conjugate(s). Are the conjugates equivalent? What are the orbits of the representations? What are the little groups?

Problem VI.4.d-5: While it is preferable to induce from the largest (invariant) subgroup, it is educational to also look at what happens when using a smaller subgroup. The octahedral group can be written

$$O=T\wedge C_2'=D_2\wedge C_3'\wedge C_2'=D_2\wedge(C_3'\wedge C_2')=D_2\wedge D_3. \qquad \text{(VI.4.d-4)}$$

Find the representations, by induction, of D_3. From them, find those of O. Discuss the differences between (the form of) these, and the ones obtained using the maximal invariant subgroup. Are there important differences in the procedure? For this there is a little group that is neither O nor the invariant subgroup [Bradley and Cracknell (1972), p. 192]; check that it is D_4.

VI.4.e Inducing from a nonnormal subgroup

The examples that we have considered, and the applications that we will study, use normal subgroups. Suppose the subgroup is not invariant. A simple example is given by S_3, which we have considered, but for which we now use nonnormal subgroup C_2 consisting of the identity E and transposition T = (12). We take the coset representatives as E and transpositions $T' = (23)$, $T'' = (13)$. So, with $P = (123)$, $P^2 = (132)$,

V	E	T	T'	T''	P	P^2
EVE	E	T	T'	T''	P	P^2
$T'VE$	T'	P^2	E	P	T''	T
EVT'	T'	P	E	P^2	T	T''
$T'VT'$	E	T''	T'	T	P^2	P
$T''VE$	T''	P	P^2	E	T	T'
EVT''	T''	P^2	P	E	T'	T
$T''VT''$	E	T'	T	T''	P^2	P
$T''VT'$	P^2	T	T''	T'	P	E
$T'VT''$	P	T	T''	T'	E	P^2 .

(VI.4.e-1)

Using the symmetric representation of C_2, we get the induced representation matrices

$$E = \begin{pmatrix} 1 & 0 & 0 \\ 0 & 1 & 0 \\ 0 & 0 & 1 \end{pmatrix}, \; T = \begin{pmatrix} 1 & 0 & 0 \\ 0 & 0 & 1 \\ 0 & 1 & 0 \end{pmatrix}, \; T' = \begin{pmatrix} 0 & 1 & 0 \\ 1 & 0 & 0 \\ 0 & 0 & 1 \end{pmatrix},$$

$$T'' = \begin{pmatrix} 0 & 0 & 1 \\ 0 & 1 & 0 \\ 1 & 0 & 0 \end{pmatrix}, \; P = \begin{pmatrix} 0 & 1 & 0 \\ 0 & 0 & 1 \\ 1 & 0 & 0 \end{pmatrix}, \; P^2 = \begin{pmatrix} 0 & 0 & 1 \\ 1 & 0 & 0 \\ 0 & 1 & 0 \end{pmatrix}. \quad \text{(VI.4.e-2)}$$

Again the matrices are not in a standard form, and are clearly reducible — there are no three-dimensional representations of S_3.

The characters, which are used for the reduction, are, for the symmetric, antisymmetric, and two-dimensional representations, (1,1,1), (1,-1,1) and (2,0,-1). The representations in the decomposition are the symmetric and the two-dimensional ones, each appearing once. The subduced representations are the symmetric, appearing twice, and one copy of the antisymmetric representation. Here, since the subgroup is not normal, the representations related by conjugation are not that of the same subgroup. Thus many of the concepts introduced for normal subgroups do not apply.

The antisymmetric representation of C_2 gives

$$E=\begin{pmatrix}1&0&0\\0&1&0\\0&0&1\end{pmatrix},\ T=\begin{pmatrix}-1&0&0\\0&0&-1\\0&-1&0\end{pmatrix},\ T'=\begin{pmatrix}0&1&0\\1&0&0\\0&0&-1\end{pmatrix},$$

$$T''=\begin{pmatrix}0&0&1\\0&-1&0\\1&0&0\end{pmatrix},\ P=\begin{pmatrix}0&-1&0\\0&0&1\\-1&0&0\end{pmatrix},\ P^2=\begin{pmatrix}0&0&-1\\-1&0&0\\0&1&0\end{pmatrix}.$$

(VI.4.e-3)

The representations in the decomposition are now the antisymmetric and the two-dimensional ones, again each appearing once.

Problem VI.4.e-1: Are there significance differences between these results and ones obtained using a normal subgroup? Why?

VI.4.f Representations of S_4 from S_3

The octahedral group, symmetric group S_4, has S_3 subgroups, for example one that acts on numbers 1,2,3. The S_3 subgroup is not normal for there are four such, acting on any three of the four numbers 1,2,3,4, and these are taken into each other by transformations of S_4. Inducing from S_3 has some interesting aspects. In particular, we can use S_3 representations in orthogonal form. What does that lead to?

The coset representatives are E, (14), (24), (34); these form a direct product of three C_2 groups. Similarity transformations of S_3 by these give S_3 groups acting on (1,2,4), (1,3,4), and (2,3,4), which are all isomorphic. We label them S_3 (123), S_3 (124), S_3 (134), and S_3 (234). S_4 is the union of these S_3 subgroups, plus the set of permutations (1234), (1324), (2134), (2143), (3124), (3142).

Induced representations of S_4 are four-dimensional blocks of matrices (there are four cosets), and are thus four- or eight- dimensional, for one- and two-dimensional representations of S_3. The irreducible representations of S_4 have dimensions 1,1,2,3,3. Hence the induced representations are reducible. The characters of the S_3 irreducible representations, (1,1,1), (1,-1,1) and (2,0,-1), are also the characters of the induced reducible S_4 representations, since these have only a single block on the diagonal. We can now use these, and the characters of the S_4 irreducible representations, to find the decompositions. The S_4 operators that belong to S_3 subgroups are in the classes of E, $8C_3$, and $6C_2$.

We will be content here with just the decomposition of the induced representations into irreducible ones. Thus we need the characters of the classes so only one element of each class. The positions of the nonzero diagonal elements are the same for all subgroup representations (although some are blocks), and we first give the members of each class that we use, with their diagonal elements, the nonzero ones being indicated by n. These classes are

$$[E:(1111)],\ [(123):(n000)],\ [(12):(n00n)],$$

$$[(12)(34):(0000)],\ [(1234):(0000)]. \qquad \text{(VI.4.f-1)}$$

We then get, for the S_3 symmetric representation, the characters η_c for each class c,

$$\eta_{[E(1111)]} = 4,\ \eta_{[(123)(1000)]} = 1,\ \eta_{[(12)(1001)]} = 2,$$

$$\eta_{[(12)(34)(0000)]} = 0,\ \eta_{[(1234)(0000)]} = 0; \qquad \text{(VI.4.f-2)}$$

for the antisymmetric representation,

$$\eta_{[E(1111)]} = 4,\ \eta_{[(123)(1000)]} = 1,\ \eta_{[(12)(-100-1)]} = -2,$$

$$\eta_{[(12)(34)(0000)]} = 0,\ \eta_{[(1234)(0000)]} = 0; \qquad \text{(VI.4.f-3)}$$

and for the two-dimensional one,

$$\eta_{[E(11111111)]} = 8, \eta_{[(123)(.5)(-1-1000000)]} = -1, \eta_{[(12)(.5)(1-100001-1)]} = 0,$$

$$\eta_{[(12)(34)(00000000)]} = 0,\ \eta_{[(1234)(00000000)]} = 0. \qquad \text{(VI.4.f-4)}$$

The first of these decomposes into the one-dimensional symmetric representation of S_4, plus the three-dimensional τ_2 representation. The second reduces to the one-dimensional antisymmetric representation of S_4, plus the three-dimensional τ_1 representation. The relationship between these reductions is as expected. The eight-dimensional representation has the decomposition into one each of the E, τ_1 and τ_2 representations. These are all inequivalent.

Problem VI.4.f-1: Finish this. Find the representation matrices for all the various subgroup representations, diagonalize them, and check that the decompositions are correct. The form of the matrices depends on the form of the S_3 matrices. If the orthogonal type is used for the latter, are the induced matrices also orthogonal?

Problem VI.4.f-2: Discuss how these results and analyses are dependent on the subgroup being noninvariant, and differ from those of cases with an invariant subgroup.

Problem VI.4.f-3: Alternating group A_5 has subgroup A_3, which is not normal (it cannot be since A_5 is simple [Mirman (1995a), pb. IV.9-5,

p. 143]). There are twenty coset representatives, so the induced representation is 20-dimensional, and of course, reducible. In it representations of A_3 occur several times; these are not only conjugate, but equivalent. The little group of (at least some of) these is A_4, providing an example of a little group larger than the subgroup, but smaller than the group. Prove this. Also work out the induced representations, and check these statements explicitly. Explain the effect of the subgroup not being normal.

Problem VI.4.f-4: A_5 also has subgroup A_4, which cannot be invariant. Is it possible to tell from the induced (or subduced?) representations that it is not normal? Is it possible to tell that A_5 is simple? Can this be done for general A_n? Are the results different for S_n? Why?

Problem VI.4.f-5: We have now obtained representations in several different ways. State explicitly the differences in the procedures, emphasizing the effects of the subgroup being, and not being, normal. What other differences are there in the subgroups? What effects do these have? Which methods seems likely to be useful in the general case of, especially, larger and more complicated groups? Why?

VI.5 PROPERTIES OF INDUCED REPRESENTATIONS

These examples should help us understand the meaning of induced representations. But before we can apply them to space groups there are various aspects still to be considered, both for special and general cases. So next we mention some of these.

We have considered the special, but important, cases with normal subgroups. While these are sufficient to give the representations of the space groups, it is useful to state results in general, not taking the subgroup normal, both to understand more fully special cases, and because of potential value elsewhere (assuming, of course, that we do not in some hidden way require the subgroup be invariant.)

VI.5.a Basis functions of induced representations

Further insight into the method can be gained by considering the induced basis functions [Inui, Tanabe and Onodera (1990), p. 82]. Group G with subgroup H is decomposed into left cosets

$$G = R_1H + R_2H + \ldots + R_kH, \quad R_1 = E. \qquad \text{(VI.5.a-1)}$$

The basis states of subgroup H are denoted by $\phi_\mu^\lambda, \mu = 1, \ldots, d$, for d-dimensional representation λ (a symbol usually suppressed). Then

$$h\phi_\nu = \sum M(h)_{\nu\mu}\phi_\mu, \quad h\varepsilon H. \tag{VI.5.a-2}$$

Next define the functions

$$\psi_{i\nu} = R_i\phi_\nu, \quad i = 1, \ldots, k; \; \nu = 1, \ldots, d. \tag{VI.5.a-3}$$

We wish to show that these form a basis of a representation of G. Any operator P of G can be expressed as a product of a coset representative and a member of H,

$$P = R_j h_l, \tag{VI.5.a-4}$$

so

$$P\psi_{i\nu} = R_j h_l R_i\phi_\nu = R_j R_k h_l\phi_\nu = \sum R_m M(h_l)_{\nu\mu}\phi_\mu$$

$$= \sum M(h_l)_{\nu\mu} R_m\phi_\mu = \sum M(h_l)_{\nu\mu}\psi_{m\mu}, \tag{VI.5.a-5}$$

for some k and m. Thus these $\psi_{m\mu}$ form a closed set under the group operators — a set of basis states.

For any $h \; \varepsilon \; H$ there is a coset representative R_j such that

$$hR_i = R_j h, \quad h = R_j^{-1} h R_i, \quad h = R_j h R_i^{-1}, \tag{VI.5.a-6}$$

giving

$$h\psi_{i\nu} = hR_i\phi_\nu = R_j h\phi_\nu = R_j \sum M(h)_{\nu\mu}\phi_\mu = \sum M(h)_{\nu\mu} R_j\phi_\mu$$

$$= \sum M(h)_{\nu\mu}\psi_{j\mu} = \sum M(R_j^{-1} h R_i)_{\nu\mu}\psi_{j\mu}. \tag{VI.5.a-7}$$

Now

$$h_l = R_j^{-1} P, \tag{VI.5.a-8}$$

so

$$h_l\psi_{i\nu} = R_j^{-1} P R_i\phi_\nu = \sum_\rho M(h_l)_{\nu\rho}\psi_{i\rho} = \sum_\rho M(h_l)_{\nu\rho} R_i\phi_\rho, \tag{VI.5.a-9}$$

which leads to

$$P\psi_{i\nu} = \sum \Gamma(P)_{i\nu,j\mu}\psi_{j\mu} = \sum_\rho M(h_l)_{\nu\rho} R_j R_i\phi_\rho, \tag{VI.5.a-10}$$

defining Γ. With $R_j^{-1} P R_i \; \varepsilon \; H$,

$$R_j^{-1} P R_i\phi_\nu = h_l\phi_\nu = \sum M(R_j^{-1} P R_i)_{\nu\mu}\phi_\mu, \tag{VI.5.a-11}$$

therefore

$$P\psi_{i\nu} = \sum M(R_j^{-1}PR_i)_{\nu\mu}R_j\phi_\mu = \sum M(R_j^{-1}PR_i)_{\nu\mu}\psi_{j\mu}, \quad \text{(VI.5.a-12)}$$

hence

$$\Gamma(P)_{i\nu,j\mu} = M(R_j^{-1}PR_i)_{\nu\mu}, \quad \text{(VI.5.a-13)}$$

giving the induced matrix for operator P. However if $R_j^{-1}PR_i$ is not in the subgroup, we take the matrix as zero since $R_j^{-1}PR_i$ then produces a basis vector that is not in the defined set of subgroup basis vectors. This indicates why the definition of induced representation was chosen.

Consider the completely symmetric representation of A_3, the invariant subgroup of S_3 (sec. VI.4.a, p. 294), and write

$$|s) = 123 + 231 + 312, \quad \text{(VI.5.a-14)}$$

(one realization of) the basis vector of a one-dimensional representation. So

$$(123)|s) = 231 + 312 + 123, \quad \text{(VI.5.a-15)}$$

$$(132)|s) = 312 + 123 + 231. \quad \text{(VI.5.a-16)}$$

Using the coset representatives E and (12), we define

$$|s,1) = |s), \quad |s,2) = (12)|s) = 213 + 132 + 321. \quad \text{(VI.5.a-17)}$$

We need consider only neighboring transpositions [Mirman (1995a), sec. IX.5.a.iii, p. 263]. Now (ignoring normalization throughout),

$$(12)|s,1) = |s,2), \quad (12)|s,2) = |s,1). \quad \text{(VI.5.a-18)}$$

Under (23) these become

$$(23)|s,1) = (23)|s) = 132 + 321 + 213 = |s,2),$$

$$(23)|s,2) = (23)(12)|s) = (23)\{213 + 132 + 321\}$$
$$= 312 + 123 + 231 = |s,1), \quad \text{(VI.5.a-19)}$$

$$(13)|s,1) = (13)\{123 + 231 + 312\} = 321 + 213 + 132 = |s,2). \quad \text{(VI.5.a-20)}$$

These two states go into themselves, thus form a basis for a representation of S_3.

What is the effect of

$$E(123)(12) = (13), \quad \text{(VI.5.a-21)}$$

which is not in the subgroup (which contains no transposition)? Its matrix element, for the induced representation, is zero. From the definition (eq. VI.3.b-2, p. 291), with $t = v = 1, j, k = 1, 2$, this is

$$\Gamma((13))_{11} = M(E(13)E) = M((13)) = 0, \quad \text{(VI.5.a-22)}$$

as there is no representation matrix in the subgroup for (13). Also

$$\Gamma((13))_{12} = M(E(13)(12)) = M((123)) = 1. \quad \text{(VI.5.a-23)}$$

For this, and the other transpositions, the only nonzero elements are off-diagonal, giving the action of these operators on the basis states in agreement with the definition of an induced representation. However subgroup element (123) is a product of (13) and (12), so has diagonal elements, both 1, as we found (sec. VI.4.a, p. 294). The induced representation is two-dimensional and reduces to the symmetric and antisymmetric representations of S_3, these given by basis states

$$|sr) = |s,1) + |s,2), \text{ and } |ar) = |s,1) - |s,2). \quad \text{(VI.5.a-24)}$$

For the E representation of A_3 we use

$$|s) = (123) + \varepsilon(231) + \varepsilon^*(312). \quad \text{(VI.5.a-25)}$$

Again

$$|s,1) = |s), \ |s,2) = (12)|s) = (213) + \varepsilon(132) + \varepsilon^*(321), \quad \text{(VI.5.a-26)}$$

so

$$(12)|s,1) = |s,2), \ (12)|s,2) = |s,1), \quad \text{(VI.5.a-27)}$$

while

$$(23)|s,1) = (23)|s) = (132) + \varepsilon(321) + \varepsilon^*(213) = \varepsilon^*|s,2),$$

$$(23)|s,2) = (23)(12)|s) = (23)\{(213) + \varepsilon(132) + \varepsilon^*(321)\}$$
$$= (312) + \varepsilon(123) + \varepsilon^*(231) = \varepsilon|s,1). \quad \text{(VI.5.a-28)}$$

Permutation (123), a subgroup member, has diagonal elements, here ε and ε^*, in agreement with what we know, giving the two-dimensional representation of S_3.

For another example we consider the basis states for induction from a nonnormal subgroup (sec. VI.4.e, p. 305), starting with the subgroup symmetric representation, which has its one basis state

$$|s) = 123 + 213. \quad \text{(VI.5.a-29)}$$

The action of the coset representatives on this is

$$|s)' = T'|s) = (23)|s) = 132 + 312, \quad \text{(VI.5.a-30)}$$

$$|s)'' = T''|s) = (13)|s) = 321 + 231. \quad \text{(VI.5.a-31)}$$

Also

$$T|s) = |s),\ T|s)' = (12)(132 + 312) = (231 + 321) = |s)'', \quad \text{(VI.5.a-32)}$$

$$T|s)'' = (12)(231 + 321) = (132 + 312) = |s)', \quad \text{(VI.5.a-33)}$$

$$T'|s)' = |s),\ T'|s) = (23)(123 + 213) = 132 + 312 = |s)', \quad \text{(VI.5.a-34)}$$

$$T''|s)'' = |s)'', \quad \text{(VI.5.a-35)}$$

$$P|s) = (123)(123 + 213) = 231 + 321 = |s)'', \quad \text{(VI.5.a-36)}$$

$$P|s)' = (123)(132 + 312) = 213 + 123 = |s), \quad \text{(VI.5.a-37)}$$

$$P|s)'' = (123)(321 + 231) = 132 + 312 = |s)', \quad \text{(VI.5.a-38)}$$

so we see that these do form a basis of a representation of S_3.

Problem VI.5.a-1: Show, using all R's and all h's, that

$$R_j h_l R_i = R_j R_k h_l \quad \text{(VI.5.a-39)}$$

can always be satisfied.

VI.5.b Conjugate representations, little groups, and orbits

With these results we can now find examples of the relationships, fundamental for the following development, between representations that we have introduced.

For the two-dimensional representation of S_3, transposition T gives the conjugate representation — it interchanges P and P^2 (and also T' and T'', though these are not in the A_3 subgroup). Thus the conjugate representations are obtained from each other by $\varepsilon \Longleftrightarrow \varepsilon^*$, this interchanges the two E representations, so these are conjugate and form an orbit. They are not equivalent since T, not being a member of the subgroup, has no representation matrix (and could not have, since it is represented in S_3 by a two-dimensional matrix, not a one-dimensional one). And the characters of these A_3 classes are different. The little group is just the subgroup.

Problem VI.5.b-1: Discuss D_4. Notice which operators of the subgroup are interchanged, and which not.

Problem VI.5.b-2: The tetrahedral group is similar.

Problem VI.5.b-3: For the octahedral group, elements (13)(24) and (14)(23) are interchanged by (12). Here (12) is also three-dimensional. Thus conjugate representations also can (but may not) be equivalent. What are the orbits of the two three-dimensional representations?

Problem VI.5.b-4: For which of these examples is the little group the subgroup induced from, and for which is it the group itself? Are there any that do not fit into these two categories?

Problem VI.5.b-5: The structures of other groups have been given (sec. I.7.a, p. 50). Use these to induced the representations of the remaining (crystallographic) dihedral groups. Can this procedure be used to obtain a general formula, or algorithm, for the representations (or at least characters) of any dihedral group?

Problem VI.5.b-6: It should be possible to obtain general formulas for the cyclic groups. How about C_∞?

Problem VI.5.b-7: Extending the procedure to improper groups that are direct products is not difficult. Is it trivial?

Problem VI.5.b-8: Group S_4 can also be written (sec. I.7.b, p. 55) as a direct sum. Can the induction procedure be used?

Problem VI.5.b-9: Induce the representations of C_{nv}. We know the relationships of some of these to their subgroups (sec. I.7, p. 50). Is it possible to find general results?

Problem VI.5.b-10: Repeat this for D_{nh}.

Problem VI.5.b-11: The only remaining group listed is T_d.

Problem VI.5.b-12: The icosahedral groups (how many?) are not crystallographic, nor has their product structure been given. Does this matter?

Problem VI.5.b-13: Not all finite groups are crystallographic. Yet it might be possible to find general results, or algorithms, for, at least some (classes) of them.

Problem VI.5.b-14: It would be useful to obtain information about A_n and S_n, for arbitrary n.

VI.5.c Conjugate subgroups

There is, for subgroups, a concept similar to that of conjugate representation (sec. VI.2.a, p. 281). For subgroup H from which the inducing is done, we define a conjugate subgroup

$$H_{-p} = s_p H s_p^{-1}, \qquad \text{(VI.5.c-1)}$$

using element s_p of G, giving an isomorphism of H [Mirman (1995a), sec. III.4.b, p. 85]. We consider in particular the set of elements giving all (distinct) cosets of H. If H is normal H_{-p} is identical to H (the isomorphism is an automorphism), but in general it is not. We need the groups each of which is the intersection of H and H_{-p} (the set of elements they have in common), for each coset p.

For example, S_3, when the inducing is done from noninvariant subgroup C_2 consisting of E and T (sec. VI.4.e, p. 305), conjugation using T' gives subgroup E and T'', conjugation by T'' gives subgroup E and T'. These three are clearly isomorphic. The tetrahedral group can be written in terms of three C_2 subgroups and a C_3 one under which these three are isomorphic.

In these examples, the intersection of each pair H and H_{-p} is just the identity.

For the octahedral group (sec. VI.4.d, p. 303),

$$O = T \wedge C_2' = D_2 \wedge C_3' \wedge C_2' = D_2 \wedge (C_3' \wedge C_2') = D_2 \wedge D_3$$

$$= (D_2 \wedge C_3') \wedge C_2', \qquad \text{(VI.5.c-2)}$$

we take subgroup D_2. The C_2 groups consisting of E plus any transposition give automorphisms of D_2. This is also true for elements of the form (123). Thus D_2 is invariant under D_3, necessarily for there is no other subgroup it can be taken into; it is the intersection.

This group, which is S_4, has subgroup S_3 that acts on numbers 1,2,3 (sec. VI.4.f, p. 306). The coset representatives are E, (14), (24), (34). Similarity transformations of S_3 by these give S_3 groups acting on (124), (134), and (234), all isomorphic. The groups acting on (123) and (124) share E and transposition (12) — group S_2, which is their intersection — an example of an intersection larger than the identity. Thus the set of intersections are the S_2 groups acting on 1,2; 1,3; 1,4; 2,3; 2,4; and 3,4, and the group common to all is the identity. However the transpositions on 1,2; 1,3; and 2,3 all belong to one S_3 subgroup (say that given by coset representative E). So there are four independent intersections, going with the four coset representatives.

Problem VI.5.c-1: Check that in the above equation, all parentheses are correct.

Problem VI.5.c-2: Prove that an intersection of groups is a group.

Problem VI.5.c-3: How are conjugate subgroups and conjugate representations related (or at least similar or different)?

VI.5.d The characters of the induced representation

As usual an essential function (over the classes) is the character [Burrow (1993), p. 77; Jansen and Boon (1967), p. 138; Lomont (1961), p. 224]. Having the representation matrices, we are now able to obtain the characters of the induced representation ζ from those of the subgroup representation η by summing diagonal elements [Mirman (1995a), sec. VII.4, p. 190]. These are related by

$$\zeta(g) = \sum_{j} \eta(r_j g r_j^{-1})\Delta_{jj}; \qquad \text{(VI.5.d-1)}$$

the sum is over all coset representatives r_j such that $r_j g r_j^{-1} \varepsilon$ H. The characters of the elements for which this does not hold are 0, which gives the definition

$$\Delta_{jj} = 1, \quad \text{for } r_j g r_j^{-1} \varepsilon H,$$

$$= 0, \quad \text{otherwise.} \qquad \text{(VI.5.d-2)}$$

To obtain this we denote the order of class l of G by c_l, and the set of elements of this class that are in H (the intersection of the class with H) by C_l — this consists of complete classes of H, and for H normal also complete classes of G. The order of class m (a member of C_l), which may occur several times in C_l, is $c_{l(m)}$, labeling it to show that it is the order of class m of H obtained from class l of G — the number of elements of H in class m. We get from the expression of the induced matrix, summing over the diagonal elements to relate characters,

$$c_l\zeta_l = \sum_{g\, \varepsilon \text{ class } l} \zeta(g) = \sum_{j=1}^{o(G)/o(H)} \sum_{g\, \varepsilon \text{ class } l} \eta(r_j g r_j^{-1})\Delta_{jj}; \qquad \text{(VI.5.d-3)}$$

the sum is over all elements of class l but their contribution is zero unless $r_j g r_j^{-1}$ is in C_l. The upper limit of the sum, the ratio of the orders of the groups, is the number of cosets. So, again summing over the elements of class l,

$$c_l\zeta_l = \sum_{j=1}^{o(G)/o(H)} \sum_{l(m)} c_{l(m)}\eta_{l(m)} = \frac{o(G)}{o(H)} \sum_{l(m)} c_{l(m)}\eta_{l(m)}. \qquad \text{(VI.5.d-4)}$$

This gives

$$\zeta_l = \frac{o(G)}{o(H)c_l} \sum_{l(m)} c_{l(m)}\eta_{l(m)}, \qquad \text{(VI.5.d-5)}$$

expressing the character of class l of the group in terms of the characters of the subgroup obtained from this class. Even though the character for the group is the same for all members of class l, the different classes of the subgroup obtained from l can have different characters, since the matrices of an irreducible representation of the group may be reducible for the subgroup, and the irreducible components can have different characters. The sum of the characters of the irreducible components of the representation matrix of an element of G is, of course, the same for all members of the class, although the terms in the sum may not be.

This is general. The subgroup need not be normal nor its representation irreducible. It shows that the representation induced from a reducible representation of H is equivalent to a sum of representations induced from its irreducible components, as expected; if the representation of H is reducible, so is the induced one.

If the subgroup consists solely of identity E, there are nonzero matrix elements only for

$$r_k g r_j^{-1} = E. \quad \text{(VI.5.d-6)}$$

So, expressing the group representation matrix Γ in terms of the subgroup matrix M (sec. VI.3.b, p. 291),

$$\Gamma(g)_{kt,js} = M(r_k g r_j^{-1})_{ts} = \delta_{ts}, \text{ if } r_k g r_j^{-1} = E. \quad \text{(VI.5.d-7)}$$

These are the matrix elements of the regular representation, which can be considered as the representation induced from the identity (using the subgroup one-dimensional completely symmetric representation to avoid repetitions). From the character formula

$$\eta(h) = g, \text{ if } h = E$$
$$= 0, \text{ otherwise}, \quad \text{(VI.5.d-8)}$$

as must be true. This provides a check, but not a useful means of finding the representations. Larger subgroups are needed.

Problem VI.5.d-1: Why does the intersection of a class of G with H consist of complete classes of H, but only if H is normal does it consist of complete classes of G [Jansen and Boon (1967), p. 136]?

Problem VI.5.d-2: The induced representation has been found using a particular coset decomposition. Of course the characters are independent of the choice of coset representative. But this should be proven [Altmann (1977), p. 144]. Clearly then induced representations found using different coset representatives are equivalent (although perhaps not identical in form).

Problem VI.5.d-3: Induction is transitive: if L is an invariant subgroup of G, H an invariant subgroup of L, and a representation of G is found by induction from a representation of L, which is found by induction from a representation of H, inducing directly from that representation of H gives the same representation of G [Bradley and Cracknell (1972), p. 182]. State the formulas for the matrices of G found by the two methods. How are the characters related? Compare this to pb. VI.2-1, p. 281. Is distributive [Mirman (1995a), pb. XIII.2.c-1, p. 371] meaningful here?

VI.5.e The Frobenius reciprocity theorem

There is a relationship (of equality) between the number of times representation R^s is subduced from an irreducible representation, and the number of times that irreducible representation is contained in the reducible representation induced from R^s [Altmann (1977), p. 148; Bradley and Cracknell (1972), p. 175; Burrow (1993), p. 80; Chen (1989), p. 38; Coleman (1968), p. 86; Inui, Tanabe and Onodera (1990), p. 83; Jansen and Boon (1967), p. 138; Janssen (1973), p. 142; Lomont (1961), p. 226; Sternberg (1994), p. 110]. This is given by the Frobenius reciprocity theorem.

Consider group G and (not necessarily normal) subgroup H, with (perhaps reducible) representation R of G induced from irreducible representation R^s of H. In the decomposition of R, irreducible representation R_p of G appears N_p times. Now R^s is subduced from R_p whose decomposition into irreducible representations of H contains n_s copies of R^s (n_s blocks of R^s). Then the theorem states that

$$N_p = n_s. \tag{VI.5.e-1}$$

So the number of times the decomposition of an irreducible representation of the group into irreducible representations of the subgroup contains representation R^s equals the number of times that the group irreducible representation appears in the reducible representation induced from subgroup representation R^s. The value is

$$N_p = n_s = \frac{1}{o(H)} \sum_{l(m)} c_{l(m)} \eta^s_{l(m)} \zeta^p_l, \tag{VI.5.e-2}$$

summed over the classes of H, where ζ and η are the characters of the representations of G and H respectively, and $c_{l(m)}$ is the order of the class of H contained in the intersection of H and (any) class l of G (a complete class of H, if H is normal).

To show this [Jansen and Boon (1967), p. 139] we use the expression for the induced character, which can be for a reducible representation. Then we reduce that into irreducible representations p of G, these appearing N_p times, which we know [Mirman (1995a), eq.VII.6.a-4, p. 200] in terms of the reducible characters ζ_l, and irreducible ones ζ_l^{p*}, summing over the classes,

$$N_p = \frac{1}{o(G)} \sum_l \zeta_l \zeta_l^{p*} = \frac{1}{o(H)} \sum_{l(m)} c_{l(m)} \eta^s_{l(m)} \zeta_l^{p*}. \tag{VI.5.e-3}$$

The reducible representation, R^s subduced from irreducible representation R_i of G [Jansen and Boon (1967), p. 135] decomposes into ν

(perhaps equal 1) irreducible representations, labeled s, each appearing n_s times. This gives, for each class of H, the reducible characters as a sum over irreducible ones,

$$\zeta_l^p = \sum_{s=1}^{\nu} n_s c_{l(m)} \eta_{l(m)}, \qquad \text{(VI.5.e-4)}$$

giving the reducible character for subduced representation s as a sum of the irreducible characters into which the subduced representation decomposes, using the fact that the irreducible character ζ_l^p of G equals the reducible character of H, (since the representation matrices are the same). The n_s are the dot (scalar) products of the character vectors, and of course are nonnegative and real (fortunately). Substituting the formula for the character (eq. VI.5.d-5, p. 315), we get,

$$\zeta_l^p = \frac{o(G)}{o(H)c_l} \sum_{l(m)} c_{l(m)} \eta_{l(m)}^s = \sum_{l(m)} \sum_{s=1}^{\nu} n_s \eta_{l(m)}^s. \qquad \text{(VI.5.e-5)}$$

The terms in the sum are nonzero only if the class of H with index $l(m)$ is contained in the class l of G; then the sum is equal to $o(G)/c_l$ (being only over the classes $l(m)$ of H that form the intersection of the class of G with H). The character of irreducible representation s for the class is $\eta_{l(m)}^s$. We now multiply the first and last terms of this equation by $c_l \zeta_l^{p*}$, sum over l, and using the character orthogonality formula [Mirman (1995a), sec. VII.5.c, p. 195] obtain

$$n_s = \frac{1}{o(H)} \sum_{l=1}^{\nu} c_{l(m)} \eta_{l(m)}^s \zeta_l^{p*}. \qquad \text{(VI.5.e-6)}$$

This is the same as N_p (eq. VI.5.e-3), giving the Frobenius reciprocity theorem.

A consequence of this is that the same R (up to equivalence) is induced by all members of an orbit of P_j. This means that we can consider the representation induced by an orbit — it does not matter which member we take. This is perhaps not trivial since the representations in the orbit are not equivalent.

To see this we induce representations R and R' of G using inequivalent conjugate irreducible representations of H, and decomposing into a sum of representations of H we get

$$R^i = \sum c_i^s P_i, \qquad \text{(VI.5.e-7)}$$

$$R^{i'} = \sum c_i^{s'} P_i, \qquad \text{(VI.5.e-8)}$$

so, by Clifford's theorem (sec. VI.2.f, p. 288),

$$c_i^s = c_i^{s'}, \tag{VI.5.e-9}$$

thus R and R' are equivalent, since they contain the same representations the same number of times.

Problem VI.5.e-1: By taking the class of the identity we find relationships between the dimensions of the representations, and also between these and the decomposition coefficients. Further since the n_s are non-negative integers, show that

$$n_s \neq 0, \quad \text{any s.} \tag{VI.5.e-10}$$

So every irreducible representation of H appears at least once in the decomposition of the set of induced representations. This is not surprising. Why?

VI.6 IRREDUCIBILITY AND COMPLETENESS FOR ARBITRARY SUBGROUPS

Since we (almost) always want irreducible representations, we want induced representations to be irreducible. Are they? Of course the representation of the subgroup from which they are induced must be irreducible, otherwise we can just induce from each of its irreducible components. Are there other conditions [Bradley and Cracknell (1972), p. 181]? Thus we now turn to the irreducibility of the induced representations.

When is an induced representation irreducible? How do we know? The necessary and sufficient condition that induced representation R^i be irreducible is that no pair of intersections of H and H_{-p}, for each pair p, p' giving cosets, can have representations subduced from R^i in common [Jansen and Boon (1967), p. 140; Lomont (1961), p. 225].

If representation R (whose symbol is suppressed), with character ζ, of group G, which has O(G) elements, is irreducible then [Mirman (1995a), eq.VII.6.b-3, p. 201]

$$\sum_g |\zeta(g)|^2 = O(G), \tag{VI.6-1}$$

where the sum is over all operators of G. We get, from the expression for the characters of the induced representation (sec. VI.5.d, p. 314) the requirement that

$$\sum |\zeta(g)|^2$$

$$= \sum_j \sum_g |\eta(r_j g r_j^{-1})|^2 \Delta_{jj}^2 + \sum_{j,j' \neq j} \sum_g \eta(r_j g r_j^{-1}) \eta(r_{j'} g r_{j'}^{-1}) \Delta_{jj} \Delta_{j'j'}$$

$$= \frac{o(G)}{o(H)} o(H) + \sum_{j,j' \neq j} \sum_g \eta(r_j g r_j^{-1}) \eta(r_{j'} g r_{j'}^{-1}) \Delta_{jj} \Delta_{j'j'} = o(G), \quad \text{(VI.6-2)}$$

thus for irreducibility,

$$\sum_g \sum_{j,j' \neq j} \eta(r_j g r_j^{-1}) \eta(r_{j'} g g_{j'}^{-1}) \Delta_{jj} \Delta_{j'j'} = 0. \quad \text{(VI.6-3)}$$

The only contributing terms in the second sum are for those elements g that lie in the intersection of the subgroups

$$H_{-j} = r_j H r_j^{-1}, \text{ and } H_{-j'} = r_{j'} H r_{j'}^{-1}; \quad \text{(VI.6-4)}$$

these conjugate to subgroup H by r_j and $r_{j'}$. With

$$w = r_{j'} g r_{j'}^{-1}, \quad \text{(VI.6-5)}$$

$$r_j g r_j^{-1} = (r_j r_{j'}^{-1}) w (r_j r_{j'}^{-1})^{-1}. \quad \text{(VI.6-6)}$$

Summing over $w \in$ H we use the restriction that

$$(r_j r_{j'}^{-1}) w (r_j r_{j'}^{-1})^{-1} \in \text{H}. \quad \text{(VI.6-7)}$$

If we keep r_j fixed, the sum over all $r_j \neq r_{j'}$ gives a complete set of coset representatives, except for the identity, although not necessarily identical to the previous one. Thus, using unitarity [Mirman (1995a), sec. VII.5.a, p. 193],

$$\sum_g \sum_{j,j' \neq j} \eta(r_j g r_j^{-1}) \eta(r_{j'} g r_{j'}^{-1}) \Delta_{jj} \Delta_{j'j'}$$

$$= \frac{o(G)}{o(H)} \sum_{j=2}^{o(G)/o(H)} \sum_g \eta(g) \eta^*(r_j g r_j^{-1}), \quad \text{(VI.6-8)}$$

where g belongs to the intersection, and the original set of coset representatives has been used since the character does not depend on this choice. If

$$r_j g r_j^{-1} = r_{j'} g r_{j'}^{-1}, \quad \text{(VI.6-9)}$$

then as the terms in the sum are absolute values, they are all positive, so for the sum to equal zero, each must.

We denote a general element of the intersection (subgroup) by τ so matrices $M(\tau)$ form a representation of this (intersection) subgroup, which is subduced from representation R^s of H (since the intersection is a, not necessarily proper, subgroup of H). Likewise matrices $M(r_j\tau r_j^{-1})$, for fixed r_j, form a representation, with matrices defined by

$$M'(\tau) = M(r_j\tau r_j^{-1}), \qquad \text{(VI.6-10)}$$

for all τ in the intersection.

Thus a necessary and sufficient condition that induced representation R^i of G be irreducible is that

$$\sum_g \eta(g)\eta^*(r_jgr_j^{-1}) = 0, \qquad \text{(VI.6-11)}$$

for all g in the intersection, and all coset representatives r_j. From the character orthogonality theorem [Mirman (1995a), sec. VII.5.c, p. 195] this means that representations M and M', which are reducible representations of the intersection, subduced from the irreducible representation of H, contain no irreducible representations in common for all $j = 2, \ldots, \frac{o(G)}{o(H)}$. Then, and only then, the cross-term in the expression for the square of the absolute value of the character is 0. (This form of the condition for irreducibility has been called Johnston's irreducibility criterion [Bradley and Cracknell (1972), p. 174]).

The representations of S_3 induced from noninvariant subgroup C_2 are seen to be reducible (sec. VI.4.e, p. 305). The intersection is the identity, whose representation, the number 1, subduced from the representations of S_3, is in fact common for all subgroups. In this case the characters are all 1, and the sum over them cannot be 0.

For the representations of S_4 induced from S_3 (sec. VI.4.f, p. 306), there are irreducible representations shared by intersections, those of C_2. And the induced representations are reducible.

Problem VI.6-1: Check some other cases, trying to find irreducible induced representations.

VI.7 INDUCING FROM A NORMAL SUBGROUP

The process of inducing is simpler and more revealing if, rather than being arbitrary, the subgroup is invariant, as in most of the pertinent examples [Altmann (1977), p. 185; Bradley and Cracknell (1972), p. 176; Jansen and Boon (1967), p. 147; Kocinski (1983), p. 33; Streitwolf (1971), p. 44]. Since the space groups have invariant subgroups, this, though special, is the relevant case. One reason for considering the general case is that there is at least one important class of groups for which this simpler procedure cannot be used, the simple groups [Mirman (1995a), sec. IV.9, p. 143]; these have no invariant subgroups. And comparison elucidates how invariance affects the analysis. Nevertheless, since this special case is generally the important one, it is useful to restate the procedure and formulas for it.

Problem VI.7-1: Since H is a normal subgroup of G, it is a normal subgroup of each of its little groups.

VI.7.a Proof of irreducibility of the induced representation

Once we have found the representations we would not be finished if they were reducible. But we do not need to reduce them, the representations found by this procedure are irreducible. First we outline one proof before returning the method used above.

To show that an induced representation is irreducible we use Schur's lemma [Mirman (1995a), sec. VII.3, p. 186], finding any matrix M that commutes with all those of the representation and show that it must be a multiple of the unit matrix. Now the matrices of the induced representation, R(h), for elements belonging to the invariant subgroup, are block-diagonal with the blocks being representation matrices of H for these form a representation of it which we take as completely reduced. The desired matrix M must be similarly block-diagonal since the representations of normal subgroup H are completely reduced (which means that there are such matrices M with each block proportional to the unit matrix); considering just the group operators belonging to subgroup H is sufficient for obtaining an M. However proportionality constants can have different values for each of the non-equivalent blocks — conjugate representations can have different constants, and it is this that we must show does not occur, that the values for all blocks are the same, so M is a multiple of the unit matrix, for if it is — and it is — then the induced representation is irreducible.

The equality of these constants follows from the commutation relations between M and those matrices representing elements not in H, but in the little group. Because it is the little group, and H is a normal subgroup of it, M must commute with its elements. But among these are elements linking different representations of H, with at least one linking any two such representations. However M can commute with the element linking representations only if their proportionality constants are the same. This applies to each pair of constants, thus they are all the same, M is a multiple of the unit matrix, and the induced representation is irreducible.

Problem VI.7.a-1: Why must the representation of normal subgroup H (only?) be completely reduced? Also M commutes with the elements of the little group. Why?

Problem VI.7.a-2: This is the outline of the proof. Write it out explicitly [Streitwolf (1971), p. 45].

VI.7.b Irreducibility of representations induced from normal subgroup

We have a general condition for irreducibility (sec. VI.6, p. 319) of the induced representation but if the subgroup is normal it is not only simpler, but its meaning is clearer.

Representation R of G induced from representation R^s of invariant subgroup H, with the groups having orders O(G) and O(H) respectively, is irreducible if, and only if, the different irreducible representations of H conjugate to R^s generated by the $\frac{O(G)}{O(H)}$ coset representatives are inequivalent. If there were several equivalent irreducible representations we would expect that they generate several components of the representation of G, which would thus be reducible. However for only a single irreducible representation, there is just one component, and the representation is irreducible. This means that for irreducibility of R, the orbit of H containing R^s must have $\frac{O(G)}{O(H)}$ members, its maximum number, and any such orbit gives an irreducible representation. It follows from this that the necessary and sufficient condition for irreducibility is that the little group of R^s be H.

In general this is not true. For a representation of the subgroup there is a little group, larger than H, whose representations induce irreducible representations of G, so we need a procedure for finding the irreducible induced representations for arbitrary normal subgroups.

This requirement follows from the proof for an arbitrary subgroup (sec. VI.6, p. 319), but H_{-j} and H are now the same, since H is normal [Jansen and Boon (1967), p. 147]. Then the sum required to be zero (eq. VI.6-8, p. 320) is the product of characters (from the same class) of conjugate representations. This must be zero, and it is by orthogonality [Mirman (1995a), sec. VII.5.c, p. 195], unless the representations are equivalent. So we have that the necessary and sufficient condition that induced representation R^i of G be irreducible is that conjugate representations M and M' contain no equivalent irreducible representations of the intersection in common for all $j = 2, \ldots, \frac{O(G)}{O(H)}$. But for an invariant subgroup, the intersection of conjugate subgroups is just H. This means that all coset representatives generate inequivalent, but conjugate, representations of H. The orbit must be as large as possible (each coset representative gives an inequivalent member of the orbit), which is true, only, if the little group of any representation of this orbit is just

H, for if the little group were larger there would be in the orbit conjugate representations that are equivalent. Generally then an irreducible representations must be induced not from H, but from the little group.

For S_3, we saw that the induced representations are irreducible for the cases in which the little group is normal subgroup A_3, but reducible when it is S_2 (sec. VI.4.a, p. 294). This is emphasized by the four-dimensional form.

Problem VI.7.b-1: Check this for the other examples above of representations induced from a normal subgroup. Does it work for nonnormal subgroups? If not, how does it break down?

VI.7.c Allowable representations of the little group

The induction process uses representations of the little group, however not all, for that could give induced representations appearing more than once. Thus we must specify a set that does not give duplications, but does give all irreducible representations (up to equivalence) of group G. Note that it is the representations of the little group, not those of H, that we use to get the irreducible representations of G.

The little group, $L(R_s)$, of irreducible representation R_s of H, is that group under which the conjugate representations of the orbit of R_s are equivalent (sec. VI.2.c, p. 284). Now take an irreducible representation P_j^s of the little group that subduces a (perhaps reducible) representation of H containing R_s. By Clifford's theorem (sec. VI.2.f, p. 288) P_j^s subduces a multiple of R_s since R_s is the only member of that orbit obtained by regarding H as a subgroup of this little group (though not necessarily the orbit obtained by taking H as a subgroup of G).

An allowable (acceptable, allowed, permitted) representation of the little group, $L(R_s)$ of irreducible representation R_j of subgroup H invariant in G, is any irreducible representation P_j^s of $L(R_s)$ that subduces a multiple of P_j^s [Birman (1974), p. 56; Jansen and Boon (1967), p. 149; Janssen (1973), p. 139; Kocinski (1983), p. 31]. That is all representations in the decomposition of the subduced representation are equivalent [Streitwolf (1971), p. 44]. For this it is sufficient that R_j occurs in the representation subduced from $L(R_j)$, since R_j is self-conjugate under L.

To find the induced representations, only allowable representations of the little group are used. If H itself is the little group, all representations are allowable.

Rather than considering this in general we return to it when we study space-group representations (sec. VII.4.a, p. 353).

Problem VI.7.c-1: Explain why all these requirements for an allowable representations are really the same.

VI.7.d All representations are given by the allowable representations

The representations obtained using induction are irreducible. Do we get all representations of the group by using only the allowable representations; is the set of representations complete? Fortunately it is [Kocinski (1983), p. 34; Streitwolf (1971), p. 46]. It can be seen from the definition (sec. VI.3.b, p. 291) that every representation of a group can be written as an induced representation of any subgroup (although not necessarily usefully). But we have to show that the restriction to allowable representations does not eliminate any needed ones. This is done by showing that every irreducible representation is (equivalent to one) induced from an allowable representation.

Take irreducible representation R of G, with subduced representation R^s of H decomposed. In this decomposition consider one representation, which we label block 11, and its little group L, and write G in terms of its cosets with respect to L,

$$G = \sum l_i L. \tag{VI.7.d-1}$$

The conjugate representations generated by l_i,

$$R(h)_i = R(l_i^{-1} h l_i), \tag{VI.7.d-2}$$

for all h of H, belong to one orbit, since conjugate representations obtained from a coset l_iL are equivalent. There is an

$$h' = l_k h, \tag{VI.7.d-3}$$

belonging to H, for some l_k of this coset such that

$$R(h)_k = R(l_k^{-1} h l_k) = R(h')^{-1} R(l_i^{-1} h l_i) R(h') = R(h')^{-1} R(h)_i R(h'). \tag{VI.7.d-4}$$

All R_i are block-diagonal. So, with the subscripts denoting the blocks, and l belonging to l_iL,

$$R(l^{-1} h l)_{rs} = \sum_{tv} R(l^{-1})_{rt} \delta_{tv} R(l)_{vs}, \tag{VI.7.d-5}$$

or

$$R(h)_{ss} R(l)_{sr} = R(l)_{sr} R(l^{-1} h l)_{rr}. \tag{VI.7.d-6}$$

By Schur's lemma, block $R(l)_{sr}$ is nonzero only if the representations are equivalent,

$$R(l_s^{-1} h l_s) \sim R(l_r^{-1} l^{-1} h l l_r), \tag{VI.7.d-7}$$

giving

$$R(h) \sim R(l_r^{-1} l^{-1} l_s h l_s^{-1} l l_r). \tag{VI.7.d-8}$$

This means that $l_s^{-1} l l_r \varepsilon$ L, the little group, is the condition that block $R(l)_{sr}$ not vanish. This condition can hold for a fixed l only for a single block in each row or column — the representation conjugate under a fixed element is unique. Conjugate representations obtained using a coset representative of the little group are equivalent, though not (necessarily) using a coset representative of the subgroup — indicating the importance of the little group in the procedure.

Transforming the matrices of the representation of G with the matrix

$$U_{sr} = \mathrm{R}(l_r)_{s1}\delta_{sr}, \tag{VI.7.d-9}$$

we get representation matrices (of H) with the nonvanishing blocks whose positions are unchanged. This gives the transformed blocks, R', as

$$\mathrm{R}(l_s)'_{s1} = \sum_{kt} \mathrm{R}(l_k)^{-1}_{k1}\delta_{sk}\mathrm{R}(l_s)_{kt}\mathrm{R}(l_1)_{11}\delta_{t1}$$

$$= \mathrm{R}(l_s)^{-1}_{s1}\mathrm{R}(l_s)_{s1}\mathrm{R}(l_1)_{11} = E, \tag{VI.7.d-10}$$

the unit matrix, since

$$\mathrm{R}(l_1)_{11} = E, \tag{VI.7.d-11}$$

with $l_1 = 1$. The nonzero blocks then are of the form

$$\mathrm{R}(l)'_{sr} = R(l_s h l_r^{-1})'_{sr} = \sum_{kt} \mathrm{R}(l_s)'_{sk} R(h)'_{kt} \mathrm{R}(l_r)'^{-1}_{tr}$$

$$= \sum_{kt} \delta_{k1} R(h)'_{kt}\delta_{t1} = R(h)'_{11}, \tag{VI.7.d-12}$$

using the unitarity of the R' matrices. Thus the transformed representation R' is induced from $R(h)'_{11}$ — every representation can be obtained by transforming from a representation induced from this.

We must show that the matrices $R(h)'_{11}$ form an allowable irreducible representation of little group L. However this follows from R'_{11} being a representation of little group L since the matrices of the representation subduced from R' are block-diagonal with only the 11 block nonzero. So the representation is allowable, and must be irreducible since if it were reducible then R would be, contrary to assumption.

Thus every allowed irreducible representation of a little group induces an irreducible representation of the group, and all irreducible representations are obtained by inducing from these. The form of the representation obtained by induction in this way is the standard form.

Problem VI.7.d-1: That every irreducible representation of group G can be obtained as an induced representation of this form can be verified [Altmann (1977), p. 191; Bradley and Cracknell (1972), p. 181] by checking that the sum of squares of the dimensions of the representations so found equals the order of the group [Mirman (1995a), pb. VI.b-1, p. 175]. Thus we have all.

Problem VI.7.d-2: This discussion shows that all representations of a group are obtained by induction from a normal subgroup. We do not consider whether this is true for an arbitrary subgroup, or whether, and how, the procedure must be modified; if so that would emphasize the value of an invariant subgroup to induce from. This is an interesting exercise for the reader. As a brief introduction to this problem it is useful to look at the nonnormal subgroups discussed above, and see whether all their representations are obtained. What general rules might be guessed from these examples?

VI.7.e Obtaining only nonequivalent representations

We do not want representations that are equivalent to others, so are redundant. Summarizing the rules we have formulated to give each representation, and only once, makes these explicit.

Different orbits clearly give inequivalent representations. We need therefore one from each orbit. The little groups of the representations of a single orbit are isomorphic, and induced representations that they give are equivalent, although perhaps with rows and columns of each block shuffled. Thus we use only one representation of an orbit. From the little group, we can take different allowed representations. The representations that two of these induce are equivalent if they are, inequivalent if the little-group representations are, so we do not use equivalent representations of the little group. Different coset representatives give equivalent induced representations, thus a definite choice must be made.

Problem VI.7.e-1: Check that after these choices are made there is no freedom left, so the resultant induced representation appears but once. Also check that all these choices are necessary.

VI.7.f Finding the induced representation using the little group

The procedure for obtaining the induced representation has now been discussed and hopefully justified. It is useful to review it, emphasizing for which cases it is complicated and the details of the method for such situations [Jansen and Boon (1967), p. 151].

We start by finding a normal subgroup H of G, and all its irreducible representations; the value of the method is based on this being (hopefully) relatively easy. These are sorted into different orbits by finding the sets conjugate under the elements of G. One representation is picked (arbitrarily) from each orbit. The little groups of each of these are found, and then their allowable representations. From each such representation, the representations of G are induced.

The procedure must, in general, go through the little group. We wish an irreducible representation R of G that subduces a multiple of an orbit of normal subgroup H, this containing R_j. We induce R from any allowable representation of the little group of R_j, $L(R_j)$, not from H. The set of induced irreducible representations of G is the same (up to equivalence) for all little groups of the members of the orbit, and is the same as the set induced from the allowable representations of $L(R_j)$. This is the full set of irreducible representations of G corresponding to the orbit — these are the irreducible representations contained in the decomposition of the reducible representation induced from the representation of H. This set contains one member if H is the little group, else not. Thus any of these representations can be taken as the required induced irreducible representation.

That is we take irreducible representation R_i from each orbit of H; the irreducible representations of G induced from the allowable representations of the little group of R_i, for all these subgroup representations, are all irreducible representations of G, each appearing once.

Problem VI.7.f-1: Clearly the method is not too useful unless each of these steps is relatively simple and straightforward — unless they can be computerized, of course. Or perhaps, in the relevant cases (that we consider), this is not necessary.

VI.7.g The matrices of the induced representations

The method gives an induced representation — the matrices of the representation. While the matrices follow from the definition, the procedure does not directly use that. Explicitly how are these given?

We start with a member R_j of some orbit of H; the other members of the orbit induce the same representation of G (up to equivalence). Take an allowable representation P^L of $L(R_j)$. Since the little group is a subgroup of G, we can decompose G in terms of its left cosets, with coset representatives written t_m ($t_1 = E$),

$$G = t_1L(R_j) + t_2L(R_j) + \ldots + t_mL(R_j), \quad m = \frac{O(G)}{O(L)}. \qquad \text{(VI.7.g-1)}$$

This representation of G induced by P^L consists of blocks, as before.

The representation of G is induced from an allowable representation of a little group, L_j. Denote the set of basis vectors of this allowable representation (its carrier space) by $V(L_j)$. The set of basis vectors of the representation of G is the direct sum over all coset representatives of $t_m V(L_j)$. These sets are disjoint (spaces for different coset representatives have no vectors in common), by Clifford's theorem (sec. VI.2.f, p. 288). Then it follows from the definition (sec. VI.3.b, p. 291) that the direct sum is the carrier space for the induced representation.

Having the basis vectors we can compute the matrices by determining the action of those operators of G not in H on these (the matrices for H are already known).

Problem VI.7.g-1: Discuss explicitly why this does give the correct matrices [Jansen and Boon (1967), p. 151].

VI.7.h Induced representations of direct product groups

While the case of an invariant subgroup is special, and simpler, there are special cases of it, for which the procedure is even simpler, or can be replaced by another that is.

First let group G be a direct product of two (of course, invariant) subgroups, H and K [Jansen and Boon (1967), p. 153]. Then all inequivalent irreducible representations of G are Kronecker products (direct products of the matrices) between the inequivalent irreducible representations of H and of K. A Kronecker product of matrices A and B is matrix C, with elements related by

$$C_{ij,kl} = A_{ik}B_{jl}. \qquad \text{(VI.7.h-1)}$$

The Kronecker product of representations (direct product of representations) R and S, is the representation whose matrices are the Kronecker product of those of R and S.

In this case therefore not only need we not induce representations from each of the subgroups, we can not, since the required little groups are all just G — to construct the allowable representations of the little groups we need those of G, which is what we want to find. The reason that the little groups are all just G is that conjugate representations are mutually equivalent under all of G's elements. But we get all the representations of the group from the Kronecker products.

Problem VI.7.h-1: It should be clear that the number of classes of G is the product of those of the subgroups, so the number of inequivalent, irreducible representations of G is also the product of those of the subgroups [Mirman (1995a), sec. VII.6.c, p. 202].

Problem VI.7.h-2: It immediately follows that all inequivalent, irreducible representations of G are Kronecker products of those of its subgroups.

Problem VI.7.h-3: Character tables can be regarded as square matrices. So the character table of G is the Kronecker product of those of its subgroups.

Problem VI.7.h-4: Since we have already given much information about the product structure of groups (sec. I.7.a, p. 50), representation matrices and character tables, this can be checked. Try it for the four-group.

Problem VI.7.h-5: Other direct-product groups are C_6 and D_2.

Problem VI.7.h-6: Among the improper point groups that are direct products are D_{6h}, T_h and O_h.

Problem VI.7.h-7: Groups C_{3h}, C_{2v} and D_{2d} are of this form, but there are differences. Do these matter?

Problem VI.7.h-8: Is it relevant whether the group (which group?) is Abelian or not?

VI.7.i Semi-direct product groups

Perhaps the most important case in which induction is used is for semi-direct product groups [Jansen and Boon (1967), p. 157], these including the Poincaré group and the space groups. Here G is (isomorphic to) the semi-direct product of subgroup S and normal subgroup N. For these there are two special — though relevant — subcases for which we can be especially explicit in giving the induced representations. They are the cases for which the induced representations can be found as a Kronecker product, and when the normal subgroup has index two. For them it is clearer that the resultant simplified procedure gives irreducible representations and all, in particular that the induced representation of a semi-direct product group (finite although it is directly extendable to continuous groups) is irreducible and that the complete set of unitary irreducible representations are found, only once, by choosing one point in each orbit and constructing the representation matrices based on each corresponding irreducible subgroup representation.

Problem VI.7.i-1: We wish to induce a representation for the semi-direct product of R and the invariant, Abelian T. The subgroup of R that forms the little group for element t is L. For $M(r)$ a representation of R, and $K(t)$ of T, the representation of G is

$$\Gamma(tr) = K(t)M(r). \qquad \text{(VI.7.i-1)}$$

Prove this, and show that the representation is unitary if M is, and K is real.

VI.7.i.i *Cases for which induced representations can be found as Kronecker products*

There are many semi-direct product groups, including the symmorphic space groups, for which the induced representations can be obtained as Kronecker products. As this simplifies the procedure we consider it first. If the little group of representation R_s of normal subgroup N is the full group,

$$L(R_s) = G, \tag{VI.7.i.i-1}$$

this method gives all irreducible representations of G that subduce R_s, although again (as with direct-product groups) the general procedure does not work (as the little group is G itself). These are called the associate irreducible representations of G with respect to R_s. Also if $L(R_s)$ is the semi-direct product of N and some subgroup of S the following procedure is used to find the allowable representations of $L(R_s)$, and the irreducible representations of G are induced from these allowable representations.

We denote the matrices of representation R_s of normal subgroup N by M^R, which we take unitary. The little group of R_s is

$$L(R) = N \wedge S = G. \tag{VI.7.i.i-2}$$

The representation of subgroup S has matrices $\Delta(s)$. All representations of N conjugate to R_s are mutually equivalent, since $L(R_s) = G$, so for each s_p,

$$M^R(h)' = M^R(s_p^{-1}hs_p) = \Delta(s_p)^{-1}M^R(h)\Delta(s_p), \tag{VI.7.i.i-3}$$

for all $h \varepsilon$ N. Since the $\Delta(s_p)$ are taken unitary, they are defined up to a phase — a unitary matrix is one that leaves the absolute value of a vector invariant, so can multiply it by a phase, but nothing more, hence is itself undetermined up to a phase, but nothing more. We must check that the $\Delta(s_p)$ do form a representation, that is that the phases can be chosen consistently. Now

$$M^R(h)'' = M^R((s_ps_{p'})^{-1}h(s_ps_p')) = \Delta(s_{p'})^{-1}M^R(s_p^{-1}hs_p)\Delta(s_{p'})$$

$$= \Delta(s_{p'})^{-1}\Delta(s_p)^{-1}M^R(h)\Delta(s_p)\Delta(s_{p'}) = \Delta(s_ps_{p'})^{-1}M^R(h)\Delta(s_ps_{p'}), \tag{VI.7.i.i-4}$$

for all $s_ps_{p'} \varepsilon$ S, and all $h \varepsilon$ N. This means that $\Delta(s_ps_{p'})$ and $\Delta(s_p)\Delta(s_{p'})$ can differ by at most a phase. If it is possible to find a set of phases, and there is arbitrariness in them, for which these matrices form a representation then the following procedure applies, otherwise not.

Problem VI.7.i.i-1: Prove that there are two types of semi-direct products for which the method can be employed. One is if N is Abelian and

S arbitrary, as for space groups that are semi-direct products — symmorphic ones. In this case all representations of N are one-dimensional and the $\Delta(s_p)$ are taken as 1 for all p. The other case is for S cyclic, and N arbitrary. Then there is a single generator s of S, and

$$s^n = E. \tag{VI.7.i.i-5}$$

The phase is chosen so that

$$\Delta(s)^n = 1. \tag{VI.7.i.i-6}$$

Problem VI.7.i.i-2: When the little group is G, what goes wrong with the general induction procedure?

Problem VI.7.i.i-3: If the phases cannot be chosen so that the Δ's form a representation, then they would seem to form a projective representation. Do they? Always? Can the procedure then be generalized?

VI.7.i.ii *The representations that are of the form of Kronecker products*

Having mentioned when the procedure works, we next describe it, using $M^j(h)$, the matrices of irreducible representation j of normal subgroup N, and $\Delta^i(s_p)$, the matrices of the i'th irreducible representation of S. Now for the little group of representation j of N

$$L(j) = G, \tag{VI.7.i.ii-1}$$

we have that all representations of N conjugate to representation j are mutually equivalent. Thus there are a set of matrices $U(s)$ giving similarity transformations between any pair of these representations. These do not necessarily form a representation of S, therefore we consider only those cases for which there are a set of matrices $U(s)$ that do form a representation. Then we define matrices

$$\Gamma^{ij}(g = s_p h) = \Delta^i(s_p) \odot U(s_p) M^j(h), \tag{VI.7.i.ii-2}$$

where $\odot$ is the Kronecker product, for all $g = s_p h \in$ G. If $U(s_p)$ is the unit matrix for all p, this reduces to the definition of the irreducible representation of a direct-product group.

We have to show that if these do form a representation of S, the Γ^{ij} form a representation of the semi-direct product group G = N $\wedge$S, and that is irreducible. Also these, for all i, are all associate with respect to representation R_j of N, and are distinct for different i. They are the complete set of associates with respect to R_j. That is they all subduce R_j, and are all the irreducible representations of G that do so.

First

$$\Gamma^{ij}(s_p h)\Gamma^{ij}(s_{p'}h') = \Gamma^{ij}(s_p h s_{p'} h'), \quad \text{all } s_p, s_{p'} \in \mathrm{S}, \quad \text{all } h, h' \in \mathrm{N}, \tag{VI.7.i.ii-3}$$

for as these are Kronecker products,

$$\Gamma^{ij}(s_p h)\Gamma^{ij}(s_{p'}h') = \Delta^i(s_p) \odot U(s_p)M^j(h)\Delta^i(s_{p'}) \odot U(s_{p'})M^j(h')$$

$$= \Delta^i(s_p)\Delta^i(s_{p'}) \odot U(s_p)M^j(h)U(s_{p'})M^j(h')$$

$$= \Delta^i(s_{p''}) \odot U(s_p)M^j(h)U(s_{p'})M^j(h'), \tag{VI.7.i.ii-4}$$

where

$$s_{p''} = s_p s_{p'}, \tag{VI.7.i.ii-5}$$

which is always true for a semi-direct product. Also

$$\Gamma^{ij}(s_p h s_{p'} h') = \Gamma^{ij}(s_p s_{p'}(s_{p'}^{-1} h s_{p'})h')$$

$$= \Delta^i(s_{p''}) \odot U(s_p s_{p'})M^j(h)U(s_{p'})M^j((s_{p'}^{-1} h s_{p'})h')$$

$$= \Delta^i(s_{p''}) \odot U(s_p)M^j(h)U(s_{p'})M^j(h'). \tag{VI.7.i.ii-6}$$

The two expressions for $\Gamma^{ij}(s_p h s_{p'} h')$ are the same, so these do form a representation of G.

To show completeness we use the characters. Then

$$\sum_{g \in \mathrm{G}} |\zeta(g)^{ij}|^2 = \sum_{s(p) \in \mathrm{S}} \sum_{h \in \mathrm{N}} |\zeta(s_p h)^{ij}|^2$$

$$= \sum_{s(p) \in \mathrm{S}} \sum_{h \in \mathrm{N}} |\rho(s_p)^i|^2 |\zeta(U(s_p)d(h)^j)|^2, \tag{VI.7.i.ii-7}$$

with ρ^i the character of representation $\Delta^i(s_p)$ of S. The character can be written in terms of the matrix elements, so

$$\sum_{h \in \mathrm{N}} |\zeta(U(s_p)d(h)^j)|^2$$

$$= \sum_{h \in \mathrm{N}} \sum_{\mu\nu} \sum_{\mu'\nu'} U(s_p)_{\nu\mu} U(s_p)^*_{\nu'\mu'} M(h)^j_{\mu\nu} M(h)^{j*}_{\mu'\nu'}$$

$$= \frac{\mathrm{O(N)}}{n_j} \sum_{\mu\nu} U(s_p)_{\nu\mu} U(s_p)^*_{\nu'\mu'} = \left(\frac{\mathrm{O(N)}}{n_j}\right) n_j = \mathrm{O(N)}, \tag{VI.7.i.ii-8}$$

where O(N) is the order of N, using the orthogonality relations for the elements of the matrices of the irreducible representations of N [Mirman (1995a), sec. VII.5.a, p. 193], and the unitarity of the U's. Thus

$$\sum_{g \in \mathrm{G}} |\zeta(g)^{ij}|^2 = \sum_{s(p) \in \mathrm{S}} |\rho(s_p)^i|^2 \mathrm{O(N)} = \mathrm{O(G)}, \tag{VI.7.i.ii-9}$$

since $\Delta^i(s_p)$ is an irreducible representation of S whose index in G is $\frac{O(G)}{O(N)}$. So the Γ^{ij} do form all irreducible representations of G.

Next consider subduced representation $M^{(s)ij}$ of N. Then

$$M^{ij}(g = eh) = \Delta^i(e) \odot U(e)M^j(h), \tag{VI.7.i.ii-10}$$

where $\Delta^i(e)$ is the unit matrix of dimension equal to the dimension of the irreducible representation $\Delta^i(s_p)$ of S, and $U(e)$ is the unit matrix of dimension n_j. Thus the representation Γ^{ij} subduces $M^{(s)ij}$ a number of times equal to the dimension of $\Delta^i(s_p)$. That means that these, for different i, are all associate representations of G with respect to $M^{(s)ij}$. And if $\Delta^i(s_p)$ and $\Delta^i(s_p)'$ are not equivalent irreducible representations of S, Γ^{ij} and $\Gamma^{ij}{}'$ are also not equivalent, which follows from the orthogonality of the characters.

Finally we must show that the Γ^{ij}, for all i, form a complete set of associates of G with respect to $M^{(s)ij}$ of N. With k_{ij} the number of times the irreducible representation of G subduces M^{si} of N, from the Frobenius reciprocity theorem (sec. VI.5.e, p. 317), we get

$$\sum_i k_{ij}^2 = \frac{O(L(M^{(s)ij}))}{O(N)} = \frac{O(G)}{O(N)}, \tag{VI.7.i.ii-11}$$

the last equation holding because

$$L(M^{(s)ij}) = G. \tag{VI.7.i.ii-12}$$

To prove that these are all associates of G with respect to $M^{(s)ij}$ we note that Γ^{ij} subduces $M^{(s)ij}$ a number of times n_i, which is the dimension of Δ^i. So

$$k_{ij} = n_{ij} \tag{VI.7.i.ii-13}$$

for M^i. Summing over all irreducible representations Γ^{ij} gives

$$\sum_i k_{ij}^2 = \sum_i n_{ij}^2 = \frac{O(G)}{O(N)}, \tag{VI.7.i.ii-14}$$

since $\frac{O(G)}{O(N)}$ is the order of subgroup S. The set therefore gives all associates.

If N is an Abelian normal subgroup, this procedure is the same as that for direct-product groups. This case includes the 73 symmorphic space groups, where N is the translation subgroup, the Abelian normal subgroup of a symmorphic space group (however it is not a subgroup of a nonsymmorphic space group, which is why the procedure does not work for these).

VI.7.i.iii *Subgroups with indices 2*

This process is also simpler if the index of N in G is 2 for then it must be normal [Jansen and Boon (1967), p. 161; Mirman (1995a), pb. IV.7.a-3, p. 134]. This includes, as we have seen, the symmetric groups, and many crystallographic point groups, as well as groups with inversion or time reversal (sec. VIII.2.c, p. 398). All that is required here is to carry out the simplifications of the general method for semi-direct product groups, just given, and then apply the results.

Problem VI.7.i.iii-1: Group G is the set consisting of (necessarily) invariant subgroup N, and the product of it with (of course, only) one coset representation s, so

$$G = \{N, sN\}. \qquad \text{(VI.7.i.iii-1)}$$

Factor group

$$F = G/N \qquad \text{(VI.7.i.iii-2)}$$

consists of two elements, identity e, and f, with

$$f^2 = e, \qquad \text{(VI.7.i.iii-3)}$$

which can be identified with the form of G in the last sentence.

Problem VI.7.i.iii-2: The necessary and sufficient condition that G be (able to be written as) a semi-direct product,

$$G = N \wedge S, \qquad \text{(VI.7.i.iii-4)}$$

is that there is an element s' belonging to sN such that

$$s'^2 = e. \qquad \text{(VI.7.i.iii-5)}$$

Then

$$s = \{e, s'\}. \qquad \text{(VI.7.i.iii-6)}$$

Explain this in general, and check it for symmetric group S_n. Try it also for some point groups, finding not only examples for which it applies, but ones for which it does not.

Problem VI.7.i.iii-3: The little group of an irreducible representation of N is either N or G. Thus an orbit has either one or two members. Try this for the groups in the last problem.

Problem VI.7.i.iii-4: Consider an irreducible representation of G in the form

$$\Gamma = \{R^i(h), R^i(sh)\}, \qquad \text{(VI.7.i.iii-7)}$$

where $R^i(h)$ is a representation of N. Is this form ever possible? Or is it always true (here)? Then

$$\Gamma' = \{R^i(h), -R^i(sh)\} \qquad \text{(VI.7.i.iii-8)}$$

is also an irreducible representation (that is the signs of half the matrices can be changed). These are distinct associate representations of G if and only if $R^i(h)$ is an irreducible representation of N, that is if and only if Γ and Γ' subduce once an irreducible representation of N having a little group which is G. These are the only two associate representations with respect to this irreducible representation of N.

Problem VI.7.i.iii-5: With the character of Γ^i written as

$$\zeta^i = \{\zeta^i(h), \zeta^i(sh)\}, \qquad \text{(VI.7.i.iii-9)}$$

verify that representations Γ^i and Γ^j are distinct if and only if $\zeta^i(sh)$ is nonzero for some $h \varepsilon$ N. If $\zeta^i(sh) = 0$ for all N, then these are equivalent and Γ^i is self-associate. Why? If it is, then it subduces once an orbit of N containing the irreducible representation R^i. Check that it is the only associate representation of G with respect to this orbit. Thus there are two different cases. In the first

$$L(R^i) = G. \qquad \text{(VI.7.i.iii-10)}$$

Then there is one conjugate non-equivalent representation of N. The irreducible representation R^i has two distinct associates

$$\Gamma = \{R^i(h), R^i(sh)\} \text{ and } \Gamma' = \{R^i(h), -R^i(sh)\}, \qquad \text{(VI.7.i.iii-11)}$$

which subduce R^i once, so Γ, Γ' have the same dimension as R^i. In the other case

$$L(R^i) = N. \qquad \text{(VI.7.i.iii-12)}$$

Here there are two conjugate non-equivalent representations of N. Irreducible representation R^i has only one associate

$$\Gamma = \{R^i(h), R^i(sh)\} \text{ with } \zeta^i(sh) = 0 \text{ for all } h \varepsilon \text{N}. \qquad \text{(VI.7.i.iii-13)}$$

It subduces R^i, and its non-equivalent conjugate once, so the dimension of Γ is twice that of R^i. Explain the dimensions in these two cases.

Chapter VII

Representations of Space Groups

VII.1 FINDING THE REPRESENTATIONS

Although in crystallography and solid-state physics groups themselves provide more information than in many other applications — for example in classifications — knowledge of their representations is still essential. But besides applications, representations of point and space groups are interesting because they require significant new concepts in group theory, especially as nonsymmorphic groups link the two types of transformations, the homogeneous (point-group) ones, and the (inhomogeneous) translations. How does this linkage affect the representations, and how do we find representations when it is present? What are the representations, how do they differ for symmorphic and nonsymmorphic groups (and for nonsymmorphic and other types of groups in general), how do they reflect the properties of the groups themselves, including their inhomogeneity, and the objects they describe? Why are nonsymmorphic groups possible, why do they occur? And, why are the representations of these groups relevant, what applications do they have, what do they teach us about mathematics, physics, group theory, about crystals and other objects? To study these questions we have introduced useful concepts, and are now ready to see their suitability, and how they are used.

The representations of the space group of free space, the Poincaré group, are products of those of the translation subgroup and the homogeneous subgroup, the Lorentz group. But there are crystals with symmetry transformations that are not products. Thus their space groups, besides coming from richer point and translation groups, can have more complicated relations to these. The representations of the space groups we then expect are richer — in some sense — than the Poincaré ones, even though these groups are subgroups of the Poincaré group. We study this, but as not everything is known about the Poincaré group, we do not try to understand the relative complexity, although this discussion might provide hints of productive lines of thought for the future.

For representations we need basis functions, the eigenfunctions of a set of group operators whose eigenvalues label these eigenfunctions, which operators we must also find — the state-labeling problem [Mirman (1995a), sec. I.6, p. 25]. To specify the (form of the) representations we first have to decide which operators are diagonal. We take these to include all momentum operators, thus limiting the types of representations studied [Mirman (1995c), sec. 1.1.1, p. 2], and do not consider whether this limitation is important, or whether other types of representations may have interesting mathematical or physical value.

Space groups, particularly nonsymmorphic ones, have an interesting property. Certain transformations move some special points on the boundary of the Brillouin zone to equivalent ones that are displaced by a reciprocal lattice vector. Thus the product of two transformations is represented, not by the product of the representation matrices of these transformations, but by this times a multiplicative factor — the product does not obey the multiplication rules of the group, as a representation is defined to, but obeys rules that are almost, but not quite, like those of the group. We require then a generalization of the concept of representation, to projective representation, thus illustrating application of these (sec. V.4, p. 259).

The representations of space groups provide the foundation for the study of the behavior of objects, like electrons or holes, in crystals (sec. XI.2, p. 569). And these illustrate the applications of, and further illuminate, the representations. Thus it is useful to review this chapter when studying material such as that as a way of gaining deeper understand of both subjects.

VII.2 CONCEPTS FOR THE REPRESENTATIONS

Having prepared the foundations (to a reasonable extent), we now turn to the development of representations of space groups [Altmann (1977), p. 196; Birman (1974), p. 47; Bradley and Cracknell (1972); Burns (1977), p. 300; Burns and Glazer (1978), p. 193; Chen (1989), p. 474; Cornwell (1969), p. 88, 165; Cornwell (1984), p. 222; Davies and Dirl (1994a,b);

Falicov (1966), p. 151; Heine (1993), p. 273; Herring (1942); Inui, Tanabe and Onodera (1990), p. 248; Jansen and Boon (1967), p. 246; Janssen (1973), p. 130; Joshi (1982), p. 91; Kocinski (1983), p. 29; Koster (1957), p. 211; Lax (1974), p. 240; Lomont (1961), p. 236; Ludwig and Falter (1988), p. 203; Lyubarskii (1960), p. 91; Streitwolf (1971), p. 78; Zak, *et al* (1969)].

How do we find these representations, and why do we use this procedure? A space group has two factor groups [Mirman (1995a), sec. IV.8.c, p.138], not necessarily subgroups, a point group, and an invariant translation group. Its representations are found from those of its factor groups, for one reason because these are known so we can build on work already done. The method then is that of induced representations. Having considered the representations of the factor groups our analysis continues with a study of how those of space groups are related to them. Combining these representations is the step that presents difficulty, and, often, novelty, so its study is useful not only because of its present application, but as an introduction to a valuable technology (sec. VI.3, p. 288).

Problem VII.2-1: Although using the procedure of induced representations is (apparently) the easiest way, is it the only possible one?

VII.2.a The type of space group representations studied here

The representations of the space groups — that we consider here — are those with the translation operators diagonal. The best known representations of the Poincaré group, the transformation group of space [Mirman (1995a), sec. II.3.h, p. 45; (1995b), sec. I.4.1, p. 9; (1995c)], have diagonal translation operators; however there are other representations, for which (not all) translation, or momentum, operators are diagonal, with properties not obvious from knowledge of more familiar ones [Mirman (1995c), sec. 1.1.1, p. 2], and it is possible that space groups have analogs, although if they do their significance is not clear. Thus this analysis may not be entirely general. Among the questions that we leave open are whether the representations found here form a complete set, whether other types of representations exist, and if so can they be expanded in terms of those we consider, and are they relevant to anything, or provide useful information not given by the present ones?

VII.2.b Induced representations in the terminology of space groups

The theory of induced representations has been developed in the last chapter, now we outline it, but in the terminology used for space groups. First we need a subgroup, or factor group, with known representations. There are two, a point group, and a translation group. We start with the latter, this being normal [Mirman (1995c), sec. 2.3, p. 15]. Every space group has a normal subgroup, the translation group, of index 2 or 3 [Streitwolf (1971), p. 83]. Since there are two factor groups, the space group has ones with index 2. However some point groups can also be written in terms of factor groups, the normal one of pure rotations, with the second coset these rotations times the inversion. Writing the space group this way gives it in terms of its normal subgroup of index 3. But this is not necessary, so we regard it as having index 2.

Space group representations (that we are considering) are based on Bloch's theorem (sec. IV.2.e, p. 193) to determine the translation group representations, so using the Brillouin zone (sec. IV.2.c, p. 189) — which gives a representation label, momentum vector $\underline{k}$, determined by the periodicity — and the Bloch functions, labeled u, these part of the basis vectors. We thus need the Brillouin zones (here always the first) for the space groups, the translation representations for them, and assignment of the proper one to each space group. With these zones we have the momentum vectors, thus the translation representations. Vectors of a Brillouin zone are related by its point-group symmetry transformations. To prevent redundancy we take a minimum set of $\underline{k}$'s, the ones of the representation domain (sec. IV.2.c, p. 189); the others are given by the point-group transformations.

We also need the representation matrices and basis states for the corresponding operators of the point (factor) group, and the representation matrices for the operators consisting of both rotations and translations (for those of nonsymmorphic groups that are not products), and eigenvalues labeling representations and their states. And we have to combine the representations, with combinations restricted because they are related.

Having these, we form a coset decomposition with respect to the subgroup and the little groups going with it (sec. VI.2.c, p. 284). The little group determines a set of equivalent representations, the orbit of any one — the meaning of these have to be stated in the present context, and they have to found.

The determination of the representations of space groups proceeds thus in a series of steps. A large part of the work has already been done, all that is left (in principle) is to put it together, not trivial, but at least instructive. For symmorphic groups, constructing the representations from those of point and translation groups is straightforward, and the representations are, as expected, the products — since symmorphic space groups are products of point and translation groups. For nonsymmorphic groups, point-group operations and translations are connected — these groups are not products, so representations are not products and finding them is less elementary, thus more stimulating and educational. Nevertheless we can hope to build on our understanding of simpler cases.

VII.2.c Orbits

Transformations acting on an element of translation group T (and in general any vector) give a set of vectors that depends on that element. Thus rotations acting on a momentum vector yield momenta in all directions, but all with the same magnitude (their ends fill the surface of a sphere). For the point on a sphere at the tip of a radius vector, the set of points resulting from the transformations of SO(2) trace out a curve on the surface of the sphere (for a rotation around the center, a great circle). The curve depends on the starting point. This suggests the proper nomenclature.

The set of points or elements produced by the action of the operators of a group acting on point (or element) P is called the orbit of P [Cornwell (1984), p. 121; Janssen (1973), p. 142; Lomont (1961), p. 221; Senechal (1990), p. 28]. If it does not take up the entire space (as it would take up the entire surface of the sphere if we considered SO(3)), it depends on P.

Poincaré transformations take points inside the light cone to others inside, ones outside to outside ones, and those on the light cone to others on the cone [Mirman (1995c), sec. 1.1, p. 1]. The orbits then are the subspaces inside, outside and on the light cone (light cones for the proper Poincaré group). Orbits, despite the terminology, need not be lines — these are surfaces and volumes, for crystals they are sets of discrete points, not surprisingly.

If the group operators acting on a point in the space generate the entire space, the space is transitive under the group. Four-dimensional space-time (the set of vectors from the origin) is not transitive under the Lorentz group (or Poincaré group), but consists of three subspaces that are (or five for the proper group, outside the light cone, the two light cones, and the two interiors). However the Euclidean three-dimensional subspace is transitive under the inhomogeneous rotation group.

Besides the definition of the orbit of a vector, we also need one of a representation, which is distinct (sec. VI.2.b, p. 283).

Problem VII.2.c-1: How are the two definitions of orbit related?

VII.2.d The star of a vector

Space group representations are labeled, in part, by the representations of the (discrete) translation group, the vectors k (sec. IV.2.c, p. 189). However not all are needed, they form sets, and we need only one of each; the others going with it are redundant. How are they related? To any k apply the members of the point group to obtain a set of vectors — the star of k [Altmann (1977), p. 202; Birman (1974), p. 53; Bradley and Cracknell (1972), p. 148; Chen (1989), p. 487; Cornwell (1969), p. 90; Cornwell (1984), p. 228; Falicov (1966), p. 152; Heine (1993), p. 274; Inui, Tanabe and Onodera (1990), p. 85; Jansen and Boon (1967), p. 255; Janssen (1973), p. 133; Joshi (1982), p. 276; Kocinski (1983), p. 37; Lax (1974), p. 242; Lomont (1961), p. 237; Ludwig and Falter (1988), p. 203; Lyubarskii (1960), p. 92; Streitwolf (1971), p. 79; Tinkham (1964), p. 279]. It is the orbit of k under the point-group operators. The star of a vector is the set of vectors generated by applying to any one all operators of the point group that is the factor group of the space symmetry group of the crystal.

As can be seen from any of the diagrams containing stars, the star of a vector does indeed look like a star.

To induce a representation we start with one of the subgroup, and find those equivalent to it. The representation here is that of the translation group, and is labeled by k; those equivalent to it are found using the transformations of the little group (sec. VII.2.g, p. 343), and are labeled by the k vectors given by these transformations — which take k to others in its star. Hence the vectors of a star label the set of equivalent representations, those of different stars label inequivalent ones. We choose a minimum set of k's — only one of each star, more would give representations appearing several times — and for each find its star.

VII.2.e Classification of positions of Brillouin zones

Finding the stars also classifies the points of the Brillouin zone, the vectors of the reciprocal space (sec. IV.2.c.ii, p. 190); there are several types. A general point is one for which the set containing that vector has one distinct member for each element of the point group — the number of members of the star equals that of the point group. However there are points — special points (more generally special positions) — for which some point-group operations take the vector into itself (sec. IV.6, p. 239) [Falicov (1966), p. 155]. Thus the set containing such a vector (the star of that vector) has fewer members than the point group. This is true, for example, for a vector that lies on a line of symmetry — a rotation around this line leaves it invariant. It is also true for some vectors on a zone boundary; the stars of these have members related by reciprocal-lattice vectors, and ones so related are regarded as identical.

Problem VII.2.e-1: Show that the star of every vector is invariant under the space group — an element of it takes a vector in the star to another vector in the same star (though the space group is defined over the points of the crystal, not the $\underline{k}$'s).

VII.2.f The cosets formed from the translations

The representations of space group G are found from those of invariant translation group T, so we write G in terms of its left coset representatives [Mirman (1995a), sec. IV.6, p. 128] acting on T (sec. III.4.d.i, p. 153),

$$G = \{R_1|\underline{w}_1\}T \oplus \{R_2|\underline{w}_2\}T \oplus \ldots \oplus \{R_h|\underline{w}_h\}T$$

$$= T \oplus \{R_2|\underline{w}_2\}T \oplus \ldots \oplus \{R_h|\underline{w}_h\}T, \qquad \text{(VII.2.f-1)}$$

with

$$R_1 = E, \ \underline{w}_1 = 0. \qquad \text{(VII.2.f-2)}$$

A coset representative is the element of a coset that multiplies the (set of elements of the) factor group in this sum over cosets making up a group, here $\{R_i|\underline{w}_i\}$. Thus G is obtained by taking the translations, a subset of its operators, and applying to these point-group operation R_2 and simultaneously translation $\underline{w}_2$ (G may be nonsymmorphic) to get a second set of translations, the second coset, R_3 and $\underline{w}_3$ to get a third set, and so on. For a symmorphic group (only) all $\underline{w}_i$ can be taken zero.

The point group operations that take a vector into another of the same star are the R's giving the coset decomposition. This decomposition assembles vectors $\underline{k}$ into sets, each the star of any vector in it.

The cosets then are sets of translations. As we saw, the coset representatives, the operations picked one from each coset, need not form a group (and do only for symmorphic groups), but the set G/T is a group, the factor group, and is isomorphic to the point group consisting of point-group operations $R_1, R_2, \ldots$ (sec. III.4.d, p. 152).

Problem VII.2.f-1: Is it possible to have more than one coset decomposition, and obtain more than one star for $\underline{k}$? Why?

VII.2.g Little groups and reciprocal lattice vectors

A little group of a vector from the reciprocal lattice is the set of transformations leaving that vector invariant (sec. VI.2.c, p. 284; sec. VI.3.a,

p. 289). However by periodicity of the lattice,

$$exp(2\pi i\underline{K} \bullet \underline{a}) = 1, \qquad \text{(VII.2.g-1)}$$

where $\underline{K}$ is a sum (with integral coefficients) of primitive reciprocal lattice vectors, and $\underline{a}$ is likewise an integral sum of primitive lattice vectors. Thus the addition of a reciprocal lattice vector takes a vector to a corresponding one in another Brillouin zone, which we (thus must) regard as identical. And since these are the basis vectors of the translation subgroup representation, adding a reciprocal lattice vector does not change the space-group basis vector. So the little group has to be defined as the (still incompletely specified) group that leaves a vector of the reciprocal space invariant — but up to the addition of a reciprocal lattice vector. It can be seen from this that the little group is the set of transformations mixing the equivalent translation representations.

Problem VII.2.g-1: Previous definitions of "little group" (sec. VI.2.c, p. 284) are stated somewhat differently than here. Relate the various definitions [Jansen and Boon (1967), p. 251].

Problem VII.2.g-2: The little group of the translation representation labeled by a general point is just the translation symmetry group of the lattice [Jansen and Boon (1967), p. 252; Streitwolf (1971), p. 80].

Problem VII.2.g-3: Show that every element of the little group of $\underline{k}$ transforms a Bloch function labeled by vector $\underline{k}$,

$$|\underline{k}) = exp(i\underline{k} \bullet \underline{x})u(x), \qquad \text{(VII.2.g-2)}$$

with u having the translational symmetry of the lattice, into another Bloch function also labeled by $\underline{k}$. How do these functions differ?

Problem VII.2.g-4: Describe completely the action of space group operators (especially nonsymmorphic ones) on Bloch functions [Bradley and Cracknell (1972), p. 144]. Are there features equivalent to those of the previous problem? Why?

VII.2.h Little co-groups and little groups

This procedure for inducing space-group representations is based on little groups. However for a nonsymmorphic space group there are two such groups, the point group that is the factor group, and a space group that is a subgroup. Since equivalent vectors in different Brillouin zones are taken the same, these both satisfy the definition of an invariance group, but the little group means the largest such — the largest group mixing equivalent representations — and this is a space group. However it is the operations and the representations of point groups that are known, thus (to induce representations of the space groups) we need a group that is a point group.

A transformation that leaves $\underline{k}$ invariant either leaves it unchanged or adds a reciprocal lattice vector. Because of the latter, the definition of little group differs from that used for, say, the Poincaré group [Mirman (1995c), sec. 2.2, p. 12] for which invariance means, and can mean, only the former. A crystal translation by a reciprocal lattice vector gives a point physically equivalent to the original; the translation eigenfunction is

$$|\underline{k}> = exp(i\underline{k} \bullet \underline{t}), \qquad (VII.2.h-1)$$

and is invariant for

$$\underline{k} \Rightarrow \underline{k} + \underline{K}, \qquad (VII.2.h-2)$$

where

$$\underline{K} \bullet \underline{t} = 2\pi, \qquad (VII.2.h-3)$$

for all lattice vectors $\underline{t}$. This broader definition of invariance is the one (necessarily) used.

In the representations obtained from the decomposition of the space group into cosets, representative j acts on the translation basis state, so, from the definition of space-group operation (eq. III.2.a-1, p. 134),

$$\{R_j|\underline{w}_j\}T \Rightarrow \{R_j|\underline{w}_j\}exp[i\underline{k} \bullet \underline{t}] = exp[i\underline{k} \bullet (R_j\underline{t} + \underline{w}_j)]$$

$$= exp[i(\underline{k}+\underline{K}) \bullet \underline{t} + i\underline{k} \bullet \underline{w}_j] = exp[i(\underline{k}+\underline{K}) \bullet \underline{t}]exp[i\underline{k} \bullet \underline{w}_j], \qquad (VII.2.h-4)$$

since we can consider R either to act on $\underline{t}$ or its reciprocal to act on $\underline{k}$. It is only for nonsymmorphic operations and special positions on the boundary of the Brillouin zone that the $\underline{w}_j$ and $\underline{K}$ terms appear. When they do the space group operation results in a change of momentum, and a phase.

The coset representatives in the decomposition of space group G into left cosets with respect to its translation subgroup are those elements that acting on the translation factor group give the cosets, and we chose them so that their point transformations correspond to the factor group G/T (sec. VII.2.f, p. 343). These however are not homogeneous operations (rotations, reflections or inversions); each coset representative contains, for nonsymmorphic groups, a translation, so they form a space group. The group formed by the set of these coset representatives that leave $\underline{k}$ invariant is the little group of $\underline{k}$ — it is that group giving representations equivalent to the one determined by $\underline{k}$ (sec. VI.2.c, p. 284).

The point group that is the factor group of a space group consists of homogeneous operations and acting on $\underline{k}$ gives a set of equivalent vectors (not all necessarily distinct; rotations around $\underline{k}$, say, if any, leave it invariant). The factor (point) group under which a vector is invariant is the little co-group of that vector [Altmann (1977), p. 205; Bradley and Cracknell (1972), p. 149; Burns and Glazer (1978), p. 194; Chen (1989), p. 475]. These point-group transformations are not (necessarily) symmetries of the crystal; they are obtained from the symmetries in this manner.

So the little co-group — a point group, but not necessarily a group of symmetries of the crystal — is the set of factor-group elements that is isomorphic to that point (factor) group of the space group that leaves $\underline{k}$ invariant; the little group, a space group, is the corresponding set of coset representatives whose homogeneous parts form a group isomorphic to the little co-group. The little co-group of $\underline{k}$ is (regarded here as) a factor group of its little group, not of the full space group G. Adding to $\underline{k}$ another reciprocal lattice vector results in a function equivalent to that labeled by $\underline{k}$. There is a difference between symmorphic and nonsymmorphic groups, leading to the distinction between little groups and little co-groups — for symmorphic groups little groups are point groups so coincide with the little co-groups. The method of finding the representations of G starts then with the little co-groups since their representations are what we know.

Little group $G(\underline{k})$ can be factorized into left cosets with respect to T, the translation subgroup of G,

$$G(\underline{k}) = \{R_1|\underline{w}_1\}T \oplus \{R_2|\underline{w}_2\}T \oplus \ldots \oplus \{R_n|\underline{w}_n\}T, \qquad \text{(VII.2.h-5)}$$

with

$$\{R_1|\underline{w}_1\} = 1; \qquad \text{(VII.2.h-6)}$$

the R's are the point-group transformations of the little co-group, $\underline{w}_i$ the smallest translation going with R_i, and n the number of cosets. For a symmorphic group, all $\underline{w}_i$ are zero (for proper choice of the origin); a nonsymmorphic group has at least one nonzero (sec. III.4.d, p. 152). We can write the product of space-group operations in a similar way, the product of a space group operation with a translation. For each i and j there is an R_l and $\underline{w}_l$ such that (pb. III.4.c.i-4, p. 151)

$$\{R_i|\underline{w}_i\}\{R_j|\underline{w}_j\} = \{R_iR_j|\underline{w}_i + R_i\underline{w}_j\} = \{E|\underline{t}_{ij}\}\{R_l|\underline{w}_l\}, \qquad \text{(VII.2.h-7)}$$

where $\underline{w}_l$ is that translation going with the product of space group operations $\{R_i|\underline{w}_i\}$ and $\{R_j|\underline{w}_j\}$ if that product is nonsymmorphic, $\{E|\underline{t}_{ij}\}$ is a translation and

$$\underline{t}_{ij} = \underline{w}_i + R_i\underline{w}_j - \underline{w}_l, \quad R_l = R_iR_j. \qquad \text{(VII.2.h-8)}$$

Note the relationship of this to the Frobenius congruence [Senechal (1990), p. 69],

$$\underline{w}_l \equiv \underline{w}_i + R_i\underline{w}_j - \underline{w}_l \text{ (mod T)}. \qquad \text{(VII.2.h-9)}$$

Problem VII.2.h-1: There might be a slight problem with discussing the little group of a vector here. One way of defining a little group gives a group determined by a representation (sec. VI.2.c, p. 284). However almost every point group (here the one defining the star of a vector) has several representations, so this little group, and little co-group, of the vector could depend on the point group representation. Thus the little group that leaves the star of a vector invariant may give several little groups, these taking the different point and space group representations to equivalent ones. Can this occur, or is the little group for a representation determined uniquely by the little group for the vector?

Problem VII.2.h-2: Show that the symmetry group (of what?) can be written as a (set) sum of cosets, starting with the little co-group of a vector, and the number of cosets equals the number of vectors in its star. What are the coset representatives? Thus the vector whose star it is can be chosen to lie within, or on the surface of, the representation domain of the Brillouin zone (sec. IV.2.c, p. 189), and this domain contains one, and only one, vector from each star.

Problem VII.2.h-3: Prove that the little co-group of every vector in the star of $\underline{k}$ is isomorphic to that of $\underline{k}$ [Bradley and Cracknell (1972), p. 149]. The groups of the vectors of a star then form a set of conjugate subgroups (sec. VI.5.c, p. 313). What are they subgroups of, and what are the operations under which they are conjugate? Similar results hold for little groups. Are there differences?

VII.3 REPRESENTATIONS OF LITTLE GROUPS AND LITTLE CO-GROUPS

The representations of space groups are found from those of the little groups, and these are found from the representations of the little co-groups. What are these, how are they found, and how, explicitly, does it help to use them. The discussion of these is thus our next task.

VII.3.a Small representations, and why they help

The Bloch functions labeled by the vectors forming the star of $\underline{k}$ are the basis vectors of a representation of the little group of $\underline{k}$ since they are generated using its operators. This representation is clearly irreducible since every one of its states is generated by applying the little group operators to any of these (properly chosen) states. Irreducible representations of this little group are called small representations [Altmann (1977), p. 176; Bradley and Cracknell (1972), p. 152; Burns and Glazer (1978), p. 194; Lyubarskii (1960), p. 93].

So finding space group representations involves nothing new except for the representations of the little groups (the small representations), which are themselves space groups. Does it lead anywhere, except in circles? It does give one nice (and hopefully useful) thing. In the representations of space group G, the translation operators $\{E|\underline{t}\}$ are diagonal — because we have chosen to consider these types of representations (sec. VII.2.a, p. 339); we leave open the question whether this leads to all space-group representations since it need not for the Poincaré group [Mirman (1995c), sec. 2.7, p. 28]. In the representation of $G(\underline{k})$, the little group of $\underline{k}$, these translations are also represented by a diagonal matrix, but in addition they have all entries the same, $exp(-i\underline{k} \bullet \underline{t})$ is the same for each; the basis state is a scalar,

$$|\underline{k}>= exp(-i\underline{k} \bullet \underline{t})\, I, \qquad \text{(VII.3.a-1)}$$

where I is the unit matrix of dimension that of the representation. Why does this extra restriction occur and how does it help?

For space group G translation operators are diagonal, but can have different diagonal elements — for a nonsymmorphic group, not all translations can be set to zero simultaneously, these give the diagonal elements, so there has to be at least two different ones. The little group has been chosen (wisely) to be that of a single $\underline{k}$, thus all translation eigenvalues are the same. The difference in the $\underline{k}'$s has been transferred to the cosets in the expansion for the space group (eq. VII.2.f-1, p. 343). Thus what this method does is split the procedure for finding the space group representations into two parts, one finds the representation of the little group, but this has all translation eigenvalues the same, the other combines the representations for the different orbits. However, in some cases, the representations for different orbits differ in phase. It is because of this that nonsymmorphic space group representations are not simply products, and further steps are required.

VII.3.a.i *Translation representations are scalars in little groups*

First we have to see why all diagonal elements of the translation matrices of the little-group representation are the same [Bradley and Cracknell (1972), p. 155]. The little group of $\underline{k}$ is first expanded as a sum (of sets) of cosets with respect to the translation factor group (eq. VII.2.h-5, p. 346). The basis functions of T are

$$\{E|t\} = exp(-i\underline{k} \bullet \underline{t}). \qquad \text{(VII.3.a.i-1)}$$

So acting on these

$$\{R_i|\underline{w}_i\}exp(-i\underline{k} \bullet \underline{t}) = exp(-i\underline{k} \bullet (\underline{w}_i + R_i\underline{t})), \qquad \text{(VII.3.a.i-2)}$$

but by definition of the little group all representation matrices of the terms in the coset decomposition have the same exponential which can be factored out, that is the diagonal elements are all the same (note that the vectors are equivalent,

$$\underline{k}R_i \equiv \underline{k}). \qquad \text{(VII.3.a.i-3)}$$

It is this equivalence which leads to the translation representations being scalars. In the corresponding coset decomposition of the space group itself (eq. VII.2.f-1, p. 343), such an equivalence does not hold, so all $\underline{k}$'s cannot be replaced by one, and the diagonal matrix elements are not all the same. The little group is purposely defined to give this equivalence, so identical diagonal elements (and of course, off-diagonal ones which are zero by restriction to momentum-diagonal representations).

VII.3.a.ii *Representations can be obtained from the little co-group*

That a matrix of the little group representing a translation is not only diagonal but a scalar allows the little-group representation matrices to be factorized into representation matrices of translation representations and of point-group representations. The point group is the little co-group; this is a point group, so its representations are known, and we immediately obtain those of the little group, from which we find the space group representations.

The problem is thereby simplified because we now have only the factor group of the little group, rather than of space group G, and that is (isomorphic to) the little co-group of $\underline{k}$ — its operators are given by the coset representatives, which are elements of this point group, the little co-group. Thus the space-group representation matrix is a matrix-valued function of the elements of the factor group of the little group with respect to the translations (sec. VI.3.b, p. 291), $G(\underline{k})/T$ — the elements of the factor group labeled by the coset representatives are each matrices representing the (discrete) translations [Bradley and Cracknell (1972), p. 155]. A representation matrix of the little group is just a product of that of the translation representation, the exponential, and a point group matrix. It is a function of the translations — each matrix element is a block representing the translations — so the matrix itself, labeled by its blocks, is independent of the translations, thus is a point-group representation, and substituting the (known) representation matrices of the little co-group in the matrix-valued functions we then obtain the representation matrices of the little group and thus space group G. So we need not find the representations of the (space) little group, but only those of the (point) little co-group, which is why this procedure is useful.

We now consider this explicitly.

VII.3.a.iii *The induced representation matrices*

Knowing that a matrix of the little group representing a translation is not only diagonal, but a scalar allows us to obtain the induced representation of space group G. Here we obtain the expression for the induced representation matrices and show that they do form a representation (perhaps projective). These matrices are block matrices, with the blocks known representation matrices.

Write the little group of $\underline{k}$ as a sum (of sets) of cosets with respect to the translation factor group (eq. VII.2.h-5, p. 346). With the matrices for representation p (giving the representation of the little co-group) of the little group of $\underline{k}$ (the small representation) designated $\Gamma_p^{\underline{k}}(\{R_i|\underline{w}_i\})$, we get

$$\Gamma_p^{\underline{k}}(\{R_i|\underline{w}_i\})\Gamma_p^{\underline{k}}(\{R_j|\underline{w}_j\}) = exp(-i\underline{k} \bullet (\underline{w}_i + R_i\underline{w}_j - \underline{w}_l))\Gamma_p^{\underline{k}}(R_l|\underline{w}_l), \tag{VII.3.a.iii-1}$$

since the elements of the cosets are products and for each i and j there is such an R_l and $\underline{w}_l$ (eq. VII.2.h-7, p. 346); also $\underline{t}_{ij}$ is a translation and basis states of $\{E|\underline{t}_{ij}\}$ are of this form.

We write these matrices more explicitly as $\Gamma_p^{\underline{k}}(\{R|\underline{t}_n + \underline{\tau}\})$ [Burns and Glazer (1978), p. 194], which exhibits the vector left invariant, $\underline{k}$, the semisimple part, operator R, lattice translation vector $\underline{t}_n$, a sum of primitive lattice vectors with coefficients n_1, n_2, n_3, abbreviated n, and $\underline{\tau}_j$ the nonsymmorphic translational part associated with R_j in the decomposition of the little group into left cosets. Now

$$R_i^{-1}\underline{k} = \underline{k} + \underline{K}_i, \tag{VII.3.a.iii-2}$$

using

$$R^{-1}\underline{k} \bullet \underline{t} = \underline{k} \bullet R\underline{t}, \tag{VII.3.a.iii-3}$$

for any $\underline{t}$, since R is orthogonal. The symmorphic part of $\underline{w}_i + \underline{w}_j - \underline{w}_l$ is zero by definition of $\underline{w}_l$, so the product becomes

$$\Gamma_p^{\underline{k}}(\{R_i|\underline{w}_i\})\Gamma_p^{\underline{k}}(\{R_j|\underline{w}_j\}) = exp(-i\underline{K} \bullet \underline{\tau}_j)\Gamma_p^{\underline{k}}(R_l|\underline{w}_l). \tag{VII.3.a.iii-4}$$

We label the representation matrix of the induced representation as $D_p^{\underline{k}}(\{R_i|\underline{w}_i\})$; it is the representation matrix of space group G induced from little group representation $\Gamma_p^{\underline{k}}$, with the labels $\underline{k}$ giving the little group, and p giving its representation. These are (eq. VI.3.b-2, p. 291),

$$D_p^{\underline{k}}(\{R_l|\underline{w}_l\})_{ij} = \Gamma_p^{\underline{k}}(\{R_i|\underline{w}_i\}\{R_l|\underline{w}_l\}\{R_j|\underline{w}_j\}^{-1}),$$

$$\text{if } \{R_i|\underline{w}_i\}\{R_l|\underline{w}_l\}\{R_j|\underline{w}_j\}^{-1} \varepsilon\, G(\underline{k}), \text{ else } 0. \qquad \text{(VII.3.a.iii-5)}$$

There are two cases, first for symmorphic groups and for interior points the origin can be picked so the translation part of every space-group element $\underline{w}_l$ is zero. Then the matrices of G are isomorphic to those of the little co-group, a point group — the space group representation matrices are just the product of the point group matrices times an exponential.

However on the boundary, point group operations can take special points to equivalent ones, and if the space group operation is nonsymmorphic, it is not equivalent to the identity, so adds a fraction of a lattice vector. But matrices $D_p^{\underline{k}}$ are the same for all members of any fixed coset $\{R_i|\underline{w}_i\}$T in the decomposition of $G(\underline{k})$, since the Γ's must be — they do not depend on the translation.

Problem VII.3.a.iii-1: Find the matrices explicitly and show that there is however a difference in the exponential on the two sides, thus the representation is projective.

VII.3.a.iv *The point group part of the matrices*

The space group representations are thus derived from those of the little co-group, a point group. However this group is restricted, and there are few that are possible. Vectors $\underline{w}$ and $\underline{k}$ are (proportional to) lattice vectors of the crystal, and of the reciprocal lattice. The product $\underline{k} \bullet \underline{w}$ that appears in the exponential thus depends on only the angle between the two, this is limited for lattices thus limiting the point groups that need be considered, giving

$$exp(-i\underline{k} \bullet \underline{w}) = exp(\frac{2\pi in}{m}), \qquad \text{(VII.3.a.iv-1)}$$

where

$$m = 1, 2, 3, 4, \text{ or } 6, \text{ and } 0 \le n \le m-1. \qquad \text{(VII.3.a.iv-2)}$$

VII.3.b The central extensions of the little co-group

As some representations of space groups are projective representations (sec. V.4, p. 259), we need their central extensions (sec. V.5, p. 270), and factor systems. The factor system, as we know (sec. VII.3.a.iii, p. 350), is

$$\omega(R_i, R_j) = exp(-i\underline{K}_i \bullet \underline{\tau}_j), \qquad \text{(VII.3.b-1)}$$

where

$$R_i^{-1}\underline{k} = \underline{k} + \underline{K}_i. \qquad \text{(VII.3.b-2)}$$

Since

$$R_l = E \text{ implies } \underline{K}_i = 0, \qquad \text{(VII.3.b-3)}$$

we get

$$\omega(E, R_j) = 1, \qquad \text{(VII.3.b-4)}$$

for all j. Because $\underline{\tau}_j$ is a fractional part of a translation (a fractional part because translations are restricted to the first Brillouin zone), $\omega(R_j, R_l)$ is of the form of this exponential, with the order of the rotations in the little co-group, a crystallographic point group, limited to 1,2,3,4 or 6. So to find the irreducible representations of a space group it is only necessary to find the representations of a few groups of small order, the extensions given by these factor systems, in addition to the representations of the point groups.

We limit $\underline{k}$ to be as short as possible, that is in, or on the boundary of, the first Brillouin zone. If it is in the zone there is no vector equivalent to it so

$$\underline{K}_i = 0, \qquad \text{(VII.3.b-5)}$$

and the representation of the little group is given by a vector representation of a point group, times a phase. It is only if it is on the boundary, for a nonsymmorphic group, that $\underline{K}_i$ can be nonzero, which thus is the only case needing further consideration.

Problem VII.3.b-1: Show that, for a nonsymmorphic space group, with $\underline{k}$ on the boundary of the Brillouin zone, a way of writing the factor system, other than that given previously (sec. V.5.c, p. 277), is

$$\omega(R, R') = exp[-i(\underline{k} - R^{-1}\underline{k}) \bullet \underline{\tau}_{R'}], \qquad \text{(VII.3.b-6)}$$

that is this obeys the requirements for a factor system, and in particular it gives the correct representation of the translation parts of the operators [Janssen (1973), p. 135].

VII.3.c Symmorphic space groups

If the space group is symmorphic all $\underline{w}_i$ are equivalent to 0, and $D_p^{\underline{k}}$ forms a (not necessarily faithful) representation of the little co-group. It also forms such a representation for a general space group if for all $\underline{k}$,

$$R_i\underline{k} \equiv \underline{k}; \qquad \text{(VII.3.c-1)}$$

that is

$$\underline{K}_i = 0. \qquad \text{(VII.3.c-2)}$$

This is the case for $\underline{k}$ an interior point of the Brillouin zone — because then the only point of the Brillouin zone equivalent to $\underline{k}$ is $\underline{k}$ itself.

Problem VII.3.c-1: In these cases the irreducible representations of $\Gamma_p^{\underline{k}}$ implies those of $D_p^{\underline{k}}$. We know the representations of the little co-groups, these being point groups, thus we have now found the representations for all symmorphic space groups [Cornwell (1984), p. 228]. Give these representations explicitly [Streitwolf (1971), p. 81]. Show that these are actually representations, and irreducible. Also for nonsymmorphic groups, we have found the little groups for all $\underline{k}$'s, except for a special subset. Finding the representations for these special cases completes the task of obtaining all representations of nonsymmorphic groups, thus of all space groups.

VII.4 INDUCING REPRESENTATIONS OF NONSYMMORPHIC SPACE GROUPS

The methods for finding space group representations have been developed, except for the few special cases. So to these more interesting instances we now turn, making the general prescriptions specific. The theory of induced representations (sec. VI.3, p. 288) finds perhaps its most important use in the development of those of space groups [Kocinski (1983), p. 33; Streitwolf (1971), p. 78]. For symmorphic groups the development is straightforward, and explicit knowledge of the theory of induced representations is not needed — it can be applied almost intuitively; also they are semi-direct products which have already been discussed (sec. VI.7.i, p. 330). For nonsymmorphic groups however this theory does provide a valuable tool, and these groups provide valuable illustrations.

VII.4.a Representations, allowable and not

Only a subset of the representations of the little group of $\underline{k}$ (the small representations) are used, those whose matrices $\Gamma_p^{\underline{k}}$ satisfy the condition

$$\Gamma_p^{\underline{k}}(\{E|\underline{t}\}) = exp(-i\underline{k} \bullet \underline{t}_n)\Gamma_p^{\underline{k}}(\{E|0\}), \qquad \text{(VII.4.a-1)}$$

where $\Gamma_p(\{E|0\})$ is the unit matrix, for every $\{E|\underline{t}\}$ of the translation subgroup, that is for all

$$\underline{t}_n = n_1\underline{t}_1 + n_2\underline{t}_2 + n_3\underline{t}_3, \qquad \text{(VII.4.a-2)}$$

for which the n's are integral; the $\underline{t}_i$ are the three primitive lattice vectors. These are called the appropriate, allowable, allowed, permitted or relevant representations [Altmann (1977), p. 177; Birman (1974), p. 56;

Cornwell (1969), p. 169; Cornwell (1984), p. 239; Falicov (1966), p. 174; Jansen and Boon (1967), p. 251; Ludwig and Falter (1988), p. 206; Streitwolf (1971), p. 79]. Since the sum of the squares of the dimensions of these representations equals the order of the (factor) point group of $\underline{k}$, which is the requirement for the complete set of representations [Mirman (1995a), sec. VI.2.b, p. 175], not all irreducible representations of the little group of $\underline{k}$ are allowable. The others must thus be equivalent to the allowable ones and using them would give multiple appearances of induced representations.

It may seem obvious that a pure translation is represented this way, in fact must be so represented. And for a general point this is the only possibility — all representations are allowable. But for special positions (in momentum — reciprocal — space), there are point-group transformations under which the momentum is invariant. So $\Gamma_p^{\underline{k}}(\{E|\underline{t}\})$ and $\Gamma_p^{\underline{k}}(\{R|\underline{t}\})$, where R is a point-group transformation leaving special point $\underline{k}$ invariant, give the same translation. Thus we have a choice of of the form of the representations labeled by these points — for special points of nonsymmorphic groups, there are several translations that give equivalent representations. Using all would result in representations appearing more than once (perhaps in different, but equivalent, forms). A choice must be made to eliminate extra appearances, and this condition is it. This means that the character table of the factor group is not (in general) square.

An example of a representation that does not satisfy, thus is not "allowable" is the identity representation for $\underline{k} \neq 0$.

The number of nonequivalent allowable-irreducible representations of the little group of $\underline{k}$ is the same as the number of irreducible representations of its little co-group, indicating both the need for only allowable representations, and that the representations of the little co-group, a point group, give the irreducible representations of the space group.

Problem VII.4.a-1: How is this definition of allowable related to that given previously (sec. VI.7.c, p. 324).

Problem VII.4.a-2: Prove that the sum of the squares of the dimensions of these representations equals the order of the (factor) point group of $\underline{k}$ [Cornwell (1984), p. 239].

VII.4.b Inducing representations of symmorphic space groups

We first summarize the simpler symmorphic case. All these groups are (isomorphic to) semi-direct (so disjoint) products of a point group and an invariant translation subgroup [Cornwell (1984), p. 226]. Translations, remain translations under point-group operations, so form an invariant subgroup [Mirman (1995a), sec. XIII.4.b, p. 382; (1995b), sec. 6.1, p. 108]. Symmorphic operations are defined as those that are products of (independent) point-group and translation transformations, unlike nonsymmorphic operations — nonsymmorphic space groups are not semi-direct products. Since the operations have this form, all representations of the little group are allowable, because the little group is just the little co-group, times translations.

The little groups are thus isomorphic to subgroups of the point groups so all their representations are known, as are representations of the translation groups, hence the theory of induced representations gives the representations as products of these two [Jansen and Boon (1967), p. 253], so the space group representations for symmorphic groups are immediate.

Problem VII.4.b-1: Apply the theory of semi-direct product groups (sec. VI.7.i, p. 330) to find the representations explicitly. Show that the allowable representations are all.

VII.4.c Representations of nonsymmorphic groups

With this background we are ready to to induce representations of the space groups [Cornwell (1984), p. 235]. The translation group is infinite (but discrete) so the properties that we have found for finite groups may not hold. We therefore visualize a crystal as having cyclic boundary conditions, the Born-von Karman boundary conditions [Birman (1974), p. 12; Jansen and Boon (1967), p. 258; Streitwolf (1971), p. 76], which means that the crystal repeats after a translation of $N_i \underline{t}_i$, an integral number of translations by a primitive lattice vector in any of the three directions (sec. XI.2.f, p. 577). Thus the translation subgroups of the space group in these three directions are of orders N_1, N_2, N_3.

Take any allowed vector $\underline{k}$ — a sum of the primitive reciprocal lattice vectors, with integral coefficients n_i/N_i, with the n's any integers (of course less than the N's otherwise repetitions occur). The little group of this vector is the set of transformations $\{R|\underline{\tau}_R + \underline{t}\}$, with $\underline{\tau}_R$ the nonsymmorphic translation going with R, $\underline{t}$ an arbitrary lattice vector, and

$$R^{-1}\underline{k} \equiv \underline{k} + \underline{K}_m; \quad \text{(VII.4.c-1)}$$

$\underline{K}_m$ is any (including zero) lattice vector of the reciprocal lattice, a sum, with three integral coefficients labeled by the m's, of primitive reciprocal lattice vectors. The group, acting on the lattice in real space, is defined so that its homogeneous factor group, a factor group of the space group, is that set of operators taking reciprocal-space vector $\underline{k}$ into equivalent vectors, and is labeled by this vector. It contains the group of pure primitive translations (as a subgroup). Thus $\{R|\underline{\tau}_R + \underline{t}\}$ is contained in the left coset, labeled by $\{R|\underline{\tau}_R\}$, in the decomposition of the space group with respect to the little group of $\underline{k}$. That is, translation by a lattice vector gives the same transformation of the space group, and so of its subgroup. These coset representatives can thus be chosen of the form $\{R|\underline{\tau}_R\}$ which label the set by the point group transformations since $\underline{\tau}_R$ is determined by R.

This gives the space group representations. We take any allowed vector $\underline{k}$ of the Brillouin zone, its little group, and the decomposition of the space group with respect to this group, with coset representatives $\{R_j|\underline{\tau}_j\}$, $j = 1, 2, \ldots, O_{\underline{k}}$, this being the number of cosets

$$O_{\underline{k}} = \frac{g_o}{g_o(\underline{k})} = \frac{g}{g(\underline{k})}. \tag{VII.4.c-2}$$

Here the orders are g for the space group and g_o for the (factor) point group, with the $\underline{k}$'s indicating the relevant subgroups of these, and with

$$N = N_1 N_2 N_3, \tag{VII.4.c-3}$$

the order of the translation subgroup,

$$g = g_o N \text{ and } g(\underline{k}) = g_o(\underline{k}) N, \tag{VII.4.c-4}$$

since the set of translations is independent of $\underline{k}$. Because cyclic boundary conditions are used the translation subgroup of the space group has only a finite number of terms, thus the orders of it, so also of the space group, are finite. However this is a device to allow these calculations, and if the orders appear in the final result something is wrong.

Next take a unitary irreducible representation, labeled p, of the little group of $\underline{k}$, of dimension d_p (sec. VII.3.a.iii, p. 350), restricted to be allowable so with matrix elements, for every $\{E|\underline{t}\}$ of the translation subgroup, given by eq. VII.4.a-1, p. 353. Then an induced unitary irreducible representation of the space group, of dimension $d_p O_{\underline{k}}$, has matrices

$$D_p^{\underline{k}}(\{R|\underline{\tau}_R + \underline{t}_n\})_{ir,js} = \Gamma_p^{\underline{k}}(\{R_i|\underline{\tau}_i\}\{R|\underline{\tau}_R + \underline{t}_n\}\{R_j|\underline{\tau}_j\}^{-1})_{rs}$$

$$= exp(-iR_i^{-1}\underline{k} \bullet \underline{t}_n)\Gamma_p^{\underline{k}}(\{R_i|\underline{\tau}_i\}\{R|\underline{\tau}_R\}\{R_j|\underline{\tau}_j\}^{-1})_{rs}, \tag{VII.4.c-5}$$

if $\{R_i|\underline{\tau}_i\}\{R|\underline{\tau}_R\}\{R_j|\underline{\tau}_j\}^{-1}$ is in the little group of $\underline{k}$, otherwise 0, with

$$i, j = 1, \ldots, O_{\underline{k}},\ r, t = 1, \ldots, d_p. \tag{VII.4.c-6}$$

The labels of the representation are $\underline{k}$, the representation label of the translation subgroup, the point in the Brillouin zone from whose little group we induce the space group representation, and p the set of labels of the representation of the point factor group, the little co-group. The little co-group (so its representation) varies within the Brillouin zone. The representation basis states are thus functions of $\underline{k}$. These relate the different little groups and their representations.

This gives every inequivalent irreducible representation of space group G using all inequivalent irreducible allowable representations of the little group of $\underline{k}$, taking one $\underline{k}$ for each star, and every star within the representation domain (sec. IV.2.c, p. 189) of the first Brillouin zone.

VII.4.c.i *These matrices form a representation*

First we have to show that these matrices form a representation, and a unitary one. That they do follows from the general theory for induced representations. All that needs to be done is to identify the symbols here with those in the definition (sec. VI.3.b, p. 291). Group G is here the space group, and the subgroup is G($\underline{k}$), the little group of $\underline{k}$, with orders g and $g_{(}\underline{k})$. The coset representatives are $\{R_1|\underline{\tau}_1\},\ldots,\{R_{O_{\underline{k}}}|\underline{\tau}_{O_{\underline{k}}}\}$. Matrices $\Gamma_p^{\underline{k}}$ (here) form a d_p-dimensional representation of the little group. The other notational changes are obvious. Thus with change of notation, it is clear that these matrices do form a representation.

VII.4.c.ii *The representations are irreducible*

That these representations are irreducible follows from the irreducibility of the representations of the little group. From the expression for the matrix elements (eq. VII.4.c-5, p. 356), that for the characters ζ of the space group, in terms of the characters η of the little group, is

$$\zeta_p^{\underline{k}}(\{R|\underline{\tau}_R+\underline{t}_n\}) = \sum exp(-iR_j\underline{k}\bullet\underline{t}_n)\eta_p^{\underline{k}}(\{R_j|\underline{\tau}_j\}\{R|\underline{\tau}_R\}\{R_j|\underline{\tau}_j\}^{-1}), \quad \text{(VII.4.c.ii-1)}$$

where the summation is over all coset representatives fulfilling the condition on the product being in the subgroup, the little group of the vector. Now, summing over the (direct) lattice vectors of a unit cell, using the Born-von Karman conditions and eq. VII.4.c-3, p. 356, we get [Cornwell (1984), p. 209],

$$\sum_{\underline{t}_n} exp(i\underline{k}\bullet\underline{t}_n) = N, \text{ if } \underline{k}=\underline{K}_m$$

$$= 0, \text{ otherwise}, \quad \text{(VII.4.c.ii-2)}$$

and similarly

$$\sum_{\underline{k}} exp(-i\underline{k} \bullet \underline{t}_n) = N, \text{ if } \underline{t}_n = 0,$$

$$= 0, \text{ otherwise}, \qquad \text{(VII.4.c.ii-3)}$$

where the sum is over the allowed $\underline{k}$ vectors. A $\underline{K}_m$ is an integer sum of reciprocal lattice vectors, and $\underline{t}_n$ of direct lattice vectors, with the subscripts indicating the coefficients. The two nonzero values arise as all terms in the sum are the same, under the stated conditions, and there are $N = N_1N_2N_3$ terms. Otherwise for each term there is its negative because all lattices have inversion symmetry. This gives [Cornwell (1984), p. 237]

$$\sum |\zeta_p^{\underline{k}}(\{R|\underline{\tau}_R + \underline{t}_n\})|^2 = N \sum |\eta_p^{\underline{k}}(\{R_j|\underline{\tau}_j\}\{R|\underline{\tau}_R\}\{R_j|\underline{\tau}_j\}^{-1})|^2, \qquad \text{(VII.4.c.ii-4)}$$

where the sum on the left is over n, on the right over j. Thus

$$\sum |\zeta_p^{\underline{k}}(\{R|\underline{\tau}_R + \underline{t}_n\})|^2 = \sum\sum\sum |\eta_p^{\underline{k}}(\{R_j|\underline{\tau}_j\}\{R|\underline{\tau}_R + \underline{t}_n\}\{R_j|\underline{\tau}_j\}^{-1})|^2, \qquad \text{(VII.4.c.ii-5)}$$

the sum on the left is now over the coset representatives, those on the right over the coset representatives of the factor group of the space group and over the $\underline{t}_n$, and j.

Using the condition on the product of the coset representatives and interchanging summations (their ranges are finite, because of the Born-von Karman conditions) we get

$$\sum |\zeta_p^{\underline{k}}(\{R|\underline{\tau}_R + \underline{t}_n\})|^2 = \sum\sum \eta_p^{\underline{k}}(\{R|\underline{\tau}_R + \underline{t}_n\})|^2 = O_{\underline{k}}g(\underline{k}) = g, \qquad \text{(VII.4.c.ii-6)}$$

with the first sum on the right from 1 to $O_{\underline{k}}$, while the second is over the coset representatives of the group of the vector. Using the expression in terms of the characters for the number of times a representation appears in the decomposition of a reducible representation shows that these representations are irreducible [Mirman (1995a), sec. VII.6.a, p. 199].

It then follows, by the character orthogonality theorem, that two representations are inequivalent if the vectors are in different stars or if they are obtained from inequivalent representations of the (factor) point groups, since they are sums of products of characters of the representations of the factor groups, which are then zero.

Problem VII.4.c.ii-1: That a representation induced from an allowable irreducible representation of the little group is itself irreducible can also be shown using Schur's lemma [Streitwolf (1971), p. 44].

VII.4.c.iii *All inequivalent representations are obtained*

Does this give all inequivalent representations? If so the sum of the squares of the dimensions of all these representations, using the allowable little group representations, and taking one $\underline{k}$ for each star, equals the order of the space group, g [Mirman (1995a), sec. VI.2.b, p. 175]. First for each star we note that the sum of the squares of the dimensions of the allowable inequivalent irreducible representations of the group for that star is $g_o(\underline{k})$, so these give all its representations. Then summing over all stars gives

$$\sum g_o(\underline{k})O_{\underline{k}}{}^2 = \sum g_o O_{\underline{k}} = Ng_o = g, \qquad \text{(VII.4.c.iii-1)}$$

as required. So the allowable little group representations are necessary and sufficient — every irreducible representation can be induced from an allowable representation of a little group, and each only once [Streitwolf (1971), p. 46].

VII.4.d The allowable representations of the little group

To find the representations of a space group we use representations of the little group, that of vector $\underline{k}$, but only the allowable ones. How do we determine which representations are allowable [Jansen and Boon (1967), p. 252]?

VII.4.d.i *Obtaining the allowable representations from representations of the little co-group*

First [Cornwell (1984), p. 239] are the cases for which

$$exp[-i\underline{k} \bullet (R\underline{\tau}_R{}' - \underline{\tau}_R{}')] = 1, \qquad \text{(VII.4.d.i-1)}$$

for every pair of transformations $\{R|\underline{\tau}_R\}$ and $\{R'|\underline{\tau}_R{}'\}$ of the little group of $\underline{k}$, $G(\underline{k})$, with little co-group $S(\underline{k})$. Using

$$R\underline{k}^{-1} \equiv \underline{k} + \underline{K}_m, \qquad \text{(VII.4.d.i-2)}$$

it follows that

$$exp(-i\underline{\tau}_R{}' \bullet \underline{K}_m) = 1, \qquad \text{(VII.4.d.i-3)}$$

which means the representations are allowable (sec. VII.4.a, p. 353). To obtain the representation, take the matrices, defined for all $\{R|\underline{\tau}_R + \underline{t}_n\}$ of this group,

$$\Gamma_p^{\underline{k}}(\{R|\underline{\tau}_R + \underline{t}_n\}) = exp[-i\underline{k} \bullet (\underline{\tau}_R + \underline{t}_n)]M_p^{\underline{k}}(\{R|0\}), \qquad \text{(VII.4.d.i-4)}$$

where $M_p^{\underline{k}}(\{R|0\})$ belongs to a unitary irreducible representation of little co-group $S(\underline{k})$. These matrices form an allowable, unitary irreducible representation of little group $G(\underline{k})$. All such representations are obtained by using all unitary irreducible representations of $S(\underline{k})$. Thus in this case we have obtained the required representations, and those of the space group, G, follow using induction.

Why is this true? First it is necessary to show that this is a representation of the little group. For any two members of the little group, $\{R|\underline{\tau}_R + \underline{t}_n\}$ and $\{R'|\underline{\tau}_R' + \underline{t}_n'\}$, this gives

$$\Gamma_p^{\underline{k}}(\{R|\underline{\tau}_R + \underline{t}_n\})\Gamma_p^{\underline{k}}(\{R'|\underline{\tau}_R' + \underline{t}_n'\})$$

$$= exp[-i\underline{k} \bullet (\underline{\tau}_R + \underline{t}_n + \underline{\tau}_R' + \underline{t}_n')]M_p^{\underline{k}}(\{RR'|0\}). \qquad \text{(VII.4.d.i-5)}$$

Also

$$\Gamma_p^{\underline{k}}(\{R|\underline{\tau}_R + \underline{t}_n\}\{R'|\underline{\tau}_R' + \underline{t}_n'\})$$

$$= exp[-i\underline{k} \bullet (\underline{\tau}_R + \underline{t}_n + R\underline{\tau}_R' + R\underline{t}_n')]M_p^{\underline{k}}(\{RR'|0\}). \qquad \text{(VII.4.d.i-6)}$$

Since the little group is that under which representations labeled by (the star of) $\underline{k}$ are equivalent, using eqs. VII.4.d.i-1,2,3, we get

$$exp(-i\underline{k} \bullet R\underline{t}_n') = exp(-iR^{-1}\underline{k} \bullet \underline{t}_n') = exp(-i\underline{k} \bullet \underline{t}_n'); \qquad \text{(VII.4.d.i-7)}$$

the two expressions for the products are the same, so these matrices form a representation.

That $\Gamma_p^{\underline{k}}$ is unitary and irreducible follows from $M_p^{\underline{k}}$ being so. Also it is immediate that this representation is allowable. We can also see that the sum of the squares of the dimensions of the inequivalent, irreducible representations of the little co-group is equal to the order of this group, $g_o(\underline{k})$; this must also be true of the allowable, irreducible representations given by this procedure, so there can be no further ones. Hence all required representations have been found.

For a symmorphic space group $\underline{\tau}_R = 0$, for all $\{R|0\}$, so the condition holds for every allowed value of $\underline{k}$, thus the procedure gives every allowable, irreducible representation of the little group of $\underline{k}$, and is the same as above.

Problem VII.4.d.i-1: Explain why the condition is sufficient, and show that the representations are unitary and irreducible. Is it necessary? How is it related to the requirement determining allowable representations (eq. VII.4.a-1, p. 353)?

Problem VII.4.d.i-2: Check that the statement about the sum of the squares, and the ones about its implications, are correct.

VII.4.d.ii *The cases of internal vectors*

This method works, except in a few special cases (the ones for which projective and induced representations are needed). If $\underline{k}$ is a vector inside the Brillouin zone (that is not on the boundary) it is applicable, as well as for general points on the boundary, and of course for symmorphic groups, and these include almost all vectors. In such cases all representations are allowable since none of these points is equivalent to another in the same zone, thus the condition (eq. VII.4.a-1, p. 353) is satisfied.

This condition can be written for these cases as

$$exp[-iR^{-1}\underline{k} \bullet \underline{\tau}_R'] = exp[-i\underline{k} \bullet \underline{\tau}_R']. \qquad \text{(VII.4.d.ii-1)}$$

Since

$$R\underline{k}^{-1} \equiv \underline{k} + \underline{K}_m, \qquad \text{(VII.4.d.ii-2)}$$

which follows from $\{R^{-1}|0\}$ being in the little co-group, and if $\underline{k}$ is an internal point $\underline{K}_m = 0$, so the requirement for an allowable representation is met.

Because the little co-group is a point group, all of whose representations are known, and all representations are allowable, the (parts of the) representations of the space groups for which these conditions hold now follow by induction.

If the condition does not hold then the product of two matrices equals the matrix of the product of the two group elements, times $exp[-i\underline{k} \bullet (\underline{\tau}_R + R\underline{\tau}_R')]$. The representation is then projective, and this is what we have to consider next.

VII.4.d.iii *Special positions on the boundary*

The only cases remaining then are for special positions on the boundary of the Brillouin zone [Jansen and Boon (1967), p. 253]. For these the little group is not isomorphic to the little co-group. Group operations connect them to other boundary positions, so equivalent representations occur more than once, there is duplication, not all representations are allowable, and the allowable ones must be determined [Cornwell (1984), p. 241]. And representations of the little group are projective (sec. VII.3.a.iii, p. 350) since transformations between equivalent vectors add phases.

The translation group of wave vector $\underline{k}$, $T(\underline{k})$, the set of all $\{E|\underline{t}_n\}$ such that

$$exp(-i\underline{k} \bullet \underline{t}_n) = 1, \qquad \text{(VII.4.d.iii-1)}$$

is an invariant subgroup of the little group of $\underline{k}$ — $T(\underline{k})$ is a subgroup of translation group T, which is a subgroup of little group $G(\underline{k})$, thus

$T(\underline{k})$ is. Let $\{R|\underline{\tau}_R+\underline{t}_n\}$ belong to $G(\underline{k})$, and $\{E|\underline{t}_n'\}$ belong to $T(\underline{k})$. So, since $\underline{\tau}_R+\underline{t}_n$ is invariant under R,

$$\{R|\underline{\tau}_R+\underline{t}_n\}\{E|\underline{t}_n'\}\{R|\underline{\tau}_R+\underline{t}_n\}^{-1}=\{E|R\underline{t}_n'\}. \qquad \text{(VII.4.d.iii-2)}$$

Now $\{E|R\underline{t}_n'\}$ is also a member of $T(\underline{k})$ since

$$exp(-i\underline{k}\bullet\underline{R}t_n')=1. \qquad \text{(VII.4.d.iii-3)}$$

Thus the subgroup is invariant.

To determine the space-group representation matrices for these positions we construct factor group

$$F(\underline{k})=G(\underline{k})/T(\underline{k}). \qquad \text{(VII.4.d.iii-4)}$$

The elements of this are taken as left cosets of the form $\{R|\underline{t}_R\}T(\underline{k})$; the identity of the factor group is $\{E|0\}T(\underline{k})$. We take a unitary irreducible representation of $G(\underline{k})/T(\underline{k})$, with matrices $M_p^{\underline{k}}$, satisfying

$$M_p^{\underline{k}}(\{E|0\}T(\underline{k}))=exp(-i\underline{k}\bullet\underline{t}_n)M_p^{\underline{k}}(\{E|0\})T(\underline{k}), \qquad \text{(VII.4.d.iii-5)}$$

for every coset $\{E|\underline{t}_n\}T(\underline{k})$ formed from primitive translations. Then the matrices defined by (eq. VII.4.d.i-4, p. 359)

$$\Gamma_p^{\underline{k}}(\{R|\underline{\tau}_R+\underline{t}_n\})=exp[-i\underline{k}\bullet(\underline{\tau}_R+\underline{t}_n)]M_p^{\underline{k}}(\{R|0\}T(\underline{k})), \qquad \text{(VII.4.d.iii-6)}$$

for every element of the little group, form a unitary irreducible representation of $G(\underline{k})$ satisfying the condition

$$\Gamma_p^{\underline{k}}(\{E|\underline{t}\})=exp(-i\underline{k}\bullet\underline{t}_n)\Gamma_p^{\underline{k}}(\{E|0\}). \qquad \text{(VII.4.d.iii-7)}$$

A representation that obeys this equation is an allowable representation of $G(\underline{k})$, which not every irreducible representation of the group is, in general.

Substituting these into eq. VII.4.c-5, p. 356, we get the expression for the space group representation matrix

$$D_p^{\underline{k}}(\{R|\underline{\tau}_R+\underline{t}_n\})_{ir,js}=exp[-i(\underline{k}+\underline{K})\bullet(\underline{\tau}_R+\underline{t}_n)]M_p^{\underline{k}}(\{R|0\})_{rs}, \qquad \text{(VII.4.d.iii-8)}$$

giving the matrices in terms of those of the (point) little co-group, which are known.

To show that every allowable irreducible representation of $G(\underline{k})$ can be obtained in this way we first show that these matrices do form a representation of $G(\underline{k})$. For any two members of the little group, $\{R|\underline{\tau}_R+\underline{t}_n\}$ and $\{R'|\underline{\tau}_R'+\underline{t}_n'\}$,

$$\Gamma_p^{\mathbf{k}}(\{R|\underline{\tau}_R+\underline{t}_n)\Gamma_p^{\mathbf{k}}(R'|\underline{\tau}_R'+\underline{t}_n')$$

$$= exp[-i\underline{k}\bullet(\underline{\tau}_R+\underline{t}_n)]exp[-i\underline{k}\bullet(\underline{\tau}_R'+\underline{t}_n')]M_p^{\mathbf{k}}(\{R|0\}T(\underline{k})\{R'|0\}T(\underline{k}))$$

$$= exp[-i\underline{k}\bullet(\underline{\tau}_R+\underline{t}_n)]exp[-i\underline{k}\bullet(\underline{\tau}_R'+\underline{t}_n')]M_p^{\mathbf{k}}(\{R|0\}\{R'|0\}T(\underline{k}))$$

$$= \Gamma_p^{\mathbf{k}}(\{R|\underline{\tau}_R+\underline{t}_n\}\{R'|\underline{\tau}_R'+\underline{t}_n'\}), \qquad \text{(VII.4.d.iii-9)}$$

as required for a representation, using the fact that the product of a coset representative acting on an invariant subgroup, times another coset representative acting on that subgroup, is the same as the product of the coset representatives acting on the subgroup [Cornwell (1984), p. 31].

That it is irreducible is shown using the characters [Mirman (1995a), sec. VII.6.a, p. 199]. We have, with $o(\underline{k})$ the order of $T(\underline{k})$,

$$\sum|\eta_p^{\mathbf{k}}(\{R|\underline{\tau}_R+\underline{t}_n\})|^2 = o(\underline{k})\sum|\eta_p(\{R|\underline{\tau}_R+\underline{t}_n\}T(\underline{k}))|^2. \qquad \text{(VII.4.d.iii-10)}$$

The summations are over all elements of the groups indicated by the subscripts. Now $M_p^{\mathbf{k}}$ is taken irreducible, the sum on the right side equals the order of $G(\underline{k})/T(\underline{k})$, so that on the left must equal the order of $G(\underline{k})$, which means that the representation is irreducible.

It also follows from eq. VII.4.d.iii-7, p. 362, that the representation is allowable, satisfying eq. VII.4.a-1, p. 353. Conversely if $\Gamma_p^{\mathbf{k}}$ satisfies this condition for being allowable every member of $T(\underline{k})$ in it is represented by a unit matrix, so all members of a coset of $T(\underline{k})$ are represented by the same matrix, and $\Gamma_p^{\mathbf{k}}$ must have this form.

For symmetry points, axes and planes the order of $G(\underline{k})/T(\underline{k})$ is less than that of $G(\underline{k})$, so its irreducible representations can be found more easily, the whole point of this procedure.

Problem VII.4.d.iii-1: Check the statements of the last three paragraphs.

Problem VII.4.d.iii-2: Explain why the product of coset representatives acting on an invariant subgroup, equals the product of the two sets (what does this mean?) obtained by having each act on that subgroup.

Problem VII.4.d.iii-3: If the normal subgroup of the space group has index 2 (sec. VII.2.b, p. 340; pb. VIII.2.c-8, p. 400) [Streitwolf (1971), p. 83], there are two cosets, and the order of the orbit is either 1, so the representation of the invariant subgroup (which is what?) is self-conjugate, and the dimension of the space group is equal to that of the invariant subgroup, or the order of the orbit is 2. Then the subgroup has pairs of conjugate irreducible representations, and the matrices of the space-group representations have dimension twice that of the invariant subgroup, with those belonging to this subgroup (chosen) block-diagonal, the others with two off-diagonal blocks. Work this out explicitly. We shall meet this form later in a different context (sec. VIII.7.d, p. 437).

VII.5 WHAT THE PROCEDURE IS, AND WHAT IT MEANS

Having outlined the procedure for inducing representations of a space group, symmorphic or nonsymmorphic, from its little co-groups and little groups, we next review it, and then make it more explicit by considering examples. What have we obtained? What are the components of the representations, what do they mean? Space groups, being inhomogeneous, are quite different from point groups, or the rotation groups. And they are discrete. Thus their representations are different, and the concepts familiar from point and rotation groups may not hold. So we briefly mention the corresponding ones, if they are — or seem to be — different, and why they are the relevant ones.

VII.5.a The procedure in essence

First we summarize the procedure for finding the representations of space group G [Altmann (1977), p. 216; Birman (1974), p. 77; Bradley and Cracknell (1972), p. 151, 181]. Given a crystal, the first step, of course, is to determine its unit cell, Brillouin zone, and symmetry space group G. Finding the representations starts with a choice of vector $\underline{k}$ from a representation domain of the first Brillouin zone (sec. IV.2.c, p. 189), the first being sufficient since a sum, with integral coefficients, of reciprocal lattice vectors can be added bringing $\underline{k}$ into the first zone. The little co-group of this vector, a (not necessarily proper) subgroup of the point (factor) group of the space group, is determined and from it we construct the little group, a subgroup of G. We also find the star of $\underline{k}$, or equivalently, write G in terms of its left cosets with respect to the little group of $\underline{k}$, these being in one-to-one correspondence with the vectors of the star (sec. VII.2.d, p. 342).

Then (from the list of these known representations) an irreducible representation of the little group (a small representation) is picked, but only one whose basis vectors are Bloch functions labeled by wave vector $\underline{k}$ such that the eigenfunctions of $\{E|\underline{t}\}$, of translation group T, are

$exp(-i\underline{k} \bullet \underline{t})$ — an allowable representation. The set of basis vectors of this representation and the vectors obtained by having the left coset representatives act on them (and their sums) gives a vector space irreducible under G so the basis of a representation of it.

Finding these representations is repeated for each small representation of the little group of $\underline{k}$ whose basis elements are the Bloch functions, giving all representations of G determined by the vectors in the star that includes $\underline{k}$. The entire process is repeated for each $\underline{k}$ in the first Brillouin zone belonging to distinct stars (each vector in the representation domain) to get all representations of G.

These representations are labeled by the star of $\underline{k}$ (that is by $\underline{k}$), and by the labels of the small representations of its little group. Each star, and each (allowable) small representation, gives a (different) representation of the space group.

If the point-group operations and the translations are independent, as for a symmorphic group (sec. III.3, p. 137), then a representation of the space group is a semi-direct product of the representations of the point group and of the translations. The representations are then immediate. Since the crystal translations are represented by phases which are roots of unity (sec. IV.2.a, p. 186), the representations of one of these space groups is the product of a point group representation with a phase.

VII.5.b What operators are diagonal?

Representations and states are labeled by the eigenvalues of functions of group operators — the state-labeling procedure [Mirman (1995a), sec. I.6, p. 25]. These operator functions are diagonal which is required by the assumption that they have eigenvalues on states of a representation, and on individual states within a representation. The form of the representation depends on which operators are taken diagonal. For the rotation group SO(3), the diagonal operators are (identified as) the total angular momentum, J^2, and its z component, J_z, labeling the representation and state, respectively. However inhomogeneous groups, like the inhomogeneous rotation group, the Poincaré group, and space groups, are richer and more complicated, thus there are choices of which operators to take diagonal.

For the inhomogeneous rotation group the diagonal operators can again be taken as J^2 and J_z, so the basis states are spherical harmonics. The inhomogeneous operators, the translations, change the origin, the point about which the spherical harmonics are defined, thus acting on one representation of the rotation group produce a sum of states of different representations. If we take the angular momentum about a different point, the total angular momentum is different, as it is for a moving particle. Another choice is to take the diagonal operators to be the momentum operators (as is done in studies of space groups, such as the present one). The basis states are exponentials, which can be expanded in terms of an infinite number of spherical harmonics (the representations of the homogeneous part, the rotation group). These representations and basis states are quite different. These aspects are similar for the Poincaré group, although because the homogeneous part, the Lorentz group, is noncompact (some of the transformations are given not by trigonometric functions, but by hyperbolic ones, so the ranges of some parameters are infinite), there are more types of representations [Mirman (1995c)].

What we are doing here seems to contradict these statements. The operators that are diagonal are the momenta, but rather than each space-group representation containing, as for the rotation group, an infinite number of representations, it contains, and is labeled by, only one of the point group. However the spaces on which these sets of operators act are different. The transformations on positions within the crystal, changing one to another, form a group for which the representations are labeled by the momenta. But each basis vector also contains a Bloch function, and it is these that the point group acts on. Thus the Bloch functions should really contain indices given the point-group basis state, which has been suppressed. A group transformation acts on several spaces simultaneously (as with an electron in free space, for which the rotation group acts on objects from different spaces, the momentum, and also on the spin state, changing both together). The operators diagonal on one space need not be diagonal on others, and here cannot be since momenta do not act on the Bloch functions (or on the basis state giving the spin of an electron). It is important to know what the spaces are, and what operators are diagonal on them.

Problem VII.5.b-1: The momentum operators need not be taken diagonal. Consider space group representations in which the point group, acting on positions of lattice points, labels the representations. For these point-group diagonal (momentum non-diagonal) representations, momentum operators mix different point group representations (of which there are only a few), and the space group representation contains (presumably) all representations of the point group. Are Brillouin zones still relevant? Is reciprocal space? Why? How? Work this out for a few space groups, particularly nonsymmorphic ones. Does it give anything interesting, and relevant? Might these have applications? Do they provide information about the crystal and the objects within it?

VII.5.c The meaning of space group representations

In summary what is the meaning of a representation of a space group representation (sec. XI.2.c.ii, p. 573)? A representation of the inhomogeneous rotation group is a set of functions — of the coordinates since this is an inhomogeneous group — that are intermixed by the homogeneous subgroup, the rotation group. Inhomogeneous operators, the exponentiations of the momenta, take a function at one point (of direct space) to that at another. The representations (considered here) are taken with the momentum operators diagonal, so it is their eigenvalues that label the states. The representations are labeled by the magnitude of the momentum. Each representation then consists of a set of basis states, each an exponential function of position, all with the same momentum magnitude, and these states are mixed by the rotation operators (acting on momentum).

For a space group, the homogeneous subgroup is not the same at all points of reciprocal space. The Bloch functions are point-group basis vectors, and the group of which they are basis states differs at different special positions, and these groups also are different from that at a general position. The space group representation basis vector then is a product of an exponential, containing the momentum, with a Bloch function, a state of the point-group representation, and this group, and its representation, can change with the momentum. But the representations at different momenta are not independent, and this is a fundamental aspect of the properties of space-group representations, and of the physical properties of a crystal, and objects within it.

Physically if the object whose statefunction the basis vector is, has its momentum changed slightly, there will be very little change in its properties. The dimension, say, of the point group representation cannot change with this very small variation of momentum. Space group representations are subject to continuity conditions, so are required to change smoothly with momentum, even though the point group changes abruptly. These representations can then be (actually must be) combined into sets, each member labeled by a different momentum value, and whose basis states are continuous functions of momentum. There are different sets of representations, these labeled by the point group representations at some specified point of reciprocal space, such as the origin. Our notation for the representation matrices indicates this (sec. VII.3.a.iii, p. 350).

Thus we can regard a space group representation as being given by a set of functions that at each momentum value forms a representation of the point group at that momentum. However the sets are different at different values, so a function belonging to a representation of the point group at one momentum belongs to a representation of a different point group at another — with the other functions in the representations different at different points; the members of the sets that it belongs to are different. The sets of degeneracies differ at different points, as do the functions that are degenerate.

The groups of the special points at the ends of a line of symmetry are different from that of the line and from each other (sec. XI.2.c, p. 571), and that of the line is a subgroup of those at the endpoints. Consider a small displacement from a point with the larger symmetry group, so going to one with less symmetry. All symmetry operations here must also have been so at the point. Thus the smaller symmetry group is a subgroup of the larger, the degeneracy at the point is split by the displacement, the irreducible representation of larger-symmetry group becomes reducible when restricted to the subgroup, and the irreducible representation of the smaller group must be in its decomposition. This is possible only for some sets of representations, giving the name, compatibility relations (sec. XI.2.c.i, p. 572), of the lists of representations of the point groups at each pair of special positions lying on the same special line or plane, having states that (with the proper change of momentum value) can belong to both.

A function then must be a basis function of both larger groups at the endpoints and of the representation of the subgroup of the line to which these reduce. However there are only certain cases in which representations of these larger groups reduce to the same ones of the smaller group that is a subgroup of both. The point group representations for the various special positions are only those for which this is possible — the entire set must be consistent, giving a condition on which sets of space group representations are possible.

Thus if we start, say, with the point of highest symmetry, the origin in reciprocal space, $\underline{k} = 0$, and consider all its representations, then for each there are representations at the other special positions giving a consistent set, and each set is part of a representation of the space group of the crystal. The point group representations are thus related.

The simple cubic structure (sec. IV.6.a.i, p. 240; sec. XI.2.h.i, p. 579), for example, has at point Γ, the center ($\underline{k} = 0$), the full cubic symmetry group, O_h. On the k_x axis the symmetry group is C_{4v}. This has four one-dimensional representations, and one two-dimensional one (tbl. E.2-1, p. 657). The three-fold degenerate T_{1u} representation of O_h thus splits into one A_1 and one E representation. So the representations A_1 and E of C_{4v} with $\underline{k}$ on the k_x axis are compatible of the symmetry T_{1u} of O_h for $\underline{k} = 0$.

These compatibility relations are needed for the endpoints of all special lines (and likewise bounding lines for special planes) of a Brillouin zone. They are straightforward to find, and there are many tables of various ones [Heine (1993), p. 278; Joshi (1982), p. 288; Tinkham (1964), p. 287; Wherrett (1986), p. 277].

Problem VII.5.c-1: Is it possible to have space group representations for quasicrystals [Jaric (1989); Lifshitz (1997); Senechal (1995)]? What would they be like? One possible way of studying this is to consider space group representations for higher-dimensional spaces (an interesting possibility in itself) and project [Coxeter (1973), p. 236; Senechal (1995), p. 54] into three-dimensional space (pb. II.4.h-6, p. 97).

VII.5.d What is the dimension of the space group representation?

A fundamental propery of a representation is its dimension. Yet for the space group the dimension — the number of basis vectors mixed by the point group — depends on the particular value of $\underline{k}$, the particular point of the Brillouin zone, since the point group does. What is its significance? How are representations at different points (so of different point groups) related [Birman (1974), p. 85]? Within a Brillouin zone there is a set of functions (of $\underline{k}$), each of which retains its identity as $\underline{k}$ is varied, and which belongs to a set forming a representation of the point group for each $\underline{k}$. And the dimension of this point-group representation varies over the zone (as does the relevant point group (sec. XI.1.d, p. 567) — the site-symmetry group).

Thus we can regard the dimension the space group as the number of functions (of $\underline{k}$) that are mixed by the point group — there are a set of basis functions, mixed by the point group operators, and this is the number of them. What happens at special points, where the dimension is lower?

Let us put the crystal in an external field, breaking the symmetry. The basis functions then have different energies, so are no longer degenerate, and may not be time-independent — the system can go from a state to others (of different energy). However they are still basis functions of a single representation [Mirman (1995b), sec. 5.3.1, p. 101; (1995c), sec. 6.2, p. 98]. Consider, at any time, two observers, related by a space group transformation, for example two relatively rotated objects [Mirman (1995b), sec. 5.1.1, p. 86; (1995c), sec. 6.3.7, p. 109]. The basis functions that they see are different of course, but they are related by the operators of the transformation. Thus for the rotation group the basis states seen by the observers are related by the angles of rotation [Mirman (1995a), sec. X.4, p. 277]. A basis state of one observer is a linear combination of the basis states of the other, but the terms in this sum are all of the same representation, and all observers see states of the same representation. The rotation-group transformed states of total angular momentum J are functions of the untransformed states — of the same J. The states of each J value, for one observer, depend on, and only on, the states of the same J of the other observer. So it is with all geometrical transformation groups (which need not be symmetry groups). That is why, even though the states have different energies, say, they still belong to a group representation, and the same one as that for all energy differences zero. This is one way of giving the meaning of a group representation.

Now with the crystal in an external field, the states can have different energies — the degeneracy is broken. But at the special points, some degeneracy remains.

At a general point the number of states is the dimension of the irreducible point group representation, at that point. The group operators mix all these states, so the representation cannot be reduced. At special points the point group becomes smaller, as do representations of it (obviously not all, for example one-dimensional representations). But the number of states remains the same — and we regard the dimension of the space group as this number of states, even though they belong to a reducible point-group representation. Thus the point group representation becomes reducible, but the states must be such as to smoothly go into the states of the special points as the $\underline{k}$ value is moved to these.

VII.5.e Representations can contain more than one momentum magnitude value

For the Poincaré group the magnitude of the momentum,

$$p^2 = p_o^2 - \sum p_i^2 = m^2, \qquad \text{(VII.5.e-1)}$$

is an invariant, the mass of the object, as is the distance between two points

$$d^2 = x_o^2 - \sum x_i^2. \qquad \text{(VII.5.e-2)}$$

Thus each representation of the Poincaré group contains only a single magnitude of momentum, which is one of its representation labels. The translation and momentum operators of course leave the momentum invariant, and the square of the momentum is invariant under the semisimple subgroup (for the Poincaré group, the Lorentz group). For nonsymmorphic space groups, there is a difference: a representation can contain more than one momentum magnitude. It is this which makes these representations special, and requires that additional procedures be developed to find them.

The reason is that for some points — special points of the boundary of the Brillouin zone — addition of a reciprocal lattice vector gives an equivalent point. Basis states are invariant under this transformation — but as it takes one vector and adds to it another, it changes the magnitude of the vector. That there are such equivalent points is in part a consequence of what the boundary of the Brillouin zone is (sec. IV.2.c.iii, p. 191). This is fundamental in determining the properties of representations of nonsymmorphic groups, and how they differ from those of symmorphic ones, and of the objects these describe (sec. XI.2.c.ii, p. 573).

For a symmorphic space group, all translation vectors attached to group elements can be chosen zero, so translation basis states can be factored out of the space-group basis states. For nonsymmorphic groups however, not all translations (in the space of the crystal, real space) can be transformed away, so a translation by a reciprocal lattice vector leaves the lattice invariant, sending the basis state into another (of the same representation, of course) but for one transformation, at least, the operator includes a translation. At least one translation vector of the space group cannot be set to zero; not all basis states can be so factored. It is a property of glides and screws making up a space group that the reciprocal-space vector going with the associated translation connects two special points on the Brillouin zone boundary. Taking the Fourier transform, to go to reciprocal space, we get from this a momentum vector that cannot be set to zero — this is the vector of the representation whose magnitude is different from that of at least one other of its vectors. There are nonprimitive transformations that take a basis vector into another of the same representation, but are of the form of a semisimple (point-group) operator times the addition of a lattice vector, so such point-group operators take a special point on the boundary of the Brillouin zone to an equivalent one, but with a different momentum value. Acting on the Fourier transform of the basis vector they take it into another basis vector times an exponential in $\underline{K}$, which gives the Fourier transform of the term added by the translation. As can be seen from eq. VII.2.h–4, p. 345, the space group operation results in a change of momentum, and an additional phase. So some basis states of single representations of nonsymmorphic groups have different values of the magnitude of $\underline{k}$.

From eq. VII.4.d.iii–8, p. 362, we see that the point group transformation takes a state with momentum $\underline{k}$ to one with momentum $\underline{k} + \underline{K}$, which has a different magnitude (even though the exponentials they give have the same values). For this, there must be at least one τ that is nonzero.

VII.6 THE SQUARE AS AN EXAMPLE

To illustrate stars, general and special points, and point-group transformations that add a reciprocal lattice vector connecting points on the boundary of a Brillouin zone, we return to the plane square lattice (sec. III.3.a, p. 138; sec. IV.2.c.iv, p. 192) of side a, whose Brillouin zone is also a plane square, with side $2\pi/a$ [Falicov (1966), p. 158; Heine (1993), p. 273; Jansen and Boon (1967), p. 259; Joshi (1982), p. 277]. Its space group is symmorphic, and after studying it we consider two-dimensional nonsymmorphic groups [Heine (1993), p. 287], specifically this with the addition of glides to make the group nonsymmorphic.

VII.6.a The reciprocal lattice vectors of the square

Since the reciprocal lattice of a square is a square, with lattice points at the corners, the reciprocal lattice vectors are those connecting corners, so are the sides of the reciprocal square, plus all integral sums of them, including the square's diagonals.

Notice that a reciprocal lattice vector from an arbitrary point in the square, or almost all on its boundary, goes to a point distinct from the first — the points are placed differently with respect to the square. However there are some points on the boundary for which a reciprocal lattice vector gives equivalent points, ones that bear the same relationship to the square.

VII.6.b General vectors and vectors giving symmetry

Vector $\underline{k}$ is a general vector if every element of the point group acting on it gives a distinct vector (sec. VII.2.e, p. 342). The number of vectors in its star equals the number of point group elements. If the number is less than the number of point group elements, then point $\underline{k}$ (that point to which vector $\underline{k}$ goes) is in a position of symmetry — it is a special point of the Brillouin zone.

What are these different vectors in the representation domain [Falicov (1966), p. 160; Joshi (1982), p. 277]? With

$$0 \leq k_x, k_y \leq \frac{\pi}{a}, \qquad \text{(VII.6.b-1)}$$

the three points, the origin

$$\Gamma : \underline{k} = (0,0), \quad X : \underline{k} = (\frac{\pi}{a}, 0) \text{ and } M : \underline{k} = (\frac{\pi}{a}, \frac{\pi}{a}) \qquad \text{(VII.6.b-2)}$$

are special. The special points on the three lines of symmetry are

$$\Delta : \underline{k} = (k_x, 0), \; Z : \underline{k} = (\frac{\pi}{a}, k_y), \; \Sigma : \underline{k} = (k_x, k_x), \qquad \text{(VII.6.b-3)}$$

as we see from

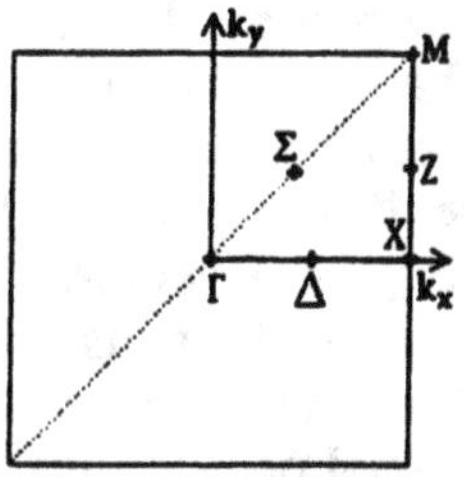

Figure VII.6.b-1: SPECIAL AND GENERAL VECTORS OF THE SQUARE.

Problem VII.6.b-1: Check that, with rotations about the z axis perpendicular to the square through a lattice point, the space group is that with operators (times translations),

$$\{E|0\}, \{C_4|0\}, \{C_4{}^{-1}|0\} = \{C_4{}^3|0\}, \{C_4{}^2|0\} = \{C_2|0\},$$

$$\{\sigma_x|0\}, \{\sigma_y|0\}, \{\sigma_d|0\}, \{\sigma_d{}'|0\}, \qquad \text{(VII.6.b-4)}$$

which defines the symbols, and explains their meaning. Why are there two diagonal mirrors? What is the point (factor?, sub?) group?

Problem VII.6.b-2: Why are these three points, and these three lines, special, and only these? Check the bounds on the components of the $\underline{k}$'s. Apply all point group operators (which are what?) to these points and lines, and to a general point, and find the points and lines to which they are taken. Check the number of members for each set obtained.

VII.6.c The stars of the points of the square

What are the stars of the vectors for the square,

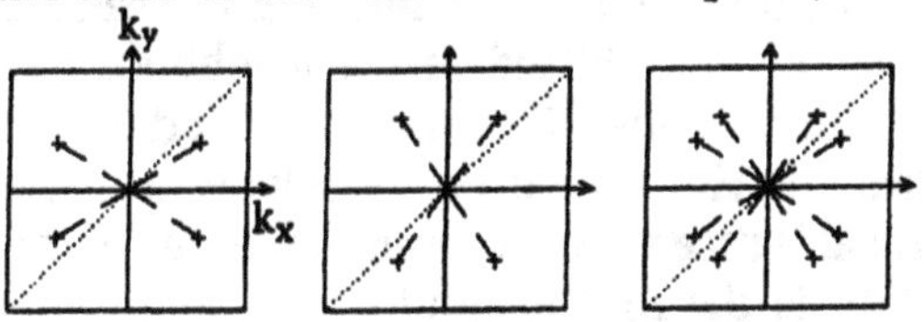

Figure VII.6.c-1: STARS OF GENERAL VECTORS OF THE SQUARE?

A general k has eight vectors in its star produced by the eight symmetry transformations of the square: rotations about the center of $0, \pi/2, \pi, 3\pi/2$; reflection in the two lines bisecting the sides; reflection in the two diagonal lines; inversion through the center — not all distinct. On the left is the square with four general vectors (drawn dashed) produced from any one by the reflections in bisecting lines. The point (given by k) of the Brillouin zone is at the end of the vector, so points, and vectors to them from the origin, are equivalent. Next the vectors are reflected in the diagonals (one of which is drawn, dotted); these are superposed, giving eight vectors, all distinct. Here vectors end inside the Brillouin zone so there is no reciprocal lattice vector that added to any of the star gives another of the star.

As an example of the special position for the square, we have

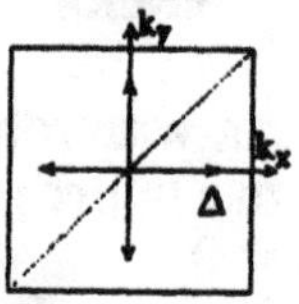

Figure VII.6.c-2: STAR FOR A SPECIAL POSITION,

with a vector to arbitrary point Δ (and similarly Σ) on the axis, are the vectors obtained from the first, four altogether, by the symmetry rotations (and also reflections in the diagonals). There are only four vectors in its star — since k lies on a symmetry line, reflections in that (and similarly those obtained from it by symmetry rotations) leave k unchanged. It is clear that no reflections through orthogonal lines bisecting the edges of the square, nor through a diagonal, one shown dotted, give other vectors.

Next is the vector to point X, the midpoint of the boundary of the Brillouin zone, and that obtained from it by a $\pi/2$ rotation — there are two vectors in the star. Reflection in a diagonal gives the same vector. The vector lies on a symmetry line (plane in three dimensions), reducing to four the number of vectors obtained with point group transformations. These form two pairs, the members of each differing by a reciprocal lattice vector, but shifted to the neighboring quadrant and back to the first Brillouin zone. Rotations of π and $3\pi/2$ give further vectors, drawn dashed. However they are related to the other two by addition of reciprocal lattice vectors, from the center to the sides of the square. On the right is the first Brillouin zone, and under it another. The point at the end of the dashed vector is the same as that reached by one of the vectors of the star, but starting at the center of the second cell (drawn heavily), which is displaced by a reciprocal lattice vector,

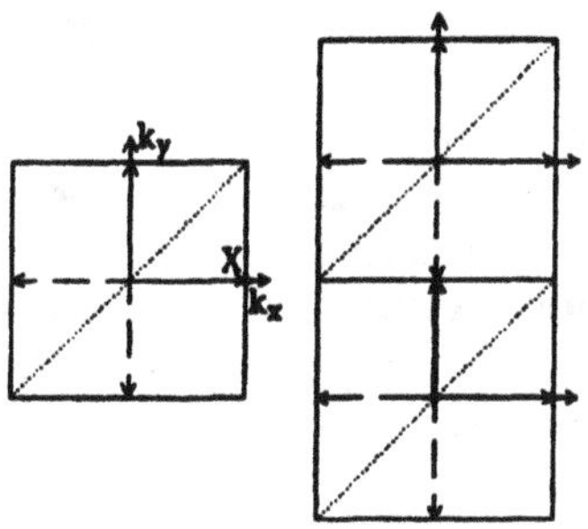

Figure VII.6.c-3: VECTOR TO THE MIDPOINT OF BOUNDARY.

Therefore these vectors are equivalent; the dashed one is equivalent to that (heavy one) from the center of the first cell to its top — it is not a distinct vector, and similarly for those along the other axis.

Only for a few special points does shifting back to the first Brillouin zone gives equivalent points.

Problem VII.6.c-1: Show that point Z likewise has only four vectors in its star, because four of the eight vectors generated by the point group transformations are equivalent to the other four. Point M has only one vector in its star, the group generates four vectors, to the four corners of the square, but these all differ from M by a reciprocal lattice vector — the side of the square. Here "reciprocal lattice vector" is singular. Why? Point Γ also has only one vector, but here since its length is zero, all symmetry operations leave it invariant.

Problem VII.6.c-2: It is important that for some points, the others in the star are reached both by point group transformations (giving the definition of the star) and also by reciprocal lattice vectors. Find these for the square. We see the role this aspect of space groups plays in representations and applications.

Problem VII.6.c-3: For the square, for each special vector, draw the eight vectors in its star and the reciprocal lattice vectors that make vectors in subsets equivalent to each other. Verify that these points actually are equivalent, and that it is impossible to make general points equivalent — the points obtained by shifting back to the first Brillouin zone are different.

Problem VII.6.c-4: Write the little groups for these vectors (the smallest point groups leaving the figure invariant), using the nomenclature for point-group transformations (tbl. B-2, p. 634). Name these little groups [Joshi (1982), p. 279], remembering that in an even-dimensional space, like that of 2 or 3+1 dimensions, an inversion is equivalent to a π rotation [Mirman (1995c), sec. 4.2.6, p. 60]. Give the orders of these groups and notice that *all* orders times the number of vectors in the corresponding star are equal (to what?). Why?

Problem VII.6.c–5: Consider for example point X. Write the translation eigenfunctions, taking the Bloch function arbitrary, and show that these are invariant under the point group operations, and also under the addition of the reciprocal lattice vector going with this point, that is vectors shown as equivalent do actually give equivalent representations. Take other points not related by reciprocal lattice vectors, and show that such addition does not give equivalent representations.

Problem VII.6.c–6: Repeat this analysis for the nonsymmorphic space group (sec. III.4.f, p. 156).

VII.6.d The square and the rectangle

The meaning of special positions is emphasized by comparing the rectangle and the square. The rectangle, with general and special points, is then

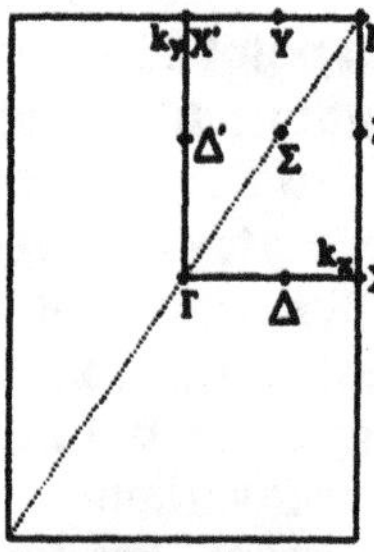

Figure VII.6.d–1: RECTANGLE WITH GENERAL AND SPECIAL POINTS.

This is like fig. VII.6.b–1, p. 373, with the object stretched along y, except there is now three more special points labeled Y, X', Δ'. For the square there is no difference between Z and Y — as we can see they are in the same star, so only one is considered. Here they are in different stars. The reason that there are more points is that there is less symmetry. The special points — considered — are those of the representation domain, the smallest part of the crystal from which the rest can be found using the point symmetry operations. Special positions related to those in this domain by a point group operation are not used — that would give redundancy. For the square the representation domain is an eighth of the square. But for the rectangle it is necessary to use a quarter. Hence three special points must be added to the Brillouin zone — they are no longer related to other points.

VII.6.e Little groups of general, and of special, vectors

The little group of a general vector (which factorizes into the little co-group times the translations) consists only of the identity times the translations — since all point group elements change the vector, the only group that leaves it invariant is the identity. If the vector goes to a special point, or to a point on a line or plane of symmetry (to a point in a special position), the little co-group, and little group, are larger.

Problem VII.6.e-1: Explain why the order of the little group times the number of vectors in the star equals the order of the point group. Check that this is true for the square.

Problem VII.6.e-2: The groups of k, the subgroups of the symmetry point group of the square, that leave points invariant [Joshi (1982), p. 279], are

Point	q	group	$o(\underline{k})$
general	8	C_1	1
Σ	4	C_{1h}	2
Z	4	C_{1h}	2
Δ	4	C_{1h}	2
X	2	C_{2v}	4
M	1	C_{4v}	8
Γ	1	C_{4v}	8

Table VII.6.e-1: GROUPS OF k,

where q is the number of points in the star, and $o(\underline{k})$ is the order of the group of k. Verify the correctness of the table. For each of these groups find the characters (tbl. E.2-1, p. 657) of all representations [Falicov (1966), p. 163]. This requires stating all symmetry transformations, and their classes.

VII.6.f Representations of the space group of the square

Having the point groups, for each vector of the Brillouin zone, their representations (sec. V.2, p. 247), and those of the translation group, we are now ready to put them together and induce the representations of the space group from the translation subgroup [Jansen and Boon (1967), p. 259]. Since the group is symmorphic this step is straightforward.

VII.6.f.i *The little groups and the representations*

For general point k, the little group is just the translation group, and the allowable representation of the little group is the translation group representation from which we are inducing.

Problem VII.6.f.i-1: Explain why. Since the point group has eight elements, the induced representation has dimension 8. The matrices follow from the definition of induced representation (sec. VI.3.b, p. 291), and are

$$D_{pq}(\{R_m|\underline{t}_n\}) = exp(iR_p^{-1}\underline{k} \bullet \underline{t}_n), \ \ R_q = R_pR_m,$$
$$= 0, \ \text{otherwise}. \qquad \text{(VII.6.f.i-1)}$$

So to find the matrices we write those for the regular representation [Mirman (1995a), chap. VI, p. 170] of the point group, replacing the single nonzero entry of each row and column with $exp(iR_p^{-1}\underline{k} \bullet \underline{t}_n)$. Justify these statements.

Problem VII.6.f.i-2: More interesting are the special points. For Δ (fig. VII.6.b-1, p. 373) explain why the little group is {T,σ_xT}, where T is the translation group, and σ_x the reflection in the x axis. It is thus larger than T (twice as large). Draw the square and show that the star of this $\underline{k}$ consists of the four vectors (as given above) $\underline{k}$, $C_4\underline{k}$, $C_2\underline{k}$, $C_4{}^3\underline{k}$. This — the set of four vectors — is the orbit of $\underline{k}$, and the four representations given by these are equivalent. Explain why the little group of these representations can be written

$$L(\underline{k}) = \{E, \sigma_x\} \wedge T, \qquad \text{(VII.6.f.i-2)}$$

where ∧ denotes the semi-direct product. Might it be a direct product? Why? Group $\{E, \sigma_x\}$ has two one-dimensional representations, for one the representative of σ_x is 1, for the other -1. There are therefore two representations of $L(\underline{k})$, both one-dimensional. Why are both allowable? The matrix elements of the members of T, and of the coset σ_xT, are $exp(i\underline{k} \bullet \underline{t}_n)$ for the first representation, and $exp(i\underline{k} \bullet \underline{t}_n)$ and $-exp(i\underline{k} \bullet \underline{t}_n)$, respectively, for the second. Verify that this agrees with the formula defining the induced matrix elements (sec. VI.3.b, p. 291). Thus the basis vectors of the space group labeled by $\underline{k}$ for the first representation are both $exp(i\underline{k} \bullet \underline{t}_n)u(x,y)$, with u a Bloch function, and an even function (of what?). Write the basis vectors for the second representation, and explain why u is an odd function for this. The other basis vectors — labeled by the members of the star of $\underline{k}$ — are found by having the elements of the group of $\underline{k}$ act on these. This gives two four-dimensional representations, whose matrices should be immediate. Verify that they are irreducible. Notice that these basis vectors are more restricted — there are more relations — than those for a general point.

Problem VII.6.f.i-3: For point Γ ($\underline{k}$ = 0), the center, the little group is the space group of the square. Why? The point group (which is what?), thus the space group (why?), has five irreducible representations of dimensions 1,1,1,1 and 2. Since the star consists only of $\underline{k}$ = 0, the basis functions of the space group representations are any Bloch functions (having the periodicity of the lattice).

Problem VII.6.f.i-4: The other special points are similar to these two (except that those on the boundary can be equivalent to others under reciprocal lattice translations). Repeat the analysis for them.

VII.6.f.ii *What determines the little groups?*

The little group is that set of operations under which conjugate representations of the subgroup, here the translation subgroup, are equivalent. Transformations by elements of the point-group leave a general point invariant — there are no points equivalent to it. Thus a translation group representation, labeled by this point, is invariant under these operations, and therefore equivalent to all representations to which it is conjugate — itself. So the little group is the translation group.

A point on Δ is taken into an equivalent point (itself) by reflection in the mirror on which it lies. The little group is the translation group, and it multiplied by the reflection; representations labeled by these two are equivalent under them, and under no other elements of the space group. The representations of the little group are all allowable (sec. VI.7.c, p. 324) since there are no equivalent representations to give redundancy — all translation representations subduced from a little group representation are equivalent, as there is only one.

VII.6.g The square with nonsymmorphic glides

More intricate are nonsymmorphic groups. If the square crystal has glides [Falicov (1966), p. 168], its group can be nonsymmorphic. An example is shown in fig. III.4.f-2, p. 156. There is also a corresponding rectangle (fig. III.4.f-1, p. 156). Since the Brillouin zone is determined by the holohedry group of the lattice (but not the representations), that for this crystal is the same as the square with symmorphic glides (fig. VII.6.b-1, p. 373). The special positions are the center of the Brillouin zone, always, and clearly the center of a side, plus the special lines — the diagonal and the side, this being equivalent to the opposite side by a reciprocal lattice vector.

Problem VII.6.g-1: Are there other glides not drawn? Are there glides that appear here, but not for the rectangle (fig. III.4.f-1, p. 156)?

Problem VII.6.g-2: Check that the space group of this square has generators (times translations)

$$\{E|0\}, \{C_4|0\}, \{C_4{}^3|0\}, \{C_2|0\}, \{\sigma_x|\frac{1}{2}a\}, \{\sigma_y|\frac{1}{2}a\}, \qquad \text{(VII.6.g-1)}$$

where a is the side of the square. Are there others? Compare this to the symmorphic group. Explain the differences. What is the point (factor?, sub?) group?

Problem VII.6.g-3: Find the Bloch functions and space group representations for the special positions.

VII.6.h A nonsymmorphic group of the rectangle

Another example [Falicov (1966), p. 168] is a rectangle with glides that make its group nonsymmorphic shown in fig. III.4.f-1, p. 156. The glide is labeled g, and there is a mirror, drawn dotted. The origin, taken at O, displaced a/4 from the mirror, is at a C_2 (also written C_{2z}) axis. The magnitudes of the primitive reciprocal lattice vectors are thus

$$k_1 = \frac{2\pi}{a}, \quad k_2 = \frac{2\pi}{b}, \qquad \text{(VII.6.h-1)}$$

along x and y. Its unit cell (containing two points) is

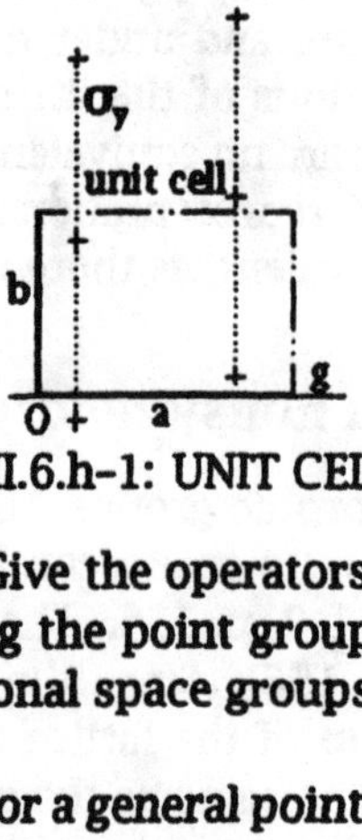

Figure VII.6.h-1: UNIT CELL WITH GLIDES.

Problem VII.6.h-1: Give the operators of the symmetry space group of this object, including the point groups at the various points. From the list of two-dimensional space groups (appendix D, p. 650), find the name of the group.

Problem VII.6.h-2: For a general point there are four basis functions,

$$\psi_1(\underline{k}) = exp(i\underline{k} \bullet \underline{r})u(\underline{r}), \qquad \text{(VII.6.h-2)}$$

$$\psi_2(\underline{k}) = exp(iC_2\underline{k} \bullet \underline{r})u(C_2\underline{r}), \qquad \text{(VII.6.h-3)}$$

$$\psi_3(\underline{k}) = exp(i\sigma_y\underline{k} \bullet \underline{r} + \frac{1}{2}i\underline{k} \bullet \underline{a})u(\sigma_y\underline{r} + \frac{1}{2}\text{a}), \qquad \text{(VII.6.h-4)}$$

$$\psi_4(\underline{k}) = exp(i\sigma_x\underline{k} \bullet \underline{r} + \frac{1}{2}i\underline{k} \bullet \underline{a})u(\sigma_x\underline{r} + \frac{1}{2}\text{a}), \qquad \text{(VII.6.h-5)}$$

showing the point group operators, and half-lattice vectors; $\underline{r}$ is the vector whose components are x and y. Explain this, including the inhomogeneous terms ($\frac{1}{2}$a), especially that $\frac{1}{2}$a (specifically the $\frac{1}{2}$) appears because there are glides — the group is nonsymmorphic — and glides require $\frac{1}{2}$. Check that the glides shown do lead to these basis states. Why does b not enter? The space of these four functions is invariant under the space group (why?) and irreducible. To reduce it would require sums, but these have different $\underline{k}$'s, so are not irreducible representation basis functions. Why? This gives a four-dimensional representation of the space group, and is the same as for a symmorphic group, as expected. Why?

Problem VII.6.h-3: Verify that the Bloch functions for a general point are the same as for the square without glides. For these points the glide does not affect the representations.

Problem VII.6.h-4: The interesting cases are for $\underline{k}$ at special points. Then the star of $\underline{k}$ has less than four members, so the four terms given by the point group operations are not all distinct. The four-dimensional representation can be reducible. Give the elements of the space group, G, the little group of $\underline{k}$, $G(\underline{k})$, the translation group T, and the factor groups G/T and $G(\underline{k})/T$. Do this for each special point (although several might give the same results), as well as a general point. Find the cases in which G/T is not a subgroup, and similarly $G(\underline{k})/T$. There are cases in which the space group operator acting on a Bloch function, $\{R|\underline{\tau}\}u(x,y)$, gives a phase of the form $exp(i\underline{k}\bullet\underline{\tau})u(x,y)$. Find these, and show that it is impossible to remove all by a shift of origin, or other redefinition. The representation of the space group is a ray representation.

Problem VII.6.h-5: One way of analyzing this representation [Falicov (1966), p. 172] is to use a smaller translation group $T_{\underline{k}}$, that translation subgroup whose members $\{E|\underline{t}_{\underline{k}}\}$ include only those translations, defined by this equation, for which

$$\underline{k}\bullet\underline{t}_{\underline{k}} = 2\pi m, \quad m \text{ is an integer, so } exp(i\underline{k}\bullet\underline{t}_{\underline{k}}) = 1. \qquad \text{(VII.6.h-6)}$$

Since $T_{\underline{k}}$ is smaller, the factor group $G(\underline{k})/T_{\underline{k}}$ is larger. It is an extension of $G(\underline{k})/T$. Show that these translation groups for the points given by the subscripts are, for the set of all integer n's,

$$T_{\Gamma} = T = \{E|n_1\underline{a} + n_2\underline{b}\}, \quad Tx = \{E|2n_1\underline{a} + n_2\underline{b}\},$$

$$T_M = \{E|2n_1\underline{a} + 2n_2\underline{b}\}. \qquad \text{(VII.6.h-7)}$$

Give these groups for the other special points.

Problem VII.6.h-6: Show that every element of a coset of $G(\underline{k})/T_{\underline{k}}$ acting on the Bloch function $u_{\underline{k}}(x,y)$ going with the translation basis state

$$\psi_{\underline{k}}(r) = u_{\underline{k}}(x,y)exp(ik_x x + ik_y y), \qquad \text{(VII.6.h-8)}$$

gives the same state. For $\{R|\underline{\tau}\}$ in $G(\underline{k})$, explain why a coset of $G(\underline{k})/T_{\underline{k}}$ is determined by

$$\{R|\underline{\tau}\}\underline{t}_{\underline{k}} = \underline{t}_{\underline{k}}\{R|\underline{\tau}\}. \qquad \text{(VII.6.h-9)}$$

Why does the equality hold? With the element $\{E|\underline{t}_{\underline{k}}\}\{R|\underline{\tau}\}$ of this coset, and defining $\psi_{\underline{k}}'(r)$, check that

$$\{E|\underline{t}_{\underline{k}}\}\{R|\underline{\tau}\}\psi_{\underline{k}}(r) = \{E|\underline{t}_{\underline{k}}\}\psi_{\underline{k}}'(r) = exp(i\underline{k}\bullet\underline{t})\psi_{\underline{k}}'(r) = \psi_{\underline{k}}'(r). \qquad \text{(VII.6.h-10)}$$

So the entire coset of operators is found by using all allowed $\underline{t}_{\underline{k}}$. Prove that it is independent of $\underline{t}_{\underline{k}}$. This action of the basis vector is thus the same for all elements of the coset.

Problem VII.6.h-7: The representations are induced from the allowable representations of the little group. So for example at point X

$$\underline{k} = \frac{1}{2}\underline{K}_1, \qquad \text{(VII.6.h-11)}$$

where $\underline{K}_1$ is a primitive lattice vector. Explain why. Now T_X is the translation group not only for $\frac{1}{2}\underline{K}_1$, but also for $\underline{K}_1$, $\frac{3}{2}\underline{K}_1$, $2\underline{K}_1\ldots$; why? However check that $\frac{1}{2}\underline{K}_1$, $\frac{3}{2}\underline{K}_1,\ldots$, all label the same representation of T, so are equivalent, and similarly for $\underline{K}_1$, $2\underline{K}_1$, and so on, which are all equivalent to $\underline{k}=0$ (point Γ). Why does this mean that $G(\underline{k})/T_X$ has two types of representations? Lattice translations can be represented by $exp(i\underline{k}\bullet\underline{t})$, and also $exp(in\underline{k}\bullet\underline{t})$ for all integer n. But these are equivalent, so we allow only $exp(i\underline{k}\bullet\underline{t})$ times the unit matrix for all these translations, thus obtaining the allowable representations, in agreement with the definition of allowable (sec. VII.4.a, p. 353; sec. VII.4.d, p. 359). Also the lattice translations not giving allowable representations have the form $\{E|(2n_1+1)\underline{a}+n_2\underline{b}\}$, and for these

$$exp(i\underline{k}\bullet\underline{t}) = exp[i\frac{1}{2}\underline{K}((2n_1+1)\underline{a}+n_2\underline{b})]$$

$$= exp[in_1\underline{K}_1\bullet\underline{a}+\frac{1}{2}i\underline{K}_1\bullet\underline{a}] = -1. \qquad \text{(VII.6.h-12)}$$

Explain why this means that the translations of the representations not in T_X are represented by the negatives of the ones that are.

Problem VII.6.h-8: For point X, show that $G(\underline{k})/T_X$ has the eight cosets (the elements of this factor group) [Falicov (1966), p. 175; Heine (1993), p. 289]

$$\{E|2n_1\underline{a}+n_2\underline{b}\},\ \ \{E|(2n_1+1)\underline{a}+n_2\underline{b}\},$$

$$\{C_{2z}|2n_1\underline{a}+n_2\underline{b}\},\ \ \{C_{2z}|(2n_1+1)\underline{a}+n_2\underline{b}\},$$

$$\{\sigma_{\mathbf{y}}|(2n_1 + \frac{1}{2})\underline{a} + n_2\underline{b}\}, \ \{\sigma_{\mathbf{y}}|(2n_1 + \frac{3}{2})\underline{a} + n_2\underline{b}\}$$

$$\{\sigma_{\mathbf{x}}|(2n_1 + \frac{1}{2})\underline{a} + n_2\underline{b}\}, \ \{\sigma_{\mathbf{x}}|(2n_1 + \frac{3}{2})\underline{a} + n_2\underline{b}\}. \qquad \text{(VII.6.h-13)}$$

Find the group table for these elements and check that this factor group is isomorphic to C_{4v}(4mm), but is not the same as it. Why? It has five representations (tbl. E.2-1, p. 657), four one-dimensional and the last two-dimensional. From these it follows that the representation matrix of $\{E|(2n_1 + 1)\underline{a} + n_2\underline{b}\}$ is the negative of that for $\{E|2n_1\underline{a}+n_2\underline{b}\}$. Why? Show that this is true, but only for the two-dimensional representation, so it is the only one giving a representation of the space group. Why? This gives a two-fold degeneracy and is called sticking together of bands (sec. XI.2.i.i, p. 586).

VII.6.i What determines the space group and its representations?

To analyze the space group, and how the glide affects its representations we first note its generators, which are

$$\{E|0\}, \{C_2|0\}, \{\sigma_{\mathbf{y}}|\frac{1}{2}\underline{a}\}, \{\sigma_{\mathbf{x}}|\frac{1}{2}\underline{a}\}; \qquad \text{(VII.6.i-1)}$$

the rotation is determined by the choice of the origin, and the other two operators are the mirror and glide.

Problem VII.6.i-1: Show that the form of the transformation involving the mirror is obtained from that at the origin [Falicov (1966), p. 169] by the transformation $t\{\sigma_{\mathbf{y}}|0\}t^{-1}$, where t is a translation (what translation)?

Problem VII.6.i-2: That the expression for the glide is correct can be checked by noting its action on a point in the diagram. The expressions for the mirror and glide have the same form. Why then is one a mirror, the other a glide?

Problem VII.6.i-3: The operations listed and drawn are not all, but if the generators are correct, the rest can be expressed as products of them. Show that none of the generators can be written as a product of others (if any can, remove it from the list of generators), and all other symmetry transformations can be written as a product of these (or add any necessary generators).

VII.6.i.i *The representation basis functions*

To find the space-group representation basis functions of a general point (pb. VII.6.h-2, p. 380), we start with the first, the general form of the statefunction, and operate on it with the generators. These act on the coordinates of the direct lattice, but since all are their own inverses (more generally represented by orthogonal matrices), they can be taken to act on the coordinates of the reciprocal lattice.

Problem VII.6.i.i-1: Work this out explicitly to find all the basis vectors.

VII.6.i.ii *How the glide affects basis functions*

The glides, as we can see, are obtained from mirrors by displacing points. How does this affect the statefunctions at special points? Consider point X, which is on the k_x axis. For a point on the x axis $\psi(x,y) = \psi(x,0)$, so a reflection in the x axis leaves it unchanged, thus

$$\{\sigma_{\mathbf{x}}|\frac{1}{2}\underline{a}\}\psi(X,y) = \psi(X+\frac{1}{2}\underline{a},y). \quad \text{(VII.6.i.ii-1)}$$

Also

$$\psi(X,0) = exp(ik_xX)u(X,0). \quad \text{(VII.6.i.ii-2)}$$

So the addition of displacement $\underline{a}$ gives a phase factor,

$$\{\sigma_{\mathbf{x}}|\frac{1}{2}\underline{a}\}\psi(X,0) = exp(ik_xX + ik_x\frac{1}{2}\underline{a})u(X,0)$$

$$= exp(ik_x\frac{1}{2}\underline{a})\psi(X,0). \quad \text{(VII.6.i.ii-3)}$$

A transformed basis vector belongs to a different representation, for a rotation takes it to a nonequivalent point, unless the point is on the boundary of the Brillouin zone. Then the addition of a reciprocal lattice vector brings it back, but still leaves the basis vector with a phase, resulting in two basis states at the point, identical except for phase.

Problem VII.6.i.ii-1: The remaining step is to find the representation matrices, which is done by induction. To work this out we use the method outlined above (sec. VII.4.d.iii, p. 361).

Problem VII.6.i.ii-2: Repeat this for all other special positions.

VII.7 THE CUBIC AND DIAMOND STRUCTURES

To show how space group representations are found in three dimensions we take two related examples [Birman (1974), p. 342; Bradley and Cracknell (1972), p. 161; Cornwell (1969), p. 98; Cornwell (1984), p. 231, 242; Cracknell (1975), p. 80; Herring (1942), p. 538; Inui, Tanabe and Onodera (1990), p. 260; Janssen (1973), p. 144; Jones (1975),

p. 155; Koster (1957), p. 228; Ludwig and Falter (1988), p. 204; Shubnikov and Koptsik (1974), p. 207, 217; Streitwolf (1971), p. 88], the cubic close-packed structure (rock-salt, NaCl, is an example) with group $Fm3m(O_h{}^5)$, No. 225, this being symmorphic [Burns and Glazer (1990), p. 154], and the diamond structure (sec. III.5, p. 159) with nonsymmorphic space group $Fd3m(O_h{}^7)$, which is space group No. 227 [Burns and Glazer (1990), p. 157; Zak, *et al* (1969), p. 267]. (Here to maintain conformity with space-group notation the International notation is used with the Schoenflies in parentheses). There are different space groups based on O_h, and these are labeled by their superscripts. Both are for the face-centered cubic Bravais lattice (sec. II.6.b, p. 105), and have lattice translations

$$\underline{t}_1 : \frac{a}{2}(0,1,1),\ \underline{t}_2 : \frac{a}{2}(1,0,1),\ \underline{t}_3 : \frac{a}{2}(1,1,0). \qquad \text{(VII.7-1)}$$

For both cases the factor group with respect to the translations is the full cubic group $m3m(O_h)$. The Wigner-Seitz cell cell, with the general and special points marked, is shown in fig. IV.6.a.iii-1, p. 243. One interesting question is how the addition of nonsymmorphic operations changes the representations; those of these two groups should otherwise be closely related.

Another example is the close-packed hexagonal structure (sec. XI.2.j, p. 590).

Problem VII.7-1: Summarize how the addition of nonsymmorphic operations changes the representations. Does this seem reasonable? Why? After finishing this chapter (and book), return to this problem. Does the answer seem different?

Problem VII.7-2: Give the action of each operator of O_h on an arbitrary translation [Bradley and Cracknell (1972), p. 162], on the lattice vectors and on the reciprocal lattice vectors. Is there a difference in the last two?

Problem VII.7-3: The reciprocal lattice vectors are easily found,

$$\underline{b}_1 : \frac{2\pi}{a}(-1,1,1),\ \underline{b}_2 : \frac{2\pi}{a}(1,-1,1),\ \underline{b}_3 : \frac{2\pi}{a}(1,1,-1). \qquad \text{(VII.7-2)}$$

Problem VII.7-4: It would be interesting to build a model of this Wigner-Seitz cell, either physically or as a computer model, and check the answers to the problems with that. Also it can be checked that this is the correct Wigner-Seitz cell for these crystals.

Problem VII.7-5: In these tables are the coordinates, multiplicity (number of points in the star), whether an internal (I) or surface (S) point, and the point group (little co-group), for the points, lines and planes (with the stated equations) are [Birman (1974), p. 346; Bradley and Cracknell (1972), p. 162; Cornwell (1969), p. 101; Cornwell (1984), p. 232; Janssen (1973), p. 145; Streitwolf (1971), p. 88]. Check and finish the tables. Place each point, line and plane on a diagram of the Brillouin zone. Are all special? Find the coordinates of the points of the stars for each [Birman (1974), p. 347]. These coordinates are with respect to the k_x, k_y, k_z axes. Give them with respect to $\underline{b}_1$, $\underline{b}_2$, $\underline{b}_3$,

Point	Coordinates	Multiplicity	Group
Γ	$(0,0,0)(1/a)$	1*I*	O_h
X	$(2\pi,0,0)(1/a)$	3*S*	D_{4h}
L	$(\pi,\pi,\pi)(1/a)$	4*S*	D_{3d}
W	$(2\pi,0,-\pi)(1/a)$	6*S*	D_{2d}
m	$(0,\pi,\pi)(1/a)$	?	D_{4h}
r	$(\pi,\pi,\pi)(1/a)$	?	O_h
x	$(0,0,\pi)(1/a)$	?	D_{4h}
H	$(0,0,2\pi)(1/a)$	?	O_h
n	$(0,\pi,\pi)(1/a)$	?	D_{2h}
p	$(\pi,\pi,\pi)(1/a)$	?	T_d

Table VII.7-1: PARAMETERS FOR POINTS OF $O_h{}^5$ AND $O_h{}^7$,

Line $0<\kappa<1$			
Δ	$(\kappa,0,0)(1/a)$	6*I*	C_{4v}
Λ	$(\kappa,\kappa,\kappa)(1/a)$	8*I*	C_{3v}
S	$(\kappa,\kappa,1)(1/a)$	12*S*	C_{2v}
Z	$(2\pi,0,\kappa)(1/a)$	12*S*	C_{2v}
Σ	$(\kappa,\kappa,0)(1/a)$	12*I*	C_{2v}
T	$(\kappa,1,1)(1/a)$	?	C_{4v}
Σ	$(0,3\kappa/2,3\kappa/2)(1/a)$	?	C_{2v}
$0<\kappa\le 1$			
q	$(1-\kappa,1,1+\kappa)(1/a)$	?	C_2
S	$(\frac{1}{2}\kappa,\frac{1}{2}\kappa,2)(1/a)$	?	C_{2v}
$0<\kappa\le 1$			
Z	$(0,\kappa,2)(1/a)$	?	C_{2v}
Δ	$(2\kappa,0,0)(1/a)$	?	C_{4v}
Σ	$(0,\kappa,\kappa)(1/a)$	?	C_{2v}
D	$(\kappa,1,1)(1/a)$	?	C_{2v}
F	$(1-\kappa,1-\kappa,1+\kappa)(1/a)$	?	C_{3v}
g	$(0,1-\kappa,1+\kappa)(1/a)$	?	C_{2v}
G	$(2\pi,\kappa,0)(1/a)$	12	?
M	$(\kappa_1,\kappa_1,\kappa_2)(1/a)$	24	?
Q	$(\kappa,\pi,2\pi-\kappa)(1/a)$	24*S*	?
N	$(\kappa_1,\kappa_2,0)(1/a)$	24	?
P	$(\kappa_1,\kappa_2,2\pi)(1/a)$	24	?
A	$(\kappa_1,\kappa_2,\kappa_3)(1/a)$	48*I*, *S*	C_1

Table VII.7-2: PARAMETERS FOR LINES OF $O_h{}^5$ AND $O_h{}^7$,

plane	equation		
ΓMX	$k_x = 0$	?	$G(\underline{k})$
ΓRM	$k_y = k_z > k_x$	?	C_s
ΓRX	$k_y = k_z < k_x$	?	C_s
MRX	$k_y = (\pi/a)$	?	C_s
ΓKWX	$k_x = 0$	?	$G(\underline{k})$
ΓKL	$k_y = k_z > k_x$	?	C_s
ΓLUX	$k_y = k_z < k_x$	?	C_s
WUX	$k_y = (\pi/a)$	?	C_s
ΓHN	$k_x = 0$	?	$G(\underline{k})$
ΓNP	$k_y = k_z > k_x$	?	C_s
ΓHP	$k_y = k_z < k_x$	?	C_s
HNP	$k_y = (\pi/a)$	?	C_s

Table VII.7-3: PARAMETERS FOR PLANES OF O_h^5 AND O_h^7.

Problem VII.7-6: The little groups for some of these points are symmorphic, for the others nonsymmorphic. Which are which [Streitwolf (1971), p. 92]?

Problem VII.7-7: The groups listed are the abstract groups. For each give its elements, defined with respect to the relevant axes, lines and points of the crystal [Streitwolf (1971), p. 88; Zak, *et al* (1969), p. 242]. Assign these to classes. Explain the geometrical significance of the classes, with reference to the crystal. Is there a difference between the symmorphic and nonsymmorphic cases? Why?

Problem VII.7-8: For $Fm3m(O_h^5)$ all rotation operators of $m3m(O_h)$ are associated with pure lattice translations only (it being symmorphic), so an appropriate decomposition of this group into left cosets with respect to the translation group T is [Bradley and Cracknell (1972), p. 163; Ludwig and Falter (1988), p. 205]

$$Fm3m(O_h^5) \Rightarrow \sum_R \{R|000\}T, \qquad \text{(VII.7-3)}$$

the sum running over all elements R of point group $m3m(O_h)$. Explain what this schematic statement means. Does each group element give a distinct coset; if not which do? It should be checked that this gives the complete group, and is completely defined. The form of the space group elements in three dimensions is $\{R|\underline{t}_x, \underline{t}_y, \underline{t}_z\}$, showing the point group operator, R, and the translations along the three axes.

Problem VII.7-9: For nonsymmorphic group $Fd3m(O_h^7)$, only the operators belonging to subgroup $\bar{4}3m(T_d)$ are associated with pure lattice translations (why?) so a decomposition into left cosets is

$$Fd3m(O_h{}^7) \Rightarrow \sum_R \{R|000\}T + \sum_R \{RI|\frac{1}{4},\frac{1}{4},\frac{1}{4}\}T, \qquad \text{(VII.7-4)}$$

as for $Fm3m(O_h{}^5)$, but with a second sum of cosets. The sums run over all (?) elements R of $\bar{4}3m(T_d)$; I is the inversion. Explain this. Is this the only way of writing such a decomposition? Why? Show that the translation part of the coset representatives for the elements of $m3m(O_h)$ not in $\bar{4}3m(T_d)$ can be taken as

$$\underline{v} = \frac{1}{4}(\underline{t}_1 + \underline{t}_2 + \underline{t}_3) = \frac{1}{4}(a, a, a), \qquad \text{(VII.7-5)}$$

when referred to axes $Oxyz$ (along the sides of the cube). The little group (what little group?) G_l can be written in terms of left cosets with respect to translation subgroup T. For $Fm3m$ (obviously?),

$$G_l = \sum_S \{S|000\}T, \qquad \text{(VII.7-6)}$$

with the sum over all elements S of the little group. For $Fd3m$, check that

$$G_l = \sum_P \{P|000\}T + \sum_Q \{Q|\frac{1}{4},\frac{1}{4},\frac{1}{4}\}T; \qquad \text{(VII.7-7)}$$

the first sum is over elements P common to G_l and $\bar{4}3m(T_d)$, while the second is over the remaining elements Q of the little group. So the first set of elements are pure point-group operators, while the second are nonsymmorphic operators — they can include glide translations and screw rotations. Use of $\bar{4}3m(T_d)$ separates the little group elements into these two types. While the statements of this problem are generally regarded as correct, no proof seems known, so they should not be used without being carefully checked.

Problem VII.7-10: If the group is nonsymmorphic, it can be written as a sum over two cosets — it has index 2 (sec. VII.2.b, p. 340), as has just been done. Show that one coset is a group, the other not, which is necessary for it to be so written. The first must be a point group (why?), so is one that has been considered, and with representations known. For diamond, which group is it? Why?

Problem VII.7-11: While it is generally believed that a nonsymmorphic group can be written as a sum over two cosets, this means that its elements are in two sets, symmorphic, and nonsymmorphic, and these have the same number of elements. If this were not true, then most of our considerations apply only to special cases. Thus a rigorous proof that these sets have equal numbers of elements would be useful. And if this should not be true in general, a rule or algorithm to determine for which cases it is true, and which elements are symmorphic, which nonsymmorphic, would be necessary. Check that it actually is true for diamond.

VII.7.a The representations of the factors for $Fm3m$

Obtaining the representations of the two subgroups is the next step. For each $\underline{k}$ we need the projective representations of the little group of $\underline{k}$ with factor system $exp(-i\underline{\tau}_j \bullet \underline{w})$, with $\underline{\tau}_j$ the nonsymmorphic part added to the reciprocal lattice vector by S_j^{-1}, and $\underline{w}$ the translation in the coset decomposition. All $\underline{\tau}$'s are zero — this is a symmorphic group. Thus the factor system consists only of units and belongs to the class corresponding to the identity element of the multiplicator (sec. V.4.b.i, p. 262) of the little group. The little group is a point group so its representations are known.

The small representations $\Gamma_p^{\underline{k}}$ of the little group are now immediately related to those of the point group,

$$\Gamma_p^{\underline{k}}(\{S|\underline{t}\}) = exp(-i\underline{k} \bullet \underline{t})M_p^{\underline{k}}(S). \qquad \text{(VII.7.a-1)}$$

In this case it is only necessary to give the various little co-groups (the point groups) for the different reciprocal lattice vectors $\underline{k}$.

Problem VII.7.a-1: Check these statements. Are they all correct?

Problem VII.7.a-2: Find the representations, and explicit matrices, for a general point, and for several special points, several lines and several planes [Streitwolf (1971), p. 90].

VII.7.b The representations of the factors for $Fd3m$

This group, that of the structure of diamond, is more interesting as it is nonsymmorphic [Birman (1974), p. 494; Koster (1957), p. 230].

Problem VII.7.b-1: In its decomposition into left cosets (eq. VII.7-7, p. 388) for symmorphic elements P, the translational part is zero. Which elements of the point group are these? For the nonsymmorphic set Q [Streitwolf (1971), p. 92], which should be the remaining elements (which are ?), but of course not of a subgroup (why?) the translational part is given by

$$\underline{\tau}_j : \frac{1}{4}(a, a, a). \qquad \text{(VII.7.b-1)}$$

Explain [Bradley and Cracknell (1972), p. 163]. Show that the factor system, $exp(-i\underline{\tau}_j \bullet \underline{w}_j)$, consists entirely of units if

$$\underline{\tau}_j = 0, \quad \text{for all } j, \tag{VII.7.b-2}$$

or if there are no elements of set Q in the little group. The former holds if $\underline{k}$ is an internal point of the Brillouin zone, the latter if the little group is composed entirely of elements in $\bar{4}3m(T_d)$; find these elements. For these the projective representations are just the ordinary representations of the point group which is the little group. Of course, the type of representations depends on the vector $\underline{k}$ whose little group we are finding. Verify, and complete, these statements. We must use a complete set of these $\underline{k}$'s, so have to consider the different types of little groups and correspondingly the different types of factor systems.

Problem VII.7.b-2: Note the difference from the preceding symmorphic case for which

$$\Gamma_p^{\underline{k}}(\{P|\underline{t}\}) = exp(-i\underline{k} \bullet \underline{t})M_p^{\underline{k}}(P), \tag{VII.7.b-3}$$

but check that here

$$\Gamma_p^{\underline{k}}(Q|\underline{t} + \underline{v}) = exp(-i\underline{k} \bullet (\underline{t} + \underline{v}))M_p^{\underline{k}}(Q), \tag{VII.7.b-4}$$

where $\underline{v}$ depends on Q.

Problem VII.7.b-3: The little groups for points X, L, W, for any on the lines containing Z, S, Q, and the plane WUX are nonsymmorphic [Bradley and Cracknell (1972), p. 164; Cornwell (1984), p. 242]. These are special positions, and some are on the boundary. Find the stars for them, and show that in the relevant ones of these stars there are equivalent points connected by reciprocal lattice vectors, that is there are point group transformations that take members of the stars to others so related.

Problem VII.7.b-4: Explain why the necessary and sufficient condition for a nonsymmorphic little group is that the point be special and on the boundary of the Brillouin zone. Use these objects to make the argument concrete and visualizable.

VII.7.b.i *Representations at Point* Γ

The first point to consider is the center of the Brillouin zone, Γ. For it $\underline{k} = (0,0,0)$; its (symmorphic) little group is the complete space group G, the translation subgroup is T, the factor group is thus O_h. We know the characters of the representations as well as other information about them (pb. V.2.d-6, p. 253). Since the translations are represented by the unit matrix for this point, the representations of the space group at Γ are those of O_h.

VII.7.b.ii *Points with nonsymmorphic little groups*

The little co-group of point $\underline{k}$, for which the little group $G(\underline{k})$ is nonsymmorphic, can be written as a sum over two cosets — it has index 2 (sec. VII.2.b, p. 340, pb. VII.7-10, p. 388) — $G_o(\underline{k})$, plus one other. Elements of the second have the form $\{Q|\underline{\tau}\}$, where Q is an operation not in T_d, and $\underline{\tau}$ a translation, necessary for nonsymmorphic groups.

The allowable irreducible representations of $G(\underline{k})$ are

$$\Gamma_p^{\underline{k}}(\{R|\underline{t}\}) = exp(-i\underline{k} \bullet \underline{t})M_o^{\underline{k}}(R), \tag{VII.7.b.ii-1}$$

where $M_o^{\underline{k}}(R)$ is any irreducible representation of $G_o(\underline{k})$. A little cogroup element that takes a point to an equivalent one related by a reciprocal lattice vector $\underline{\tau}$ gives an equivalent representation, so to prevent this multiplicity we consider only allowable representations. We must determine the orbits of the representations — find the cases for which

$$\Gamma_p^{\underline{k}}(R) \sim \Gamma_p^{\underline{k}}(R_oRR_o^{-1}). \tag{VII.7.b.ii-2}$$

Problem VII.7.b.ii-1: The groups $G_o(\underline{k})$ for L, S, Q, WUX have only self-conjugate representations [Streitwolf (1971), p. 93]. Why? So they each have two irreducible representations (why?), with R_o the coset representative,

$$D_\pm(\{R|\underline{t}\}) = exp(-i\underline{k} \bullet \underline{t})M_o^{\underline{k}}(R), \text{ for } R \,\varepsilon\, G_o(\underline{k}) \tag{VII.7.b.ii-3}$$

$$D_\pm(\{R|\underline{t} + \underline{\tau}\}) = \pm exp(-i\underline{k} \bullet \underline{t})\mu M_o^{\underline{k}}(R_o^{-1}R), \text{ for } R \text{ not in } G_o(\underline{k}), \tag{VII.7.b.ii-4}$$

and we must determine μ. In these cases we can choose $R_o^2 = E$, since R_o belongs to the coset for the inversion; check that this gives

$$\mu^2 = D_\pm(R_o^2) = D_\pm(\{E|R_o\underline{t}+\underline{\tau}\}) = exp(-i\underline{k}\bullet(R_o\underline{\tau}+\underline{\tau}))I, \tag{VII.7.b.ii-5}$$

where I is the unit matrix. Also

$$D_o(R) = \Gamma_p^{\underline{k}}(R_o^{-1}RR_o) = \mu^{-1}D_o(R)\mu, \tag{VII.7.b.ii-6}$$

so μ commutes with $\Gamma_p^{\underline{k}}$, thus by Schur's lemma is a multiple of I, giving

$$\mu = exp(-\frac{i}{2}\underline{k} \bullet (R_o\underline{\tau} + \underline{\tau}))I, \tag{VII.7.b.ii-7}$$

a phase. As R_o is not in T_d, there is a coset in the decomposition of O_h that contains R_o. Thus

$$R_o\underline{\tau} + \underline{\tau} = \underline{t}, \tag{VII.7.b.ii-8}$$

that is the element R_o, acting on lattice vector $\underline{\tau}$ to which is added $\underline{\tau}$ to bring the vector back to the first Brillouin zone, equals a translation that depends on the coset. Explain.

Problem VII.7.b.ii-2: We next need the representations of the little group, and the little co-group. Since R_o commutes with all elements of G^T_{ok}, the group that is the unit element of the factor group of little group G_{ok} of point $\underline{k}$ (why?), we can write

$$G_{ok} = G^T_{ok} \otimes \{E, R_o\}. \qquad \text{(VII.7.b.ii-9)}$$

Why? The character table of G_{ok} then is a 2×2 block, headed by G_{ok} and IG_{ok}, with the two entries in the first row identical, both being the character table of G_{ok}, in the second row the first entry is the same, the second is its negative. Hence the $\pm$ of $D_\pm(\{R|\underline{t}\})$ can be included in $M^{\mathbf{k}}_o T$, the representation of G^T_{ok}, and we just use irreducible representations $M^{\mathbf{k}}_o$ of G_{ok}. So we get

$$\Gamma^{\mathbf{k}}_p(\{R|\underline{t}\}) = exp(-i\underline{k} \bullet \underline{t})M^{\mathbf{k}}_o(R), \quad \text{for } R \in G^T_{ok} \qquad \text{(VII.7.b.ii-10)}$$

$$\Gamma^{\mathbf{k}}_p(\{R|\underline{t} + \underline{\tau}\}) = exp(-i\underline{k} \bullet \underline{t})\mu M^{\mathbf{k}}_o(R_o^{-1}R), \quad \text{for } R \text{ not in } G^T_{ok}, \qquad \text{(VII.7.b.ii-11)}$$

with

$$\mu = I, \quad \text{for points L and Q,} \qquad \text{(VII.7.b.ii-12)}$$

$$\mu = exp(-\frac{i}{2}\underline{k} \bullet \underline{t}), \quad \text{for S and WUX.} \qquad \text{(VII.7.b.ii-13)}$$

We can then use the character table for $D_6 = D_3 \otimes C_2$.

Problem VII.7.b.ii-3: Check that G_{ok} contains the inversion for X and L. For which other points is the inversion included? For which not? Is there a general rule?

Problem VII.7.b.ii-4: Finish this. What remains to be done?

Problem VII.7.b.ii-5: One point for which the little co-group is neither Abelian nor contains the inversion is W. Find the representations for it [Streitwolf (1971), p. 96].

Chapter VIII

Spin and Time Reversal

VIII.1 MORE COMPLICATED CRYSTALS

Crystals are sets of physical objects arranged on a lattice. A lattice itself is a set of points; to get a crystal we attach to each a collection of objects, which themselves can be sets of points. But at these points (as we refer to them) there are really atoms, ions, molecules, perhaps sets, thus with structure. So next we take a further step toward reality, endowing the points with an additional label, besides atomic or molecular species, one that corresponds to the physical property of angular momentum. With angular momentum (including spin) crystals, and the groups, become more complicated, so more interesting.

There are, besides angular momentum representations of the rotation group, other aspects of physical laws that determine the properties of crystals. One is that laws and properties are time-reversal invariant. What effect does that have, what limitations does it place, on representations and on our physical entities? And what does it mean — we cannot make time run backwards? We investigate these two extensions next, and together. Spin is a label, but more, a physical property, and it is (especially half-odd-integral) angular momentum that accentuates the importance of time reversal — and this is connected to complex conjugation, which goes to the foundation of quantum mechanics [Mirman (1995b), sec. 2.2, p. 35].

VIII.2 TIME REVERSAL

In classical physics the dynamical law,

$$\sum \underline{F} = m\underline{a} = m\frac{d^2\underline{x}}{dt^2}, \tag{VIII.2-1}$$

is of second order in the time. Thus if the sign of the time is reversed the law is unchanged. There is no way to tell if a motion picture is being run forward or backwards (although the initial conditions for one direction may be very unusual, but not for the other — which is irrelevant here, since we consider only the laws). Classical physics is the limit of quantum mechanics so we would expect the laws of quantum physics to also be invariant under time reversal — at least, for those systems that have a classical limit. But unfortunately Schrödinger's equation,

$$i\frac{d\psi(t)}{dt} = H\psi(t), \tag{VIII.2-2}$$

is of first order in the time (as is Dirac's). It does not look invariant. How then can Newton's second law, the classical limit of Schrödinger's equation [Mirman (1995b), sec. 3.3, p. 118], be time-reversal invariant? This equation is not a dynamical law, or an equation of motion (an undefined concept), despite general belief, but merely (?) a definition of the Hamiltonian, which (exponentiated) is the time-translation operator. Thus it is not a physical law. The dynamics, the physical assumptions, the physical description of the system, appear in the form of the Hamiltonian.

The time-reversed equation is

$$i\frac{d\psi(-t)}{d(-t)} = H\psi(-t) = -i\frac{d\psi(-t)}{d(t)}, \tag{VIII.2-3}$$

where it is assumed that Hamiltonian H is invariant under time reversal (as it is, for an entire system, if there is a classical limit). As we go on we examine this more closely. Now the complex conjugate of Schrödinger's equation,

$$-i\frac{d\psi^*(t)}{dt} = H\psi^*(t), \tag{VIII.2-4}$$

taking H as real, shows that $\psi^*(t)$ obeys the same equation as $\psi(-t)$. The equation is invariant under time reversal, with these assumptions about H, if we simultaneously reverse the sign of the time, and change ψ to its complex conjugate. Hence the equation is invariant, not under the reversal of the sign of the time, but under the combined operation of this reversal and complex conjugation (that it is this combined operation is the reason that physical theories must be invariant under the product of time reversal, parity and particle-antiparticle interchange, the TCP theorem, rather than under TP alone [Mirman (1995c), sec. 2.6, p. 60]).

There is time-reversal invariance of the formalism of quantum mechanics, but expressed differently from that of classical mechanics [Birman (1974), p. 159; Bradley and Cracknell (1972), p. 605; Burns (1977), p. 332; Cornwell (1984), p. 158; Cracknell (1975), p. 76; Elliott and Dawber (1987), p. 369; Evarestov and Smirnov (1993), p. 57; Falicov (1966), p. 183; Heine (1993), p. 164, 290; Inui, Tanabe and Onodera (1990), p. 291; Jansen and Boon (1967), p. 281; Janssen (1973), p. 169; Joshi (1982), p. 184, 318; Koster (1957), p. 249; Lax (1974), p. 275; Lomont (1961), p. 175; Ludwig and Falter (1988), p. 186; Lyubarskii (1960), p. 327; Streitwolf (1971), p. 176; Tinkham (1964), p. 141; Wigner (1959), p. 325]. Classically all functions and quantities are real, so the classical limit cannot have complex conjugation. More important here, time reversal in quantum mechanics differs from the types of invariance that we are familiar with in which a transformation leaves the statefunction invariant (or more generally mixes a set of statefunctions). All basis vectors met up to this point were related only by the group transformations. Here one basis vector is sent into another, but that is also obtained by a different operation, complex conjugation.

The transformation group of space then, including time reversal, is $E(3) \otimes C_2$, the direct product of the Euclidean group, $E(3)$, with this (realization of the) cyclic group of two elements [Janssen (1973), p. 218]. Here we are interested in subgroups, the point and space groups with time reversal.

That the quantities of classical physics are time-reversal invariant while in quantum mechanics the extra operation $\psi \Longleftrightarrow \psi^*$ is required should not be surprising. The probability, which gives the relevant classical values, is proportional to $\psi^*\psi$, so is unchanged by complex conjugation and time reversal. But quantities, like momenta and magnetic fields for example, do change sign when the sign of the time is changed. Before considering the effect of this, since time-reversal is a new type of operator, we have to study its properties and how to handle it.

Problem VIII.2-1: Since Schrödinger's equation is its nonrelativistic limit, we would expect Dirac's equation [Mirman (1995b), sec. 3, p. 114] to be invariant under time reversal, would we not? Is it really?

VIII.2.a Antilinear and antiunitary operators

The operators met up to now are linear — acting on a vector they give a sum of vectors. And they could all be made unitary — all group representations considered are equivalent to unitary ones. The time-reversal operator θ is different, it is antilinear and antiunitary.

A linear operator L is one satisfying

$$L\sum(c_i\psi_i) = \sum c_i L(\psi_i), \tag{VIII.2.a-1}$$

where the c's are scalars (not affected by the group operations, thus belonging to the scalar representation), that is complex numbers. If in addition L leaves invariant $|\psi|^2$, it is unitary.

But time reversal is an antilinear operator,

$$A\sum(c_i\psi_i) = \sum c_i^* A(\psi_i), \tag{VIII.2.a-2}$$

changing the c's to their complex conjugates. If the c's are real there is no difference, but in quantum mechanics complex numbers are essential. An antilinear operator that leaves $|\psi|^2$ invariant is then antiunitary.

As we see later (pb. VIII.7.e-2, p. 441), all representations of all finite — and although we do not consider them, compact (semisimple) — groups are equivalent to ones in which every operator is either unitary or antiunitary, a generalization of this well-known property [Mirman (1995a), sec. VII.2.a, p. 182]. But what is the effect on a representation of having operators in the group that are antilinear?

VIII.2.b The general form of an antilinear operator

An antilinear operator A differs from a linear one only in taking numbers to their complex conjugates. This implies that it might be written as a linear operator L times the operation of complex conjugation C.

Problem VIII.2.b-1: Show that the product of two antilinear operators is linear and that of two antiunitary operators is unitary; the product of a linear operator and an antilinear operator is antilinear; the inverse of an antilinear operator is antilinear. The rules for unitary and antiunitary operators are the same. Also complex conjugation is antilinear and antiunitary.

Problem VIII.2.b-2: The product of an antilinear operator with complex conjugation C is linear,

$$AC = L. \tag{VIII.2.b-1}$$

Also since

$$C^2 = 1, \tag{VIII.2.b-2}$$

$$A = LC, \tag{VIII.2.b-3}$$

$$A = CL, \tag{VIII.2.b-4}$$

as we expected. Check that similar equations hold for antiunitary operators. The antiunitary operators in which we are interested act on statefunctions and applied twice to a statefunction give back the same statefunction, up to a possible phase — this having no effect on the probability density. Thus, for operator U, we define V, ϕ and c,

$$V^2 = UU^* = e^{i\phi}E = cE; \tag{VIII.2.b-5}$$

E is the unit matrix. But since U is unitary,

$$c^2 = 1, \tag{VIII.2.b-6}$$

so

$$c = \pm 1. \tag{VIII.2.b-7}$$

Explain the last sentence. In general applied to a statefunction,

$$V^2\psi = \pm\psi. \tag{VIII.2.b-8}$$

If ψ is a scalar then

$$V\psi(t) = a\psi(-t), \tag{VIII.2.b-9}$$

where

$$a = \pm 1 \quad \text{or} \quad \pm i. \tag{VIII.2.b-10}$$

But it is possible that ψ could have more than one component and that V also mixes them. Why that would occur and its effect is perhaps of the most interest physically.

Problem VIII.2.b-3: It should be clear that time reversal commutes with the coordinates,

$$\theta x_i = x_i\theta, \tag{VIII.2.b-11}$$

and anticommutes with the momenta

$$p_i = \frac{dx_i}{dt}, \tag{VIII.2.b-12}$$

$$\theta p_i = -p_i\theta, \tag{VIII.2.b-13}$$

and since orbital angular momentum can be constructed from coordinates and momenta it also anticommutes with it,

$$\theta L_i = -L_i\theta; \tag{VIII.2.b-14}$$

that is, a movie of a projectile or a merry-go-round run backwards would show the objects moving in the reverse direction. From this show that acting on spherical harmonics [Mirman (1995a), sec. XI.4.c, p. 322],

$$\theta Y_m^l = (-1)^l Y_{-m}^l. \tag{VIII.2.b-15}$$

Interpret the change in sign of m in terms of the merry-go-round. What about spin? Explain the $(-1)^l$. Since the product of two spin-$\frac{1}{2}$ statefunctions transforms as a sum of a scalar plus an angular-momentum 1 state, the effect of θ can be to only multiply the statefunction by $\pm i$, or multiply it and interchange the two spin states. What actually happens, and why? Is there any freedom? What is the effect of θ on the Pauli σ spin matrices [Mirman (1995a), sec. II.4.g, p. 56]? What is the commutation relation of θ and a magnetic field? An electric field?

VIII.2.c The general form of the time reversal operator

Time-reversal operator θ has been specified by its action on the various objects of quantum mechanics. This is as far as we can go — in general. There is no way of writing a general operator in terms of anything else that reverses, say, time t or momentum p (although maybe with some skill an artificial expression can be given). The four Pauli σ matrices (including the identity) form a complete set of two-by-two matrices. Any transformation acting on them is also a two-by-two matrix. Thus it should be expressible in terms of these four; time reversal acting on a scalar cannot be.

Problem VIII.2.c-1: Check that the product of orbital and spin vectors, $\underline{L} \bullet \underline{\sigma}$, is a scalar under rotations, and is (up to constants and powers) the only scalar formed by these two vectors. (Note that these are both vectors, although the representations on whose basis vectors they act are different — in what way, and why then are they both vectors?) Since we have to construct scalars from these vectors, for example for a Hamiltonian that includes spin-orbit coupling, the σ's must transform under θ in such a way that $\underline{L} \bullet \underline{\sigma}$ is also invariant under time reversal. Since L anticommutes with θ, so does σ. Of course it is not surprising that σ, an angular momentum operator like L, has the same transformation properties as it. But that does not mean it must — it is incompletely defined by its transformation properties under the rotation group; how it transforms under all relevant groups, including all inversions, also has to be specified. The need for (existence of) invariants determines how it transforms (matrices are just sets of numbers; transformation properties are added requirements).

Problem VIII.2.c-2: Writing

$$\theta = UC, \tag{VIII.2.c-1}$$

we find [Wigner (1959), p. 331], from the anticommutation of θ with the σ's, that U anticommutes with σ_x and σ_z, these having real elements, and commutes with σ_y, since its elements are imaginary. Explain. If a state consists of several particles (that is, it is a product of states), we consider the operators, s, that are the products of σ's, one for each particle, and this result holds similarly for the s's. Verify this. What operator has these commutation relations with the σ's? Of course,

$$U = \sigma_y. \quad \text{(VIII.2.c-2)}$$

Problem VIII.2.c-3: Matrix σ_y is hermitian. Is it also unitary? Is this the only possible operator? It can be multiplied by a constant without affecting its commutation relations, but we wish it unitary — so it does not change $|\psi|^2$. Thus the only constant is a phase. Since the phase is free we make the obvious choice and set the phase of U to 1.

Problem VIII.2.c-4: Any other operator that satisfies these commutation relations must be a sum of σ_x and σ_z. Why? How about including the unit matrix? Show that there is no such operator and that, except for a phase, the time reversal operator acting on spins is unique.

Problem VIII.2.c-5: We then find, for a system of n (spin-$\frac{1}{2}$) particles,

$$\theta = C\Pi_i(\sigma_y)_i; \quad \text{(VIII.2.c-3)}$$

the time reversal operator is the product of the matrices σ_y for each particle (labeled i), times complex conjugation [Inui, Tanabe and Onodera (1990), p. 293]. Check that this is antilinear and antiunitary, that

$$\theta^2 = 1, \quad n \text{ even (for integral angular momentum)}, \quad \text{(VIII.2.c-4)}$$

$$\theta^2 = -1, \quad n \text{ odd (for half-integral angular momentum)}, \quad \text{(VIII.2.c-5)}$$

and that it has the proper commutation relations with all σ's. This time reversal operator is an off-diagonal matrix. Thus it not only multiplies the statefunction by a factor (with an implied change of the sign of t on which the statefunction depends), but also interchanges the two components of a spin-$\frac{1}{2}$ statefunction. The factor means that the representation is a ray (projective) representation (sec. V.4, p. 259).

Problem VIII.2.c-6: Write an expression for θ, in terms of the σ's, that commutes instead of anticommutes, with all σ's. What happens to $\underline{L} \bullet \underline{\sigma}$ under it? Is the expression reasonable?

Problem VIII.2.c-7: Given a statefunction $\psi(r_i, m_i, t)$, where the r's are the coordinates of the n particles and the m's label their z components of spin, show that

$$\theta\psi(r_i, m_i, t) = (-i)^{2M}\psi(r_i, -m_i, -t), \quad \text{(VIII.2.c-6)}$$

with

$$M = \sum m_i. \quad \text{(VIII.2.c-7)}$$

So

$$\theta^2\psi = \psi, \ n \text{ even}, \quad \theta^2\psi = -\psi, \ n \text{ odd}. \quad \text{(VIII.2.c-8)}$$

As ψ is a multi-component function, this is really an abbreviation. Rewrite it correctly. Is there a general expression for θ for every (each?) angular momentum value?

Problem VIII.2.c-8: Consider a group containing an antiunitary operator, like time reversal. Show that it also contains unitary operators [Jansen and Boon (1967), p. 170], and that the latter form an invariant subgroup of index two [Mirman (1995a), sec. IV.6.b, p. 131]. How are the numbers of unitary and antiunitary operators related? Ray representations imply that there is a larger group for which they are ordinary representations, so the appearance of this group is not surprising.

Problem VIII.2.c-9: The groups that we are considering, ones with invariant subgroups of index two (sec. VI.7.i.iii, p. 335), belong to a general class, defined by that (pb. VII.4.d.iii-3, p. 363). It is useful to list some properties of these groups, which if not obvious, can easily be demonstrated [Jansen and Boon (1967), p. 161]:

1. With G the group, and H the invariant subgroup,

$$G = \{H, sH\}, \tag{VIII.2.c-9}$$

with s any coset representation of H in G; that is the group consists of two cosets.

2. The factor group, $F = G/H$, has only two elements, the identity e, and f, with

$$f^2 = e. \tag{VIII.2.c-10}$$

3. G is a semi-direct product [Mirman (1995a), sec. III.5.c, p. 102] of the form

$$G = H \wedge S, \tag{VIII.2.c-11}$$

with S invariant, if and only if there is an element v within the coset sH, such that

$$v^2 = e, \tag{VIII.2.c-12}$$

(why?) and then

$$S = \{e, v\}. \tag{VIII.2.c-13}$$

4. An irreducible representation of G is either an irreducible representation of H, or contains two such representations, related of course by operators not in H. What are the orbits, the little groups, in these cases?

5. With $D(h)$ an irreducible (?) representation matrix of an element of H, and $D(sh)$ that of an element of the second coset, we can write a representation of G as

$$\Gamma = \{D(h), D(hs)\}; \tag{VIII.2.c-14}$$

then

$$\Gamma' = \{D(h), -D(hs)\}, \qquad \text{(VIII.2.c-15)}$$

is also an irreducible (?) representation of G.

6. With $\xi(h)$ the character of element h, and

$$\xi = \{\xi(h), \xi(sh)\}, \qquad \text{(VIII.2.c-16)}$$

Γ and Γ' are distinct representations if and only if $\xi(sh)$ is nonzero for some h. If

$$\xi(sh) = 0, \text{ all } h, \qquad \text{(VIII.2.c-17)}$$

then Γ, Γ' are equivalent — Γ is then called self-associate. Otherwise Γ, Γ' are associate representations. What are the little groups for these two cases? Summarizing, the representations of G are either self-associate, or there are pairs (but no more than two) which are associate representations; so the representation of G has dimension equal to that of H, in the former case, or twice that of H, for the latter. Representation Γ subduces representation D (sec. VI.2, p. 281). How many times for these two cases? What are the orbits (sec. VI.2.b, p. 283)? Restate these results for the special cases in which the coset representative is complex conjugation, and for which it is time reversal.

Problem VIII.2.c-10: The antiunitary operators considered here are time reversal, and products of it with the unitary operators, perhaps giving the impression that the invariant unitary subgroup is fully determined. However these concepts are abstract. Take a few point groups, both proper and improper, and divide each into two sets in all possible ways, with one a subgroup and the other a coset of that with a representative, denoted by s, of order two, and show that each of these pairs of subsets satisfy the conditions of the preceding problems, in particular the subgroup is invariant, and can be taken unitary, while the other coset can be taken to consist of antiunitary operators. Thus, for group G (of course only for ones with invariant subgroups of index two), it is necessary to specify the invariant subgroup. Different choices give different realizations of the antiunitary operators, and have different representations. This means in particular that there can be several different crystals (with different symmetry groups) having objects with spin that give the same crystal (with a single symmetry group) if the objects are replaced with (otherwise) identical ones, but without spins. For each invariant subgroup of each point group, take a crystal with symmetry given by the point group and place spins in such a way to give a crystal with symmetry given by the group with that invariant subgroup. Also the groups we find are not antiunitary groups (ones containing antiunitary operators), they are point and space groups, but become antiunitary when some idempotent operator [Mirman (1995a), pb.VIII.5.a-1, p. 233] is identified as time reversal. (Why idempotent?) Thus it is the physical interpretation, and its implications like the form of the representations, that distinguish these groups from purely abstract ones.

Problem VIII.2.c-11: Show that for a group with r generators [Mirman (1995a), sec. III.3, p. 82], the maximum number of subgroups of index 2 [Jaswon and Rose (1983), p. 170] is

$$n_s = 2^r - 1. \qquad \text{(VIII.2.c-18)}$$

Calculate this maximum for all (crystallographic) point groups, including improper ones (tbl. B-1, p. 633).

VIII.2.d Kramer's Theorem

If there is a symmetry in a system we would expect degeneracy — several states with the same energy. Is this true for systems invariant under time reversal? The answer is given by Kramer's theorem: All energy levels containing an odd number of half-integral spin particles must be at least doubly degenerate if the system is time-reversal invariant [Burns (1977), p. 184; Elliott and Dawber (1987), p. 373; Hamermesh (1962), p. 118; Heine (1993), p. 169; Inui, Tanabe and Onodera (1990), p. 232; Jansen and Boon (1967), p. 287; Janssen (1973), p. 173; Joshi (1982), p. 188; Lax (1974), p. 285; Ludwig and Falter (1988), p. 95; Tinkham (1964), p. 78, 143; Tsukerblatt (1994), p. 249]. If there are external magnetic fields the meaning of this statement must be examined more closely; here we assume no such fields.

Problem VIII.2.d-1: For an odd number of half-integral spin particles [Koster (1957), p. 251]

$$\theta^2\psi = -\psi, \qquad \text{(VIII.2.d-1)}$$

so $\theta\psi$ and ψ are orthogonal. Why? Show in particular that

$$(\theta\psi, \psi) = -(\psi, \theta\psi), \qquad \text{(VIII.2.d-2)}$$

and relate this to θ being antiunitary. If the system is time reversal invariant $\theta\psi$ and ψ have the same energy, showing that the system is doubly degenerate. Restate this in terms of the Hamiltonian. Of course, if there is further symmetry ψ may consist of a set of degenerate states so the degeneracy can be higher than two.

Problem VIII.2.d-2: What is the degeneracy for one electron (for example)? For three? For n?

VIII.3 COMPLEX CONJUGATE REPRESENTATIONS

Group representation matrices, including orthogonal ones, can have complex entries [Mirman (1995a), sec. V.2.a.ii, p. 148; (1995b), sec. 2.1.1, p. 25; sec. 7.2, p. 124; Wigner (1959), p. 285]. If we take their complex conjugates we obtain another set of matrices. Are these also representation matrices of the group? If so what is their relationship to the original representation; are they equivalent: sometimes; always; never? What physically is the significance of this equivalence, or lack of it, and of the matrices containing complex entries? For what groups, and physical contexts, are representations (necessarily) complex, and for which are they not, perhaps cannot be, and why? A complex number can be written as the sum of two real ones, and from a complex number and its conjugate we obtain, by taking sums, two real numbers. Is there anything comparable for matrices, particularly representation matrices?

It is not surprising that the complex-conjugate matrices also give a representation. From

$$M(R)M(S) = M(RS), \qquad \text{(VIII.3-1)}$$

we obtain

$$M^*(R)M^*(S) = M^*(RS), \qquad \text{(VIII.3-2)}$$

so the conjugate matrices obey the same product rules as the original, and as the operators of the group. The answer to the question of whether a representation and its complex conjugate are equivalent is sometimes, so this needs detailed discussion.

Problem VIII.3-1: That both M and M^* are simultaneously either reducible or irreducible should be clear. Show that if matrix D reduces representation M, then D^* reduces M^*.

VIII.3.a When are conjugate representations equivalent?

There is a simple way of telling whether a representation and its conjugate are equivalent. They are if all their characters are real, else not. The characters of conjugate matrices are obviously complex conjugates of each other. And two representations are equivalent if, and only if, for each class, the two characters are equal [Mirman (1995a), sec. VII.4.a, p. 191].

One might suspect that whether representations and their conjugates are equivalent is related to the structure of the group. In fact, if every operator of the group belongs to the same class as its reciprocal (the group is ambivalent [Mirman (1995a), pb. IV.5-15, p. 121]), then all representations are equivalent to their conjugates. (As seen in other ways, there are relationships between the properties of the classes and those of the representations [Mirman (1995a), sec. VII.6.c, p. 202]). If the representation matrices are unitary then

$$M(R^{-1}) = M^{+}(R), \tag{VIII.3.a-1}$$

where M^{+} is the conjugate transpose. Since characters depend only on diagonal elements, M^{+} and M have complex conjugate characters, so in general the unitary matrices for a transformation and its reciprocal are complex conjugates of each other. All elements of the same class have the same character, so if R and R^{-1} are in the same class, for all R, then their characters are real. However we have to be careful of the converse. If they belong to different classes, those classes might have the same characters for some representations (the scalar, for example). We also have to consider whether all representations can be equivalent to their conjugates if the group is not ambivalent.

Problem VIII.3.a-1: Prove that if all operators have the same characters as their reciprocals, for all representations, then all operators belong to the same classes as their reciprocals. In this case, all characters are real. Further if all characters are real, for all representations, the group is ambivalent.

Problem VIII.3.a-2: Check that the three-dimensional rotation group has only real characters, as is also true of SU(2) — but it is not true of SO(2) [Mirman (1995a), pb. X.3.a-3, p. 274; sec. X.4.a.ii, p. 278; eq. X.5.a-1, p. 285].

VIII.3.b Operators mixing representations and their conjugates

If a representation is equivalent to its conjugate then there is a matrix U performing a similarity transformation from one to the other,

$$M^{*} = UMU^{-1}. \tag{VIII.3.b-1}$$

It is unlikely that any matrix can do this. What are the restrictions on U needed to make it a candidate for such a role? It turns out that U is either symmetric or antisymmetric [Lax (1974), p. 294; Wigner (1959), p. 286],

$$U_{ij} = \pm U_{ji}. \tag{VIII.3.b-2}$$

Of course given the M's, U is uniquely determined, up to a constant.

Problem VIII.3.b-1: As both M and M^{*} are unitary, U can (must?) also be taken as unitary. Then

$$UU^{*}M^{*} = UMU^{*} = M^{*}UU^{*}, \tag{VIII.3.b-3}$$

follows from the relations between M and M^*. If M (so M^*) is irreducible, then since UU^* commutes with it, $UU^* \sim I$, the unit matrix. Since U is unitary,

$$U^t U^* = I; \qquad \text{(VIII.3.b-4)}$$

U^t is the transpose. This gives

$$U = \pm U^t, \qquad \text{(VIII.3.b-5)}$$

as stated. Also check that if the matrix taking a representation to its complex conjugate is symmetric (antisymmetric), so is the matrix taking any equivalent representation to its conjugate. If the matrix relating one representation of a group to its complex conjugate is symmetric, need this be true for the matrices for all representations?

VIII.3.c Classification of groups under complex conjugation

Groups can be classified into three types depending on the behavior of their representations under complex conjugation [Burns (1977), p. 336; Cornwell (1984), p. 127; Heine (1993), p. 171; Jansen and Boon (1967), p. 125], complex, potentially real and pseudo-real. The classification is the same under time reversal, since it is complex conjugation times a unitary operator, and this does not distinguish these types. But time reversal illustrates the physical meaning of the kinds of groups, and why they fall into these classes.

Determining the type from the representation matrices can be difficult, but there is a test, the Frobenius-Schur test, which we shall discuss when we consider other (closely related) applications for it (sec. VIII.7.k, p. 449).

VIII.3.c.i *Potentially real, and pseudo-real, representations*

Two classes of representations are equivalent to their conjugates, those for which the similarity transformation is symmetric, these called potentially real, and those for which it is antisymmetric, called pseudo-real. Potentially real representations are really real; if U is symmetric they can be transformed into a form in which all matrix elements are real numbers. For pseudo-real representations, all characters are real, but there is no form in which all matrix elements are. If a representation is not equivalent to its complex conjugate, we call it essentially complex. Most representations of unitary groups SU(n), for n > 2, which are not discussed here, are of this form.

To show that if U is symmetric there is a similarity transformation that makes all matrix elements of all M's of the representation real numbers, we note that if U is both unitary (which we can always require [Mirman (1995a), sec. VII.2.a, p. 182]) and symmetric, then its eigenvectors can be taken as real [Wigner (1959), p. 287] (of course they need not be; we can always multiply all components by a complex number).

Problem VIII.3.c.i-1: Prove this.

Problem VIII.3.c.i-2: Show that the necessary and sufficient condition for a set of representation matrices to be equivalent to a set with all real entries, is that U be symmetric. Thus for U symmetric we can take the matrices as all real, which seems like the most reasonable thing to do. However if it is antisymmetric the representation has to be looked at more closely.

Problem VIII.3.c.i-3: Show that every pseudo-real irreducible representation has even dimension [Cornwell (1984), p. 128; Lax (1974), p. 297]. Why should this be?

Problem VIII.3.c.i-4: Prove that if a representation is real and reducible, than any pseudo-real irreducible representation that appears in its reduction does so an even number of times [Cornwell (1984), p. 129]. If an essentially complex irreducible representation is contained in it then its complex conjugate is also, and the same number of times. Explain the reasons for these results.

Problem VIII.3.c.i-5: Double cyclic groups (sec. V.3.a, p. 257) have some representations that are real, with others pseudo-real. Prove, and explain (sec. VIII.7.k, p. 449). Ordinary crystallographic point groups have no pseudo-real representations, and double-valued representations (those of the double group) are real only if they are one-dimensional, all others being either pseudo-real or complex [Lax (1974), p. 299]. Interpret this physically, after checking that it is correct, remembering that the double group describes systems that have particles with (half-odd-integral) spin.

VIII.3.c.ii *Ambivalent groups and their character sets*

Characters can tell us much about representations, and often about the groups themselves. If a group is ambivalent there is a strong restriction on all its characters — they are real (sec. VIII.3.a, p. 403). Is the converse true — if all characters are real, for all representations, is the group ambivalent? If the group is ambivalent, all elements have the same characters are their reciprocals (every element is in the same class as its inverse). Suppose this is true; is the group ambivalent?

To investigate these questions we use the orthogonality theorems for the characters (of unitary representations) [Mirman (1995a), sec. VII.

5.c, p. 195] which state the product of the character vectors of two representations summed over the classes is zero if the representations are inequivalent, and the product of character vectors of two classes summed over the representations is zero if the classes are different. Here the products are of one character vector with the complex of the other.

Now if the representation matrices are unitary, the character of an element and that of its inverse are complex conjugates. If all elements have the same characters as their reciprocals, the (relevant) product of the character vectors is nonzero, so the classes they belong to are the same for all, and the group is ambivalent. And if all characters are real, again the product cannot be zero, giving again that the group is ambivalent.

VIII.3.c.iii *Why physically are groups so classified?*

Groups with real representations include the symmetric groups. Whether the state of particles is symmetric or antisymmetric (or is a basis vector of another representation) does not change in time, so will not vary if time runs backwards. Hence under time reversal, it acts like a scalar, unchanged, so real.

For a spin-$\frac{1}{2}$ particle, the up and down states, along a line given by angles θ and ϕ, are [Mirman, (1969b), eq. 2]

$$|u) = \begin{pmatrix} exp(\frac{i\phi}{2})cos\frac{\theta}{2} \\ iexp(\frac{-i\phi}{2})sin\frac{\theta}{2} \end{pmatrix}, \quad \text{(VIII.3.c.iii-1)}$$

$$|d) = \begin{pmatrix} iexp(\frac{i\phi}{2})sin\frac{\theta}{2} \\ exp(\frac{-i\phi}{2})cos\frac{\theta}{2} \end{pmatrix}. \quad \text{(VIII.3.c.iii-2)}$$

So

$$|u,\theta,\phi)^* = |u,\theta,-\phi), \quad \text{(VIII.3.c.iii-3)}$$

and similarly for $|d)$. Thus the time-reversed state is given by the negative angles. Why? Let the angular momentum of the particle be up along z at $t = 0$. Then after a time it will be in direction θ, ϕ. However if we let time run backwards, then after the same interval of time it points in direction $\theta, -\phi$, as we would expect a time-reversed state to do. The matrix taking the state at $t = 0$ to this has complex entries, since the initial vector has real ones, the final complex. The representations of the group, SU(2), are pseudo-real, although this does not rule out some being potentially real as we see next.

An example of potentially real representations is given by the spherical harmonics [Mirman (1995a), sec. XI.4.c, p. 322], which for $l = 1$ are (up to normalization),

$$Y_1^1 = exp(i\phi)sin\theta,\ Y_0^1 = cos\theta,\ Y_{-1}^1 = exp(-i\phi)sin\theta. \quad \text{(VIII.3.c.iii-4)}$$

These are sums of $cos\theta, sin\theta cos\phi$ and $sin\theta sin\phi$, which are all real. The matrices acting on the first form are complex but can be written in terms of these acting on the second which are real.

Multiplying these by a constant r gives another form for the basis vectors of the $l = 1$ representation, x, y, z. And this indicates why these representations are potentially real. A particle with orbital angular momentum traces a path through space. Thus its statefunction must be a function of the coordinates. Since it transforms according to a representation of the rotation group, it must transform as a representation formed from the powers of the coordinates, these giving all irreducible representations, and every representation can be expanded as a sum of this complete set of irreducible representations [Mirman (1995a), sec. XI.4.a.ii, p. 317; sec. XIV.1.b, p. 404]. Statefunctions describing orbital angular momentum and transforming under an irreducible representation of the rotation group are then (up to constants) just the set of polynomials in the coordinates that transform according to the same representation, or some linear combination of them. Coordinates are real, thus any integer angular-momentum representation of the rotation group has real basis vectors, so real representation matrices. Any other irreducible representation (giving orbital angular momentum) must be equivalent to one of these — potentially real.

However the representation describing orbital angular momentum is only potentially real, it is actually written in terms of complex basis vectors, and so matrices. Why not take it as real? It describes the motion of a particle, say in a plane given by θ, which has moved through an angle ϕ in time t. It has to be written in a way that properly describes the motion, which includes giving the behavior if t goes backwards. From the proceeding equation,

$$Y_{\pm 1}^1(\theta,\phi)^* = Y_{\pm 1}^1(\theta,-\phi) = Y_{\mp 1}^1(\theta,\phi). \quad \text{(VIII.3.c.iii-5)}$$

That is if time is reversed, the plane in which the particle moves is unaffected, but instead of moving to angle ϕ, it moves to $-\phi$. Or we can take the particle, still in the same plane, as moving in the opposite direction, so its angular momentum is reversed.

Thus although the basis vectors can be taken as real, to properly display the physical behavior, the correct form is complex, so they can describe the motion and the behavior under time reversal. They are only potentially real.

Problem VIII.3.c.iii-1: Check all this using the representation matrices for these cases, including the spinor [Mirman (1995a), sec. XI.5, p. 329].

Problem VIII.3.c.iii-2: Explain why the Y_0^1 state is unaffected by time reversal.

VIII.3.c.iv *Reality and the rotation group*

The rotation group and its subgroups furnish the most important applications (here). To study these [Wigner (1959), p. 288] we find the matrix U that transforms representation M^j to its complex conjugate M^{j*},

$$UM^j = M^{j*}U, \qquad \text{(VIII.3.c.iv-1)}$$

Any representation matrices can be used, and the simplest ones are for a rotation about z through angle β, for which the matrix elements are [Mirman (1995a), eq. XI.4.d-3, p. 324]

$$M^j_{lm} = \delta_{lm}exp(im\beta), \; M^{j*}_{lm} = \delta_{lm}exp(-im\beta), \qquad \text{(VIII.3.c.iv-2)}$$

which must hold for all β's, giving for the matrix that transforms one to the other,

$$U^j_{nm}M^j_{mk} = M^{j*}_{nm}U^j_{mk}, \qquad \text{(VIII.3.c.iv-3)}$$

so

$$U^j_{nk}exp(ik\beta) = U^j_{nk}exp(-in\beta), \qquad \text{(VIII.3.c.iv-4)}$$

and the exponents must be equal, resulting in

$$U^j_{nk} = u^j_n\delta_{n,-k}. \qquad \text{(VIII.3.c.iv-5)}$$

Thus U has nonzero elements only on the anti-diagonal, the line running from upper right to lower left (and orthogonal to the diagonal). What are these elements? Using this equation for an arbitrary group element, we have for the j, m element of the representation matrices

$$u^j_nM^j_{-n,k} = M^{j*}_{n,-k}u^j_k, \qquad \text{(VIII.3.c.iv-6)}$$

and substituting, for a general rotation, the parts of the matrix elements [Wigner (1959), p. 166] that differ,

$$u^j_j\sqrt{\binom{2j}{j-m}} = (-1)^{j-m}u^j_m\sqrt{\binom{2j}{j+m}}, \qquad \text{(VIII.3.c.iv-7)}$$

we get

$$u^j_m = (-1)^{j-m}u^j_j. \qquad \text{(VIII.3.c.iv-8)}$$

Fortunately (?) $j - m$ is always an integer, whether j is integral or half-integral. Since there is one free constant for each representation we choose

$$u^j_j = 1. \qquad \text{(VIII.3.c.iv-9)}$$

So U is a matrix having all zero elements except on the anti-diagonal where the elements are alternately 1 and -1, starting with 1 in the upper right hand corner and ending in the lower left-hand corner with 1 if j is an integer, and -1 if j is a half (odd) integer.

Problem VIII.3.c.iv-1: Verify the last sentences.

Problem VIII.3.c.iv-2: Check that U is symmetric for integral j. These representations are potentially real. Verify that the spherical harmonics can be brought into real form. For half-integral j, U is antisymmetric. Thus the spin-$\frac{1}{2}$ representations (and representations of an odd number of spin-$\frac{1}{2}$ particles) are pseudo-real. Compare these results with those of sec. VIII.3.c.i, p. 405.

Problem VIII.3.c.iv-3: For consistency the direct product of two potentially real representations must be potentially real, the product of two pseudo-real ones must be potentially real also, and the product of a potentially real and pseudo-real one must be pseudo-real. Why? Verify that this is true for the rotation group.

Problem VIII.3.c.iv-4: This should give for the representation matrix of rotation R,

$$M^{j*}_{m'm}(R) = (-1)^{m-m'} M^j_{-m',-m}(R); \qquad \text{(VIII.3.c.iv-10)}$$

check that the matrices transforming spherical harmonics do have this property. Write the basis vectors for the spin-$\frac{1}{2}$ representation (the vectors that the Pauli spinors [Mirman (1995a), sec. II.4.g, p. 56] act on), and verify that they also obey this equation. Of course in doing this, we use the fact that the components of the basis vectors are functions of the angles [Mirman (1969)]. Thus the statefunctions for a system with an odd number of spin-$\frac{1}{2}$ particles provide the pseudo-real representations — showing that they do exist.

VIII.3.d Equivalent sets can be distinguishable

A basis vector of a group may also transform under a different group (as with spin and isospin). Thus a group may have many different, equivalent, sets of basis vectors, distinguished by some other property. Here the other property is complex conjugation — a distinction without a difference. Neither set can be taken as the representation with the other its complex conjugate. Rather the two sets are complex conjugates of each other.

VIII.3.d.i *Mathematical equivalence, and physical equivalence*

This emphasizes that equivalent representations may be mathematically equivalent, but they may not be equivalent physically. The coordinates, and the statefunction of a particle with unit orbital angular momentum, both transform according to the $l = 1$ representation of the rotation group. However the correct forms describing them are different, the real form for real coordinates, the complex form to describe the motion of a particle. To use the proper form for one to describe the other would be misleading, and would hide what is happening — although what is proper can vary with circumstances. There are physical reasons for the forms that we use so while these representations are equivalent mathematically — there is a similarity transformation linking them — they are different physically. It is always important to understand the physical situation and meaning, thus what the proper mathematics, and the proper form for the mathematics, are to describe the physics, and not automatically apply mathematical formalism without thinking what it means, and why.

VIII.3.d.ii *Complexity of half-integral angular momentum statefunctions*

Odd half-integral angular momentum does not arise from a particle moving through space, so the statefunctions do not depend on coordinates, therefore need not transform as coordinates. They can be complex, and have to be complex, being basis vectors, not of the orthogonal group on real numbers, but the unitary one on complex numbers [Mirman (1995a), sec. II.3.d, p. 43; (1995b), chap. 7, p. 122]. They are thus pseudo-real.

As we come across different groups we should examine their behavior under complex conjugation, and so under time reversal, and see whether there are physical reasons for their properties, and whether they, and the forms in which their representations are given, correctly describe the physics of the systems whose transformation groups they are.

Problem VIII.3.d.ii-1: Discuss the case of a particle with both spin and orbital angular momentum, and also the various cases of several particles.

VIII.4 COLOR GROUPS

Crystals, though we have are thought of them as (restricted) collections of points, really consist of objects, electrons, holes, atoms, ions, molecules, ..., and these have structure, in particular angular momentum (producing magnetic fields). How does this affect the symmetry, how does having a collection of, say, spinning particles change the group that describes the (spinless) system?

A picturesque, although not completely similar, way of envisioning this is to take the crystal and color different regions of it — starting with black and white. This lowers the symmetry. However it also introduces a new operation, interchange of colors. So we come then to color (or magnetic) groups [Bradley and Cracknell (1972), p. 569; Burns (1977), p. 274; Hamermesh (1962), p. 63; Inui, Tanabe and Onodera (1990), p. 306; Jansen and Boon (1967), p. 355; Janssen (1973), p. 218; Jaswon and Rose (1983), p. 48; Joshua (1991), p. 127; Kocinski (1990), p. 22; Lifshitz (1997); Lockwood and Macmillan (1983), p. 59, 195; Ludwig and Falter (1988), p. 26, 91; Megaw (1973), p. 180; Senechal (1988); Senechal (1990), p. 74; Shubnikov and Belov (1964), p. xx, 112; Shubnikov and Koptsik (1974), p. 263; Tinkham (1964), p. 299; Tolédano and Tolédano (1987), p. 308]. Here unfortunately they will not be very colorful — we consider only two colors.

This is the next step toward reality. We started with lattices, sets of points. Then on each we placed another set, identical for all lattice points, forming a crystal, in general lowering the symmetry of the system — the groups of most crystals are not holohedry groups. Here each of these points is itself replaced by an object with further structure, viewed as angular momentum, say. Of course, real objects, especially at nonzero temperature, are more complicated; for example, instead of considering two colors, or angular momentum states, we can consider more, as with larger angular momentum, so giving other groups, reasonably named polychromatic groups [Bradley and Cracknell (1972), p. 677; Lockwood and Macmillan (1983), p. 206; Shubnikov and Belov (1964), p. 228]. Thus here we take the next step, but only a step.

The groups we now study are found by assigning to each point a two-valued variable, spin-$\frac{1}{2}$ or color — black and white. However there can also be higher values of angular momentum, so a variable with more than two values, thus more than two colors is needed, leading to many more point and space groups (than we can consider here), although these are less of a generalization than the present one which introduces a new type of operator, an antiunitary one. Other cases might also be handled by redefinitions, such as changing the unit cell.

While we study only two colors we do advance a step. At each stage we gain information, and place restrictions. It is always possible, given the *time* and *energy*, to go further toward the actual physical objects that we deal with in life, though not necessarily in theories, or books. But time and energy are limited, thus so must these considerations be, and we shall consider only these color groups, but with the realization that we have started, not finished, the study of actual objects. The real world requires more time, more energy — and more concepts.

VIII.4.a The conditions on color groups

The operator that interchanges black and white will (usually) be interpreted as time reversal; we denote it by θ. Clearly,

$$\theta^2 = E. \qquad \text{(VIII.4.a-1)}$$

What is the symmetry group of the colored crystal? Not all transformations of the uncolored one are symmetries with the addition of color; rotations interchanging white and black segments are not. The point group has elements, denoted by S, that are rotations, reflections and the inversion (and do not interchange colors), and in addition symmetry transformations of the form θS — the product of such an uncolored transformation and θ. For this, the rotation (reflection or inversion) moves black to white, and white to black, and then θ interchanges the colors, leaving the crystal invariant. Thus the symmetry group consists of a subset, clearly a group, of (non-color changing) point-group transformation, plus another subset, clearly not a group, of the form θS.

What limitations does this place on the larger (color) group, the one with θ?

Suppose that θ were a symmetry of the object. Then a change of black and white would leave the object the same — obviously impossible. Or interpreting θ as time reversal, it interchanges the direction of currents, also spin up and spin down, and magnetic fields, and if it were a symmetry all currents and fields would be zero. But we want to generalize to cases for which they are not. So operator θ is not a member of the symmetry group. This means for groups with elements θS, that S cannot be of odd order, for then there would be a power such that

$$(\theta S)^n = E, \qquad \text{(VIII.4.a-2)}$$

the identity. It therefore cannot have elements like θC_3 or θS_3. Also if element S_1 is in the subgroup then θS_1 is not a member of the symmetry group, for if it were,

$$\theta S_1 S_1^{-1} = \theta \qquad \text{(VIII.4.a-3)}$$

would be as well. If S_1 is in the group, θS_1 is not, and conversely.

Problem VIII.4.a-1: *This assumes that θ and S commute. Why do they? Why must the set of S form a group? Is it a subgroup? Invariant?*

Problem VIII.4.a-2: Why, mathematically and physically, is the set θS not a group? Why is the sum of sets S and θS a group?

Problem VIII.4.a-3: Explain why the sets of elements S form one of the thirty-two point groups (here we are considering only elements S that are rotations, reflections or inversions).

Problem VIII.4.a-4: How then do we find the point groups, G, containing θ? The set of uncolored symmetries form a subgroup H of index 2 (pb. VIII.2.c-8, p. 400; sec. VIII.7, p. 434). So the group can be written

$$G = H \oplus VH, \qquad \text{(VIII.4.a-4)}$$

where H is a maximal subgroup, not containing an antiunitary operator, and V an element of G not in H (and G is the sum of the sets, it is the set consisting of subsets H and VH). Showing θ explicitly, write G as

$$G' = H \oplus \theta SH, \qquad \text{(VIII.4.a-5)}$$

where θ is the antiunitary operator and θS an element of G' not in H, prove that G' is a group. Thus both G and G' are groups, and both are one of the 32 point groups (sec. I.6.d, p. 42). Why? That is G is obtained from G' by replacing θ with the identity E. We would expect this to be true since if black faded, or the vector representing angular momentum became zero, and G' is a symmetry no matter how light the color or how short the arrow, then G' would become G. The effect of θ would be the same as the identity. Show that H is an invariant subgroup. Also the number of elements in G, and in G', is twice that in H (of course?). Therefore G and G' each have only two cosets. Further show that $H \oplus (G' - H)$, the set sum of H and all elements of G' not in H, is a point group.

Problem VIII.4.a-5: List all elements of all point groups (tbl. B-2, p. 634). Which give elements of the form θS that can be an element of a color point group?

Problem VIII.4.a-6: Among the groups previously considered [Mirman (1995a), chap. II, p. 29; chap. III, p. 67], say the symmetric groups, the water group, and so on, which are possible choices for such subgroups of color point groups? Why?

VIII.4.b Point groups with time reversal

The point groups containing, say, time (or more graphically color) reversal θ, are thus found from the ordinary 32 point groups, those which do not contain θ. To do so, for each find a maximal subgroup, and construct the two cosets, but for the second replace coset representative h by θh. The (direct) sum of the two cosets is the group containing θ. What are these groups [Janssen (1973), p. 220; Tinkham (1964), p. 299]?

There are 32 ordinary point groups, 32 gray point groups, and 58 black-and-white point groups, giving (in three dimensions) a total of

$$32 + 32 + 58 = 122 \qquad \text{(VIII.4.b-1)}$$

color (or magnetic) point groups [Bradley and Cracknell (1972), p. 570; Burns (1977), p. 274; Joshua (1991), p. 186; Lockwood and Macmillan (1983), p. 202; Nowick (1995), p. 81, 219; Shubnikov and Koptsik (1974), p. 266]. To the description of these we now turn.

We denote the color group with ', and show both notations, Schoenflies and International. To construct the color group the elements of the ordinary group are divided into two equinumerous sets, one an invariant subgroup, the elements of the other are then multiplied by θ. The set sum of these two forms the color group.

Problem VIII.4.b-1: Find the magnetic point groups in one and two dimensions [Lockwood and Macmillan (1983), p. 196; Shubnikov and Belov (1964), p. 222; Shubnikov and Koptsik (1974), p. 270].

VIII.4.b.i *Cyclic color groups*

The simplest groups are the cyclic ones [Mirman (1995a), sec. III.2.a, p. 68]. What are their generalizations to color groups [Jaswon and Rose (1983), p. 49]? Cyclic groups are generated from a single operator, so there is a unique decomposition (pb. VIII.2.c-11, p. 402). Only ones of even order can give color groups. Take $\mathrm{C_n(n)}$ the cyclic group of order n (with n elements), which is necessarily even, and $\mathbf{C_n}$ an element of it (for example a rotation through $\frac{2\pi}{n}$), which of course obeys

$$\mathbf{C_n}^n = E. \qquad \text{(VIII.4.b.i-1)}$$

We then get the color groups of order $2n$,

$$\mathrm{C_2}'(2') = \mathbf{C_2} + \theta\mathbf{C_2} = \mathbf{C_2}(E+\theta), \qquad \text{(VIII.4.b.i-2)}$$

$$\mathrm{C_4}'(4') = \mathrm{C_2}(2) + \theta\mathrm{C_2}(2) = \mathrm{C_2}(2)(E+\theta), \qquad \text{(VIII.4.b.i-3)}$$

$$\mathrm{C_6}'(6') = \mathrm{C_3}(3) + \theta\mathrm{C_3}(3) = \mathrm{C_3}(3)(E+\theta), \qquad \text{(VIII.4.b.i-4)}$$

$$\ldots,$$

displaying the coset structure. Also

$$(\theta\mathbf{C_n})^n = (\theta)^n\mathbf{C_n}^n = E, \qquad \text{(VIII.4.b.i-5)}$$

for n even. Thus the n elements of $\mathrm{C_n(n)}$ become $2n$ of $\mathrm{C_n(n')}$, each element becoming a pair, one the same as that in $\mathrm{C_n(n)}$, the other the same (rotation, say) with color switch.

This might seem vacuous. Consider $C_2(2)$ and an object on which it acts, say a circle with the top half white, the bottom black. A rotation of C_2 around the center interchanges these, as does θ, so why does the group have four elements? The number of representations of $C_2(2)$ is 2, but of $C_2'(2')$ is 4. If the object were two atoms with one having spin up, the other down, there would be four representations describing it, but only two for atoms with no spin — the symmetric (the proper one for bosons) and the antisymmetric (for fermions). However if there were another quantum number both could be relevant for fermions, thus the objects could be protons and neutrons, with isospin; there are correct statefunctions with the space part symmetric, the isospin antisymmetric, and also the reverse. There might be selection rules, so whether the representation is symmetric or antisymmetric would determine allowed transitions — the type of representation has physical meaning. For $C_2'(2')$ statefunctions have parts giving the spins. There are then four possibilities, both spin and space parts of the statefunction symmetric, both antisymmetric, or one symmetric the other antisymmetric (giving two states if, say, the particles were a proton and neutron, but still giving two were the distinction to disappear). Even for electrons the space-symmetric, spin-antisymmetric state is not the same as that of space-antisymmetric, spin-symmetric. There are important physical distinctions among these various states. Here there are four states because there are two products of two states, the ones on which the sets $\{E, C_2\}$ and $\{\theta E, \theta C_2\}$ act. If the system is, for example, that of two particles of opposite spin displaced from each other, these states are symmetric under rotations, antisymmetric under time reversal (thus under spin reversal), antisymmetric and symmetric, both symmetric, and both antisymmetric. Also, while θ and C_2 may have the same effect on a two-particle state, they do not on the system's environment, and the system's relation to its environment, so are physically different, and have different physical consequences.

VIII.4.b.ii *Dihedral color groups*

Next is the generalization of dihedral groups to their color versions [Jaswon and Rose (1983), p. 50]. These are generated by two elements [Mirman (1995a), sec. III.2.h, p. 79] so there are a maximum of three decompositions (pb. VIII.2.c-11, p. 402). With R (say) the axial mirror reflections (through the mirror perpendicular to the principle axis), we get two decompositions for $\{nm\}, n = 2,4,6$,

$$\{2mm\} = C_2 + RC_2 = \{E, C_2, R, RC_2\}, \quad C_2^2 = R^2 = E, \quad \text{(VIII.4.b.ii–1)}$$

$$\{2mm\} = \{m\} + C_2\{m\} = \{E, R, C_2, C_2R\}, \quad C_2R = RC_2. \quad \text{(VIII.4.b.ii–2)}$$

That is, we can take the unitary invariant subgroup as either (the identity plus) the two-fold rotation, or the mirror reflection. The corresponding color groups are, with the primes indicating which elements become antiunitary,

$$\{2m'm'\} = C_2 + \theta R C_2 = \{E, C_2, \theta R, \theta R C_2\}, \qquad \text{(VIII.4.b.ii-3)}$$

$$\{2'mm'\} = \{m\} + \theta C_2\{m\} = \{E, R, \theta C_2, \theta C_2 R\}, \qquad \text{(VIII.4.b.ii-4)}$$

where $\theta R, \theta R C_2$ are color reflections and θC_2 a color rotation. Likewise

$$\{4m'm'\} = (4)(E + \theta R), \qquad \text{(VIII.4.b.ii-5)}$$

$$\{4'mm'\} = \{2mm\}(E + \theta C_4), \qquad \text{(VIII.4.b.ii-6)}$$

$$\{6m'm'\} = (6)(E + \theta R), \qquad \text{(VIII.4.b.ii-7)}$$

$$\{6mm'\} = \{3m\}(E + \theta C_6). \qquad \text{(VIII.4.b.ii-8)}$$

There is only one decomposition for each of the others (the maximum is not always reached),

$$\{1m'\} = \{1\}(E + \theta R), \qquad \text{(VIII.4.b.ii-9)}$$

$$\{3m'\} = \{3\}(E + \theta R). \qquad \text{(VIII.4.b.ii-10)}$$

Problem VIII.4.b.ii-1: Are all these statements and expressions correct? Explain why some groups have fewer than the maximum number of decompositions. Illustrate this with diagrams. Is the interpretation of R relevant? This question has (at least) two meanings. Are the structures of the color groups different for different interpretations? The other is: Are the symmetries different? And this has various interpretations, such as geometrical, as with colors, or physical, as with spins or fields [Tolédano and Tolédano (1987), p. 311]. Diagrams would help answer (at least) the last question. Give answers for the different interpretations. Are there interesting distinctions?

Problem VIII.4.b.ii-2: For these expressions the International notation seems best. Restate them in Schoenflies notation.

Problem VIII.4.b.ii-3: Give physical meaning to this. Explain the meaning of elements becoming antiunitary. Why are these the only ones possible? What are the geometrical, and for those interpretations physical, differences between the various color groups obtained by different decompositions from each dihedral group?

Problem VIII.4.b.ii-4: Find the color groups obtained from the cyclic and dihedral groups with improper elements. For these, since there are additional generators, orders of the proper groups need no longer be even. Need they? Describe the symmetries given by these color groups, especially for the ones obtained from any ordinary groups that could not give color groups without the adjunction of improper elements. Do all the resultant color groups have improper elements? Why? What effect do such elements have? What are the maximum numbers of decompositions for these groups? How do these numbers compare with those for groups without improper elements? Are the maximum numbers reached? Why? What are the interpretations of the different color groups obtained from the different decompositions?

VIII.4.b.iii *Gray groups*

Antiferromagnetic crystals (ones with spins of neighboring atoms equal and opposite) [Inui, Tanabe and Onodera (1990), p. 306; Joshua (1991), p. 123; Tinkham (1964), p. 301; Tolédano and Tolédano (1987), p. 310] provide examples of the second set of 32 point groups — for these the product $\theta\tau$ is a symmetry operation; translation τ connects opposite spins, θ reverses spins. Thus besides the ordinary 32 point groups, we have another 32 groups, often called the gray point groups [Burns (1977), p. 275]. Here each point (or in the antiferromagnetic case, small region) has for example one atom with each spin direction, one up, one down, so there is an extra variable, with two values (up and down). Hence time reversal which appears as an element of these 32 groups, unlike the ordinary 32, leaves the point, and crystal, invariant. Each point is colored black and white simultaneously, giving the name gray group. The difference between an ordinary point group and the corresponding gray group is that the latter has twice the number of elements — it consists of the sum of H and θH, where H is the point group and θ is time reversal (in general an antiunitary operator), and θ and H commute, which distinguishes this class of gray point groups from the next class, the magnetic or black and white point groups. Thus for gray groups G,

$$G = H \oplus \theta H, \qquad \text{(VIII.4.b.iii-1)}$$

where the $\oplus$ indicates the direct sum of the two sets, H and θH. The direct sum is the set containing all members of the subsets, and the members of different subsets commute.

A gray group is a direct product of two invariant subgroups, that of H, and the two-element group consisting of the identity E and θ. These groups commute, which it is why it is a direct product. It is an extension of one group by the other, specifically a central extension (sec. V.5, p. 270).

What is the difference between coloring a point gray and coloring it white; why does introducing the extra operator have an effect? If each point of the crystal is replaced by two objects of opposite spin on top of each other (say two current loops, with the currents opposite; these being interchanged by time reversal), the symmetry of the crystal is not affected. But, as discussed with cyclic groups, statefunctions are basis vectors of the symmetry group. By enlarging it with the adjunction of an operator, though without changing the crystal symmetry, we can enlarge the representation basis space. Thus each basis vector is replaced by either one or two, in the latter case the basis vectors behave the same under the point group, but differently under the antiunitary operator. This then introduces (additional) degeneracy, as required by Kramer's theorem (sec. VIII.2.d, p. 402), and, for example, changing the density of states, on which transition probabilities depend. This gives content to the gray groups. Also for the corresponding space groups if the star of momentum $\underline{k}$ includes -$\underline{k}$ (which is independent of the presence of spin), then the states they label are degenerate (sec. VIII.9, p. 456). However for those gray-group cases in which this is not true, but there is time reversal invariance, there must still be this degeneracy, one imposed by the group.

Gray groups describe paramagnetic crystals, whose atoms have permanent magnetic moments, but oriented randomly, so with net moment zero. Diamagnetism is the result of magnetic moments due to current loops induced by magnetic fields external to the atom, perhaps from the rest of the crystal, and again randomly oriented. Thus it can be considered (here) together with paramagnetism.

Problem VIII.4.b.iii-1: A gray group is both a direct product, and a direct sum. How are these related?

VIII.4.c Magnetic groups

There are crystals that are changed by a combined translation and reversal of spins. For example if other atoms were added to an antiferromagnetic crystal, then even though $\theta\tau$, time reversal plus a transformation from one spin to the opposite, is a symmetry of the antiferromagnetic part it is not of the rest, so not of the entire crystal. Also [Joshua (1991), p. 117; Nowick (1995), p. 80] ferromagnetic and ferrimagnetic crystals (respectively, those with all spins aligned in a domain, and those with some aligned in one direction, some in the other, but without complete cancellation, so a net moment) do not have $\theta\tau$ as a symmetry. These are not the only possibilities for magnetic crystals — life can be complicated — but these are all we consider [Bradley and Cracknell (1972), p. 600].

So to study such cases we consider first point groups with antiunitary elements, but not containing θ (it is not a symmetry) — the (58) black-and-white (magnetic) point groups. Magnetic point groups are defined so point group M obtained from ordinary point group G is the (set) sum of H and $\theta(G-H)$, where H is an invariant unitary subgroup of G of index two (sometimes called the halving subgroup), and $G-H$ is the (equinumerous) set of the elements of G not in H. Black-and-white point group M, then is

$$M = H \oplus \theta(G-H), \qquad \text{(VIII.4.c-1)}$$

where the two subsets need not commute. What are these 58, and how do we get them?

In essence, we take each point group G in turn, find all its invariant point subgroups of index 2, and for each construct magnetic group M. The remaining elements, equal in number to those of H, are converted into antiunitary ones, and these plus those of H form magnetic group M. There are more than 32 of these — almost twice 32 — because some point groups have more than one (nonisomorphic) invariant factor group.

Like a gray group, a magnetic group is an extension of one group by the other, by the two-element group (E, θ), again a central extension (sec. V.5, p. 270); it has an invariant factor group. However it is not a direct product, the two groups need not commute, but a semidirect product. The two groups, G and M, have the same order (while a gray group has twice the order), but, when interpreted this way, are not isomorphic.

Problem VIII.4.c-1: Are there relations between the gray, magnetic, and double groups (sec. I.8, p. 59; sec. V.3, p. 256) constructed from a point group? From any groups? All? What? Why? Might any of these be of technological or economic value?

Problem VIII.4.c-2: It is interesting to speculate about other types of magnetism. For example we might consider that the spin varies along a helix (perhaps in a solid, perhaps in a liquid crystal), that is as we move along a line the spins of the atoms along it trace out a helix [Kocinski (1990), p. 269]. This line might be a rotation axis, or perhaps more intriguing, a screw. The spins along different lines may turn in the same, or opposite, senses. Or we might take a glide plane with the spins of consecutive atoms related by the glide parallel, or even antiparallel. Thus we could put spins in various ways on (some of) the atoms of say a diamond structure (sec. VII.7, p. 384). For each of these (infinite number of?) possibilities, find, or design, a crystal with that structure, or prove that such is impossible. What are the point groups, at each point, and the symmetry space groups (can there be any?) of those crystals that are possible?

VIII.4.d An example of a magnetic group

Group C_{2h} has elements E, C_2, σ_h and I, each with unit square. The gray group obtained from it has operators E, C_2, σ_h, I, θE, θC_2, $\theta\sigma_h$ and θI. What are the magnetic groups? There are three invariant subgroups with elements (E, C_2), (E, σ_h), (E, I).

Problem VIII.4.d-1: Check that these are all invariant subgroups and that there are no others. Find which, if any, commute with the elements outside it.

Problem VIII.4.d-2: Verify that E, C_2, $\theta\sigma_h$, θI form a group, a point group, and obey the restrictions for a crystal symmetry group (sec. I.5, p. 23). Find the other two magnetic groups obtained from C_{2h} and verify that these also are correct. Can any of these in addition be interpreted as gray? Why? How are these groups related? What are the physical differences between them?

VIII.4.e Construction of a magnetic group

An example of the construction of a magnetic group and an illustration of the reason a point group leads to more than one magnetic group is given by the color group of the square [Bradley and Cracknell (1972), p. 571; Burns (1977), p. 276; Joshua (1991), p. 129; Lifshitz (1997), p. 1194; Senechal (1990), p. 80; Shubnikov and Belov (1964), p. 121]. The uncolored square (sec. III.3.a, p. 138; sec. IV.2.c.iv, p. 192; sec. VII.6, p. 372) has symmetry group $C_{4v}(4mm)$ with elements E, C_{4z}, $C_{4z}{}^{-1}(= C_{4z}{}^{3})$, $C_{2z}(= C_{4z}{}^{2})$, σ_x, σ_y, σ_{da} and σ_{db}, where z is the axis perpendicular to the square [Mirman (1995a), sec. II.2.f, p. 37]. That is, the square is invariant under (respectively) rotations of $\frac{\pi}{2}$ and π around z in both positive or negative directions, as well as reflections in lines bisecting the sides of the square (the x and y axes) and the lines through its diagonals.

From its center to each corner draw a line, dividing the square into four parts, and color two opposite quarters black, leaving the other two white, which decreases the symmetry. Then E, C_{2z}, σ_x, σ_y remain symmetries, while C_{4z}, $C_{4z}{}^{-1}$, σ_{da} and σ_{db} also interchange black and white. With θ the operation interchanging black and white, operators θC_{4z}, $\theta C_{4z}{}^{-1}$, $\theta\sigma_{da}$ and $\theta\sigma_{db}$ are also symmetries. Thus θC_{4z} rotates the square by $\frac{\pi}{2}$, so changing black to white, and then θ changes the color back, and the square is unchanged. Hence the colored square also has an eight element symmetry group, but a different one, as we see on the left in

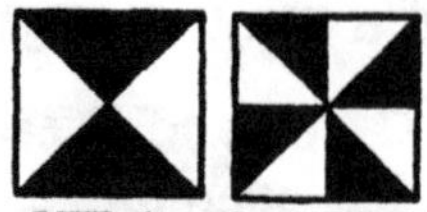

Figure VIII.4.e-1: COLORED SQUARE.

This group (the magnetic group of the square) is {4′mm′}, with the primes referring to the antiunitary operators.

Problem VIII.4.e-1: Find the group tables of the groups, {4mm} and {4′mm′}, and verify that they are different. To what (ordinary) point group is {4′mm′} (abstractly) isomorphic?

Problem VIII.4.e-2: What are the other ways of dividing the square into quarters? What symmetry groups do these give? Are they the same?

VIII.4.e.i *The square has other magnetic groups*

Suppose that instead of this coloring we divided the square into eight parts, by bisecting each of the four of the previous example giving four black strips and four white ones, as on the right of the diagram. How does the symmetry group change? Why?

Problem VIII.4.e.i-1: Show that the symmetry group, now {4m′m′} (why?), has elements E, C_{4z}, C_{4z}^{-1}, C_{2z}, $\theta\sigma_x$, $\theta\sigma_y$, $\theta\sigma_{da}$ and $\theta\sigma_{db}$. Find the group table and check that these three groups of the square are nonisomorphic. Also if for the first case we take, instead of the division there, one obtained by rotating the axes by $\frac{\pi}{4}$ (so the lines between the segments lie not on the diagonal, but on the bisectors to the sides) show that we get a group, isomorphic to {4′mm′}, with elements E, θC_{4z}, θC_{4z}^{-1}, C_{2z}, $\theta\sigma_x$, $\theta\sigma_y$, σ_{da} and σ_{db}. To what (ordinary) point groups are these (abstractly) isomorphic?

Problem VIII.4.e.i-2: Give the three groups (white, gray, and black and white) of an equilateral triangle: group elements, multiplication table, and a diagram showing objects having these groups [Ludwig and Falter (1988), p. 26].

Problem VIII.4.e.i-3: Go through the list of 32 point groups (sec. I.6.d, p. 42; tbl. B-2, p. 634) and for each construct all its magnetic groups. Show that there are 58 [Bradley and Cracknell (1972), p. 573; Burns (1977), p. 278; Tinkham (1964), p. 304], giving a total of 122 (colored) point groups altogether (sec. VIII.4.b, p. 414).

VIII.4.e.ii *Why the square has these magnetic groups*

Looking at the square could we guess its magnetic groups? The elements of its uncolored symmetry group are in five classes: E; C_{2z}; C_{4z}, C_{4z}^{-1}; σ_x, σ_y; σ_{da}, σ_{db}. To find the magnetic groups the elements are divided into two sets, with one an invariant subgroup, so it contains only complete classes. The sets have equal numbers of elements since there are equal numbers of black and white regions (the antiunitary operator interchanges black and white and could not give products that are symmetries unless the numbers of regions are equal). The invariant subgroup is a symmetry of each of the two sets of regions alone.

The subgroup then contains classes E and C_{2z}, and any one of the others, giving three possibilities. One has three rotations, the other two a single rotation and two reflections. Since the product of a rotation and reflection is a reflection of the same type (through a bisector, or along a diagonal), these two subgroups are isomorphic. But the multiplication table of rotations is not the same as that of reflections, or reflections and rotations. Thus one subgroup is not isomorphic to the other two.

Two of the magnetic groups then are isomorphic, the third is not. Of course its group table can be relabeled, rearranging the rows and columns (so that those of the subgroup no longer take up the first four) to make this group isomorphic to the other two. What determines then whether magnetic groups obtained from the same point group are isomorphic is not their group tables, since these are actually the same as that of the point group — all these magnetic groups are (abstractly) isomorphic in this sense — but the choice of the invariant subgroup.

Looking at the square we see that it is divided into eight parts that are mixed by the symmetry operations. Half are black. Which? We can color alternate parts black, giving eight distinguishable regions, two neighboring ones, giving four such regions, or four neighboring ones giving two regions, say the upper half black, the lower white. However the last is not symmetric under a magnetic group because the number of elements that leave regions of the same color invariant is only two. For the other two cases it is four, as it need be.

Thus we get three patterns, one of eight regions, and two of four, these triangular in two cases, square in the other. It is not surprising that the magnetic group of one is nonisomorphic to those of the other two, while they are isomorphic to each other. But how can we be sure of that just by looking at the square?

Problem VIII.4.e.ii–1: One coloring was not drawn. Draw it.

Problem VIII.4.e.ii–2: Why are all these magnetic groups (abstractly) isomorphic?

Problem VIII.4.e.ii–3: Note first that the shape of the colored regions does not matter. The corners (and midpoints) can be labeled (perhaps by putting spinning particles on them), and the symmetry group then acts by shuffling these labels. Do this and find the action of the group, and then of the three magnetic groups derived from it, on these sets of labels. Show that the action of two of the groups makes them isomorphic, that of the third is different, so that group is nonisomorphic to the other two. In particular note the action of rotations and of reflections. Check that it does not matter whether the reflections are in lines through the midpoints or along the diagonals, nor is this relevant to the action of the rotations on the lines. Also check, both from this labeling and from a diagram of a square, that the product of two reflections differs from the product of a reflection and a rotation. Looking at the two diagrams with eight, and with four, regions distinguished, we see that the first is invariant under all four symmetry rotations of the square, while the second is not, this being invariant under one of the rotations and two reflections, and these two differ in the two cases. However groups containing all rotations are not isomorphic to ones containing both rotations and reflections. Hence the magnetic group for the eight-region diagram cannot be isomorphic to those of the other two, and a slightly closer look shows that these are isomorphic to each other.

Problem VIII.4.e.ii-4: The magnetic groups, although isomorphic as abstract groups, are taken as distinct because they are physically different. In particular the colors are not really of interest, they just represent the directions of spin of the particles. Reinterpret the colors in terms of spins. Find models of the eight-region and the two four-region cases. Are they physically different (in some important way)? Does it make sense to take one magnetic group nonisomorphic to the other two, with these mutually isomorphic? Groups can be isomorphic abstractly, but not physically. and it is important to understand why, and what a reasonable definition of isomorphism is, in any situation.

Problem VIII.4.e.ii-5: Repeat this analysis for a rectangle.

Problem VIII.4.e.ii-6: Repeat it also for a hexagon.

VIII.4.f The orthorhombic magnetic crystal

How is a three-dimensional object colored? The orthorhombic unit cell has symmetry group $D_{2h}(\frac{222}{mmm})$ with elements E, C_{2x}, C_{2y}, C_{2z}, σ_x, σ_y, σ_z and I. The cell on the left of the next diagram is invariant under rotations of π about any axis perpendicular to a pair of faces and through their centers (one such line is shown, dashed), and under reflections in planes perpendicular to, and bisecting, a pair of faces (indicated by a glazed plane), as well as inversion through the center (pb. II.4.c-3, p. 87; sec. II.6.a, p. 103). If the structure were a ferromagnet [Tinkham (1964), p. 303], so that the atoms at each of the eight corners had spins, taken as along a side of the lattice (all in the same direction, parallel to z, since it is ferromagnetic), the symmetry would be reduced. How? Operations C_{2x} and C_{2y}, as well as σ_x and σ_y, reverse the spins;

C_{2z} and σ_z, and $\mathcal{I}$, leave them unchanged — angular momenta are axial (pseudo-) vectors, so are unchanged by these reflections and inversions, as we see,

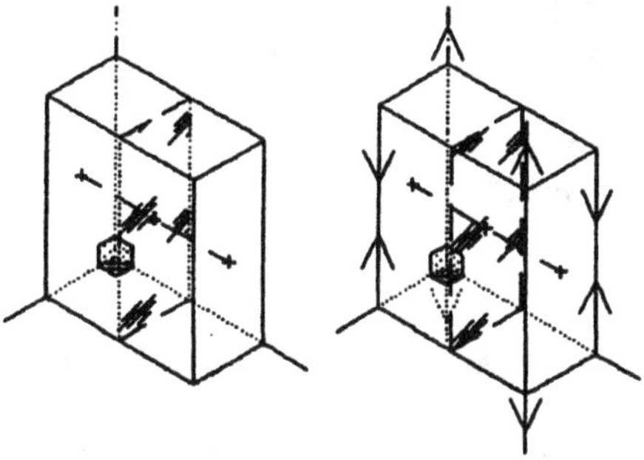

Figure VIII.4.f-1: ANTIFERROMAGNETIC ORTHORHOMBIC CELL.

The group containing E, C_{2z}, σ_z and $\mathcal{I}$ is $C_{2h}(\frac{2}{m})$.

Problem VIII.4.f-1: Verify that the symmetry group of this ferromagnetic orthorhombus has the elements E, C_{2z}, σ_z and $\mathcal{I}$, θC_{2x}, θC_{2y}, $\theta\sigma_x$, $\theta\sigma_y$, and that this satisfies all requirements for a magnetic point group. Check also if it obeys the product rules of $D_{2h}(\frac{222}{mmm})$, with the unitary subgroup being C_{2h}. The group is $D_{2h}'(\frac{222}{m'm'm})$, which is non-Abelian so the two sets, H and $\theta(G-H)$, do not commute. Which pairs of elements from each of the two sets do not commute? How do the commutation relationship compare to that of D_{2h}? Show this also using a model.

Problem VIII.4.f-2: Construct a match box (a right prism whose face is a parallelogram), and show that its symmetry group is $C_{2h}(\frac{2}{m})$ [Shubnikov and Koptsik (1974), p. 237]. Give the permutations (of what?) to which its operations correspond. What gray, and what magnetic, groups can be obtained from this?

Problem VIII.4.f-3: Group D_{2h} allows two other crystals, $(\frac{222}{mmm'})$, with unitary subgroup C_{2v} and $(\frac{222}{m'm'm'})$ with unitary subgroup D_2, this being antiferromagnetic. Thus three magnetic (black and white) groups are obtained from it [Burns (1977), p. 279]. Verify this, giving the elements of the unitary subgroups in these two cases, and draw diagrams showing the spin directions, or construct models.

Problem VIII.4.f-4: On the right of the figure is an orthorhombic antiferromagnet with the spins shown. What is its symmetry group? Can it be replaced by a cell having, not spins, but volumes that are colored, and with the same symmetry group?

Problem VIII.4.f-5: The lines in the diagram are not exactly straight. To what extent does this affect the symmetry? We consider crystals as sets of points placed so that translations take them to another point for which the environment is exactly the same — the positions (or in quantum mechanics the statefunctions) are invariant, to infinite precision, under these translations. Of course, in reality this is never true. Even ignoring defects, and the impossibility of placing physical objects at exact positions, crystals are at finite temperatures, so the objects within them are in constant motion. Is there still symmetry? In what sense? Is the waviness of the lines a good representation of a crystal at a finite temperature? Why? Is it a good representation of a crystal with defects?

VIII.4.g What determines which groups are different?

The physical meanings, and the realizations, of the different magnetic groups are different, so they are regarded as different, even though abstractly they are not. The number of magnetic point groups is greater than the number of allowed abstract crystallographic groups, so their distinction must result, not only from the differences between abstract groups, but from the way they are defined, used and realized.

Problem VIII.4.g-1: There are point groups that describe symmetries of ferromagnetic crystals, but only some [Burns (1977), p. 279]. Why? Which are these? There are others for antiferromagnetic crystals. Is there a difference for ferrimagnetic crystals? Determine which groups go with which crystals, and, perhaps, which can be used for none.

Problem VIII.4.g-2: Here we have been using the International notation more than usual. Which notation more clearly indicates the symmetry here, this or the Schoenflies?

VIII.5 MAGNETIC BRAVAIS LATTICES

Space groups and their representations are built upon Bravais lattices. For color space groups we need such lattices again, but with changes of color, the set is enlarged and enriched. When points (regions) have different colors, the number and (in a certain sense) types of lattices are greater. And there is also a new class of operators, the color translation operators (connecting points of different color) [Jaswon and Rose (1983), p. 90]. This leads to color (or magnetic) Bravais lattices [Jaswon and Rose (1983), p. 96; Lockwood and Macmillan (1983), p. 203; Ludwig and Falter (1988), p. 38; Tolédano and Tolédano (1987), p. 312]. Understanding these lattices and the crystals that have the symmetries given by them [Janssen (1973), p. 222; Joshua (1991), p. 133, 211] is our next task although all we do here is outline and indicate some of the questions.

The magnetic Bravais lattices consist of the fourteen uncolored lattices, which we have considered (sec. II.5, p. 98), plus a set of two interpenetrating lattices, one black, the other white, twenty-two in number, giving thirty-six altogether [Bradley and Cracknell (1972), p. 583; Jaswon and Rose (1983), p. 90; Joshua (1991), p. 211; Shubnikov and Belov (1964), p. 176; Shubnikov and Koptsik (1974), p. 299]. Members of each set may themselves consist of interpenetrating lattices. The reason that there are more such lattices than for the uncolored case is that they have an additional primitive lattice vector, that connecting the lattices of different color, reminiscent of centered lattices. By definition, the environment of a point is unchanged by translation by a primitive lattice vector. Here we are doing something new, leaving the environment unchanged except for a spin flip. This primitive vector then connects points that are the same except for a spin flip (or color change).

Magnetic lattices are combinations of the uncolored lattices, so we find them by considering all such unions of uncolored lattices going with the same symmetry group — all are constructed from the same primitive lattice. Each colored primitive lattice contains one black, and one white, primitive lattice, and except for color these are identical.

Colored lattices must remain lattices if black fades to white (or the spin become zero) — if the distinction disappears. This determines the possible lattices. Each of the two unit cells becomes unit cells of the uncolored lattice. Thus it contains two units cells — each of its units are half that. Magnetic lattices are therefore defined such that a primitive translation takes a point to another with the same environment, except that there is spin flip. A colored lattice is of the same lattice system as the uncolored one from which it is constructed, and identical to it, differing only in that points are so distinguished.

Problem VIII.5-1: Is it possible to combine two different lattices into a colored lattice (so that they do not form a lattice if the color fades) and still get a lattice? Does the answer depend on whether the lattices go with the same point group? What would happen if they do but one was primitive, the other not (it has structure)?

Problem VIII.5-2: A lattice is a group (pb. II.2.a-3, p. 68). Is this true for a colored lattice also? If so, what type of group is it? Are there (important) distinctions between colored and uncolored lattices when considered as groups?

Problem VIII.5-3: There are various extensions of this that might be interesting to consider, such as finding all lattices with k colors, for arbitrary k, and all in n dimensions, for arbitrary n (including spaces with different signatures, if possible). Perhaps computer programs could be useful. Since lattices are groups, these provide further types of groups (although ones of different k and n might be isomorphic, which is something to look at). Does this help classify groups? What information does it provide about them? is it a useful means of studying groups?

VIII.5.a Properties of the magnetic Bravais lattices

We leave it as a problem for the reader to find these lattices only noting several of their properties [Shubnikov and Belov (1964), p. 176], and briefly describing them. First, the sum of two consecutive color translations (going from one lattice point to a point with the other color, so connecting lattices of different colors) is equivalent to an uncolored translation (connecting two points of the same lattice) in the same direction. The length of a shortest color lattice translation is thus half that of the shortest uncolored lattice translation in the same direction. Also lattice points of opposite sign (color) can be only in the center of an edge, face or cell of one of the fourteen uncolored Bravais lattices. That is, the only way of getting a colored lattice is to take one of these fourteen and put a dot of opposite color at one of these points, and draw the same primitive lattice vectors as the uncolored lattice starting from that.

Problem VIII.5.a-1: Prove these statements.

Problem VIII.5.a-2: Additional lattice points of opposite sign cannot be at $\frac{1}{4}$ of a plane diagonal when there is an uncolored centering of a face, or of $\frac{1}{4}$ of a space diagonal when there is uncolored body centering. Why?

Problem VIII.5.a-3: If the face of a cell of a Bravais lattice is centered, additional lattice points of opposite sign at $\frac{1}{4}$ of both plane diagonals of this face give a two-fold reduction of the shortest translations along these edges which have previously been chosen shortest. Likewise having such lattice points on one diagonal is also impossible. If there is a centered face in the lattice it is rectangular and parallel even if a two-fold axis or normal to the symmetry plane is along an edge. In the first case there is a rotation that reverses the oblique translation to bring it along a second diagonal, so again halving the shortest uncolored translation. For the second case, a mirror reflects the colored translation to bring it along a second diagonal.

Problem VIII.5.a-4: Show that these conditions also are required with a screw axis and with a glide plane.

Problem VIII.5.a-5: Draw diagrams for each lattice, illustrating these.

Problem VIII.5.a-6: How is the procedure for finding colored lattices the same as, and different from, that for finding centered lattices? To what extent are the results the same, and to what extent are they different? Why?

VIII.5.b Two-dimensional magnetic Bravais lattices

To give a brief description we consider first the two-dimensional colored Bravais lattices [Jaswon and Rose (1983), p. 92, 181; Shubnikov and Belov (1964), p. 211]. For each of the five uncolored lattices (sec. II.3.a, p. 71) we have to add, if we can, a second to give a colored one.

For the oblique parallelogram we put the second lattice in its center, or at the midpoints of opposite sides. The first is ruled out because a lattice must be invariant (at least) under rotations of π about any of its lattice points. The second placement gives a colored lattice.

A rectangular lattice gives a colored one by putting colored points at the centers of an opposite pair of sides, both opposite pairs of sides, and at the center of the cell. There are three colored lattices of this system. The centered rectangular lattice is similar. For the square a colored point is put at the center of the cell, so there is one colored lattice for this uncolored one. (Putting points in the middle of the edges gives smaller lattices, so nothing new.)

There are thus ten two-dimensional colored lattices.

Problem VIII.5.b-1: Compare this with centering; explain the similarities and differences.

Problem VIII.5.b-2: Draw, and (so) check, these.

Problem VIII.5.b-3: Since we now know how the Bravais lattices are related to the axioms of geometry (pb. II.4.h-6, p. 97; pb. II.7.f-10, p. 123; pb. III.4.c.ii-2, p. 152; pb. IV.3.h-8, p. 211), it should be simple to relate these (?) axioms to the colored lattices.

Problem VIII.5.b-4: What are the Wigner-Seitz cells for these lattices? How does coloring affect them?

VIII.5.c Three-dimensional magnetic Bravais lattices

Next we briefly turn to the three-dimensional lattices [Shubnikov and Belov (1964), p. 180].

For the triclinic, if another lattice were put in the center, or the center of a face, it would reduce under fading to a lattice which is not possible. Hence the only colored triclinic lattices have the corners of one lattice at the midpoints of four of the lines joining the corners of the other lattice. Any set of four is possible, but all resultant lattices are the same. Making the colors identical gives a triclinic lattice with unit cell half the size, but otherwise identical to the unit cell from which we constructed the colored lattice. The triclinic thus has two magnetic lattices, one uncolored, and another with black displaced from white along the edges to their midpoints.

The monoclinic system, whose unit cells have two sides orthogonal, has two uncolored lattices, the primitive and one with a unit cell of three points, with two in the centers of opposite faces. How can colored lattices be obtained from them? For the primitive lattice colored points can be put at the midpoints of four parallel lines, which if color were ignored would give a unit cell with edges half the length of the original uncolored cell. There are two pairs of such parallel lines (distinguished by whether the angles are $\frac{\pi}{2}$ or not), so two colored lattices. Another way of seeing this is that colored lattices are formed from two intersecting uncolored ones (of different colors); here there are two colored ones as the pairs of faces on which the two lattices intersect are different. In addition, colored points can be put in the centers of two of the faces of the unit cell. Hence from the primitive lattice, three colored ones can be constructed. In addition since the monoclinic allows face centering, a face-centered black lattice can be placed so that the corners of its unit cell are at the midpoints of four of the edges of the white lattice and there can also be face-centered white lattices with the black lattices interpenetrating so its six corners are at the midpoints of the six edges. The monoclinic system gives seven magnetic lattices, the two uncolored ones, and five black and white ones.

The orthorhombic system has eighteen lattices, of which five are uncolored, the primitive, the body-centered, two two-face-centered, and the all-face-centered. The uncolored primitive lattice gives five colored ones. For the first two, colored points are placed at the centers of four parallel edges giving a lattice which would be identical to the primitive one if color is ignored, but with half the height. The next two have colored points at the centers of a pair of opposite faces and go to the two-face-centered nonprimitive lattice if the color fades. Finally a colored point can be put at the center of the unit cell, going to the body-centered nonprimitive cell which gives a colored lattice by the emplacement of colored points in the centers of one pair of opposite faces, and four at the midpoints of four parallel lines. It reduces to a two-face-centered nonprimitive cell, of smaller size, when the colors become the same. There are two of these latter uncolored cells. They give six colored ones. Each gives a cell consisting of identical black and white parallel cells, displaced from each other by half the length of the uncolored cell. Another pair of cells has eight black points at the midpoints of the four edges of each of a pair of opposite faces. Another pair has four black points at the centers of the four faces not occupied by the centering white points. The all-face-centered cell gives one colored cell, in which black points are placed in the middle of all twelve edges and at the center of the cell.

The tetragonal system has six lattices, two uncolored, a primitive, and a body-centered, one. The primitive gives three colored cells, again one with four colored points at the centers of four parallel edges, at the centers of one pair of opposite faces, and at the center of the cell. The body-centered one gives one colored cell, with colored points at the centers of four parallel edges, and a pair at the centers of the two opposite faces perpendicular to these edges.

For the hexagonal (and rhombohedral or trigonal) set there are four lattices, two uncolored (the primitive and the rhombohedral) and two colored, one obtained from the primitive again by centering four edges and one obtained from the rhombohedral, by putting a point at the center of the cell.

There are finally five cubic lattices, three uncolored and only two colored. One colored lattice has a colored point at the center of the cell, reducing (when color fades) to the body-centered lattice, while the other obtained from the face-centered cell has colored points at the centers of all edges and of the cell. It reduces to a primitive cubic lattice.

The total number of magnetic lattices is then thirty six, the fourteen ordinary Bravais lattices, and twenty-two black and white ones.

Problem VIII.5.c-1: Check these, with diagrams, models, computer programs, and analytically. Are all these lattices really different (whatever this means), or are they just obtained in different ways?

Problem VIII.5.c-2: Compare this with centering; explain the similarities and differences.

Problem VIII.5.c-3: How are these results related to the axioms of geometry? Are there fundamental differences for spaces of other dimensions? Are there corresponding results for spaces that are not Euclidean?

Problem VIII.5.c-4: What are the Wigner-Seitz cells for these lattices? How does coloring affect them? If we use a finer classification, distinguishing lattices also by their Wigner-Seitz cells (sec. IV.4, p. 212), how many lattices are there? What are the relationships between these and the uncolored lattices? How are the classifications for the colored and uncolored lattices related? How does one type of Wigner-Seitz cell go into another as the parameters of the lattice are varied, and as color fades?

VIII.6 MAGNETIC SPACE GROUPS

We are now ready for the color (magnetic) space groups [Bradley and Cracknell (1972), p. 574; Evarestov and Smirnov (1993), p. 237; Inui, *Tanabe and Onodera (1990), p. 306; Jaswon and Rose (1983), p. 144;* Joshua (1991), p. 134, 215; Lifshitz (1997), p. 1196; Ludwig and Falter

(1988), p. 38]. As with the uncolored groups, these are constructed from the point and translation groups, here the colored point groups and colored Bravais lattices. While there are three types of colored point groups, combining them with colored lattices gives four types of colored space groups. We only list the four types, leaving the discussion of these groups, and their relationships to the colored lattices, to the references, and the reader.

VIII.6.a Types of colored space groups

The magnetic space groups, M, are constructed from the ordinary space groups G, its halving group H (sec. VIII.4.c, p. 419), and antiunitary operator θ. There are four types:

I: The ordinary (colorless) space groups of which there are 230 (the Fedorov space groups):

$$M = G. \tag{VIII.6.a-1}$$

II: The gray space groups of which there are (clearly also) 230:

$$M_g = G \oplus \theta G. \tag{VIII.6.a-2}$$

III: The black and white space groups based on the ordinary Bravais lattices of which there are 674, defined similarly to the black and white point groups (sec. VIII.4.c, p. 419):

$$M_{bwo} = H \oplus \theta(G - H). \tag{VIII.6.a-3}$$

IV: The black and white space groups based on the black and white (colored) Bravais lattices of which there are 517:

$$M_{bwc} = G \oplus \tau\theta G, \tag{VIII.6.a-4}$$

where τ is the color translation operator introduced by the colored Bravais lattices, which is needed since there has to be a translation connecting colored and uncolored points (sec. VIII.5.a, p. 428). There are thus a total of

$$230 + 230 + 674 + 517 = 1651 \tag{VIII.6.a-5}$$

space groups (Shubnikov space groups, sometimes Heesch-Shubnikov groups [Cracknell (1975), p. 6]; with corresponding terms for point groups). These are considered in depth in various places [Shubnikov and Belov (1964), p. 175; Shubnikov and Koptsik (1974), p. 269]. The uncolored point and space groups are also known as Fedorov groups.

The ordinary space groups have been discussed, and the gray ones are constructed from the gray point groups in the same way that the ordinary space groups are constructed from the ordinary point groups.

While type III space groups are given exactly as the corresponding magnetic point groups, with the symbols now referring to the space groups, the definition of a type IV group is more comparable to that of the gray point group,

$$M = G \oplus \theta\{E|\tau\}G, \tag{VIII.6.a-6}$$

where G is an ordinary point group, and now there is a translation associated with the antiunitary operators. Types III and IV differ in that the former has no translation, the latter does.

VIII.6.b A space group obtained from the primitive cubic lattice

As an example, consider the primitive cubic lattice and color the corners black. Then add a white point in the center. The resulting lattice consists of two interpenetrating lattices — one black, one white — each an ordinary primitive cubic lattice. The symmetry group then consists of the symmetries of the ordinary primitive cubic lattice, plus another set. Here τ is the vector that is half the body diagonal of the cube — it connects a corner to the center. This translation moves the black points to the white ones and white to black. Then θ interchanges black and white (or spin direction), returning the lattice to the initial one. Any of the cube's symmetry operations still leave the lattice unchanged. This gives the second set of symmetries.

Problem VIII.6.b-1: Check that the translation, the color reversal and the set of cubic symmetries all commute, so it does not matter in which order we perform these operations. Verify all these statements.

VIII.6.c Implications and extensions

While it is obvious that this discussion is a mere mention of these groups, even in the references the understanding of them is not complete (nor ever will be). It is therefore useful to consider some possible extensions. That they are interesting is clear, that it is possible to actually develop them is much less so.

Problem VIII.6.c-1: This assumes (probably) that the point groups are the ordinary point groups. How would it change if double point groups are used? What are the magnetic space groups then? Are there significant differences?

Problem VIII.6.c-2: Is there any space, of n dimensions, that has 1651 ordinary space groups?

Problem VIII.6.c-3: How many color groups are there in an n-dimensional space?

Problem VIII.6.c-4: Can there be such groups in pseudo-Euclidean space — one with nonpositive signature (eq. IV.3.h-1, p. 211), in particular with dimension 3+1? Can there be such groups in non-Euclidean spaces? How many?

Problem VIII.6.c-5: A lattice is a group. Space groups are also infinite discrete groups (pb. III.9.c-7, p. 184). To what extent do magnetic space groups (help) classify such groups? Are all infinite discrete groups (magnetic) space groups in the proper dimension, or might there be others that cannot be so classified? Can these be used to classify and study lattices in arbitrary dimensions? How do the answers differ from those for ordinary space groups? For groups based on double point groups?

VIII.7 REPRESENTATIONS OF GROUPS WITH ANTILINEAR OPERATORS

The aspect of group theory essential to applications is (usually) that of representations. We have now introduced a new type of group (realization), one with antilinear (so antiunitary) operators — here specifically time reversal and products with other group elements. Do such groups have representations; how are these (if any) affected by the new operators? If they do not have representations, how can they be used, say for physical applications? And how are the group and physical properties affected by the presence of (realizations by) antiunitary operators?

To see that there are actually representations of groups with antiunitary operators (fortunately), and what these mean, we start with the unitary operators S_μ of group G, these forming an invariant unitary subgroup (the subgroup with only unitary operators) of index 2 — this has representations — plus the operators θS_μ; the group consists of these two sets. Then, with the matrices determined by the action of these operators denoted by Δ (not — yet? — representation matrices),

$$S_\mu \psi_i = \sum \Delta(S_\mu)_{ik}\psi_k, \qquad \text{(VIII.7-1)}$$

$$\theta S_\mu \psi_i = \sum \Delta(\theta S_\mu)_{ik}\psi_k = \theta \sum \Delta(S_\mu)_{ik}\psi_k = \sum \Delta(S_\mu)^*_{ik}\theta\psi_k, \qquad \text{(VIII.7-2)}$$

since both θ and S acting on a member of the set of basis vectors of the unitary subgroup gives a sum over the members of that set. The matrices for the S subset, the invariant subgroup, are familiar; the θS set is new and must be explored.

Problem VIII.7-1: Any set of vectors with a scalar product can be orthonormalized (for example by the Gram-Schmidt procedure [Arfken (1970), p. 437; Courant and Hilbert (1955), p. 4, 50; Margenau and Murphy (1955), p. 298]). Assuming that the ψ's are orthonormal — they are the basis vectors of the invariant subgroup [Mirman (1995a), sec. VII.5, p. 193] — show that the matrices Δ are unitary. However the Δ's do not form a representation of G. For them to do so, they must obey the same product rules as the group operators. Now

$$(\theta S_\mu)S_\nu\psi_i = \sum \Delta(\theta S_\mu S_\nu)_{il}\psi_l = \sum(\Delta(S_\nu)\Delta(\theta S_\mu))_{il}\psi_l$$

$$= \sum \Delta(\theta S_\mu)_{ij}\Delta(S_\nu)^*_{jl}\psi_l, \qquad \text{(VIII.7-3)}$$

because the Δ's have complex entries (in general) which cannot be eliminated. Why? Also

$$\theta S_\mu S_\nu = S_\mu S_\nu \theta, \qquad \text{(VIII.7-4)}$$

since θ commutes with the unitary operators. But check that

$$\Delta(\theta S_\mu)\Delta(S_\nu) = \Delta(S_\mu)^*\Delta(\theta S_\nu) = \Delta(\theta S_\mu S_\nu)^* \neq \Delta(\theta S_\mu S_\nu). \qquad \text{(VIII.7-5)}$$

Thus the matrices do not have the same product as group G. Groups with antilinear operators do not (seem to) have representations, in the usual sense. Since representations are needed for physical applications, and certainly groups containing the time reversal operator have physical applications, we have a problem, and we must find a way of dealing with it. What? Is this reminiscent of any previous discussion?

VIII.7.a Corepresentations

What are the entities for these groups that are the equivalent of representations? They are corepresentations — the irreducible representations of the nonunitary (antiunitary) groups, ones with antiunitary operators [Birman (1974), p. 180; Bradley and Cracknell (1972), p. 610; Evarestov and Smirnov (1993), p. 241; Inui, Tanabe and Onodera (1990), p. 294; Jansen and Boon (1967), p. 162; Janssen (1973), p. 223; Joshua (1991), p. 135; Kocinski (1983), p. 86; Kocinski (1990), p. 37; Ludwig and Falter (1988), p. 91, 222; Tolédano and Tolédano (1987), p. 313; Wigner (1959), p. 335]. What are they, how do they differ, if they do, from representations, are their properties so different that we have to construct a whole new theory, or can we just change some language and symbols?

We want the corepresentations of antiunitary group G, with invariant unitary subgroup H, of index 2. Taking A as any antiunitary operator of it (time reversal, say, or the product of it with an element of H), the group is a direct sum of cosets,

$$G = H \oplus AH, \quad \text{(VIII.7.a-1)}$$

with elements of H, denoted by R, unitary, those of AH, whose members we write as AR, antiunitary. It is a direct sum because the subgroups H and AH commute (a group commutes with itself, although its elements may not commute); clearly time reversal, for example, commutes with those elements that are independent of it (the ones in H).

Problem VIII.7.a-1: Explain why any element of AH, any coset representative, can be chosen as A.

Problem VIII.7.a-2: Why would we expect that these are projective representations (of what?) (sec. V.4, p. 259)?

VIII.7.b Multiplication rules for corepresentations

Representations are defined as matrices whose products are the same as those of the group. However matrices of the corepresentations obey the multiplication rules

$$D(R_1)D(R_2) = D(R_1R_2), \quad \text{(VIII.7.b-1)}$$

$$D(R_1)D(A) = D(R_1A), \quad \text{(VIII.7.b-2)}$$

$$D(A)D(R_1)^* = D(AR_1), \quad \text{(VIII.7.b-3)}$$

$$D(AR_1)D(AR_2)^* = D(AR_1AR_2). \quad \text{(VIII.7.b-4)}$$

We now have to find matrices satisfying these rules, in particular we have to relate the D matrices of the corepresentation to the representation matrices, Δ, of the unitary subgroup.

Problem VIII.7.b-1: The multiplication rules can easily be found. Check that acting on a set of basis vectors, ψ_i, the transformations have the representation matrices

$$R_1R_2\psi_k = \sum[D(R_1)D(R_2)]_{kj}\psi_j, \quad \text{(VIII.7.b-5)}$$

$$RA\psi_k = \sum[D(R)D(A)]_{kj}\psi_j, \quad \text{(VIII.7.b-6)}$$

$$AR\psi_k = \sum[D(A)D(R)^*]_{kj}\psi_j, \quad \text{(VIII.7.b-7)}$$

$$A_1A_2\psi_k = \sum[D(A_1)D(A_2)^*]_{kj}\psi_j. \quad \text{(VIII.7.b-8)}$$

These show again that the representations are not ordinary representations. In particular A and R commute, but the actions of AR and RA differ.

VIII.7.c Representation Spaces

To find what corepresentations are, and why, we start by asking how the representation space of an antiunitary group differs from that of a unitary one, if it does [Bradley and Cracknell (1972), p. 609]?

The basis vectors of a representation of the unitary subgroup are denoted by ψ_i^r, with these transforming under matrices $\Delta^r(R)$, where r labels the representation (which we can suppress). From ψ_i^r we obtain the set of vectors

$$\theta\psi_i^r = \phi_i^r, \qquad \text{(VIII.7.c-1)}$$

in one-to-one correspondence with the ψ's. How do the ϕ_i^r transform under the unitary subgroup? We have

$$R\phi_i^r = R\theta\psi_i^r = \theta R\psi_i^r = \theta\sum\Delta(R)_{ik}\psi_k^r = \sum\Delta(R)_{ik}^*\phi_k^r. \qquad \text{(VIII.7.c-2)}$$

Thus the ϕ's transform according to the complex conjugate representation; being of different representations (if these are) the ϕ's and ψ's are mutually orthogonal. Combining these two sets (if they are distinct) we have the basis vectors of a representation of the whole group, which we now consider in detail.

VIII.7.d Constructing the corepresentations

To find the corepresentations of group G with invariant unitary subgroup H, which has operators R, we need only one, any one, of the antiunitary operators which we call A [Wigner (1959), p. 340]. For a representation of H of dimension d, the basis vectors are written as

$$|\psi> = |\psi_1, \psi_2, ..., \psi_d>, \qquad \text{(VIII.7.d-1)}$$

where $|\,>$ indicates a set, and for element R of H,

$$R\psi_i = \sum_1^d \Delta(R)_{ij}\psi_j, \qquad \text{(VIII.7.d-2)}$$

with Δ the representation matrix, which can be abbreviated

$$R|\psi> = \Delta(R)|\psi>. \qquad \text{(VIII.7.d-3)}$$

The interesting question is the effect of antilinear operator A. We define the functions

$$|\phi> = |\phi_1, ..., \phi_d>, \qquad \text{(VIII.7.d-4)}$$

by

$$|\phi> = A|\psi>. \qquad \text{(VIII.7.d-5)}$$

If it is not obvious that the space $|\phi>, |\psi>$ — the direct sum of the two spaces (unless they are identical), thus of twice the dimension of each — is invariant under G, we consider why it should be, thereby getting the representation matrices on this entire space. We have to find the action of unitary element R on $|\phi>$, which is

$$R|\phi>= RA|\psi>= A(A^{-1}RA)|\psi>$$

$$= A\Delta(A^{-1}RA)|\psi>= \Delta(A^{-1}RA)^*|\phi>, \qquad \text{(VIII.7.d-6)}$$

since $A^{-1}RA$ belongs to H, and A is antilinear (eq. VIII.7.c-2, p. 437). This gives (assuming the representation is complex), with the first row acting on the first term in $|>$, the second row on the second,

$$R|\psi,\phi>= \begin{pmatrix} \Delta(R) & 0 \\ 0 & \Delta(A^{-1}RA)^* \end{pmatrix} |\psi,\phi>, \qquad \text{(VIII.7.d-7)}$$

supplying the matrices for all elements of H, and we now need those for the antiunitary coset. Consider some member of this set

$$B = AR; \qquad \text{(VIII.7.d-8)}$$

we have

$$B|\psi>= D(B)|\psi>, \qquad \text{(VIII.7.d-9)}$$

where $D(B)$ is the representation matrix to be determined, and

$$B|\psi>= AR|\psi>= A\Delta(R)|\psi>= \Delta(R)^*|\phi>= \Delta(A^{-1}B)^*|\phi>, \qquad \text{(VIII.7.d-10)}$$

so

$$B|\phi>= BA|\psi>= \Delta(BA)|\psi>, \qquad \text{(VIII.7.d-11)}$$

as BA is a member of H. Combining gives

$$B|\psi,\phi>= \begin{pmatrix} 0 & \Delta(BA) \\ \Delta(A^{-1}B)^* & 0 \end{pmatrix} |\psi,\phi>, \qquad \text{(VIII.7.d-12)}$$

which shows that this space $|\psi,\phi>$ goes into itself under all operators of the group.

Thus the representation matrices for all group transformations are

$$D(R) = \begin{pmatrix} \Delta(R) & 0 \\ 0 & \Delta(A^{-1}RA)^* \end{pmatrix}, \quad \text{all } R \text{ in } H, \qquad \text{(VIII.7.d-13)}$$

$$D(B) = \begin{pmatrix} 0 & \Delta(BA) \\ \Delta(A^{-1}B)^* & 0 \end{pmatrix}, \quad \text{all } B \text{ in } AH. \qquad \text{(VIII.7.d-14)}$$

This is the representation of G obtained from that one of H given by the Δ matrices. These can be seen to obey the relevant multiplication rules. The representation space has dimension either equal to, or twice that of, the representation space of H, being that space plus the one obtained from it by the action of any antiunitary operator, with the vectors of the two subspaces in one-to-one correspondence. The last part of the sentence might seem vacuous, since they have the same dimension. However each of these vectors bears labels given by the group H, and its subgroups, and is transformed into other vectors of the same subspace by H. It is this which gives the statement content.

How does this argument change if the representation spaces of D and Δ are the same, so the matrices have the same dimension? Then

$$|\phi>= A|\psi>= |\psi>, \tag{VIII.7.d-15}$$

remembering that $|\,>$ is a set which A might shuffle. So

$$R|\phi>= RA|\psi>= A(A^{-1}RA)|\psi>= A\Delta(A^{-1}RA)|\psi>$$
$$= \Delta(A^{-1}RA)^*|\phi>= \Delta(A^{-1}RA)^*|\psi>, \tag{VIII.7.d-16}$$

which requires that (ignoring possible reordering)

$$\Delta(A^{-1}RA)^* = \Delta(R). \tag{VIII.7.d-17}$$

Thus representation $\Delta(R)$ is equivalent to its conjugate, here denoted by $\Delta(A^{-1}RA)^*$. Also

$$B|\psi>= BA^{-1}|\phi>= D(B)|\psi>= AR|\psi>= A\Delta(R)|\psi>$$
$$= \Delta(R)^*|\psi>= A^{-1}\Delta(B)^*|\phi>= \Delta(A^{-1}B)^*|\phi>$$
$$= \Delta(A^{-1})\Delta(B)|\psi>= \Delta(A^{-1}B)^*|\psi> . \tag{VIII.7.d-18}$$

From

$$B|\phi>= BA|\psi>= D(BA)|\psi>, \tag{VIII.7.d-19}$$

we get

$$D(BA) = D(ARA) = \Delta(A^{-1}B)^* = \Delta(R)^*. \tag{VIII.7.d-20}$$

We thus have all matrices of the corepresentation, starting from those of the unitary subgroup.

Problem VIII.7.d-1: It can be seen that these obey the multiplication rules for the D's (sec. VIII.7.b, p. 436).

Problem VIII.7.d-2: Above we stated that the group does not have representations (in the ordinary sense), yet here we (may) have found them, by doubling the set of basis vectors of the unitary subgroup. How does this avoid the difficulty?

VIII.7.e Transformations of the corepresentations

An essential property of representations is reducibility. To study this for ones with antiunitary operators, we have to consider how they are acted on by transformations of basis vectors. Labeling the basis vectors as

$$|\xi>=|\psi,\phi>, \tag{VIII.7.e-1}$$

we perform unitary transformation U on them,

$$|\xi>' = U|\xi>, \tag{VIII.7.e-2}$$

giving

$$R|\xi>' = D'(R)|\xi>' = UD'(R)|\xi>= RU|\xi>, \tag{VIII.7.e-3}$$

so

$$D'(R)|\xi>= D(U^{-1}RU)|\xi>, \tag{VIII.7.e-4}$$

and the representation matrices of the elements of H transform as

$$D'(R) = U^{-1}D(R)U. \tag{VIII.7.e-5}$$

Also

$$B|\xi>' = D'(B)|\xi>' = UD'(B)|\xi>= BU|\xi>= U^*D(B)|\xi>, \tag{VIII.7.e-6}$$

(eq. VIII.7.b-7, p. 436) and for this,

$$D'(B) = U^{-1}D(B)U^*, \tag{VIII.7.e-7}$$

which is somewhat different from those of H.

This leads to the definition of two corepresentations as unitarily equivalent [Bradley and Cracknell (1972), p. 611] if there is a unitary matrix U such that

$$D'(R) = U^{-1}D(R)U, \quad D'(B) = U^{-1}D(B)U^*, \tag{VIII.7.e-8}$$

for all R in H and B in AH. There is thus a difference between the definitions of unitary equivalence for representations and for corepresentations, but the definitions of course are not arbitrary.

Suppose that instead of A we used as coset representative antiunitary operator

$$A' = VA, \tag{VIII.7.e-9}$$

for some V in H. That would have given the basis $|\psi,\phi'>$, with

$$|\phi'>= A'|\psi>= VA|\psi>= V|\phi>= |\phi>\Delta(A^{-1}VA)^*, \tag{VIII.7.e-10}$$

(eqs. VIII.7.d-6, p. 438; VIII.7.d-16, p. 439). Matrix

$$W = \begin{pmatrix} 1 & 0 \\ 0 & \Delta(A^{-1}VA)^* \end{pmatrix}, \qquad \text{(VIII.7.e-11)}$$

can be seen to be unitary, therefore the two corepresentations with these bases are unitarily equivalent. Thus the choice of the coset representative (for a fixed coset) does not matter; the representations obtained are unitarily equivalent.

Problem VIII.7.e-1: Two corepresentations are equivalent if and only if the corresponding representations, the subduced representations — the representations of the unitary subgroup from which the corepresentations are induced (sec. VI.3.a, p. 289) — are equivalent [Jansen and Boon (1967), p. 174]. Prove and explain this. Why are induced representations relevant?

Problem VIII.7.e-2: Prove that all corepresentations of a finite group are unitarily equivalent to completely reduced ones. Likewise they are equivalent to ones in which all representation matrices are either unitary, or antiunitary (as for ordinary representations [Mirman (1995a), sec. VII.2.a, p. 182]).

VIII.7.f Reducibility of corepresentations

A representation is reducible (and likewise decomposable) if it is unitarily equivalent to one in reduced (or decomposed) form [Mirman (1995a), sec. VII.2.c, p. 185]. We next consider this for corepresentations, for which the definitions are the same. We have found corepresentations; are these reducible [Inui, Tanabe and Onodera (1990), p. 296; Wigner (1959), p. 335]?

A corepresentation of G is constructed from two of subgroup H and we now need the relationship of representations $\Delta(R)$ and $\Delta(A^{-1}RA)^*$ to determine if a corepresentation is irreducible. The corepresentation obtained from Δ is equivalent to that obtained from $\Delta(A^{-1}RA)^*$ (eqs. VIII.7.d-6, p. 438; VIII.7.d-16, p. 439). Even if these two representations of the unitary subgroup are inequivalent, the corepresentations that they give are equivalent. To study irreducibility we have to consider then two cases, when these subgroup representations are inequivalent, and when they are equivalent, with the latter having two subcases.

VIII.7.g The types of corepresentations

To construct the corepresentations we first classify them; for this we use the representations, Δ, of the unitary subgroup, as implied by the preceding argument. What are the possibilities? First to simplify the notation we define, for any fixed antiunitary operator A,

$$\Delta(R)^{+} = \Delta(A^{-1}RA)^{*}. \qquad \text{(VIII.7.g-1)}$$

There are clearly three types of representations:

1. Δ and Δ^* are equivalent, and equivalent to some real irreducible representation (that is with all matrix elements real).
2. Δ and Δ^* are inequivalent.
3. Δ and Δ^* are equivalent, but there is no representation with matrices having real entries to which they are equivalent.

It follows that there are three types of corepresentations, denoted by D, going with the three types of subgroup representations. We now construct them from the unitary subgroup representations.

Problem VIII.7.g-1: Of course that these are the only types does not mean that all are actually realized, does it? Is this related to any classification given previously?

VIII.7.h The case of inequivalent subgroup representations, Δ and Δ^+

This is the easier case, so we start with it [Wigner (1959), p. 340]. The question is whether corepresentation D is irreducible; is there a unitary matrix U that reduces it? Since $D(B)$ is an off-diagonal block matrix, U must be (block) off-diagonal to reduce it. If there is no such U, then D is irreducible. The representation matrices are given above (sec. VIII.7.d, p. 437), and we use them to analyze this.

Problem VIII.7.h-1: It is elementary [Bradley and Cracknell (1972), p. 612] to show that, with U the (presumed) matrix reducing the corepresentation (sec. VIII.7.e, p. 440), because $\Delta(R)$ and $\Delta(A^{-1}RA)^*$ are inequivalent subgroup representations, the two block off-diagonal elements of U are zero, by Schur's lemma [Mirman (1995a), sec. VII.3, p. 186]. Otherwise they would give a similarity transformation relating these representations. Thus a corepresentation obtained from two inequivalent subgroup representations is irreducible.

Problem VIII.7.h-2: We can take $\Delta(R)$ as unitary since all representations of finite groups are equivalent to unitary ones [Mirman (1995a), sec. VII.2.a, p. 182]. Check that we can also take $\Delta(A^{-1}RA)^*$ unitary, so the corepresentation is equivalent to a unitary one.

Problem VIII.7.h-3: Show that these results are independent of the choice of A. Considering some crystals with spin, check physically that this is correct and explain the meaning of the different choices.

VIII.7.i Equivalent subgroup representations

The more difficult case is that of the two subgroup representations, Δ and Δ^+, equivalent [Bradley and Cracknell (1972), p. 613; Wigner (1959), p. 340].

As we saw introducing antiunitary operators either doubles the dimension of the representation, or leaves it unchanged. Thus either D and Δ have the same dimension — the same number of rows and columns, in which case, for the operators of the unitary subgroup,

$$D(R) = \Delta(R), \tag{VIII.7.i-1}$$

or D has twice the dimension, and we first want

$$D(R) = \begin{pmatrix} \Delta(R) & 0 \\ 0 & \Delta(R) \end{pmatrix}. \tag{VIII.7.i-2}$$

To reduce the representation, if possible, we use unitary block matrix

$$U = \begin{pmatrix} 1 & 0 \\ 0 & W^{-1} \end{pmatrix}, \tag{VIII.7.i-3}$$

to bring $D(R)$ from the form of eq. VIII.7.d-13, p. 438,

$$D(R) = \begin{pmatrix} \Delta(R) & 0 \\ 0 & \Delta(A^{-1}RA)^* \end{pmatrix}, \tag{VIII.7.i-4}$$

to

$$D(R)' = \begin{pmatrix} 1 & 0 \\ 0 & W \end{pmatrix} \begin{pmatrix} \Delta(R) & 0 \\ 0 & \Delta(A^{-1}RA)^* \end{pmatrix} \begin{pmatrix} 1 & 0 \\ 0 & W^{-1} \end{pmatrix} = \begin{pmatrix} \Delta(R) & 0 \\ 0 & \Delta(R) \end{pmatrix}, \tag{VIII.7.i-5}$$

a block unit matrix. That is with Δ and Δ^+ equivalent, there is a unitary matrix W such that

$$\Delta^+(R) = \Delta(A^{-1}RA)^* = W^{-1}\Delta(R)W, \tag{VIII.7.i-6}$$

or

$$\Delta(R) = W\Delta(A^{-1}RA)^*W^{-1}. \tag{VIII.7.i-7}$$

It remains to find $D(B)$, for antiunitary operator B. Since the square of B is unitary (thus an element of H), and B commutes with all elements of H,

$$\Delta(B^2) = W\Delta(A^{-1}B^2A)^*W^{-1} = W\Delta(B^2)^*W^{-1}. \tag{VIII.7.i-8}$$

Also

$$\Delta(R)^* = W^*\Delta(A^{-1}RA)W^{*-1}, \tag{VIII.7.i-9}$$

so (with $R = A^2$, for any element of H),

$$\Delta(A^{-1}RA)^* = W^*\Delta(A^{-2}RA^2)W^{*-1} = W^*\Delta(A^{-2})\Delta(R)\Delta(A^2)W^{*-1}. \tag{VIII.7.i-10}$$

Thus combining,

$$\Delta(R) = WW^*\Delta^{-1}(A^2)\Delta(R)\Delta(A^2)W^{*-1}W^{-1}, \tag{VIII.7.i-11}$$

giving that $WW^*\Delta^{-1}(A^2)$ commutes with all matrices of the irreducible representation of the unitary subgroup, so is a multiple, λ, of the unit matrix over the representation space of $\Delta(R)$ by Schur's lemma. This, and its complex conjugate, give

$$\Delta(A^2) = \frac{WW^*}{\lambda}, \tag{VIII.7.i-12}$$

$$\Delta(A^2)^* = \frac{W^*W}{\lambda^*}. \tag{VIII.7.i-13}$$

We have another relationship between these two (eq. VIII.7.i-8), so together

$$\frac{WW^*}{\lambda} = \frac{WW^*WW^{-1}}{\lambda^*} = \frac{WW^*}{\lambda^*}, \tag{VIII.7.i-14}$$

which means that

$$\lambda = \lambda^*. \tag{VIII.7.i-15}$$

The matrices of the unitary subgroup are, of course, unitary, as is W, so

$$|\lambda| = 1, \tag{VIII.7.i-16}$$

and

$$\lambda = \pm 1. \tag{VIII.7.i-17}$$

Thus either

$$WW^* = \Delta(A^2), \tag{VIII.7.i-18}$$

or

$$WW^* = -\Delta(A^2). \tag{VIII.7.i-19}$$

There are two cases, not surprisingly, and these are considered separately — the results for the reducibility of D differ for the two signs.

VIII.7.i.i *Transforming to reduced form*

This gives the restrictions on the transformation, and we now have to write it explicitly. Since B is antiunitary, the corresponding matrices are transformed by

$$D(B)' = U^{-1}D(B)U^*, \text{ with } D(R)' = U^{-1}D(R)U, \tag{VIII.7.i.i-1}$$

giving

$$D(B)' = \begin{pmatrix} 1 & 0 \\ 0 & W \end{pmatrix} \begin{pmatrix} 0 & \Delta(BA) \\ \Delta(A^{-1}B)^* & 0 \end{pmatrix} \begin{pmatrix} 1 & 0 \\ 0 & W^{-1*} \end{pmatrix}$$

$$= \begin{pmatrix} 0 & \Delta(BA)W^{-1*} \\ W\Delta(A^{-1}B)^* & 0 \end{pmatrix}, \qquad \text{(VIII.7.i.i-2)}$$

and in particular

$$D(A)' = \begin{pmatrix} 1 & 0 \\ 0 & W \end{pmatrix} \begin{pmatrix} 0 & \Delta(A^2) \\ 1 & 0 \end{pmatrix} \begin{pmatrix} 1 & 0 \\ 0 & W^{-1*} \end{pmatrix}$$

$$= \begin{pmatrix} 0 & \Delta(A^2)W^{-1*} \\ W & 0 \end{pmatrix}. \qquad \text{(VIII.7.i.i-3)}$$

Now D is reducible if, but only if, $D(R)$ and $D(A)$ can be written in the same block-diagonal form simultaneously and $D(R)$ has already been written in reduced block-diagonal form. So we have to find a unitary transformation W that will block-diagonalize $D(B)$, while not changing $D(R)$ — one that commutes with it. If there is none, the corepresentation is irreducible. We write it as

$$W = \begin{pmatrix} c_{11}I & c_{12}I \\ c_{21}I & c_{22}I \end{pmatrix} = c \times I, \qquad \text{(VIII.7.i.i-4)}$$

where I is a unit matrix of the same dimension as Δ, and c the matrix of the c's. It is clear that since $D(R)$ is in block-diagonal form, W must be in this corresponding block form (by Schur's Lemma). Because W is unitary,

$$det(W) \neq 0, \qquad \text{(VIII.7.i.i-5)}$$

which gives

$$c_{11}c_{22} \neq c_{12}c_{21}. \qquad \text{(VIII.7.i.i-6)}$$

For W to reduce $D(B)$ requires that the diagonal terms vanish. Also since A commutes with all R, we can take

$$D(A) = \begin{pmatrix} W & 0 \\ 0 & W \end{pmatrix} = I_2 \times W, \qquad \text{(VIII.7.i.i-7)}$$

where I_2 is a 2×2 block matrix. Then from eq. VIII.7.d-14, p. 438,

$$D(B)' = \begin{pmatrix} c_{11}I_2 & c_{12}I_2 \\ c_{21}I_2 & c_{22}I_2 \end{pmatrix} \begin{pmatrix} 0 & \Delta(BA)W^{*-1} \\ W\Delta(A^{-1}B)^* & 0 \end{pmatrix} \begin{pmatrix} c_{11}I_2 & c_{21}I_2 \\ c_{12}I_2 & c_{22}I_2 \end{pmatrix}$$

$$= \begin{pmatrix} c_{11}I_2 & c_{12}I_2 \\ c_{21}I_2 & c_{22}I_2 \end{pmatrix} \begin{pmatrix} c_{12}\Delta(BA)W^{*-1} & c_{22}\Delta(BA)W^{*-1} \\ c_{11}W\Delta(A^{-1}B)^* & c_{21}W\Delta(A^{-1}B)^* \end{pmatrix}$$

$$= \begin{pmatrix} T_{11} & T_{12} \\ T_{21} & T_{22} \end{pmatrix}, \qquad \text{(VIII.7.i.i-8)}$$

where

$$T_{11} = c_{11}c_{12}[\Delta(BA)W^{*-1} + W\Delta(A^{-1}B)^*], \qquad \text{(VIII.7.i.i-9)}$$

$$T_{12} = c_{11}c_{22}\Delta(BA)W^{*-1} + c_{21}c_{12}W\Delta(A^{-1}B)^*, \qquad \text{(VIII.7.i.i-10)}$$

$$T_{21} = c_{21}c_{12}\Delta(BA)W^{*-1} + c_{11}c_{22}W\Delta(A^{-1}B)^*, \qquad \text{(VIII.7.i.i-11)}$$

$$T_{22} = c_{21}c_{22}[\Delta(BA)W^{*-1} + W\Delta(A^{-1}B)^*]. \qquad \text{(VIII.7.i.i-12)}$$

We now use eqs. VIII.7.i-18, VIII.7.i-19, p. 444, and the condition from the nonvanishing determinant and find that the reduction is possible only if

$$c_{11}c_{22} = -c_{12}c_{21}. \qquad \text{(VIII.7.i.i-13)}$$

This requires

$$WW^* = -\Delta(A^2). \qquad \text{(VIII.7.i.i-14)}$$

There is enough freedom in W to take

$$c_{11}c_{12} = c_{21}c_{22} = \frac{1}{2}. \qquad \text{(VIII.7.i.i-15)}$$

So the case with the plus sign is for an irreducible representation, with the minus sign, the representation can be reduced.

VIII.7.i.ii *The case with the sign positive*

The first possibility for the dimension of D is that it is equal to the dimension of Δ, which is true if the sign is positive. We can take

$$D(A) = W^*, \qquad \text{(VIII.7.i.ii-1)}$$

since

$$\Delta^+(R) = \Delta(A^{-1}RA)^* = W^{-1}\Delta(R)W, \qquad \text{(VIII.7.i.ii-2)}$$

which gives the result because in this case D and Δ must be the same, up to a multiplicative constant — there being no other choice as Δ and Δ^+ are equivalent, and D reduces to Δ for the unitary subgroup. Hence

$$W^* = W. \qquad \text{(VIII.7.i.ii-3)}$$

This picks out the plus sign, so for it the representation is irreducible.

Specifically W is a symmetric unitary matrix, and so can be written [Wigner (1959), p. 287]

$$W = r^{-1}wr, \qquad \text{(VIII.7.i.ii-4)}$$

where w is diagonal, and r real and orthogonal. Using the similarity transformation given by this we reduce the representation matrices. As $D(R)$ is diagonal, and transformed by a real orthogonal matrix, it remains diagonal. However $D(A)$ is diagonalized, with the diagonal terms perhaps differing by an irreducible constant, depending on the elements of w. For it we use eq. VIII.7.d-13, p. 438, so getting two diagonal representation matrices. The entries are equivalent and it is irrelevant which is taken.

Then the representation matrices are,

$$D(R) = \Delta(R), \qquad \text{(VIII.7.i.ii-5)}$$

$$D(B) = \Delta(BA^{-1})W. \qquad \text{(VIII.7.i.ii-6)}$$

This matrix of a product cannot be written as a product of matrices as Δ is a representation matrix of the unitary subgroup, but neither of the terms in the product belongs to it, although the product does. We now have the matrices, D, of the corepresentation in terms of those of the unitary subgroup, Δ.

Problem VIII.7.i.ii-1: Since $\Delta(R)$, so $D(R)$, are unitary it remains only to check that we can also take $D(A)$ antiunitary, so the corepresentation is equivalent to a unitary one (defined this way).

VIII.7.i.iii *The case with the minus sign*

Next is the case for which the dimension of D is twice that of Δ.

To show that

$$\Delta(BA) = W\Delta(A^{-1}B)^*W^*, \qquad \text{(VIII.7.i.iii-1)}$$

we first note that from eq. VIII.7.d-20, p. 439, and eq. VIII.7.i-9, p. 443,

$$\Delta(A^{-1}B)^* = \Delta(R)^* = W^*\Delta(A^{-1}RA)W^{*-1}, \qquad \text{(VIII.7.i.iii-2)}$$

and taking the case of the minus sign,

$$WW^* = -\Delta(A^2), \qquad \text{(VIII.7.i.iii-3)}$$

$$WW^*\Delta(A^{-1}RA)W^{*-1} = -\Delta(A^2)\Delta(A^{-1}RA)W^{*-1}, \qquad \text{(VIII.7.i.iii-4)}$$

so

$$W\Delta(A^{-1}B)^*W^* = WW^*\Delta(A^{-1}RA)W^{*-1}W^*$$

$$= WW^*\Delta(A^{-1}RA) = -\Delta(A^2)\Delta(A^{-1}RA) = \Delta(ARA) = \Delta(BA), \quad \text{(VIII.7.i.iii-5)}$$

as required. The diagonal block matrix is

$$\Delta(BA)W^{*-1} + W\Delta(A^{-1}B)^* = 2W\Delta(A^{-1}B)^*. \quad \text{(VIII.7.i.iii-6)}$$

This gives the reduced form

$$D(A)' = \begin{pmatrix} W & 0 \\ 0 & W \end{pmatrix}, \quad \text{(VIII.7.i.iii-7)}$$

and

$$D(B)' = \begin{pmatrix} \Delta(BA^{-1})W & 0 \\ 0 & -\Delta(BA^{-1})W \end{pmatrix}. \quad \text{(VIII.7.i.iii-8)}$$

Problem VIII.7.i.iii-1: As $\Delta(R)$ is unitary it remains merely to check, for each case, that we can also take $D(B)$ antiunitary, so the corepresentation is equivalent to a unitary one.

Problem VIII.7.i.iii-2: Compare the discussion here with sec. VIII.3.c, p. 405. Compare it also to the discussion of induced and subduced representations, and conjugate, nonequivalent, representations.

VIII.7.j A simple corepresentation

As an example of a corepresentation [Jansen and Boon (1967), p. 172] consider the two-element group e and s, with (of course)

$$s^2 = e. \quad \text{(VIII.7.j-1)}$$

Then the matrices, obeying the corepresentation multiplication rules (sec. VIII.7.b, p. 436)

$$D(e) = 1, \; D(s) = exp(i\chi), \; \text{any } \chi, \quad \text{(VIII.7.j-2)}$$

form a one-dimensional corepresentation — these matrices are merely numbers. The invariant unitary subgroup H is just $\{e\}$, and realizing s as complex conjugation,

$$\phi = exp(\frac{-i\chi}{2})f(y), \quad \text{(VIII.7.j-3)}$$

is a basis vector of the corepresentation, where χ, f and (variable) y are real, as can be seen from the action of $D(s)$ on ϕ. The group has also two one-dimensional representations,

$$D(e) = 1, \; D(s) = 1; \;\; D(e) = 1, \; D(s) = -1; \quad \text{(VIII.7.j-4)}$$

s can be realized as a reflection, inversion, or rotation of π. The basis vectors are

$$\phi = f(y), \tag{VIII.7.j-5}$$

with f symmetric for the first representation, antisymmetric for the second, under s. While the corepresentation is one dimensional, it can be regarded as a two-dimensional ordinary representation. The representation matrices are

$$D(e) = \begin{pmatrix} 1 & 0 \\ 0 & 1 \end{pmatrix}, \quad D(s) = \begin{pmatrix} 0 & 1 \\ 1 & 0 \end{pmatrix}. \tag{VIII.7.j-6}$$

Notice that the dimension of the representation space has been doubled.

Problem VIII.7.j-1: Find functions that form basis vectors of the two-dimensional corepresentation. How are they related to the basis vectors of the one-dimensional one? Why can this be regarded as a corepresentation, rather than as a representation?

Problem VIII.7.j-2: There are only two possible matrices for $D(s)$ for the representation, but a continuous infinite number for the corepresentation, in one-dimensional form, and one (given) for the two-dimensional form. Explain.

Problem VIII.7.j-3: Why is the one-dimensional corepresentation not a representation? Check that the two-dimensional form does give a representation. Why can it be regarded as an ordinary representation? Abelian groups, which this clearly is, have only one-dimensional representations — over the complex numbers (pb. V.4.c.i-2, p. 268). But they can have two-dimensional representations if limited to real numbers. How is this related to the present results? In developing the theory of corepresentations we have been using an invariant subgroup of index 2. Is that, and the form and properties of corepresentations, relevant to this property of Abelian groups? Are these relevant to the discussion of this section?

VIII.7.k Which representations are which?

The three types of corepresentations (sec. VIII.7.g, p. 441) have now been studied. To determine the type of a particular one we could construct its representation matrices, and find whether or not there are equivalent representations that can be made real. This could be difficult, especially the last step; a simpler way would be preferable.

The Frobenius-Schur test evaluates the sum of the characters of the *squares* of the h group elements [Birman (1974), p. 168; Bradley and Cracknell (1972), p. 617; Burns (1977), p. 336; Cornwell (1984), p. 161;

Falicov (1966), p. 196; Inui, Tanabe and Onodera (1990), p. 298; Jansen and Boon (1967), p. 168; Kocinski (1983), p. 53; Lax (1974), p. 298; Streitwolf (1971), p. 183; Tinkham (1964), p. 146]. This actually determines how representations and their complex conjugates are related, but for reasons that should be clear from the forms we have discussed, this is immediately applicable to corepresentations. Here we ignore spin, and return to that below (sec. VIII.8.c, p. 455). Then, with the classification of sec. VIII.7.g, p. 441,

$$\Sigma\eta^r(R^2) = \begin{matrix} h, & \text{type } 1, \\ 0, & \text{type } 2, \\ -h, & \text{type } 3, \end{matrix} \qquad \text{(VIII.7.k-1)}$$

summed over the group elements; η^r is the character of class (of elements) r. (The class of elements should not be confused with the type of a representation given here.) To show this we start with

$$\sum_R \eta^r(R^2) = \sum_R \sum_{ij} \Delta^r(R)_{ij}\Delta^r(R)_{ji}. \qquad \text{(VIII.7.k-2)}$$

By the orthogonality theorem [Mirman (1995a), sec. VII.5.a, p. 193],

$$\sum_R \Delta^{a*}(R)_{ij}\Delta^b(R)_{kl} = (\frac{h}{d})\delta^{ab}\delta_{ik}\delta_{jl}, \qquad \text{(VIII.7.k-3)}$$

for representations a, b with d the dimension of the representation. The first type has $\Delta = \Delta^*$ so

$$\sum_R \eta^r(R^2) = \sum_R \sum_{ij} \Delta^{a*}(R)_{ij}\Delta^a(R)_{ji} = \sum \delta_{ij}(\frac{h}{d}) = h. \qquad \text{(VIII.7.k-4)}$$

For the second type the representations are inequivalent so $\sum \eta^r(R^2) = 0$ by the orthogonality theorem. This sum is defined by this equation, otherwise it is meaningless when applied to different representations. Since for the third type Δ and Δ^* are equivalent,

$$\Delta^* = U\Delta U^{-1}, \qquad \text{(VIII.7.k-5)}$$

with U unitary, so

$$\sum_R \eta^r(R^2) = \sum_R \sum_{ij} U_{ik}\Delta^{a*}(R)_{kl}U^{-1}_{lm}\Delta^a(R)_{mi}$$

$$= \sum_R \sum_{ij} U_{ik}\Delta^{a*}(R)_{kl}U^*_{ml}\Delta^a(R)_{mj} = (\frac{h}{d})Tr(UU^*). \qquad \text{(VIII.7.k-6)}$$

Taking the complex conjugate,

$$\Delta = U^*\Delta^*U^{*-1} = U^*U\Delta U^{-1}U^{*-1}, \qquad \text{(VIII.7.k-7)}$$

gives

$$U^*U = \pm 1. \qquad \text{(VIII.7.k-8)}$$

Problem VIII.7.k-1: Prove that [Birman (1974), p. 172] for

$$U^*U = 1, \qquad \text{(VIII.7.k-9)}$$

Δ can be made real (sec. VIII.7.i.ii, p. 446). Finally when the Δ's cannot be made real,

$$U^*U = -1. \qquad \text{(VIII.7.k-10)}$$

Then (sec. VIII.7.i.iii, p. 447)

$$Tr(U^*U) = -d, \qquad \text{(VIII.7.k-11)}$$

giving the result. This proves the test for all three classes.

VIII.7.l The Herring test

This rule distinguishing the representations given by the Frobenius-Schur test requires a sum over group elements, whose number is infinite for a space group. However since a space-group representation is determined by a small representation of the little group, we can use the simpler Herring test [Birman (1974), p. 173; Bradley and Cracknell (1972), p. 639; Cornwell (1969), p. 151, 187; Falicov (1966), p. 198; Inui, Tanabe and Onodera (1990), p. 300; Joshi (1982), p. 320; Streitwolf (1971), p. 180; Tinkham (1964), p. 297]. This says

$$\sum \eta^r(P^2) = \begin{matrix} n, & \text{type 1,} \\ 0, & \text{type 2,} \\ -n, & \text{type 3,} \end{matrix} \qquad \text{(VIII.7.l-1)}$$

summed over those n space-group elements P that take $\underline{k}$ into -$\underline{k}$, a subgroup of the little group of $\underline{k}$. If this little group contains the inversion then elements P form the little group. Otherwise the set IP found by multiplying them by the inversion does. Thus instead of using the representation of the space group, this uses only the small representation of the little group (sec. VII.3.a, p. 347), actually just the little co-group.

To show this [Lax (1974), p. 305] we express the character of group element S in the little group (sec. VII.4.c.ii, p. 357), as

$$\eta^{\underline{k}}(S) = \sum_i \Delta_{ii}^{\underline{k}}(S), \qquad \text{(VIII.7.l-2)}$$

the sum of the diagonal matrix elements of S, labeled by i, while the character of the space group representation matrix (of which the little group representation matrix is a block) requires in addition a sum over the star of $\underline{k}$,

$$\eta(S) = \sum_{\underline{k}} \eta^{\underline{k}}(S). \tag{VIII.7.1-3}$$

A space group element S (with a homogeneous operator that depends on the translation for a glide or screw) can be written (sec. III.2.a, p. 134)

$$S = \{R|\underline{v}(R) + \underline{t}\} = \{E|\underline{t}\}S_0, \tag{VIII.7.1-4}$$

$$S_0 = \{R|\underline{v}(R)\}, \tag{VIII.7.1-5}$$

and its square is

$$S^2 = \{E|\underline{t} + R\underline{t}\}S_0^2 = exp[-i\underline{k} \bullet (\underline{t} + R\underline{t})]S_0^2, \tag{VIII.7.1-6}$$

from eq. VII.3.a.i-1, p. 348, where the last form holds for a space-group basis state taken at $\underline{k}$, the relevant basis state for the little group. Thus, summing over all these space group elements gives

$$\sum_{\underline{t}}\sum_{R} \eta(S^2) = \sum_{\underline{k}}\sum_{R} \eta^{\underline{k}}(S^2) \sum_{\underline{t}} exp[-i(\underline{k} + R^{-1}\underline{k}) \bullet \underline{t}]. \tag{VIII.7.1-7}$$

The sum on the arbitrary translation $\underline{t}$ has for each value another for a translation that gives the negative of the term, so is zero unless

$$\underline{k} + R^{-1}\underline{k} = 0, \text{ or } \underline{K}, \tag{VIII.7.1-8}$$

$$R^{-1}\underline{k} = -\underline{k} + \underline{K}; \tag{VIII.7.1-9}$$

$\underline{K}$ is a reciprocal lattice vector. So the only elements R giving nonzero terms are those for which R^{-1}, thus R, take $\underline{k}$ to -$\underline{k}$, or an equivalent point. Then the sum over $\underline{t}$ equals N, the number of lattice points, using the Born-von Karman cyclicity condition (sec. VII.4.c, p. 355; sec. XI.2.f, p. 577), and if the number of elements of the point group is h, then the order of the space group is

$$g = Nh, \tag{VIII.7.1-10}$$

and the Frobenius criterion becomes

$$\frac{1}{g}\sum \eta(S^2) = \frac{1}{h}\sum_{\underline{k}}\sum_{Q} \eta^{\underline{k}}(Q^2), \tag{VIII.7.1-11}$$

where the sum on Q is only over those elements taking $\underline{k}$ to -$\underline{k}$. The (artificial) value N does not appear. All vectors of a star are equivalent so the sum over them, for each star, is equal to the number in it, m, times the function for any one vector, giving, from eq. VIII.7.k-1, p. 450,

$$\frac{1}{g}\sum \eta(R^2) = \frac{1}{h}\sum_{\underline{k}}\sum_{Q}\eta^{\underline{k}}(Q^2) = \pm 1 \quad \text{or} \quad 0, \qquad \text{(VIII.7.l-12)}$$

where

$$h(k) = \frac{h}{m}, \qquad \text{(VIII.7.l-13)}$$

is the order of $G_{\underline{k}}/T$, the number of elements of $P(\underline{k})$.

Problem VIII.7.l-1: Put everything together, and finish this.

VIII.7.m Are corepresentations representations?

We have developed a structure for antiunitary groups that is equivalent to representations. Are these essentially different from ordinary representations, or are they just disguised? Note that a complex number differs from a real one in that it is a pair of real numbers. It can be written as a sum of real and imaginary parts — both real numbers. This is similar to what we are doing for corepresentations.

The representation space is unaltered or doubled by the introduction of the antiunitary operator, and that mixes the two parts in the latter case. Then the representation matrices for a corepresentation of a group whose invariant subgroup has representation matrices of dimension n, are of dimension $2n$. These are of blocks, each of dimension n. Each representation matrix has two such such blocks, either on the diagonal or on the anti-diagonal. For the latter case, the antiunitary operator, time reversal say, mixes the two blocks.

What has been done, because the multiplication rules involve the complex conjugate matrices, is to enlarge the representation matrices to include these. The entire set, this set of larger matrices, of dimension $2n$, then forms a representation of the group. So using the larger representation space, antiunitary groups do have representations, with the same form as for ordinary groups.

Problem VIII.7.m-1: If the dimension of the representation space is unaltered, what can be said about the reality of the (relevant) representations? Which representations are potentially real, pseudo-real, and essentially complex (sec. VIII.3.c, p. 405)?

Problem VIII.7.m-2: Are the corepresentations the same as ordinary representations? Does this disagree with the argument that groups with antiunitary operators have no representations?

VIII.8 SPIN AND COREPRESENTATIONS

The types of corepresentations, and their properties, in particular the degeneracy they require [Lax (1974), p. 301; Tinkham (1964), p. 146], depend on whether the spin is integral or half-integral. Here we consider this for the magnetic point groups [Jansen and Boon (1967), p. 289; Wigner (1959), p. 344].

VIII.8.a Integral spin

The analysis of these representations is in two parts, the first for integer spin, so (pb. VIII.2.c-5, p. 399)

$$\theta^2 = 1. \tag{VIII.8.a-1}$$

For the first type (sec. VIII.7.k, p. 449) the basis states, the ψ's, are real; if they were not we could find two real sets by taking sums of their real and imaginary parts. These sets are not mixed by the transformations, including θ, as can be seen from its action (eq. VIII.7.c-1, p. 437). Thus they both form sets of basis vectors of the same representation of the group, and since the group elements commute with the Hamiltonian (taking it as time-reversal invariant), these states have the same energy. Up to equivalence then, they are the same. For this type, θ does not introduce additional degeneracy. The second type has the two representations inequivalent, so their basis vectors are orthogonal. Thus the representation space of the group is twice as large as that of the unitary subgroup. The two subspaces are mixed by θ, hence it doubles the degeneracy (for a Hamiltonian invariant under θ).

Problem VIII.8.a-1: Since for the third type, matrix elements are of the form $(\psi^+, S\psi)$ and cannot be made real, verify that neither can ψ or ψ^+. For this, the unitary subgroup has two sets of basis vectors for each representation, complex conjugates of each other, equivalent under this subgroup, but mixed by θ. Thus here again time reversal invariance doubles the degeneracy.

VIII.8.b Half-integer spin

If the system has an odd number of half-integer spins (pb. VIII.2.c-5, p. 399),

$$\theta^2 = -1. \tag{VIII.8.b-1}$$

Then for the first type (sec. VIII.7.k, p. 449), using the two real combinations u and v,

$$\theta u = iv, \tag{VIII.8.b-2}$$

and

$$\theta v = -iu; \tag{VIII.8.b-3}$$

for this the degeneracy is doubled.

It is clear that for the second type since the degeneracy is doubled for $\theta^2 = 1$, it is also doubled for $\theta^2 = -1$, because the change in sign does not cause the two distinct sets to become equivalent, and θ can do no more than double the degeneracy.

The third type is the most interesting — it has no further degeneracy. The reason is that the statefunction can be taken as a product of one for integral spin times a spin-$\frac{1}{2}$ statefunction, and we need consider only the latter whose two states of spin (up and down) are mixed by θ, but they are also mixed by the group of rotations, so there is no additional degeneracy because of θ.

Problem VIII.8.b-1: Why can we consider only the spin-$\frac{1}{2}$ statefunction (and can ignore spin-$\frac{3}{2}$, and so on)?

Problem VIII.8.b-2: Would you expect degeneracy doubling for the first type from Kramer's theorem (sec. VIII.2.d, p. 402)? How about the other two?

VIII.8.c Classifying the corepresentations

With spin the type of corepresentation is determined by the generalization of the Frobenius-Schur test (sec. VIII.7.k, p. 449), with the sign dependent on whether the number of fermions (particles with half-odd-integral spin) is even or odd [Bradley and Cracknell (1972), p. 648; Kocinski (1983), p. 55; Wigner (1959), p. 344]. The test uses the sum of the characters of the *squares* of the h elements of the magnetic little cogroup $M^{\underline{k}*}$, the little group of the magnetic space group (sec. VIII.4.c, p. 419). With $\Gamma_p^{\underline{k}}$ a small representation (sec. VII.3.a, p. 347) of little group $G^{\underline{k}}$, which has character $\eta_p^{\underline{k}}$, and $\Delta_p^{\underline{k}}$ the irreducible representation of space group G induced from $\Gamma_p^{\underline{k}}$, using the classification of sec. VIII.7.g, p. 441, the type of corepresentation of M induced from $\Delta_p^{\underline{k}}$ is given by

$$\sum \eta_p^{\underline{k}}(\{S_{\underline{k}}|w_{\underline{k}}\}^2) = \begin{array}{ll} \gamma h, & \text{type 1,} \\ 0, & \text{type 2,} \\ -\gamma h, & \text{type 3,} \end{array} \tag{VIII.8.c-1}$$

summed over the group elements $S_{\underline{k}}$, restricted to those point group members $S_{\underline{k}}$ of the set $(M^{\underline{k}*} - G^{\underline{k}*})$ for which $S_{\underline{k}}\underline{k}$ is equivalent to $-\underline{k}$.

Then

$$\gamma = \Delta(\theta^2) = 1; = -1, \qquad \text{(VIII.8.c-2)}$$

for an even, and odd, number of fermions, respectively. For

$$M = G + \theta G, \qquad \text{(VIII.8.c-3)}$$

this is the same as the Herring criterion (sec. VIII.7.l, p. 451).

Problem VIII.8.c-1: Verify, and if necessary correct, the statements of this section.

VIII.9 APPLICATION OF COREPRESENTATIONS TO MAGNETIC SPACE GROUPS

Although we have studied corepresentations in principle, we have to relate them specifically to the magnetic point [Bradley and Cracknell (1972), p. 622] and space groups, and consider their physical applications. This emphasizes the effect of time reversal on the properties of a crystal, in particular its degeneracies, and the differences between crystals consisting of particles with spin, and those for which the particles are purely point-like. First we review the latter.

VIII.9.a The translation operators

We start, using the elements $\{E|\underline{t}\}$ of the translation subgroup, by writing those of the translation subgroup of unitary subgroup H, thus of magnetic group G, as

$$H \Rightarrow \{R|\underline{v}_R\}, \qquad \text{(VIII.9.a-1)}$$

and for coset θH,

$$\theta H \Rightarrow \theta\{S|\underline{w}_S\}, \qquad \text{(VIII.9.a-2)}$$

where R and S are point group operators, thus of subgroup H; $\underline{v}_R$ and $\underline{w}_S$, determined by R and S respectively, are either zero (or lattice translations equivalent to zero), or non-lattice translations (for nonsymmorphic groups). Thus

$$A\{R|\underline{v}_R\} = \{S|\underline{w}_S\}. \qquad \text{(VIII.9.a-3)}$$

This is true for all four types of space groups, the three of sec. VIII.7.g, p. 441, plus the unitary ones (for which the equations involving arbitrary antiunitary operator A are vacuous).

VIII.9.b Corepresentations of the magnetic space groups

The corepresentations of the magnetic space groups are found using the method of induced representations (sec. VI.3.a, p. 289), generalizing the method used for unitary groups; how is this procedure affected by the appearance of time reversal [Bradley and Cracknell (1972), p. 638, 646; Evarestov and Smirnov (1993), p. 244]?

VIII.9.b.i *The procedure for unitary groups*

First, for ordinary space groups, we find the little co-group of vector $\underline{k}$ of the Brillouin zone (sec. VII.2.h, p. 344). An irreducible representation of translation subgroup T is given by $\underline{k}$ and has the representation matrix

$$F^{\underline{k}}(\{E|\underline{t}\}) = exp(-i\underline{k} \bullet \underline{t}). \qquad \text{(VIII.9.b.i-1)}$$

Then (eq. VII.4.d.iii-2, p. 362)

$$\{R|\underline{v}\}^{-1}\{E|\underline{t}\}\{R|\underline{v}\} = \{E|R^{-1}\underline{t}\}, \qquad \text{(VIII.9.b.i-2)}$$

so

$$F_R^{\underline{k}}(\{E|\underline{t}\}) = F^{\underline{k}}(\{E|R^{-1}\underline{t}\}) = exp(i\underline{k} \bullet R^{-1}\underline{t}) = exp(iR\underline{k} \bullet \underline{t}) = F^{R\underline{k}}\{E|\underline{t}\}. \qquad \text{(VIII.9.b.i-3)}$$

The little co-group (sec. VII.2.h, p. 344) of $F^{\underline{k}}(\{E|\underline{t}\}), G^{k*}$ (with the * indicating the co-group), is the set of all point group operations R such that

$$F^{R\underline{k}} = F^{\underline{k}}, \qquad \text{(VIII.9.b.i-4)}$$

for those R's that satisfy

$$R\underline{k} = \underline{k} + \underline{K}, \qquad \text{(VIII.9.b.i-5)}$$

where

$$exp(-i\underline{K} \bullet \underline{t}) = 1, \qquad \text{(VIII.9.b.i-6)}$$

for all $\underline{t}$. The vectors $\underline{K}$ are reciprocal lattice vectors, and vectors $\underline{k}+\underline{K}$ are those equivalent to $\underline{k}$. Thus this equation means that $R\underline{k}$ is equivalent to $\underline{k}$. The little group $F^{\underline{k}}$ in G, the group of $\underline{k}$, is then written as a set sum,

$$G^{\underline{k}} = \bigoplus \{R|\underline{v}_R\}T, \qquad \text{(VIII.9.b.i-7)}$$

summed over all elements of the little co-group and

$$G = \bigoplus \{R|\underline{v}_R\}G^{\underline{k}}, \qquad \text{(VIII.9.b.i-8)}$$

summed over the little groups. The set of representations $F^{R\underline{k}}$ is formed from the star of $F^{\underline{k}}$.

VIII.9.b.ii *How the procedure is changed for magnetic space groups*

This is for ordinary space groups; how is it modified for magnetic ones? We know that, with θ time reversal and A any antiunitary operator,

$$A\{R|\underline{v}\} = \theta\{S|\underline{w}\}; \qquad \text{(VIII.9.b.ii-1)}$$

the effect of an antiunitary operator on a unitary space group operation is the same as another unitary operator, with different point and translation group parts, times time reversal. As θ commutes with all space group operations, these acting on coordinates,

$$\theta\{S|\underline{w}_S\}\{E|\underline{t}\}\theta\{S|\underline{w}_S\}^{-1} = \{E|S^{-1}\underline{t}\}, \qquad \text{(VIII.9.b.ii-2)}$$

with $\underline{t}$ a lattice translation; that is conjugation with a product of a nonsymmorphic operator with time reversal of a translation gives the translation obtained by the action of the inverse of the point group operator part of the nonsymmorphic operator. Thus, just as for ordinary space groups,

$$F_S^{\underline{k}}(\{E|\underline{t}\}) = F^{S\underline{k}}(\{E|\underline{t}\}). \qquad \text{(VIII.9.b.ii-3)}$$

Now S is an antiunitary operator in the coset obtained by time reversal, and since that reverses the direction of the momentum, for S to belong to the magnetic little co-group, it is necessary that

$$(F^{\underline{k}})^{-1} = F^{-\underline{k}}. \qquad \text{(VIII.9.b.ii-4)}$$

So the magnetic little co-group consists of all R belonging to unitary subgroup G plus all S such that $S\underline{k}$ is equivalent to $-\underline{k}$ — giving an extra condition for these groups. If there is no such S, then the magnetic little co-group is just the little co-group obtained from G. The magnetic little co-group of momentum $\underline{k}$, $M^{\underline{k}*}$, can thus be written

$$M^{\underline{k}*} = \bigoplus\{R|\underline{v}_R\}T + \bigoplus\{S|\underline{w}_S\}T, \qquad \text{(VIII.9.b.ii-5)}$$

where the first sum is over all R in $G^{\underline{k}*}$, the second over all S in $M^{\underline{k}*} - G^{\underline{k}*}$.

The effect of θ on $\underline{k}$ is

$$\theta^{-1}exp(-i\underline{k}\bullet\underline{r})\theta = exp(i\underline{k}\bullet\underline{r}), \qquad \text{(VIII.9.b.ii-6)}$$

since $\underline{k}$ is a momentum, not a coordinate, so reverses sign under reversal of time (the direction of motion is reversed).

Thus the procedure for finding magnetic-group corepresentations is almost identical to that for representations of ordinary space groups, with this extra sum in the expression for the little co-group and the requirement that the star of $\underline{k}$ include $-\underline{k}$.

VIII.9.c Projective corepresentations of magnetic groups

An important aspect of space groups is that some have projective representations. This implies that when we adjoin time reversal we obtain magnetic space groups with projective corepresentations [Janssen (1973), p. 229]. It is interesting to see how these differ, if they do in any essential way, from the projective representations of ordinary space groups, and from ordinary corepresentations and from ordinary representations.

If unitary transformations R commutes with the Hamiltonian — are symmetries of the system — when there is no spin, what operators commute with it if the particles have spin? These are

$$V_S = S; \quad S \text{ unitary, integral spin,} \tag{VIII.9.c-1}$$

$$V_S = CS; \quad S \text{ antiunitary, integral spin,} \tag{VIII.9.c-2}$$

$$V_S = u(S)S; \quad S \text{ unitary, odd half-integral spin,} \tag{VIII.9.c-3}$$

$$V_S = \sigma_2 Cu(S)S; \quad S \text{ antiunitary, odd half-integral spin,} \tag{VIII.9.c-4}$$

where C is complex conjugation, $u(S)$ the SU(2) representation matrix for S, and σ_2 the Pauli matrix. The first equation defines the symbols, the second follows from our previous considerations, since for it CS is unitary. We have to discuss the last two. To transform the Hamiltonian we transform the coordinates and momenta, these by S, and the spinor statefunctions, these by $u(S)$. This explains the third equation, the standard expression for the operations of SU(2). Why does σ_2, which is anti-hermitian, appear for antiunitary operators and odd half-integral spin? Time reversal interchanges spin up and down states; for orbital angular momentum the direction in which a particle moves is reversed, so this is the requirement for spin also. Complex conjugation, which gives θ, requires σ_2 to interchange states, and to get the proper phase (multiplication by i) the operator is σ_2C.

These operators, unitary and antiunitary, belonging to the group, have representation matrices which satisfy

$$V_R V_S = \omega(R,S) V_{R,S}, \tag{VIII.9.c-5}$$

where the phase factors obey

$$\omega(R,S) = 1; \text{ integral spin,} \tag{VIII.9.c-6}$$

$$\omega(R,S) = \omega_s(r,s); \text{ odd half-integral spin, } R \text{ or } S \text{ unitary,} \tag{VIII.9.c-7}$$

$$\omega(R,S) = -\omega_s(r,s); \text{ odd half-integral spin, } R \text{ and } S \text{ antiunitary,} \tag{VIII.9.c-8}$$

with r, s the homogeneous parts of R, S and ω_s a spin factor system (sec. V.4, p. 259), which we must determine (or at least limit). The sign comes from the multiplication rules for corepresentations (sec. VIII.7.b, p. 436).

Such representations, having both unitary and antiunitary operators and with multiplication rules determined up to a phase, are projective corepresentations. To find them we note that the subgroup of unitary operators is homomorphic to a group of unitary operators. Thus the methods are essentially the same as before (sec. V.4, p. 259). The representation of the unitary subgroup that is a subrepresentation of the projective corepresentation can be diagonalized. This diagonal subrepresentation is then a direct sum of irreducible representations of the subgroup, these given by a vector $\underline{k}$ of the Brillouin zone. With the transformation R written as

$$V_R \Rightarrow (\{R|\underline{t}\}, \varepsilon); \tag{VIII.9.c-9}$$

$$\varepsilon = 1 \text{ for unitary operators,} \tag{VIII.9.c-10}$$

$$\varepsilon = -1 \text{ for antiunitary operators,} \tag{VIII.9.c-11}$$

then for a basis vector ψ of the representation labeled by $\underline{k}$, $(\{R|\underline{t}\}, \varepsilon)\psi$ belongs to a representation determined by

$$\underline{k}' = \varepsilon R \underline{k}. \tag{VIII.9.c-12}$$

This gives the value of ε for antiunitary operators, these reversing the sign of the momentum. The representation space labeled by $\underline{k}$ is invariant under the group of $\underline{k}$ defined by the set

$$M^{\underline{k}} = \{w | \varepsilon R \underline{k} = \underline{k}\}, \tag{VIII.9.c-13}$$

where w is a member of M, the magnetic space group. Now M can be written as a sum of cosets

$$M = M^{\underline{k}} \oplus w_2 M^{\underline{k}} \oplus w_3 M^{\underline{k}} \oplus w_4 M^{\underline{k}} \oplus \ldots \oplus w_s M^{\underline{k}}; \tag{VIII.9.c-14}$$

the w's belong to M. We can write the basis of the representation given by $\underline{k}$ as $\psi_{11}, \ldots, \psi_{1d}$. Then the set

$$\psi_{\mu i} = (\{R_\mu|\underline{t}\}, \varepsilon)\psi_{1i}; \quad \mu = 1, \ldots, s; \; i = 1, \ldots, d, \tag{VIII.9.c-15}$$

forms a basis for the sd-dimensional representation of M under consideration. We therefore have the relations between this representation and those of the subgroup to which it reduces,

$$D(\{R|\underline{t}\}, \varepsilon)\psi_{\mu i} = D(\{R|\underline{t}\}, \varepsilon) D(\{R_\mu|\underline{t}\}, \varepsilon)\psi_{1i}$$

$$= [\frac{\omega(R,R_\mu)}{\omega(R_\nu,R_h)}]D(\{R_\nu|t\},\varepsilon)D(\{R_h|t\},\varepsilon)\psi_{1i}$$

$$= [\frac{\omega(R,R_\mu)}{\omega(R_\nu,R_h)}]\sum_j D_k(R_h)_{ji}\psi_{\nu j}; \quad (\{R_\nu|\underline{t}\},\varepsilon) \text{ unitary,}$$

$$= [\frac{\omega(R,R_\mu)}{\omega(R_\nu,R_h)}]\sum_j D_k(R_h)^*_{ji}\psi_{\nu j}; \quad (\{R_\nu|\underline{t}\},\varepsilon) \text{ antiunitary,} \quad \text{(VIII.9.c-16)}$$

with

$$(\{R|\underline{t}\},\varepsilon)(\{R_\mu|\underline{t}\},\varepsilon) = (\{R_\nu|\underline{t}\},\varepsilon)(\{R_h|\underline{t}\},\varepsilon) \quad \text{(VIII.9.c-17)}$$

which belongs to $(\{R_\nu|\underline{t}\},\varepsilon)M^{\underline{k}}$.

So the problem has been reduced to finding each irreducible corepresentations of those subgroups $M^{\underline{k}}$ whose restriction to the unitary subgroup is the direct sum of d representations labeled by $\underline{k}$.

On a representation space of $M^{\underline{k}}$ define an operator P_R, for each R of the point group of $M^{\underline{k}}$,

$$P_R = exp(-i\underline{k}\bullet\underline{t}_R)(\{R|\underline{t}_R\},\varepsilon). \quad \text{(VIII.9.c-18)}$$

These operators form a projective corepresentation of $M^{\underline{k}}$ with factor system

$$\omega(R,R')' = \omega(R,R')exp[i(R^{-1}\underline{k}-\varepsilon_k\underline{k})\bullet\underline{t}_{R'}]. \quad \text{(VIII.9.c-19)}$$

Thus the sd corepresentation of M is found by induction from a corepresentation of $M^{\underline{k}}$, and this is determined by the irreducible projective corepresentation of the point group with this factor system.

This system reduces to the factor system of pb. VII.3.b-1, p. 352, for a spin-0 particle, for which

$$\omega = 1, \quad \text{(VIII.9.c-20)}$$

and for a spin-$\frac{1}{2}$ particle,

$$\omega = \pm 1. \quad \text{(VIII.9.c-21)}$$

It is in the sign that a spinless, and a spin, factor system differ [Janssen (1973), p. 162].

Problem VIII.9.c-1: Show that P_R does not depend on the particular $\{R|\underline{t}_R\}$.

Problem VIII.9.c-2: Why does the effect of time reversal on orbital angular momentum give its action on spin?

Problem VIII.9.c-3: These results were stated, not proved. Either prove them, rigorously, or find counterexamples.

Problem VIII.9.c-4: Both projective representations, and corepresentations, are ordinary representations of larger groups (group extensions). What are projective corepresentations?

VIII.9.d Time reversal invariance and gray space groups

Time reversal invariance gives, in some cases, extra degeneracy, specifically doubling it (sec. VIII.2.d, p. 402), and requires that the eigenstates and eigenvalues of the Hamiltonian be the same for $\underline{k}$ and $-\underline{k}$ since it reverses the direction of motion. We now see how this affects the representations of the gray space groups [Bradley and Cracknell (1972), p. 650]. The study of space group representations is based on little groups each given by the momentum vector it leaves invariant (sec. VI.3.a, p. 289). Here there are two such groups for each vector, one a subgroup of the invariance group, M, the magnetic little group, the other of unitary subgroup G of M.

With these groups we can choose A (sec. VIII.7.d, p. 437) to be the time reversal operator — this commutes with the elements of the unitary subgroup, these acting on coordinates — so

$$\Delta(R)^+ = \Delta(A^{-1}RA) = \Delta(R)^*. \tag{VIII.9.d-1}$$

For a gray space group (eq. VIII.6.a-2, p. 432),

$$M = G \oplus \theta G, \tag{VIII.9.d-2}$$

the magnetic little group for momentum $\underline{k}$, $M^{\underline{k}}$, has one of three forms,

$$M^{\underline{k}} = G^{\underline{k}}, \tag{VIII.9.d-3}$$

$$M^{\underline{k}} = G^{\underline{k}} + \theta G^{\underline{k}}, \tag{VIII.9.d-4}$$

$$M^{\underline{k}} = G^{\underline{k}} + BG^{\underline{k}}, \tag{VIII.9.d-5}$$

where B is an antiunitary element which is a product of θ and some space group element other than a translation, that is

$$B \neq \theta\{E|\underline{t}\}. \tag{VIII.9.d-6}$$

While M is a gray group the little group $M^{\underline{k}}$ need not be, for the antiunitary element can be, not θ, but some product of it with a space group element. Thus this classification follows from the results of pb. VIII.4.a-4, p. 414.

VIII.9.d.i *Unitary magnetic little groups*

For the first of these cases there are no elements in G that transform $\underline{k}$ into $-\underline{k}$; that is $-\underline{k}$ is not in the star of $\underline{k}$ — if these vectors were in the same star, its invariance group would contain time reversal, since that reverses the direction of motion. For this case, $M^{\underline{k}}$ contains no antiunitary operators and is the same as $G^{\underline{k}}$, and so has no extra degeneracies at $\underline{k}$. However there is degeneracy between the spectra (the set of Hamiltonian eigenvalues) at $\underline{k}$ and $-\underline{k}$, since the little corepresentation of $M^{\underline{k}}$ must contain both little corepresentations of $G^{\underline{k}}$ and $G^{-\underline{k}}$.

VIII.9.d.ii *Magnetic little groups with antiunitary operators*

The other two cases have elements of G that transform $\underline{k}$ into $-\underline{k}$ so that these both appear in the same star. For the second case the elements that do this are in $G^{\underline{k}}$, but these, by definition of the little group, leave $\underline{k}$ invariant so $\underline{k}$ and $-\underline{k}$ are equivalent. This can occur only for points of symmetry for which $\underline{k}$ is zero, or half a reciprocal-lattice vector, since

$$\underline{k} \equiv -\underline{k}. \tag{VIII.9.d.ii-1}$$

For the third case $\underline{k}$ and $-\underline{k}$ are in the same star, but not equivalent. As all vectors in a star have the same spectra for these last two cases, the spectra at $\underline{k}$ and $-\underline{k}$ are the same, even without time-reversal invariance. Thus additional degeneracies come only from elements of θG that give degeneracies for different values of $\underline{k}$.

VIII.9.d.iii *Degeneracies depend on the magnetic corepresentation*

Whether there are extra degeneracies depends on the type of corepresentation of $M^{\underline{k}}$, according to the classification of sec. VIII.7.k, p. 449. For one type there are none, for the other two the degeneracies are doubled; in one of these case the degeneracy is between two sets of eigenvalues belonging to the same representation of the relevant little group, for the other, it is between the eigenvalues of different representations.

Problem VIII.9.d.iii-1: In the last paragraph, which cases are which?

Problem VIII.9.d.iii-2: Prove, and explain, the results of this section.

VIII.9.e The physical meaning of the representations

It is useful to look at a group in order to emphasize what this means physically, taking a double point group.

Problem VIII.9.e-1: For D_2 (sec. V.3.b, p. 258), prove that some representations are real, others pseudo-real. Interpret this physically remembering that the double group describes systems of particles with spin. What is the type of the two-dimensional representation? Why?

What are the types of the representations of D_2; of the new representations obtained in going to $D_2{}^*$? Is this related to the fact that the former group describes the arrangement of point particles, while for the latter the particles have spin, so are "rotating"?

Problem VIII.9.e–2: The magnetic groups can also be interpreted as color groups, for which there is no time dependence. Yet the basis vectors are affected by the time reversal operator. Is it possible to write basis vectors if the color group interpretation is used? Are there other interpretations of these groups besides that of the description of spinning particles? What would be the effect of time reversal for such interpretations?

VIII.10 REPRESENTATIONS OF DOUBLE SPACE GROUPS

The representations of double space groups [Bradley and Cracknell (1972), p. 454; Cornwell (1969), p. 205; Evarestov and Smirnov (1993), p. 79; Inui, Tanabe and Onodera (1990), p. 256; Janssen (1973), p. 165; Ludwig and Falter (1988), p. 218; Streitwolf (1971), p. 167] are projective representations of the corresponding space groups, where the phase is only a sign. There are two types of such representations, as with the double point groups (sec. V.3, p. 256), one for which the representation matrices are related by

$$\Gamma(\bar{E}|t) = \Gamma(E|t), \qquad \text{(VIII.10–1)}$$

corresponding to integral angular momentum representations of the rotation group, and the other for which

$$\Gamma(\bar{E}|t) = -\Gamma(E|t), \qquad \text{(VIII.10–2)}$$

corresponding to half-integral angular momentum (spin) representations. It is interesting therefore, besides their usefulness in applications in spin-orbit coupling, to consider these since they give simple examples of projective representations. We just mention a few general results [Lax (1974), p. 250].

For the interior of the Brillouin zone, and everywhere for symmorphic (double) space groups, the irreducible representations of the double little group of $\underline{k}$ can be obtained from those of the corresponding little group by multiplying the matrix representation or character of R of the double point group of the space group by a phase factor $exp(-i\underline{k} \bullet \underline{u})$ to get the representation or character of $\{R|u\}$.

To show this we recall that if the operators of the single point group obey

$$RS = T, \qquad \text{(VIII.10–3)}$$

then those of the double point group (sec. I.8, p. 59) satisfy

$$RS = \mu(RS)T, \tag{VIII.10-4}$$

where

$$\mu(RS) = \pm 1 \tag{VIII.10-5}$$

depending on which SU(2) operator these correspond to. Similarly for the space group,

$$\{R|u\}\{S|v\} = \{RS|u + Rv\}, \tag{VIII.10-6}$$

becomes

$$\{R|u\}\{S|v\} = \mu(RS)\{RS|u + Rv\}. \tag{VIII.10-7}$$

Likewise if the operators

$$O(R) = exp(i\underline{k} \bullet \underline{u})\{R|u\} \tag{VIII.10-8}$$

obey (in a projective representation)

$$O(R)O(S) = \lambda(RS)O(RS), \tag{VIII.10-9}$$

then for the corresponding double group operators

$$O(R)O(S) = \mu(RS)\lambda(RS)O(RS). \tag{VIII.10-10}$$

Now

$$\lambda(RS) = 1, \tag{VIII.10-11}$$

except possibly on the zone boundary for nonsymmorphic groups, because all representations are ordinary, except for some ray representations of nonsymmorphic groups on the zone boundary. Thus, except for these,

$$O(R)O(S) = \mu(RS)O(RS), \tag{VIII.10-12}$$

and the operators $O(R)$ realize the algebra of the double point group and have the same representations and characters.

The representations of the double group of primitive translations [Cornwell (1969), p. 206] are similar to those of the single group. The statefunction is replaced by a two-component — spinor — statefunction, which obeys Bloch's theorem in the same form as for the single group.

The representations of the double space groups are found in a manner similar to those of the ordinary space groups [Cornwell (1969), p. 207; Jansen and Boon (1967), p. 272]. The orbits of the translation subgroup in the double group are the same as the orbits of the translation group in the ordinary space group, and are classified by the stars of the vectors in the same way. The irreducible representations of the double group are again classified by the star of $\underline{k}$ and the allowable representations of its little group.

Essentially then we find the representations of the double space groups by finding those of the little double group and using the same procedure as for the ordinary space groups. Knowing the representation matrices of the double little group, and knowing how the representation matrices of the ordinary little group and space group are related, we can immediately find the matrices for the double space group. Where necessary, we have to use the factor group as the little group; in these cases the space group is not a semi-direct product.

Chapter IX

Tensors, Groups and Crystals

IX.1 MACROSCOPIC PHYSICAL PROPERTIES OF CRYSTALS

The crystalline symmetries that we have been studying are those given by the unit cell. But a crystal is (also) a macroscopic object. With unit cells aligned (as they are by definition of a lattice, and hopefully by careful crystal growth), microscopic symmetry should manifest itself in the macroscopic shape. In addition objects, say solids, here particularly crystals, have macroscopic properties that make them interesting physically, and often economically: magnetic or electric moments (giving ferromagnetic, ferrimagnetic, antiferromagnetic, and ferroelectric materials); electrical and Hall conductivitities; piezoelectricity (distortion in electric fields and production of an electric field by stress); effects on light, and so on [Cracknell (1975), p. 37, 166; Heine (1993), p. 304; Jerphagnon, Chemla and Bonneville (1978); Jones (1990), p. 78; Lomont (1961), p. 146; Lovett (1990), p. 126; Nowick (1995); Nye (1992), p. 289; Shubnikov and Koptsik (1974), p. 307]. How are these determined, and limited, by the underlying symmetry of the crystal?

The description and behavior of an object is independent of the reference frame from which it is viewed, so are unchanged by the operations of its symmetry group. The sets of functions giving this description thus form bases of representations of the rotation, and the symmetry, groups. To understand the properties, their symmetries, and their relationships to the symmetry of the crystal and the object, we have to study the functions, sets of numbers, moments, polarizabilites and the like — the group basis vectors — that determine their occurrence, nature and description. These sets, tensors, can be analyzed and restricted using group theory, hence so can the physical object itself and the properties described by these basis vectors (tensors).

Many of the concepts considered above, space groups, unit cells, Bravais lattices, Brillouin zones, Wigner-Seitz cells, play little if any part here. These generally are related to microscopic quantities, certainly the magnitudes relevant to them are microscopic, but the present discussions are for macroscopic quantities. Also many properties here are static — time is not involved. The properties that we generally consider in this book involve time, although often not explicitly. Those subjects are, say, motions of electrons or vibrations of atoms. The Brillouin zone, and so the Wigner-Seitz cell, for example, describe inverse space, that of the momentum — a fundamental quantity of representations of space groups. But most objects studied here have no momentum, and even when it appears it is not relevant to the quantities we are discussing, such as the number of components, and their relationships.

Group theory, and symmetry, regard physical quantities as representation basis vectors, not distinguishing different quantities that transform the same. Since our interest is in group theory, other properties of these quantities, essential for full understanding and application, are skimmed over. However they are needed, among other reasons, to determine the representations to which these basis vectors belong. To help readers who wish a deeper understanding we suggest problems requiring more knowledge than can be presented here. It is implied that they will use other sources, such as books on solid-state physics and crystals and on the science of materials, to obtain the needed information [Aris (1989); Bhagavantam and Venkatarayudu (1951), p. 201; Jones and March (1985); Lovett (1990); Nye (1992); Wooster (1973)]. The topics we consider provide interesting examples of the physics of basis vectors. But there is too much physics to allow discussion of anything more than basis vectors.

IX.2 TENSORS FOR THE ROTATION GROUPS

The physical quantities are rotation-group basis vectors and are often written as symbols with subscripts, the number giving the rank of the tensor. One with 0 subscripts, S, is a scalar (a zeroth-rank tensor), a rank-1 tensor, V_μ, is a vector, $T_{\mu\nu}$ is a second-rank tensor, and so on. However this, though convenient, is misleading. There are two rotation groups, SO(3) and O(3), and the number of indices does not give the transformation properties under inversion. But there is something perhaps more important.

IX.2.a The number of representations symbolized by a tensor

Tensor T_{ij} is not a single quantity, but three — it belongs to three irreducible representations, it actually represents three different physical quantities — and similarly for higher-rank tensors [Mirman (1995a), sec. XI.4.b.ii, p. 321]. Thus we may use a symbol to show that a crystal has some physical property, but the symbol actually implies that it has several — correctly or not. There may be reasons, either from symmetry, or other physical considerations, why one, two, or all three of these physical quantities must be (or at least are) zero. Also the number of independent components found from the indices is the maximum number. These components may be related, and one of the uses of group theory is to determine what the relations are, and find what symmetry alone requires. It is essential to remember that a physical entity is described by an irreducible representation, and that a symbol transforming according to a reducible representation gives several physical ones, although not all irreducible representations in its decomposition need have physical meaning — in which case there are excess, and perhaps misleading, terms. So a symbol like κ_{ij}, with nine (or if traceless, eight) components actually represents three (or two) different — independent — entities (which are not mixed by the group transformations), this being hidden by the use of the same symbol for all. It is like writing momentum with four components. Actually this denotes a (three) vector, the (three) momentum, and a scalar, the energy, so two different quantities, with different definitions and physical interpretations. There is actually a group of which the (three-dimensional) rotation group is a subgroup — the Lorentz group, the (pseudo-)rotation group of four space — and for it the (four-)momentum is a vector. If the different irreducible parts of a tensor are written with one symbol then the writer should be prepared to either find a group of transformations mixing them, or clarify any confusion that might result.

The dangers of counting indices must be emphasized. Too often physicists (usually unconsciously) assume that it is correct to set equal quantities because their numbers of indices are equal. However they can only be set equal if they transform the same way — as (perhaps different realizations of) the same basis state of the same representation, and (apparent) equality of the number of indices is neither necessary nor sufficient to determine this. Thus for the rotation group SO(3) (but not for O(3))

$$V_k = T_{ij} - T_{ji}, \qquad \text{(IX.2.a-1)}$$

where (ijk) equals (123), (231) or (312), both terms in the equation do transform the same, as a vector, though their numbers of components look different. And for most groups there are many different representations with the same number of states, for example SU(3) has two 10-dimensional representations. The states of these cannot be set equal because they transform differently, unless the group is completely irrelevant, and then it is meaningless to assign the objects to representations. Also these groups are semisimple. But as we see, many of the groups that we have to deal with are not, many are inhomogeneous. For these, not only must all states in an equation transform the same under the homogeneous part — the semisimple part — but also under the inhomogeneous part, they must be (perhaps different realizations of) the same basis state of this — in the cases considered here, of the translation part. For physical equations to be consistent all terms in it must transform as the same basis state of the entire (relevant) group.

It is often useful (and wise) to use representation basis vectors of irreducible representations, these often called symmetry coordinates [Nowick (1995), p. 46].

Problem IX.2.a-1: Show that the symmetric part of a second-rank tensor

$$T^s_{ij} = \frac{1}{2}(T_{ij} + T_{ji}), \qquad \text{(IX.2.a-2)}$$

the antisymmetric part,

$$T^a_{ij} = \frac{1}{2}(T_{ij} - T_{ji}), \qquad \text{(IX.2.a-3)}$$

and the trace

$$T = \sum T_{ii}, \qquad \text{(IX.2.a-4)}$$

all go into themselves under rotations (symmetric terms remain symmetric, antisymmetric ones stay antisymmetric, and the trace is unchanged — for orthogonal groups; how is this used?). Thus these are basis vectors of three different representations, so give three independent physical properties. This holds for SO(3) and O(3). The SO(3) reducible representation to which T_{ij} belongs breaks up (with the total number of components remaining nine) into three irreducible ones, a scalar, a vector (the antisymmetric part), and one with five (independent) components — this is traceless (why?) so that the independent diagonal components are T_{11} and T_{22} (or any other two). Show that the antisymmetric part transforms as a vector (under SO(3)); relate the three components of the vector to the components of the tensor. Note that a vector has two independent components; under a rotation its magnitude is constant, so specifying two components determines the third. Verify that the Lie-algebra angular-momentum generators expressed in terms of coordinates and momenta (derivatives with respect to coordinates) provide an example of this decomposition [Mirman (1995a), sec. X.7.c, p. 295]. Extend these results to higher rank tensors.

Problem IX.2.a-2: Check that the completely antisymmetric symbol, ε_{ijk}, which is zero if two indices are equal, 1 is they are in order 123, and -1 for order 132, is a tensor, for any orthogonal group (only, why?). That is, the transformed object has the same properties. Extend this definition to larger groups whose indices have larger ranges. Likewise the symbol ε_{ijkl}, which is zero for two indices equal, 1 for an even permutation of 1234, and -1 for an odd permutation, is a tensor. This is true for any number of indices. For SO(3) these other symbols are zero. Why? If the set of indices is a basis state of the relevant symmetric group, belonging to a representation other then the completely symmetric or completely antisymmetric ones [Mirman (1995a), sec. IX.4, p. 257], is the symbol a rotation group tensor?

Problem IX.2.a-3: The quantity

$$V_i = \sum \varepsilon_{ijk} T_{jk}, \qquad \text{(IX.2.a-5)}$$

summed over the dummy (repeated) indices, looks like a vector. Prove that it is (for SO(3)). Check that the angular momentum operators (up to constants) expressed as coordinates and their derivatives can be written in this form.

Problem IX.2.a-4: The transformation matrix of a second-rank tensor is given by the direct product of matrices for the defining representation (for the rotation groups its basis states are called vectors) so

$$V_i' = a_{im} V_m, \quad T_{ij}' = a_{im} a_{jn} T_{mn}; \qquad \text{(IX.2.a-6)}$$

this can be seen by realizing T as the product of two vectors. Another realization is that of an object transforming one vector to another. Prove that this transformation rule is correct for these realizations, by finding the transformation of these objects — of course the transformation cannot depend on the realization [Wooster (1973), p. 328].

Problem IX.2.a-5: A third-rank tensor, U_{ijk}, has, in three dimensions, 27 components. Using the characters of the rotation group [Mirman (1995a), sec. IV.5.d, p. 125; pb. X.4.c-3, p. 284; sec. XI.4.d, p. 324] show that it decomposes into one irreducible representation, a scalar (l = 0), three irreducible representations with l = 1 (vectors), two with l = 2, and one with l = 3, these having 1, 3 × 3 = 9, 2 × 5 = 10 and 7 independent components, for a total of 27. Also decompose the product of two vector representations and then decompose the product of a vector with each term in the first decomposition and verify that this gives (with multiplicity) the same decomposition as that of a three-indexed symbol. While the products of two representations of SO(3) are simply reducible, those of more than two are not; in the decomposition of the product of two representations each representation occurs (no more than) once, but for three representations there can be more than one occurrence, as seen here. The symbol U_{ijk} thus stands for seven different physical quantities. Perform the symmetrization and extraction of traces and find the independent components of these seven quantities. Check that

$$S = \sum \varepsilon_{ijk} U_{ijk}, \tag{IX.2.a-7}$$

is a scalar. Also (with the repeated index summed over)

$$V_i = \sum U_{ijj}, \; V_j' = \sum U_{iji}, \; V_k'' = \sum U_{iik}, \tag{IX.2.a-8}$$

transform as vectors. Show that no rotation mixes these three so there are three independent vector representations in the decomposition. Note that the completely symmetric traceless part has seven independent components, so transforms as the $l = 3$ representation. Find the sums that are the basis vectors of the two $l = 2$ representations, and show that they are not mixed by rotations, so describe two distinct physical quantities. Nor, of course, are they mixed with sums from other representations. These objects can be symmetrized to give symmetric-group irreducible basis states, thus for example,

$$W_{ijj} = U_{ijj} + U_{jij} + \ldots. \tag{IX.2.a-9}$$

What rotation-group representations does this give?

Problem IX.2.a-6: The character of a representation that is the product of two representations is

$$\chi(R) = \chi_1(R)\chi_2(R), \tag{IX.2.a-10}$$

giving the character of a product of a representation with itself,

$$\chi(R) = \chi_w(R)^2. \tag{IX.2.a-11}$$

What are the characters of the symmetric and the antisymmetric representations? How do they compare with the character of the rotation-group vector representation? How about the trace (scalar) representation?

IX.2.b Pseudo-tensors

A representation irreducible under SO(3) is irreducible under O(3), and conversely. But representations distinct under O(3) need not be so under SO(3). Inversion invariance places conditions on tensors and physical properties. Here we consider how inversion distinguishes representations. Coordinates change sign under inversion, and there are other vectors (say proportional to coordinates) that do also. These are polar vectors — they point to poles which are interchanged by reflections; one such describes electrical polarization, which requires two poles, that is opposite charges, with the polarization vector going from one charge to the other. There are also vectors that do not change sign; these are (not quite vectors, so are called) pseudo-vectors or (since they determine an axis of rotation) axial vectors. Angular momentum and angular velocity (about an axis) are examples.

Problem IX.2.b-1: Why physically do these two not change sign?

Problem IX.2.b-2: Why is a representation irreducible under SO(3) irreducible under O(3), and conversely?

IX.2.b.i *Tensors formed from products of vectors*

A second-rank tensor (which can be) formed from the product of two vectors (as is a magnetic field, or angular momentum — which can result from the spatial motion of an object) behaves the same under the proper and the improper rotation groups; it does not change sign under inversion. The square of a reflection is 1 so a second-rank tensor is invariant under inversion whether it is a product of two vectors, or two pseudo-vectors, but reverses sign if it is the product of a vector and a pseudo-vector.

Second-rank tensors exist whose behavior under the groups differs from ones such as those that transform the same as tensors that are realized as products of vectors (of course the transformation properties are independent of the realization, but that is often a useful means of determining the representation). Coefficients relating two vectors are second-rank tensors,

$$F_i = \sum \sigma_{ij} G_j. \qquad \text{(IX.2.b.i-1)}$$

If vectors $\underline{F}$ and $\underline{G}$ transform the same under the improper rotation group, both polar, or both axial (like angular momentum $\underline{J}$ and a magnetic field $\underline{B}$), then σ does not change sign under inversion. However if they transform oppositely (as does a current and a magnetic field [Harter (1993), p. 489]), it does change sign. Tensors relating electric and magnetic fields, or currents and magnetic fields, are examples of pseudo-tensors.

Under inversion a scalar does not change sign, a pseudo-scalar does, a vector changes sign, a pseudo-vector does not, and so on. A rank n tensor is a tensor (polar tensor) if it changes sign for n odd, and does not for n even, it is a pseudo-tensor (axial tensor) for the reverse. The product of a vector and a pseudo-vector is a pseudo-scalar. Rotational symmetry alone does not completely determine the tensor — whether it is polar or axial must also be specified.

Problem IX.2.b.i–1: Give physical examples of pseudo-scalars.

IX.2.b.ii *Restrictions on crystal properties*

A crystal with a center of symmetry cannot have a property described by a tensor that changes sign under inversion. The existence of such a tensor thus indicates that the crystalline group does not contain an inversion. And tensors that do not change sign provide no information about the presence of this element, so do not distinguish between groups differing only in its presence. These are examples of physical properties following from the restrictions imposed by inversion.

Problem IX.2.b.ii–1: Why can a crystal with a center of symmetry not have a property described by a tensor that changes sign under inversion?

Problem IX.2.b.ii–2: Find the number of nonzero components of a second-rank symmetric tensor for groups $C_{2v}(mm2)$, $S_4(\bar{4})$, $D_{2d}(\bar{4}2m)$ [Wooster (1973), p. 257, 259]. Repeat this for a second-rank symmetric pseudo-tensor. How do the results compare? Explain any differences.

IX.2.c Tensors relating vectors

One way that tensors enter the theory of solids is as operators transforming one vector to another. Both the electric field $\underset{\sim}{E}$ and the current density $\underset{\sim}{J}$ are ordinary (polar) vectors, and are related by

$$\underset{\sim}{J} = \sigma \underset{\sim}{E} \Rightarrow J_i = \sum \sigma_{ij} E_j, \qquad \text{(IX.2.c-1)}$$

where the conductivity σ is a second-rank tensor, a 3×3 matrix (in three-space), so with nine components, although perhaps not all independent, or nonzero. In general, in a solid the current is not parallel to the electric field. This can be visualized, although not in a serious way, by regarding the solid as made up of tubes with the current able to flow only along the tubes, thus not generally along the field. So σ is *not* merely a number that determines the value of the field from that of the current, it also changes the direction of one vector to give that of the other.

If a tensor relates two vectors

$$\underline{J} = \sigma \underline{E}, \tag{IX.2.c-2}$$

then under a rotation

$$\underline{J}' = R\underline{J}, \quad \underline{E}' = R\underline{E}, \tag{IX.2.c-3}$$

where R is a rotation operator,

$$\sigma' = R\sigma R^{-1}, \quad \sigma'_{ij} = R_{ik}R_{jl}\sigma_{kl}, \tag{IX.2.c-4}$$

using the orthogonality of R. This means that σ is a representation basis vector, but of a product representation of the vector representation with itself, therefore reducible. Since the realization is irrelevant a second-rank tensor is a reducible-representation basis vector, and symmetrization and taking traces is the standard method of reduction (pb. IX.2.a-1, p. 470).

IX.2.d Required symmetry in indices

The tensors that form bases of irreducible representations have symmetry in their indices. There are often reasons, physical or mathematical, why a tensor must be symmetric or antisymmetric. One is because of the nature of a particular tensor itself. If a tensor is a product, say of vectors and these are identical such as the product of coordinate vectors, the antisymmetric term must be zero. In general a second-rank tensor actually gives three quantities, but in specific cases, such as this, some do not appear.

There can be physical reasons for a particular tensor, often the antisymmetric part, being zero; this representation does not appear in the description of the phenomenon. An example is the dielectric tensor κ_{ij} that relates the electric field and polarization [Burns (1977), p. 65; Lovett (1990), p. 72; Nye (1992), p. 68; Shubnikov and Koptsik (1974), p. 313; Wooster (1973), p. 61],

$$P_i = \kappa_{ij}E_j. \tag{IX.2.d-1}$$

The product of these two vectors is an energy [Lax (1974), p. 115; Nye (1992), p. 74], as might be seen by studying a capacitor,

$$W = \sum P_i E_i, \tag{IX.2.d-2}$$

so

$$E_i = \frac{dW}{dP_i}, \quad \kappa_{ij} = \frac{d^2W}{dE_i dP_j}, \tag{IX.2.d-3}$$

and as the derivatives commute κ must be symmetric.

Problem IX.2.d-1: The tensor κ_{ij} gives the polarization of an object in a field, thus gives the direction in which its constituents move when placed in the field. From its definition there is no reason for it to be symmetric, there is no reason why the displacement in the x direction when the field is in the y direction should equal the displacement in the y direction when the field is along x. This results from an extra condition which is (apparently) arbitrarily imposed. Yet it is symmetric. Why?

IX.3 TENSORS AND SYMMETRY

How does the symmetry of a crystal determine the tensors describing its properties? What restrictions does it place upon them? And what assumptions (or is it requirements?) underlie the application of group theory to the analysis of the macroscopic properties of crystals? Those considered here are of course perfect crystals, each a single lattice. A real physical object usually consists, perhaps unfortunately, perhaps not, of a set of different lattices with relative orientations, often random, with defects and impurities, and with particles displaced by thermal motion. It is always understood that these aspects are not considered, that our analyses, so results, are merely part of the beginning of the study of actual physical objects.

This is true in other ways. Tensors give relations between physical quantities — other tensors. They are written as if proportional, thus the current density $\underline{J}$ is related to the electric field $\underline{E}$ by the conductivity σ,

$$J_i = \sum \sigma_{ij} E_j; \qquad \text{(IX.3-1)}$$

σ is a second-rank tensor (with two indices) because the current and field need not be parallel. This is Ohm's law. In general these are not proportional, $\underline{J}$ depends on $\underline{E}^n$ for all n, so is actually given by a Taylor's expansion. For small $\underline{E}$ only the first term is needed (unless there is a singularity, which should be checked if, say, the symmetry or the Wigner-Seitz cell changes, or if to first order $\sigma = 0$, if these are possible, perhaps at some values). Higher-order terms do not affect the analysis here, they just give more tensors to be considered, these treated the same way. Lowest order illustrates the methods, all we are (able to be) interested in. However when applying results, it should be understood if they are true in general, or just for the first term in an expansion. So should we find that a quantity must be zero by symmetry, it may be that this is for lowest order. In many cases only a slight generalization of the argument is needed for all orders. Also tensors for different orders give different physical properties of the crystal. Knowing the lowest-order tensor provides no information (in general) about others; different crystals have different sets, and these are also different functions (say of temperature). It is important, as we must emphasize, to know what physical properties, and how many and why, we are dealing with.

Another aspect that could be relevant in studies of actual materials, but not considered here for it does not change the analysis, is hysteresis. The magnetization for example can depend not on the external field now, at time t, but on its history, on what it was at time t_p. Thus we might have

$$M_i(t) = \sum \mu_{ij} B_j(t_p), \qquad \text{(IX.3-2)}$$

or perhaps more likely

$$M_i(t) = \int \sum \mu_{ij}(t, t_p) B_j(t_p) dt_p; \qquad \text{(IX.3-3)}$$

it depends on the history of the external field (and perhaps on the history of M). Also the susceptibility could vary with time. Here we are concerned only with the transformation properties of these objects, so such possibilities are included, although not shown explicitly.

IX.3.a Foundations of the tensor analysis of crystal properties

In considering physical tensors, we have to relate these macroscopic quantities to the symmetry of the microscopic unit cell, and also to two groups, the rotation group, of which these are basis vectors, and the symmetry group of the crystal. There are two linkages so two principles on which the analysis is based, one for each.

IX.3.a.i *Neumann's principle*

The first is Neumann's principle [Burns (1977), p. 63; Burns and Glazer (1990) p. 167; Cracknell (1975), p. 11; Kocinski, (1983), p. 79; Lovett (1990), p. 31; Nowick (1995), p. 64; Nye (1992), p. 20; Wooster (1973), p. 15]. It states that all macroscopic quantities describing a crystal have — at least — the symmetry of the point group of the crystal [Lax (1974), p. 111]. This implies that some quantities can have greater symmetry — why? Consider a second-rank tensor symmetric in its indices. It can be diagonalized giving three nonzero components. For a cubic crystal the number of conditions is such that all three must be equal, hence the tensor has not merely symmetry under the group of the cube, the octahedral group, but under the full rotation group. It has greater symmetry than the crystal. Besides the symmetry of the point group there may be other conditions, as symmetry of the indices, giving greater symmetry for the tensor.

Neumann's principle applies to tensors specifying properties of the crystal. It does not apply to other tensors which we consider, for example the stress (due to an external force) is determined by the imposed force, not by the properties of the object on which it is imposed; this object cannot place restrictions on it.

In general, the number of restrictions imposed by the symmetry of a crystal is not related to the dimension of the representation to which a tensor belongs; it can even be greater. There can be enough requirements — coming from the symmetry, the nature of the tensor (such as symmetry in indices as with products or resulting from commuting derivatives, say), or the limit on the number of basis vectors of the representation — to make a tensor zero, or to relate components, giving it greater symmetry than the crystal. It can have greater symmetry than given by the crystal point group but not less. This provides explanations for many zero tensors (the nonappearance of certain physical properties) or for relations among components (the reduced freedom to specify properties of objects) which might not be apparent otherwise.

Problem IX.3.a.i-1: Take the coordinate axes along the edges of the cube and assume that the tensor then has nonzero components T_{xx}, T_{yy}, T_{zz}; it is clear that since all three edges are identical these are equal. But the tensor need not be diagonal in this coordinate system (need it?). Show however that no matter which system the tensor is diagonal in, an equivalent argument holds, and these are always equal.

IX.3.a.ii *Is Neumann's principle obvious?*

Neumann's principle should be obvious — almost. If the crystal has a symmorphic space group than it is symmetric under the point group of the space group, thus so are all its properties. But if the space group is nonsymmorphic then it is noninvariant under the point group, so why should its — macroscopic — properties be invariant?

Rigorously this follows from properties of the representations of space groups: since tensors are invariant under translations, they belong to the $k = 0$ representation, so are invariant under the point group [Lax (1974), p. 244]; the object the tensor describes, the crystal, has no momentum, although objects within it do. There is also a simple physical argument, and luckily it gives the same result as the rigorous mathematical one. A crystal, so the tensor, is invariant under a space-group operation, a point-group operation followed by a translation through a fraction of a lattice vector. For a nonsymmorphic space group it is not invariant under the point group, so we cannot conclude from space-group symmetry that the tensor is invariant under the point group. The tensor obtained by a point-group operation can be found by applying a space-group operation, under which the tensor is invariant, followed by a translation through a fraction of the lattice spacing, which might change the tensor. But how much is this change? Clearly for a macroscopic property such a small translation will give a change too minute to be detected. Thus the tensor is (effectively) invariant under the point group.

And what is a macroscopic property? Here it is one that varies so slowly that its change under such a translation is smaller than the precision of the experiment being done. But since the tensor is invariant under a translation of a lattice spacing the argument breaks down only if the experimental precision is smaller than this spacing.

It is because of Neumann's principle that here we consider only point groups, not space groups.

Problem IX.3.a.ii-1: This argument seems to imply that the space group operation actually changes the tensor, only the change is unnoticeable. But rigorously there is no change. How are these arguments related?

IX.3.a.iii *Why tensor components are point-group scalars*

The second fundamental requirement is that each component of the tensor be a scalar (transform under the identity representation) of the point group of the crystal. This may seem strange — how can a component of a tensor be a scalar? However the groups, and the operations, are different, and this emphasizes the importance of defining the meaning of the transformations, carefully.

Take a component of say a vector, V_x. Under (relative) rotations of the crystal with respect to the axes drawn in space (the word "fixed" is too confining here), this component is mixed with others. However under the operations of the invariance group of the crystal it — each component separately — is a scalar. It is not changed by an operation that leaves the crystal invariant — that takes each point to an equivalent point.

Why is there a difference between the invariance group of space, for which the tensor can belong to any representation, and the invariance group of the crystal, for which it must be a scalar? Consider a vector, a magnetic field produced by the crystal for example. Different (external) observers have different views of it — its components in different coordinate systems are different. However it is along some line in the crystal, which can be picked as an axis, so symmetry transformations of the crystal cannot move it with respect to that line — its components along these axes are fixed, thus it must be a scalar under the group. This is true no matter what rotation-group representation it transforms under. A rotation of the crystal changes components with respect to external axes, but not with respect to axes fixed in the crystal.

A cubic crystal and a vector describing it that lies along a diagonal illustrates this. If the x axis in space is drawn along the diagonal, the vector has one nonzero component, V_x. Changing the orientation of the crystal with respect to the axes — an operation of the rotation group — changes these components. However an operation of the crystal invariance group, say one interchanging diagonals, leaves V_x invariant. The component (of the tensor under O(3)) is a scalar under group O_h.

Problem IX.3.a.iii-1: Actually this implies there is, in this case for example, not one vector, but one along each diagonal. These are all equivalent and are interchanged by O_h. But since there is no distinction between the vectors before and after the transformations of O_h, they have to be taken as scalars under this group. While this provides a clear and simple example illustrating this point (what point?), it should be obvious that there can be no vectors along a diagonal describing physical properties of a cubic crystal. Can there be any such vectors at all?

Problem IX.3.a.iii-2: Pick a crystal that can have a vector property, and explain why each component is a scalar. Repeat this for a higher-rank tensor.

Problem IX.3.a.iii-3: Why is a tensor under the rotation group a scalar under its subgroup (which all point groups are)? Also a group is (here) a set of transformations. What are the operations of these groups?

IX.3.b Equilibrium and non-equilibrium properties

Although time does not enter our discussion, there is one aspect to which it is relevant [Cracknell (1975), p. 44; Nowick (1995), p. 20]. All quantities here are (taken as) time-independent, all are steady-state, but not all are static. An electric field produces a polarization (in general), and this is static, nothing changes, there is no change in entropy. The field also produces a current, and this is constant in time (always assuming the field, and the conductivity tensor are). But here there is a change, a flow of charge, so an increase in entropy. This conductivity tensor is an example of a transport property, describing the transport of something; non-equilibrium thermodynamics [de Groot and Mazur (1984); Haase (1990)] provides much of the foundations for studies of these. For most of the discussion this is irrelevant, but it is an essential distinction when considering time reversal (with operator denoted by θ). There are quantities that must be zero because, and only because, of it. This must always be checked when magnetism is involved.

Neumann's principle holds for time-reversal symmetry — for static properties [Nowick (1995), p. 83], and may place restrictions on these tensors. However this extension cannot be made for transport properties.

Problem IX.3.b–1: Show that if it were extended the principle would require that the conductivity be identically zero [Nowick (1995), p. 84]. Explain how the extension fails.

IX.3.c Magnetic tensors and the effect of time reversal

A magnetic field is reversed by reversal of time — it is produced by a current (perhaps due to spin) whose direction is reversed. What restrictions are placed on tensors for magnetic properties by time-reversal invariance [Cracknell (1975), p. 40; Nowick (1995), p. 85]? Obviously only groups containing θ need be considered. There are two types of tensors, one invariant under this transformation like coordinates, electric fields and electric polarizations, the other changes sign, for example currents and magnetic fields.

For gray groups (sec. VIII.4.b.iii, p. 418), tensors relating two objects that are both time-symmetric, or both time-antisymmetric, are time-symmetric so the presence of θ does not affect them. If one is symmetric, the other antisymmetric, the tensor relating them is time-antisymmetric and must be identically zero for these groups as θ is a symmetry.

The other groups are the black-and-white ones (sec. VIII.4.c, p. 419). Again if two physical tensors are both time-symmetric, or time-antisymmetric, the tensor relating them is time-symmetric, and time reversal puts no restrictions on it.

The interesting case is if one of these is time-symmetric, the other antisymmetric. Then they are related by a time-antisymmetric tensor, and it is a basis vector of a black-and-white group. The general theory for these representations can be used (sec. VIII.7, p. 434) to find the structure of the tensor, however it can usually be found more simply in actual physical cases by using the properties of the specific system.

IX.3.d The number of independent components

A tensor is a rotation-group basis vector and the number of its components is determined by the representation. However it describes the properties of a crystal, thus is restricted by its symmetry group. This relates components, so the number of independent ones is generally less than that of the rotation-group representation. If the number is 0, a crystal with that symmetry cannot have the property. We need to determine the number of independent components of a tensor describing an attribute of a crystal in order to study it [Bhagavantam and Venkatarayudu (1951), p. 210; Ludwig and Falter (1988), p. 198; Wooster (1973), p. 243].

One application of group theory is finding this number, and the specification of which components are related. This has been worked out for low-rank tensors for all point groups and the results tabulated [Bhagavantam and Venkatarayudu (1951), p. 208; Lovett (1990); Nowick (1995), p. 18; Nye (1992)]. However one of the purposes of the problems below is to find some of these results, which should be done before looking at the tabulations.

Consider a crystal with a vector property, a dipole moment, say. If the group leaving the crystal invariant includes operators that move every axis, there is no direction in which this vector can point. Hence it can have no property given by a vector. If there is some axis invariant under all group operations such a property is possible (but of course not necessary). The function describing the property, while a vector under the rotation group, is an invariant under the crystal symmetry group — the vector is along the invariant axis (whose direction is changed with respect to axes in space by rotations).

If irreducible rotation-group representation l is reducible under the point group, the rotation-group character for element R is

$$\chi_l(R) = \sum n_i \chi_i^p(R); \tag{IX.3.d-1}$$

n_i is the number of times point-group representation i, with character χ_i^p, appears in rotation-group representation l [Mirman (1995a), sec. VII.6.a, p. 199]. Using the orthonormality of characters [Mirman (1995a), sec. VII.5.c, p. 195],

$$n_i = \frac{1}{g}\sum_R \chi_l(R)\chi_i^p(R)^*, \tag{IX.3.d-2}$$

summing over all g elements of the point group. Now each tensor component transforms as the point-group identity representation all of whose characters are 1. So the number of times, N, the point-group identity representation appears in representation l is

$$N = \frac{1}{g}\sum_R \chi_l(R). \tag{IX.3.d-3}$$

Each appearance is due to an independent tensor component, therefore this is the number of its independent components.

Problem IX.3.d-1: Why is each appearance due to an independent component?

Problem IX.3.d-2: Show that the characters of the O(3) group [Mirman (1995a), sec. IV.5.d, p. 125, pb. X.4.c-3, p. 284], in the defining representation, for the class given by angle θ , are

$$\Gamma^{+} = 1 + 2cos\theta, \tag{IX.3.d-4}$$

$$\Gamma^{-} = \pm(1 + 2cos\theta), \tag{IX.3.d-5}$$

for representations even, and odd, under inversion. How do these differ from the characters of SO(3); why? Then the characters of the point group elements (that of R^{-1} is the same as of R) are [Lax (1974), p. 71], with the * indicating an improper element,

	1	2	3	4	6	1*	2*	3*	4*	6*
	E	C_2	C_3	C_4	C_6	I	σ	S_6	S_4	S_3
Γ^+	3	−1	0	1	2	3	−1	0	1	2
Γ^-	3	−1	0	1	2	−3	1	0	−1	−2 .

(IX.3.d-6)

Explain why the character of R^{-1} is the same as that of R. Check that all point-group elements are in this table, or are products of these.

Problem IX.3.d-3: Verify that the number of independent components of a symmetric second-rank tensor is determined not by the group, but by the crystal system — by the holohedry group — and all its subgroups allow the same number of components. Is this true in general? Why? Give the number, and indices, of the independent components of a second-rank tensor for each system [Lax (1974), p. 117].

Problem IX.3.d-4: For group D_3(32), show that the number of independent constants for a vector is 0 — if a crystal has this symmetry it can have no vectorial property [Wooster (1973), p. 246].

Problem IX.3.d-5: For point group C_{3v}(3m), check that the number of independent components of a vector is 1 — a crystal with this symmetry can have vector properties but the symmetry relates the three components [Wooster (1973), p. 247].

Problem IX.3.d-6: How many independent components for a vector are there for the group $T_d(\bar{4}3m)$ [Wooster (1973), p. 247, 257]?

Problem IX.3.d-7: For a second-rank tensor (one with two indices) show that for point group $D_{2h}(\frac{222}{mmm})$ the number of independent components is 3 [Wooster (1973), p. 249]. If the tensor can be diagonalized this means the three diagonal elements are independent. Why might the tensor not be diagonalizable? Must the tensor be symmetric in its two indices?

Problem IX.3.d-8: How many independent components are there for a third-rank tensor for groups $S_2(\bar{1})$ and T(23) [Wooster (1973), p. 249, 257]? Does it matter if the tensor is symmetric or antisymmetric in its two indices? This tensor gives three physical properties. What does the number of components indicate about these properties?

Problem IX.3.d-9: For group $D_3(32)$ show that a third-rank tensor has two independent components [Wooster (1973), p. 251]. How many properties does this tensor give? What can we conclude from the number of components?

IX.3.e The meaning of the number of independent components

For a tensor of rank n we can find, as in these examples, the number of independent components for different groups. It is important to note whether this number gives the number of independent basis vectors of an irreducible representation or the number for different, and independent, representations.

Thus for a symmetric second-rank tensor the number in some cases is three, these being the diagonal elements. However they represent two physical quantities. This means that the group places no restriction on these two tensors. If the number were two then the independent quantities might be the two (independent) components of a vector representation, and the other (scalar) does not appear — it does not describe (anything related to) the physical property, or there might be two representations with the vector having only one independent component. These two cases are physically different. If there is one independent component, then one of the two representations does not occur, and the choice is not arbitrary, and if the scalar is the one absent then the vector representation has only one independent component. Similar results holds for other tensors.

IX.3.f Requirements imposed by symmetry elements on tensors

Certain symmetry elements and groups are incompatible with the existence of some tensors; the presence of an element requires such tensors to be zero. For example if the symmetry group contains an inversion all vector properties are zero. Thus if (it is known that) a crystal has a property given by a particular tensor we can rule out all symmetry elements and groups for it that do not allow such a tensor (which can be experimentally useful in studying the crystal). Also conflicting restrictions due to the presence of more than one element lead to components, or entire tensors, being zero. The greatest symmetry, so the strongest restrictions, come from the holohedry group of a crystal system. An experimentally determined violation of a requirement shows that the symmetry is broken, the crystal has lesser symmetry than that of the holohedry group. Nonzero properties give information about the crystal symmetry. However these may not be easy to discern experimentally; measurements may distort the crystal decreasing its symmetry, and properties may not be zero but have values below experimental limits. In the real world these issues are essential, but since we study only principles we must ignore them. Here we list some requirements from symmetry.

Problem IX.3.f-1: Under O(3) a vector transforms as

$$V_i' = a_{ij}V_j = V_i; \qquad \text{(IX.3.f-1)}$$

the right equation is true if there is symmetry. For an inversion

$$a_{ij} = -\delta_{ij}, \qquad \text{(IX.3.f-2)}$$

so the vector is zero, if inversion is a symmetry. A crystal with a symmetry group that includes the inversion can have no vector property. Prove that in general it can have no properties given by tensors odd under inversion; this does not depend on the number of indices, or any other representation label. In particular scalars are not limited by symmetry, although pseudo-scalars are excluded by a center of symmetry — these tensors are odd under inversion. Thus for crystal classes containing the inversion, even-rank axial tensors, and odd-rank polar tensors must vanish [Kocinski, (1983), p. 80].

Problem IX.3.f-2: Check that a two-fold symmetry axis, say along z, for which

$$a_{ij} = -\delta_{ij}, \; i,j = 1,2, \quad a_{33} = 1, \qquad \text{(IX.3.f-3)}$$

requires that any crystal vector be parallel to z [Wooster (1973), p. 20]. How many independent components does this vector have? The same holds for three-fold, four-fold and six-fold axes. A plane of symmetry perpendicular to z requires that the vector lie in that plane. Find the components if the plane of symmetry is parallel to z and bisects the angle between the x and y axes [Wooster (1973), p. 21, 23]. Thus for vectors we have found all conditions imposed by the symmetry elements: they are ruled out by inversion, required to be along a symmetry axis, and in a plane of symmetry, and the components in that plane can be related by its orientation. If a crystal has a symmetry axis and a plane of symmetry not parallel to that axis (so a symmetry plane and more than one axis) it can have no vector property. These provide examples in which symmetry requires certain tensors, say the antisymmetric part of a rank-two tensor, which transforms like a vector under SO(3), to be zero.

Problem IX.3.f-3: A tensor with two indices transforms as the product of two vectors (a direct product), which does not mean that it can be written as such a product; the transformation matrix is the (direct) product of two rotation matrices [Mirman (1995a), sec. XII.2, p. 340]. Thus verify that a two-fold axis (along z) imposes no requirement — the transformation matrix acts as the identity, and in general gives no requirement on a tensor with an even number of indices. For a tensor with an odd number, the requirement is that a component be zero unless the number of x and y indices is even. State the requirement for the symmetry axis in an arbitrary direction. A vector can be taken as an object with a single index, or as the antisymmetric part of one with two. Check that these both give the same result. Note that here the number of indices does not fully determine the representation (for two indices there are three representations). However this restriction on the nonzero components is determined by the number of indices — it applies to all representations with the same value of this label.

Problem IX.3.f-4: It should be clear that the result is the same for a symmetry plane. State it for an arbitrary orientation of the plane. How are the nonzero components related by the plane?

Problem IX.3.f-5: For a four-fold axis, components of a vector in the plane perpendicular to the axis are ruled out by the minus sign introduced by the rotation. So a tensor with an odd number of indices is also excluded. For three- and six-fold axes, the vector component in the perpendicular plane is changed, so the component of a tensor bearing such an index is zero. State this result carefully, and for axes at an arbitrary angle to the x, y, z coordinates. Give the relations for the nonzero components.

Problem IX.3.f-6: For each crystal system determine if a vector property is allowed by its holohedry group (tbl. B-1, p. 633). Suppose one not so allowed were found. What would be the largest subgroup that the crystal could be invariant under? Could the vector be used to determine what the symmetry group actually is? How?

Problem IX.3.f-7: If a crystal has a property given by a second-rank tensor what can be said about the crystal system to which it belongs, and its symmetry group? Would it help in answering the latter question if the crystal system were known? Could an experimental study of the tensor yield further information about the group?

Problem IX.3.f-8: How many tensors, and of what type, are needed to determine the crystal system to which an object belongs? How about its symmetry group? Answer this question for both knowing, and not knowing, the crystal system.

Problem IX.3.f-9: Which of the 32 point groups allows physical properties given by third-rank tensors, and by third-rank pseudotensors? Do these answers differ? Why?

IX.3.g How group theory provides information about tensors

One of the uses of group theory is to determine the number of distinct physical quantities needed to describe a system. Thus without a knowledge of the rotation group it would be quite difficult to see an important physical fact: a second-rank tensor, say, gives not one, but three different properties. It would be like not knowing whether electric and magnetic dipole moments are physically different quantities. And without group theory it would be also be hard — even impossible — to know how many numbers are needed to specify a quantity, why that many are needed, and how this is related to a particular crystal. There would be little understanding of the reason very different physical quantities are given by the same number of components, why they have the same relations among components, why the indices of nonzero components are what they are, why these answers are the same for many different crystals, and why they differ when they do. For many other applications of group theory we can obtain information, perhaps with greater difficulty and less understanding, by solving equations of motion, but it is difficult to see how the distinct physical quantities here could be found, or how they could be understood, without using group theory. That it is needed is not surprising — these facts are determined by symmetry. And this is emphasized by the many discussions of tensor properties that use little or no group theory, and fail to indicate an understanding of even the number of quantities being considered — thus losing much physical understanding, and even resulting in misunderstanding.

IX.4 RANK-1 TENSORS — ELECTRIC AND MAGNETIC DIPOLE MOMENTS

Having discussed the general properties of these tensors, we next review some examples starting with properties that are described by the lowest-rank tensors (aside from the trivial scalar), vectors [Lax (1974), p. 112; Nowick (1995), p. 93, Wooster (1973), p. 19]. These include dipole moments, electric and magnetic. Group-theoretically the cause of dipole moments, and their physical interpretations, are irrelevant, they are merely basis vectors of the adjoint (regular) representation

[Mirman (1995a), chap. VI, p. 170] of the rotation group. There is no difference between electric and magnetic dipole moments for the proper rotation group — it has only one vector representation. However there is a difference, due to inversion, for the improper group. The electric dipole moment is proportional to the displacement vector from one charge to another so changes sign under inversion, as do coordinates. The magnetic dipole moment is proportional to an angular momentum — it might be due to a charge moving in a circle so of the form of a coordinate times a momentum, $r \times p$, and both change sign, so it does not. Electric dipole moments are vectors (polar vectors), angular momenta and (so) magnetic dipole moments are pseudo-vectors (axial vectors — angular momentum has an axis). Which of these are possible for a crystal depends on whether it has inversion as a symmetry element. If it does not, then it can have both types of moments.

Properties described by axial vectors include the magnetocaloric effect, and its converse, the pyromagnetic effect [Nowick (1995), p. 6, 95].

For a crystal with symmetry group C_3 (3) there are no improper elements (reflections or inversions). The three-dimensional representation of the rotation group becomes a reducible three-dimensional representation of C_3; the characters are (3,0,0), so $N = 1$ (eq. IX.3.d-3, p. 482). There is one independent component as we expect, the value of the moment along the axis that is left invariant, the threefold axis.

For group C_{3v}(3m), the characters are (3,0,1) and (3,0,-1), for the vector and pseudo-vector representations. Hence the number of independent constants for the electric dipole case is again 1, but for the magnetic case it is 0. Since the crystal is invariant under a reflection, while a magnetic moment is not, there cannot be a magnetic moment. Such a crystal can be ferroelectric but not ferromagnetic.

Problem IX.4-1: Check the statements of these two paragraphs.

Problem IX.4-2: If a crystal has a plane of symmetry through z that is at equal angles to x and y, what are the possible nonzero components of a vector property [Wooster (1973), p. 21, 23]?

Problem IX.4-3: We can now list the point groups that allow ferroelectric crystals, those that allow ferromagnetism, and those that allow both. Check that the only crystals that can be ferroelectric are those with point groups C_1, C_2, C_3, C_4, C_6, C_{1v}, C_{2v}, C_{3v}, C_{4v}, C_{6v} [Lax (1974), p. 113]. A crystal can be ferromagnetic only if its point group is C_1, C_2, C_3, C_4, C_6, C_i, C_{2h}, C_{4h}, C_{6h}, S_6; or one isomorphic to C_n, that is $C_s \sim C_2$, $S_4 \sim C_4$, and $C_{3h} \sim C_6$ (pb. I.6.d-9, p. 43). Does the International notation give further insight about the groups allowing these properties? The second set is obtained by adjoining the inversion to the members of the first set. There are thus some crystals that are allowed to be both ferromagnetic and ferroelectric, others that can be only one.

Are there any that can be neither? For each of these crystals explain why each component of the dipole moment is a scalar under the point group. This discussion does not consider time reversal. Would that change these groups [Cracknell (1975), p. 42]?

IX.5 SECOND RANK TENSORS

Dipole moments provide examples of first-rank tensors — vectors. Tensors that relate vectors provide examples of second-rank ones. These give three irreducible representations, the ones antisymmetric and symmetric in the indices, and the (invariant) trace.

Vectors can usefully be visualized (and it is no more than that) as arrows. A symmetric (part of a) second-rank tensor, T_{ij}, can also be represented geometrically, though not so usefully, by a surface, the representation surface, or representation quadric [Aris (1989), p. 271; Lovett (1990), p. 27; Nowick (1995), p. 195; Nye (1992), p. 16; Wooster (1973), p. 25]. The equation

$$\sum T_{ij}x_ix_j = 1, \qquad \text{(IX.5-1)}$$

defines a surface in real space which shows some of the properties of the tensor. If all T's are nonnegative it is an ellipsoid, else it is hyperbolic (or in extreme cases, parabolic). The axes for which T is diagonal are the principle axes of the figure [Nye (1992), p. 41]. If there is circular symmetry it becomes a surface of revolution. For spherical symmetry the ellipsoidal representation surface reflects the symmetry, becoming a sphere. The best way of describing the maximally symmetric hyperboloid (there cannot be spherical symmetry) is as a (light) cone.

Problem IX.5-1: The tensor of an irreducible representation has to be traceless. What is the significance of this for the representation surface?

Problem IX.5-2: In studies of examples below, and in other situations in which second-rank tensors are met, it would be useful to consider which representation surfaces are ellipsoidal, and which need (or must) not be, and whether any can be paraboloidal, and the physical reasons for, and implications of, these. It would also be nice to design crystals for which each property has a paraboloidal surface, or show that there cannot be any.

IX.5.a Thermal conductivity

A temperature gradient, the derivative of the temperature with respect to coordinates thus a vector, causes heat flow. The flow of heat is the (heat) energy crossing a surface per unit time, thus is also a vector (actually a vector function of the angle of orientation of the surface) with direction that of the normal to the surface.

The greater the temperature gradient, the greater the heat flow, and we take them proportional (to lowest order), though for a crystal not necessarily parallel. Thus, with temperature gradient $\frac{dT}{dx_j}$, and heat flow $\frac{dQ}{dx_i}$,

$$\frac{dQ}{dx_i} = K_{ij}\frac{dT}{dx_j}; \qquad \text{(IX.5.a-1)}$$

K is the thermal conductivity, a second-rank tensor [Lovett (1990), p. 36; Nowick (1995), p. 20; Nye (1992), p. 195].

By the second law of thermodynamics heat flows from a point of higher temperature to one of lower, so all components of K are positive, and the representation surface is an ellipsoid.

Problem IX.5.a-1: Define these quantities carefully, in particular "angle of orientation" is undefined.

IX.5.a.i *The meaning of the symmetric and antisymmetric parts*

Since K_{ij} stands for three independent physical quantities we have to see what these are.

Problem IX.5.a.i-1: Show that the symmetric part of K gives the heat flow along straight lines, the antisymmetric part part gives heat flow along a spiral — emphasizing that these refer to different physical properties of the crystal [Nye (1992), p. 205; Wooster (1973), p. 26]. Design a crystal for which heat flows only linearly, in which it also flows spirally, and one in which it flows only spirally (the tensor is antisymmetric [Nowick (1995), p. 21]), or prove that these are impossible — so that there are physical reasons why this tensor is symmetric or antisymmetric in its indices. Would they have technological value? The symmetric part is actually two tensors, the trace, a scalar, and the traceless part. What are their physical meanings? An instructive example is given by a circular disk heated at the center whose edge is kept a constant temperature. This is circularly symmetric. Can heat move along a spiral? Circular symmetry requires an equal probability for each orientation of the spiral. However the initial conditions are never exactly symmetric; the source or sink is never precisely at the center. Thus heat (presumably) can — if there is no other reason to prevent it — start moving in an arbitrary direction, and then spiral outward. Since K is a property of the crystal, once we have determined its properties using this simplified setup, circular symmetry, the properties hold in general, and the temperature distribution does not matter, does it? We do not consider these interesting possibilities but limit the discussion to crystals for which the thermal conductivity is a symmetric, second-rank tensor — one independent of position and temperature (although since we are considering only lowest-order it is also independent of the temperature gradient). Discuss what — essential — aspects, that are here relevant, would change if each restriction did not hold (and they likely are not exact in reality). Interesting extensions might be found by making some of these time dependent.

IX.5.a.ii *The physical quantities given by the symmetric parts*

The thermal conductivity, taken as symmetric, still represents two distinct physical quantities. What are these? Can either be zero, but not the other? Since K is independent of the thermal gradient we can take that as the simplest possible — spherically symmetric — and write the gradient as dT/dr. Then

$$\frac{dQ}{dx_i} = \sum K_{ir}\frac{dT}{dr}, \qquad \text{(IX.5.a.ii-1)}$$

for each i. Let

$$s = \sum K_{ii}, \qquad \text{(IX.5.a.ii-2)}$$

denote the invariant trace. It is easiest to visualize this first in two dimensions. The two physical quantities are s and the traceless tensor with components

$$K_{12},\ K_{11} - \frac{s}{2},\ K_{22} - \frac{s}{2}. \qquad \text{(IX.5.a.ii-3)}$$

We pick coordinates to diagonalize the tensor, so

$$K_{12} = 0. \qquad \text{(IX.5.a.ii-4)}$$

Hence there are two quantities, s and

$$\Delta = K_{11} - K_{22}. \qquad \text{(IX.5.a.ii-5)}$$

The total heat flow, integrated over angle, is

$$\Delta Q = \int d\theta\left(\frac{dQ}{dx_1} + \frac{dQ}{dx_2}\right) \sim s\frac{dT}{dx}. \qquad \text{(IX.5.a.ii-6)}$$

Thus s gives the total heat flow, Δ the difference between the flow along the two principle axes (for a spherically symmetric temperature gradient). These are two distinct properties of the crystal. It is clear that there can be objects for which the flow is symmetric, so $\Delta = 0$; the traceless, symmetric second-rank tensor is zero.

This emphasizes that symbol K_{ij} gives not one quantity, but three, here the spiral heat flow, the total linear heat flow, and the asymmetry of the latter. Some may be zero, indeed may have to be zero. But the statement that one is zero gives a physical property, following perhaps from symmetry, or (also) other physical reasons.

Problem IX.5.a.ii-1: Design a crystal for which $\Delta \neq 0$, or prove that such is impossible.

Problem IX.5.a.ii-2: Is it possible to imagine a crystal for which the heat flowing along one principle axis (for this temperature distribution) is the negative of that along the other, giving $s = 0$, so the total heat flow is zero (but Δ is not)? Would it be interesting? Could it be used for anything?

Problem IX.5.a.ii-3: For three dimensions the situation is the same, though slightly more complicated. Give, and describe the meaning of, the physical quantities in this case. Does the antisymmetric part still give a (perhaps impossible) spiral or can the curve (surface?) be more interesting? Find, or invent, crystals for which these various quantities are nonzero, alone or in combinations, or prove that particular ones are impossible (using what physical assumptions?). Find uses for them.

IX.5.b Thermal expansion

If an object is heated its size and shape changes [Lovett (1990), p. 106; Nowick (1995), p. 6; Nye (1992), p. 106; Wooster (1973), p. 36]. There is an atom (at least hypothetically) at some point that does not move (we do not consider whether there can be more than one such point, which might be the case if the expansion took the form of a wave); that point we take as the origin. Then an atom (or classically, a small region of the object) at point x_i is displaced by dl_k. The displacement is proportional (again to first approximation) to the distance from the origin, and the temperature, T. So

$$\frac{dl_k}{dx_i} = s_{ki}T, \qquad \text{(IX.5.b-1)}$$

where the s's are the coefficients of thermal expansion. We take them independent of x and of T (although higher powers of T, which is a scalar, do not affect the discussion — they would only give more tensors of the same type; however an expansion depending on the gradient of T gives higher-rank tensors). We regard the tensor as symmetric, and ignore the question of whether there can be antisymmetric ones.

Some materials might expand so much in one direction that they have to contract in others. Then different components of s_{ij} would have different signs, and the representation surface would be a hyperboloid. While an ellipsoid is bounded, a hyperboloid is not. If the object expands along all axes the displacement, and the inverse displacement, is always bounded (the thermal expansion tensor is defined inversely to the tensor in the equation for the representation surface (eq. IX.5-1, p. 489). However if it expands along one axis, and this displacement is bounded, and contracts along another, then the distances of all atoms from the origin along this axis (assuming the coefficients do not change) become zero. The properties of the hyperboloid are relevant, and perhaps serve as a warning to consider whether the assumptions used always hold.

Problem IX.5.b-1: Define the inverse displacement in terms of the displacement.

Problem IX.5.b-2: Explain physically — a model would be nice — why thermal expansion is (generally) a tensor (not a scalar). Are there objects for which this tensor can have an antisymmetric part? If so, what are their properties, if not, why? What physical property would an antisymmetric tensor describe?

Problem IX.5.b-3: If the tensor components have both signs, there must be some directions of zero expansion. Could you tell this from the representation surface being a hyperboloid? Why? A simple model (even in two dimensions) of an object that expanded in one direction and contracted in others should emphasize the meaning of representation surface, and show which assumptions used here should be checked when the change of shape becomes too great. Could the surface ever be a paraboloid? Why?

Problem IX.5.b-4: A tensor, symmetric in its indices and independent of coordinates, describes a homogeneous expansion: straight lines remain so, angles are unchanged (so parallel lines also remain so), but distances are varied. The transformations given by the tensor are affine (sec. III.2.b, p. 134). Under what conditions would a tensor describe transformations that curved lines, changed angles, or caused parallel lines to meet (changing the geometry of the object from Euclidean to non-Euclidean)? Is it possible for this tensor, or ever? If there are materials that had such properties would they be of economic value?

Problem IX.5.b-5: The symmetric part of s gives two independent physical quantities. What are they? When would one, but not the other, be zero? The symmetric part can be diagonalized by transforming to principle axes. What is the significance of these axes? The significance of the diagonal components? What do they tell about the crystal, how are they related to it (and its symmetry group)?

IX.5.c Stress and strain

A force is a vector, and acting on a point particle we need only specify its magnitude and direction. Of course this is all that can be given for a force. But to find its effect on an extended object the orientation of the object with respect to the force must be known. The behavior of the object is determined by the integral, over its surface, of the force acting on each small area of the surface, times the area, thus by the stress tensor

$$c_{mn}(x) = f_m(x)a_n, \tag{IX.5.c-1}$$

with $\underline{f}$ the force at x, and $\underline{a}$ the vector normal to the surface [Harter (1993), p. 471; Joshi (1982), p. 305; Lovett (1990), p. 56; Nowick (1995), p. 199; Nye (1992), p. 82; Wooster (1973), p. 55]. The total force is the integral of this over the surface.

The stress tensor is not a property of the crystal, but is imposed externally. Therefore the symmetry of the crystal does not give any conditions on it.

Problem IX.5.c-1: Show that if the stress tensor is not symmetric it results in a rotation of the object [Lax (1974), p. 124] (and if the antisymmetric part is a function of the coordinates a rotational distortion). Under what conditions (on the external forces, and on the crystal) does it give rotation of the object as a whole, and when does it give relative rotation of the parts of the object with respect to each other? The symmetric part determines the effect of a force on an extended object, the antisymmetric part the effect of a torque. What are the possible representation surfaces; do they provide useful information?

Problem IX.5.c-2: The symmetric part gives two distinct physical objects. What is their significance? Names for them would be helpful. Under what conditions would one, but not the other, be zero? Give examples. This part can be diagonalized by transforming to principle axes. What is their significance? Are they related to the properties of the crystal?

Problem IX.5.c-3: A stress produces a strain, an expansion, contraction, or other deformation of the object [Burns (1977), p. 66; Lovett (1990), p. 50; Nowick (1995), p. 200; Nye (1992), p. 93; Wooster (1973), p. 53]. This is described by a second-rank tensor s (sec. IX.5.b, p. 492),

$$\frac{dl_k}{dx_i} = s_{ki}(x). \tag{IX.5.c-2}$$

An atom at point x is displaced by vector with components dl_k, and since the object is distorted, not moved, this displacement varies in the neighborhood of the point, and the amount it changes is direction dependent. Like the stress, the strain is not (only) a property of the crystal; it depends on the external stress. However the response to a given stress — the strain tensor, giving the strain — is a property of the object. Taking it symmetric, we diagonalize by rotating to principle axes. What is their significance? What do they tell about the crystal? Are they related to the symmetry axes? What are the possible representation surfaces, and their significance (if any)? Does the crystal system, or the symmetry group, impose conditions on these axes? What do the diagonal values mean, and what do they say about the crystal? Does the system, or symmetry group, give conditions on diagonal values (or on any properties of the strain tensor)? What is the physical significance of the two independent tensors, the trace, and the traceless part (the shear strain [Lax (1974), p. 125; Nye (1992), p. 103])? Why this name? What information do they provide about the crystal?

Problem IX.5.c-4: What is the significance of the antisymmetric part?

Problem IX.5.c-5: The principle axes for the two tensors, the stress and the strain, need not be the same. Why? Describe the physical situation if they are, and if they are not. Is it useful if they are? Can it be arranged so that they are?

IX.5.d Second-rank tensors for group C_{3v}

To illustrate the application of group theory to the study of these tensors we consider $C_{3v}(3m)$. This group has classes E, $2C_3$, $3\sigma_v$, with characters (3,0,1), for the three-dimensional reducible representation acting on coordinates, so the product of this representation with itself, acting on products of coordinates (second-rank tensors), has characters (9,0,1), and the number of independent parameters is therefore (eq. IX.3.d-3, p. 482)

$$\frac{1}{6}(9+0+3(1)) = 2. \qquad \text{(IX.5.d-1)}$$

What are these the two components, and why only two? A general three-by-three symmetric traceless matrix (the trace, being invariant, has been subtracted off) has five independent components, three off-diagonal and two on it. This implies that for C_n the off-diagonal elements are zero, and the two independent ones are on the diagonal. Since a crystal with this group has an axis of symmetry, taken as z, we would expect the x and y components to be equal, but differ from the z component. That is the independent components are

$$T_{xx} = T_{yy}, \text{ and } T_{zz}. \qquad \text{(IX.5.d-2)}$$

Why are the off-diagonal terms zero? First T_{xz} and T_{yz} cannot be invariant under rotations around z. And the two-by-two matrix labeled by x and y commutes with group transformations since it must be invariant. But by Schur's lemma it is a multiple of the unit matrix, this multiple being

$$T_{xx} = T_{yy}. \tag{IX.5.d-3}$$

Problem IX.5.d-1: The antisymmetric part has one independent component, and as this vector is invariant under the group it must be along the axis of transformation C_n, the z axis, so the component is

$$T_{xy} = -T_{yx}, \text{ with } T_{xz} = T_{yz} = 0. \tag{IX.5.d-4}$$

IX.5.e Nonzero components for different crystal systems

Besides specific groups, the crystal systems provide information about the tensors. What restrictions are placed on second-rank tensor k by the crystal system to which an object belongs [Wooster (1973), p. 30]?

Problem IX.5.e-1: The monoclinic system has a two-fold symmetry axis, taken along z. Show that this requires

$$k_{iz} = k_{zi} = 0, \quad i = x, y. \tag{IX.5.e-1}$$

Of course it does not require that the other components be nonzero. State the requirement if the orientation of the axis is arbitrary.

Problem IX.5.e-2: For the orthorhombic system, there are three mutually perpendicular symmetry axes. Check that only the diagonal components can be nonzero.

Problem IX.5.e-3: The tetragonal system has a tetrad (four-fold) axis, again labeled z. And again the only components that can be nonzero are the diagonal ones. Observe that this also applies to the rhombohedral and hexagonal systems. For the cubic system this is also true, but now these three must in addition be equal. Repeat this for arbitrary orientation. Thus the only second-rank tensor allowed by octahedral group O(432) is isotropic — one invariant under all rotations [Lax (1974), p. 131]. Does this depend on which of the three irreducible rotation-representations the tensor belongs to? Why?

Problem IX.5.e-4: Find the general second-rank tensor for a crystal with C_n symmetry and principle axis along z, and show that a reflection plane containing z requires that the off-diagonal elements be zero [Lax (1974), p. 116, 131]. Is this also true for general orientation of the axes?

Problem IX.5.e-5: For all these systems, does the presence of other symmetry elements gives further restrictions? Explain. Suppose the symmetry of a crystal were less than that of the holohedry group of the system. Would that weaken, or strengthen, any of these requirements? Why?

Problem IX.5.e-6: Describe the representation surfaces for the different crystal systems. Would the answers be different for a crystal with less than holohedral symmetry; for any crystal system; for all? What information be obtained about a crystal knowing the representation surface? Could any (all) of these systems have representation surfaces that are ellipsoidal, or even spherical; hyperboloidal; paraboloidal?

IX.6 THIRD-RANK TENSORS

For physical properties of crystals that are described by third-rank tensors the approach is the same, though perhaps more challenging in determining which of the irreducible tensors given by such a symbol are nonzero, and what their significance is. A few examples will help prepare the reader to analyze these in real life.

IX.6.a Piezoelectricity

Certain crystals develop electric moments when subject to mechanical stress, and conversely in an electric field they deform [Burns (1977), p. 68; Joshi (1982), p. 313; Lovett (1990), p. 109; Nowick (1995), p. 7; Nye (1992), p. 110; Wooster (1973), p. 64]. This is the piezoelectric effect ("piezo" from the Greek word to press or squeeze), in which pressing on an object produces electricity. Thus a stress described by tensor c_{ij} produces a dipole moment (per unit volume) given by a vector with components P_k, so

$$P_k(x) = \sum d_{klm}(x)c_{lm}(x); \qquad \text{(IX.6.a-1)}$$

the third-rank tensor d_{klm} gives the direct piezoelectric effect (charge separation in an object by a stress).

Problem IX.6.a-1: Since this is a third-rank tensor, the presence of a center of symmetry for a crystal requires it to be zero. Why? The occurrence of the piezoelectric effect for a material thus limits its possible crystal forms.

Problem IX.6.a-2: Taking c as symmetric, we see that d is symmetric in its last two indices. However further symmetry beyond that is an extra physical requirement. Give the representations in the decomposition of the reducible representation of the rotation group of which d is a basis vector, and express their basis vectors as sums over the components of d. Are there reasons why some of these must (or must not) be zero? What is the physical significance of the nonzero ones? What do they tell about the crystal?

Problem IX.6.a-3: For each value of the first index, d is a symmetric second-rank tensor. It can be diagonalized. What is the meaning of these diagonal elements? What physical properties of the crystal do they give? How are the principle axes related to the crystal? Are they the same for all values of the first index? Does the crystal system, or the symmetry group, impose any conditions on these axes or diagonal elements? There is a representation surface (eq. IX.5-1, p. 489) for each value of the first index. What are these like? Why? Need they be related? Why? Need they all have the same type? Why?

Problem IX.6.a-4: For the cubic system, the only crystals that can show piezoelectricity are those whose symmetry group lacks a center of symmetry. Show that of these subgroups, O(432) only, requires all components to be zero [Wooster (1973), p. 69]? Why? Find any conditions on, or relations between, components for the other groups of this system.

Problem IX.6.a-5: The converse piezoelectric effect is the production of a strain by an electric field. Define the tensor describing it [Nye (1992), p. 115; Wooster (1973), p. 80]. What are the symmetry requirements on it? Reduce it to a sum of basis vectors of irreducible representations and interpret them, stating which, if any (representations and basis vectors), must (or must not) be zero. Relate the tensors of the two effects, both in reducible and in irreducible form, and discuss how their interpretations compare. Does one give information about the object that the other does not? Why? What are the possible representations in the sum? Why? What do they tell about the crystal?

Problem IX.6.a.-6: Explain why a piezoelectric crystal must have at least one transition that is allowed for both infrared (sec. X.7.a, p. 554) and Raman transitions [Burns (1977), p. 120].

IX.6.b The Hall effect

A conductor in an electric field $\underline{E}$ (giving a current) which is placed in magnetic field $\underline{H}$, gains an additional current, perpendicular to that due to the electric field, the Hall current, $\underline{I}$, given by

$$I_\mu = \sum \sigma_{\mu\nu\lambda} E_\nu H_\lambda. \qquad \text{(IX.6.b-1)}$$

Since $\underline{I}$ and $\underline{E}$ are vectors, but $\underline{H}$ is a pseudo-vector, σ is a pseudo-tensor. The occurrence for a material of the Hall effect [Lax (1974), p. 118; Lovett (1990), p. 99; Nowick (1995), p. 25; Wooster (1973), p. 121], described by a third-rank tensor, rules out the presence of a center of symmetry.

Problem IX.6.b-1: The restrictions on the tensor, and the effect, due to crystal symmetry are similar to that for piezoelectricity, but since it is a pseudo-tensor it is necessary to check whether further ones are imposed by inversion. Are there point groups that allow one effect but not the other, in particular ones allowing the Hall effect but not piezoelectricity?

Problem IX.6.b-2: This tensor should be reduced to give irreducible basis vectors of the rotation group, and the physical significance of these (how many?) irreducible representations stated. A magnetic field causes charges to flow in curved paths, unlike the straight lines due to an electric field. Of these irreducible tensors which must be zero? How does this action of the magnetic field affect the answer? How do these results differ from those for piezoelectricity? Explain. What are the physical manifestations of these differences?

Problem IX.6.b-3: The tensor $\sigma_{\mu\nu\lambda}$ is restricted by the Onsager relation [Cracknell (1975), p. 44; de Groot and Mazur (1984), p. 35, 100; Haase (1990), p. 90; Jones and March (1985), p. 684; Lavenda (1990), p. 25; Lax (1974), p. 118, 287; Nowick (1995), p. 21; Nye (1992), p. 207]

$$\sigma_{\mu\nu\lambda}H_\lambda = -\sigma_{\nu\mu\lambda}H_\lambda, \tag{IX.6.b-2}$$

a consequence of time-reversal invariance (sec. VIII.2, p. 394). If the magnetic field is reversed the Hall current is given by the tensor with these indices interchanged. Why? How does this affect the number of components of σ?

Problem IX.6.b-4: Prove that if the material is isotropic, the tensor has only one independent component (why?) σ_o, and

$$\underline{J} = \sigma_o(\underline{E} \times \underline{H}), \quad \sigma_{\mu\nu\lambda} = \sigma_o\varepsilon_{\mu\nu\lambda}, \tag{IX.6.b-3}$$

where $\varepsilon_{\mu\nu\lambda}$ is the completely antisymmetric symbol (pb. IX.2.a-2, p. 471).

Problem IX.6.b-5: Find the number of independent components of the Hall tensor [Lax (1974), p. 122], and then list those that can be nonzero, first for a reasonable set of axes then in general, for $C_{3v}(3m)$. List the representations of this group to which the nonzero tensor belongs.

Problem IX.6.b-6: For each point group find the number of independent components of the tensor [Wooster (1973), p. 124], and then list those that can be nonzero, first for a reasonable (?) set of axes, then in general. Are there point groups that rule out a Hall effect? What information about a crystal can be obtained by studying this effect?

IX.6.c Optical activity

In a material object there are two electric fields, the external $\underline{E}$, and the measured one $\underline{D}$, the displacement vector. They are (taken as) proportional (at least to lowest order), but not (necessarily) in the same direction, so expanding $\underline{E}$ in plane waves $\sim exp(i\underline{k} \bullet \underline{r})$,

$$D_\mu = \sum \varepsilon_{\mu\nu}(\underline{k})E_\nu, \qquad \text{(IX.6.c-1)}$$

for each term, with the dielectric constant $\varepsilon_{\mu\nu}(\underline{k})$ (to the extent that it is) dependent on wave vector $\underline{k}$. Thus the electric field, that is the direction of polarization, of the wave is rotated when traversing a crystal that has optical activity. The angle of rotation per unit length is, generally, dependent on the direction in which the wave travels, as is the index of refraction, giving birefringence.

Problem IX.6.c-1: By time-reversal invariance [Lax (1974), p. 131],

$$\varepsilon_{\mu\nu}(\underline{k}) = \varepsilon_{\nu\mu}(-\underline{k}); \qquad \text{(IX.6.c-2)}$$

explain this. Does it look familiar? To first order (long wavelength) it should be clear that

$$D_\mu = \sum \varepsilon_{\mu\nu}(0)E_\nu - i\varepsilon'_{\mu\nu\lambda}\frac{dE_\nu}{dx_\lambda} + \ldots, \qquad \text{(IX.6.c-3)}$$

where

$$\varepsilon'_{\mu\nu\lambda} = \frac{d\varepsilon(\underline{k})_{\mu\nu}}{dk_\lambda}, \qquad \text{(IX.6.c-4)}$$

and is antisymmetric in $\mu\nu$ (why?), so giving optical rotation — the rotation of the polarization vector of the wave (why?) [Lovett (1990), p. 91; Nowick (1995), p. 16; Nye (1992), p. 260; Wooster (1973), p. 114]. Here then is another third-rank tensor. How many different physical quantities does it give? What physical significance do they have? Is there a difference — in terms of symmetry — between it and the Hall tensor? How about the set of point groups that allow it? What is the next term of the expansion, and its rank? Does it lead to physical effects? How many? Might they be of economic interest? How do the point groups allowing it compare with those that allow the first-order term? What does the appearance, and nonappearance, of these terms experimentally tell about the crystal? Can an algorithm be given for determining which higher-order terms are allowed by any given point group? Would it be useful in studying a crystal? Compare this information with that obtained by studying the Hall effect. Would it be helpful to study both?

IX.6.d Groups that can, and cannot, have these tensor properties

After these examples it is useful to consider the tensors as purely representation basis vectors, and find their properties in general.

Problem IX.6.d-1: For a crystal with a two-fold axis of symmetry (taken along z) show that the only components that can be nonzero have one or three subscripts "z", $d_{zzz}, d_{zij}, d_{izj}, d_{ijz}$. Give the result for the symmetry axis in an arbitrary direction.

Problem IX.6.d-2: Show that if there is a plane of symmetry perpendicular to y, the number of "y" subscripts must be even [Wooster (1973), p. 67, 73]. Give the result for the symmetry axis in an arbitrary direction.

Problem IX.6.d-3: The orthorhombic system has three mutually perpendicular diad axes (along the coordinates). Check that only d_{xyz}, d_{yxz}, d_{zxy} can be nonzero. Give the result for general orientation of the axes.

Problem IX.6.d-4: Find the components that are allowed to be nonzero for the tetragonal system, first [Wooster (1973), p. 68] for symmetry axes along the coordinate axes (which are what?), then in general.

Problem IX.6.d-5: For a third-rank tensor, for group D_3 (32) there are two independent components (pb. IX.3.d-9, p. 484). What would be a reasonable choice for these [Wooster (1973), p. 251]? How is this choice related to the structure of the crystal? Explain geometrically why there are two and why these two. Are pseudo-tensors different?

Problem IX.6.d-6: For each crystal system, does the presence of other symmetry elements give further restrictions? Explain the result. Suppose the symmetry of a crystal were less than that of the system's holohedry group. Would that weaken any requirements? Why?

Problem IX.6.d-7: Which point groups allow physical properties described by third-rank tensors? Which allow third-rank pseudo-tensors?

IX.7 FOURTH RANK TENSORS

Second-rank tensors often arise as coefficients relating one vector to another. And second-rank tensors are themselves often related, giving thus fourth-rank tensors. The principles involved in application of symmetry to them are the same as for lower rank, but they provide further examples of the physical meaning and value of these concepts (as well as more stimulating exercises).

IX.7.a Relating stress and strain

The strain of an object is a function of the stress on it (sec. IX.5.c, p. 494) and (to first order)

$$c_{ij}(x) = r_{ijmn}(x)s_{mn}(x), \qquad \text{(IX.7.a-1)}$$

so the response of the material to an applied force is given by the elastic coefficients, a fourth-rank tensor. Thus

$$f_i(x)a_j = r_{ijmn}(x)\frac{dl_m}{dx_n}. \tag{IX.7.a-2}$$

The inverse of this, giving the strain directly in terms of the stress is,

$$s_{mn}(x) = t_{mnij}(x)c_{ij}(x), \tag{IX.7.a-3}$$

and

$$\frac{dl_m}{dx_n} = t_{mnij}(x)f_i(x)a_j, \tag{IX.7.a-4}$$

relating the variation with distance in the n direction of the extension (or compression) in the m direction to the component i of the force at the point crossing the surface with normal in direction j. These are generalizations of Hook's law and Young's modulus [Burns (1977), p. 68; Lovett (1990), p. 57; Nowick (1995), p. 202; Nye (1992), p. 131].

Problem IX.7.a-1: Express tensor t in terms of tensor r.

Problem IX.7.a-2: A model made of hard clay or wax (especially using a nonisotropic material, perhaps by inserting toothpicks) should clarify the meaning of this. See what the deformation is, as a function of position, and how it varies with direction, as the orientation of the external force is changed, and as it is applied to different surfaces.

Problem IX.7.a-3: Find the potential energy of an object under stress [Lax (1974), p. 126; Nye (1992), p. 136; Wooster (1973), p. 86], relate r_{ijmn} to the derivatives of it with respect to displacement, and from the commutativity of derivatives, show that r is symmetric in its first two indices, and also in its last two (sec. IX.2.d, p. 475). This does not imply symmetry between the pairs. How many components does a fourth-rank tensor have, and how many does one with such symmetry have?

Problem IX.7.a-4: Using this symmetry, how many independent components are there for an isotropic crystal [Nye (1992), p. 142]?

Problem IX.7.a-5: Is it possible that there can be objects for which the derivatives have to be replaced by covariant ones, so the tensor is not symmetric?

Problem IX.7.a-6: Tensor r is a basis vector of a reducible representation of the rotation group. Give the representations in its decomposition, and express their basis vectors as sums over the components of r. First do this for a general fourth-rank tensor, then one with this symmetry. Are there reasons why some of these must (or must not) be zero? In general? With this symmetry? What is the physical significance of the nonzero ones? Are there cases in which the crystal symmetry group requires one of these representations to be zero, but others are allowed? The tensor can be diagonalized in either pair of indices. What is the significance of these diagonal elements, for the two cases? What physical properties of the crystal do they give? How are the two sets of principle axes related, and related to the crystal? Under what conditions, if any, does one set of axes simultaneously diagonalize the tensor in both sets of indices? Does the crystal system, or the symmetry group, impose conditions on these axes or diagonal elements? For each set of fixed values of one pair of indices there is a representation surface (eq. IX.5-1, p. 489) for the other pair. What are these like? Need they be related? Must the types of surfaces be the same for both pairs? Are hyperbolic surfaces possible?

Problem IX.7.a-7: Give the restrictions imposed by various symmetry elements of the crystal — axes, mirrors, and center of symmetry — on this tensor. Note that the symmetry affects the strain, but not the stress — this is due to an arbitrary external force, hence it can only give limitations on some indices. Of course the strain is also determined by the external force, so is not only a property of the object.

Problem IX.7.a-8: For group $C_{3v}(3m)$, find the number of independent components, and give them, first in a well-chosen coordinate system, then in general [Lax (1974), p. 129].

Problem IX.7.a-9: For each crystal system find the components of the tensor that can be nonzero, and also any relations between these components, first for intelligently chosen axes, then in general [Lovett (1990), p. 62; Nye (1992), p. 137; Wooster (1973), p. 89]. Are there further, or fewer, restrictions or conditions if the symmetry is less than that of the holohedry group of the system?

Problem IX.7.a-10: Devise expressions for the number of independent constants for each of the irreducible tensors. Give the total number for group $C_{4h}(\frac{4}{m})$ [Wooster (1973), p. 252, 258]. What are reasonable ones? What is their physical significance? Explain these geometrically.

IX.7.b Photoelasticity

Some materials rotate the polarization of light — a vector — passing through them. Such a medium has a property given by a quantity relating the initial and final vectors — a second-rank tensor (sec. IX.6.c, p. 499). If the medium is subject to stress, given by a second-rank tensor (sec. IX.5.c, p. 494), its optical properties are varied; this is photoelasticity [Lovett (1990), p. 111; Nowick (1995), p. 13, 159; Nye (1992), p. 241; Wooster (1973), p. 131], and is described by a tensor which relates two second-rank tensors, a fourth-rank tensor. The details that relate to symmetry are similar to other fourth-rank tensors — except that these can, or must, have different symmetry in their indices; the optics is not relevant here.

Problem IX.7.b-1: Nevertheless it is useful to think about the physical meaning of these quantities. What symmetry of the indices does the tensor have? Why? The tensor is a basis vector of a rotation-group reducible representation whose decomposition should now be known. What is the physical significance of each representation with nonzero basis vectors — why do these representations, and only these, occur, and what physical properties do they describe? How do they differ from those of other fourth-rank tensors with different symmetry in their indices? What restrictions on the appearance of these representations are imposed by the crystal system, or symmetry group? What information do they provide about the crystal? Those parts that are symmetric in their indices can be diagonalized. What is the meaning of these diagonal elements, for the different sets of symmetric indices (if more than one)? What physical properties of the crystal do they give? How are the principle axes related to the crystal? How are different sets of principle axes related? Under what conditions, if any, does one set of axes give the tensor simultaneously diagonal in other sets of indices? Does the crystal system, or the symmetry group, impose conditions on these axes or diagonal elements? Discuss the various (possible) representation surfaces.

Problem IX.7.b-2: This fourth-rank tensor is reducible, and some irreducible representations appear more than once. Reduce it, finding the sums of components giving the basis vectors of each irreducible representation, and give the physical meaning of each irreducible component, unless it can be proven zero. In cases, if any, in which an irreducible representation appears more than once, explain the physical distinctions between them. What physical properties do each describe? Design a crystal with each property, first for the cases for which the properties given by the other irreducible components can occur, then find crystals for which only a single property is possible — or prove that these are impossible. What do such occurrences tell about the crystal? About its symmetry group? Why? Would these objects be of use?

Problem IX.7.b-3: Give expressions for the number of independent components for each irreducible tensor. Find the total number for group T(23) [Wooster (1973), p. 252, 258]. What would be reasonable ones? What is their physical significance? Explain these results geometrically, and physically.

Problem IX.7.b-4: While the conditions given by the crystal symmetry group are the same for all fourth-rank tensors (checking though for any differences between a tensor and a pseudo-tensor due to, say, a center of symmetry), different irreducible representations may, for other reasons, be zero, or nonzero (if allowed by symmetry), for different tensors. For each crystal system, and symmetry group — taking into account all other physical principles affecting this tensor including any that impose requirements on symmetry of indices — which irreducible representations can have nonzero basis vectors? How does this compare with the tensor relating stress and strain? Explain the similarities and differences, mathematically and physically.

Problem IX.7.b-5: As with the tensor that relates stress and strain, the photoelasticity tensor refers to several irreducible representations, some of which may appear more than once giving different properties of the crystal. Compare the results of the previous problem for these two tensors.

Problem IX.7.b-6: For each crystal system, does the presence of other symmetry elements give further restrictions? Explain. Suppose the symmetry of a crystal were less than that of the system's holohedry group. Would that weaken, or strengthen, requirements? Why?

Problem IX.7.b-7: Which point groups allow physical properties described by fourth-rank tensors? Fourth-rank pseudo-tensors?

IX.8 THE EFFECT OF IRREDUCIBILITY ON THE PHYSICS OF TENSORS

Tensors are usually written as symbols with indices, these being reducible representations of the relevant groups. But the physical quantities are given by irreducible representations. Thus for the rotation group T_{ij} is a sum of terms from three irreducible representations, S_0, S_1 and S_2, where the subscript gives the "angular-momentum quantum number", j; the dimension of the representation is $2j+1$. Third-rank tensor U_{ijk} transforms like the product V_iT_{jk}, which is a sum of products V_1S_0, V_1S_1, V_1S_2; a vector transforms under the $j=1$ representation (all symbols are schematic). Reducing these terms gives R_1; R_0', R_1', R_2'; R_1'', R_2'', R_3'', with the R coming from V_1S_0, the R' from V_1S_1, and the R'' from V_1S_2. Notice that the triple product is not simply reducible; there can be more than one term going with a representation, that is more than one sum of terms of V_iT_{jk} transforming under each representation. Here the vector representation, R_1, appears three times, representation R_2 occurs twice. This non-simple-reducibility holds for tensors with more indices, giving larger numbers of representations.

The realization is irrelevant so that U_{ijk}, whether a product or not, transforms like one thus is a sum of terms of different irreducible representations. State m of irreducible representation l is therefore a sum of the U's (schematically),

$$V^{r}_{l,m} = \sum C(l, m; ijk; r) U_{ijk}, \qquad \text{(IX.8-1)}$$

where the C's are the Clebsch-Gordan coefficients [Mirman (1995a), sec. XII.2, p. 341], the sum is over all ijk, and r is the multiplicity index, needed as representations can occur several times — there can be several different sums of the U's giving the same state, m, of the representation, l, to which V belongs.

It is the irreducible representations that give the physical quantities. Thus U_{ijk} is not the symbol for a physical property of an object, but rather for seven of them. For a given system the number of these quantities that are allowed to be nonzero can be anywhere from zero to seven. The irreducible representation is not, generally, sufficient to specify a physical quantity since it can occur more than once. Further the equation

$$V_i = U_{ijk} T_{jk}, \qquad \text{(IX.8-2)}$$

actually indicates the object found by taking the term transforming as a vector of the Clebsch-Gordan decomposition of the twenty-one products of seven times three representations (for the U and for the T). Of these twenty-one representations, the $R_0'S_0$ term does not contribute. Thus this expression gives not one vector, but several. They are all different, they could all appear (unless there were rules for eliminating them) and they are for different physical properties of the object. It may be that only some of these representations, or only some occurrences of multiply-occurring ones, have physical meaning, thus only these appear. That is not all components of U need be independent, there may be relations or some may be zero leading to sums that are zero — all sums giving the states of a representation that cannot appear are zero (which requires relationships among the components of U). Also the symmetry group may require relations among the components of U, or among the components of representation V in its decomposition, so on those of the components of U. (And there can be restrictions on the components of T). Such symmetry requirements are one of the factors determining the representations in the decomposition of U for a particular physical situation and particular symmetry group. It is important thus to know how many independent, nonzero components of U there actually are, and not assume a number obtained from counting indices.

Here we take relations between objects to lowest order, getting tensors with several indices. But usually there are higher-order terms, and for enough precision these must be considered. They have more indices, and terms of different order give different physical properties. Thus higher-order terms represent larger and larger numbers of physical properties, with the number increasing rapidly — unless the restrictions on them also increase.

Why should an object have two properties that do the same thing, connect the same two tensors, and transform the same way? Consider say piezoelectricity, and a crystal consisting of two interpenetrating lattices of different atoms. The deformation of a lattice leads to a potential; depending on the masses of their atoms and their interactions, either or both of the lattices could deform, independently, and each could effect the deformation of the other. Moreover the lattices could have different symmetry, thus the point and space groups could place different demands on them, allowing a tensor for one to be nonzero, but not for the other. The deformation tensors of the lattices are different physical properties. Thus an observation that only measured the total tensor, the sum of the two, would throw away important information about the crystal. There can be other factors determining properties than those considered. Thus the strain depends on the crystal structure, and on the stress, but might also depend on the orientation of the crystal surface to the crystal symmetry axes and planes. These possibilities must all be considered.

A crystal can consist of many such lattices, but there are only a few tensors, two for example, that give its properties. Why is there a limitation? To find the tensor, which is a representation basis vector of the point group, we consider the product of the representations of the point groups for each of the lattices, if these groups are the same, or the product of the groups (keeping in the decomposition only those groups that are point groups, if not all need be), and find the particular representation of the specified group that describes the crystal. Terms in these products can occur more than once, but the number is limited. It can be no greater than the number for the rotation group, of which the point groups are subgroups. Hence the physical properties are given by the different terms in the decomposition, and these are determined by the properties of individual lattices. But it is the combination of these latter properties that fixes the properties of the crystal as a whole. There is, in general, more than one such combination, but there cannot always be as large a number as that of the individual terms, the number of lattices say. Of course here we are assuming one fourth-rank tensor and finding the number of irreducible tensors it gives. But there could be several fourth-rank tensors, so multiple occurrences of irreducible tensors, all giving different physical properties.

If there are multiple lattices, the crystal as a whole might have a structure (a crystal system) that differs from each when alone (compare magnesium and spinel; sec. III.5.d, p. 166). Or it might be invariant under a subgroup of the groups of each. However some point groups share distinct subgroups (as we saw). Thus the crystal as a whole can be of several different forms, and these might coexist as domains. The tensor could be different for the different domains, though belonging to the same rotation-group representation. This is another example of why the crystal can have different tensors that transform the same, but are physically different. Which are, or are not, zero, depends on which domains are present, and to what extent.

How would these different tensors show up experimentally?

Consider a light wave passing through a crystal. It is actually two waves of opposite circular polarization. The paths of the two, and the amount by which their polarization is changed, can differ (birefringence or double refraction) [Lovett (1990), p. 71; Nowick (1995), p. 120; Nye (1992), p. 235]. Hence the tensors that give these, and that transform the same, are different, they have different values, and perhaps different nonzero components (though they do not come from a decomposition, being only of second order). One could be zero, the wave might be undeviated (or for that tensor, its polarization unchanged), the other not zero. And these tensors might be affected by stress (a photoelastic effect, which provides another example of a tensor) [Lovett (1990), p. 111; Nowick (1995), p. 14; Nye (1992), p. 243], perhaps differently for the polarizations. Thus the different occurrences of an irreducible representation in the decomposition of the fourth-rank tensor relating optical activity to stress could be different; one relevant to left-handed polarization, the other right. Moreover a stress might cause a wave of a single polarization to break up and travel in two paths. The rotation of polarization might be different for these paths. In this case the irreducible tensor relating the stress tensor to the optical rotation tensor (giving the optical activity) would connect two tensors that form the same basis state, and would itself belong to only one representation and basis state. However there would be two (or more) such tensors, one for the optical activity for one path, the other for the other path. Under the rotation group these would be the same, but physically, in their physical effects and relevance, they would be different. Thus this crystal would have more than one tensor, relating the same two basis states with (essentially) the same meaning, which transformed according to the same representation, but which are distinct. They are completely independent.

Again this emphasizes the value of group theory. Without it there seems no way of determining the number of distinct physical properties given by a symbol, or understanding why they differ.

Problem IX.8-1: The reader can undoubtedly think of more interesting cases — and should when studying objects, else important information could be lost, or errors made. Think of other reasons why there can different tensors giving different properties. Also there are systems that might seem capable of having several tensors, but the number is actually limited by the groups. In such cases it should be understood how the limitation actually reduces the number of physical properties (if that is what happens).

Problem IX.8-2: Write two different sums (and explain how they differ) of products of components of three rank-1 tensors, these thus transforming under the $j = 1$ representation (the product is then a third-rank tensor) — and which have the same value of the magnetic quantum number (eigenvalue of the labeling operator). Show that they cannot be transformed into each other by any rotation. Repeat for the $j = 2$ representation.

Problem IX.8-3: Could the above be a correct description of the photoelastic tensor?

Problem IX.8-4: In the decomposition of tensor $U_{ijkl...}$, representation V appears. Suppose that the symmetry group imposes conditions on U, requiring components to be zero, or relating them. Can it also impose other conditions on V, or are all the requirements on V obtained from those on U? Can those on U be obtained from the ones on V?

Problem IX.8-5: We have not considered direct products of point groups; one way of doing so is to consider each as a subgroup of a symmetric group, find the product of that with itself [Mirman (1995a), sec. IV.3.d.ii, p. 113], and then find the subgroup that is the product of the point groups. The outer product of symmetric groups is not simply reducible, which means that there are several distinct sums of basis states transforming under the same symmetric-group representation. Is this true also for any point groups? If we take the product of several symmetric groups, there are many such distinct sums. However if the resultant group is a subgroup of the rotation group its basis states must be states assigned to an occurrence in the decomposition of a product of rotation-group representations, resulting in conditions relating the sums (or requiring some to be zero), so that their number is equal to the number of occurrences of the rotation representation in the decomposition. Try this for a few cases. Are there other ways of finding these products. How are products found for space groups?

Problem IX.8-6: Here the rotation group was considered. Discuss the extent to which crystallographic groups place restrictions on these physical quantities, such as requiring some to be zero, and whether they can distinguish between objects that transform under the same rotation-group representation.

Problem IX.8-7: A fourth-rank tensor is a basis vector of a reducible representation, thus a sum of irreducible-representation basis vectors. However here there is an additional point. Many such tensors relate two second-rank tensors, and these can also be reduced. Thus we have a product of each irreducible representation in the decomposition of the fourth-rank tensor, with each such representation in the decomposition of a second-rank tensor, and this gives a reducible representation. We wish to project out the irreducible representation corresponding to the irreducible component of a second-rank tensor; that is we have a set of Clebsch-Gordan decompositions [Mirman (1995a), sec. XII.2, p. 340]. Give the list of these for this case. Now we can get a particular irreducible representation from the product of different irreducible representations — a specific irreducible representation can come from products of two different physical quantities. Thus for a given irreducible-representation stress tensor, and a given irreducible-representation strain tensor, there can be several elastic tensors belonging to different irreducible representations (in the decomposition of the fourth-rank tensor) that link them. These would then give different physical properties of the crystal. Moreover since there might be several stress, and several strain, tensors, belonging to different irreducible representations, there could be several elastic-coefficient tensors, each connecting different pairs of irreducible representations, and each of which can consist of tensors of different irreducible representations. Thus in this case, the set of elastic coefficients can give many physical properties. Make a list of all these elastic-coefficients irreducible representations, some, or all, of which may thus appear several times, using in particular the different second-rank tensors describing different physical properties. Give for each the physical property of the crystal that it describes, or that determines it, or explain why it cannot occur. Of course, these equations are not all for direct products, since there are sums over indices. What does this lead to? The Clebsch-Gordan decomposition is that of the rotation group. However the symmetry is reduced to that of point groups. For each point group, tell which entries on this list must be zero; it may be that some, but not all, occurrences of an irreducible representation must be zero, thus reducing the number of times it appears. For each nonzero term give the set of independent components. Explain physically the meaning of, and the reason for, the occurrence of each irreducible representation, and each nonzero component, the reason for zero terms, and for the relationships among coefficients; include explicitly the information given by the point groups. Design crystals for which all of these tensors, but one, are zero, for each entry on the list (for each point group), unless it is impossible, perhaps in certain cases, for only one to be nonzero. Repeat for several nonzero. Are any of these crystals likely to have technological interest? Why? Try this for the other such physical tensors.

Problem IX.8-8: Repeat this for the third-rank tensors considered.

Problem IX.8-9: For each tensor considered, or found elsewhere, using its definition (perhaps as a product, or a tensor relating other tensors), find the maximum number of independent representations it belongs to. Which of these can actually occur? Why? What is their physical significance?

IX.9 THE USEFULNESS OF TENSORS IN ANALYZING CRYSTALS

Knowing the symmetry of a crystal tells us much about its tensor properties. So studying these properties gives information about the crystal. Here each type has been considered separately, but it is useful to look at them all at once.

Problem IX.9-1: Suppose you have a crystal about which you wish to learn as much as possible. Develop an experimental program for studying its tensor properties that provides the maximum amount of information (most efficiently).

Problem IX.9-2: We have been discussing crystals as if they were infinite in extent, with properties the same at all points, with rotation axes and symmetry planes placed at one point parallel to those placed at any other. Yet it is very difficult, really impossible, to grow — or find in nature — perfect crystals. Even so there are objects found, indeed often difficult to avoid, with symmetry, and with properties limited by it, with electric and magnetic dipole moments, with thermal conductivity and expansion, with stress-energy tensors, with piezoelectricity, and so on, and these are governed by, and reflect, the symmetry of the crystal — imperfect though the crystal may be. Why? To what extent do the restrictions and results obtained here, for each physical property, and each crystal symmetry, depend on the perfection of the (macroscopic) crystal? Which remain if it consists of regions in which the symmetry elements, axes and planes, are oriented differently? Why? Might there be reasons why these are, or are not, so oriented? Would defects change these properties? Crystals are not merely interesting, and helpful in illustrating group theory, but are useful — in many ways, including technological and economic. How do these imperfections, and their effect on the crystal properties, influence the value of these materials? How might they increase it? To what extent does group theory provide a guide to developing such systems and to their value? Is group theory economically useful?

Chapter X

Groups, Vibrations, Normal Modes

X.1 WHAT GROUPS TELL US ABOUT MOLECULES AND CRYSTALS

Symmetry groups of crystals and molecules have now been introduced, providing information about groups and about these objects, so illustrating the applicability and usefulness of group theory in the study of real objects. There is a vast amount of information and applications still, too vast to consider fully here. But to at least provide some indication of their applications and to gain experience with these methods and the insight they can supply we next briefly discuss some further examples. Here we start with vibrations, specifically of molecules.

We have an object — for concreteness a molecule (but it can be viewed more broadly) — whose parts vibrate relative to each other; for what purpose do we study these vibrations? They are determined (in part) by the physical characteristics of the object, so can be used as probes — to learn about objects. Here however we are studying them in a particular way, using group theory; why should it, and symmetry, be relevant? How can they be used as probes? Objects change states, radiating or absorbing energy, affecting their environment, and are affected by it, as is energy flow to and from the environment. This, part of the most fundamental aspect of the physical universe, must be understood. It is through radiation that we learn of our surroundings, of the universe, and that we alter it. It is through radiation that parts of the universe interact with each other, scatter, cause changes of state. Vibrations, radiation that carries information, what these tell us about the object and the environment, their interactions and effects, all are of fundamental interest.

What are these vibrational states, what about them are we trying to learn, to understand, how do these give us insight into the objects and into nature? Coordinates of the parts (more properly, since this is quantum mechanical, their expectation values) are functions of time, and can be written as Fourier expansions, as sums of exponentials. The relevant, and necessary, entities then include the eigenfunctions of the Hamiltonian, their eigenvalues (the energies), and knowledge of how these are determined by, and determine, the properties of the system. Changes of states, and radiation produced, are limited, and we have to know these are related to the properties of the system, and how the radiation tells us about these. Of course we wish to view such questions as ones about group theory — if there are groups that allow us to do so. How are these properties and objects affected by, and how do they show, the symmetry? For concreteness, we also wish pictures for each state of the motion of the parts (the atoms; for complex molecules perhaps sets of atoms).

Problem X.1-1: These are what we want to learn. What do we want to understand?

X.2 VIBRATIONAL STATES AND SYMMETRY

The rotational states of a system are properties of the rotation group, and this (not necessarily a symmetry group) provides information, and constrains the properties of the system, including energies and transition probabilities. But besides rotations, systems — say, atoms in crystals and molecules [Goldstein (1953), p. 333] — undergo vibration [Altmann (1977), p. 318; Bhagavantam and Venkatarayudu (1951), p. 95; Bishop (1993), p. 164; Burns (1977), p. 78; Califano (1976); Cornwell (1984), p. 165; Cotton (1990), p. 304; Elliott and Dawber (1987), p. 106; Hargittai and Hargittai (1987), p. 201; Harris and Bertolucci (1989), p. 93; Harter (1993), p. 205, 286; Heine (1993), p. 229; Inui, Tanabe and Onodera (1990), p. 220; Kettle (1995), p. 191; Kocinski (1983), p. 80; Lax (1974), p. 134; Leech and Newman (1969), p. 35; Lomont (1961), p. 96; Ludwig and Falter (1988), p. 139; Lyubarskii (1960), p. 103; Pauling and Wilson (1967), p. 282; Schensted (1976), p. 98; Sternberg (1994), p. 94; Tinkham (1964), p. 234; Tsukerblatt (1994), p. 355; Wherrett (1986), p. 98; Wilson, Decius and Cross (1980); Wooster (1973), p. 262]. There are energies and transitions due to this motion also. Yet there may seem no *symmetries to govern vibrational states, nothing equivalent* to the rotation group giving the rotational levels, suggesting that their properties lie outside the purview of group theory, that these, unlike rotational states, are determined by other conditions and aspects of systems and nature. However there are constraints on the forms, energies and transition probabilities for these states. In fact, there is symmetry, and it often plays a key role. Why? How does it lead to such constraints? And how do we extract information about vibrations from symmetry (of what?) and group theory. The treatment of point groups provides the foundations — for reasons we still do not know — and having studied them, we are now ready to turn to these questions.

A related topic, which unfortunately we cannot study, is the of vibrations of quasicrystals [Gazeau (1997); Jaric (1989); Lifshitz (1997); Quilichini and Janssen (1997)].

Problem X.2-1: Why are point groups (the) relevant (mathematical) objects?

X.2.a Normal modes

For a single particle undergoing — simple — harmonic motion, the eigenstates of the Hamiltonian are (schematically)

$$x = Aexp(i\omega t + \phi), \qquad \text{(X.2.a-1)}$$

where frequency ω is the energy (we ignore constants), the Hamiltonian eigenvalue [Mirman (1995b), sec. 3.4.2, p. 53], and there are a set of allowed frequencies. A system of several interacting particles, an object whose parts vibrate with respect to each other, say each undergoing simple harmonic oscillations, has eigenstates, with the subscript labeling the particles,

$$x_i = A_i exp(i\omega t + \phi), \qquad \text{(X.2.a-2)}$$

where ω and ϕ are the same for all particles. These are, as eigenstates of the Hamiltonian, the stationary states; if the object is in one — with a single frequency — it remains in that. These stationary states — the normal modes — are a property of the system, so provide information about it. In general the system is in a superposition of normal modes, essentially a wavepacket.

Perhaps the here most relevant definitions of the word "normal" are "standard" and "typical", these modes being typical of the system. It apparently goes back to the Proto-Indo-European word for know — and they provide much knowledge of the system — through a Latin word one of whose meanings is rule; they rule the system and provide rules for obtaining knowledge of it. The word mode comes from the Latin term whose meanings include manner, and also measure, size, harmony and melody. A cognate is model. The modes tell the manner in which, provide models for the way, the system oscillates, they measure it. The normal modes are the standard, the typical, manners, measures, of vibrations. These mathematical functions provide models for the physical actions.

When an object is vibrating in a normal mode, all coordinates have the same time dependence. Thus we need not know all to find the behavior of the object; from a few we get the others. We can therefore reduce the number of coordinates to a minimum, this minimum set being the normal coordinates, the ones which are each a function of a single frequency. The system is governed by a partial differential equation, in non-relativistic quantum mechanics Schrödinger's equation; using normal coordinates this separates into a set of ordinary differential equations, one for each normal coordinate. In general, coordinates are sums of normal coordinates; for a normal mode — an energy eigenstate — only one term appears in the sum.

In these discussions for concreteness — and because of Fourier's theorem — we often think of the simple harmonic oscillator [Goldstein (1953), p. 318]. But nothing that we do requires this. An anharmonic oscillator can still respect symmetry. So these investigations are more general. When amplitudes are calculated the form of the force law is important, but its details need not be for symmetry, nor for the applications that we consider here, such as selection rules.

If the motion is not simply harmonic, the eigenstates are more complicated. However if it does not deviate too much from being so, if the anharmonic terms are small (to a good approximation), we can start with these states and then find the deviations from them, or saying it another way, we can assume the system is in a normal mode and regard the anharmonic terms as causing transitions between them (so the system moves among these modes). Thus, unless there is strong anharmonicity, it is the set of normal modes, the coordinates as functions of time, and the corresponding energies (frequencies), that we want.

Our aim thus is to find the normal modes and coordinates, to understand how these are determined by the object (and its symmetry), and what transitions between them are possible — which are allowed by the symmetry of the system — and what the allowed transitions tell us.

Problem X.2.a-1: The object as a whole moves and its parts move relative to each other. However, provided it is not accelerated, we can always go to a frame in which the object is at rest, the center-of-mass frame (actually center-of-momentum frame), and throughout we consider only that system, so only the relative motion of the parts. What is the physical reason that we can always take the center-of-mass at rest?

Under what conditions is the center-of-mass frame equivalent to the center-of-momentum frame? Why? Are the differences relevant?

Problem X.2.a-2: Why can we always perform a Fourier expansion? A simple harmonic oscillator has such normal modes. For an anharmonic oscillator these modes — single exponentials — are not eigenfunctions of the Hamiltonian, so do not have definite energy, nor are they time independent (which means what?). If the coordinates in a stationary state, an energy eigenstate, are sums of terms of different frequencies, what is the energy of the system — how is it related to these frequencies?

X.2.b The simple harmonic oscillator

Vibrations imply an equilibrium point (in configuration space), that set of values for which the system remains unchanged in time if all initial velocities are zero. The coordinates are taken with respect this origin. Often, to a good approximation, the system's Hamiltonian is quadratic in these coordinates — the displacements from equilibrium — and can be replaced by a phenomenological one with the forces pictured as due to springs. Since the Hamiltonian is invariant under the symmetry group (the essential condition here), it is a function of invariants formed from the group operators. Thus the force constants, introduced by this set of equivalent springs, are related and their number determined, at least in part, by the symmetry group [Lax (1974), p. 150, 153]. This language is classical, while our view is quantum mechanical, but it is useful in reading this discussion to consider classical analogs, including restatements of such results as relations among constants.

Problem X.2.b-1: The simple harmonic oscillator (the usual example of a quadratic Hamiltonian) is studied in (almost) all books on quantum mechanics [Merzbacher (1967), p. 51; Pauling and Wilson (1967), p. 67; Schiff (1955), p. 60, 353]. Its energy E_n is proportional to n, the principle quantum number, which runs over all nonnegative integers and its statefunctions, the Hamiltonian eigenfunctions, are Hermite polynomials. More interesting and here relevant, is the three-dimensional simple harmonic oscillator [Bohm (1951), p. 351; Elliott and Dawber (1987), p. 486; Pauling and Wilson (1967), p. 100]. The Hamiltonian is

$$H = \sum \frac{d^2}{dx_i^2} + \frac{1}{2}\sum x_i^2, \qquad \text{(X.2.b-1)}$$

suppressing unnecessary constants [Mirman (1995b), sec. A.1.1, p. 177], with eigenfunctions determined by

$$H\psi_n = E_n\psi_n. \qquad \text{(X.2.b-2)}$$

Write this equation in spherical coordinates, and separate it into two [Schiff (1955), p. 70], one for the spherical harmonics, the other (related to) that of the one-dimensional oscillator (how are the constants related?). Solve these. The Hamiltonian eigenfunctions are given by spherical harmonics and Hermite polynomials. How? We would expect that the one-dimensional oscillator is modified by the appearance of centripetal (and Coriolis?) forces. How do these affect the statefunctions, and their meaning? Explain qualitatively how such effects can be seen from the statefunctions. Each energy level is given again by $E_n \sim n$, for all nonnegative integers n, but now each consists of a set of states, these with different total angular momentum (and different z components). What are the allowed values of the angular momentum for each n? Give a physical explanation for this dependence of the angular momentum on n.

Problem X.2.b–2: States with different angular-momentum z-components must all have the same energy by rotational invariance of space. However for the simple harmonic oscillator (but not for anharmonic ones) states of different total angular momentum also have the same energy (analogous to the non-relativistic hydrogen atom); the energy depends only on n. This is an "accidental degeneracy", but it implies that there is another symmetry, besides rotational invariance, for this system (this particular potential). Find a group, of which the rotation group is a subgroup, under which the Hamiltonian is invariant [Harter (1993), p. 426]. Show that it requires that the energy be independent of the total angular momentum. Explain what goes wrong when there are anharmonic terms. Can a simple explanation be given for this extra symmetry? For an oscillator in four, or more, dimensions, would there also be extra symmetry? (Generalize angular momentum to the larger number of dimensions [Harter (1993), p. 418].) Is it possible to find the (largest) symmetry group of a k-dimensional oscillator? Can an explanation be given why it should be this group?

X.2.c Group theory of the one-dimensional simple harmonic oscillator

While these systems are studied using group theory, there is one system, the paramount one here, the one-dimensional simple harmonic oscillator, that is not. This does not mean that it cannot, or should not be. Actually the group theory of this system is fundamental in many areas of physics. All we do is explain why we do not study it. The reason is that the group is a Lie group, and here we consider only discrete groups, except for the rotation group. The states of this and many another system are given by creation and annihilation operators, a^* and

a, and these form a nilpotent Lie algebra, which exponentiates to a Lie group, known as the simple harmonic oscillator group, or the Heisenberg group [Miller (1972), p. 397]. Since this brings in new concepts, not used elsewhere in this book, and since these are not necessary for this chapter, we, regrettably, do not include it.

X.2.d The linear triatomic molecule

To illustrate how symmetry affects vibrational states we start with a classical example. Perhaps the simplest molecular vibration is that of the linear triatomic molecule [Goldstein (1953), p. 333] with two atoms of mass m attached by bonds which can be taken as springs, with spring constant K, to an atom of mass M half-way between them, as shown in

Figure X.2.d-1: LINEAR TRIATOMIC MOLECULE.

A related, but not quite identical, example of this is nitrous oxide, N_2O [Leech and Newman (1969), p. 43].

Since this is standard we just summarize. The displacements of the atoms from their equilibrium positions are denoted by ξ_i (which are different from the coordinates of the atoms, denoted by x, these all measured from the same origin), and we wish these displacements as a function of time. They can be Fourier analyzed giving

$$\xi_i = \sum a_{il} exp(i\omega_l t); \qquad \text{(X.2.d-1)}$$

these must obey the equations of motion so there are a limited number of ω's that appear in this expansion. We choose the initial conditions so that only one ω appears. The general solution is a sum of these single frequency ones. What then are the allowed ω's, and for initial conditions chosen to give only one ω in the expansion, how are the a's related — how do the atoms move with respect to each other?

For each l, a_{il} forms a column vector and the kinetic and potential energies are taken as matrices acting on these vectors. The condition determining the ω's is

$$|V - \omega^2 T| = 0, \qquad \text{(X.2.d-2)}$$

the secular equation. There are three ω's,

$$\omega_1 = 0, \ \omega_2 = \sqrt{\frac{k}{m}}, \ \omega_3 = \sqrt{\frac{k}{m}(1 + \frac{2m}{M})}. \qquad \text{(X.2.d-3)}$$

The first gives uniform translation of the entire molecule, while the a's for the other two cases are

$$a_{12} = \frac{1}{\sqrt{2m}}, \quad a_{32} = -\frac{1}{\sqrt{2m}}, \quad a_{22} = 0; \tag{X.2.d-4}$$

$$a_{33} = a_{13} = \frac{1}{\sqrt{2m(1+\frac{2m}{M})}}, \quad a_{23} = \frac{-2}{\sqrt{2M(1+\frac{2M}{m})}}. \tag{X.2.d-5}$$

For a_{i2} the two outer atoms are moving with the same speed, but in opposite directions, while the center atom remains at rest. Coordinates a_{i3} describe the two outer atoms moving together with the same speed, while the center atom moves in the opposite direction.

To what extent could we have determined these results using symmetry and group theory? The two outer atoms are taken as identical, and space is taken as uniform (which it might not — appear to — be if the molecule were inside a material, where it might thus be used as a probe). Hence the motion can be decomposed into the motion of the center of mass, the $\omega = 0$ case, and the motion with a fixed center of mass. The motions of the two identical outer atoms are essentially identical; they have the same speed, with velocities either parallel or antiparallel. The motion of the center atom, with the total momentum zero, follows from this, and the a's then follow from normalization. But why must the motions of the outer atoms be related? The system (for fixed center-of-mass) is invariant under the two-element group of the identity and inversion. Thus there are two basis vectors, given by the a's, the symmetric and antisymmetric ones. So from symmetry and group theory — not a lot, this being a very simple system — we can find the relationship between the a's of the outer atoms, and using conservation of momentum, a consequence of the translational invariance of space, the a for the center atom, thus solving the problem.

From these examples we begin to guess the relevance of symmetry and group theory to the determination of the normal modes. We next work this out explicitly.

Problem X.2.d-1: What symmetry requirements are necessary for the decomposition that separates the motion of the center of mass? Since space is uniform, momentum is conserved. Check that in these three cases it is actually conserved. Give a rigorous explanation of why the center-of-mass motion is given by (only) $\omega = 0$. It is possible to have motion in which the speeds of the two outer atoms differ. Explain why, and how, and why this does not show that the argument is wrong, and what the relevance (and value) of the argument is.

Problem X.2.d-2: How would these arguments change if space were not symmetric (say for an object in a material)? Would symmetry still be useful in any way?

Problem X.2.d-3: Nitrous oxide can be taken to differ from this example by regarding it as having one nitrogen atom at an end, another in the middle, with the oxygen at the other end. Is this relevant? Why?

Problem X.2.d-4: Place two identical beads on a horizontal string [Schensted (1976), p. 105]. When these are moved vertically by (small) distances y_1 and y_2, there is a restoring force with a potential energy

$$V = \frac{k}{2}(y_1^2 + y_2^2) + by_1y_2, \qquad \text{(X.2.d-6)}$$

where k and b are constants depending on the string. Explain why a cross term has to be included. Describe the symmetries of the system, and using the invariance group show that the normal modes are

$$q_1 = y_1 + y_2, \quad q_2 = y_1 - y_2. \qquad \text{(X.2.d-7)}$$

Are these expected? Why? Describe the motion given by the two normal modes. This is classical. How is the system described quantum mechanically? What are the statefunctions? Are they reasonable?

Problem X.2.d-5: Analyze, using group theory, the (classical) system consisting of two masses, constrained to move on a line, connected by a spring, and each connected by a spring to one of the two opposite walls [Lomont (1961), p. 99, 109]. Consider the cases in which the masses and spring constants are different, and then in which there are various equalities linking these. What is the quantum mechanical analog? Do the descriptions of motions differ between classical and quantum physics? Why?

X.3 WHY, AND HOW, GROUP THEORY IS RELEVANT TO VIBRATIONS

The appearance of group theory in the study of vibrations is due, partially, to invariance under rotations, and the identity of particles. That we can interchange particles, with no change (in the physical properties) of the object, and that we can rotate them into each other, requires relationships among vibrational energies (frequencies), amplitudes, phases, and transitions, these given through group theory. In principle this is (obviously?) clear; but how do these invariances impose constraints on vibrations? While a molecule may have symmetry at rest, its vibrations break that symmetry, yet still seem to be governed by it. How could this be?

The symmetry group is that of the system in equilibrium — but we are interested in displacements from equilibrium. Need these also be (physically) invariant under the (same) group — if the system with all displacements zero is invariant, must this be true for arbitrary vibrations? Why, and in what sense?

What is fundamental in determining if there is symmetry, and which, is the invariance of the Hamiltonian under the group, not the configuration at any time. So it is not the object — the configuration at any time — that is invariant, but the Hamiltonian (classically the sum of the kinetic and potential energies) governing it. This is the time-translation operator: it takes the statefunction at any time, say when the system is at equilibrium, and gives the statefunction at a later time, as when the parts are displaced. The statefunction at an arbitrary time is the basis vector obtained from the decomposition of the product of the Hamiltonian with the equilibrium statefunction. But the Hamiltonian is a scalar, so the statefunctions at both times transform as the same basis vector, therefore their difference does, requiring the functions describing the oscillations to transform as that basis vector. Thus if a system has symmetry, so interchange of (nonvibrating) particles gives a physically indistinguishable object, the operations of displacing a particle from its equilibrium position in the object and displacing the corresponding one in the transformed object are also indistinguishable, which means that the resultant systems are — the object is invariant in equilibrium and the (constant) Hamiltonian is, so the system is invariant at all times, during all parts of the vibrational cycle. Classically, the sets of positions of the parts at all times realize the same basis vector, thus their differences, so velocities and momenta, do.

Group theory is relevant, why is it useful? The Hamiltonian being invariant, must contain group operators only in combinations that are invariants, so its eigenvectors, giving the normal modes, are symmetry-group irreducible-representation basis vectors, the energies of these modes are the Hamiltonian eigenvalues, and these describe the behaviors of the object. Actually as we will see (sec. X.5.b.v, p. 543), these arguments still work even if the object, or the configuration, is not symmetric, as long as the Hamiltonian is.

Normal modes, oscillations with a single frequency, functions of only a single exponential, have time dependence $\sim exp(i\omega t)$. Quantum mechanically, states that vary as $exp(i\omega t)$ are the eigenstates of the Hamiltonian; these we find by diagonalizing it. Using normal coordinates greatly simplifies, in fact solves, the problem since the time variation is (known as) exponential and we can easily determine the parameters. The real challenge then is to find (and interpret physically) the normal coordinates. How, and how does symmetry help? In general, each pair of points oscillates about its center of mass with a frequency unrelated to those of other pairs. But with symmetry, if we can interchange pairs of points, their frequencies must be the same. Thus with a smaller number of distinct eigenvalues, diagonalizing the Hamiltonian is simplified and physical interpretation of solutions is easier and clearer. Fortunately there often is symmetry, partly from physics — the symmetry of space (say) — and partly from limitations on the systems that we are considering.

Also, transitions — of the system from one state to another — must maintain the symmetry (the Hamiltonian is a scalar), so the possible transitions are limited, and information can be obtained about them using the properties of the group required by symmetry, and conversely from them we gain information about the group, so the object.

These are some aspects we wish to demonstrate, and describe how to utilize.

Problem X.3-1: Discuss the linear triatomic molecule using the terminology and formalism implied here.

Problem X.3-2: What is the relevance of the Hamiltonian being the time translation operator to $q \sim exp(i\omega t)$? Why do normal modes, only, each depend on a single frequency? How are the classical and quantum-mechanical views related? What is the meaning of normal mode in quantum mechanics?

X.3.a Symmetry and normal coordinates

Eigenvectors — the normal modes — of the (invariant) Hamiltonian are irreducible-representation basis vectors with eigenvalues (having physical meaning as) the energies. Each set of normal coordinates, those of the normal modes, describes the system in a definite energy state. Transformations mix coordinates, normal or general; transformations commuting with the Hamiltonian cannot mix coordinates for different energies. Hence nondegenerate normal coordinates go into only themselves or their negatives under a symmetry transformation — they form a one-dimensional representation. Degenerate coordinates, those whose vibrations have the same energy, can be intermixed, so are basis vectors of larger representations.

To find the normal coordinates we thus find the irreducible representations. We first take the group operations to act on a set of coordinates, say Cartesian coordinates for each particle, which gives a reducible representation (called here the "coordinate representation"); reducing this, which diagonalizes the potential energy matrix, gives a transformation from the original, to the normal, coordinates.

The number of vibrational degrees of freedom is the number of degrees of freedom minus the number, three, for translation of the molecule, minus the number of rotational degrees of freedom. The number of basis vectors of the coordinate representation is not in general equal to the number of vibrational degrees of freedom of the atoms in the molecule (as should be clear?). Thus how are the normal coordinates related to these basis vectors? Do all irreducible representations give (distinct) normal coordinates? How many different normal coordinates correspond to each irreducible representation? How does degeneracy affect these considerations and answers?

Problem X.3.a-1: What is the number of vibrational degrees of freedom of a linear molecule [Bishop (1993), p. 184; Cotton (1990), p. 305]? A planar one?

X.3.a.i *Kinetic energy, potential energy and normal coordinates*

The Hamiltonian, in classical physics, is a function of displacements (potential energy) and momenta (kinetic energy), so is a sum of different functions of different variables, in quantum mechanics it is a sum of operators for these; can it always be diagonalized? Might the coordinates that diagonalize one term leave the other not diagonal? Can the kinetic and potential energies always be simultaneously diagonalized? Actually "always" is wrong; except for constants (whose irrelevance we must show and then ignore), the simple-harmonic-oscillator Hamiltonian (for each value of the number of particles) is unique.

Consider N point objects (say atoms in molecules) in 3-space (although clearly the dimension is not of much importance); these have masses m_α, for particle α, orthogonal coordinates $x_{i\alpha}$ (with the same origin for all particles), displacements $\underline{q}_\alpha$ measured from equilibrium, and momenta $\underline{p}_\alpha$. They interact with each other through forces which we take (to a first approximation) as simple harmonic. The Hamiltonian is the sum of kinetic and potential energies,

$$T = \frac{1}{2}\sum m_\alpha\left(\frac{dx_{i\alpha}}{dt}\right)^2, \quad V = \frac{1}{2}\sum k_{ij}x_{i\alpha}x_{j\alpha}; \qquad \text{(X.3.a.i-1)}$$

where the potential energy is zero at equilibrium so its first-order term is zero, and (possible) higher-order terms are, in the "simple harmonic" approximation, not considered (or at least wished away). With N particles, the sum is over the $3N$ coordinates — the $3N$ degrees of freedom (for particles restricted to a plane or line, obvious numbers that are smaller). That T is of this form is a property of the kinetic energy (actually coming from the Poincaré group, being the lowest-order term in its expression for energy), while the expression for V is merely a Taylor series, which requires no more (?) than that V be well-behaved, and that the displacements be small (compared to what, and why they should be so, we do not consider). The reason this Hamiltonian is so useful, and simple, is that we can transform coordinates to diagonalize V — while leaving T diagonalized [Wilson, Decius and Cross (1980), p. 14]. The coordinates, $q_{i\alpha}$, in which T and V are simultaneously diagonal are normal coordinates. A vibration can have all — but one — normal coordinates constant in time (zero). This is not possible with the x's; if one varies, others must.

Problem X.3.a.i-1: Can there be a zeroth-order term (a constant) in V? Might T ever have cross terms? Would these matter?

Problem X.3.a.i-2: To show that V can be diagonalized, while leaving T diagonal, perform the transformation to normal coordinates in two steps; first define "mass-weighted" coordinates

$$w_{i\alpha} = \sqrt{m_\alpha} x_{i\alpha}, \qquad (X.3.a.i-2)$$

giving

$$2T = \sum (\frac{dw_{i\alpha}}{dt})^2, \quad 2V = \sum d_{ij} w_{i\alpha} w_{j\alpha}, \qquad (X.3.a.i-3)$$

where T is still diagonal. Then express the q's, which diagonalize V, in terms of the w's. Show that the latter is an orthogonal transformation, but the transformation from the x's to the w's is not. The transformation from the w's to the q's is a rotation. In what space, and what is the significance of this space? The transformation from the x's to the w's, and from the x's to the q's involve a "stretching" or "contraction" of the axes, so is not orthogonal — these are not similarities (pb. III.8-2, p. 173). Do these transformations form groups? Why? If so, which? Thus for the simple harmonic oscillator — in any dimension — the kinetic and potential energies can be simultaneously diagonalized, giving an immediate solution to the problem.

Problem X.3.a.i-3: It might be thought that as the x's are sums of normal coordinates, the coefficients can be so chosen so that all but one are zero. Explain why this cannot be. Show that it is always possible to pick initial conditions such that all normal coordinates but one are zero and remain so in time. The normal coordinates are therefore not only useful in solving the problem but they are related to, and express, the properties of the system. Explain how. What do they tell about the system?

Problem X.3.a.i-4: That the solution of the simple harmonic oscillator is so simple is a consequence of the existence of coordinates that simultaneously diagonalize the kinetic and potential energies. Why is this possible? Show that there are coordinates that diagonalize the potential energy while leaving the kinetic energy still diagonal because they are both of second order and the potential is (must be — why?) symmetric in the variables. This allows the transformation to be orthogonal. Why is this relevant? Explain why this procedure breaks down when higher-order (anharmonic) terms appear, and why this spoils the simplicity of the solution. The solutions for the simple harmonic oscillator form a complete set (in quantum mechanics, what about classical mechanics?) — any (well-behaved) function of the coordinates can be expanded as a sum of them. Discuss how this can be used to study anharmonic oscillators. Suppose that the Hamiltonian of a multi-dimensional oscillator were written in spherical coordinates. Would this prevent T and V from being simultaneously diagonalized? Why? Are there coordinate systems that would?

X.3.a.ii *Proof of the existence of normal coordinates*

For a system of several particles undergoing (simple-harmonic) vibrations with respect to each other, is it always possible to find normal coordinates — those for which we can take all but one as zero, and which do remain so in time? Starting from the expressions for the kinetic and potential energies in terms of the mass-weighted coordinates and (obviously)

$$d_{ij} = d_{ji}, \qquad \text{(X.3.a.ii-1)}$$

we define $\underline{q}$'s,

$$w_{i\alpha} = \sum a_{il} q_{l\alpha}, \qquad \text{(X.3.a.ii-2)}$$

so

$$2T = \sum a_{il} a_{im} \left(\frac{dq_{l\alpha}}{dt}\frac{dq_{m\alpha}}{dt}\right), \quad 2V = \sum d_{ij} a_{il} a_{jm} q_{l\alpha} q_{m\alpha}, \qquad \text{(X.3.a.ii-3)}$$

with sums over repeated indices. The requirements are

$$\sum a_{il} a_{im} = \delta_{lm}, \quad \sum d_{ij} a_{il} a_{jm} = \lambda_l \delta_{lm}. \qquad \text{(X.3.a.ii-4)}$$

The first says that the a's form an orthogonal matrix. The second diagonalizes the d matrix. They can both be satisfied if d can be diagonalized by an orthogonal transformation. Since it is symmetric and real, it can [Gel'fand (1989), p. 132; Lomont (1961), p. 15, 97; Margenau and Murphy (1955), p. 316]. This is analogous to the moment-of-inertia tensor which has cross terms in arbitrary coordinates, but can be brought to diagonal form — written in terms of its principle axes — by a rotation [Goldstein (1953), p. 153]. The rotation keeps T diagonal since, in mass-weighted coordinates, it is "spherically symmetric". The q's are the eigenvectors of the d matrix (giving the potential energy), one eigenvector for each particle (why?), with the components being the (normal) coordinates of that particle. Thus these are the eigenvectors of the matrix giving the potential energy, expressed in the "mass-weighted" coordinates, the w's — as might be expected since the form of T is fixed. It is the potential energy then that distinguishes the particular object (how?), and that gives its normal modes.

Problem X.3.a.ii-1: Justify the last few sentences.

Problem X.3.a.ii-2: What is the relevance of the eigenvectors (which are what?) and the corresponding eigenvalues?

Problem X.3.a.ii-3: Another way of simultaneously diagonalizing T and V [Schensted (1976), p. 98] is to start with both arbitrary (including cross terms for both) and write each coordinate as

$$q_l = a_l exp(i\omega t); \qquad \text{(X.3.a.ii-5)}$$

then the equations of motion give a set of algebraic conditions for the a's. For there to be a solution requires that the determinant formed by the coefficients appearing in T and V be 0 — this is the secular equation (eq. X.2.d-2, p. 518) [Goldstein (1953), p. 120, 154, 321], which gives the eigenvalues of the matrix obtained from the Hamiltonian that are the allowed frequencies. Work this out and show that the results are the same. Must the determinant in the secular equation always be 0? Might there be physical situations for which it could (conceivably) not be — for example an oscillator immersed in a nonisotropic medium, like a crystal, for which the kinetic energy might not be diagonal, or diagonalizable? Explain.

Problem X.3.a.ii-4: This implies that normal coordinates do not exist for anharmonic oscillators. Take the above Hamiltonian and add a term [Harris and Bertolucci (1989), p. 106] to the potential of higher order in the displacements (small if necessary). Check whether such coordinates can be found for it. How does this lead to a difference in the way harmonic, and anharmonic, oscillators are described? Can there ever be such terms that allow normal coordinates? Explain why, or why not.

X.3.b Finding normal coordinates

To a large extent the problem of describing vibrations, at least that part to which group theory is relevant, is solved when stated in normal coordinates [Bishop (1993), p. 164; Wilson, Decius and Cross (1980), p. 19, 71]. How are these found? Let the displacements of the objects from their equilibrium positions be given by arbitrary mass-reduced (or mass-weighted) coordinates $w_{i\alpha}$. We need a transformation, S, from the w's to normal coordinates $q_{l\alpha}$,

$$q_{l\alpha} = \sum S_{li,\alpha} w_{i\alpha}. \qquad \text{(X.3.b-1)}$$

The set of symmetry transformations acting on the w's mixes them — they form a basis of a reducible representation of the symmetry group. Acting on the q's however only subsets are intermixed — these form sets of irreducible representations. Thus the transformation from the w's to the q's reduces that representation whose bases the w's are. But we know how to reduce representations, using characters for example [Mirman (1995a), sec. VII.6.a, p. 199]. This gives the number of times each irreducible representation appears in the decomposition of the reducible one.

Problem X.3.b-1: It is, in addition, necessary to find the explicit matrices giving the reduction. Methods for doing this have been implied by the discussion above. Make them explicit. Once these have been determined the transformation from the w's to the normal coordinates follows immediately. Hence the key is the reduction of the reducible representation to which the coordinates belong.

Problem X.3.b-2: Should there be a subscript α on S? Why?

X.3.b.i *Valence bond lengths and interbond angles*

Choosing coordinates properly often simplifies a problem, and an understanding of the physical (and chemical) context can help. Such a choice is often the internal coordinates [Kettle (1995), p. 193; Wooster (1973), p. 273], those giving the relative positions of the points, so eliminating the extraneous translations and rotations. These may be of different kinds, angles and lengths say, and symmetry transformations can only mix ones of the same kind. So dividing coordinates into — physically distinct — sets immediately leads to simplifications.

Thus the displacement of an atom can be specified by giving the change in the distance between it and a neighbor — the change of the valence bond length — and also the change of the angle between two of its bonds. Symmetry transformations can only mix distances among themselves, and angles among themselves, but not intermix them.

For ammonia, NH_3, instead of twelve coordinates (for four atoms), there are only six internal ones, and these come in two sets, three distances and three angles (sec. X.5.b, p. 537). So instead of solving the secular equation by diagonalizing a 12 x 12 matrix, with a proper choice of coordinates we need only diagonalize two 3 x 3 matrices, and the symmetry of the molecule can yield further reductions.

X.3.b.ii *Coordinates that are not independent*

While the $x_{i\alpha}$ (and perhaps various internal coordinates) lead to a simple analysis of the oscillator they may not be optimal. We are interested in vibrations so wish to ignore translations and rotations of the object as a whole. This can be done by fixing two points (which give the rotation axis) reducing the number of coordinates from the $3N$ $x_{i\alpha}$ to $3N-6$ (fewer for planar and linear objects), with N the number of particles. But more convenient are often internal normal coordinates. What are these, why, and how are they related to the molecule and its symmetries? And why cannot we pick internal coordinates arbitrarily?

For objects on the corners of a triangle one set of coordinates are the vertex angles. However they have a problem, as emphasized by the completely symmetric representation; under it all three angles vary identically. But all angles cannot change together — only two are independent. This leads to trouble in two ways, first it messes up the formalism somewhat, but more it can lead to incorrect results if it is overlooked. It is important that all coordinates be independent, or if not then the analysis be properly modified.

Problem X.3.b.ii-1: One way of handling this is to remove one occurrence of the completely symmetric representation from the set in the reduction of the coordinate-reducible-representation into representations giving the vibrational representations. This is an *ad hoc* procedure and we expect that it comes out of the formalism [Wilson, Decius and Cross (1980), p. 142]. How should we expect this to occur? Show that it does.

Problem X.3.b.ii-2: Find a general (group-theoretical) method for eliminating non-independent coordinates even if they do not transform according to the symmetric representation (there are also geometrical methods, for example, as with the obvious one for the angles of a triangle) [Wilson, Decius and Cross (1980), p. 143]. Prove its correctness and generality. Compare with the method of Lagrange multipliers [Goldstein (1953), p. 41]. Is this relevant here? Why? Does it provide any (other, physical) information?

X.4 CHARACTERS AND COUNTING

Since characters are central to the investigation of vibrations, simple ways of obtaining them help. And such methods aid interpretation as we shall see. They can be found by taking the traces of the representation matrices. This is general, but it requires that we know these; there is a simpler way for the coordinate representation (sec. X.3.a, p. 522), which also has the advantage of being directly related to the symmetry of, and the effect of transformations on, the object [Leech and Newman (1969), p. 41; Tsukerblatt (1994), p. 185; Wilson, Decius and Cross (1980), p. 102; Wooster (1973), p. 269]. For this representation, the method is the same as that for the characters of a point-group representation, which (here) need not be crystallographic (sec. V.2.a, p. 248). It is to consider each atom, find the character for that atom, and then sum these over all atoms of the molecule.

Reflections and inversions leave particles fixed, or interchange two (which contributes 0 to the character). An inversion changes the sign of the three coordinates of a fixed particle, adding -3 to the character; a reflection changes the sign of one coordinate, the other two are unchanged, so adds 1. The contribution of C_n^k for each rotated particle is $1+2cos(\frac{2\pi k}{n})$. The character of the identity transformation is the total number of coordinates, which is three times the number of particles. This is the dimension of the (reducible) representation. Each coordinate representation appears the same number of times as there are atoms in the molecule, since it is the same for each, and the coordinate representation for the molecule is their direct sum. These contributions to the character χ_R of transformation R in the coordinate representation for each unshifted particle are thus known (tbl. V.2.d-1, p. 252; tbl. E.2-1, p. 657) [Wilson, Decius and Cross (1980), p. 105].

Thus we get the rule that the character of a transformation is given by the factor (in the table) going with the transformation times the number of particles not interchanged by it. For internal coordinates two particles are fixed, these determining the axes, so to find the character of a rotation multiply the factor for the rotation by the number of unshifted particles minus 2.

These are the characters for the reducible representation defined by its basis vectors which are the (three) coordinates for each atom. It is this representation that we reduce to find the normal modes.

Problem X.4-1: The dimension of this coordinate representation is larger than that of the regular representation (what regular representation?). Why; what is its significance? Can there be cases for which it is not?

X.5 EXAMPLES OF APPLICATION OF SYMMETRY TO VIBRATIONS

Having now outlined the general theory of vibrations, we are ready to illustrate by considering (or actually letting the reader consider) some molecules, starting with two in depth, not only to see how the theory applies, but more, what we learn about them, about molecules, about the theory of groups. These provide exercises in the theory and application of point groups, usually all that is necessary, except — of course — for finding the correct group.

X.5.a Group theory, symmetry and the vibrations of water

It is helpful that the likely most important molecule, at least on Earth, is also good for illustrating many facets of physics and chemistry, and group theory, the water molecule, H_2O [Brown (1975), p. 247; Burns (1977), p. 81; Cotton (1990), p. 319; Elliott and Dawber (1987), p. 115; Hargittai and Hargittai (1987), p. 203; Harris and Bertolucci (1989), p. 138; Jones (1990), p. 85; Kettle (1995), p. 10; Lomont (1961), p. 117; Tinkham (1964), p. 243; Tsukerblatt (1994), p. 357; Wilson, Decius and Cross (1980), p. 18, 114]. The molecule — at equilibrium — is taken planar and triangular, the oxygen atom at the apex, the two hydrogens at the other vertices with bonds to the oxygen. About equilibrium they undergo (small) vibrations. What are these, and why are they what they are?

The water molecule,

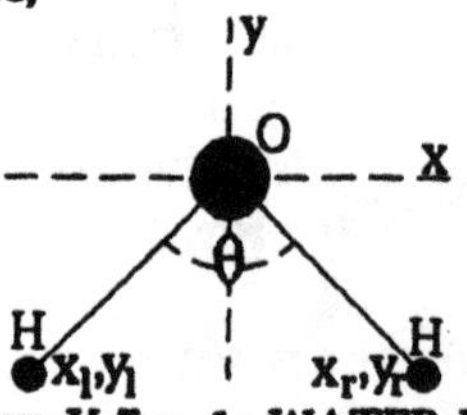

Figure X.5.a-1: WATER MOLECULE,

with the origin at the oxygen atom, z perpendicular to the plane of the molecule, x parallel, y perpendicular, to the line joining the hydrogens, has coordinates x_O, y_O, z_O, for oxygen, and $x_l, \ldots, x_r, \ldots$, for the left and right hydrogen atoms. The molecule has symmetry under reflections in the plane holding it (taking the atoms as points), as well as in the perpendicular plane through the oxygen, bisecting the angle between the two hydrogen bonds and thus the line connecting them. The intersection of these two planes forms a line, the y axis, which is a two-fold axis of symmetry: a rotation of π interchanging the two hydrogens leaves the molecule unchanged (actually indistinguishable from the first, though it would be distinguishable if one of the atoms were deuterium). The symmetry group of water is thus C_{2v}, with four operations, the identity E, C_2, a rotation of π about the y axis, $\sigma_v(yz)$, a reflection through this line, and $\sigma_v(xy)$ a reflection in the plane in which the three atoms lie. The reflection is labeled by the plane of the mirror.

Water has three atoms, so the character of the identity is 3×3. The rotation leaves the oxygen atom fixed (contributing -1 to the character) and interchanges the two hydrogen atoms (so contributes 0 for each) as does one of the reflections, while the other leaves all atoms fixed. Hence the character table of reducible coordinate representation Γ is

$$\begin{array}{ccccc} & E & C_2 & \sigma_V(yz) & \sigma_V(xy) \\ \Gamma & 9 & -1 & 1 & 3 \end{array} \tag{X.5.a-1}$$

Problem X.5.a-1: Using this model write the group table of C_{2V} and compare it with previous discussions (chap. I, p. 1; sec. I.7.b, p. 55) [Mirman (1995a)]. Check that it has four representations (sec. V.2, p. 247), all one-dimensional (labeled A_1, A_2, B_1 and B_2), and that the character table, with the first line giving the characters of the coordinate representation, is

$$\begin{array}{c|cccc} & E & C_2 & \sigma_V(yz) & \sigma_V(xy) \\ \Gamma & 9 & -1 & 1 & 3 \\ \hline A_1 & 1 & 1 & 1 & 1 \\ A_2 & 1 & 1 & -1 & -1 \\ B_1 & 1 & -1 & 1 & -1 \\ B_2 & 1 & -1 & -1 & 1 \end{array} . \tag{X.5.a-2}$$

Does it matter for water which plane is taken as yz? Check geometrically that all operations commute, so each is in a class by itself, and that all (except the identity) are of order two. From this it is clear that the decomposition of the coordinate representation is

$$\Gamma = 3A_1 \oplus A_2 \oplus 2B_1 \oplus 3B_2. \tag{X.5.a-3}$$

The subgroups are the cyclic groups C_2, consisting of the identity and the rotation, and two groups C_S with operators E and σ, for each of the σ's. They are all isomorphic so have the same representations and character tables; one representation is symmetric so both characters are 1, the other antisymmetric so its characters are 1 and -1.

Problem X.5.a-2: Show that the three translational degrees of freedom for C_{2V} transform as A_1 (for y), B_1 (for z) and B_2 (for x). The three rotational degrees of freedom transform as R_X, R_Y, R_Z, which here become B_1, A_2, B_2. Subtracting these gives for the vibrations,

$$\Gamma_{vib} = 2A_1 \oplus B_2. \tag{X.5.a-4}$$

X.5.a.i *The vibrational representations*

The number of times, n^γ, irreducible representation γ appears in a reducible representation, in terms of the characters χ_j^γ of classes j with

g_j elements, for a group with g elements, is [Mirman (1995a), eq. VII.6.a-4, p. 200]

$$n^{\gamma} = \frac{1}{g}\sum g_j\chi_j^{\gamma *}\chi_j^{\gamma}, \qquad \text{(X.5.a.i-1)}$$

so

$$n^{A_1} = \frac{1}{4}(9-1+1+3) = 3;\ n^{A_2} = \frac{1}{4}(9-1-1-3) = 1, \qquad \text{(X.5.a.i-2)}$$

$$n^{B_1} = \frac{1}{4}(9+1+1-3) = 2; n^{B_2} = \frac{1}{4}(9+1-1+3) = 3; \qquad \text{(X.5.a.i-3)}$$

nine states altogether. We must subtract the six translational and rotational states. Rotations of π around y leave y unchanged while changing the signs of x and z, a reflection in the plane perpendicular to the atom through the y axis, changes the sign of x, leaving y and z fixed. So y is invariant under all transformations thus belongs to A_1; z and x belong to B_1 and B_2. Translations transform as the axes, so as A_1, B_1 and B_2. Rotations around y are invariant under C_2, but being pseudo-vectors do not change sign under reflections, so transform as A_2. The signs of rotations around x and z are changed by C_2 and one of the reflections; they transform as B_1 and B_2. Subtracting these we then get the decomposition of the vibrational representation,

$$\Gamma_{vib} = 2A_1 \oplus B_2. \qquad \text{(X.5.a.i-4)}$$

The representation matrices are

$$E = \begin{pmatrix} 1 & 0 & 0 \\ 0 & 1 & 0 \\ 0 & 0 & 1 \end{pmatrix},\quad C_2 = \begin{pmatrix} -1 & 0 & 0 \\ 0 & 1 & 0 \\ 0 & 0 & -1 \end{pmatrix},$$

$$\sigma_V(yz) = \begin{pmatrix} -1 & 0 & 0 \\ 0 & 1 & 0 \\ 0 & 0 & 1 \end{pmatrix},\quad \sigma_V(xy) = \begin{pmatrix} 1 & 0 & 0 \\ 0 & 1 & 0 \\ 0 & 0 & -1 \end{pmatrix}, \qquad \text{(X.5.a.i-5)}$$

giving the transformations of the coordinates,

$$C_2\begin{pmatrix} x_l \\ y_l \\ z_l \\ x_r \\ y_r \\ z_r \\ x_O \\ y_O \\ z_O \end{pmatrix} = \left(\begin{array}{ccc|ccc|ccc} 0 & 0 & 0 & -1 & 0 & 0 & 0 & 0 & 0 \\ 0 & 0 & 0 & 0 & 1 & 0 & 0 & 0 & 0 \\ 0 & 0 & 0 & 0 & 0 & -1 & 0 & 0 & 0 \\ -1 & 0 & 0 & 0 & 0 & 0 & 0 & 0 & 0 \\ 0 & 1 & 0 & 0 & 0 & 0 & 0 & 0 & 0 \\ 0 & 0 & -1 & 0 & 0 & 0 & 0 & 0 & 0 \\ 0 & 0 & 0 & 0 & 0 & 0 & -1 & 0 & 0 \\ 0 & 0 & 0 & 0 & 0 & 0 & 0 & 1 & 0 \\ 0 & 0 & 0 & 0 & 0 & 0 & 0 & 0 & -1 \end{array}\right)\begin{pmatrix} x_l \\ y_l \\ z_l \\ x_r \\ y_r \\ z_r \\ x_O \\ y_O \\ z_O \end{pmatrix} = \begin{pmatrix} -x_r \\ y_r \\ -z_r \\ -x_l \\ y_l \\ -z_l \\ -x_O \\ y_O \\ -z_O \end{pmatrix} \qquad \text{(X.5.a.i-6)}$$

and

$$\sigma_v(yz)\begin{pmatrix} x_l \\ y_l \\ z_l \\ \\ x_r \\ y_r \\ z_r \\ \\ x_O \\ y_O \\ z_O \end{pmatrix} = \begin{pmatrix} 0 & 0 & 0 & | & -1 & 0 & 0 & | & 0 & 0 & 0 \\ 0 & 0 & 0 & | & 0 & 1 & 0 & | & 0 & 0 & 0 \\ 0 & 0 & 0 & | & 0 & 0 & 1 & | & 0 & 0 & 0 \\ \\ -1 & 0 & 0 & | & 0 & 0 & 0 & | & 0 & 0 & 0 \\ 0 & 1 & 0 & | & 0 & 0 & 0 & | & 0 & 0 & 0 \\ 0 & 0 & 1 & | & 0 & 0 & 0 & | & 0 & 0 & 0 \\ \\ 0 & 0 & 0 & | & 0 & 0 & 0 & | & -1 & 0 & 0 \\ 0 & 0 & 0 & | & 0 & 0 & 0 & | & 0 & 1 & 0 \\ 0 & 0 & 0 & | & 0 & 0 & 0 & | & 0 & 0 & 1 \end{pmatrix} \begin{pmatrix} x_l \\ y_l \\ z_l \\ \\ x_r \\ y_r \\ z_r \\ \\ x_O \\ y_O \\ z_O \end{pmatrix} = \begin{pmatrix} -x_r \\ y_r \\ z_r \\ \\ -x_l \\ y_l \\ z_l \\ \\ -x_O \\ y_O \\ z_O \end{pmatrix}. \tag{X.5.a.i-7}$$

Problem X.5.a.i-1: Verify these. Also draw a diagram and check that they are correct. State the effect of the other two matrices. What are the characters of these four matrices?

Problem X.5.a.i-2: To study transitions we need the decomposition of the direct product of representations. It should be clear from the character table that

$$A_1 \times A_1 = A_1,\ A_1 \times A_2 = A_2,\ A_1 \times B_1 = B_1,\ A_1 \times B_2 = B_2,\ A_2 \times A_2 = A_1,$$

$$B_1 \times B_1 = A_1,\ B_2 \times B_2 = A_1,\ B_1 \times B_2 = A_2,\ A_2 \times B_1 = B_2,\ A_2 \times B_2 = B_1. \tag{X.5.a.i-8}$$

Note the symmetry of the decomposition, as required (why?).

X.5.a.ii *The symmetry coordinates*

To find the normal modes we use the idempotents (projection operators) [Mirman (1995a), pb. VIII.5.a-1, p. 233] which, summed over the group elements τ, are

$$\chi^{\alpha} = \frac{1}{4}\sum \chi_{\tau}^{\alpha}\tau; \tag{X.5.a.ii-1}$$

only one index is needed for the representation since all are one-dimensional, and each matrix is the character. These acting on the coordinates, give the sums of them that are the basis vectors of the irreducible representations, the normal modes. So

$$4\chi^{A_1} = E + C_2 + \sigma_v(yz) + \sigma_v(xy), \tag{X.5.a.ii-2}$$

$$4\chi^{B_2} = E - C_2 - \sigma_v + \sigma_v'. \tag{X.5.a.ii-3}$$

We now get for A_1, using the above matrices,

$$4\chi^A = \begin{pmatrix} x_l \\ y_l \\ z_l \\ \\ x_r \\ y_r \\ z_r \\ \\ x_O \\ y_O \\ z_O \end{pmatrix} + \begin{pmatrix} -x_r \\ y_r \\ -z_r \\ \\ -x_l \\ y_l \\ -z_l \\ \\ -x_O \\ y_O \\ -z_O \end{pmatrix} + \begin{pmatrix} -x_r \\ y_r \\ z_r \\ \\ -x_l \\ y_l \\ z_l \\ \\ -x_O \\ y_O \\ z_O \end{pmatrix} + \begin{pmatrix} x_l \\ y_l \\ -z_l \\ \\ x_r \\ y_r \\ -z_r \\ \\ x_O \\ y_O \\ -z_O \end{pmatrix} = \begin{pmatrix} -2x_r + 2x_l \\ 2y_r + 2y_l \\ 0 \\ \\ 2x_r - 2x_l \\ 2y_r + 2y_l \\ 0 \\ \\ 0 \\ 4y_O \\ 0 \end{pmatrix}. \quad \text{(X.5.a.ii-4)}$$

Since the molecule is planar z = 0, and for A_1 only the H atoms move, there is only one pair of coordinates $(x_l - x_r, y_l - y_r)$, so they move together. For A_1 the coordinates are

$$q_{1H} = x_l - x_r = Q_1 exp(i\omega t), \quad q_{2H} = y_l + y_r = Q_2 exp(i\omega t), \quad \text{(X.5.a.ii-5)}$$

for the hydrogens, and

$$q_{1O} = 0, \quad q_{2O} = y_O = -Q_O exp(i\omega t), \quad \text{(X.5.a.ii-6)}$$

for the oxygen, with amplitude Q_1 an initial condition, and the other Q's depend on Q_1 since the center-of-mass is fixed. All displacements leave the (vibrating) object unchanged, so the three displacements are related. The hydrogen atoms move with the same speed, in phase along y, but with velocity x-components opposite. There are two ways of doing this: when the oxygen is moving in the y direction, the hydrogens can move toward it, or away; these differ, at all times, in the sign of the hydrogen-velocity x-component. The velocity vectors are thus either parallel to the hydrogen-oxygen bonds (shown as solid lines), or perpendicular, so there are two A_1 representations. The O moves, as it must, along y, with velocity opposite to that of the y component of the hydrogens. Therefore one A_1 representation (with the arrows not to scale, in particular that of the O atom is too large) is,

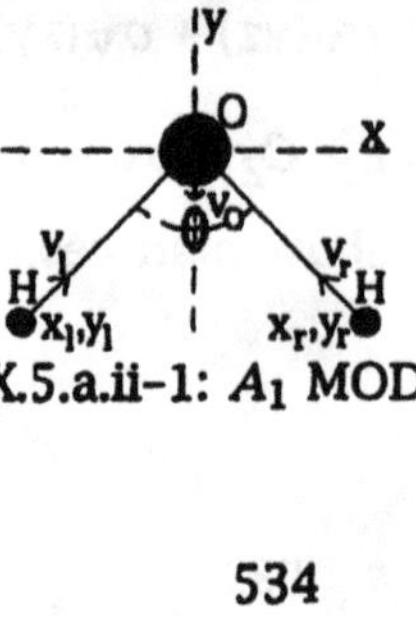

Figure X.5.a.ii–1: A_1 MODE OF WATER.

The second A_1 mode is,

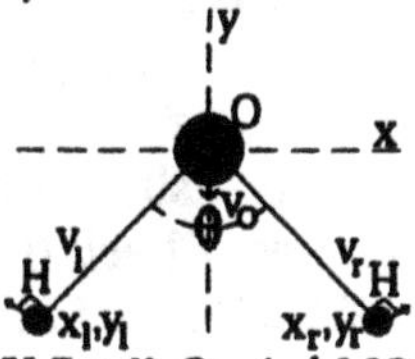

Figure X.5.a.ii-2: $A_1{}'$ MODE OF WATER.

Likewise for B_2,

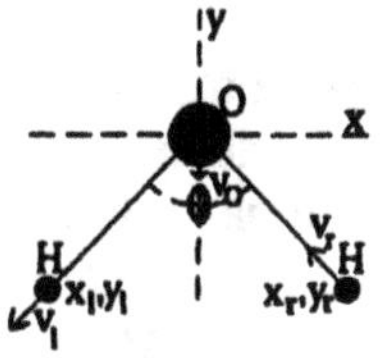

Figure X.5.a.ii-3: B_2 MODE OF WATER.

For this mode the coordinates are x_O and $(x_l - x_r, y_l + y_r)$. The velocity x-components of the hydrogens are equal, the y-components opposite; the oxygen, again as it must for fixed center-of-mass, moves along x. This we have from

$$4\chi^B = \begin{pmatrix} x_l \\ y_l \\ z_l \\ \\ x_r \\ y_r \\ z_r \\ \\ x_O \\ y_O \\ z_O \end{pmatrix} - \begin{pmatrix} -x_r \\ y_r \\ -z_r \\ \\ -x_l \\ y_l \\ -z_l \\ \\ -x_O \\ y_O \\ -z_O \end{pmatrix} - \begin{pmatrix} -x_r \\ y_r \\ z_r \\ \\ -x_l \\ y_l \\ z_l \\ \\ -x_O \\ y_O \\ z_O \end{pmatrix} + \begin{pmatrix} x_l \\ y_l \\ -z_l \\ \\ x_r \\ y_r \\ -z_r \\ \\ x_O \\ y_O \\ -z_O \end{pmatrix} = \begin{pmatrix} 2x_r + 2x_l \\ 2y_r + 2y_l \\ 0 \\ \\ 2x_r + 2x_l \\ 2y_r + 2y_l \\ 0 \\ \\ 4x_O \\ 0 \\ 0 \end{pmatrix}. \quad \text{(X.5.a.ii-7)}$$

In the B_2 mode the interchange of the two hydrogens, by either a rotation or a reflection, changes the sign of the statefunction, so these two atoms vibrate together, along their bonds — the line connecting them to the oxygen — and in the same direction, and the oxygen must vibrate along the line parallel to the hydrogens.

Problem X.5.a.ii-1: Find the coordinates for the other representation. Describe the motion.

Problem X.5.a.ii-2: Here are the x's and y's the mass weighted coordinates or cartesian coordinates?

Problem X.5.a.ii-3: Although there are nine coordinates, only four have been given for each representation. The others are determined by the requirements that the translation and rotation of the molecule both be zero. Using these, give equations for the time dependence of each of the nine (cartesian) coordinates for all representations and check that the statements about the motions of the atoms are correct. Then assume that the molecule is both translating and rotating, write general equations for the nine coordinates. Check the decomposition of the coordinate representation (eq. X.5.a.i–4, p. 532) using these results. Find the angles that the velocities of the H atoms make with the x axis for the two vibrational representations. What are the ratios of the speeds of hydrogen and oxygen? Why?

Problem X.5.a.ii–4: Reasonable internal coordinates are the angle between the two bonds and the bond lengths. Express these in terms of both the cartesian and the symmetry (normal-mode) coordinates, and give the time variation of the internal coordinates. Which frequencies go with which irreducible representations? Explain this connection.

X.5.a.iii *Internal coordinates for water*

Properly chosen internal coordinates not only simplify the problem, but help make the solutions clearer and more understandable. We use as the three internal coordinates the two bond lengths (from the oxygen to the two hydrogens, the interaction of the latter being neglected) and the interbond angle. We picture this by fixing the center of mass, which eliminates translations, and take the three rotational axes through the oxygen. The bonds are interchanged by C_2 and one σ, while all three coordinates are left invariant by the reflection in the plane in which the molecule lies. Hence the normal coordinate given by the interbond angle transforms according to A_1, as does one of those going with a bond-length normal-coordinate, the other transforming as B_2, as we found.

So for one normal coordinate, the two hydrogen atoms move toward and away from each other, their bonds opening and closing like a scissors. For the other A_1 mode, the two bonds lengthen and shorten together, and for the center-of-mass to be fixed, the oxygen atom must move opposite to them. For the B_2 mode, the hydrogens move opposite to each other along their bond lengths and for fixed center-of-mass the oxygen moves parallel to the base of the triangle, opposite to the direction of the atom moving away from it. The velocities depend on the masses and force constants (the interatomic forces), not on symmetry, so are not considered here.

Problem X.5.a.iii–1: Thus two ways of finding, and picturing, the normal modes have been used. Derive each from the other.

Problem X.5.a.iii–2: There is one further vibrational mode we might consider: the hydrogen atoms moving perpendicular to the plane, so that one moves toward the front while the other moves toward the rear, and then the reverse. Presumably this — if it occurs — is a combination of the rotational and vibrational modes already discussed. Which? How does this mode transform under the symmetry group? Is this consistent with the combination of motions just found?

Problem X.5.a.iii-3: Hydrogen sulfide, H_2S, is similar to water, but with a different bond angle (92°, compared to 104°) [Brown (1975), p. 102]. Does this change the analysis?

Problem X.5.a.iii-4: Another molecule with three atoms is ozone, O_3, but all its atoms are identical. Repeat the analysis for it [Heine (1993), p. 233].

X.5.b Ammonia

The ammonia molecule, NH_3, is a pyramid, an N atom at the top and a base which is an equilateral triangle, each corner an H atom [Cornwell (1984), p. 173; Cotton (1990), p. 333; Elliott and Dawber (1987), p. 116; Harter (1993), p. 205; Heine (1993), p. 242; Kettle (1995), p. 123; Lax (1974), p. 135; Schensted (1976), p. 113; Wherrett (1986), p. 98; Wilson, Decius and Cross (1980), p. 107]. We take the z axis through the N and perpendicular to the base, the x and y axes thus in the basal plane:

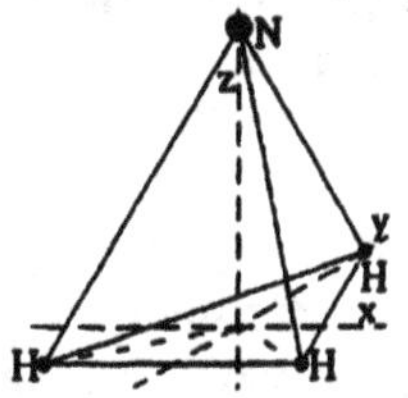

Figure X.5.b-1: AMMONIA.

The point group is C_{3v} which has three types of transformations, implying — correctly — three classes, the first containing E, the second $C_2{}^1$ and $C_2{}^2$, the $\frac{\pi}{3}$ rotations about z, and the third σ_v, σ_v', and σ_v'', reflections in the three planes (through the z axis) holding an H and the N (their intersections with the base are dashed lines). The z axis is invariant under the group, the other two are intermixed. The three translations are along, so transform as, these axes; that along z then belongs to the symmetric representation, A_1, the other two being mixed, belong to the two-dimensional one, E. Rotations leaving N fixed are invariant under rotations, odd under reflections, so belong to A_2. There is also a rotation about an axis in the basel plane, and since there are two axes mixed by rotations, this belongs to an E representation.

Problem X.5.b-1: Check that the group, and the action of the transformations, are correct. There are four atoms so the character of E is 12, that for the rotations 0, and for the σ's 2. Verify the correctness of the set of characters (and that the group is a subgroup of the tetrahedral group — why?). Find the characters of the irreducible representations and show that the decomposition of the twelve-dimensional representation Γ_{12} is

$$\Gamma_{12} = 3A_1 \oplus A_2 \oplus 4E; \qquad (X.5.b-1)$$

the symmetric representation appears three times, A_2 once, and the two-dimensional representation E four times, giving a 12 x 12 matrix.

X.5.b.i *The vibrational representations*

Now that we have found this, what do we learn? What can we say about the vibrations of the NH_3 molecule? There are 12 coordinates but three give translations and three rotations; only six are vibrational. We know the translational and rotational representations; the remaining terms in the decomposition of the twelve-dimensional representation give the vibrations, so for these

$$\Gamma_{12}^{v} = 2A_1 \oplus 2E, \qquad (X.5.b.i-1)$$

correctly giving six vibrational coordinates. From this we see that there are four frequencies, two, those going with A_1, nondegenerate, while each of the other two is doubly degenerate. That is there are two normal modes each with its own frequency, while there are two pairs, again each with its own frequency, but with the two normal coordinates of each pair mixed by the group transformations (thus their frequencies are the same); the frequencies of the two pairs are not the same — unless there is "accidental degeneracy", a degeneracy due to some "accident", or more likely a symmetry of a particular case that we have overlooked. This would occur for a molecule with this same geometry, but with masses and force constants that had the *exact values* necessary to give the frequencies equal. The group requires that certain frequencies be equal but does not prevent others, on which it places no requirements, from being so.

Group theory thus gives the (maximum) number of distinct frequencies, and their degeneracies. The frequencies can be found by solving the secular equations (notice the plural — there is one for each frequency); these depend on the masses and force constants. The potential and kinetic energies are expressed in terms of the normal coordinates, and the secular equations are then solved to find the frequencies. To do this we have to find the normal coordinates (using group theory).

Problem X.5.b.i-1: Repeating the analysis of H_2O, find the symmetry coordinates for NH_3 as functions of the cartesian coordinates of its atoms. Check the decomposition of the coordinate representation. Discuss the behavior of the atoms for each normal mode [Wherrett (1986), p. 98].

X.5.b.ii *Displacements in the normal modes*

How do the atoms move in the normal modes [Lax (1974), p. 139, Wherrett (1986), p. 103]? Their behavior can be understood, to some extent, from the properties of the group, especially its characters. This qualitative discussion, as is often the case, is more revealing than a derivation, and a similar one for other molecules can often help simplify such derivations. But that this can be done here does not imply that an equivalent analysis (completely) works in more complicated cases.

Since the center-of-mass remains fixed there are three general possibilities. One is for the nitrogen to have zero velocity, and the vibrations confined to the triangle formed by the three hydrogens, with the triangle's center-of-mass fixed. Next the triangle can move along the line (taken as z) between its center and the nitrogen atom, that moving in the opposite direction with velocities related by the requirement of fixed molecular center-of-mass. Finally the velocities of the center of the triangle and that of the nitrogen can be antiparallel (of course in the xy plane). Clearly these modes are independent; if the triangle and the N atom are relatively displaced along z there will be a force causing vibration, but there will be no force giving vibration or rotation of the triangle (perpendicular to the z axis), and conversely. This we can guess, but how do we determine it from the symmetry group?

The scalar representation, A_1 appears twice. Since it is a scalar, all group operators must leave the (appearance and properties of the) molecule unchanged at all times, both at equilibrium and when the atoms are displaced, so must leave the velocities unchanged, or interchange them to give the same configuration. What set of velocities allows this? All hydrogen atoms move toward and away from their equilibrium positions, with the same speed and all in phase, along the lines from the center of the triangle they form, with the nitrogen fixed. This is a "breathing" mode, with the equilateral triangle expanding and contracting, and always equilateral. The center-of-mass of the molecule, and of the triangle, are stationary. Another case is for the triangle to remain fixed in size, but move along the line between its center and the nitrogen (the z axis), while the latter moves in the opposite direction, with speed determined by the center-of-mass being fixed. Again, somewhat similar to water, there are two ways of doing this, thus two A_1 modes, but here one is a translation, giving then two one-dimensional vibrational modes, as found. These are (the arrows are not to scale)

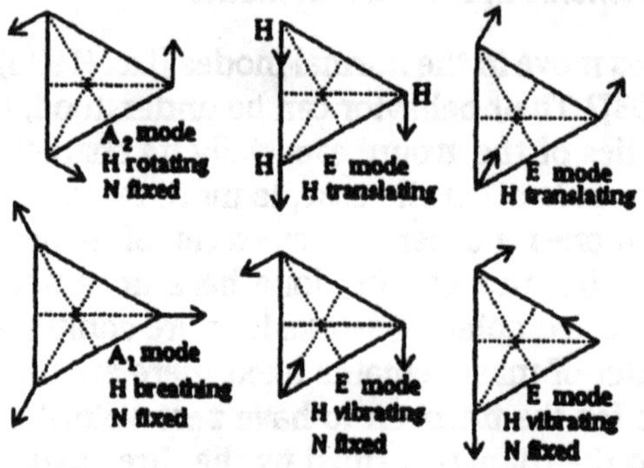

Figure X.5.b.ii-1: AMMONIA MODES WITH NONVIBRATING N, and illustrating the A_1 modes with moving nitrogen,

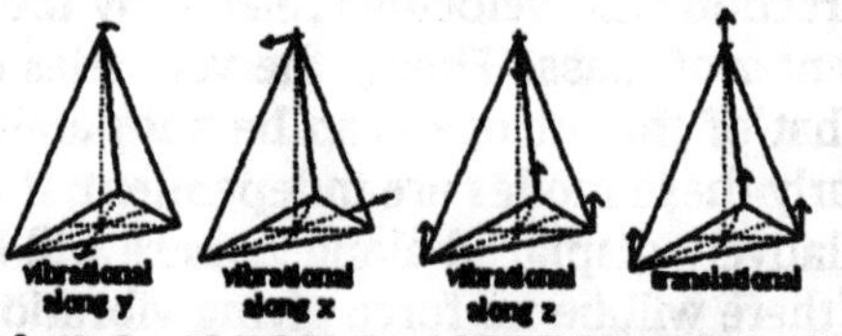

Figure X.5.b.ii-2: AMMONIA MODES WITH MOVING NITROGEN.

Might the eigenstates of the Hamiltonian be two mutually orthogonal sums of these modes? Suppose we start from the molecule at equilibrium and displace the atoms to give one of these normal modes, for example keep the nitrogen fixed and expand the triangle, keeping it equilaterial. How will the atoms move — what will be the effect of the Hamiltonian on the statefunction that initially describes this displacement? Since the Hamiltonian is a scalar acting on a state of one representation, it cannot give any others (the product of a scalar with a basis state gives the same basis state) so this normal mode is not mixed with those of any other representations. But might these two A_1 normal modes be mixed together? It is clear that with this displacement of the atoms there will be no force on the nitrogen, so it will not start to move — the other A_1 normal mode will not be excited. Likewise if the hydrogens and the nitrogen are displaced vertically there are no forces changing the size of the triangle — these two normal modes are eigenstates of the Hamiltonian. If the initial displacement gives one, the molecule will remain in that mode for all time. This indicates why there are two A_1 representations, and that different representations have meaningful physical differences.

For the A_2 representation the character of the rotations is 1, of the reflections, -1. This means that the three hydrogen atoms have velocity vectors that are taken into each other by a $\frac{2\pi}{3}$ rotation, and reflection in a plane through any of the H's gives the reversed motion. This representation is that of a pure rotation.

The basis states of these two representations are realized as nine-dimensional (or 12 dimensional if the nitrogen is included) vectors, containing the three velocity components of the three atoms. They are however each a single vector since these representations are one-dimensional. For the A_1 mode (the scalar) this basis vector is invariant, so it must be such a function of the velocities that under the group transformations (mixing their components by a rotation, say, and taking them into each other) the basis vector goes into itself, giving the velocities described.

Problem X.5.b.ii-1: Show that this basis vector, with the described velocities, is a scalar.

Problem X.5.b.ii-2: Can there be a scalar Hamiltonian that mixes the various A_1 modes? Give one, or explain why not. Is it possible to construct a molecule, with this symmetry group, for which these two normal modes are mixed? If so, what is its Hamiltonian? What linear combinations of the A_1 modes are the eigenstates of the Hamiltonian? Why?

X.5.b.iii *The two-dimensional E modes*

One two-dimensional vibrational representation, E, has its two basis vectors each realized as a nine-dimensional object, v, going with the velocity (or momentum) vectors $\underline{v}_1, \underline{v}_2, \underline{v}_3$ of the hydrogen atoms, all of equal magnitude, taken as 1, and similarly w, for nitrogen. The character of rotation C_3 is -1, of reflection σ_v is 0. Thus the two basis vectors are interchanged by reflections, while each C_3 is a rotation of $\frac{2\pi}{3}$ in the two-dimensional basis vector space. The operators are then

$$C_3 = \begin{pmatrix} cos\frac{2\pi}{3} & -sin\frac{2\pi}{3} \\ sin\frac{2\pi}{3} & cos\frac{2\pi}{3} \end{pmatrix}, \quad \sigma = \begin{pmatrix} 0 & 1 \\ 1 & 0 \end{pmatrix}. \qquad \text{(X.5.b.iii-1)}$$

Now

$$\underline{v}_1 + \underline{v}_2 + \underline{v}_3 = 0, \qquad \text{(X.5.b.iii-2)}$$

so each velocity is at angles $\frac{2\pi}{3}$ to the others (not surprisingly).

Why does symmetry impose requirements on the velocities in this case? The basis vectors are realized, simultaneously, in two different ways, as vectors in the two-dimensional representation space, and as nine-dimensional velocity vectors. These must be consistent. Under C_3, with $\underline{v}$ denoting the velocity (or momentum) of a particle,

$$\underline{v}_1 \Rightarrow \underline{v}_2 \Rightarrow \underline{v}_3 \Rightarrow \underline{v}_1. \qquad \text{(X.5.b.iii-3)}$$

Under σ, with the mirror through particle 1, taken along the y axis, with the x axis parallel to the line through particles 2 and 3,

$$v_{1x} \Rightarrow -v_{1x}, \quad v_{1y} \Rightarrow v_{1y}, \quad v_{2x} \Rightarrow -v_{3x}, \quad v_{2y} \Rightarrow v_{3y},$$

$$v_{3x} \Rightarrow -v_{2x}, \quad v_{3y} \Rightarrow v_{2y}. \qquad \text{(X.5.b.iii-4)}$$

This gives relations among the components of the velocities of the particles, restricting the functions of these that the basis vectors are, thus determining (although not necessarily completely) the motion of the atoms when the vibration of the molecule is described by a basis vector of this representation.

There is one other vibrational E mode, with the N and the triangle moving oppositely, along either x or y, these interchanged by rotations, so the representation is two-dimensional (in the diagram vibrations along two altitudes, rather than two orthogonal axes, are shown).

Problem X.5.b.iii-1: Work out the algebra and describe the motion as given by the two basis vectors. How do these differ?

Problem X.5.b.iii-2: While it is not surprising that the velocity vectors are at $\frac{2\pi}{3}$ to each other it is not inevitable (using only the group). Why?

X.5.b.iv *Internal coordinates for ammonia*

What are the (best) internal coordinates for ammonia [Wilson, Decius and Cross (1980), p. 110]? There are two reducible representations, each of dimension three, implying the bond extensions and the bond angles, as these coordinates. The character of the identity is 3 for both, 0 for rotations — it interchanges atoms so has no diagonal elements — and 1 for reflections. Each symmetry plane contains one bond, so a reflection changes the sign of two coordinates, for both extensions and angles, giving diagonal elements 1, 1, and -1, for a character of 1.

Problem X.5.b.iv-1: This gives

$$\Gamma = 2A_1 \oplus 2E, \qquad \text{(X.5.b.iv-1)}$$

as (twice) before, but now it is simpler.

Problem X.5.b.iv-2: Check the diagrams to verify that they behave as stated under the group transformations.

Problem X.5.b.iv-3: The atoms of the ammonia molecule can be taken to interact via forces due to charge separation, which classically can be modeled by springs. A model using this picture should show that the two A_1 modes are not mixed — and give a qualitative understanding of how the reasons are related to the structure of the molecule. A proper quantum-mechanical analysis, to a reasonable approximation, would involve the solution of Schrödinger's equation (taking account of the

Pauli principle) and should also give this result, undoubtedly after more work. It would also give different frequencies for the two E modes, indicating why they are different. To what extent is the Pauli exclusion principle relevant (group-theoretically)? Why?

X.5.b.v *Why is group theory relevant here?*

That group theory is applicable to the motion of the H atoms, alone, is clear; the set of these is invariant under their interchange. But we have also been using it to study the motion of the N atom, and this is different. How then could it be so used? Replace the triangle by a single atom, giving a linear diatomic molecule, with the atoms interchanged by the group C_2, but not invariant under it if they are different. This has two representations, the symmetric giving vibrations, and the antisymmetric, giving translations. Is there symmetry? Using velocities no, but using momenta there is — for vibrations these are equal and opposite, because of the symmetry of space, even though the molecule itself is not symmetric. This indicates the importance of the proper choice of variables, and also that groups can be used to study objects that are not invariant under their transformations — if there is other symmetry, here of space.

Taking the molecule as linear hides the other motions, but these are clear if the atoms have structure, as when one is replaced by a triangle, showing the presence of rotations and translations perpendicular to the line. What about oscillations perpendicular to the line? If, at some time, the momenta are tangent to (any of the great circles on) the spheres centered at the center-of-mass, then the motion is purely rotational. If there is a component along a radius, there will be in addition vibrations, and this we have considered. There can be no oscillations perpendicular to this line because the atoms do not exert forces on each other in such a direction. However for NH_3, if the N atom moves parallel to the triangle then the sum of the forces due to the three H atoms has a parallel component, so there are vibrations involving N, not only along the line joining it to the triangle, but also perpendicular to this line.

Problem X.5.b.v-1: Why does the symmetric representation give vibrations, and the antisymmetric translations?

Problem X.5.b.v-2: How do the vibrations of NH_3 differ from those of a molecule with the same geometry, but with all four atoms identical?

Problem X.5.b.v-3: The language here is classical; restate this in quantum-mechanical terms.

X.5.c How vibrational spectra depend on the molecule

The spectrum of a molecule gives information about it — how is this related to a molecule's structure? One good way of studying this is to compare different molecules. There are many examples of molecules with various symmetry groups [Hargittai and Hargittai (1987), particularly sec. 3-5, p. 84], and since discussions of them are widely available it is unnecessary to give full details. Here we list some problems that help explore how, and to what extent, symmetry determines the vibrations. To answer these questions state how the group transformations affect each molecule, and give the points, lines and planes invariant under every transformation. Describe the normal modes — the degeneracy and the expressions for the symmetry coordinates in terms of (some reasonably defined) orthogonal coordinates. Determine internal coordinates for each molecule that best express the symmetry. Also find other molecules for each symmetry group, and discuss how the vibrations of the different molecules with the same symmetry differ, and why.

Problem X.5.c-1: The symmetry group of S_2Cl_2 is C_2 [Hargittai and Hargittai (1987), p. 84]. Analyze its vibrations using symmetry, to the extent possible.

Problem X.5.c-2: Repeat this for molecules with groups C_3, C_4, ... To what extent is knowledge of the molecule, not merely its group, needed? If you knew that a molecule had a cyclic symmetry group, but did not know which, how much information about the group could be obtained from the spectrum of vibrations? How much information could be obtained from just a quick look?

Problem X.5.c-3: An interesting example of a molecule is the square planar XeF_4 [Harris and Bertolucci (1989), p. 145].

Problem X.5.c-4: The acetylene molecule, C_2H_2, is linear, with two carbon atoms each flanked by a hydrogen [Brown (1975), p. 5, 95; Cornwell (1984), p. 175, 178, 197; Pauling (1988), p. 164; Tinkham (1964), p. 244]. Analyze its vibrations using symmetry. The symmetry group is the infinite-dimensional dihedral group $D_{\infty h}$; why infinite, why dihedral, why this one? Does the infinite dimensionality affect the vibrational behavior? How? Another linear molecule is carbon dioxide, CO_2 [Harris and Bertolucci (1989), p. 142]. Compare the behavior of these molecules.

Problem X.5.c-5: Tetraarsene, As_4 [Hargittai and Hargittai (1987), p. 99], and methane, CH_4 [Cotton (1990), p. 335; Elliott and Dawber (1987), p. 212, 278; Ladd (1989), p. 225; Wherrett (1986), p. 113; Wilson, Decius and Cross (1980), p. 80; Wooster (1973), p. 275], are both regular tetrahedra,

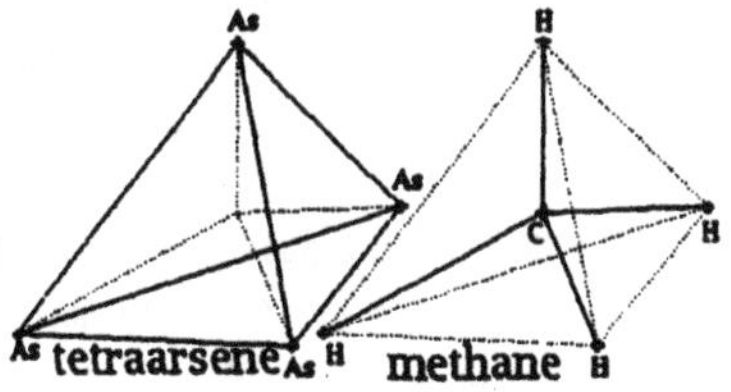

Figure X.5.c-1: TETRAARSENE AND METHANE,

(with what symmetry group?). For the former (with three As atoms in the planes, and one out of it) the edges of the figure are chemical bonds, shown as solid lines, connecting every pair of As atoms, for the latter they are not, these (the solid lines) going from the hydrogens to the central carbon (which lies out of the plane of the page, this containing the four hydrogens). The dotted lines are put in the left part of the diagram to show how the two structures are related. Analyze their vibrations. Describe the normal modes, and internal coordinates. How do these differ for the two molecules; why? Representations can occur multiple times; what is the effect of this [Wooster (1973), p. 280]?

Problem X.5.c-6: The octahedral molecule [Wooster (1973), p. 289] sulfur hexafluoride, SF_6, has cubic symmetry [Cotton (1990), p. 337; Hargittai and Hargittai (1987), p. 94; Wherrett (1986), p. 106], so point group O_h. Describe its vibrational spectrum. Compare it with tetrahedral molecules. Suppose one fluorine atom were a different isotope than the others. What would the effect be on the spectrum?

Problem X.5.c-7: Benzene, C_6H_6, is a planar molecule with the six carbons in a ring and a hydrogen attached to each carbon [Hargittai and Hargittai (1987), p. 95; Wherrett (1986), p. 67; Wilson, Decius and Cross (1980), p. 240]. In this diagram each line represents a double bond (sec. XI.7.c, p. 610),

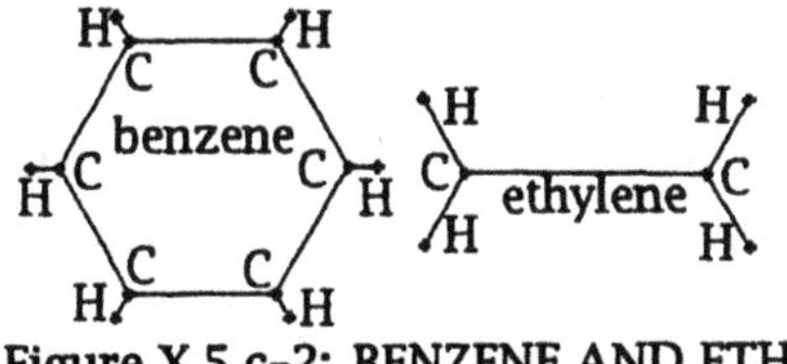

Figure X.5.c-2: BENZENE AND ETHYLENE.

What is its symmetry group (pb. I.6.a.ii-7, p. 31)? Analyze its vibrations. Besides its own importance, other molecules are constructed from it by substitution of various side chains in place of hydrogen atoms. Substitute one side chain and discuss the effect on the vibrations. Continue with two, and so on, now considering the different emplacements of these chains and cases for which side chains are both the same and are different. Determine both how the normal modes and frequencies change from those of benzene, and what ones are added, or subtracted, for each substitution (and this includes cases in which the atoms of the side chain vibrate, but the rest of the molecule does not move, if possible, and in which there are vibrations of several side chains). To what extent could the positions of these side chains be obtained from the spectra? Another way of studying these is to substitute deuterium which is about twice as massive as hydrogen, for some atoms (giving broken symmetry) and then finding the frequencies and normal modes (experimentally, but this also requires they be found theoretically). There are, of course, differences in frequencies if a deuterium nucleus is substituted instead of a side chain. Are there other differences? Why? It is also interesting to compare the behavior for some side chain containing hydrogen, and the corresponding chain with deuterium. And some carbon atoms can be replaced by different isotopes. In particular if a pair of opposite carbon atoms are replaced by a different isotope, the molecule has more symmetry than if only one were [Wherrett (1986), p. 116]. Could such isotopic replacements be determined by studying the vibrational spectra? How? Suppose that a molecule were constructed with identical side chains attached to each of the six identical carbon atoms, but the side chains were not completely known. To what extent can these substitutions be used to gain information about them? Suppose that different types of side chains were attached. Would the vibrational spectrum of the molecule, and of these substitutions, help in telling how many different chains there were, and how they differed?

Problem X.5.c-8: Ethylene (ethene), C_2H_4 [Brown (1975), p. 95; Elliott and Dawber (1987), p. 212, 278; Harris and Bertolucci (1989), p. 276; Kettle (1995), p. 63; Pauling (1988), p. 229; Wilson, Decius and Cross (1980), p. 128], a planar molecule with the carbons joined and each linked to two hydrogen atoms with bonds at a nonzero angle to the line between the carbons (fig. X.5.c-2, p. 545); answer the questions of the previous problem for it.

Problem X.5.c-9: The molecule ethane, C_2H_6 [Brown (1975), p. 95; Hargittai and Hargittai (1987), p. 80, 110; Pauling (1988), p. 222], consists of two joined tetrahedra (each reminiscent of methane). Write its structural formula. Is it possible to determine its structure (completely) using (experimental) knowledge of its vibrational spectrum. Isotopic substitutions, if helpful, are assumed not to vary the structure. Would other substitutions help? Why?

Problem X.5.c-10: Answer these questions for propane, C_3H_8 [Brown (1975), p. 122; Pauling (1988), p. 222]. This, like many molecules, has more than one configuration (for it, two), known as the eclipsed and the staggered. Draw a diagram showing these. How do the spectra differ, if they do?

Problem X.5.c-11: Methanol (methyl alcohol) is derived from methane by substitution of OH for hydrogen giving COH_4 [Pauling (1988), p. 242]. What can be learned of its structure from its vibrations? Would a comparison with the spectrum of methane be helpful? Why? Is this the best way of writing its formula? A related compound is methylamine in which an amine group, NH_3, is substituted for hydrogen giving CNH_5. Repeat the analysis for this and compare results. A related molecule is methyl chloride, CH_3Cl [Leech and Newman (1969), p. 42], with C_{3v} symmetry. Compare the spectra of these molecules.

Problem X.5.c-12: Analyze the vibrational spectrum: find the symmetry group, degeneracies, normal modes, and internal coordinates of a molecule that consists of four identical atoms at the corners of a square [Duffey (1992), p. 80, 88; Elliott and Dawber (1987), p. 124, 277; Inui, Tanabe and Onodera (1990), p. 223]. Describe the motions for the normal modes. How would the spectrum differ if the molecule were in the shape of a rectangle? What would it be if the molecule where a rhombus [Wherrett (1986), p. 113]? Repeat the analysis for a molecule with two different types of atoms [Burns (1977), p. 103]. Do it both for bonds (springs) along the edges of the square, and for the case in which, in addition, there are bonds along the diagonals. Discuss the difference in the vibrational spectra, and the representations of the relevant groups.

Problem X.5.c-13: The planar molecule boron trifluoride, BF_3 [Elliott and Dawber (1987), p. 124, 278; Ladd (1989), p. 224], has the three fluorine atoms at the vertices of an equilateral triangle, and the boron at the center; its symmetry group is D_{3h}. Why? Repeat the previous problem for it.

Problem X.5.c-14: Planar molecule BCl_3 also has D_{3h} symmetry [Harris and Bertolucci (1989), p. 147]. The formulas of it and BF_3 are very similar. What is the easiest way of finding the correct structures from their vibrational spectra?

Problem X.5.c-15: Repeat this for phosphorous, P_4 [Bhagavantam and Venkatarayudu (1951), p. 108; Elliott and Dawber (1987), p. 212, 278; Pauling (1988), p. 168], with atoms at the vertices of a regular tetrahedron (and compare the results with those for other tetrahedral molecules). Compare its group with that of PCl_5 [Pauling (1988), p. 245].

Problem X.5.c-16: A set of atoms can be arranged in several ways, giving molecules with very different properties. For example [Hargittai and Hargittai (1987), p. 76] $C_2H_4O_2$ in its incarnation as $HCOOCH_3$ is a molecule (which we take as planar) with an O in the middle of a line and C atoms on either side. To one carbon is attached three hydrogens, one on the COC line, one above and the other below. To the other carbon is attached a hydrogen below the line and an oxygen above. In the form CH_3COOH the oxygen replaces the hydrogen on the second carbon, and to it is attached this hydrogen atom. As $HCOCH_2OH$, the two carbons are bonded to each other and on the same line is the oxygen attached to the second carbon and to the oxygen, lying on the same line, is a hydrogen, with an oxygen above this line and a hydrogen below bonded to the first carbon. The last two hydrogens are attached to the middle carbon, one above and one below the line, with the line between them perpendicular to the CCO line. All oxygens attached to one carbon have a double bond. Draw these, find (and compare) their symmetry groups, and analyze their vibrations. To what extent can the spectra distinguish the molecules? Would it be possible to determine, by studying these, if the molecules were actually planar? Do the spectra fully determine the molecular structure (perhaps knowing the atoms)? Why?

Problem X.5.c-17: For a molecule with tetrahedral symmetry, find the vibrational spectrum, normal modes, and internal coordinates. Do this for molecules with each of the symmetry groups of the Platonic solids (sec. I.4.b, p. 8). How much information can be obtained, particularly about internal coordinates, from just the symmetry group, without knowledge of the constituents of the molecule? Can the symmetry group be completely determined from the spectra? Would it help, be necessary, to include rotational spectra?

Problem X.5.c-18: Is it possible to tell by studying the vibrations whether the symmetry group is cyclic or not? How easy would it be?

Problem X.5.c-19: The normal coordinates (or symmetry coordinates (sec. IX.2.a, p. 469)) are the basis vectors of the representations of the symmetry group of the system. This is a subgroup of a symmetric group. For NH_3 (also try other molecules, such as PH_3 or CH_4 [Harris and Bertolucci (1989), p. 150, 508]) the symmetry group has subgroup S_3 which permutes the three hydrogens. The statefunctions of S_3 are known [Mirman (1995a), sec. VIII.6.c, p. 244]. There are two two-dimensional representations (so each carries two indices), these being in the same equivalence class. Thus there are four statefunctions with the members of each of the two pairs intermixed (such multiplicity holds for all finite groups). In discussing the normal modes we have been (conveniently) overlooking this point. Does it matter?

Problem X.5.c-20: Which of the following sets of degeneracies might go with real molecules (note tbl. E.2-1, p. 657): (1), (1,1), (2), (2,1), (2,2), (3), (3,1), (3,2), (1,1,1), (2,1,1), (2,1,1,1), (3,2,1), (3,2,2), (3,3), (3,3,1), (3,3,2,1)? Why? For each construct (a three-dimensional model of) a molecule.

Problem X.5.c-21: Develop a procedure for determining (as much as possible of) the symmetry group of a molecule (there are an infinite number of molecules — and so symmetry groups?) from the set of degeneracies of its frequencies (the experimental data from which we could reasonably wish to find the molecules). Try it for a few (arbitrary) sets of degeneracies. Do these all lead to symmetry groups? Why? Could a set be found experimentally that would not? What would be the significance of such?

Problem X.5.c-22: Is there an algorithm (implemented as a computer program) for finding (much information about) molecules having any given set of degeneracies? What would be needed, besides this set, to fix (completely) the molecule?

Problem X.5.c-23: Can general statements be made about the relationship between a molecule's structure and the information that can be obtained from an experimental study of its vibrations?

X.5.d Breaking of symmetry

The molecules studied here are taken to be in free space — an essential, though hidden, assumption is that their symmetry groups are subgroups of the rotation group. Of course they really are not. While the environment may provide merely a small correction, which an experimenter has to be cautious about, we might also use the effect to advantage. The disagreement between the observed vibrational spectrum and what is calculated can serve as a probe of a molecule's environment.

If the symmetry is broken, say if the molecule were in a crystal [Cotton (1990), p. 341], then the group describing the system (not the molecule) is a subgroup of the symmetry group of the molecule [Wherrett (1986), p. 106]. Representations irreducible for the full group are not (necessarily) so for the subgroup. Thus the degeneracy can decrease. And we know how an irreducible representation becomes reducible under a subgroup. So by studying the spectra of the molecule both when it is free, and when it is under the influence of its environment, we can gain information about the latter (and how it affects the molecule). Also, as we saw in examples in the previous section, broken symmetry, say due to changing a molecule to a less symmetric one, can be used to study the molecule itself.

Problem X.5.d-1: Is it possible to use water as a probe of its environment? What information would it give? Could this be medically useful? How about ammonia?

Problem X.5.d-2: Suppose ammonia were placed in a crystal of unknown symmetry. How much information can be found about this symmetry from the vibrational spectrum of the molecule? Would the molecule's orientation matter?

Problem X.5.d-3: For which of the molecules discussed previously can the vibrational spectrum be used to study the molecule's surroundings? What information would be obtained? Which (types) would be most useful, or efficient? Why?

Problem X.5.d-4: A solid is known to be a crystal, but its structure is unknown. Can the vibrational spectrum of an embedded water molecule be used to find the crystal structure? Which molecule, if any, would be best for determining this structure? Why? Does the best molecule depend on the crystal structure?

Problem X.5.d-5: A crystal has symmetry, and the implication here is that it is smaller (or perhaps different) than that of the molecule. Suppose the environment had a higher symmetry; if the material were amorphous, say, the environment could be spherically symmetric. Or the environment may have no symmetry. Why? Discuss the effect on the spectrum (and the rotational spectrum could be included) if the environment were symmetric under a subgroup of the symmetry group of the molecule, under a group of which the molecular symmetry group were a subgroup, under a group having neither subgroup relationship, if it were spherically symmetric, and if it had no symmetry. For which of these cases could the spectra be used to obtain information about the molecule? About the environment? About both? To what extent? How? Why?

X.6 MULTIPLE EXCITATIONS

Given the normal modes, the frequencies and the linear combinations of coordinates that form the symmetry coordinates, and (what is emphasized here) their degeneracies, we still lack complete information (and in real life, always will). There are two sets of numbers (relevant here) still to be given, amplitudes and phases. What are the amplitudes and how do they affect the analyses? A simple harmonic oscillator has classically a continuous range of amplitudes each giving a different energy, quantum mechanically it has discrete energies. Classically, knowing a motion, say one of the normal modes of NH_3, and the frequency, leaves the amplitude still undetermined. Quantum mechanically, the system has sets of energy levels, each set going with a mode (so frequency), and the level to which the system is excited corresponds to the classical amplitude — the frequencies give the energy levels, but does not tell which is occupied. The ground states of the sets are the fundamentals. Excited levels of a single set are overtones — a level can be multiply excited, having a larger energy (thus classically a larger displacement of the parts) than the fundamental excitation. States having excitation from more than one set are combination levels [Bhagavantam and Venkatarayudu (1951), p. 89; Bishop (1993), p. 192; Harris and

Bertolucci (1989), p. 107, 164; Heine (1993), p. 252; Tinkham (1964), p. 245; Wilson, Decius and Cross (1980), p. 35, 151].

A state of the system is a group-representation basis state. When more than one is excited the state is given by the product of the basis states going with each mode, reduced into states of irreducible representations.

Thus for water (sec. X.5.a, p. 530) there is a normal mode in which the two hydrogens are moving toward the oxygen, and it moves toward the middle of the line between them. This does not fully describe the motion. Classically, while the velocities are related, their values still need be given, and the larger they are, the greater the energy. Quantum mechanically, the amplitudes of the basis vectors of which the statefunction consists are not determined by knowing the normal modes, and the larger those of higher energy are, the greater the energy of the system.

Problem X.6-1: If a single normal mode is excited its phase is meaningless (or should it be "does not matter" — if there is a difference?). For water take two different normal modes excited. Write the cartesian coordinates as sums over the symmetry coordinates, putting in the time dependence, and include relative phases (and amplitudes). Are the phases relevant? The amplitudes? Repeat this for one mode, but doubly excited. Next take two different modes, one doubly excited, and one triply excited. What effect do the phases have? Are phases measurable experimentally? Classically? Quantum mechanically? How? Why? Would they be interesting? Does this depend on the environment? Could there be a (measurable) phase if only one mode is excited, but multiply so? How, experimentally, would two different modes be excited, with specified amplitudes and phases? Is the particular molecule relevant to these question? What would be the effect of anharmonicities? Try this for all combinations of normal coordinates excited, thus for multiple excitations.

Problem X.6-2: Discuss multiple excitations and anharmonicities for ammonia (sec. X.5.b, p. 537). Is there a difference in usefulness between this and water, perhaps in studying their environments, if they are studied when they are not isolated gases? Why?

Problem X.6-3: Work this out for acetylene (pb. X.5.c-4, p. 544). What is the spectrum of energy levels for this molecule taking multiple excitations into account?

Problem X.6-4: The carbon dioxide molecule, CO_2 [Hargittai and Hargittai (1987), p. 215; Schensted (1976), p. 107; Wherrett (1986), p. 87], can be taken as linear with the carbon atom midway between the two oxygens (pb. X.5.c-4, p. 544). Find the group giving its symmetries. Show that it is Abelian so its representations are one-dimensional. Which give the vibrational states of the molecule? Describe the motion of the atoms for the normal modes. Would you have guessed these without using the group? Why? What are the statefunctions? Is oxygen a boson or fermion? Does this affect the normal modes and statefunctions? How do these answers change if the molecule contains O^{17} instead of O^{16}? What would be the effect if the two atoms were different isotopes of oxygen? Could the vibrations be used to test experimentally whether they are? How? Discuss the effects of multiple excitations and anharmonicities, and their usefulness. Another possibility is one O^{16} and one O^{18}. Why is this different? Would the answers change for this case?

Problem X.6-5: Formaldehyde, H_2CO [Harris and Bertolucci (1989), p. 280; Pauling (1988), p. 171], is a molecule with carbon and oxygen joined and the two hydrogens attached to the carbon, above and below the CO line. Repeat the analysis of the previous problem, and compare the results. How does the vibrational spectrum of this planar molecule differ from that of the linear CO_2?

Problem X.6-6: Thus besides the states of excitation of a single normal mode (discuss what this — including that *states* is plural — means both classically and quantum mechanically), we can consider more than one mode excited simultaneously. What does group theory say about such multiple excitations? Would studying this give us more information about the system? What? Is this information useful, or merely redundant? Why? Now suppose that the frequencies of different normal modes happen "accidentally" to be the same. What effect would this have? Would it be possible to guess from the spectrum whether this equality was truly accidental, or perhaps a sign of another symmetry? How would (small) anharmonicities affect these spectra [Heine (1993), p. 256; Tinkham (1964), p. 246]? How useful could they be?

X.7 TRANSITIONS AND SELECTION RULES

Our knowledge of the universe is determined by the study of transitions between states. Macroscopically this may be less obvious than for transitions found using quantum mechanics, but certainly the state of a light beam, at least, must be changed for us to gather information about the object it illuminates (and so, though we may not be conscious of it and the effect may be small, the state of the observed object must also change — as must our state, the state of the observer). In microscopic physics the centrality of transitions is so intuitive that it is often not thought about — but very much used. What does symmetry, and group theory, tell us about transitions, about which are possible, and which not (selection rules), and why? Can it provide us with other information, and what can it not provide?

Although to find completely the states of a system and the transitions between them requires full knowledge of all interactions among all parts, symmetry alone can provide much information. In particular there are transitions that cannot occur because they would change the symmetry of the system which is impossible if the interaction is itself symmetrical, or change it in a way not allowed by the symmetry of the interaction. This leads to selection rules, and we consider these here.

The number of (inequivalent) basis states of a (finite) symmetry group is finite, but the number of states that can be excited might (depending on the potential) be infinite — these having different energies (different diagonal matrix elements of the — perhaps noninteraction part of the — Hamiltonian). Thus an excited state is in general a sum of states of all modes each with different numbers of quanta, that is different amplitudes. The state of the system is then found by taking the product, for each mode, of the basis state for that mode, the number of times equal to the number of quanta, then taking the product of all these, and reducing (with care; the reduction of a product of states of different energy can give a sum which has no physical meaning). This gives a sum over the group basis states. The question now is what are the allowed transitions between two basis states? For arbitrary energies, we have to consider the transitions between each basis state of one level, and each basis state of the other. Some of these transitions may be forbidden, but if not all are there are transitions between the two energy levels. Here when we refer to a state we mean a basis state, and do not consider which energy level it belongs to, or whether there are others for that level. However in studies of experimental data, say, this is quite relevant.

In deciding that transitions are forbidden, we assume that the symmetry is exact. We have been treating atoms as points, molecules as having no neighbors that break their symmetries, and so on. Effects due to the failure of such idealizations cause forbidden transitions to actually, but usually weakly, occur. Whether group theory can be used to find these depends on whether they also have symmetry, probably less than that of the object, but enough so that they can be described by basis states of the object's symmetry group.

And it may be that while H_{int}, the interaction part of the Hamiltonian, has no simple symmetry it can be written as a sum of increasingly smaller terms, each with some symmetry. Then the transition may be forbidden to some order, but allowed beyond it. Although group theory does not give the value of the amplitude, here it would imply that it is smaller (by the order of the expansion parameter to the lowest power giving a nonzero term) then other amplitudes that are allowed to lower order.

Often giving the representation basis state does not determine it completely. There can be other quantum numbers, so many states that transform as the same basis state of the same representation, these differing by some quantum number not related to the symmetry (being considered). The amplitude for the transition depends on these other quantum numbers, a reason for it being outside this discussion. Here we ignore this point, considering only those state labels that come from the symmetry group (that we are concerned with).

To find the amplitude for the transition between states $|1\rangle$ and $|2\rangle$ [Birman (1974), p. 271; Bishop (1993), p. 186; Burns (1977), p. 112; Cotton (1990), p. 324; Harris and Bertolucci (1989), p. 130; Hochstrasser (1966), p. 215; Kettle (1995), p. 220; Leech and Newman (1969), p. 91; Tsukerblatt (1994), p. 223; Wherrett (1986), p. 193; Wilson, Decius and Cross (1980), p. 146], we need (schematically) the matrix element, $\langle 1|H_{int}|2\rangle$, of the interaction Hamiltonian (to lowest order perturbation theory; these results may not be exact, though violations are hopefully "small"). Assume that H_{int} transforms as state i of representation α of the symmetry group of the system, and $|1\rangle$ as state j of representation l_1 (these labels may be sets, if several are needed). Then for the amplitude to be able to be nonzero — since the symmetry does not give its value we cannot rule out that it could "accidently" be zero — the state $|2\rangle$ (of representation l_2 of the symmetry group of the object) must occur in the Clebsch-Gordan decomposition of $|\alpha, i\rangle|l_2, j\rangle$. Otherwise the matrix element would be a product of two different states, and so 0 by the orthogonality of basis vectors [Mirman (1995a), sec. VII.5.b, p. 195]. Thus here we have to consider different transformation properties of H_{int} and the transitions that it allows.

Without knowledge and use of symmetry — group theory — certain transitions would just "accidently" happen to come out zero, with no way of understanding why, or finding general patterns, or developing simple, systematic rules for determining these zeros; each case would have to be calculated separately, each calculation, each integration, would have to be done again and again, rather than, as here, bypassed.

X.7.a Transitions due to electric dipole moments

A molecule can have an electric dipole moment if atoms are charged because electrons move from one to another or are displaced by interactions. A dipole, the result of charge separation, is proportional to the vector from one charge to the other, so transforms as a coordinate (the only aspect relevant here). Interactions with electromagnetic fields also produce charge separation, so multipole moments. For interactions with light, transition matrix elements contain a plane wave, that of the light; this is expanded (in the spirt of the expansion of H_{int} just discussed) and the lowest-order term gives, so transforms as, a dipole (higher multipoles are important but not now considered). The radiation is usually in the infrared region, thus the result of "infrared transitions" [Bhagavantam and Venkatarayudu (1951), p. 82; Bishop (1993), p. 186; Brown (1975), p. 239; Cotton (1990), p. 327; Harris and Bertolucci (1989), p. 93; Heine (1993), p. 245; Inui, Tanabe and Onodera (1990), p. 227; Ludwig and Falter (1988), p. 154; Schiff (1955), p. 246; Tsukerblatt (1994), p. 362; Wilson, Decius and Cross (1980), p. 39, 156]. The amplitude is then schematically of the form $(1|x_i|2)$, $i = 1,2,3$. What are its selection rules?

In order that symmetry not require that the transition be impossible, the representation (here of the rotation group) l_1 (with state $|1)$) must appear in the Clebsch-Gordan decomposition [Mirman (1995a), sec. IX.5, p. 259; sec. XII.2, p. 340] of the product $x|2)$, that is in the decomposition of the product of representation l_2 with the $l = 1$ representation to which coordinates, so dipoles, belong.

Problem X.7.a-1: There is symmetry in this Clebsch-Gordan decomposition so this condition can be given in three other ways: l_2 must appear in the product of representations l_1 and $l = 1$; the $l = 1$ representation must appear in the product of the l_1 and l_2 representations; the scalar representation ($l = 0$) must appear in the decomposition of the triple product of the l_1, l_2 and $l = 1$ representations. For the rotation group this can (and should) be shown using the fact that the representations l appearing in the decomposition of the product of the l_a and l_b representations are

$$|l_a - l_b| \leq l \leq l_a + l_b, \qquad \text{(X.7.a-1)}$$

and also using the characters of the group representations. However the relevant group is the object's symmetry group. Explain why this result follows from that for the rotation group.

Problem X.7.a-2: Work out the allowed transitions for both ammonia and water [Elliott and Dawber (1987), p. 120].

Problem X.7.a-3: The water molecule has C_{2v} symmetry (sec. X.5.a, p. 530), and coordinates transform as A_1, A_2 and B_2. The vibrational modes are A_1 and B_2. Using the decomposition of the product representations we can analyze its infrared transitions. Note that the decompositions of the product of these two representations with themselves, and with each other, contain only these two, thus are the only representations we need be concerned with. Check that since the y component of the dipole operator belongs to the A_1 representation it can only cause transitions between two A_1 or two B_2 representations. The z component belongs to the B_1 representation so can give no transitions. The x component belongs to B_2, so gives absorption, causing a transition from the A_1 to B_2 modes, but not from an A_1 to an A_1, or B_1 to a B_1 representation. Knowing the motion of the atoms in these modes, do these results make sense? Why? Is it clear that they must hold?

X.7.b Polarizability and the Raman effect

The Raman effect is the scattering of electromagnetic radiation by a molecule due to its polarization by the incoming field, raising it to a higher state from which it falls producing the scattered radiation [Bhagavantam and Venkatarayudu (1951), p. 93; Bishop (1993), p. 189; Elliott and Dawber (1987), p. 121; Harris and Bertolucci (1989), p. 93; Heine (1993), p. 248; Tinkham (1964), p. 248; Tsukerblatt (1994), p. 364; Wilson, Decius and Cross (1980)]. Thus the scattering amplitude depends on how easy it is to polarize the molecule, that is on its polarization tensor. This is defined by [Wilson, Decius and Cross (1980), p. 157, 359]

$$\underline{\mu} = \alpha\underline{E}, \tag{X.7.b-1}$$

where $\underline{E}$ is the electric field of the incoming wave, $\underline{\mu}$ the induced dipole field and α the polarization tensor. Since $\underline{\mu}$ and $\underline{E}$ need not be in the same direction it must have nine components,

$$\mu_i = \alpha_{ij}E_j. \tag{X.7.b-2}$$

However it is symmetric,

$$\alpha_{ij} = \alpha_{ji}. \tag{X.7.b-3}$$

Problem X.7.b-1: Why is the polarization tensor symmetric?

X.7.b.i *Raman scattering is of second order*

There is another argument (these are rough since we are not interested in the reasons for the effect, but merely the limitations imposed by symmetry — which are dependent only on the basis states). In Raman scattering a photon is absorbed and then another (which can be of different frequency) emitted. The process is then of second order, so the transition amplitude is given by

$$T \sim \sum (2|\mathrm{H}_{int}|s)(s|\mathrm{H}_{int}|1), \tag{X.7.b.i-1}$$

summed over a complete set of states, and since each term transforms as a coordinate, this transforms as the product. Thus α_{ij} transforms as $x_i x_j (= x_j x_i)$. The decomposition of the product of the $l = 1$ representations with itself contains representations $l = 2,1,0$. However here the product is symmetric so with only six components, thus containing only the $l = 2$ and $l = 0$ representations.

Problem X.7.b.i-1: Show that

$$V_i = \sum \varepsilon_{ijk} T_{jk}, \quad \text{(X.7.b.i-2)}$$

transforms as a vector, where ε is the completely antisymmetric symbol (pb. IX.2.a-2, p. 471). If T_{jk} is symmetric then V is zero.

X.7.b.ii *The angular-momentum selection rules*

The $l = 0$ representation, the scalar, gives transitions only between states of the same angular momentum (we do not care whether it gives transitions between states with different values of other quantum numbers, such as the principle quantum number). We consider solely transitions changing angular momentum, so only the $l = 2$ term. (Note that the $l = 0$ and $l = 2$ terms give different classes of transitions; the pairs of states in a transition are different.) The rule then is that the only transitions that can be nonzero are those between states of representations l_1 and l_2, where l_1 occurs in the Clebsch-Gordan decomposition of the product of l_2 and $l = 2$. Using this, the selection rules for Raman scattering follow immediately.

Problem X.7.b.ii-1: Show that this is equivalent to requiring that the l_2 representation occur in the decomposition of the product of the l_1 and $l = 2$ representations, and also equivalent to the requirement that the $l = 2$ representation appear in the decomposition of the product of the l_1 and l_2 representations, and further that the scalar representation appears in the decomposition of the triple product of the l_1, l_2 and $l = 2$ representations. Prove that the equivalent statement is true for all values of l_1, l_2, l.

Problem X.7.b.ii-2: For water, x^2, y^2, z^2, transform according to the scalar (A_1) representation, yz according to the B_1 representation, and so on. Hence there can be Raman transitions from any ground state to any excited state.

Problem X.7.b.ii-3: What are the allowed transitions for ammonia (sec. X.5.b, p. 537)? Compare them with those for water. How do the selection rules for these molecules compare with those for their electric dipole transitions?

Problem X.7.b.ii-4: The molecules methane, CH_4 (pb. X.5.c-5, p. 544) and monodeutromethane, CH_3D, differ by one hydrogen isotope. How can their two types of spectra be used to distinguish them [Bishop (1993), p. 190].

Problem X.7.b.ii-5: How would these analyses change if the incoming wave induced a higher-order multipole moment? Is this possible?

Problem X.7.b.ii-6: Could magnetic moments be induced for these molecules? What selection rules would they give? Do the differences in the selection rules for electric and magnetic cases seem reasonable physically? Do they provide useful information?

Problem X.7.b.ii-7: Can selection rules be obtained, for any of these cases, in general, not considering specific molecules?

Problem X.7.b.ii-8: A large number of molecules have been considered above. For at least some, find which can have infrared spectra, which Raman spectra, which both, and which neither — if any. Give all selection rules. Explain, physically, the results.

X.7.b.iii *Exclusion rule*

There is a relationship between which objects can have infrared transitions and which can have Raman absorption.

Problem X.7.b.iii-1: The frequency of the light emitted (or absorbed) is related to the difference of the energies of the levels. Show that for a centrosymmetric molecule or crystal (one with symmetry under inversion) a frequency cannot appear in both infrared and Raman transitions; if there can be a transition between two levels by one process, there cannot be by the other. This is called the exclusion rule [Birman (1974), p. 295; Cotton (1990), p. 338; Tinkham (1964), p. 249] or (unfortunately) the exclusion principle [Burns (1977), p. 119]. Explain this using the parity of the transition operator (whether it is even or odd under inversion). This is not true for a noncentrosymmetric molecule or crystal. A piezoelectric crystal (sec. IX.6.a, p. 497) must have at least one transition that is allowed for both processes [Burns (1977), p. 120].

X.7.b.iv *Raman scattering for crystals*

Crystals as well as molecules can give Raman scattering [Birman (1974), p. 271]. The tensor that determines the amplitude has components x^2, xy, ..., the components of a second-rank tensor which is inherently symmetric — it is a product of coordinates, and coordinates commute. Further since it is so realized it is a polar tensor. The meaning and realization of the tensor are irrelevant in determining which crystals can have nonzero values for it, only the representation it transforms under plays a part. Since this is the same as cases we have considered, it is unnecessary to study this further.

Problem X.7.b.iv-1: Is the reason that it is symmetric a physical, or a mathematical, one?

Problem X.7.b.iv-2: The tensor can be diagonalized. What is the significance of the principle axes? This tensor stands for two physical quantities, the trace

$$r^2 = x^2 + y^2 + z^2, \qquad (X.7.b.iv-1)$$

and the traceless part, $x^2 - \frac{1}{3}r^2, \ldots$. What is the physical significance of these two tensors? Under what conditions is one, but not the other, zero? What information do they each provide?

X.8 HOW REASONABLE ARE THE APPROXIMATIONS?

In treating vibrations of molecules we have made approximations and used idealizations — or perhaps only language — that is not strictly correct in the real world. The extent to which these are reasonable depends on the context. But there are (often) sound rationales for them and here we try — only — to make some of these explicit.

Thus for water, the hydrogen atoms are of course (taken as) identical, so the molecule is invariant under their interchange — this effects their vibrational modes; using one deuterium would have an effect, though perhaps small, as would relative orientation of angular momenta, and differing states of excitation (as would certainly occur for some molecules in a gas being studied). There are other factors. The oxygen atom is (taken as) spherically symmetric; with, say, a magnetic moment it might distinguish the hydrogens. Also we assume that the points of space occupied by the hydrogens are identical. If there are electric or magnetic fields (for example due to nearby atoms) this discussion could fail, perhaps wildly. Water is a constituent of many materials. These considerations need not be true for such cases such as one — of some importance — for which the points of space are not identical, as in a (nonisotropic) crystal or the cells of an organism (though if they do not fail badly the disagreement could serve as a probe of the environment). Here we assume maximum symmetry. Another assumption is that there is a single equilibrium position for each particle and that displacements are small. If there are several with (relatively large) displacements, then vibrations are about each, although consideration must then be given to the effect of this multiplicity.

Problem X.8-1: The analyses here rest (perhaps somewhat uneasily) on the assumption of linearity — the potential energy is quadratic in the coordinates (thus the problem is said to be linear?). For a molecule the potential is that between atoms, but this varies inversely as a power of the distance, $V \sim r^{-n}$, or perhaps a sum of such terms. Show however that, in both one and three dimensions, V does vary quadratically in the displacements from equilibrium — for these small. State any assumptions used. (With care, we can excite vibrations in such a way that they are small — the displacement is small — so in this sense the method is general, unless, say, the required displacement is less than the size of the object, or perhaps other aspects, like rotation, prevent this.) How is the (group-theoretical) analysis affected by higher-order terms? Explain why, and even show rigorously that, the states of the system are not changed, though their energies (frequencies) are. Is this always true — under what conditions? Here states means the basis states of the symmetry group. But as we know for an oscillator (sec. X.2.b, p. 516) there are a set of levels distinguished by a principle quantum number n; each has (for three dimensions) a set of substates that are the basis states. Nonlinearities (of the right kind) do not mix different basis states, but they do mix states of different n; the states mixed are then different realizations of the same representation basis state — they are different functions (say sets of Hermite polynomials or spherical harmonics) realizing the same basis state. This is for V a scalar (of what group?). Can V not be a scalar? How? What would happen?

Problem X.8-2: Potentials need not have only the simple form given by the expansion of r^{-n}. Take the atoms to have (electric or magnetic) dipole moments. Add to the expression for the potential obtained in the previous problem a (small) term due to the interaction of two such moments. Also consider one of the atoms replaced by an ion, and the interaction of the field of the ion with the moment of the other atom. What is the effect of these interactions on the (group-basis) states of the system? Are the eigenfunctions of the Hamiltonian (the steady states) still representation basis states? Why? If not what are they, and how are they related to the group?

Problem X.8-3: Construct models for the water molecule (and similarly others) for which the oxygen has a magnetic moment that introduces a (small) interaction distinguishing the two hydrogens. Analyze, group theoretically, the vibrations in this case, and discuss how they differ from those of the most symmetric molecule.

Problem X.8-4: Molecular motions are studied (here) by separating off the translational and rotational motions leaving the pure vibrations. By Lorentz invariance (which might not appear to hold inside a crystal), we can go into a frame in which the object is at rest, so translational motion can be ignored. How about rotations? The total force on an atom includes, besides the harmonic restoring force we are using, the centripetal and Coriolis forces [Goldstein (1953), p. 135; Wilson, Decius and Cross (1980), p. 273, 361]. (Is the latter — which can be thought of as the correction needed by an observer in a rotating frame — relevant here? Why?) These depend on distances between atoms — which vary as the molecule vibrates. Thus there really is in the Hamiltonian a nonlinearity linking vibrations and rotations. Usually energy differences between rotational states and those between vibrational states are so dissimilar that each set can be considered separately — as a good approximation; in quantum mechanics this is the Born-Oppenheimer approximation [Inui, Tanabe and Onodera (1990), p. 228; Tinkham (1964), p. 210, 234; Wherrett (1986), p. 152] (sec. XI.1.a, p. 565). Discuss the effect on the symmetry and on the Hamiltonian eigenstates when there is both rotation and vibration, and the centripetal and Coriolis (?) forces are small, but nonnegligible [Tinkham (1964), p. 257, 265]. In particular, for only rotations, the eigenstates are rotation-group basis states. Do vibrations mix these? How? What are the constraints on mixing? What is the effect on the eigenstates of the vibrational part of the Hamiltonian when the molecule is rotating? Discuss also the effect on selection rules. Would the change in the spectra be interesting? Could these methods be used if the centripetal and Coriolis forces were large? Why? Might these corrections be useful as a probe? Of what?

Problem X.8-5: Another assumption is that the molecules are at zero temperature, although of course they are not. This should not affect the analysis, or should it [Kettle (1995), p. 201]?

X.9 HOW DIFFERENT GROUPS GIVE DIFFERENT VIBRATIONAL SPECTRA

The symmetry of a molecule is highly influential in determining its allowed vibrational states and the transitions between them. To clarify this it is helpful to study together different approximations, and different symmetry groups, and compare results. While higher-order approximations (usually) give only small corrections [Wilson, Decius and Cross (1980), p. 39], this is not important here, for we are interested in learning how to apply group theory, not in actual calculations.

Problem X.9-1: The next term in the expansion of the plane wave gives electric quadrupole moments. Repeat these analyses for it, including the study of specific molecules.

Problem X.9-2: Magnetic moments transform like the corresponding electric moments under SO(3), but have opposite parity under O(3). How would the analyses differ for them? How would transitions due to these arise physically? Are there circumstances, or objects, for which there are transitions due to one moment, but not the other? Can there be cases that allow neither? How about both?

Problem X.9-3: Are selection rules obtainable, for any of these cases, in general, not considering specific molecules?

Problem X.9-4: Can general algorithms be given (and be computer implemented) telling, for any moment, electric or magnetic, of any order, what transitions are possible, and what molecules (or crystals) — point groups — allow them?

Problem X.9-5: Pick molecules that have the symmetry of each point group [Hargittai and Hargittai (1987), p. 84], such as those listed next, most of which are noncrystallographic, and give the set of allowed vibrational states describing their normal modes and the motions of each of the atoms. Find reasonable internal (symmetry) coordinates and describe the motions using these. Draw each molecule and find the symmetry transformations. Also give the allowed infrared and Raman transitions, and compare them. An analysis of higher-order approximations would also be useful. For these molecules, full information about their structure may not be easily found — only the symmetry group. How important is this? If necessary, assume (or even look up) reasonable structures. There are many examples of molecules with different numbers of atoms, but the same symmetry group. Does the number of atoms play a role? Check the statements about the symmetry elements.

a) The group C_1 has no symmetry elements. One of the many examples is HCBrClF.

b) Examples with C_2 symmetry include S_2Cl_2 and $(CH_2)_2O_3$.

c) Construct (if that would be useful) molecules with symmetry C_3, C_4, C_5, ..., C_n, ..., C_∞.

d) Molecules with group C_i having a center-of-symmetry include $(HCBrCl)_2$, this giving (at least) two molecules.

e) The group C_s includes one symmetry plane; examples are C_9H_9, $(H_2C)_2S_2PCl$, and SCl_2O.

f) A molecule with a four-fold mirror-rotation axis, having group S_4, is $(H_3C)_4O_4Si$.

g) A molecule with symmetry group S_6 (which has a six-fold mirror-rotation axis — which is equivalent to a three-fold rotation axis plus a center-of-symmetry) is $[(CH_2)CH]_3CC[CH(CH_2)_2]_3$. Check that both sets of symmetry elements are present.

h) The group C_{2h} has a two-fold rotation axis and a perpendicular mirror; PF_4 is an example.

i) Continue this for C_{3h}, ..., C_{nh}, ..., $C_{\infty h}$ (if this is possible).

j) Two perpendicular mirrors whose intersection is a two-fold rotation axis give group C_{2v}. Water is one example, HBF_2 is a less familiar one.

k) For C_{3v} there are three symmetry planes; the angle between two is $\frac{2\pi}{6}$, and the intersection of the three planes is a three-fold rotation axis. Ammonia and POF_3 are examples.

l) Group C_{4v} has two inequivalent pairs of perpendicular mirrors, these pairs rotated by $\frac{\pi}{4}$ with respect to each other and their intersection is a four-fold rotation axis. A molecule with this group is WOF_4.

m) Continue for groups $C_{5v}, \ldots$, describing the groups and finding (or inventing) molecules to serve as examples.

n) Examples of molecules with $C_{\infty v}$ include HCl, HBS and HC_4Cl.

o) For D_2 there are three mutually perpendicular two-fold rotation axes; a molecule called twistane (for good reason) is an example.

p) Describe, and find examples for, groups $D_3, \ldots, D_\infty$. Is there a limit on n?

q) Give a description of group D_{2d}; B_2Cl_4 has such symmetry.

r) Do the same for $D_{3d}, \ldots, D_{\infty d}$ (if there is no limit on the groups); give examples.

s) The group D_{2h} has three mutually perpendicular mirrors; their three lines of intersection are two-fold rotation axes, and their point of intersection is a center of symmetry. An example is Al_2Cl_6.

t) This can be continued for $D_{3h}, \ldots, D_{\infty h}$. Are there interesting differences between the results for D_{nh} and D_{nv}? How about D_∞, $D_{\infty d}$ and $D_{\infty h}$?

u) An example for the tetrahedral group T is $Pt(PF_3)_4$. Describe the group and relate this description to the structure of the molecule.

v) Adding symmetry planes to T (how many, where?) gives group T_d. CCl_4 and CH_4 are examples [Lax (1974), p. 167].

w) Instead if a center-of-symmetry is added to T, the group T_h is obtained.

x) The final molecular groups (listed here) are O and O_h. For the latter, examples are a molecule called (not surprisingly) cubane, and SF_6. Another example [Harter (1993), p. 286; Lax (1974), p. 167] is the economically (and politically) important molecule uranium hexafluoride, UF_6.

y) The polyoma virus [Hargittai and Hargittai (1987), p. 407], which is more than a molecule, has icosahderal symmetry. Could this be used to study it in any useful way?

Problem X.9-6: Suppose you were given a molecule of unknown structure and you were able to study its vibrational transitions. How much could you learn about it from these [Kettle (1995), p. 174]? How would you do so (most efficiently)? If necessary, consider also rotational spectra. Would it help to irradiate the molecule? Why?

Chapter XI

Bands, Bonding, and Phase Transitions

XI.1 WHY IS SYMMETRY RELEVANT?

A crystal is the environment for objects composing it, collections of electrons, holes, quasiparticles, phonons, ..., so determines much of the prescription for their behaviors. And these are a major factor in making crystals interesting and valuable, scientifically, technologically, economically. Also crystals, and such (pseudo)particles [Mirman (1995b), sec. 3.1, p. 38], supply insight into group theory because the symmetry that they see depends on where in the crystal they are.

Crystals provide an environment; they also exist in an environment. They, their symmetry, lattice types, invariance groups, can be affected, determined, by their surroundings, by temperature, pressure, and so on, imposed by objects with which they are in contact, and by history. As these vary, so does the crystal, perhaps in consequential ways. But such variations are not arbitrary, they are governed by the circumstances, and also by the crystal, by its symmetry.

So here we wish to study these, to find the effect of crystals, and their symmetry, on the objects within them, and how the crystals, their symmetries, are effected by changes in their external surroundings, temperature, pressure, for example, how symmetries and lattice types change as external variables do, how they undergo phase transitions, and what the restrictions on these are. We wish to learn about these, and use them to learn about groups.

What we consider are the groups, but the insight gained by using groups, and how they inform about the requirements on physical systems depends largely on the group, and less on the physical system (which is a reason that group theory is so powerful and general). Often with no more than a change of nomenclature, the results can be applied to very different systems. We thus use these methods, not only to study crystals, but because of the similarity of methods, having developed them, we consider how they are applied to other fields, specifically chemistry and the study of atoms, not in (idealistically) infinite crystals but in molecules, to see how, and why, they are bound, how they behave, how they change, how they interact with the external world.

Such questions provide illustrations of some, certainly not all, applications of symmetry and group theory to the study of complicated systems. They provide a start of such study, and (here, no more than) hints of how to go further.

Problem XI.1-1: As the reader goes through this material it would be useful to think how it would be modified if the crystal were not fully translationally symmetric, as with quasicrystals [Gazeau (1997); Jaric (1989); Lifshitz (1997); Quilichini and Janssen (1997)], or if the symmetry were broken in other ways. Of course finding actual materials to which these variations could be applied would also be interesting.

XI.1.a Different types of objects can be studied separately

A crystal consists of various objects, which we classify into two broad types. First are electrons, and vacancies left by them, holes, which we treat as equivalent particles, plus combinations, generically quasiparticles, and even other such things that might be thought of with imagination. Then there are nuclei. The relevant distinction is in their masses, with nuclei much the heavier. We want to study the behavior of these, but because of their mass differences, to an excellent approximation, we can consider them separately, the speed of the latter being so much less than of the former — this (sec. X.8, p. 559) is the adiabatic (Born-Oppenheimer) approximation [Inui, Tanabe and Onodera (1990), p. 228; Tinkham (1964), p. 210, 234; Wherrett (1986), p. 152]. In it we consider the behavior of the light particles, with the nuclei taken as fixed. Nuclei are later considered separately, ignoring the other objects.

Also the state of a particle is affected by those of the other particles, certainly by their Coulomb interaction. However we ignore the interaction of pairs of particles, replacing the effect of all others on a single one by an average potential (the one-electron approximation), and then need not be concerned about which part of this potential is due to interactions with nuclei and which to interactions with other electrons or holes. All we need know (here) about the potential is that is has the symmetry of the crystal. We thus start by studying the states of particles in a crystal, then consider the behavior of the nuclei, the lattice vibrations (sec. XI.4, p. 598). With these methods developed, we use the reasoning for a different problem, the behavior of electrons in molecules, to obtain the states of molecules and chemical bonding (sec. XI.7, p. 607).

XI.1.b Degeneracy, necessary and accidental

The study of states in crystals tells about crystals, but also about representations of groups. One aspect that it emphasizes is the difference between necessary degeneracy, coming from symmetry, and accidental degeneracy, coming from particular values of the constants and form of interactions. We consider only free particles (when we consider the potential function explicitly, which is rarely — most results are general), so the potential is not only periodic, compelled to be by the periodicity of the crystal, but is constant within a zone. This simple and never really correct case has states that have the same energy but whose energies differ for any other (nonconstant) potential. But there are also states whose energies are always the same, no matter what the potential, because of symmetry. And the ordering of energy levels may be different at different symmetry points — they may cross. Here accidental symmetry is actually required — they must cross at some point — but the point depends on the potential. Usually if we find two states with the same energy, not required by symmetry, a more precise measurement shows the energies different — the constants and form of the potential can never be such as to make them exactly equal. But here, since there is crossing, energies are equal to infinite precision. But the crossing point cannot be known with complete precision. (In reality, crystals have nonzero temperature, something we fortunately ignore, so the potential is slightly time-dependent, thus the energies are exactly equal at a point, but only for an infinitesimal time; the point of exact equality jiggles a little.)

Problem XI.1.b-1: If the potential is constant in a zone, and periodic in the zones, it should be constant everywhere, so can be taken zero, thus the objects are in free space. Yet (presumably) this is not what we are doing. State properly what is being done.

XI.1.c What can we know about objects in crystals?

States of a system allowed by quantum mechanics are limited; for atoms levels are discrete, in (infinite) crystals though continuous they are restricted to form bands (for actual bounded crystals, these consist of discrete, though closely spaced states). For the hydrogen atom, most clearly, but also for others, limitations and allowable statefunctions are related to the symmetries of the system and space. Is this true of crystals also? What determines the possible states of an electron in a crystal? How are these related to the invariances of the system, to its point and space groups?

Given an object, an electron, hole, quasiparticle, . . ., in a crystal, what do we want to know? Besides the allowed energies and the states they determine, the statefunctions for each, and the other quantum numbers, like angular momentum, that statefunctions depend on, there is for a crystal other aspects, and they affect these properties. An electron in an atom has discrete energies, and angular momenta; unbound it has continuous values of quantum numbers. Which is true for crystals (perhaps both?) and why? How does the size of the crystal affect these answers? In space all directions are the same, which is not (apparently) true within crystals. How does the decrease in (apparent) symmetry affect the answers to these questions? Can energy depend on the direction of motion? Is there (an analog of) angular momentum if directions differ; if so with what meaning? A crystal exists in a space in which all directions are the same (which would not seem to be true if it were in a field). Is this relevant, though all directions in the crystal may not (seem to) be identical? By definition, crystals have symmetries. What would happen, how would the answers to these questions be affected, if these were slightly, or strongly, broken?

This discussion is about group theory, not solid-state physics. If the symmetry is broken, as the symmetry of space is (apparently) broken by a crystal, is group theory relevant and helpful? Why?

XI.1.d Symmetry varies; how is it useful?

What is novel about the symmetry seen by an object in a crystal is that it depends on where in the crystal the object is — the point symmetry group (the site-symmetry group) is a function of position (sec. VII.5.d, p. 369). An electron, say, feels a potential, and this has point-group symmetry, but the point group is different at different positions. Thus while a statefunction is a basis vector of the point group representation, this group is a function of the variables on which the statefunction depends. State labels vary — energy is a function of position, and it depends on the magnitude of the momentum, but crystals are anisotropic, so it is also a function of the momentum's direction. How then do we use symmetry, and why?

To find energies we solve the eigenvalue equation for the Hamiltonian, which for spin-$\frac{1}{2}$ particles (of course, all we are interested in here) can be done by solving Dirac's equation [Mirman (1995b), sec. 6.3,

p. 114] for each state separately. Actually, since the particles are nonrelativistic this is done by solving Schrödinger's equation. An advantage of symmetry is that with it energies are equal, and statefunctions are related by symmetry transformations. However for a crystal there is — coming from that in real space — symmetry in momentum space, but only for certain points. For some positions (only) of the Brillouin zone, points, lines or planes — the special positions (sec. VII.2.e, p. 342) — point group transformations leave momentum vectors, thus statefunctions, invariant. These our studies stress, for at them we thus have restrictions that we can use. Behaviors of objects elsewhere — general positions — where there is no symmetry to constrain them, are constrained by the presence of the special positions and the requirements they lead to. Thus for momentum zero, the center of the Brillouin zone, the states are related by the point group symmetry of the crystal, with energies all equal. At other positions, not all, there is also symmetry, (generally) less than that for the center, and different for different points — the symmetry subgroup of the crystal point group varies with position in the Brillouin zone (with momentum).

Symmetries then appear only at parts of the Brillouin zone of measure zero; how do we use these — limited — ones? Consider a special position and a set of states which symmetry requires be degenerate there. Any (infinitesimal) change of momentum breaks the degeneracy. A (finite) change from this position to another special one causes a state of the degenerate set (basis state of the representation of the point symmetry group at that point) to vary in form and energy, but upon reaching the second point it again becomes one of a degenerate set, this of a different representation and of a different symmetry group (in general). States that are members of the first set and those of the second may differ (the two sets of states may be disjoint) except for this one. That the state belongs to both sets restricts the representations, leading to compatibility relations (sec. VII.5.c, p. 367; sec. XI.2.c.i, p. 572), as we will see.

We now at each special position solve Schrödinger's equation (assuming a potential) to find the states and energies, a task simplified by invariance and the resultant degeneracy. Finding these for a general point is more difficult. But we have knowledge of the statefunctions at several points, and use that for the start of an approximation procedure to find the state and energy elsewhere [Evarestov and Smirnov (1993), p. 152]. While interpolation is, for realistic cases, often difficult, knowing functions at a few points is an important, perhaps necessary, beginning. The calculational value of symmetry for a crystal may be less than for other systems, but is useful, even essential, in developing workable approximation schemes. For crystals then one important, often the important, value of group theory is as a foundation for approximation schemes.

XI.1.e Selection rules, perhaps not exact, but still productive

Groups also provide selection rules (sec. X.7, p. 552). For emission or absorption the product of the original state with the part of the Hamiltonian giving radiation must contain the final state in its decomposition — for each point with symmetry, although there are not many. However, by continuity, we would except that if a matrix element must be zero at some point, it would be (at least) small at nearby ones, providing, if not exact at least approximate, selection rules.

It is important to know and understand, certainly for applications, how a crystal interacts with its surroundings, electromagnetic fields, electrons and holes passing through its boundaries, and so on. Symmetry supplies information, and understanding; more generally we can gain (or retain) much from positions of symmetry even where symmetry is incomplete.

An object, an electron or hole say, to enter a crystal has to be in a state allowed by the crystal, having an allowed energy for its momentum. A particle in a crystal that is in an electric field has its energy and momentum changed. Unless the field is such that the energy change is sufficient to permit a transition to a different state, the particle must move along a trajectory in reciprocal space on an energy surface. However when it reaches a point at which its state becomes degenerate, it can scatter, changing its quantum numbers and move to another state, with no change of energy or momentum, and then continue along a different trajectory. Thus special positions can be important in determining the behavior of the particle.

XI.2 ELECTRON STATES IN CRYSTALS

We start then with states of objects with small mass; electrons, though the considerations are general — it is useful to substitute other names, holes, quasiparticles, ..., to see if it matters [Burns (1977), p. 293; Evarestov and Smirnov (1993), p. 144; Falicov (1966), p. 176; Heine (1993), p. 270; Inui, Tanabe and Onodera (1990), p. 259; Janssen (1973), p. 177; Jones (1975), p. 114; Joshi (1982), p. 265; Lax (1974), p. 194; Ludwig and Falter (1988), p. 244; Tinkham (1964), p. 275; Wherrett (1986), p. 257]. As usual we want statefunctions, energies, and transition probabilities. Much of what we do applies to lattice vibrations

— we are only interested in illustrating and understanding the role of group theory, not in details that depend, not only on the crystal type, but on the specific material (which determines the potential function), or even the specific behavior described by the group and by the state-functions. Aspects coming from symmetry are similar for all.

Statefunctions — of the objects in the crystal, like electrons — are space group basis states (chap. VII, p. 337), and their properties reflect the group. How are these affected by the groups being nonsymmorphic — how do they differ for symmorphic and nonsymmorphic groups? For nonsymmorphic groups basis states can have extra phase factors (sec. VII.3.b, p. 351). What affect does this have? How is the behavior of an electron different for the different types of groups? We want numbers and functions, but also understanding, and this illustrates what we can ask, what we can learn, and why we want to.

XI.2.a Labels for states are necessary

To study how objects behave we need information about their states, and the first step is labeling these — the state-labeling problem [Mirman (1995a), sec. I.6, p. 25]. The nature of these labels itself provides insight about the system. States of a hydrogen atom are labeled by the principle quantum number and the angular momentum quantum numbers, and what this reveals can be seen by comparing these labels with ones for other systems, such as those considered here. We expect similarities for an electron in a crystal, though what is the equivalent of the principle quantum number, and since there is no spherical symmetry what replaces angular momentum? Also there is an additional quantity, momentum, or its dual, position. Given all quantum numbers of an atom, the state is fully specified. But an electron in a crystal, has (as labels, the equivalent of) angular momentum, and also can be anywhere in the crystal — its statefunction is a basis state, not of a point group, but of a space group, so its energy depends on its linear momentum. The groups provide the state labels, even though for a crystal the states are not point-group representation eigenstates except at certain points. While labels may have mathematical or physical meaning only at specific points, they are completely defined, even with no simple interpretation, at general points, and are not merely useful, but essential.

XI.2.b Bands, and why they are

Electrons in crystals can have a set of energies, as in an atom or potential well, but (for an infinite crystal) these form not a discrete set (given by Dirac deltas on the energy axis), but discrete bands of finite width

[Altmann (1992), p. 80; Cornwell (1969), p. 84; Jones (1975), p. 14; Streitwolf (1971), p. 131]. What are the physical reasons for there being bands, different bands, and ones with finite, but limited, widths?

A particle in a crystal interacting with atoms sees a potential, a function of position, but with the same value at all equivalent points — the potential is the same at points related by lattice translations. The energy of a particle depends both on its momentum $\underline{k}$, and on how its statefunction is distributed over the unit cell. Consider say a particle spending most time near an atom — correctly its statefunction is small except there. Such statefunctions can be approximated, to lowest order, by atomic statefunctions; these give a set of energies. Thus for each $\underline{k}$ value there is a set of energy states, and as $\underline{k}$ varies continuously (for an infinite crystal) over a finite range, from zero to a maximum, each discrete atomic energy state is replaced by a band of finite width (determined by the range of $\underline{k}$).

Electrons in a crystal, especially in the free electron model, are pictured as detached from their atoms wandering freely through the crystal (correctly their statefunctions extend throughout the crystal). Each atom in a crystal releases electrons, and since lattice translations take an atom to an equivalent one, each releases electrons from the same energy level, but all free electrons cannot have the same energy because of the Pauli exclusion principle, so they must have sets of energies.

It is these bands, the set of ranges of energy values, that we wish to find and for each the set of space-group representations, the statefunctions, and the transitions between them.

XI.2.c What group theory tells about electron bands

From labels of states, that is (in part) the labels of the symmetry-group representations and basis states — for atoms those of the rotation group, or a subgroup — we find (in principle) statefunctions of the system to which an object belongs. Crystal statefunctions also of course are symmetry-group basis states; their labels include momentum $\underline{k}$ (giving the translation representation), each value of which gives a set of energy levels and representation basis vectors (the energy is a function of $\underline{k}$, and for each $\underline{k}$ there is a set of energies). These basis vectors, products of Bloch functions and terms from the translation subgroup, are in sets forming point-group representations. The relevant aspect is that for some $\underline{k}$ values, at special positions such as the center or boundary of the Brillouin zone, the number of states of different energy is reduced — states whose energy differs in general can be degenerate there. For these special positions of higher symmetry there are point-group operations that take a $\underline{k}$ vector into itself, so there are fewer that mix it with others. These then form a symmetry group, a subgroup of the crystal point group, at that point. At the center that symmetry group is the full point group.

XI.2.c.i *Point groups differ at different points in reciprocal space*

As the number of distinct $\underline{k}$ vectors depends on the special position, the subgroups whose representations these momenta form also differ, but state vectors are continuous functions of $\underline{k}$. Move from a special point to another along a line (or plane) of symmetry, which is less than the symmetry at either end point (or line). The statefunction is a basis vector of the symmetry subgroup for the line, and also of the subgroups, generally different, of the special points at its end (sec. VII.5.c, p. 367). The group of the line (or plane) must be a subgroup of those of the ends, and the point group representations to which the statefunction belongs at those positions must include in their reductions that representation of the symmetry group of the line (or plane) to which the statefunction on it belongs — the representation on a line (or plane) must be in the decompositions of the representations of all subgroups for the different special positions on that line (or plane). This relates, so restricts, representations for each energy function (of $\underline{k}$), for the various positions. Thus while energy bands differ at general positions, they may be degenerate at special ones, and the bands that are related by the degeneracy depend on the special position. This is possible only for certain sets of representations, thus limiting them and the physical statefunctions. The lists of these sets of representations (sec. VII.5.c, p. 367) for the subgroups of a point group are compatibility relations (connectivity relations); only some representations are compatible with others, only these can be in the same decompositions [Birman (1974), p. 230; Burns (1977), p. 311; Cornwell (1969), p. 109; Evarestov and Smirnov (1993), p. 150; Heine (1993), p. 277; Inui, Tanabe and Onodera (1990), p. 264; Jones (1975), p. 114; Joshi (1982), p. 287; Ludwig and Falter (1988), p. 232; Streitwolf (1971), p. 136; Tinkham (1964), p. 284; Wherrett (1986), p. 277].

Each set of point-group representations for which these are all satisfied give a space group representation. That representation determines a (hyper)surface over momentum space, each point on the surface giving the energy for that momentum. Different representations have different surfaces. These surfaces defined over the representation (not as part of it) depend on the Hamiltonian thus on the crystal potential which is a property of the crystal and independent of the representation — the potential is not determined by electron statefunctions (ignoring electron correlations). Knowing the representations, we find the matrix elements and selection rules, as functions of k, thus describing the physical system.

Problem XI.2.c.i-1: Is it possible that energy surfaces for some representations are actually the same; could the same energy for all points in momentum space be given by different representations?

XI.2.c.ii *How these bands illustrate the meaning of space group representations*

Representations of space groups (chap. VII, p. 337) differ in interesting ways from those of other groups, illustrating the diversity of properties of groups and representations, and emphasizing how these depend on, and may be limited by, the type of group.

Inhomogeneous rotation groups of which space groups are subgroups — with often strikingly different properties — have representations given by the magnitude of momentum k. Each radius in reciprocal space thus is a label of a representation, and the (continuously infinite number of) points on its sphere label its states; the representations and states are also labeled by the homogeneous subgroup (the rotation group) representation and state labels. For each radius there are an infinite number of representations, these given by the rotation-group representations. A space group representation is somewhat similar, except that the homogeneous operators — forming the relevant subgroup of the improper rotation group — are discrete, and finite in number. Representations with different magnitudes of k then are still inequivalent; the space group cannot change the magnitude of k (except for a few values of k for nonsymmorphic groups). However for it all values of k with the same magnitude do not label states of a single point group representation — the homogeneous operators give only a discrete number of states acting on any one.

The lengths of a rectangle along x and y are different, thus there is no symmetry-group operator (as there is for a square) that takes k_x into k_y, although there are ones, a π rotation, or an inversion, interchanging k and -k. So for a rectangle, states of different k's (except for sign) belong to different representations, as is true for a square, except that states of orthogonal k's are in the same representation.

An inhomogeneous-rotation group representation contains an infinite number of states, rather than two or four as for these; all values on the surface of a sphere label its states, with the radius labeling the representation. The star (sec. VII.2.d, p. 342) of each general point labels a representation — each magnitude k (each radius of a sphere) gives a representation. But unlike those for the inhomogeneous rotation group, these, although dense, do not fill the entire sphere on which a $\underline{k}$ resides. That sphere consists of pieces; all $\underline{k}$'s on a piece give representations, but the pieces are related by homogeneous operators. For special points each piece is larger, and their number smaller. A (point-group) representation dimension can thus depend on the value of the (momentum) label. Every energy band therefore runs over a continuous range of representations, each going with (in general) different energy, but the energies are (usually) continuous functions of the representation momentum-label.

States of a representation labeled by a star are given by the actions of the point subgroup on any one. The translation subgroup consists of translations by an integral number of lattice vectors whose effect is to multiply the state by the corresponding exponential, thus adding to $\underline{k}$ a vector equal to a reciprocal lattice vector — the states produced by the translations are given by the points in the different Brillouin zones corresponding to the state labels defined in the first zone. Fourier transforms of these statefunctions, the statefunctions in real space, are then sums over states of different representations. This is true also for the states of the representations of the translation group in space, say for a free particle; in general it is a sum over states of different momenta (a wavepacket), thus of different translation group representations, these labeled by the momentum value.

Groups with nonsymmorphic elements have transformations that are functions of a lattice vector, with Fourier transform given by a $\underline{k}$, and also other transformations that depend on a fraction of a lattice vector, so with Fourier transform given by a multiple of $\underline{k}$. That is the space group has elements of the point subgroup mixing the $\underline{k}$'s of the same star, plus translation operators taking a $\underline{k}$ from one Brillouin zone to another, plus now operators taking one $\underline{k}$ to another of the same zone with a different magnitude. Since both these sets of transformations belong to the group, in a representation there are both of these values — there are group transformations that change a basis state with one magnitude of $\underline{k}$ into a basis state with another (sec. VII.5.e, p. 370).

XI.2.d Symmetry reduction

While this discussion (or at least terminology) is restricted to potentials that are constant, it is useful to briefly compare it to that with nonconstant potentials to elucidate the potential's effect [Burns (1977), p. 296]. Potential $V(\underline{r})$ has the periodicity of the crystal lattice so can be expanded in a Fourier series in terms of reciprocal lattice vectors,

$$V(\underline{r}) = \sum_{\underline{k}_n} A_{\underline{k}_n} exp(i\underline{k}_n \cdot \underline{r}), \qquad \text{(XI.2.d-1)}$$

with

$$\underline{k}_n = n_1\underline{b}_1 + n_2\underline{b}_2 + n_3\underline{b}_3, \qquad \text{(XI.2.d-2)}$$

where the $\underline{b}$'s are reciprocal lattice vectors. To check this we translate by such a vector,

$$V(\underline{r})' = \sum_{\underline{k}_n} A_{\underline{k}_n} exp(i\underline{k}_n \bullet \underline{r}) exp(i\underline{K} \bullet \underline{r}) = V(\underline{r}), \qquad \text{(XI.2.d-3)}$$

using (sec. IV.2.e, p. 193)

$$exp(i\underline{K} \bullet \underline{r}) = 1, \qquad \text{(XI.2.d-4)}$$

so this standard Fourier series expansion does have the required periodicity.

Problem XI.2.d-1: Potential V has the periodicity of the crystal (or lattice?) — by definition; the definition of what? Is the term "crystal lattice" correct?

Problem XI.2.d-2: Show that

$$A_{\underline{k}_n} \sim \int V(\underline{r}) exp(-i\underline{k}_n \bullet \underline{r}) d\underline{r}. \qquad \text{(XI.2.d-5)}$$

What is the range of integration? What is the normalization factor?

Problem XI.2.d-3: Now also (why?)

$$\int V(\underline{r}) exp(-i\underline{f} \bullet \underline{r}) d\underline{r} = 0, \qquad \text{(XI.2.d-6)}$$

unless $\underline{f}$ is a reciprocal lattice vector and if this is true it must also be true for $V(\underline{r} + \underline{t})$, and conversely, giving

$$\sum_{\underline{k}_n} A_{\underline{k}_n} \int exp[i(\underline{k}_n - \underline{f}) \bullet \underline{r}] d\underline{r} \sim \sum A_{\underline{k}_n} \delta_{\underline{k}_n, \underline{f}}, \qquad \text{(XI.2.d-7)}$$

with, of course, the same A's. Verify this and find the constant of proportionality. Explain the physical reasons for the integral being zero. One aspect is related to the fact that the potential is expanded in terms of plane waves. We wish to understand the effect of nonconstant V. The energy, for constant potential, is

$$E(\underline{k}) = \frac{1}{2m}|\underline{k}|^2, \qquad \text{(XI.2.d-8)}$$

and its change due to the variation of the potential is given by the secular determinant of perturbation theory which here is

$$D = \begin{vmatrix} E_1 - E & V_o \\ V_o & E_2 - E \end{vmatrix}; \qquad \text{(XI.2.d-9)}$$

E_1 and E_2 are the energies for a constant potential for the two values of $\underline{k}$ differing by a reciprocal lattice vector $\underline{a}$ (that is belonging to adjacent bands), and V_o is the off-diagonal matrix element of V (schematically), say,

$$V_o = < exp(-i\underline{k} \bullet \underline{r}) exp(\frac{-2\pi ix}{a_1}) | V(\underline{r}) | exp(-i\underline{k} \bullet \underline{r}) >$$

$$= < exp(-ikx) exp(\frac{-2\pi ix}{a_1}) | V(\underline{r}) | exp(-ikx) > . \qquad \text{(XI.2.d-10)}$$

Explain why. This result says that the only nonzero off-diagonal terms, V_o, are for states whose $\underline{k}$ values differ by a reciprocal lattice vector. Why? How does this differ for an object in free space? Find the two values of E, and show that the eigenfunctions are the sum and difference of these two states. With constant potential these two are degenerate. This is removed by the variation of the potential with position, and these considerations show how. Explain how. The details of the change depend on the specific function that V is. Compare the effect of a potential on an object in a crystal with one in free space.

XI.2.e The equation governing the objects

The (nonrelativistic) equation governing electrons in crystals is that of Schrödinger whose solutions are the statefunctions, $\psi_k(r)$. Part of these we know, the exponentials, so we need the equation determining the remainder of the statefunctions, the Bloch functions $u(\underline{r})$ [Jones (1975), p. 36]. Since

$$\psi_k(\underline{r}) = u(\underline{r}) exp(i\underline{k} \bullet \underline{r}), \qquad \text{(XI.2.e-1)}$$

Schrödinger's equation, with potential $V(\underline{r})$, gives

$$\frac{\partial^2 u}{\partial x_j^2} + 2ik_j \frac{\partial u}{\partial x_j} + 2m(\epsilon - \frac{k^2}{2m} - V(\underline{r}))u = 0; \qquad \text{(XI.2.e-2)}$$

m is the mass of the electron, $\underline{k}$ its momentum, and ϵ the (energy) eigenvalue. We consider only the free electron model, $V(\underline{r}) = 0$, thus just the effects of crystal symmetry on the electrons. This mere first step in investigating behaviors of particles in a crystal, so their properties, gives results — from symmetry — which are independent of the potential, except that degeneracies among (but not within) representations need not hold for other potentials. Orderings of energies so obtained, for this special case, are illustrative, and likely differ for other cases. However the relationships between states at different points of the Brillouin zone result from symmetry so are general.

Problem XI.2.e-1: If the potential were really zero everywhere the particle would be in free space — with no way of forcing periodicity. Explain what this assumption actually means.

XI.2.f Translational symmetry and bands

We build the theory of the relationship between the structure of a crystal and its energy bands by starting with the defining symmetry, translation. Crystals are taken infinite in extent, in all directions, but real ones are bounded — how do we put in (find a way to ignore?) boundary conditions?

A crystal is invariant under translations

$$\underline{t}_n = n_1\underline{a}_1 + n_2\underline{a}_2 + n_3\underline{a}_3; \qquad \text{(XI.2.f-1)}$$

for all integral n's. If it is finite the largest translation possible along a symmetry axis is between opposite boundaries: $N_i\underline{a}_i$. We treat this by assuming the crystal is infinite, but periodic (sec. VII.4.c, p. 355). Thus

$$\underline{t}_n' = \underline{t}_n = (n_1 + l_1N_1)\underline{a}_1 + (n_2 + l_2N_2)\underline{a}_2 + (n_3 + l_3N_3)\underline{a}_3; \qquad \text{(XI.2.f-2)}$$

the l's are integers — a translation through $\underline{t}_n$ plus several complete periods is equivalent to that through $\underline{t}_n$, so the translation group is then a cyclic group (as in reciprocal space, sec. IV.2.a, p. 186),

$$\underline{t}_n = \{E|a\}^n \equiv \{E|a\}^{N+n}. \qquad \text{(XI.2.f-3)}$$

Representations of these groups are the N'th roots of unity (sec. V.2.b, p. 248) so the matrices are

$$M(\{E|\underline{a}\}) = E^{1/N} = [exp(-2\pi il)]^{1/N} = exp(\frac{-2\pi il}{N}), \qquad \text{(XI.2.f-4)}$$

where l is any integer obeying, to avoid redundancy,

$$0 \leq l \leq N-1. \qquad \text{(XI.2.f-5)}$$

For three dimensions

$$M(\{E|a\}) = exp(\frac{-2\pi il_1}{N_1})exp(\frac{-2\pi il_2}{N_2})exp(\frac{-2\pi il_3}{N_3}), \qquad \text{(XI.2.f-6)}$$

giving the translation-group representations that form part of the state-functions of the objects.

There are thus a finite number of states, so as in atoms, they are discrete, but for crystals (as ordinarily regarded) the N's are so large the number of states is also and are therefore deemed continuous. These are the allowed energy states which form (effectively continuous) bands. For a free particle, the energy is a function of momentum, which can have any value, thus so can the energy. In a crystal, the momentum is restricted to values between 0 and the edge of the Brillouin zone, so the energy is likewise restricted to a finite range. While the momentum is continuous within a Brillouin zone, if the crystal were actually bounded, the reciprocal of the momentum, the wavelength, could only assume discrete values, thus so could the momentum, and thus the energy, within the band. Letting the crystal ends go to infinity makes these continuous, the energy then becoming densely continuous within a band.

The advantage of using these periodic boundary conditions — the Born-von Karman boundary conditions (sec. VII.4.c, p. 355) — is that they replace an infinite discrete group with a finite one, which at times is easier to work with. However there can be aspects of infinite groups that do not appear for finite ones [Grossman and Magnus (1992), p. 79], and it is possible (or at least we have not shown it to be impossible) that, far from the surface of a crystal, these differences could affect its properties.

Problem XI.2.f-1: Or could they?

XI.2.g Energy eigenvalues

An object's energy depends on its momentum, but we restrict this to the first Brillouin zone thus it also depends on the lattice vectors, so the lattice symmetry. Finding energies is essential for physical understanding, but is also interesting for this reason. How does the crystal symmetry group determine possible energies?

Reciprocal lattice vectors, $\underline{B}_n$, are obtained from the primitive reciprocal lattice vectors $\underline{b}_j$ (sec. IV.2.a, p. 186),

$$\underline{B}_n = n_1\underline{b}_1 + n_2\underline{b}_2 + n_2\underline{b}_2, \qquad \text{(XI.2.g-1)}$$

with the n's any integers. For $V = 0$, the Bloch function is constant, so the periodic solution of the equation, with subscripts denoting the set of three n's, is

$$\psi_n(r) = exp[i(\underline{k} - \underline{B}_n) \bullet \underline{r}], \qquad \text{(XI.2.g-2)}$$

whose energy eigenvalue, for momentum $\underline{k}$ (in the first Brillouin zone), is immediate,

$$E_n(\underline{k}) = \frac{1}{2m}|\underline{k} - \underline{B}_n|^2; \qquad \text{(XI.2.g-3)}$$

$||$ denotes the magnitude of a vector. The energy depends on the momentum, as for a free particle, but now also on the primitive reciprocal lattice vectors. As the momentum increases the energy does until the momentum vector reaches the end of the Brillouin zone. Rather than using a larger momentum, it is decreased to return it to the first Brillouin zone, and the value of n is increased, increasing $\underline{B}_n$, to give the actual energy. Larger $\underline{k}$'s go into other Brillouin zones, but are referred back to the first, by the periodicity of the crystal (potential). This is the reason that there are different energy bands for the same momentum in the first zone — they really belong to different momenta, but these are outside the first zone.

Since the crystal has symmetry, there will be, in general, multiple values of $\underline{k}$ giving the same energy. And each momentum gives several states of different energy. But why are these not discrete, why do they form (in the limit of infinite crystals) continuous bands? The statefunction is a product of an exponential, for the momentum, and a Bloch function — which gives the distribution of the probability density within the unit cell. And that distribution, varying continuously, determines the energy; there are, for each representation, an infinite number of states having different functional forms of the Bloch function. Thus each band is given by the momentum, and the energy value within the band by the specific Bloch function. Of course a Bloch function is not a useful label, but it provides a physical reason for the width of the bands.

Problem XI.2.g-1: Above (sec. XI.2.b, p. 570; sec. XI.2.f, p. 577), aspects of this were stated differently. Compare the rationales. Which are the most reasonable (the most nearly correct?)?

XI.2.h Energies and statefunctions for cubic lattices

The three cubic lattices, the simple, the fcc and the bcc (sec. II.6.b, p. 105, sec. IV.5.a, p. 216), are important and highly symmetric, so furnish useful examples [Cornwell (1969), p. 98]. And because there are three, they allow comparison of states and energies showing how these are affected by differences in lattice types. The labels for the special positions of the Brillouin zone have been given previously (sec. IV.6.a, p. 240).

XI.2.h.i *The simple cubic lattice*

For this [Burns (1977), p. 315; Jones (1975), p. 109; Joshi (1982), p. 289], reciprocal lattice vectors are proportional to the vectors of the real-space lattice (of length a). The energies are (eq. XI.2.g-3, p. 578), defining the reduced energy ϵ and κ_j (for $j = 1, 2, 3$) by,

$$E_{\underline{k}n} = \frac{1}{2m}\epsilon_n(\underline{k}) = \frac{1}{2m}|\underline{k} - \underline{B}_n|^2 = \frac{1}{2m}\sum(\kappa_j - n_j)^2, \qquad \text{(XI.2.h.i-1)}$$

[Jones (1975), p. 118] in terms of the three components of the momentum vector along the axes of the cube, labeled by the three κ_j's which run over lengths of the first zone, and the band labeled by the n's, these running over the integers. If the potential is constant this instantly provides the statefunctions (eq. XI.2.g-2, p. 578)

$$\psi_{\underline{k}} = exp[\frac{2\pi i}{a}\sum(\kappa_j - n_j)x_j], \qquad \text{(XI.2.h.i-2)}$$

which are functions of the real space coordinates x_j. (Normalization is ignored throughout.)

We want the energies and statefunctions at the special points, and then have to relate them. We write the coordinates of the special points as

$$\Gamma : (0,0,0), \quad \Delta : (0,\kappa,0), \quad X : (0,\frac{1}{2},0). \qquad \text{(XI.2.h.i-3)}$$

These have little groups

$$\Gamma : O_h, \quad \Delta : C_{4v}, \quad X : D_{4h}. \qquad \text{(XI.2.h.i-4)}$$

First for the Δ axis, the energies at the special points are

$$\epsilon_\Gamma = n_1^2 + n_2^2 + n_3^2, \qquad \text{(XI.2.h.i-5)}$$

$$\epsilon_\Delta = n_1^2 + (n_2 - \kappa)^2 + n_3^2, \quad (0 < \kappa < \frac{1}{2}), \qquad \text{(XI.2.h.i-6)}$$

$$\epsilon_X = n_1^2 + (n_2 - \frac{1}{2})^2 + n_3^2. \qquad \text{(XI.2.h.i-7)}$$

Expressions for the other axes and points are similar. At the center, for zero momentum, and at point X, the energy depends only on the band. Along Δ it also depends on the momentum component. There are limits on κ because momentum is referred to the first Brillouin zone (sec. XI.2.g, p. 578).

As expected the lowest energy is at the center Γ — the momentum is zero. The statefunction, seen to be constant (in direct space, of course), is a basis state of octahedral group representation of type A_1 (Γ_1) (tbl. E.2-1, p. 657) — this is the representation invariant under all point group transformations. Along Δ the lowest state has energy

$$\epsilon_{\Delta,000} = \kappa^2, \qquad \text{(XI.2.h.i-8)}$$

and statefunction

$$\psi_{\Delta,000}(r) = exp(\frac{2\pi i\kappa y}{a}), \qquad \text{(XI.2.h.i-9)}$$

which is invariant under the little group of the point so belongs to representation A_1, as can be seen from the character tables. Next for point X,

$$\epsilon_{X,000} = \epsilon_{X,010} = \frac{1}{4}, \qquad \text{(XI.2.h.i-10)}$$

with statefunctions

$$\psi_{X,000}(r) = exp(\frac{\pi iy}{a}), \quad \psi_{X,010}(r) = exp(\frac{-\pi iy}{a}). \qquad \text{(XI.2.h.i-11)}$$

There are two values of n_2 giving the same energy. Since X is on the zone boundary, the two n values give equivalent points — these are connected by a reciprocal lattice vector. The points are related by operations of the little group, such as inversion, so these transform the basis functions into each other — they form a two-dimensional representation of the little group. This is reducible and is reduced by symmetrizing, giving representations A_{1g} and A_{2u} of D_{4h}, with states, respectively,

$$\psi_{X_+} = cos(\frac{\pi y}{a}), \quad \psi_{X_-} = sin(\frac{\pi y}{a}), \qquad \text{(XI.2.h.i-12)}$$

which can be seen to transform correctly under these representations (and no other).

The other state at X has statefunction that becomes along Δ

$$\psi_{\Delta,010}(r) = exp(\frac{2\pi iy(\kappa - 1)}{a}), \qquad \text{(XI.2.h.i-13)}$$

with energy

$$\epsilon_{\Delta,010} = (\kappa - 1)^2, \qquad \text{(XI.2.h.i-14)}$$

and is invariant under the little group of Δ. Thus at X there are two degenerate states (that is with the same energy), but not belonging to the same representation, so they behave differently as the momentum is changed.

These momentum vectors all lie in the first Brillouin zone. Vectors in other Brillouin zones correspond to them, but of course give larger energies. However here we consider the first zone only so need the concept of bands. For each momentum vector in the first zone there are a set of energy values; these belong to different bands. The lowest band is for momentum in the first zone, the others for the other zones, but are labeled by vectors in the first.

For point Γ, at the center, the second band has energy

$$\epsilon_{\Gamma,1} = 1, \qquad \text{(XI.2.h.i-15)}$$

and statefunctions (the indices give the set of labels, the n's),

$$\psi_{\Gamma,100}(r) = exp(\frac{2\pi ix}{a}), \quad \psi_{\Gamma,-100}(r) = exp(-\frac{2\pi ix}{a}), \qquad \text{(XI.2.h.i-16)}$$

$$\psi_{\Gamma,010}(r) = exp(\frac{2\pi iy}{a}), \quad \psi_{\Gamma,0,-1,0}(r) = exp(-\frac{2\pi iy}{a}), \qquad \text{(XI.2.h.i-17)}$$

$$\psi_{\Gamma,001}(r) = exp(\frac{2\pi iz}{a}), \quad \psi_{\Gamma,00,-1}(r) = exp(-\frac{2\pi iz}{a}). \qquad \text{(XI.2.h.i-18)}$$

These form a representation of the little group O_h of Γ, but it is reducible. As can easily be checked from the character formula [Mirman (1995a), sec. VII.6.a, p. 199], or from the transformation properties, or characters (tbl. E.2-1, p. 657), it reduces into a sum of representations A_{1g}, E_g and T_{1u}. The statefunctions that have the proper form along Δ (since the y axis is the principle rotational axis of its symmetry group) are

$$\psi_{A_{1g}}(r) = cos(\frac{2\pi x}{a}) + cos(\frac{2\pi y}{a}) + cos(\frac{2\pi z}{a}), \qquad \text{(XI.2.h.i-19)}$$

for the scalar,

$$\psi_{E_g}(r) = cos(\frac{2\pi x}{a}) - cos(\frac{2\pi z}{a}), \qquad \text{(XI.2.h.i-20)}$$

$$\psi_{E_g}(r)' = cos(\frac{2\pi y}{a}) - \frac{1}{2}[cos(\frac{2\pi x}{a}) + cos(\frac{2\pi z}{a})], \qquad \text{(XI.2.h.i-21)}$$

for the two-dimensional representation, and

$$\psi_{T_1u}(r) = sin(\frac{2\pi x}{a}), \quad \psi_{T_1u}(r)' = sin(\frac{2\pi y}{a}),$$

$$\psi_{T_1u}(r)'' = sin(\frac{2\pi z}{a}), \qquad \text{(XI.2.h.i-22)}$$

for the three-dimensional one.

Along Δ, the n values $(0, \pm 1, 0)$ give energy

$$\epsilon_{\Delta,(0,\pm 1,0)} = (1 \mp \kappa)^2, \qquad \text{(XI.2.h.i-23)}$$

and statefunctions

$$\psi_{\Delta,0-10}(r) = exp(\frac{2\pi i(1+\kappa)y}{a}), \quad \psi_{\Delta,010}(r) = exp(\frac{2\pi i(-1+\kappa)y}{a}), \qquad \text{(XI.2.h.i-24)}$$

and these are basis states of group C_{4v}, both of representation A_1, which thus appears twice. The other four cases have energy

$$\epsilon_\Delta = 1 + \kappa^2, \quad \text{(XI.2.h.i-25)}$$

and basis functions of the irreducible representations,

$$\psi_{\Delta a}(r) = exp(\frac{2\pi iy\kappa}{a})[cos(\frac{2\pi x}{a}) + cos(\frac{2\pi z}{a})], \quad \text{(XI.2.h.i-26)}$$

$$\psi_{\Delta b}(r) = exp(\frac{2\pi iy\kappa}{a})[cos(\frac{2\pi x}{a}) - cos(\frac{2\pi z}{a})], \quad \text{(XI.2.h.i-27)}$$

for the one-dimensional representations, and

$$\psi_{\Delta c}(r) = exp(\frac{2\pi iy\kappa}{a})sin(\frac{2\pi x}{a}), \quad \text{(XI.2.h.i-28)}$$

$$\psi_{\Delta c}(r)' = exp(\frac{2\pi iy\kappa}{a})sin(\frac{2\pi z}{a}), \quad \text{(XI.2.h.i-29)}$$

for the two-dimensional one. The energies of these are equal, for the special case of constant potential. With a (more realistic) nonconstant one, the degeneracies are lifted. However the two states of representation c remain degenerate, belonging to the same representation; states of different representations have different energies.

Problem XI.2.h.i-1: Check these results.

Problem XI.2.h.i-2: At the end of this line is point X. There are eight states with energy the same as that of Δ for this band for the value of κ (which is what?) giving point X [Jones (1975), p. 120]. State them, and the basis vectors for all representations. Linear combinations of basis states may be needed so that those along Δ go smoothly into the ones at X. Give the basis states. Are combinations of the Δ states or of the X states needed, or can either be chosen, or is this unnecessary? Are the basis vectors stated for Γ correct, or are combinations (of which states?) necessary so that the Δ states go smoothly into the Γ ones? Are there representations that cannot occur because the smoothness condition cannot be met? Which representations go into which — what are the compatibility relations?

Problem XI.2.h.i-3: Repeat this for all special lines and planes. For every pair of these points (lines) that lie on a special line (plane) the states of the representations for one must go smoothly to those for the other. The lists of pairs of states that can do so, for each pair of special positions, are the compatibility relations. Find all.

XI.2.h.ii *The body-centered cubic lattice*

States and energies depend on the lattice vectors, and these differ for different lattices [Jones (1975), p. 122; Joshi (1982), p. 290], as illustrated by comparison of the simple cubic with this, whose reciprocal lattice vectors (the primitive lattice vectors of the fcc lattice) are given above (sec. IV.5.a.ii, p. 216) and which has the energy function (eq. XI.2.h.i-1, p. 580),

$$\epsilon_n = (\kappa_1 - n_1)^2 + (\kappa_2 - n_2)^2 + (\kappa_3 - n_3)^2. \qquad \text{(XI.2.h.ii-1)}$$

The only allowed n values are integral ones.

The groups for positions Γ and Δ are the same as for the simple cubic lattice, and for point H the group is that of full cubic symmetry, thus the same as for Γ. Point H appears for this lattice, but not for the simple cubic, thus adding requirements on the statefunctions: these have to be degenerate at this extra point. But points P and N, both also not appearing for the simple cubic, have symmetry groups different from any of those of the simple lattice. This lattice adds conditions, and complexity.

These symmetry positions have coordinates

$$\text{H}:(0,1,0),\ \ \text{N}:(\frac{1}{2},\frac{1}{2},0),\ \ \text{P}:(\frac{1}{2},\frac{1}{2},\frac{1}{2}),\ \ \Delta:(0,\kappa,0), \qquad \text{(XI.2.h.ii-2)}$$

and energies

$$\epsilon_\Gamma = n_1^2 + n_2^2 + n_3^2, \qquad \text{(XI.2.h.ii-3)}$$

$$\epsilon_\Delta = n_1^2 + (n_2 - \kappa)^2 + n_3^2,\ \ (0 < \kappa < 1), \qquad \text{(XI.2.h.ii-4)}$$

$$\epsilon_H = n_1^2 + (n_2 - \kappa)^2 + n_3^2. \qquad \text{(XI.2.h.ii-5)}$$

$$\epsilon_N = (n_1 - \frac{1}{2})^2 + n_2^2 + (n_3 - \frac{1}{2})^2. \qquad \text{(XI.2.h.ii-6)}$$

The statefunctions, for example for $\epsilon_H = 1$, have the form

$$\psi = exp(\frac{2\pi iy}{a})exp[-\frac{2\pi i}{a}(n_1 x + n_2 y + n_3 z)]; \qquad \text{(XI.2.h.ii-7)}$$

the six n values, with the * indicating the negative, are

$$n = (011),\ (01^*1),\ (101),\ (1^*01),\ (000),\ (002). \qquad \text{(XI.2.h.ii-8)}$$

Thus we get the properly symmetrized statefunctions

$$\text{H}_1 : \psi = cos\frac{2\pi x}{a} + cos\frac{2\pi y}{a} + cos\frac{2\pi z}{a}, \qquad \text{(XI.2.h.ii-9)}$$

$$H_{12} : \psi = [cos\frac{2\pi y}{a} - \frac{1}{2}(cos\frac{2\pi x}{a} + cos\frac{2\pi z}{a}), cos\frac{2\pi x}{a} - cos\frac{2\pi z}{a}], \qquad (XI.2.h.ii-10)$$

$$H_{15} : \psi = [sin\frac{2\pi x}{a}, sin\frac{2\pi y}{a}, sin\frac{2\pi z}{a}]. \qquad (XI.2.h.ii-11)$$

The subscripts go with the representation labels in the character table for this group.

For a nonconstant potential, these three representations have different energies, but the two states of H_{12} remain degenerate, as do the three of H_{15}. This is required by symmetry. That all these states are here degenerate is an "accidental" symmetry coming the very special form of the (unrealistic constant) potential.

Problem XI.2.h.ii-1: Repeat this for all special lines and planes. Find all compatibility relations.

XI.2.h.iii *The face-centered cubic lattice*

This has two additional symmetry points, L and W (sec. IV.5.a.ii, p. 216), so is more restrictive, providing more relations on the statefunctions, and is thus perhaps more complicated [Altmann (1994), p. 130; Jones (1975), p. 127; Joshi (1982), p. 293; Wherrett (1986), p. 266]. The rest of the discussion is similar to that for the other two lattices. In finding the statefunctions for realistic potentials the knowledge of the statefunctions at the symmetry points is important, but for our purposes this illustrates the method, and how the different, but here related, lattices give different results.

Problem XI.2.h.iii-1: Nevertheless, though all aspects have appeared for the other cases, it is a useful exercise to work this out (completely), for all special positions. Find (all) compatibility relations, and compare them to those for the other cubic lattices. This lattice has additional symmetry points (positions?). What effect does that have on the (number of) compatibility relations? Are the number, and types of, degeneracies (at the symmetry positions shared by these lattices) increased or decreased (or just changed) because of the extra symmetries? Some degeneracies are due to symmetry, others are accidental, due here to the special form of the potential. Are any degeneracies changed from one kind to the other? Why?

Problem XI.2.h.iii-2: There are sets of symmetry positions with the same group, in some cases realized differently (that is the sets of vectors at the different positions are invariant under different operations, though these form isomorphic groups). For the cubic lattices find these sets, classifying them into subsets with the same groups, and ones with only isomorphic groups. For each special position shared by lattices, do these sets and subsets change because of the presence of ones not shared?

XI.2.i Energy bands for nonsymmorphic groups

Lattice vectors going with glides and screws are determined by two constants, the lattice size (essentially an irrelevant scaling factor) as before, plus one other: the magnitude of the translation going with the glide or screw (telling how wide the screw is, or the distance from a point connected by a glide and the mirror). This adds richness to the states (containing now constants), and the discussion of the physics [Jones (1975), p. 132]. Nonsymmorphic groups have more interesting representations than symmorphic ones — there are additional aspects caused by the presence of glide planes and screw axes, especially additional degeneracy.

XI.2.i.i *Sticking together of bands*

Glide planes and screw axes result in additional degeneracies — bands that are not so in their absence become degenerate at certain points of the zone surface in reciprocal space [Burns (1977), p. 322]. This is called sticking together of bands (pb. VII.6.h-8, p. 382). Why does this occur; why are there surface points at which there is two-fold degeneracy?

Of course, if there are nonsymmorphic symmetry elements there must be more than one object (atom or molecule, say) in a unit cell.

Take for example (two-dimensional) space group $p2mg$ (Pbl), with a rectangular unit cell of length a along x, b along y (fig. D-8, p. 652); it has a mirror, a center of inversion, and a glide plane [Heine (1993), p. 284]. The Brillouin zone is also rectangular with lengths $\frac{\pi}{a}$ along k_x, and $\frac{\pi}{b}$ along k_y (sec. IV.2.c.iv, p. 192), the same Brillouin zone as for this space group since the Brillouin zone is of the lattice (sec. IV.2.c.i, p. 190), not the crystal; for the square the relevant special points are the same (fig. VII.6.b-1, p. 373). Consider (here only) point

$$X = (\frac{\pi}{a}, 0), \qquad \text{(XI.2.i.i-1)}$$

and the statefunction $\psi(X,0)$ at it. With the operators of the glide (which is along x), inversion (through the center, Γ) and mirror reflection denoted by g, I and r, the functions

$$g\psi = \psi(x + \frac{a}{2}, -y), \qquad \text{(XI.2.i.i-2)}$$

$$I\psi = \psi(-x, -y), \qquad \text{(XI.2.i.i-3)}$$

$$r\psi = \psi(-x - \frac{a}{2}, y), \tag{XI.2.i.i-4}$$

and ψ, related by these symmetry transformations, are degenerate, and have the same $\underline{k}$. Thus

$$g\psi(x,y) = \psi(x + \frac{a}{2}, -y) = r\psi(-x,-y) = rI\psi = \psi(-x,-y). \tag{XI.2.i.i-5}$$

What conditions do these impose? The four functions might be proportional, but if not there is degeneracy. Since

$$I^2\psi = r^2\psi = \psi, \tag{XI.2.i.i-6}$$

$$I\psi = \alpha\psi, \quad r\psi = \beta\psi, \quad \alpha = \pm 1, \quad \beta = \pm 1. \tag{XI.2.i.i-7}$$

Then, assuming proportionality,

$$g^2\psi(x,y) = rIrI\psi = \alpha rIr\psi = \ldots = \alpha^2\beta^2\psi = \psi(x,y)$$

$$= \psi(x+a,y) = exp(ik_x a)\psi(x,y) = exp(ia\frac{\pi}{a})\psi(x,y) = -\psi(x,y), \tag{XI.2.i.i-8}$$

since k_x is on the boundary of the Brillouin zone. This gives a contradiction, so the functions cannot be proportional and there must be degeneracy.

Why? The glide, applied twice, takes a function to its negative. However we can also obtain this transformation with products of inversions and mirrors since the square of the glide is a lattice translation — a single glide is not — but these, being lattice translations, do not give a minus sign. (These lattice translations leave the lattice, including statefunctions defined over it, invariant.) Since there are then two different functions related by the symmetry group transformations, they must be degenerate.

Another way of seeing this is to recall that the glide plane introduces a phase — the representations of the space group are projective (ray) representations (sec. V.4, p. 259). This gives a minus sign preventing the basis vectors related by the group transformations from being proportional, thus requiring degeneracy.

This occurs at the zone face, only, because it is there that a point transformation gives a vector that is also equivalent to the original under a lattice translation.

XI.2.i.ii *Irreducible representations on the zone faces*

These results indicate that the only representations at the surface are two-dimensional, that is the presence of the glide plane reduces the number of representations of the space group. How does this occur? One way of handling ray representations is to enlarge the group so that they become vector representations of the larger group. However that has more representations then the symmetry group, so only a subset is relevant. And this includes only two-dimensional representations.

To illustrate consider again this space group, and point X. We start with the little group of $\underline{k}$. Translations

$$\underline{t}_{even} : (2na, mb), \quad n, m \text{ integers,} \tag{XI.2.i.ii-1}$$

are lattice translations so represented by the unit matrix —

$$exp(i\underline{k} \bullet \underline{t}_{even}) = 1. \tag{XI.2.i.ii-2}$$

But

$$\underline{t}_{odd} : ((2n+1)a, mb) \tag{XI.2.i.ii-3}$$

is not a lattice translation. It must be represented by a matrix whose square is the unit matrix, so by the negative of the unit matrix.

We thus get the representation matrices, defining an abbreviated version of the Seitz notation,

$$M(E) = M(\{E|\underline{t}_{even}\}), \tag{XI.2.i.ii-4}$$

$$M(g) = M(\{r_y|\underline{\tau} + \underline{t}_{even}\}), \tag{XI.2.i.ii-5}$$

$$M(I) = M(\{I|\underline{t}_{even}\}), \tag{XI.2.i.ii-6}$$

$$M(r) = M(\{r_x|\underline{t}_{even}\}), \tag{XI.2.i.ii-7}$$

$$N(E) = M(\{E|\underline{t}_1 + \underline{t}_{even}\}), \tag{XI.2.i.ii-8}$$

$$N(g) = M(\{r_y|\underline{\tau} + \underline{t}_1 + \underline{t}_{even}\}), \tag{XI.2.i.ii-9}$$

$$N(I) = M(\{I|\underline{t}_1 + \underline{t}_{even}\}), \tag{XI.2.i.ii-10}$$

$$N(r) = M(\{r_x|\underline{t}_1 + \underline{t}_{even}\}); \tag{XI.2.i.ii-11}$$

with the translations

$$\underline{t}_1 = (a, 0), \tag{XI.2.i.ii-12}$$

$$\underline{\tau} = (\frac{a}{2}, 0). \tag{XI.2.i.ii-13}$$

These matrices form a group, and $M(E)$ and $N(E)$, and so on, are different elements, even though

$$N(E)^2 = 1, \tag{XI.2.i.ii-14}$$

so that $N(E)$ acts like $-M(E)$. A presentation [Mirman (1995a), sec. III.3, p. 82] for it is

$$M(g)^2 = N(E),\ M(g)^3 = N(g),\ M(g)^4 = M(E) = N(E)^2,$$
$$M(r)M(\mathrm{I}) = M(g),\ M(r)^2 = N(r)^2 = M(\mathrm{I})^2 = N(\mathrm{I})^2 = M(E),\ \ldots. \quad \text{(XI.2.i.ii-15)}$$

The group is thus (isomorphic to) point group C_{4v}(4mm) whose characters we know (tbl. E.2-1, p. 657).

Not all representations listed in the character table appear;

$$N(E) = -M(E), \quad \text{(XI.2.i.ii-16)}$$

so the only allowed ones are those for which this holds. There is but one, the two-dimensional representation. Thus this is the only one relevant to electron states at this point of the Brillouin zone, for this space group. And it requires degeneracy.

Here we see that the presence of a glide plane places an extra condition on the point group. The point group is the factor group, so does not depend on whether there is a glide, thus that cannot increase the number of representations, or make representations equivalent that are not. What it does is decrease the number of allowed representations — only two-dimensional ones are now allowed.

Problem XI.2.i.ii-1: That the groups are isomorphic can be checked by showing that the corresponding transformations satisfy the same equations.

XI.2.i.iii *The little group need not be a subgroup of the point group*

For crystals with symmorphic groups the little group of wavevector $\underline{k}$, for any special position, is a subgroup of the point group of the space group, perhaps the point group itself (as at the center of the Brillouin zone: $\underline{k} = 0$), so its order is not greater than that of the crystal point group. A crystal with a nonsymmorphic group has positions at which the symmetry group is of higher order — the point group operation of a nonsymmorphic element is not an element of the point (factor) group of the space group, so the set containing it is not a group. But a power of such an element is a transformation equivalent to the identity. Thus the set of transformations including nonsymmorphic elements, and all powers up to that equivalent to the identity, forms a finite group. However since it has these powers, it has more elements (larger order) than the point group of the space group [Jones (1990), p. 139]. This we have just seen and see again next for the closed-packed hexagonal structure for which the point factor group has order 24, while for point A the little group is of order 48, twice that.

The larger group has more representations, but not all give Bloch functions for the particular wavevector — nonsymmorphic groups have projective representations (sec. V.4, p. 259), which are vector representations of an extended group; since there are more representations for that, not all are representations of the nonsymmorphic group.

XI.2.j The close-packed hexagonal structure

An illustration of the effect of nonsymmorphic elements for a three-dimensional crystal is given by the close-packed hexagonal structure [Altmann (1994), p. 83] with space group $P6_3/mmc$ ($D_{6h}{}^4$) which is also written $P\frac{6_3}{m}mc$ [Altmann (1977), p. 103, 235; Altmann and Bradley (1965); Cornwell (1969), p. 83, 173; Herring (1942), p. 530; Jones (1975), p. 141; Slater (1965), p. 74]. It is derived from point group $D_{6h}(\frac{622}{mmm})$ by replacing the sixfold rotation axis with screw displacement 6_3, and also one set of reflection planes containing the principle axis with a set of glide planes. The lattice is (of course) primitive hexagonal, so its space group is labeled with a P. The point group of the structure is then $D_{3h}(\bar{6}m2)$, of order 12. The factor group of the space group with respect to pure translations is (isomorphic to) D_{6h}, of order 24, twice that of the point group. Since the screw axis is 6_3, two lattice translations are required for a transformation equivalent to the identity (sec. III.4.a, p. 142).

The two lattice vectors in the xy plane are $\underline{a}_1$ and $\underline{a}_2$. The unit cell of this structure has two planes separated by a distance of $c/2$. The matrices giving the lattice (sec. II.8.f, p. 130) [Cornwell (1984), p. 200; Streitwolf (1971), p. 59] and the reciprocal lattice (sec. IV.2.a, p. 186),

$$A = \begin{pmatrix} a & -(1/2)a & 0 \\ 0 & (\sqrt{3}/2)a & 0 \\ 0 & 0 & c \end{pmatrix}, \quad B = 2\pi \begin{pmatrix} 1/a & (1/\sqrt{3}a) & 0 \\ 0 & (2/\sqrt{3}a) & 0 \\ 0 & 0 & 1/c \end{pmatrix}, \qquad \text{(XI.2.j-1)}$$

are clearly reciprocal (with factor 2π).

XI.2.j.i *Nonsymmorphic elements of the crystal*

A single unit cell of the crystal is shown next on the top left, below it are two fit together horizontally, and on the right, two vertically. There are two parallel hexagons displaced along the z axis, by half the length of the unit cell, and in the xy plane by 1/3 the distance between the atom and the center of the cell, these (although of course identical) are drawn in solid and dashed lines so they can be distinguished. Atoms are shown as heavy dots (only on the upper left). The upper hexagon on the right is rotated with respect the lower by $\frac{\pi}{6}$ about z. Edges of the unit cell are dotted lines, as is the z axis through the centers of the hexagons, marked by dots, and also the center of the cell, which is on the plane of the displaced hexagon. The center of the lower hexagon is the origin, and on the left the three primitive lattice vectors are drawn from it (as dotted and dashed lines), one to the center of the hexagon above,

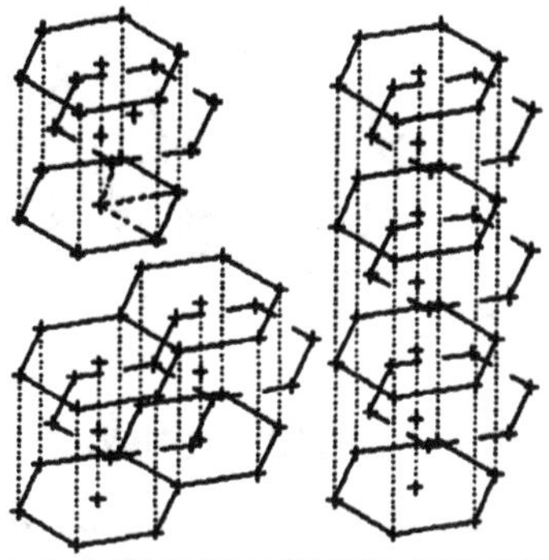

Figure XI.2.j.i-1: NONSYMMORPHIC HEXAGONAL UNIT CELL.

To go from a point on the lower hexagon, say that at the end of a lattice vector, to the corresponding one above, requires a rotation of $\frac{\pi}{6}$ about z, and a translation along it. Thus the z axis is a screw axis, and this screw is a symmetry of the system (which of course is infinite in extent, though only two sets of points are drawn).

In the next diagram is a unit cell, on the left, with one point heavier so that its path can be followed. Then next to it the cell has been rotated by $2\pi/6$, leaving the bottom hexagon unchanged (but the heavy point has been moved), however shifting the dashed one. Thus this rotation does not leave the crystal invariant. Then on the right the cell has been translated along z by half its length. It is seen that it fits onto a second copy of itself (on the extreme right this is dotted for clarity),

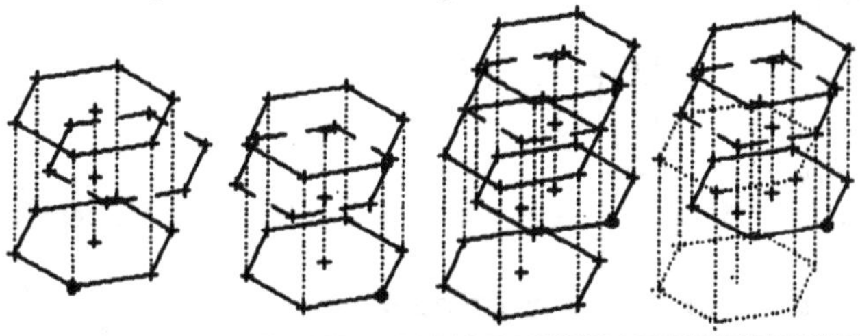

Figure XI.2.j.i-2: TRANSLATION IN THIS NONSYMMORPHIC CELL.

The combined operation of a rotation plus a translation leaves the crystal unchanged, so this screw is a symmetry of the crystal.

To give an example of a glide, the left object is a unit cell, to the right of it is the cell translated along the z axis (the vertical), shown by the dotted line connecting the centers of the bases. Note that this does not leave the crystal invariant for if the bases are brought into coincidence the hexagons above would be displaced differently, one to the left of the base, the other to the right. Also the hexagons below this point differ — lines from the point to a corner go in different directions for the two,

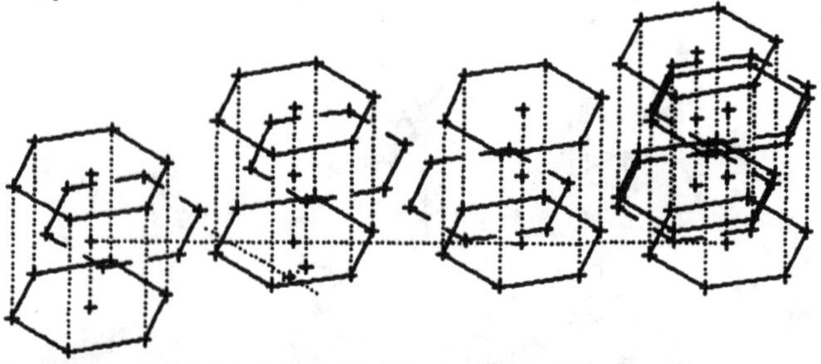

Figure XI.2.j.i-3: GLIDE IN THIS NONSYMMORPHIC CELL.

For the second figure, there is a dotted line drawn on the base which is the intercept of the vertical mirror with it. This is in the middle of the line from the corner to the projection of the corner of the hexagon above (shown by a vertical dotted line), and perpendicular to it. By reflecting in this mirror (in which the translation vector lies), the next object is obtained. That this is the same as the first is indicated in the rightmost figure, in which these two are overlayed, slightly displaced to distinguish them. The system is invariant under (only) the joint operation of the reflection and translation in the same plane — under the glide.

There are also mirrors, as we see next with a mirror through the z axis and the centers of opposite edges, indicated by a dotted line in the base. The figure is clearly invariant under this reflection. Note however the movement of the heavy corner, which shows the effect of the reflection:

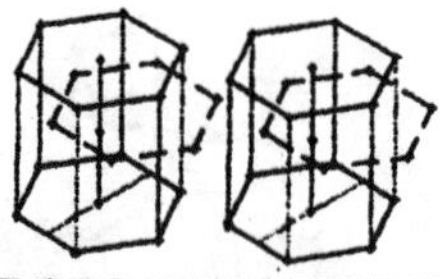

Figure XI.2.j.i-4: MIRRORS IN NONSYMMORPHIC HEXAGONAL CELL.

XI.2.j.ii *Special positions of the Brillouin zone*

Next we consider the Brillouin zone for the hexagonal lattice [Koster (1957), p. 210]. While only the subset of special positions in the representation domain (sec. IV.2.c, p. 189) need be shown, here some are drawn more than once for clarity. However only the ones in the representation domain are marked. The coordinates of the special points, listing for equivalent ones only those with positive coordinates, are [Jones (1975), p. 143]

$$\Gamma = (0,0,0),\ \ A = (0,0,\frac{\pi}{c}),\ \ M = (0,\frac{2\pi}{\sqrt{3}a},0),\ \ L = (0,\frac{2\pi}{\sqrt{3}a},\frac{\pi}{c}),$$

$$K = (\frac{2\pi}{3a},\frac{2\pi}{\sqrt{3}a},0),\ \ H = (\frac{2\pi}{3a},\frac{2\pi}{\sqrt{3}a},\frac{\pi}{c}). \tag{XI.2.j.ii-1}$$

Of course Γ has the same coordinates for all lattices. So we have

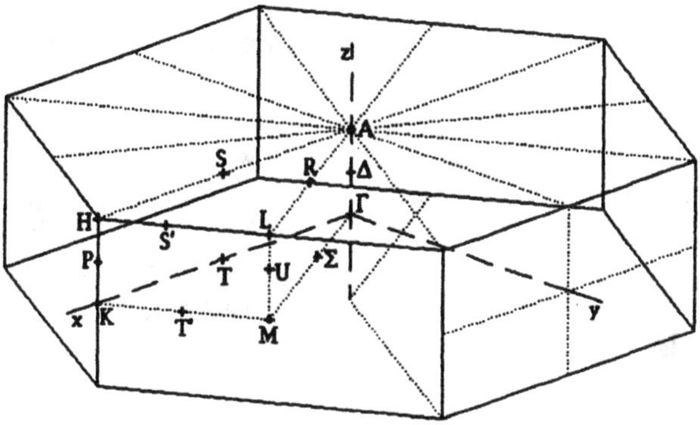

Figure XI.2.j.ii-1: BRILLOUIN ZONE FOR HEXAGONAL LATTICE.

At A the statefunction is

$$\psi_A = u_A(x,y,z)exp(\frac{\pi iz}{c}), \tag{XI.2.j.ii-2}$$

where u_A is the Bloch function for the state given by momentum

$$\underline{k}_A = (0,0,\frac{\pi}{c}). \tag{XI.2.j.ii-3}$$

Writing the glide reflection as $r\zeta$, with r the reflection of the glide, and ζ the displacement $\frac{c}{2}$ along z, we get

$$r\zeta\psi_A = u_A exp[\frac{\pi i}{c}(z+\frac{c}{2})]$$

$$= u_A exp\frac{\pi iz}{c} exp\frac{\pi i}{2}, \tag{XI.2.j.ii-4}$$

indicating the point reached with the glide, which being a symmetry leaves the Bloch function invariant, and the introduction of a phase, so

$$\zeta^2\psi_A = u_A exp\frac{\pi iz}{c} exp(\pi i) = -\psi_A. \tag{XI.2.j.ii-5}$$

Glides and screws are symmetries of the crystal, and Bloch functions are defined to be invariant under symmetries, but basis vectors of these (projective) representations need not be — although physically probabilities, given by matrix elements using these vectors and conjugates of them, have to be invariant. As ζ^2 takes this wavevector to an equivalent point in the (next) Brillouin zone it is in its little group, $(\zeta^2)^2$ is equivalent to the identity, so element ζ^2 is distinct from ζ and the adjunction of it results in a group of twice the order of the group without the glide. Also with r the reflection in the basel plane,

$$r\zeta = \zeta^{-1}r; \qquad \text{(XI.2.j.ii-6)}$$

as elements ζ and r can be seen from their effects on the crystal to not commute, ζ^2 is not the identity (r is the only element not commuting with ζ). Of course r does commute with $(\zeta^2)^2$ — it is equivalent to the identity. The group of this wavevector, with extra elements ζ^2 and products, has order 48. The reason that the group is larger than the one without the glide is that it has elements under which the crystal is invariant, but not all basis vectors of the latter group are — belonging to projective representations these are less restricted.

Problem XI.2.j.ii-1: Are all coordinates of the special points correct?

Problem XI.2.j.ii-2: Check that the only element not commuting with ζ is r, and that it does not commute.

XI.2.j.iii *Representations, and the effect of nonsymmorphic elements*

Not all listed representations (tbl. E.2-1, p. 657) belong to the little group. For these ζ^2 must be -1, and all such representations are at least two-dimensional giving, as expected, degeneracy, that is sticking together of bands. The other half of the set of representations, those for which $\zeta^2 = 1$, belong to point Γ. The groups obtained by extension of the little groups of the two points are the same, but since it has twice the number of representations it supplies half to one point, the other half to the other.

The energy is given by eq. XI.2.g-3, p. 578; in nondimensional form by eq. XI.2.h.i-1, p. 580, so with the reciprocal lattice matrix B (eq. XI.2.j-1, p. 590), the energy,

$$\epsilon_1 = (\kappa_1 - n_1)^2 + (\kappa_2 - \frac{1}{\sqrt{3}}(n_1 + 2n_2))^2 + \frac{a}{c}(\kappa_3 - n_3)^2, \qquad \text{(XI.2.j.iii-1)}$$

depends on a parameter, here a/c, since the group is nonsymmorphic. For close packing of rigid spheres [Jones (1975), p. 149], a reasonable assumption,

$$\frac{c}{a} = \sqrt{\frac{8}{3}}. \qquad \text{(XI.2.j.iii-2)}$$

Thus the energy depends not only on the symmetry and the potential, but also on the distance between atoms. Even if the potential is (taken as) constant, with the space group fixed, there is still some freedom in the positions of the points of the crystal, and these must be specified. The crystal is a lattice plus a basis and a basis requires parameters for its specification.

To illustrate how nonsymmorphic elements affect the basis vectors we consider the statefunctions for point M (using the notations for the representations in the character table) which, with the corresponding energies (and superscripts indicating whether the basis vectors are analogous to even or odd parity functions), are

$$E_M = \frac{1}{3};\ M_1^+ : \psi = cos\frac{2\pi y}{\sqrt{3}a},\ M_2^- : \psi = sin\frac{2\pi y}{\sqrt{3}a}, \qquad (XI.2.j.iii\text{-}3)$$

$$E_M = \frac{17}{24};$$

$$M_1^+ : \psi = cos\frac{2\pi z}{c}sin\frac{2\pi y}{\sqrt{3}a},\ M_2^- : \psi = cos\frac{2\pi z}{c}cos\frac{2\pi y}{\sqrt{3}a}$$

$$M_3^+ : \psi = sin\frac{2\pi z}{c}cos\frac{2\pi y}{\sqrt{3}a},\ M_4^- : \psi = sin\frac{2\pi z}{c}sin\frac{2\pi y}{\sqrt{3}a}, \qquad (XI.2.j.iii\text{-}4)$$

$$E_M = 1;\ M_1^+ : \psi = cos\frac{2\pi x}{a},\ M_2^- : \psi = sin\frac{2\pi x}{a}. \qquad (XI.2.j.iii\text{-}5)$$

These differ from those of symmorphic cases (such as ones we considered, like the cubic of sec. XI.2.h, p. 579) as they have extra phase factors, given by the z/c terms, as expected (sec. VII.3.b, p. 351). Of course these phase factors are quite relevant in computing matrix elements, such as for transitions, and selection rules, so are important in finding properties of the material, and in their experimental determination.

Problem XI.2.j.iii-1: Verify that the point groups of Γ and A are the same (isomorphic) so that the representations of the group extension fall here into two types, one giving the representations for Γ (why?), the other for A.

Problem XI.2.j.iii-2: Show that these statefunctions and energies are (hopefully) correct.

Problem XI.2.j.iii-3: Is there a maximum number of parameters needed (or possible) to specify a basis? Might the maximum number be different for different crystal systems? For spaces of different dimension?

XI.2.k What do we learn from this group-theoretical analysis?

Labels of statefunctions of atoms include the eigenvalues of the diagonal generators of the rotation group because this is a subgroup of the *transformation group* of space, even if space is not invariant under its transformations — in particular, statefunctions of individual electrons still have such labels although their environment is not rotationally invariant. But corresponding statefunctions of different observers are related by transformations of this group, so a statefunction belonging to representation j is a sum of the statefunctions seen by another observer of the same j, only. (Rotation representations are defined about an arbitrary, but fixed, origin. A rotation between observers, and change of the origin, would intermingle not only states within a representation, but representations also.)

States of a representation for an atom are thus mixed by transformations (including those) of the rotation group. A crystal though has different point groups. What is the equivalent for a crystal, how and why is it so, how are statefunctions of electrons, ..., dependent on the transformation group of the crystal? And what is that group? Why? Basis vectors of a group representation are functions of the position within the crystal, and form sets whose number depends on the group, and the representation. As space groups are inhomogeneous, not all their operators mix basis functions. However different subsets are mixed at various special positions, thus relating these functions. The sets of degeneracies then are different for different crystals, being determined by the point groups that are site-dependent subgroups of the space group — site-symmetry groups, the symmetry group of a site depends on the site [Cotton (1990), p. 342; Evarestov and Smirnov (1993); Tsukerblatt (1994), p. 187].

Problem XI.2.k-1: Answers to these questions should be clear (?) from the previous discussions. State them explicitly. How, and to what extent, are these answers useful? Does it really matter that a crystal, that points within it, or within its Brillouin zone, are invariant under certain groups? Why? However a point cannot be invariant. What is this question asking?

XI.3 THE EFFECT OF TIME REVERSAL

There is an additional symmetry, that of time reversal [Jones (1975), p. 154], with three types of behavior (sec. VIII.2, p. 394), one with additional degeneracy, thus a sticking together of bands [Heine (1993),

p. 290]. In general the effect of time reversal is different for spin-independent, and for spin-dependent, Hamiltonians [Cornwell (1984), p. 158; Falicov (1966), p. 199; Streitwolf (1971), p. 176; Tinkham (1964), p. 297].

XI.3.a The energy surfaces must have inversion symmetry

For the zero-spin case, time reversal symmetry requires the energy to obey

$$E(-\underline{k}) = E(\underline{k}); \qquad \text{(XI.3.a-1)}$$

that is there is inversion symmetry for the energy surface even if there is not at the point of the Brillouin zone since the time-reversed function is the complex conjugate. Thus for the statefunction

$$\psi_{\underline{k}} = u_{\underline{k}}(r)exp(i\underline{k} \bullet \underline{r}), \qquad \text{(XI.3.a-2)}$$

the time reversed, so degenerate, function is

$$\psi_{\underline{k}}^* = u_{\underline{k}}(r)^* exp(-i\underline{k} \bullet \underline{r}), \qquad \text{(XI.3.a-3)}$$

the same as

$$\psi_{-\underline{k}} = u_{-\underline{k}} exp(-i\underline{k} \bullet \underline{r}), \qquad \text{(XI.3.a-4)}$$

which transforms under the representation given by -$\underline{k}$, and is the statefunction for a state of the same energy as $\psi_{\underline{k}}$.

XI.3.b Degeneracy due to time reversal

Whether time reversal introduces further degeneracy depends on the relationship between $\underline{k}$ and -$\underline{k}$. There are three cases [Cornwell (1969), p. 151, 210]:

1. If -$\underline{k}$ is neither in the star (sec. VII.2.d, p. 342) of $\underline{k}$, nor equivalent to it, then (for integral spin) there is no extra degeneracy of the energy $E(\underline{k})$ — there are no additional statefunctions with this value of $E(\underline{k})$ and this momentum vector $\underline{k}$. The extra degeneracy comes from $E(\underline{k}) = E(-\underline{k})$; it is due to the equality of eigenvalues going with different momentum vectors. This can only occur if the group does not contain the inversion, because then basis functions of the representation obtained by time reversal, the complex conjugate basis functions, do not have $-\underline{k}$ as their momentum. Otherwise this degeneracy is present without time reversal. However with half-integral spin the degeneracy is doubled.

2. For -$\underline{k}$ in the star of $\underline{k}$, there is an extra degeneracy of $E(\underline{k})$. Here basis vectors of the time-reversed representation, the complex conju-

gates, correspond to momentum vector -k, as well as to k, so this representation must be equivalent to a non-time-reversed one. This belongs to either case 2 or case 3 of the Herring test (sec. VIII.7.1, p. 451). The extra degeneracy at k is then due to the equivalence of two representations of the space group — the bands "stick together".

3. If k and -k are equivalent, there is again (for integral spin) an extra degeneracy of $E(k)$ due to time reversal. Again the representations equivalent under time reversal are said to "stick together". For half-integral spin, there is no *additional* degeneracy due to time reversal invariance.

XI.3.c Degeneracy at general points

For k a general point, the little group consists only of the identity, and if present, inversion [Lax (1974), p. 308]. If there is no inversion symmetry the representation is of the real type, so no degeneracy. With spin, time-reversal symmetry doubles the degeneracy — Kramer's theorem (sec. VIII.2.d, p. 402) — thus every level is doubly degenerate. Then if inversion is in the symmetry group, spin-orbit coupling cannot break a degeneracy between up and down states that holds in the absence of such coupling.

XI.4 LATTICE VIBRATIONS

In studying motions of objects that are the constituents of a crystal we started by taking the nuclei as fixed and considering the behavior of electrons (or holes — the absences of electrons — or of combinations, like excitons). Now we turn to the motion of the nuclei, their vibrations around their equilibrium positions (we do not consider motions away from these which lead, say, to defects or melting). Vibrations — for molecules — have been studied in chap. X, p. 512, and the effect of symmetry on the behavior of electrons considered above. Much of these analyses apply here so we just summarize a few points; the details and formalism, essential but often not highly informative (about applications of group theory), are widely (though not necessarily easily) available as are various discussion of such vibrations [Bhagavantam and Venkatarayudu (1951), p. 127; Birman (1974); Burns and Glazer (1978), p. 200; Cornwell (1984), p. 218; Elliott and Dawber (1987), p. 306; Evarestov and Smirnov (1993), p. 213; Inui, Tanabe and Onodera (1990), p. 271; Janssen (1973), p. 189; Kocinski (1983), p. 80; Lax (1974), p. 324; Leech and Newman (1969), p. 81; Ludwig and Falter (1988), p. 235; Streitwolf (1971), p. 189].

Lattice vibrations are analyzed in terms of harmonic oscillators, as usual. If the Hamiltonian is linear in the interactions this gives a correct description, and is a good approximation if nonlinearities are small, otherwise it is a first approximation, the start of a more detailed study. These harmonic vibrations can be regarded as a set of "particles", phonons, with those of greater amplitude having a larger number of phonons. However it is important to remember [Mirman (1995b), sec. 3.1, p. 38] that these phonons are not particles (especially in the classical sense), but simply give a way of describing the amplitudes of oscillations — these are discrete. The language of phonons (like that for other such objects) is useful, but care must be taken that it does not become misleading.

We are interested mainly in infrared absorption and emission and Raman scattering, especially selection rules (sec. X.7, p. 552) [Altmann (1977), p. 246; Cornwell (1969), p. 102, 181; Evarestov and Smirnov (1993), p. 214; Hochstrasser (1966), p. 215; Kettle (1995), p. 220; Tsukerblatt (1994), p. 223; Wooster (1973), p. 294]. These provide information about the crystal (or molecular) structure. And if we want the crystal to radiate (a particular type of radiation) we need to know such rules. As we have seen with electrons, selection rules can be obtained for only regions of less than three dimensions (points, lines or planes) of the Brillouin zone, the ones with symmetry, thus a set of measure zero (sec. XI.1.e, p. 569). However for momentum near such a region, we expect that, except for weird and unlikely potentials (hopefully), forbidden lines, though they occur, would be weak. Thus comparing the strength of different lines should give information about the crystal space group.

The frequency (thus energy) of vibration is a function of momentum, and here we try to find restrictions on such functions (the functions depend on the potential and finding them for the entire Brillouin zone is a complicated, often not quite possible, calculation). The relation between the frequency and the wavelength (inverse momentum) is called a dispersion relation, showing the origin of the concept in the dispersion of white light by a crystal.

A wave, here of vibrations of atoms or nuclei, can oscillate parallel to the direction of propagation; these are *long*itudinal waves (they are a*long* this direction), like sound waves and are generally of (roughly) the same frequency, so are called longitudinal acoustic modes. Transverse modes are perpendicular to (cut across) the propagation direction. For short wavelengths, motions of atoms are similar to that of ions in an electromagnetic field of infrared or visible light, thus these are called optic modes. Of course, this classification is only a start; there can be, besides combinations, intermediate values as well as other types of vibrations, such as helical waves.

XI.4.a The dynamical matrix

The equation governing the system, Schrödinger's equation in the harmonic approximation — keeping only the second order term of the potential — leads for n particles, in a well-known way, to the equation for the frequencies $\nu(\underline{q})$,

$$\nu(\underline{q})^2\psi(\underline{q}) = D(\underline{q})\psi(\underline{q}), \qquad \text{(XI.4.a-1)}$$

with $\underline{q}$ the $3n$-dimensional momentum vector, the position vector in reciprocal space, and $\psi(\underline{q})$ the statefunction of the system. Here we might use the Born-von Kármán condition (sec. VII.4.c, p. 355). Matrix $D(\underline{q})$ describes the dynamics of the system, so is called the dynamical matrix. It is hermitian, and obeys

$$D(\underline{q}) = D(-\underline{q})^*, \qquad \text{(XI.4.a-2)}$$

$$D(\underline{q}+\underline{K}) = D(\underline{q}); \qquad \text{(XI.4.a-3)}$$

$\underline{K}$ is any reciprocal lattice vector.

Symmetry of the crystal leads to symmetries of $D(\underline{q})$, which we study to find those of the crystal, and properties of the vibrations. The equilibrium positions of particles are invariant under the crystal space group, however since they are vibrating, the set of instantaneous positions need not be.

Problem XI.4.a-1: Show that the dynamical matrix must be positive definite, else the crystal would not be stable [Streitwolf (1971), p. 190]. Why is it hermitian?

XI.4.b Transitions in the simple cubic lattice

The symmetry group at the center of the Brillouin zone of the simple cubic lattice (sec. XI.2.h.i, p. 579) is O_h. For infrared transitions the relevant part of the Hamiltonian is the dipole moment, which transforms like coordinates (sec. X.7.a, p. 554); for Raman scattering it is the symmetric polarizability tensor, which transforms like products of the coordinates with themselves (sec. X.7.b, p. 556). A nonzero transition requires that the representation of the final state be in the decomposition of the product of the representations of the operator and the initial state (sec. X.7, p. 552), equivalently that the decomposition of these three contain the scalar representation (pb. X.7.a-1, p. 555).

An example is the transition from a state of representation T_{2g} to A_{2u}. Operator r, a coordinate, transforms as T_{1u} and [Burns (1977), p. 403]

$$T_{1u} \otimes T_{2g} \Rightarrow A_{2u} \oplus E_u \oplus T_{2g} \oplus T_{1u}, \qquad \text{(XI.4.b-1)}$$

so the transition to this state is allowed — it is in the decomposition.

This is for a specific momentum value, but the transition is clearly allowed, not only for all nearby momenta, but for all except, perhaps, those lying on other symmetry positions. Likewise the forbidden transitions, while not strictly forbidden nearby, are, except in unusual cases, weak. Thus comparing the intensities (and their variation with $\underline{k}$) gives information about (the space group of) the crystal.

XI.5 ATOMS IN CRYSTALS AND ENERGY LEVEL SPLITTING

The states of an atom are governed by rotational symmetry. This is familiar for hydrogen, and though it may be less clear for more complicated atoms for these also rotational symmetry plays a key role. But the environment of a constituent of a crystal is not rotationally invariant — there is a symmetry, but less, being that of a finite subgroup of the improper rotation group. How does this affect the states of the constituents and especially their energy levels [Cotton (1990), p. 260; Cracknell (1975), p. 26, 48; Falicov (1966), p. 93; Janssen (1973), p. 88; Joshi (1982), p. 260; Joshua (1991), p. 70; Leech and Newman (1969), p. 84; Ludwig and Falter (1988), p. 191; Tinkham (1964), p. 67; Tsukerblatt (1994), p. 95, 255]? The methods of studying this are now (hopefully) familiar. Thus we merely mention the topic to record another example of the application of these groups, and to indicate where further information can be obtained.

XI.5.a Level splitting in crystals

What affect do atoms in a crystal have on each other? Here we only look at (or perhaps only use words referring to) states of electrons, ignoring those of nuclei. Interactions are divided (by us) into several types, and we write the Hamiltonian for an electron as

$$H = H_o + H_r + H_{so} + H_c, \qquad \text{(XI.5.a-1)}$$

where H_o is the part due to the nucleus and inner electrons, assumed spherically symmetric, H_r is that due to the repulsion by the other electrons of the atom, outside the inner shell, H_{so} gives spin-orbit coupling, and H_c the effect of the crystal — all other objects composing it except those in the atom being considered [Elliott and Dawber (1987), p. 202]. The largest part is taken as H_o — it would be difficult to consider the electron in an atom if it was not dominated by the nuclear force (minus the shielding due to the inner-shell electrons). We take Coulomb repulsion of electrons to dominate spin-orbit coupling.

Problem XI.5.a–1: The reader might wish to perform a calculation, or look one up, to see if this is (always) reasonable.

Problem XI.5.a–2: Does it matter if H_o is spherically symmetric?

XI.5.a.i *Some different possibilities*

Then we consider three limiting cases, depending on the strength of the crystal field,

I. Weak Crystal Field:	$H_c \ll H_{so} \ll H_r \ll H_o,$	
II. Intermediate Crystal Field:	$H_{so} \ll H_c \ll H_r \ll H_o,$	(XI.5.a.i–1)
III. Strong Crystal Field:	$H_{so} \ll H_r \ll H_c \ll H_o.$	

In case I, the crystal field is weak compared with spin-orbit coupling. The total angular momentum, J, of the ion considered is a (reasonably) good quantum number — its eigenstates are (approximately) the eigenstates of the Hamiltonian. For case II, both total orbital angular momentum, L, and total spin, S, of the object are good quantum numbers — the Hamiltonian eigenstates are (approximately) also eigenstates of them — but the total angular momentum, J, is not. The total angular momentum of the entire system is, of course, a constant, and labels the eigenstates of the Hamiltonian. But that of a particular ion need not be, as angular momentum can be exchanged with the crystal.

Problem XI.5.a.i–1: The strong crystal field provides a useful exercise (in explaining the physics). How does it differ from the other terms?

XI.5.a.ii *Steps in the decomposition*

Representations of the rotation group, to which statefunctions belong, are reducible under the point group so have to be reduced. The largest term results in a reduction of symmetry from the rotation group to that of some point group, the next reduces the symmetry further, to a smaller point group, and so on. Thus the computation is a series of steps, from the rotation group to the largest point group, to the next, until the group that the complete Hamiltonian is invariant under is reached. Different cases have different point groups in the series. While the final set of representations does not depend on the series — the rotation-group representation decomposes into the same set of irreducible representations of the invariance point group of the Hamiltonian — energy levels differ in ordering and energy values (although not in number). The first decomposition splits degenerate levels into a set of ones with different energy, the next splits each of these, but with less splitting, and so on. While the set of final levels is the same, they are differently arranged into shells, subshells, subsubshells, down to levels which still can be degenerate if there is remnant symmetry.

We neglect spin, taking the crystal field to act on the orbital motion of the electrons and only on spin through spin-orbit coupling which we ignore.

If the ordering of the steps is wrong, the levels will be jumbled, and important information lost.

Problem XI.5.a.ii–1: What information?

XI.5.b The octahedral group as an example

Consider an atom — its states are eigenstates of the rotation group — in a crystal, here one with octahedral symmetry [Cotton (1990), p. 260; Cracknell (1975), p. 27; Janssen (1973), p. 92; Joshi (1982), p. 260; Joshua (1991), p. 71; Tinkham (1964), p. 69; Tsukerblatt (1994), p. 111]. For the free atom, all states with the same angular momentum z-component are degenerate. What happens when it is in the crystal?

We wish to find the combinations of atomic orbitals that have the required symmetry. We have for $l = 0, 1$, the s and p states (ignoring normalization),

$$s = 1; \qquad \text{(XI.5.b–1)}$$

$$p_x = sin\theta cos\phi, \qquad \text{(XI.5.b–2)}$$

$$p_y = sin\theta sin\phi, \qquad \text{(XI.5.b–3)}$$

$$p_z = cos\theta, \qquad \text{(XI.5.b–4)}$$

the usual spherical harmonics [Mirman (1995a), sec. XI.4.c, p. 322], but taking the combinations that are real because these have the same form as coordinates — the relevant aspect.

XI.5.b.i *The intermediate case*

As usual, we find the octahedral group irreducible representations contained in each of those of the rotation group using characters [Mirman (1995a), sec. VII.6.a, p. 199], knowing those of the rotation group [Tsukerblatt (1994), p. 106]. Of course, the s state must form the identity representation of O; the three p states form the three-dimensional representation T_{1u}, so remain degenerate. However for larger rotation representations, there has to be splitting — there are no representations of group O with dimension larger then 3. Actually there can be splitting — symmetry requires the other states not split, but it cannot prevent "accidental" degeneracy. The five degenerate D ($L = 2$) states go into two representations, of two-fold and three-fold degeneracy,

$$D^2 = T_{1u} + T_{2g}, \tag{XI.5.b.i-1}$$

whose basis functions transform like

$$T_{1u} \Rightarrow (x^2 - y^2, 2z^2 - x^2 - y^2) \tag{XI.5.b.i-2}$$

$$T_{2g} \Rightarrow (yz, zx, xy). \tag{XI.5.b.i-3}$$

The seven-fold degenerate F ($L = 3$) states (capitalized since these are total angular momentum) decompose as

$$F^3 = A_{2u} + T_{1u} + T_{2g}, \tag{XI.5.b.i-4}$$

with states transforming like

$$A_{2u} \Rightarrow xyz, \tag{XI.5.b.i-5}$$

$$T_{1u} \Rightarrow (x, y, z), \tag{XI.5.b.i-6}$$

$$T_{2g} \Rightarrow (yz, zx, xy), \tag{XI.5.b.i-7}$$

giving the seven-fold level splitting into a singlet, plus two triply-degenerate levels. The same O representation, here T_{2g} occurs in the decomposition of different rotation representations. Representations of the crystal symmetry group do not completely specify the SO(3) representation they belong to.

With inversion symmetry, that is for group O_h, the S and P representations remain unsplit — the P states, having odd parity under inversion, belong to representation T_{1u}. The splitting of the D levels remains the same. For the F levels, the decomposition is now

$$F^3 = A_{2g} + T_{1g} + T_{2g}; \tag{XI.5.b.i-8}$$

the parity of the states is opposite from the previous decomposition. With spin, if the number of electrons is even, such as with filled shells, the analysis is unaffected. If odd, we have to consider the crystal double group (sec. I.8, p. 59), and the representations in the decomposition are the double-valued representations only — all degeneracies are even because of time-reversal symmetry as given by Kramer's theorem (sec. VIII.2.d, p. 402).

We can see why representations do, and do not, split [Tinkham (1964), p. 74]: take the ion to be surrounded by six ions at the vertices of a regular octahedron — this gives the required cubic symmetry. The axes pass through these ions. Then the three p orbitals, p_x, p_y, p_z, are clearly identical and so have the same energy. For the d orbitals, there are two sets, one giving the T_2 representation (yz, zx, xy), the other the E representation $(x^2 - y^2, 3z^2 - r^2)$. For E, the lobes point at the ions, for T_2 they point away. Thus the energies for these representations are different. But the states of each of the two are identical, and are mixed by cubic group operators.

Problem XI.5.b.i-1: Draw these orbitals, and using the diagram, explain the splitting [Tinkham (1964), p. 282]. How is the degeneracy reduced if the symmetry is, say if the distance between ions along one axis is different from that along the other two? If all three distances are different?

XI.5.b.ii *The case of a weak crystal field*

Since J is (reasonably well) fixed, we can consider only a single multiplet at a time. For an even number of electrons, the procedure is the same as the other cases, for an odd number we have to use the double group.

Thus for cubic symmetry, now with spin, levels $J = \frac{1}{2}$ and $J = \frac{3}{2}$ do not split, corresponding to the lack of splitting for the S and P states when spin is not considered [Janssen (1973), p. 161].

Problem XI.5.b.ii-1: Work this out explicitly.

XI.5.b.iii *Splitting of the octahedral levels*

This gives the degeneracy of the levels of an atom in an octahedrally-symmetric environment, which still has high symmetry; if the crystal is distorted (so its lattice changes) symmetry is reduced. Then there is further decrease in degeneracies. For small distortion the splitting is small, and can be used to study the crystal, for example finding (to a reasonable degree) what the lattice is, and how much it is distorted, and from what more symmetrical lattice.

Problem XI.5.b.iii-1: For the octahedral crystal distorted to tetragonal, trigonal and rhombic symmetries (with what groups?), find how each of the octahedral levels splits [Tsukerblatt (1994), p. 101].

XI.6 SPIN-ORBIT COUPLING

Electrons have spin, thus doubling the number of states within a band — the Pauli exclusion principle allows only one per state [Mirman (1995b), sec. 8.1.1, p. 147]. And electrons move, thus have orbital angular momentum. Crystal Hamiltonians couple the two angular momenta of an electron (as in an atom) changing energies (perhaps only slightly, but often interestingly). How is spin-orbit coupling included [Burns (1977), p. 326; Heine (1993), p. 293; Hochstrasser (1966), p. 265; Inui, Tanabe and Onodera (1990), p. 273; Janssen (1973), p. 161; Jones (1975), p. 256; Kibler (1997), p. 37; Streitwolf (1971), p. 171; Tinkham (1964), p. 295; Tsukerblatt (1994), p. 255]?

Space group operations now have to be applied to both space and spin parts of the statefunction, simultaneously — the crystal is not invariant under separate transformations of space and spin. Crystals are not invariant under, say, a rotation taking an electron into another with a different spin, unless spin is also acted on. Electrons (so holes) have spin $\frac{1}{2}$, thus a 2π rotation reverses the sign of the statefunction. While the point group for particles, with spin ignored, is a subgroup of the rotation group, that with spin (sec. I.8, p. 59) is a subgroup of SU(2), following from the homomorphism between O(3) and SU(2) [Mirman (1995a), sec. X.6, p. 287]. This point group is the (crystal) double group [Burns (1977), p. 326; Cornwell (1969), p. 191; Cornwell (1984), p. 147; Cotton (1990), p. 297; Falicov (1966), p. 103; Jansen and Boon (1967), p. 261; Janssen (1973), p. 155; Jones (1975), p. 256; Joshi (1982), p. 254; Joshua (1991), p. 74; Lax (1974), p. 250; Streitwolf (1971), p. 164; Tinkham (1964), p. 75; Tsukerblatt (1994), p. 245]. The corresponding space group is the double space group.

XI.6.a Spin-orbit coupling and removal of degeneracy

Schrödinger's (properly, Dirac's) equation with spin-orbit coupling is invariant, not under the single group — rotations, inversions and translations alone — but the double group [Cornwell (1969), p. 203; Joshua (1991), p. 74] since angular momentum must be transformed. With spin all states become doubly degenerate — up and down are equivalent — but spin-orbit coupling can break this degeneracy.

Representations of the double group are more numerous than those of the (single) point group. States are doubled, and degenerate states go into non-degenerate sets. We take, for a special position of the Brillouin zone, the direct product of a single-point-group representation with the spin-$\frac{1}{2}$ representation, and decompose it into a sum of representations. States in a point group representation, without spin, are degenerate, but not the different representations in the decomposition (although of course the states of each representation are), except for accidental degeneracy.

As an example, states descended from those of the p ($l = 1$) representation of the rotation group — three, which are degenerate — form, when spin is considered, two sets for which the total angular momentum, j, is $l \pm \frac{1}{2}$. One is four-fold degenerate with $j = \frac{3}{2}$, the other two-fold degenerate with $j = \frac{1}{2}$. Thus the set of three degenerate states becomes two nondegenerate sets, each consisting of states co-degenerate.

The Hamiltonian with spin-orbit coupling contains an extra term, this one, hence its eigenvalues, the energies, generally change when the term is included (for s-like states, from $l = 0$, the expectation value of the coupling is, of course, zero).

The procedure for studying a crystal including the spin of its constituents is essentially the same as for no spin, with this additional effect. Thus we merely discuss a brief example.

XI.6.b Cubic symmetry

Cubic symmetry provides a model [Cornwell (1969), p. 216; Joshua (1991), p. 76]. For example consider the representation with total orbital angular momentum $L = 3$, and multiplicity 7, the ^4F term (the F denotes $L = 3$, the superscript the multiplicity due to spin, which means that the spin is $\frac{3}{2}$). Including spin gives a direct product of rotation group representations $D^{\frac{3}{2}}$ and D^3. Using the decomposition (with the degeneracy in parentheses)

$$D^3 = A_2(1) \oplus T_{1u}(3) \oplus T_{2u}(3), \qquad \text{(XI.6.b-1)}$$

we get that the A representation splits into four (it has no choice), while the T representations, both of twelve states when spin is included, go each into four representations, two four-fold degenerate, the other two two-fold degenerate.

Computation of the level splitting requires knowledge of the state-functions, thus of the crystal potential, and is specific to the crystal, so not relevant here.

XI.7 MOLECULAR ORBITALS

In principle finding the eigenstates of the Hamiltonian of a molecule is the same as finding them for other Hamiltonians, like those that we have considered. These states have the symmetry of the molecule. Of course computationally it can be very difficult, and various methods have been developed to solve it. Here we only illustrate the role of group theory, and see whether there are differences from the role it plays elsewhere, essentially not many [Bhagavantam and Venkatarayudu (1951), p. 190; Bishop (1993), p. 197; Burns (1977), p. 199; Chen (1989), p. 405; Cotton (1990), p. 133; Evarestov and Smirnov (1993), p. 131; Falicov (1966), p. 133; Harris and Bertolucci (1989), p. 225; Harter (1993), p. 674; Heine (1993), p. 206; Hochstrasser (1966), p. 174; Inui, Tanabe and Onodera (1990), p. 183; Kettle (1995); Ladd (1989), p. 223; Lax (1974), p. 390; Lomont (1961), p. 190; Ludwig and Falter (1988), p. 159; Tinkham (1964), p. 213; Tsukerblatt (1994), p. 199; Wherrett (1986), p. 152]. One essential question is under what conditions do atoms actually bind into molecules, and when do they not?

The state of an electron is determined by its interaction with the nuclei and with other electrons, presenting a quite complicated problem. However we ignore all these, replacing the interactions by an average potential due to the nuclei and all other electrons. We do not consider actual calculations, yet even for what little we do consider approximations have to be used, at least to provide a language in which to discuss this. First we want electron statefunctions; we take the nuclei as fixed. Thus the symmetry that we use is that of the molecule with the nuclei at their equilibrium positions; vibrations (which must occur) reduce the symmetry electrons see (except, perhaps, at certain points of times).

Problem XI.7-1: It is useful to state all these approximations, but for our limited considerations not all need be necessary (and we may have included others that we do not state explicitly). In reading this it would be helpful to check where, and to what extent, each approximation enters, whether all do, and whether there are others that are hidden.

XI.7.a Relating molecular orbitals and atomic orbitals

Hamiltonian eigenstates found (or at least mentioned) here are called molecular orbitals. We have (perhaps only some) information that we can use to find them. A molecule consists of atoms, and we know (we hope) their statefunctions, thus we write molecular orbitals as sums of atomic statefunctions; this is the (approximation) method of linear combination of atomic orbitals — molecular orbitals (LCAO—MO) [Bishop (1993), p. 201; Cotton (1990), p. 134; Cracknell (1975), p. 49; Harris and Bertolucci (1989), p. 245; Jones (1975), p. 221; Ladd (1989), p. 226; Tsukerblatt (1994), p. 199; Wherrett (1986), p. 154]. Finding the coefficients in these sums is not trivial, but they are at least related by the molecular symmetry, and this is our (only) interest here.

Group basis states, at least for groups we consider, form complete sets — any (well-behaved) function of the variables on which the group acts can be expanded in terms of them. The appropriate choice of a set depends on the problem — thus if there is spherical symmetry, we use the basis states of the rotation group for these are relevant states to use to find the eigenfunctions of the Hamiltonian; if we used the basis states of the translation group, the eigenfunctions would be infinite sums, at best much less convenient.

In a molecule (or crystal) states of an atom belong to the rotation group, but these do not have the symmetry of the molecule. So to study them we need basis states of the molecular symmetry group, but we have those of the rotation group, thus the question is how are they related? States of an atom are distorted by its environment, so the eigenfunctions of the Hamiltonian of the molecule are their sums. We want these sums, but that they are such sums is alone helpful and informative, we want therefore to include in our expressions for states both the symmetry of the molecule (or crystal), and that the states come from the distortion of rotational ones.

The procedure is standard. An irreducible representation of the rotation group is a reducible representation of the molecular (rotation sub)group, and we reduce it. The questions then are what is the symmetry group of the molecule, and what representations of it are interesting, and why [Elliott and Dawber (1987), p. 282; Falicov (1966), p. 133]?

To find the Hamiltonian eigenstates, we have to obtain its matrix elements for a (reasonably) complete set of states, and diagonalize the matrix. The set of states that we use are the atomic orbitals, and the matrix elements are integrals of the Hamiltonian taken between them. Finding these integrals is usually the main part of calculating the statefunctions. However this is chemistry, so we ignore it, and simply assume that we have a set of values, and only have to diagonalize the matrix. It is here that group theory and symmetry are used to split a large matrix into smaller blocks, each for an irreducible representation; it is these blocks that have to be diagonalized, which is much easier.

XI.7.b The types of orbitals we consider

The states of an atom are angular momentum eigenstates, thus can be labeled by the total orbital angular momentum quantum number (ignoring spin-orbit coupling); the $l = 0$ states (s states), the $l = 1$ (p states), and so on. Here we regard molecular orbitals to be combinations of atomic states of only a single l value. A sum of s states of an atom is called a σ state of the molecule, that of p states is a π state, and so on. We consider σ and π states separately [Bishop (1993), p. 203; Burns (1977), p. 208; Kettle (1995), p. 227; Lax (1974), p. 397; Ludwig and Falter (1988), p. 163; Tsukerblatt (1994), p. 181]; the calculation of one set ignores the presence of the other (and of all other angular momentum states). In fact, we are mainly interested in the π electrons since these are the ones most involved in bonding the atoms and in chemical reactions. The σ electrons, as is implied by their originating from s states, have statefunctions that tend to be (reasonably) spherically symmetric — the states can have greater symmetry than the molecule, but not less.

Problem XI.7.b-1: Why, and in what sense?

XI.7.c Benzene

An example is the benzene molecule, C_6H_6 (pb. I.6.a.ii-7, p. 31; pb. X.5.c-7, p. 545), a planar molecule with six carbon atoms at the corners of a hexagon [Altmann (1994), p. 39; Bishop (1993), p. 206; Burns (1977), p. 233; Cotton (1990), p. 142; Heine (1993), p. 213; Hochstrasser (1966), p. 182, 217; Inui, Tanabe and Onodera (1990), p. 189; Lax (1974), p. 391; Lomont (1961), p. 195; Tinkham (1964), p. 228; Tsukerblatt (1994), p. 206; Wherrett (1986), p. 155]. Each contributes one π electron. Their molecular orbitals are constructed from six $2p_x$ orbitals, one from each carbon — these have principle quantum number 2, and since $l = 1$ there are three p states, the subscript indicating the ones used. The symmetry group is D_{6h} ($= D_6 \otimes C_i$), however we only need irreducible representations of subgroup D_6, since those of C_i are one-dimensional.

The six atomic orbitals form reducible representation Γ^{AO} of D_{6h}, which on reduction gives (the superscripts are the representation labels)

$$\Gamma^{AO} = \Gamma^{A_2} + \Gamma^{B_2} + \Gamma^{E_1} + \Gamma^{E_2}. \qquad \text{(XI.7.c-1)}$$

From this we find the coefficients for the reduction, so the sums of atomic orbitals forming the molecular orbitals that are the basis states of these representations. To obtain the energies, we compute the matrix elements of the Hamiltonian between these states. One important, and quite helpful, fact is that matrix elements between states of different representations are zero — the Hamiltonian transforms under the scalar representation of the symmetry group, by definition of the symmetry group.

Problem XI.7.c-1: Verify the decomposition.

Problem XI.7.c-2: Give the character of each class of D_{6h} for the representation formed by these six π orbitals. Generalize to a cyclic $(CH)_n$ molecule and show that there are n π molecular orbitals, belonging to each of the n representations of C_n [Cotton (1990), p. 144].

XI.7.d Bonding and antibonding states

Atoms are most likely to bond in a molecule if their probability densities (largely) overlap, with states thus called bonding orbitals, and resist bonding if there is small overlap, so antibonding orbitals [Cotton (1990), p. 138; Harris and Bertolucci (1989), p. 257; Kettle (1995); Wherrett (1986), p. 160]. Overlapping densities result in stronger bonding because the electrons are more likely to be between the two nuclei, where they are attracted to both, and attract both. It is the attraction between electrons and nuclei that results in the atoms being bound. Electrons that are more likely to be close to one nucleus than to both are held to that, so do not interact greatly with the other atom, giving little attraction. Then mutual repulsion of nuclei dominates, the atoms repel, so the orbitals are antibonding.

Statefunctions for bonding then should have antinodes along the lines joining atoms. These lines are permuted when identical atoms are, that is by the symmetry group. So the statefunctions must be basis states such that the set of their antinodes is invariant under the symmetry group. This picks out the states that give bonding.

XI.7.d.i *Tetrahedral carbon*

Simple, but very important, illustrations are given by molecules of carbon. For methane, CH_4 (pb. X.5.c-5, p. 544), the atoms lie on the vertices of a tetrahedron [Bishop (1993), p. 225; Burns (1977), p. 202; Lax (1974), p. 393; Tinkham (1964), p. 87; Wherrett (1986), p. 167], so exemplify the general class of tetrahedral molecules [Cotton (1990), p. 209; Evarestov and Smirnov (1993), p. 136; Kettle (1995), p. 171; Tsukerblatt (1994), p. 203]. The electronic configuration of the carbon ground state is $1s^2 2s^2 2p^2$, with the coefficients the principle quantum number, and the superscripts showing the number of electrons in each state. We ignore the $1s$ shell in considering bonding, and then have four states, s, p_x, p_y, p_z, whose combinations give the required molecular states. Inscribing the tetrahedron in a cube, and labeling the statefunctions with the direction numbers of the major lobes with respect to the axes of the cube, we obtain the states with the required symmetry,

$$\psi_{111} = \frac{1}{2}(s + p_x + p_y + p_z), \quad \text{(XI.7.d.i-1)}$$

$$\psi_{1-1-1} = \frac{1}{2}(s + p_x - p_y - p_z), \quad \text{(XI.7.d.i-2)}$$

$$\psi_{-11-1} = \frac{1}{2}(s - p_x + p_y - p_z), \quad \text{(XI.7.d.i-3)}$$

$$\psi_{-1-11} = \frac{1}{2}(s - p_x - p_y + p_z), \qquad \text{(XI.7.d.i-4)}$$

which can be seen to be orthonormal and basis states of tetrahedral group T_d (sec. V.2.f, p. 255). The representation, however, is reducible, and it is necessary to decompose it.

Along the (111) direction,

$$cos\theta = \frac{1}{\sqrt{3}}, \quad sin\theta = \sqrt{\frac{2}{3}}, \quad cos\phi = sin\phi = \frac{1}{\sqrt{2}}, \qquad \text{(XI.7.d.i-5)}$$

so, with $s = 1$,

$$\psi_{111}(111) = 2, \qquad \text{(XI.7.d.i-6)}$$

as must be true for all equivalent directions.

Problem XI.7.d.i-1: Check that these states are correct. Show that they are orthonormal and basis states of T_d.

Problem XI.7.d.i-2: It can be checked that no other combinations of these four states give a higher bond strength. Tetrahedral bonds of carbon are quite common (thus important), and this is the reason.

XI.7.d.ii *Trigonal carbon*

Carbon can also have trigonal bonds [Tinkham (1964), p. 88]. For this there are three equivalent coplanar bonds separated by $2\pi/3$. The states, which are clearly orthonormal and mixed by D_{3h}, are

$$\psi_1 = \frac{1}{\sqrt{3}}(s + \sqrt{2}p_x), \quad \psi_{2,3} = \frac{1}{\sqrt{3}}s - \frac{1}{\sqrt{6}}p_x \pm \frac{1}{\sqrt{2}}p_y. \qquad \text{(XI.7.d.ii-1)}$$

The bond strengths along the relevant directions turn out to be (once the potential is known),

$$\psi_1(\theta = \frac{\pi}{2}, \phi = 0) = 1.99, \qquad \text{(XI.7.d.ii-2)}$$

which is almost the same as that of the tetrahedral bonds. Perpendicular to the plane ($cos\theta = 1$), the only statefunction is p_z with a bond strength $\sqrt{3} = 1.73$ — much weaker.

An example with this configuration is graphite, and as these results imply, it consists of sheets with strong internal bonding, but with weaker ones between layers.

Problem XI.7.d.ii-1: Check these statements.

Problem XI.7.d.ii-2: Why is diamond (sec. III.5, p. 159) so much harder than graphite?

XI.8 CHANGE OF PHASE

Substances have various phases, often with different symmetries, indeed many are defined by their symmetry, magnetic materials for example — some solids above a certain temperature (the transition temperature) are nonmagnetic, below magnetically ordered. In both cases the object is a crystal — for both it is invariant under transformations — though with differences of (aspects of) crystal symmetry. Does one type tell about the other; how are these related? Why? Does the crystal symmetry above the transition temperature restrict that below, and conversely; what classes of structures in each region are allowed by those in the other?

The theory of phase transitions provides an example of the use of group theory in classical physics, another is that of tensor operators in crystals (chap. IX, p. 467), suggesting, not unreasonably, some relationship. While most applications involve quantum mechanics, as we see below these (seem to?) involve only classical concepts and classical properties thus are of added interest in showing that the use of group theory, and symmetry, extends (much?) beyond quantum mechanics.

The theory is presented as an illustration of the use of group theory in another context. That it has weaknesses is well known [Tolédano and Tolédano (1987), p. vii], yet it is still (often) helpful in understanding changes of phase; failings are not greatly relevant to our purpose, which is not the study of transitions, but of illuminating and developing aspects of group theory, and properties of materials, in a reasonable, if imperfect, context.

XI.8.a First and second order phase transitions

Phase transitions [Bruce and Cowley (1981), p. 2; Cracknell (1975), p. 157; Evarestov and Smirnov (1993), p. 205; Inui, Tanabe and Onodera (1990), p. 316; Kocinski (1983); Kocinski (1990); Lyubarskii (1960), p. 121; Megaw (1973), p. 434; Tolédano and Tolédano (1987)] are (limited here to) those of second order [Landau and Lifshitz (1967), p. 446]. What physically is the difference between first and second order phase transitions, and why are we here so limited?

A first-order transition involves a change of state in which the arrangements of particles in the two phases are unrelated, or for which one has (average) positions fixed, as in a solid, the other not, as for a liquid. It requires energy to move atoms from one set of positions to the other, or to free them — as for ice going to water and water to steam. The states are different but coexist at a single temperature. Energy is absorbed or given off in the change of state. In these transitions, densities (or other relevant parameters) change discontinuously at T_C, the critical temperature (at which the phase change occurs), with absorption or release of latent heat. Ice at its melting point has a smaller density than water at the same temperature, and it takes heat to melt ice. Steam has lower density than water at the boiling point; again heat is required to boil water.

For second-order (correctly continuous) transitions, there is a discontinuous change, not in quantities like density, but in their derivatives. So the specific heat, which we can think of as the rate of change of internal energy with temperature, changes discontinuously — energy is not needed to change the phase, but the effect of energy changes and discontinuously. For a continuous transition the states of the two phases are the same at the critical temperature, the functions describing the state vary smoothly in going from one to the other — the state at T_C is the limit of the states of the two sides [Tolédano and Tolédano (1987), p. 36].

Shifts in magnetic symmetry, for example, leave positions of atoms unchanged — thus the density does not vary abruptly at T_C — but orientations of spins can change, or can become random. Responses to magnetic fields then differ, and the derivatives of the response (say, the magnetic moment with respect to the field, or to temperature) are discontinuous. This is a continuous transition. As is typical for these transitions, there is a tensor that changes, such as the (magnetic) moment. So for the transition to a ferromagnetic phase, as the critical temperature is approached from above the spins become more and more aligned with each other. At critical temperature T_C, and below, they are completely aligned. No heat is needed to go from one state to the other, for at T_C they are the same. The specific heat also varies smoothly, but its derivative is discontinuous. Above energy put into the system causes (among other things) atoms to vibrate faster and also spins to oscillate. Below, the latter is (essentially) impossible so only the former contributes. Thus as a function of temperature the specific heat is different on the two sides — its derivative is discontinuous.

While the states are different for a first-order transition so their symmetries (if any) need not be related, for a continuous transition states are continuous, thus the lower-symmetry space group must be obtained from the one with higher symmetry by removing some elements — it is given by a subgroup. But existence of several subgroups allows different transitions. There is no alteration of positions of (sets of) atoms for these phase changes (in principle) thus the crystal system does not change, so the translation subgroup of the space group is the same for both phases. However there are many space groups with the same translation subgroup — belonging to the same crystal system —

hence often many possibilities for these transitions. Group theory thus places limits on continuous phase transitions, but mere limits need not determine the state of lesser symmetry, knowing the greater one.

There can be first-order transitions in which the positions of the atoms remain (almost) the same, so for which there are relations between symmetry groups; however we ignore the distinction [Tolédano and Tolédano (1987), p. 166].

Problem XI.8.a-1: This discussion implies that we are considering a crystal, not a lattice. Why?

XI.8.b Limitations on the analysis

The discussion here, based on Landau's theory of symmetry change, is limited in many ways, and is presented to show how group theory can be useful in studying — some — changes of phase. This is this most widely known theory, but there are others [Kocinski (1983), p. 160; Kocinski (1990), p. 128]. Among the basic limitations are that only continuous transitions are studied. Also positions of the atoms change little (a vague term which must be considered in each case) in the transition [Cracknell (1975), p. 157]. Thus a transition from say bcc to hcp (hexagonal-closed-packed) structures (sec. XI.2.j, p. 590) is not considered. For these no restrictions are placed (at least here) on symmetry change — this is determined by the details of the potentials in which the objects find themselves. However this discussion can be relevant to transitions such as between a paramagnetic material and an antiferromagnetic one, for which related spin orientations change, but not positions of atoms. Also for magnetostrictive or electrostrictive (piezoelectric) distortions (sec. IX.6.a, p. 497) going with such transitions, positions can vary, but little. These need not make the present considerations irrelevant.

Another limitation, if it is, requires that, in the plane (or space) of the phase diagram, the points at which the transition takes place form a line (or surface), and that there are no isolated points. Thus transition points form a line in the temperature-pressure plane, for a phase diagram containing one. The symmetry changes as the line is crossed, with the group on one side a subgroup of that on the other. If there were an isolated point, moving in the plane along a line through it, gives a symmetry change there. However the system can also be taken from the same initial, to the same final, points along a different curve, on which there is no phase (thus symmetry) change, so the symmetry would be different at the two end points, that is path-dependent. We do not consider if this is impossible, but ignore it. Likewise, for higher-dimensional phase spaces, we ignore possibilities of isolated lines and surfaces.

XI.8.c Specifying the crystal thermodynamics

To study a system we need variables and functions for it. Functions are needed to give properties of the crystal, density, internal energy, and so on; others depend on these, and on temperature, pressure, magnetic values .... A density function $\rho(x,y,z)$ gives the probability $\rho(x,y,z)dV$ of, say, finding a particle in a small region dV at x,y,z, which, for this purely classical function, is equivalent to the number of particles there [Inui, Tanabe and Onodera (1990), p. 317; Lyubarskii (1960), p. 121; Tinkham (1964), p. 309], or it might give, say, the charge or magnetic moment density. It may have indices (here suppressed) as a crystal could consist of several kinds of particles, and also particles have spin and the densities for different directions vary independently — and these are of interest. With symmetry, ρ is invariant under a set of coordinate transformations (acting also on its indices); this set forms the space group of the crystal. The question we consider is how this space group for one phase limits that of the other.

And there are functions, like free energy, that are necessary to describe the thermodynamics of the object, these functions are thermodynamic potentials, and they, and their relationships to crystal variables and functions, temperature, density, ..., and how they depend on, and give, crystal symmetry, have to be specified. Thermodynamic potentials, which ultimately depend on the potential energy functions of the crystal constituents, are not uniquely given by symmetry, so knowing the larger symmetry need not fully determine states of lesser symmetry.

All variables are intensive, they are, say, per unit mass or volume, except for those like temperature that have no meaning as such, and like density that must be so by definition.

XI.8.d Equilibrium

Systems can be complicated, with variables like temperature space and time dependent. Then concepts of phase and transition may be meaningless. Thus it is here a fundamental assumption (or limitation) that the state is one of equilibrium. Such a state is different above and below the critical temperature, there is a phase change — the critical temperature is defined by this change. The equilibrium state would be the one with the lowest energy, E, except that might violate the second law of thermodynamics. Thus it is usually the state of lowest "free energy", the energy that is "free" after taking into account the entropy S (for a collection of oscillators, say, the lowest energy state, at any temperature, might be that for which all have the same energy — a very unlikely state). The free energy is

$$\Phi = E - TS. \tag{XI.8.d-1}$$

If there are additional variables, pressure, or magnetic fields in which the crystal is immersed, say, these are included, so more generally we refer to Φ as the thermodynamic potential. Specific functions are irrelevant here; we need only know that there is a function Φ whose minimum determines the equilibrium state. We have to find how this depends on the symmetry, and how the minimum varies with temperature T — especially at the critical temperature.

Thus, with p the pressure (or generally the set of relevant variables), we have functional $\Phi(\rho, T, p)$, the thermodynamic potential, a function of p and T and function ρ (called the density, which might be, say, that of average spin, or magnetic moment), and wish to find one variable as a function of the others, $\rho(T, p)$, such that Φ is a minimum. This function, ρ, and its group of symmetries, describes the crystal. The problem of relating phase transitions to the group structure of the crystal is that of finding how the symmetries of this function change as the temperature goes through the critical value.

The continuity of the transition is fundamental in these analyses, for the symmetry group of the lower-symmetry phase is assumed to be a subgroup of that of the higher one — at the transition symmetry elements are lost — as we expect for a continuous transition, but not (generally) otherwise.

This method is used to find the possible space groups of the lower-symmetry phase knowing the larger symmetry group. The reason for going from higher to lower is that the latter phase has the order parameter (such as magnetization) whose appearance is central to this analysis [Tolédano and Tolédano (1987), p. 40]. There are ways of going in the reverse direction — knowing the symmetry group of the lower phase, finding the possible symmetry groups of the higher ones [Kocinski (1990), p. 134].

XI.8.e How symmetry change is found

On one side of the transition the crystal has symmetry group G_o, on the other it has lesser symmetry with group G', a subgroup of G_o. The question is, given G_o, which of its subgroups can describe the crystal in the lower-symmetry state? For this analysis representations of space group G_o are studied to find which can go to representations of the smaller-symmetry group. Then we find the subgroups that have such representations. These are the possible groups describing the state of lesser symmetry. Obtaining the actual one in a particular case is a different problem.

Thus this method need not be definitive, for space groups can have large numbers of subgroups. What it can do is to eliminate a considerable fraction of these as possible symmetry groups of the lower-symmetry phase, greatly reducing the work of finding the actual group [Cracknell (1975), p. 180].

XI.8.f The order parameter

The system is more ordered on one side of T_c than the other. To describe this we introduce variable χ, the order parameter [Bruce and Cowley (1981), p. 13; Cracknell (1975), p. 181; Kocinski (1983), p. 2; Tolédano and Tolédano (1987), p. 8], such that it goes to zero at T_c; the order given by χ disappears on the other side, the higher symmetry one, so there $\chi = 0$. Thus χ is a central element in the description of the transition. For a continuous transition χ, a function of temperature T, approaches zero as a power (in this theory),

$$\chi \sim (T_c - T)^{\beta}, \qquad \text{(XI.8.f-1)}$$

where β is the critical exponent for this parameter. Thermodynamic potential Φ has a minimum as a function of χ, for $\chi = 0$ above T_c, and for $\chi \neq 0$, below. The order parameter may be a set so of higher dimension, for example when there can be distortions along different axes, giving different orderings [Bruce and Cowley (1981), p. 14]. For a second-order transition the state varies continuously through critical temperature T_c — density ρ as a function of T is continuous — although the symmetry changes.

XI.8.f.i *Magnetic ordering*

Ferromagnets for which the (domains of) spins are aligned below, random above give examples. Above T_c for a transition to a magnetically-ordered state, the solid has a symmetry group, since it is a crystal. But spins of atoms (or moments of domains) are random. Below all are parallel, so the symmetry group is less, the crystal no longer being invariant (on average) under rotations changing spins. In the less-symmetrical phase the crystal is invariant under operations acting on the positions of the atoms, with the spin-state disregarded (or averaged over). In the more-symmetrical phase the crystal is still invariant under these operations, but in addition there is the operation taking the spin of one atom to that of another, under which it is also invariant. For $T > T_c$ the average correlation between spins is zero (except over a short range — beyond that the magnetic moment averages to zero). As the temperature is lowered to the critical one spins become more and more correlated, with the range increasing, becoming infinite at the critical temperature. The distance for which there is correlation approaches infinity smoothly as $(T - T_C)$ becomes zero — the transition is second-order. Positions of the atoms, thus their symmetry group do not change, but the spins align, decreasing the symmetry — the symmetry group below T_C is a subgroup of that above. Below T_C spins are ordered giving an order parameter, but being random above, the parameter is zero, to which it goes smoothly.

For an antiferromagnetic crystal (sec. VIII.4.b.iii, p. 418) — the spins of neighboring atoms are opposite — the crystal is invariant under operations taking one atom to another, with a simultaneous spin flip. This last is the extra symmetry operation introduced by the change of phase.

XI.8.f.ii *Alloys*

Alloys ordered below T_C, but not above, provide other examples of a change of symmetry, and of order parameters [Inui, Tanabe and Onodera (1990), p. 322; Megaw (1973), p. 448]. The completely-ordered alloy of copper and zinc has a cubic lattice with the zinc atoms at the vertices, say, and copper at the centers of the cubic cells [Landau and Lifshitz (1967), p. 447]; it is a simple cubic Bravais lattice. If not completely ordered, the two types of atoms each have non-zero probabilities of being at any of the sites. When probabilities of the two types at each site are equal, all sites are equivalent, and the crystal gains a new symmetry (here of the average density, not the exact positions of the atoms). There is a new lattice vector, from the vertex to the center, and the lattice becomes a body-centered cubic. Once these probabilities are equal, lowering the temperature does not make them unequal. Of course there are fluctuations, there may be either atom at any site, and these fluctuations decrease with temperature. But the probabilities remain equal.

The order parameter is nonzero in the lower symmetry phase, zero in the more symmetric one and becomes zero at the transition temperature — the reason this temperature is the one at which the transition occurs. For the CuZn alloy, this parameter is chosen as

$$\chi = \frac{(w_c - w_z)}{(w_c + w_z)}; \qquad \text{(XI.8.f.ii-1)}$$

the w's are the probabilities of finding each type of atom at a given lattice site, so χ is zero if these they are equal. As temperature T approaches the transition, χ goes continuously to zero, then remains there. However $d\chi/dT$ is nonzero above and, since χ is constant, zero below T_c, so has a discontinuity. If the derivative approached zero continuously from above it could never equal zero, contradicting the existence of a transition. It would go to zero if the rate of change of the difference in probabilities of the occupation of sites by the two kinds of atoms were to decrease to zero as the probabilities became equal. That there is a transition means that in the expansion of this rate in terms of the temperature the first term is a constant.

This provides a simple example of the relationship between crystal symmetry and phase transitions. Let the crystal consists of two types of boxes, called white and black, in a row (corresponding to the vertices and center of the unit cell of a real crystal). Assume the lowest energy state has equal probability of each box containing either atom. On average the chance of finding either atom in a box is $\frac{1}{2}$. The boxes have states of different energy because extra energy is needed if neighboring boxes have more of one atom than another; in addition there is (ignored) energy due to atomic motion and their internal excitations.

At $T = 0$, the two types of atoms are uniformly distributed. Knowing the atom in one box we know that in its neighbors, thus in all. But each box has an equal chance of holding either atom, thus there is equal probability for the two distributions. If we raise the temperature — put in energy — then on average each of the two colored boxes has an equal chance of containing either atom, but locally there will be more of one type in the white boxes, and more of the other in the black, then for the lowest-energy state. Atoms in these boxes will be forced to higher energy states. Entropy, disorder, will increase because of these local irregularities. However at some temperature, for some crystals, entropy increase is greater if the probabilities of the two types of boxes become different. That means that at some temperature

$$E_{int} - TS_e > E_{int}' - TS_u, \qquad \text{(XI.8.f.ii-2)}$$

where E_{int} and E_{int}' are the internal energies in the two cases, the S's the entropies for equal and unequal probability. The probabilities being no longer equal, there will be a phase transition. Thus the question is why do some crystals give this faster increase in entropy for unequal probabilities? How is this related to the crystal structure?

XI.8.g Expansion of the density

Critical temperature T_c depends on several variables; at it the pressure at which there is equilibrium is p_c (p is the relevant variable, or set of variables, which we refer to as pressure, but for concreteness only), the density is $\rho_c(x, y, z)$ with symmetry (space) group G_o. At a nearby

T, on the lower symmetry side of the phase transition, the equilibrium pressure is p, and the density ρ. Since the transition is second-order (and here it is used) density varies continuously, so

$$\rho(x,y,z,T) = \rho_c(x,y,z) + \delta\rho(x,y,z,T), \qquad \text{(XI.8.g-1)}$$

with $\delta\rho$ going to zero as $T \Rightarrow T_c$ (T is usually suppressed). It is a limitation of the present analysis that $\delta\rho$ is small.

We use the name density, but this is essentially any order parameter. As an example for a cubic lattice which distorted becomes tetrahedral, reducing the symmetry, the density changes, but smoothly, as an infinitesimally small distortion of a length of the cube gives a tetrahedral lattice. We can regard this difference in length as an order parameter; it becomes zero at T_c and remains so. Or density might refer to spin, say.

Now $\delta\rho$ is not invariant under symmetry space group G_o of ρ (else this discussion is meaningless). Applying its elements g we get a set of functions $\delta\rho(g^{-1}r)$, one for each g, not all distinct as there is residual symmetry; r is the point with coordinates x, y, z. These functions form a closed set, a space, a representation space of G_o, in general reducible, and we consider its irreducible components. For each we choose an orthonormal set of basis vectors, $\phi_j^{*\underline{k},t,l}$; the representation is determined by the star of $\underline{k}$, indicated by $^*\underline{k}$, that of the little group, labeled by t, with j the basis vector of that representation. A representation can appear several times, the occurrence labeled l. These symbols are often suppressed, being indicated schematically by a single one. From the completeness of these basis vectors [Mirman (1995a), sec. VI.1.b, p. 171],

$$\delta\rho(r) = \sum_{jr} C_j^{*\underline{k},t,l}(p,T)\phi_j^{*\underline{k},t,l}, \qquad \text{(XI.8.g-2)}$$

defining the coefficients, the C's [Tolédano and Tolédano (1987), p. 31]. In the more symmetrical phase, ρ is invariant under G_o, so the only term that appears is the identity (scalar) representation. Since $\delta\rho$ goes to zero continuously at the transition, so do the C's. In this expansion the identity representation is not included, that transforming as ρ_c is already accounted for.

Under the group $\delta\rho$(r) is transformed, and we can regard the group elements as acting on the ϕ's, with the C's constant, or take the ϕ's as fixed basis vectors, and transform the C's. Since it is the C's that carry information about the crystal, we transform them. Having related the change of density to the group representations, we now analyze them to find what changes of symmetry are possible, that is which representations allow nonzero C's. And why do only some representations allow this?

Problem XI.8.g-1: The crystal can also be described by order parameter χ. Find the relationship between χ and ρ for various systems, say an alloy, a ferromagnetic material, an antiferromagnetic material, crystals with different space groups, and so on.

XI.8.h Physically irreducible representations

As ρ is real, so is $\delta\rho$. If a representation is nonreal, basis vector ϕ is complex. However nonreal representations have complex conjugates and we use as basis vector $\phi + \phi^*$. Such a sum of representations is mathematically reducible but is called physically irreducible [Inui, Tanabe and Onodera (1990), p. 317; Lyubarskii (1960), p. 123; Tolédano and Tolédano (1987), p. 30]. The expansion of $\delta\rho$(r) is then over the irreducible real, and the physically irreducible, representations. Terms in the sum for the latter have the form $C\phi + C^*\phi^*$, so are real; in defining physically irreducible representations we require that coefficient C^* of ϕ^* be the complex conjugate of coefficient C of ϕ. All representations considered here are either real and irreducible, or physically irreducible.

In general, representations reducible over complex numbers need not be over the reals (pb. VI.4.a-1, p. 297) [Mirman (1995a), pb. V.5.a-4, p. 163]. Thus, as is done here, the meaning of irreducibility must be given. Usually it means over complex numbers, but this emphasizes that there are situations with ambiguity which has to be removed.

XI.8.i Symmetry restrictions on the expansion coefficients

In the more symmetric phase, ρ, thus $\delta\rho$, are invariant under G_o, but under only subgroup G' in the lower-symmetry phase. As ϕ's are intermixed so are C's, so these also are basis vectors of G_o. However thermodynamic potential Φ being invariant under G_o can be a function only of invariant polynomials in the C's — these polynomials are products of basis vectors of different representations which are expanded into terms that are irreducible-representation basis vectors, and Φ can be a function only of terms transforming under the scalar representation. The sole linear invariant term is C_o^o. In the more symmetric region ρ is constant under G_o, so

$$\rho = C_o^o \phi^o. \qquad \text{(XI.8.i-1)}$$

The C's depend on ρ, thus on T, p, and point r, $C(r, T, p)$; the ϕ's are basis vectors so are independent of the crystal properties.

The thermodynamic potential is a function of the order parameter, thus of the C's. These provide the information about the symmetry group so we have to know how Φ depends on them. But Φ is essentially arbitrary, determined by the details of the crystal, thus we expand it in a Taylor series of powers of the C's, exhibiting them explicitly. Assumptions (limitations?) of the theory are that we gain sufficient information from just the first few terms, and for a continuous transition $\delta\rho$ is small so it is reasonable to expand Φ in terms of it, thus in terms of the C's. This gives

$$\Phi(\rho,T,p) = \Phi(\rho_c,T,p)_0 + \delta\rho\Phi_1(\rho_c,T,p) + (\delta\rho)^2\Phi_2(\rho_c,T,p) + \ldots, \tag{XI.8.i-2}$$

with $\delta\rho$ evaluated at T and p. Rewriting

$$\Phi(\rho,T,p) = \Phi(\rho_c,T,p)_0 + \sum_{jr} C_j^{*\underline{\mathbf{k}},t,l}\left(\partial\Phi(\rho_c,T,p)/\partial C_j^{*\underline{\mathbf{k}},t,l}\right)_{|c}$$

$$+\frac{1}{2}\sum_{jr}\sum_{ks} C_j^{*\underline{\mathbf{k}},t,l}C_i^{*\underline{\mathbf{k}}',s,h}\left(\partial^2\Phi(\rho_c,T,p)/\partial C_j^{*\underline{\mathbf{k}},t,l}\partial C_i^{*\underline{\mathbf{k}}',s,h}\right)_{|c} + \ldots, \tag{XI.8.i-3}$$

with the derivatives evaluated at the critical point, and the dependence on the basis functions shown explicitly. We do not consider if this converges, and if not, say due to singularities if there can be such, what effect nonconvergence has, so do not rule out the possibility of materials with interesting properties arising from an incorrectness of the expansion for them. This expansion is a fundamental postulate (step?) of the Landau theory [Bruce and Cowley (1981), p. 40].

At T_c, Φ must be a scalar under G_o — it is so on one side and varies smoothly — so can depend only on invariant polynomials in the C's. The only invariants formed from the C's, to second-order, are C_o^o and $\sum_t |C_j^{*\underline{\mathbf{k}},t,l}|^2$, the product of a representation with itself summed over all representations — the only product that gives an invariant (representations are physically irreducible, requiring the absolute square). At p_c, T_c,

$$\delta\rho = 0; \tag{XI.8.i-4}$$

the transition is continuous, so all C's are 0 there, except C_o^o. As Φ_1 is the first-order term, depending linearly on the C's,

$$\Phi_1(\rho_c,T,p) = C_o^o\Phi_1(\phi_o,T,p) = 0, \tag{XI.8.i-5}$$

while Φ_2 depends on them to second order, so

$$\Phi_2(\rho_c,T,p) = \sum |C_j^{*\underline{\mathbf{k}},t,l}|^2\Phi_2(\phi_o,T,p) \equiv \sum A_s(p,T,\phi_i^s)\sum |c_j^{*\underline{\mathbf{k}},t,l}|^2. \tag{XI.8.i-6}$$

This defines the A and c coefficients; since the term involving the c's is invariant, independent of the basis vector but not the representation, we transfer the functional dependence on the system, indicated by p and T, to functions of the representations only, the A's. This sum does not include the identity representation — that is in the first term. Term Φ_1 is zero because is it the derivative with respect to the scalar, and this does not appear in the expansion of $\delta\rho$. Since ρ is a function of the basis vectors, Φ_2 must also be, and we have written it as a sum of terms over functions of those of the different representations, each depending on only a single representation. This follows from it being of second-order, and that it depends only on $\sum |C_j^{*\mathbf{k},t,l}|^2$.

XI.8.j Implications of the requirement that Φ be a minimum

The minimum of Φ gives equilibrium (because it is so chosen); thus a change of the C's increases its value. How does requiring that these coefficients give Φ a minimum limit them, thus possible symmetry changes? On the higher-symmetry side all C's, except C_o^o, are 0. From the form of Φ_1 as a sum over derivatives, not including the identity representation, and as all $\delta\rho$'s are 0 at T_c,

$$\Phi_1(p_c, T_c) = 0, \tag{XI.8.j-1}$$

and that Φ be a minimum requires that all

$$A_s(p_c, T_c) \geq 0. \tag{XI.8.j-2}$$

To see the latter assume that the symmetry is prevented from changing with temperature. This would give a larger free energy (generally, thermodynamic function) than actually occurs, thus Φ_2 is a minimum (at the point for which $\delta\rho$ is zero, that with all C's zero), and as the A's are functions of orthogonal basis vectors, the condition applies to each. The second derivatives of Φ with respect to the C's — the A's — must be positive.

This determines the sign of the derivatives of Φ, so of the A's. Not all C's can be 0 because in some neighborhood of p_c, T_c, there is symmetry change, thus $\delta\rho(r)$ is nonzero; but Φ_2 is zero. So not all A's can be strictly positive in this neighborhood; at least one $A_j(p_c, T_c)$ is 0 (with $j \neq 0$). Thus in going through the phase transition some coefficients change sign. Take only one, denoted by A_1, to do so (more would mean that two transition lines in the PT plane cross giving an isolated point — which we explicitly do not consider). This gives the line in the PT

plane at which the transition takes place,

$$A_j(p_c, T_c) = 0. \tag{XI.8.j-3}$$

On the side where A_1 is positive, the symmetry is that of point p_c, T_c and all C's are 0, so on the other side

$$\rho(r) = \rho_c(r) + \sum C_j^{*\underline{k},t,l} \phi_j^{*\underline{k},t,l}. \tag{XI.8.j-4}$$

XI.8.k Conditions from necessity of terms being zero

We find that terms of the first and third orders shift the critical temperature, so their coefficients are 0; since the thermodynamic potential has a minimum at the critical point, its first derivative with respect to the order parameter (here only one) is zero, its second derivative positive. This gives the third-order term zero, the fourth-order positive, as is also true in the neighborhood of the critical point. Thus up to fifth degree there can be no third-power terms in C. Let us consider the effect of this.

XI.8.k.i *The third-order term must be zero*

The third-order term is a product of a coefficient, a function of temperature and pressure (and perhaps other variables), with an invariant formed from the C's. A zero coefficient would give a line of critical points in the PT plane, and there would be a transition only when that intersected the other such lines we have found, that is there would be transitions only at (excluded) isolated points. So only representations that give a vanishing third-order invariant are relevant [Lyubarskii (1960), p. 125; Tolédano and Tolédano (1987), p. 42]. This is the case for a magnetic transition in which the third-order terms are required to vanish by time-reversal symmetry, which changes the sign of the magnetic moment (sec. VIII.2, p. 394), here the order parameter, but must leave the potential invariant — its expansion can contain no odd-order terms, a condition on the allowed representations.

Now the $\delta\rho$'s are basis vectors of G_o, while Φ is a scalar — it transforms under the identity representation. So $(\delta\rho)^{*\underline{k},t,l}$ transforms under a reducible representation and the Φ_k's act as Clebsch-Gordan coefficients [Mirman (1995a), sec. XII.2, p. 340] which reduce it to the representation to which Φ belongs, the identity. If the $(\delta\rho)^{*\underline{k},t,l}$ does not appear, then this product does not contain the unit representation. That is Φ is a function of the representation basis vectors and (it is assumed) can be expanded in terms of them (for point groups this is true by the Peter-Weyl theorem, but the assumption may be less rigorous for space groups). In the expansion, if the reduction of a particular product does not contain the unit representation its coefficients must be zero. Conversely, if we know that a particular product does not appear then we can assume that its decomposition does not contain the unit representation (there is a slight hole in this argument as coefficients are not completely determined by symmetry so some might be "accidently" zero, or zero for other reasons, even if the decomposition does contain the unit representation). Since no cubic term appears, the only representations allowed are those whose symmetrized cube does not contain the unit representation. This condition can be written [Mirman (1995a), sec. VII.6.a, p. 199] in terms of the characters $\chi([D(g)]^3)$ of the symmetrized cube of representation D(g) of the group with order $|G_o|$,

$$\frac{1}{|G_o|}\sum_{g}\chi([D(g)]^3) = 0. \qquad \text{(XI.8.k.i-1)}$$

This is called the Landau condition (also the stability condition) [Inui, Tanabe and Onodera (1990), p. 321; Kocinski (1983), p. 12, 55; Kocinski (1990), p. 76; Tolédano and Tolédano (1987), p. 53].

By the same argument the fifth-order term is zero.

XI.8.k.ii *The fourth-order term*

If the fourth-order term [Tolédano and Tolédano (1987), p. 44] is positive definite in the more symmetric state, then the minimum of the thermodynamic potential is given by Φ_o, and its symmetry is G_o. The necessary and sufficient condition for the existence of a fourth-order invariant is that the symmetrized fourth power of the representation contain the identity representation of G_o at least once. This gives an equivalent equation to that of the third-order term, but with the fourth power, and now the requirement is that it not be zero.

XI.8.k.iii *The requirement of spatial homogeneity*

A crystal, by definition, is spatially homogeneous, providing an additional condition on possible transitions, this requirement from spatial homogeneity is the Lifshitz condition [Inui, Tanabe and Onodera (1990), p. 329; Kocinski (1983), p. 13, 57; Kocinski (1990), p. 76; Tolédano and Tolédano (1987), p. 110]. If there were an inhomogeneity then coefficients in the expansion of the density (eq. XI.8.g-1, p. 621) would be functions of space. We shall not discuss this here but simply state that the Lifshitz criterion gives a condition on the decomposition of the antisymmetric square of a representation.

These are the two conditions that determine which representations are allowed, and so which subgroups of the symmetry group are possible. They can then be explored in depth for the various groups.

XI.8.l Halving the symmetry always allows a transition

There is one case in which the transition is always possible, that in which G' has half the number of elements as G_0. For in this case G' is an invariant subgroup of G_0 with index 2, so the factor group G_0/G' is of order 2. It has two one-dimensional representations, the completely symmetric, with both elements represented by 1, and the antisymmetric, with the identity having matrix 1, the other element, consisting of the set of elements not belonging to G', matrix -1. The latter representation excludes odd-order invariants, and being one-dimensional, also excludes the construction of an antisymmetric square representation. Thus both conditions are met [Kocinski (1983), p. 17; Tolédano and Tolédano (1987), p. 53].

XI.8.m Active and passive representations

Only with crystals of certain symmetry types is it possible to have continuous phase transitions, as we see. The representations of G_0 for which it is possible are called active representations, or acceptable representations, the others passive representations [Inui, Tanabe and Onodera (1990), p. 322; Kocinski (1983), p. 16, 53; Kocinski (1990), p. 77; Lyubarskii (1960), p. 128; Tolédano and Tolédano (1987), p. 122]. The active ones satisfy both conditions, the passive ones fail to satisfy at least one.

Thus the symmetry groups for which this type of transition is possible (in this theory, to the approximations considered) are only those having such representations.

XI.9 CLASSICAL VIEWS, QUANTUM VIEWS, REALITY

Crystals are often pictured as classical. Symmetry groups are of operations on points, with crystals depicted using ball and stick models — completely classical. But actually the atoms of crystals are governed by quantum mechanics. Should this have an effect?

Basis states are usually the quantum-mechanical statefunctions; but we have not introduced quantum mechanics, yet still have basis states.

These arise, even in classical physics, whenever there are functions that are transformed by the group operations. There are none if we consider only points making up the crystal. But we also want functions over the crystal, the density, order parameter, thermodynamic potential, and so forth. These are (necessarily) sums of (products of) basis states.

Atoms themselves are described by statefunctions; how are these related to such parameters as density? They are products of statefunctions with their complex conjugates. Thus if a statefunction belongs to an irreducible representation, the density may not. A statefunction that belongs to representation l, gives a density belonging to all representations from 0 to $2l$.

Problem XI.9-1: How is the order parameter related to the statefunctions of the crystal?

Problem XI.9-2: Would we get further restrictions on phase transitions if we used, instead of classical variables like density, quantum-mechanical ones like statefunctions? Is it possible to study phase transitions classically or is quantum mechanics needed for a full description?

Problem XI.9-3: Many of the properties of crystals, and objects in them, discussed here depend on their lattice types, and Wigner-Seitz cells. But most lattices have more than one category of Wigner-Seitz cell (chap. IV, p. 185). The dimensions of the lattice can be changed, for example by applying pressure, maybe magnetic or electric fields say (magnetostriction, electrostriction [Nowick (1995), p. 7, 12]), and perhaps some crystals can form lattices of the same type but with different relative dimensions (and if not it should be shown why not). Thus, say, the history of the crystal and its current environment can affect its category of Wigner-Seitz cell. It would be interesting to review the properties discussed here to see how they would change if the category does, as for the suggested conditions. It might be particularly stimulating to study this for regions in which the category is changing, say to find how these properties change with pressure when a small change of that gives a discrete change of Wigner-Seitz cell.

Problem XI.9-4: Crystals are nice because they are often so simple. But, like all of physics, they are idealizations. How can they be extended to real systems, to ones for which the assumptions on which the analyses of crystals are based, (start to) break down [Hargittai and Hargittai (1987), p. 395]? They are infinite in extent, and translationally invariant. But real objects have edges and faces. How can they be classified? Consider a lattice of each symmetry class, and a crystal of each space group built on it. This is infinite. Suppose we cut it to make it finite. The faces can be arbitrary, they can be irregular say. But let us do this so the resultant object has an integral number of unit cells. However there are many different choices of unit cells, so perhaps the faces are not unique. Classify, for each crystal class, the ways of putting on faces. How does this classification depend on the crystal class? One interesting question is how small the object can be, and still have (crystalline) symmetry — an aspect of mesoscale physics. So what is the minimum number of unit cells we can obtain by putting in faces? Of course the answer is one. But suppose we want the resultant object to belong to a definite crystal class, and definitely not belong to any others. What is the smallest number of unit cells that allow this? How does this number depend on the crystal class? How does it depend on the form of the unit cell (and so on what faces are put in)? Is there a relationship between these two answers? Why? We have developed many results based on the crystal symmetry, vibrational modes, electron states, and so on. How are these changed, and left invariant, by insertion of faces? How do the answers to these questions depend on the crystal class and the unit cells chosen? Why?

Problem XI.9-5: Throughout questions have been raised about physical objects, their symmetries, their properties, those of the objects of which they consist. Some have been answered in (hopefully reasonable) depth, others overlooked or ignored. It is useful, especially in reviewing this material, to consider which questions have been sufficiently discussed, which only somewhat, which overlooked, and why these choices have, accidently or purposely, been made. Answers should be found for those not treated adequately. Some, many perhaps, cannot be answered — the material here may be insufficient, the required knowledge of crystals, of physics, of group theory, of other parts of mathematics, may not exist. Or mathematics and physics may allow no definite answers. Understanding what the questions (of interest) are, which can be answered, which not, and why, serves as a spur, an inspiration, to investigation, to greater knowledge, to the development of methods useful (such as) here, and often in (seemingly) unrelated fields. The questions in this book, and in this section, emphasize that we are not dealing with reality, but only with a quite imperfect model of it. Hopefully as we have gone along our models have become more and more like the real world, although they have not gotten closer because it is impossible to get closer to infinity in a finite number of steps. But the material, the questions — and the errors, the gaps, the weaknesses — provide a start. Let us hope that the reader will understand that this is the purpose of the book, not a description of reality, but a start, and, most important, a prod to go further, to take the next small steps in the infinite journey to understanding.

Appendix A

Symbols and definitions

E Identity (operator)
ε_{ijk} completely antisymmetric symbol (pb. IX.2.a-2, p. 471)
$\underline{v}$ vector
v magnitude of vector v
$|\underline{v}|$ magnitude of a vector
$\hat{\underline{i}}$ unit vector
• dot (scalar) product of vectors
⊕ sum of sets
⊕ sum of representations
⊗ direct product (sec. III.2.e, p. 136)
∧ semi-direct product (sec. III.2.e, p. 136)
⊙ Kronecker product (sec. VI.7.h, p. 329)
⊐ subgroup $G \sqsupset L$ says that L is a subgroup of G
$\{R|t\}$ Seitz operator (sec. III.2.a, p. 134)
n_m screw axis (sec. III.4.a, p. 142), the rotation is $\frac{2\pi}{n}$, the translation $\frac{m}{n}$ of a unit cell.
$F = G/H$ factor group [Mirman (1995a), sec. IV.8.c, p. 138], where G is a group and H a subgroup, so

$$G = H \oplus g_1H \oplus \ldots \oplus g_kH, \tag{A-1}$$

where the g_i are the representatives, a set of elements of G not in H picked so the terms in this sum are distinct, and all elements of G appear in it, and once only.

Appendix B

The Point Groups

The Point Groups

The letters designating the point groups and those for their transformations are often the same. They are distinguished here by different fonts. First are listed the symbols for the transformations of the crystallographic point groups in both Schoenflies and International notations (sec. I.6.a.ii, p. 29):

Operation	Schoenflies	International
Identity	E	E
Rotation of $\frac{2\pi}{n}$ ($n = 1, 2, 3, 4, 6$)	C_n	n
Reflection, horizontal plane	σ_h	$/m$
Reflection, vertical plane	σ_v	m
Reflection, diagonal plane	σ_d	m
Inversion	I	I
Improper rotation of $\frac{2\pi}{n}$ (rotary-reflection)	S_n	
Rotary-inversion of $\frac{2\pi}{n}$		$\bar{n}$

Table B-1 gives the generators (with the identity understood) of each crystallographic point group [Yale (1988), p. 107]. Then in table B-2 is listed the crystallographic point groups [Bradley and Cracknell (1972), p. 34; Burns (1977), p. 377; Burns and Glazer (1990), p. 294; Fässler and Stiefel (1992), p. 221; Hahn (1989), p. 11; Heine (1993), p. 446; Inui, Tanabe and Onodera (1990), p. 362; Leech and Newman (1969), p. 14; Tinkham (1964), p. 61; Yale (1988), p. 112], giving both the International and the Schoenflies notations, the crystal systems to which they belong, and the space groups of which they are subgroups (or factor groups). Also listed for each crystal system are the conditions on the lattice sides, a, b, c and angles α, β, γ [Burns (1977), p. 373; Burns and Glazer (1990), p. 32; 289]. The coefficients on a symbol give the number of elements in its class. Elements appearing more than once belong to different classes; primes indicate different classes of two-fold axes, and subscripts the (other than principle) axis. These are in the Schoenflies notation, and for a few groups also in the International notation. Also listed is order $||G||$ of group G. The numbers of the space groups for which the point group is a factor group are at the end of the second row [Hahn (1987); Hahn (1989)].

Table B-1: CRYSTALLOGRAPHIC POINT GROUP GENERATORS

Triclinic	
C_1 (1)	E,
S_2 ($\bar{1}$)	$I(= C_i)$
Monoclinic	
C_2 (2)	C_2
C_{1h} (m)	σ_h
C_{2h} ($\frac{2}{m}$)	I, C_2
Orthorhombic	
D_2 (222)	C_2, C_{2y}
C_{2v} (mm2)	C_2, σ_v
D_{2h} ($\frac{222}{mmm}$)	I, C_{2y}, C_2
Tetragonal	
C_4 (4)	C_4
S_4 ($\bar{4}$)	S_4
C_{4h} ($\frac{4}{m}$)	I, C_4
D_4 (422)	C_{2y}, C_4
C_{4v} (4mm)	σ_v, C_4
D_{2d} ($\bar{4}$2m)	C_{2y}, S_4
D_{4h} ($\frac{422}{mmm}$)	I, C_{2y}, C_4
Trigonal (Rhombohedral)	
C_3 (3)	C_3
S_6 ($\bar{3}$)	I, C_3
D_3 (32)	C_{2y}, C_3
C_{3v} (3m)	σ_v, C_3
D_{3d} ($\bar{3}\frac{2}{m}$)	I, C_{2y}, C_3
Hexagonal	
C_6 (6)	C_2, C_3
C_{3h} ($\bar{6}$)	σ_h, C_3
C_{6h} ($\frac{6}{m}$)	I, C_2, C_3
D_6 (622)	C_2, C_{2y}, C_3
C_{6v} (6mm)	C_2, σ_v, C_3
D_{3h} ($\bar{6}$m2)	C_{2y}, σ_h, C_3
D_{6h} ($\frac{622}{mmm}$)	I, C_{2y}, C_2, C_3
Cubic	
T (23)	$C_2, C_3[111]$
T_h ($\frac{2}{m}\bar{3}$)	$I, C_2, C_3[111]$
O (432)	$C_4, C_3[111]$
T_d ($\bar{4}$3m)	$S_4, C_3[111]$
O_h ($\frac{4}{m}\bar{3}\frac{2}{m}$)	$I, C_4, C_3[111]$

Table B-2: THE CRYSTALLOGRAPHIC POINT GROUPS

Schoenflies notation / Elements Schoenflies	International / Elements International	Full International	Order $= \|\|G\|\|$ / Space groups
Triclinic	a, b, c distinct;	α, β, γ distinct;	
C_1	1	1	1
E			1
$S_2(C_i)$	$\bar{1}$	$\bar{1}$	2
E, I	E, I		2
Monoclinic	a, b, c distinct;	$\alpha = \beta = \frac{\pi}{2} \neq \gamma$	
C_2	2	2	2
E, C_2	$E, 2$		3 – 5
$C_{1h}(C_s)$	m	m	2
E, σ_h			6 – 9
C_{2h}	2/m	$\frac{2}{m}$	4
E, C_2, I, σ_h	$E, 2, I, /m$		10 – 15
Orthorhombic	a, b, c distinct;	$\alpha = \beta = \gamma = \frac{\pi}{2}$	
D_2	222	222	4
E, C_2, C_2', C_2'			16 – 24
C_{2v}	mm2	mm2	4
$E, C_2, \sigma_v, \sigma_v$			25 – 46
D_{2h}	mmm	$\frac{222}{mmm}$	8
E, C_2, C_2', C_2' $I, \sigma_h, \sigma_v, \sigma_v$			47 – 74
Tetragonal	$a = b \neq c$;	$\alpha = \beta = \gamma = \frac{\pi}{2}$	
C_4	4	4	4
$E, 2C_4, C_2$			75 – 80
S_4	$\bar{4}$	$\bar{4}$	4
$E, 2S_4, C_2$			81 – 82
C_{4h}	4/m	$\frac{4}{m}$	8
$E, 2C_4, C_2, I, 2S_4, \sigma_h$			83 – 88
D_4	422	422	8
$E, 2C_4, C_2, 2C_2', 2C_2''$			89 – 98
C_{4v}	4mm	4mm	8
$E, 2C_4, C_2, 2\sigma_v, 2\sigma_d$			99 – 110
D_{2d}	$\bar{4}$2m	$\bar{4}$2m	8
$E, C_2, 2C_2', 2\sigma_d, 2S_4$			111 – 122
D_{4h}	4/mmm	$\frac{422}{mmm}$	16
$E, 2C_4, C_2, 2C_2', 2C_2''$ $I, 2S_4, \sigma_h, 2\sigma_v, 2\sigma_d$			123 – 142

Schoenflies notation	International	Full International	Order $= \|\|G\|\|$
Elements Schoenflies	Elements International		Space groups
Trigonal			
(Rhombohedral)	$a = b = c;$	$\frac{\pi}{2} \neq \alpha = \beta = \gamma < \frac{2\pi}{3}$	
C_3	3	3	3
$E, 2C_3$			143 – 146
$S_6 (C_{3i})$	$\bar{3}$	$\bar{3}$	6
$E, 2C_3, I, 2S_6$			147 – 148
D_3	32	32	6
$E, 2C_3, 3C_2$			149 – 155
C_{3v}	3m	3m	6
$E, 2C_3, 3\sigma_v$			156 – 161
D_{3d}	$\bar{3}$m	$\bar{3}\frac{2}{m}$	12
$E, 2C_3, 3C_2, I, 2S_6, 3\sigma_v$			162 – 167
Hexagonal	$a = b \neq c;$	$\alpha = \beta = \frac{\pi}{2}, \gamma = \frac{2\pi}{3}$	
C_6	6	6	6
$E, 2C_6, 2C_3, C_2$			168 – 173
C_{3h}	$\bar{6}$	$\bar{6}$	6
$E, 2C_3, \sigma_h, 2S_3$			174
C_{6h}	6/m	$\frac{6}{m}$	12
$E, 2C_6, 2C_3, C_2, I, 2S_3, 2S_6, \sigma_h$			175 – 176
D_6	622	622	12
$E, 2C_6, 2C_3, C_2, 3C_2', 3C_2''$			177 – 182
C_{6v}	6mm	6mm	12
$E, 2C_6, 2C_3, C_2, 3\sigma_v, 3\sigma_d$			183 – 186
D_{3h}	$\bar{6}$m2	$\bar{6}$m2	12
$E, 2C_3, 3C_2, \sigma_h, 2S_3, 3\sigma_v$			187 – 190
D_{6h}	6/mmm	$\frac{622}{mmm}$	24
$E, 2C_6, 2C_3, C_2, 3C_2', 3C_2'', I, 2S_3, 2S_6, \sigma_h, 3\sigma_v, 3\sigma_d$			191 – 194

Schoenflies notation	International	Full International	Order $= \|\|G\|\|$
Elements Schoenflies	Elements International		Space groups
Cubic	$a = b = c$;	$\alpha = \beta = \gamma = \frac{\pi}{2}$	
T	23	23	12
$E, 8C_3, 3C_2$			195 – 199
T_h	$m\bar{3}$	$\frac{2}{m}\bar{3}$	24
$E, 8C_3, 3C_2,$ $I, 8S_6, 3\sigma_h$			200 – 206
O	432	432	24
$E, 8C_3, 3C_2$ $6C_2, 6C_4$			207 – 214
T_d	$\bar{4}3m$	$\bar{4}3m$	24
$E, 8C_3, 3C_2,$ $6\sigma_d, 6S_4$			215 – 220
O_h	$m\bar{3}m$	$\frac{4}{m}\bar{3}\frac{2}{m}$	48
$E, 8C_3, 3C_2$ $6C_2, 6C_4,$ $I, 8S_6, 3\sigma_h,$ $6\sigma_d, 6S_4$			221 – 230

Appendix C

Objects Invariant Under the Point Groups

As a way of illustrating the meaning of the point groups, a set of objects is given here that are invariant under them. These are two-dimensional projections of three-dimensional objects, so at times imagination is needed to fully understand how they behave under the group transformations, and why they are invariant. Also lines are drawn differently to clarify diagrams, but they should be the same for symmetry. Some of the lines are a little wavy. It is an interesting exercise to consider how that breaks the symmetry.

First is the most symmetric object, the cube, clearly invariant under the full cubic group, $O_h(\frac{4}{m}\bar{3}\frac{2}{m})$,

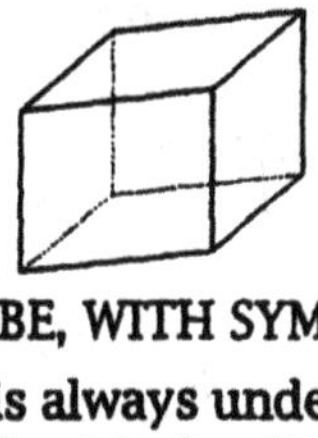

Figure C-1: THE CUBE, WITH SYMMETRY GROUP O_h.

The generators (the identity is always understood) of this group (tbl. B-1, p. 633) are $\mathcal{I}$, C_4, $C_3[111]$, with the direction of the three-fold axis, the diagonal of the cube, shown by its direction cosines (sec. II.4.a, p. 85). Its invariance under these three transformations shows that it is invariant under all group elements, these being products of the generators (by definition of generator). The symmetry of the cube can also be studied from other diagrams of it (sec. I.4.d.i, p. 17).

To reduce the symmetry, an arrow is placed on the corner of the cube,

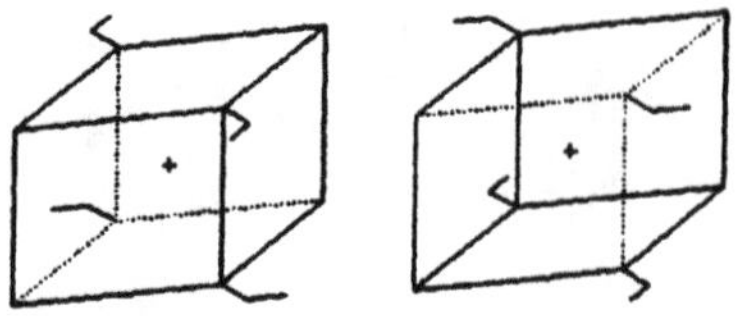

Figure C-2: BREAKING THE SYMMETRY TO O.

Arrows are placed only on half the corners to avoid excessive lines, allowing the reader to imagine them properly on all, which would give a figure invariant under all rotations of group O(432). Next to it is the object obtained by inversion through the (marked) center of the cube. The arrows are different. (The dotted lines on the left figure are the edges furthest away; these go into the nearest edges on the right.) Thus this object is an example of one with lower symmetry than the full cube, having symmetry group O(432), rather than the full $O_h(\frac{4}{m}\bar{3}\frac{2}{m})$, these differing in one generator, the inversion.

Another way of breaking this symmetry is with a question mark,

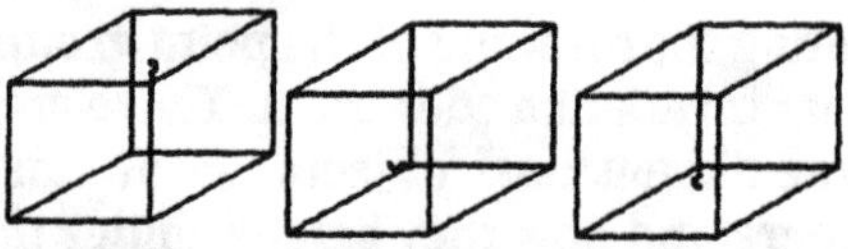

Figure C-3: A QUESTION MARK DESTROYS INVERSION SYMMETRY.

It is rotated about the center, in the middle, and inverted through the center, on the right. The question mark is at the same position in the last two, but its orientation is different. Putting question marks properly on all corners results in O(432) symmetry, but not $O_h(\frac{4}{m}\bar{3}\frac{2}{m})$.

Inversion symmetry is also broken as with the next cube that has a comma attached to the end of a diagonal, on the left. This is rotated by $\frac{2\pi}{3}$ and $\frac{4\pi}{3}$, and the resultant cubes superimposed to give the figure on its right, which is now invariant under these rotations about this diagonal. It is similarly rotated and superimposed to give the left bottom object invariant under all rotations of O. This is inverted (through its center), to give the bottom right one. These two figures are not identical as seen, say, from the set of three commas pointing out of the page, these twisted in different directions. Thus both of the bottom objects are invariant under O(432), but not under inversions, so not under $O_h(\frac{4}{m}\bar{3}\frac{2}{m})$. Superposing them, or removing the commas to give the cube in fig. C-1, p. 637, results in figures invariant under the full group,

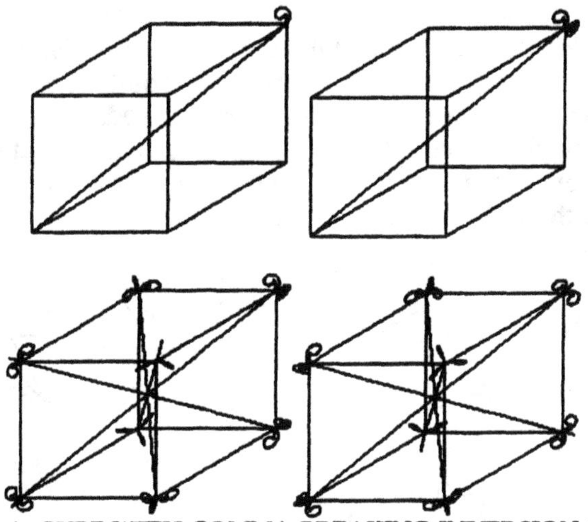

Figure C-4: CUBE WITH COMMA BREAKING INVERSION SYMMETRY.

These, simplification and complication, illustrate two ways of obtaining objects with greater symmetry, and undoubtedly crystals use both (a reason there are different properties, even with the same symmetry).

To study the tetrahedral group we start with an object, a tetrahedron which has the symmetry of the group; but of which? It is inscribed in a dotted cube to show its relationship to an object invariant under a group of which the tetrahedral groups are (proper) subgroups. This is also shown elsewhere (sec. I.7.a.ii, p. 52). Which operations leave it invariant? On its right it is inverted through the center of the cube and is clearly not invariant under inversion,

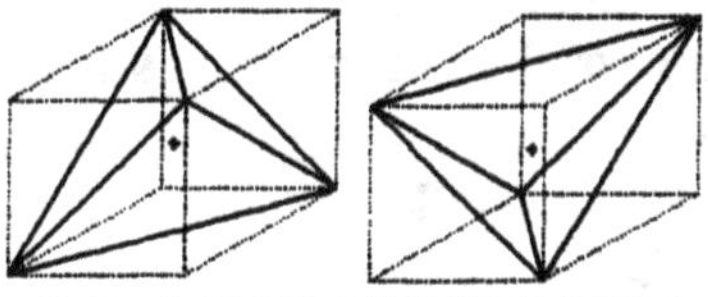

Figure C-5: A TETRAHEDRON IN A CUBE.

The next tetrahedron at top left has a three-fold axis (through the vertices, and extending slightly beyond the cube) indicated by a dashed line, and a two-fold axis (through the centers of the faces). The one to its right is obtained from it by $2\pi/3$ rotation about the three-fold axis; it is invariant under this (although the figure is not as not all axes have been drawn). Then next to it the leftmost one has been rotated by $\pi/4$ around the axis through the face, changing it. At the extreme right this rotation has been followed by a reflection in the plane of the rotation axis, giving a tetrahedron identical to the original one, the result of an

S_4 rotation-reflection, the product of the two operations. It is invariant under this product, but not the individual transformations. The bottom left figure is obtained from the first by a (C_2) rotation of π around the (drawn) axis through the face, leaving it unchanged. A diagonal mirror, bisecting the angle between two two-fold axes is indicated by a line (in the next figure) at the bottom of the cube (the mirror goes through this line, and also the top edge of the tetrahedron),

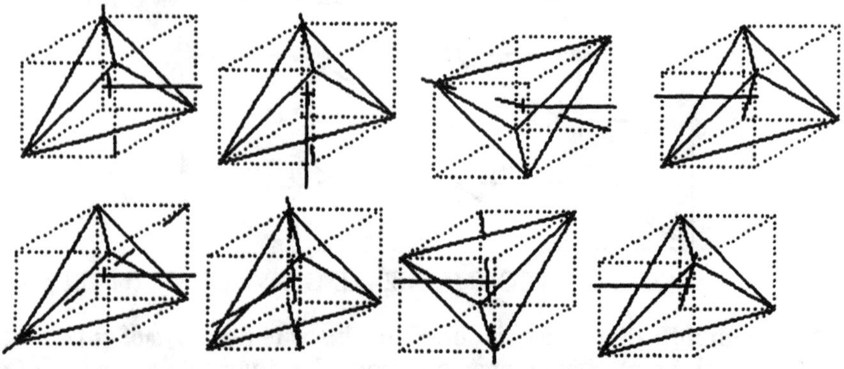

Figure C-6: OBJECT INVARIANT UNDER T_d.

The following figure shows the object after reflection in this mirror, and again except for axes, it is unchanged. The two right figures on the bottom show the object inverted through the center of the cube, on the left (changing it), and then on the right, after then being rotated through $\frac{\pi}{4}$ about the two-fold axis perpendicular to the face. The product of these two operations, inversion and rotation is rotary-inversion $\bar{4}$, and leaves the tetrahedron (although not the incompletely drawn figure) invariant, though each alone does not. Thus, with generators C_3 and S_4, the symmetry group of the tetrahedron is $T_d(\bar{4}3m)$, which does not contain the inversion.

For the next tetrahedral group

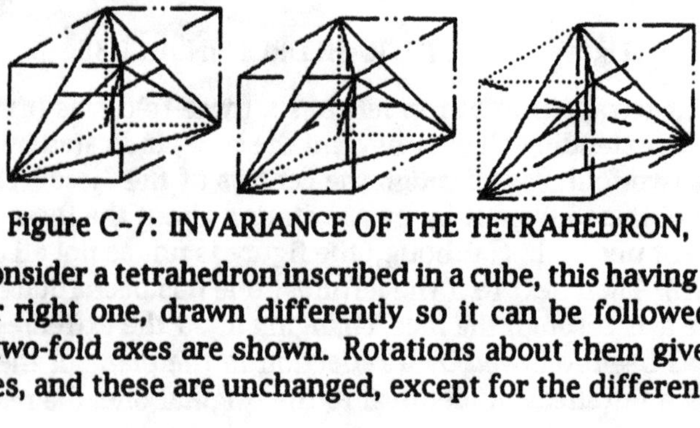

Figure C-7: INVARIANCE OF THE TETRAHEDRON,

we first consider a tetrahedron inscribed in a cube, this having one edge, the upper right one, drawn differently so it can be followed. Three-fold and two-fold axes are shown. Rotations about them give the next two figures, and these are unchanged, except for the differently drawn edges.

To consider inversions we start with the tetrahedron on the upper left

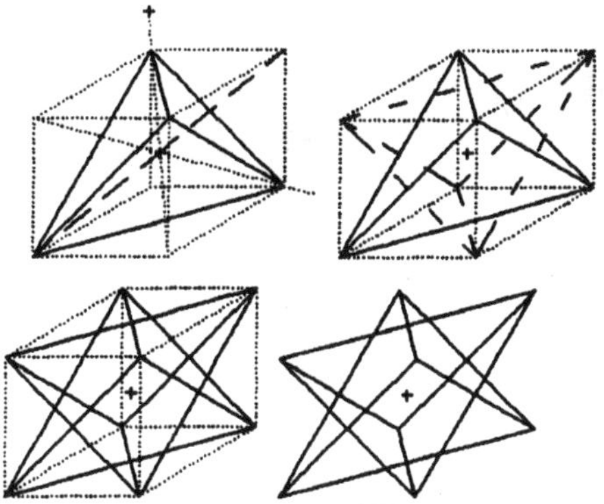

Figure C-8: OBJECTS INVARIANT UNDER T_h,

which has two of its three three-fold axes shown, as dotted lines (extending beyond the vertices of the cube), and a mirror, with a line lying on it dashed, symmetric with respect these two axes, going through the vertices; the plane of the mirror goes also through the opposite vertices of the cube, one coinciding with the vertex of the tetrahedron. The object is invariant under reflection through this mirror. These transformations lead to point group $T_h(\frac{2}{m}\bar{3})$, which contains the inversion and so is not a symmetry group [Elliott and Dawber (1987), p. 192; Inui, Tanabe and Onodera (1990), p. 172]. Then the tetrahedron is inverted through the center of the cube (a point which is not inside the tetrahedron), leaving it still invariant under the two-fold and three-fold rotations, and the two objects superposed, shown, three times for clarity, once with one tetrahedron dashed, once solid, and once without the cube, This combined object is invariant under $T_h(\frac{2}{m}\bar{3})$. But it is not convex, being essentially star-shaped (sec. IV.3, p. 196). From the way that it is inscribed in the cube, it is clear that its symmetry group is a subgroup of that of the cube. The cube with these two tetrahedra inscribed is a convex object with symmetry group $T_h(\frac{2}{m}\bar{3})$. However it also has an unusual feature. Unlike the other cases in which symmetry is reduced by making vertices different, here all are the same, but they are connected (by the lines giving the tetrahedron), and it is the set of connections (these lines) that reduce the symmetry.

There is one other tetrahedral group and to obtain it triple commas are placed at the vertices of the tetrahedron, at the left, so the resultant figure is not invariant under reflection, as it is without the commas, as seen by comparing it with that on its right, which has been reflected in this mirror (notice, say, the closest sets of commas),

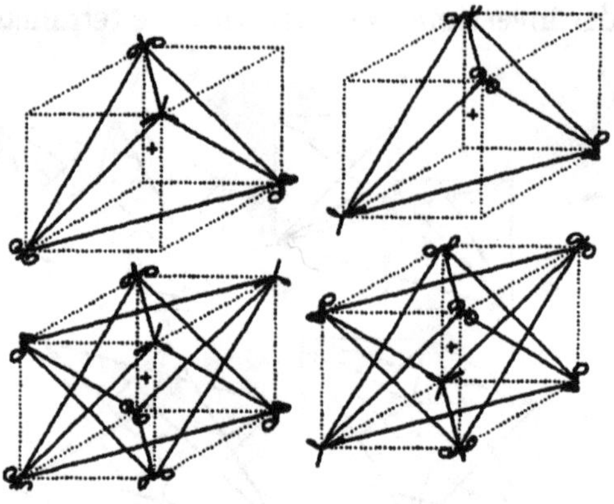

Figure C-9: TETRAHEDRON WITH SYMMETRY GROUP T.

Thus its symmetry group is T(23), and no larger one. The commas reduce the symmetry from $T_d(\bar{4}3m)$ to T(23). The bottom left object is constructed by superposing the top left one and that found by inverting the reflected (top right) object through the center of the cube (the point marked). With the commas, it is no longer inversion-invariant — the bottom figures are obtained from each other by inversion. So $T_h(\frac{2}{m}\bar{3})$ is not a symmetry group for it. The figure is clearly invariant under T(23) (only, for there are no other *generators* that it is also invariant under).

Reducing the symmetry further leads to the tetragonal system, and dihedral groups starting with $D_4(422)$, and a cube with commas, but now with two sides marked,

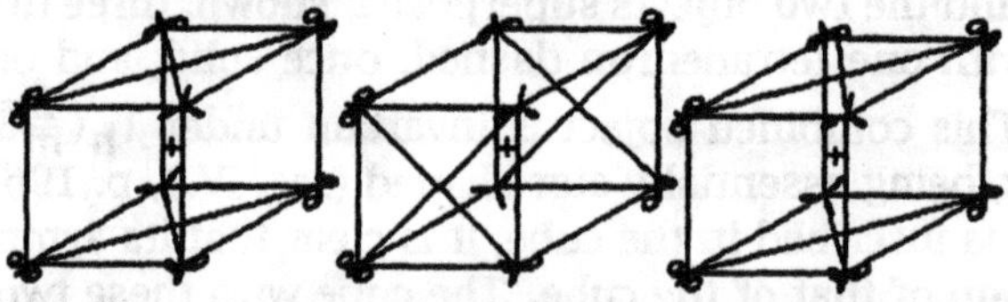

Figure C-10: A CUBE WITH COMMAS INVARIANT UNDER D_4.

It is unchanged by a rotation of $\frac{\pi}{2}$ about an axis through the centers of the top and bottom faces, and by a π rotation taking these two into each other. A $\frac{2\pi}{3}$ rotation gives the object in the middle, which is not the same as that on the left. But a π rotation taking the top and bottom faces into each other interchanges the left and right objects, which are identical. Because of the commas there is no reflection or inversion symmetry. Thus the symmetry group is $D_4(422)$.

An *object invariant* under just $C_4(4)$ is obtained from this by only marking either the top or bottom faces, distinguishing them.

To illustrate why dihedral groups are not Abelian, we have

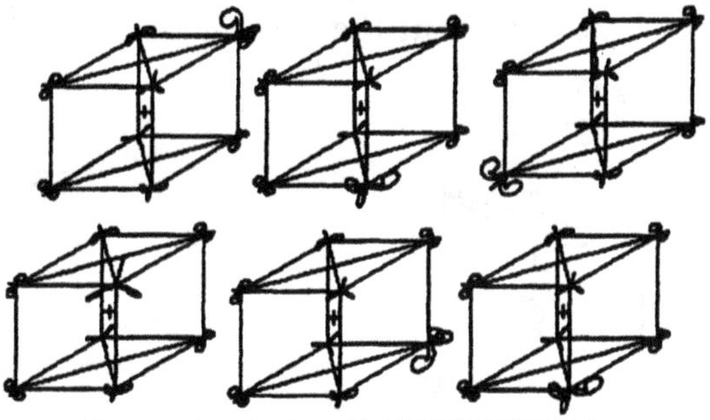

Figure C-11: D_4 IS NONABELIAN.

The top left object (with one set of commas enlarged only to visually distinguish it), invariant under $D_4(422)$, is rotated by $\frac{\pi}{2}$ about the vertical axis through the intersection of the lines on the top and bottom faces, to give the one under it, while the object to its right is that found by rotation of π about a horizontal axis through the center of the cube and the center of the rightmost face. That is then rotated by $\frac{\pi}{2}$ to give the top right one. The one at the lower center is obtained from the lower left object by the π rotation, so that it and the top right one are found from the top left object by the same rotations, but in opposite order. It is clear (for example, from the positions of the large commas) that these rotations do not commute — $D_4(422)$ is not Abelian. However the object (ignoring the fact that one set of commas is enlarged) is invariant under this group — the two transformations that are these products, in opposite order, are both in the group, and the object is invariant under each. To show the effect of the $\frac{\pi}{2}$ rotation on the lower center one the resultant object is on the lower right. It might be compared with an object found from the upper right one using the π rotation.

Now removing the commas gives the object on the left,

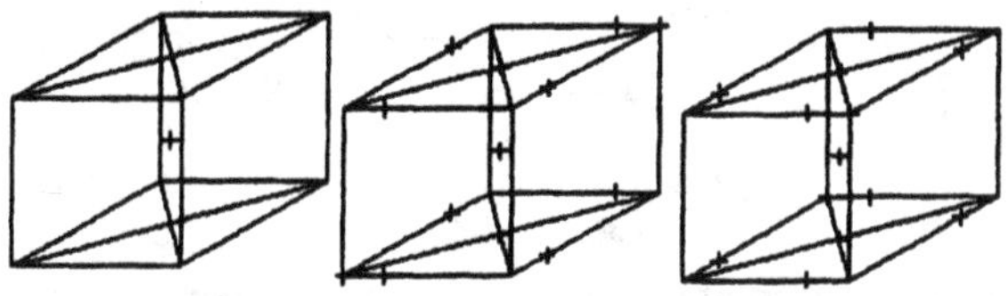

Figure C-12: OBJECTS INVARIANT UNDER D_{4h} AND C_{4h},

which is invariant under the full dihedral group $D_{4h}(\frac{422}{mmm})$, with generators the inversion, the π rotation about horizontal axes, and the $\frac{\pi}{2}$ rotation about the vertical one. To its right is the same object but with arrows (indicated, but not actually drawn), from the vertices to the marked points. On the far right is the object obtained by rotating the middle one by π around a horizontal axis, and these two are different, so the cube with arrows is not invariant under this C_2 rotation, but it is under a C_4 rotation about the vertical axis, and the inversion; the symmetry group is $C_{4h}(\frac{4}{m})$.

In the next diagram

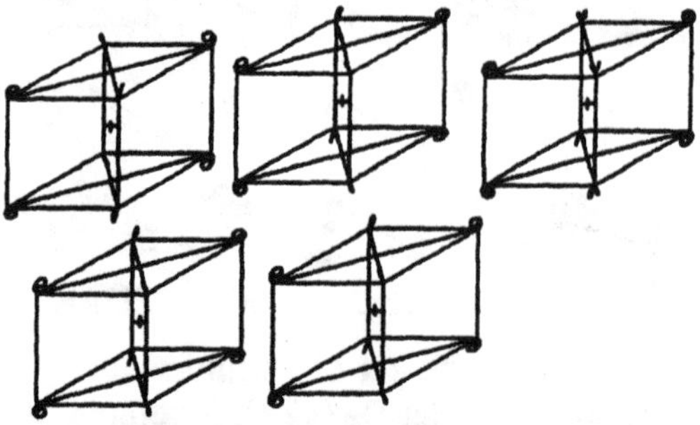

Figure C-13: OBJECTS INVARIANT UNDER D_4 AND D_{4h},

the leftmost figure of the top row is invariant only under $D_4(422)$, since it is changed by inversion or a reflection, as seen from the inverted figure on its right, with the commas different. The two objects below these are found from the first by reflection in mirrors parallel to, and perpendicular to, the faces with crossed lines, and through the (dot at the) center. These last three are identical, but different from the original (reflection in the other diagonal mirror would give another different figure). The top rightmost figure, obtained by superposing the two left ones is invariant under the full dihedral group $D_{4h}(\frac{422}{mmm})$.

There are further operations, as illustrated in

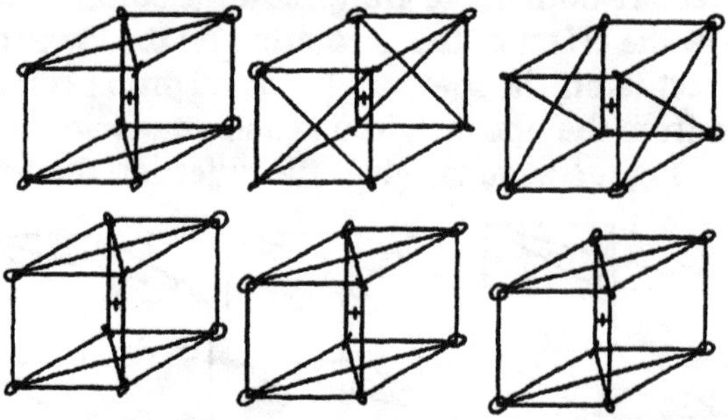

Figure C-14: TRANSFORMATIONS OF OBJECT INVARIANT UNDER D_4.

The object at the top left, now with a circle at each point, rather than commas, is constructed to be invariant under $\frac{\pi}{2}$ rotations around the vertical axis through the intersections of the diagonal lines on the faces. The two objects to its right are found with this rotation about the other two axes perpendicular to faces, and all three are different. At the bottom left the first object has been rotated about one of these axes by π, and is unchanged. To its right are the two objects obtained by reflection in the diagonal mirrors through the center of the cube and the lines on the top and bottom faces. The reflection changes the object, but the final object is the same for the two mirrors (and a horizontal mirror gives the same object). So the object is not invariant under these reflections, thus only under $D_4(422)$. These are also obtained from the first by inversion through the center, which is thus not a symmetry.

Next is group $D_{2d}(\bar{4}2m)$. In the drawing is a cube, with commas only at two corners,

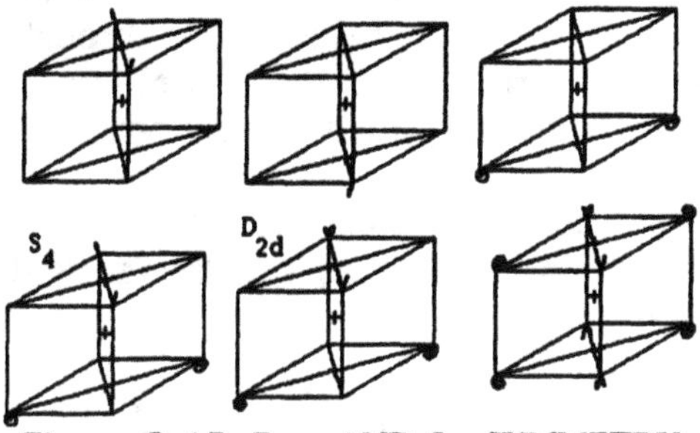

Figure C-15: D_{2d} AND S_4 SYMMETRY,

then, left to right, it rotated through π about an axis through the centers of the sides, that reflected in the vertical mirror through the center of the cube perpendicular to the faces, and this and the first cube superposed, at the bottom left. This is invariant under the π rotation about the vertical axis through the center, and an S_4 rotation-reflection with the rotation about this same axis (of course) of $\frac{\pi}{2}$, plus a reflection in the horizontal plane through the center. It is thus invariant under $S_4(\bar{4})$. The center bottom one is found by rotating this by π around the other horizontal axis and superposing the two. It is invariant under reflections in mirrors passing through both pairs of top and bottom diagonal lines, thus under $D_{2d}(\bar{4}2m)$. To obtain the object at the bottom right, a copy has been rotated by $\frac{\pi}{2}$ around the vertical axis through the center, and the two overlaid. It should be compared with the $D_{4h}(\frac{422}{mmm})$ invariant figures above.

The last group of the tetragonal set is $C_{4v}(4mm)$. This is generated by a C_4 rotation and a reflection in a vertical mirror. In

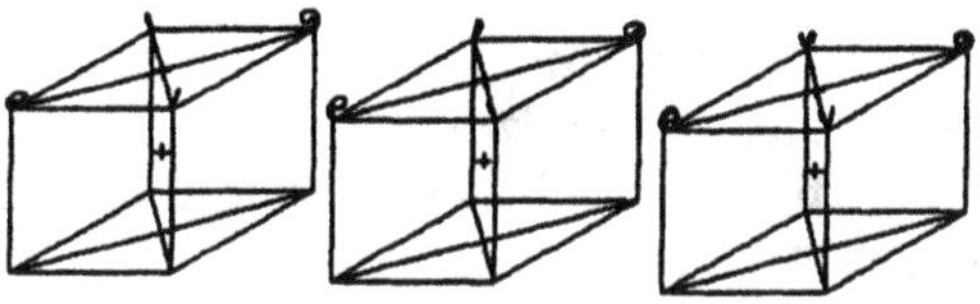

Figure C-16: EXAMPLE OF SYMMETRY GROUP C_{4v},

the cube on the left, invariant only under $C_4(4)$ is reflected through a mirror bisecting the angle between the diagonals at the top (and bottom) giving the second figure; these two are combined in the rightmost figure, which has group $C_{4v}(4mm)$.

Turning to the hexagonal system, we have at the left a hexagonal figure invariant under $D_{6h}(\frac{222}{mmm})$, the full group. That to its right has commas, thus is invariant only under $C_6(6)$. A rotation of this by π around an axis perpendicular to the principle six-fold one, going through the center of the hexagonal tube, and through the midpoint of the edge closest to the front, gives the next object. For the following one, the axis of rotation is through the midpoint of the face; clearly these are the same figures. Then overlaying the second and third (or fourth) gives an object invariant (by construction) under $D_6(622)$. It, inverted through its (marked) center, is on the far right, and clearly not inversion invariant,

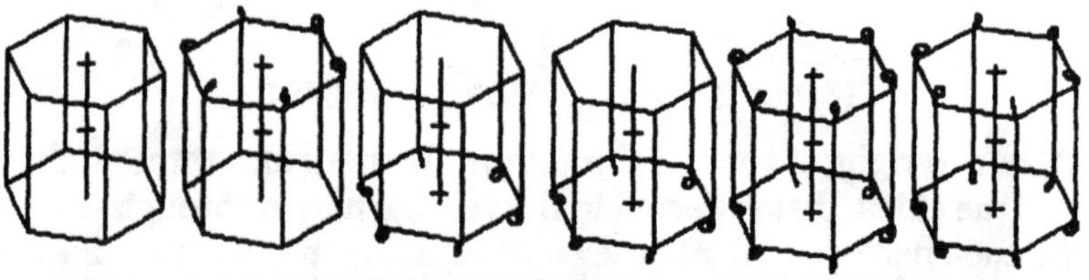

Figure C-17: SYMMETRY UNDER D_{6h} AND SUBGROUPS.

An object with only three-fold symmetry, so invariant under $C_3(3)$, is on the left,

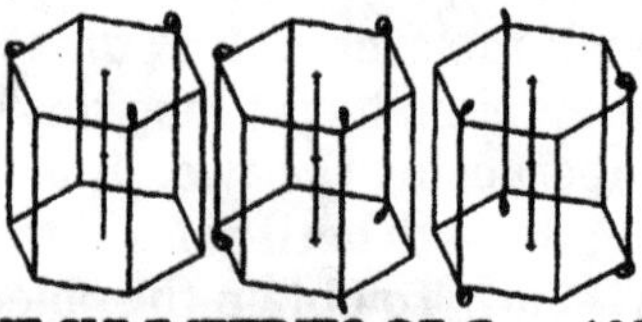

Figure C-18: THE SYMMETRIES OF C_{3h} AND SUBGROUPS.

The second object is obtained from it by reflection in a horizontal plane, and then overlaying of the two figures. Its symmetry group is $C_{3h}(\bar{6})$, by construction; it is not invariant under any larger group. This can be seen with the next one, which is it inverted; these two are quite different.

Another example with a six-fold axis, is the right object constructed by reflecting the left one in a horizontal mirror, and superposing; it is invariant under $C_{3h}(\bar{6})$. Clearly it is not invariant under inversion, so not under any larger group;

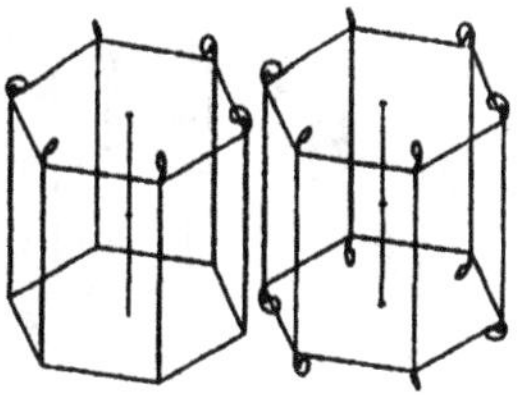

Figure C-19: OBJECT INVARIANT UNDER C_{3h}.

We have again at the left,

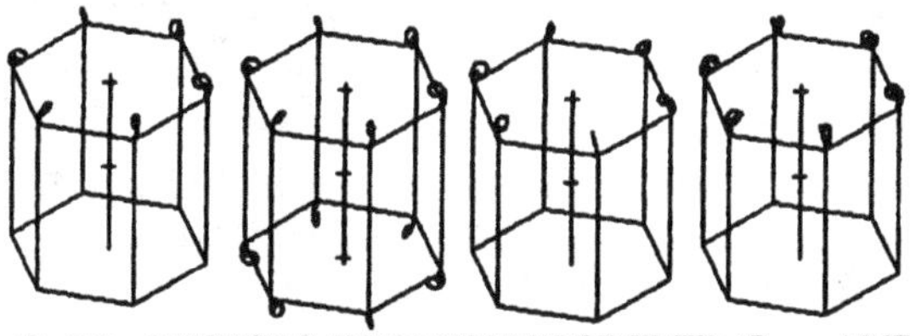

Figure C-20: OBJECTS INVARIANT UNDER C_{6h} AND C_{6v},

an object, with $C_6(6)$ symmetry, inverted about its center, and the two objects superposed giving to its right an object with symmetry of $C_{6h}(\frac{6}{m})$. The next right one is obtained by reflecting the first in a vertical mirror containing the principle (six-fold) axis through the center of a face, or the center of an edge. The rightmost object is constructed by combining this and the leftmost one and is thus invariant under $C_{6v}(6mm)$. The objects invariant under $C_{6h}(\frac{6}{m})$ and $C_{6v}(6mm)$ should be compared, and are seen to be different.

Turning to the trigonal system, we take the first object, with $C_3(3)$ symmetry, rotated about a horizontal axis, shown to its right, and the two combined, to give the top right figure, invariant under $D_3(32)$. The left bottom one is obtained by reflection in a vertical mirror, and next to it that found by overlaying the first object on this; it is invariant under $C_{3v}(3m)$. This rotated by 60° is at the bottom left, and is different,

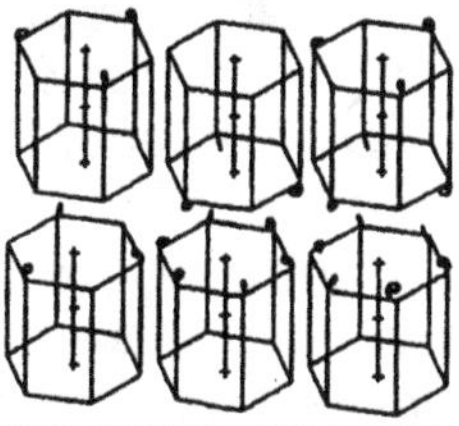

Figure C-21: OBJECTS INVARIANT UNDER D_3 AND C_{3v}.

Invariance under $C_{3v}(3m)$ does not give it under this rotation.

We start again with an $C_3(3)$ invariant object, and that combined with the one obtained by inverting it, to its right, is $S_6(\bar{3})$ invariant,

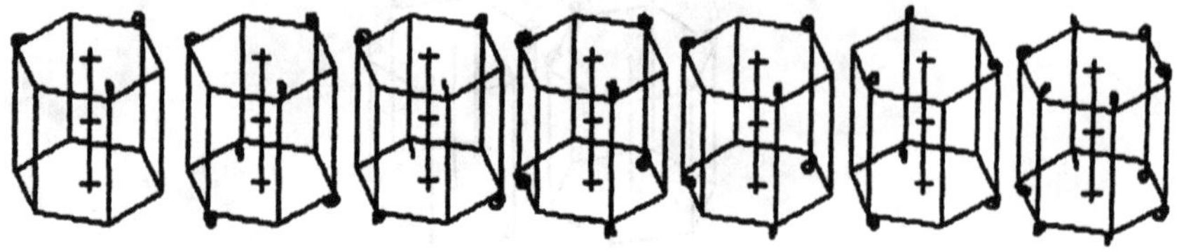

Figure C-22: OBJECTS INVARIANT UNDER D_{3h}, D_{3d} AND S_6.

The third object is found from the second using a π rotation around a horizontal axis, and these are different. The object to its right is given by the placement of the two to its left on each other, so with two commas at each occupied corner, thus an object invariant under $D_{3d}(\bar{3}\frac{2}{m})$. Next is the object find by overlaying the leftmost one on an object obtained from it by a horizontal reflection; it is invariant under $D_{3h}(\bar{6}m2)$. To the right of this is the object it gives when rotated by $\pi/6$ around a horizontal axis. These superposed are at extreme right.

These objects have been found by starting with the most symmetrical and breaking the symmetry to a lower one. It is useful to also go the other way. The object on the top left in

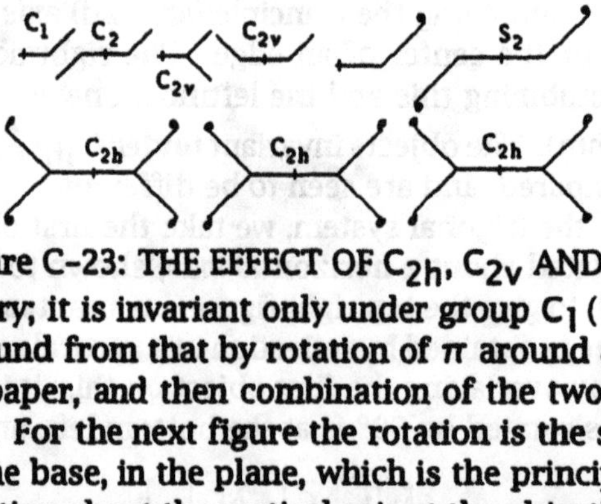

Figure C-23: THE EFFECT OF C_{2h}, C_{2v} AND S_2,

has no symmetry; it is invariant only under group C_1 (1). On its right is the object found from that by rotation of π around an axis perpendicular to the paper, and then combination of the two. Its symmetry group is C_2(2). For the next figure the rotation is the same but about an axis along the base, in the plane, which is the principle axis. Carrying out the rotation about the vertical axis at the object's end gives the following figure, similarly invariant, but for a different rotation. These last two are also invariant under reflections in mirrors on the axes of rotation. Their symmetry group is C_{2v}(mm2). This is a larger group, the object has more symmetry, so there are more generators. For the next figure on the top a comma has been put in, which leaves the group unchanged, as there was no symmetry. The object to its right found by a π rotation about a vertical axis then reflection in a horizontal mirror along the base, and superposition of the two, is invariant under $S_2(\bar{1})$ (and, necessarily [Mirman (1995c), sec.4.2.6, p. 60], also under a π rotation about an axis through its center perpendicular to the plane). An example of an object with the group $C_{1h}(m)$ is a pair of hands (as on the top left of fig. I.6.a.ii-1, p. 30). The object at the bottom left is found from the right top one with the comma by inversion through its center (this always taken as the fixed point under transformations), and then superposition; it is invariant under $C_{2h}(\frac{2}{m})$. The two rightmost figures on that line are obtained from the previous one by π rotations around an axis along the base, and one perpendicular to the plane through the center of the base. Although identical they are not the same as the first, so that figure has no higher symmetry than $C_{2h}(\frac{2}{m})$.

The top left figure (invariant under $C_2(2)$) of

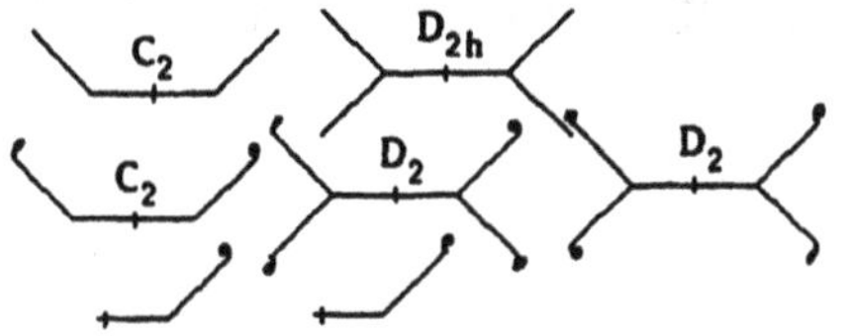

Figure C-24: THE ACTION OF D_{2h} AND SUBGROUPS,

is rotated by π around its base, and then these two combined to give the object on the right, with symmetry group $D_{2h}(\frac{222}{mmm})$, it being invariant under rotations of π about both the vertical axis through its center, in the plane, and the horizontal one perpendicular to the plane, and also mirrors along these axes. In the middle row the left figure, invariant under $C_2(2)$, is rotated by π around the axis through the base in the plane, and the objects are combined to give the center figure, which is thus invariant under π rotations around the base, and the two lines perpendicular to it. The figure to its right has been reflected through a mirror containing the base, and is different. These then have $D_2(222)$, but no higher, symmetry. The left figure on the bottom, is reflected in a mirror in the plane, giving the next object, to show the effect of this, which can be compared with the operations used to generate the other figures.

Problem C-1: The smallest groups have been illustrated differently from the others. Draw objects similar to the latter for these groups (and conversely, although this may not be possible for all, as the last two diagrams are only two-dimensional).

Problem C-2: Several groups have been illustrated more than once. These should be compared. Do some illustrations carry more information than others? Why?

Problem C-3: In all these cases the objects are obtained using the generators of the group given in tbl. B-1, p. 633. It should be verified that they are invariant under all the other group transformations given in tbl. B-2, p. 634.

Appendix D

Two-Dimensional Space Groups

There are multitudinous examples of patterns with the symmetries of the two-dimensional space groups, many quite interesting, and more artistic; those provided below, quite uninteresting, may be useful in that most are closely related. But many examples should be studied and compared. For each of the ones here, a unit cell is given with symmetry elements indicated, and also as part of a lattice, a few unit cells placed together, produced by a repeated motif. The symbols used (which when placed in the illustration are, of course, not part of the object), and the object used to devise the motif for most patterns are

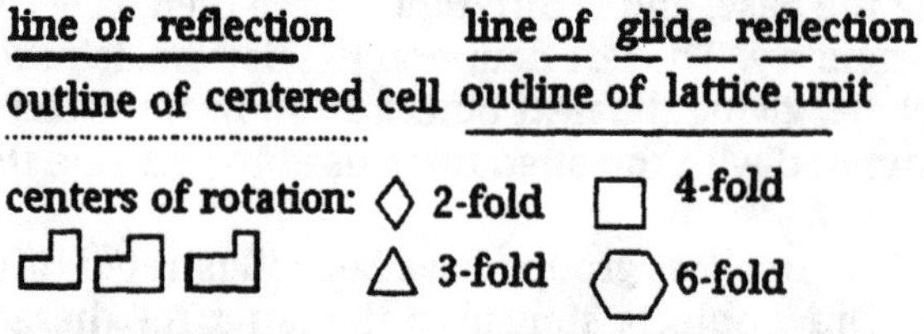

Figure D–1: SYMBOLS AND MOTIF FOR WALLPAPER GROUPS.

Using these we can construct objects invariant under the groups [Schattschneider (1978)]. The first is

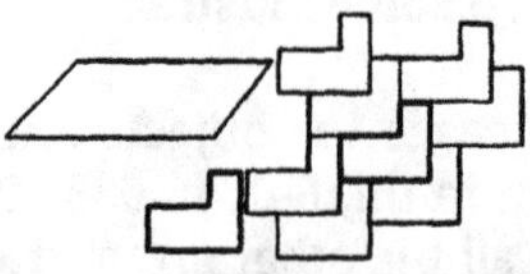

Figure D–2: WALLPAPER GROUP $p1$ ($p1$).

An example of the next one looks like

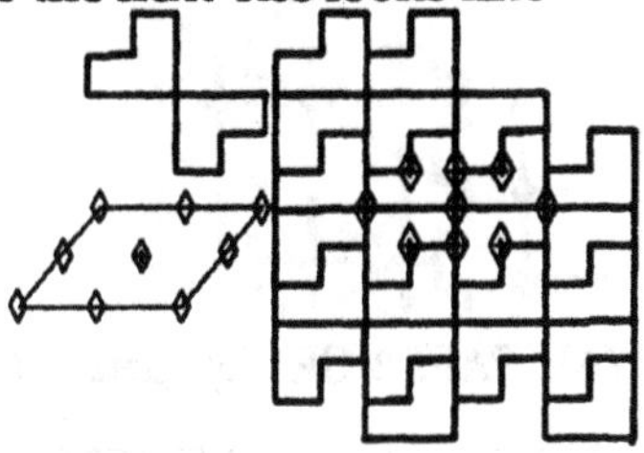

Figure D-3: WALLPAPER GROUP $p2$ ($p211$).

Then we have a crystal with a mirror,

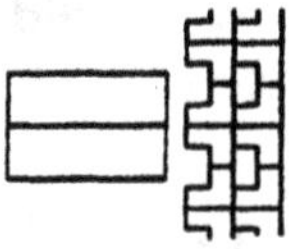

Figure D-4: WALLPAPER GROUP pm ($p1m1$).

With a glide we get a simple pattern like

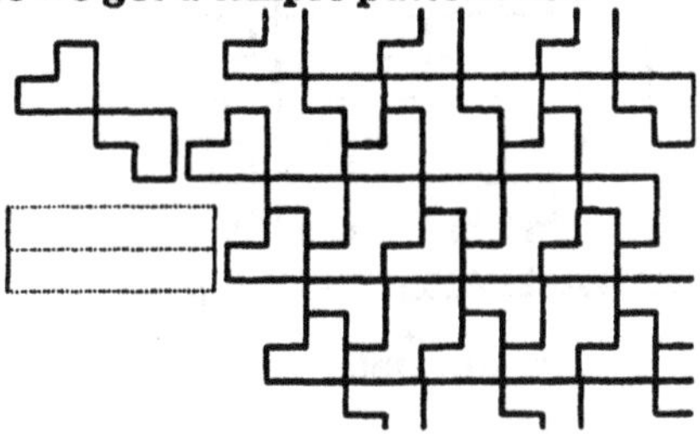

Figure D-5: WALLPAPER GROUP pg ($p1g1$).

Then we have a centered cell with glides and mirrors, and a symmorphic group. In the diagram one mirror is marked, with an m, and a glide with a g. The pattern reflected in the line marked g must be translated along this line for invariance; this line is a glide, not a mirror. The cell is rectangular, so while it has a horizontal mirror, it has no vertical one,

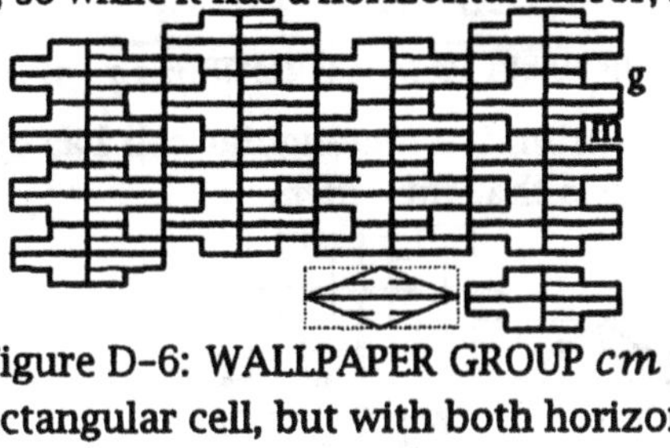

Figure D-6: WALLPAPER GROUP cm ($c1m1$),

Another rectangular cell, but with both horizontal and vertical mirrors, is

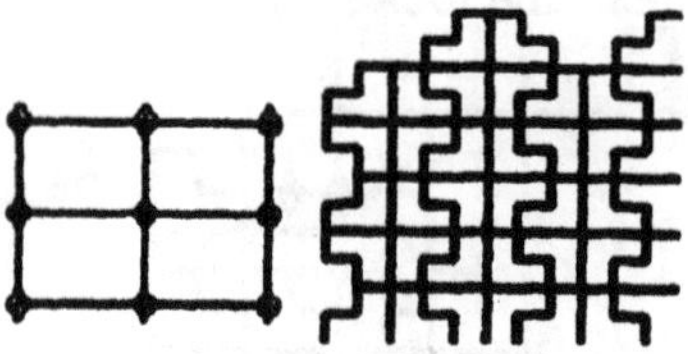

Figure D-7: WALLPAPER GROUP *pmm* (*p*2*mm*).

An illustration of a pattern with both mirrors and glides, here orthogonal to each other [Burns (1977), p. 323] is given next, and it should be noted that, as indicated by the symbol for the group, the glide makes it nonsymmorphic, for reasons that should be clear from the diagram. This serves to show the difference between symmorphic and nonsymmorphic glides,

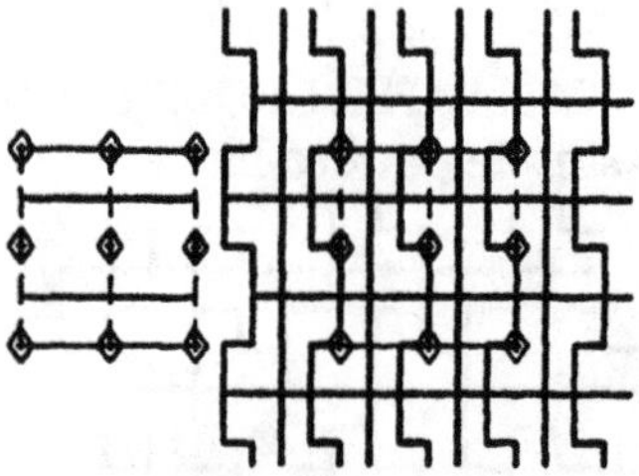

Figure D-8: WALLPAPER GROUP *pmg* (*p*2*mg*).

A figure with glides, here mutually orthogonal, but no mirrors, is that invariant under *p*2*gg*,

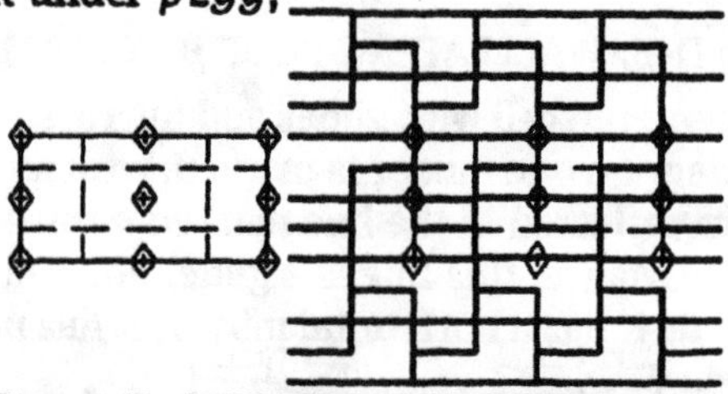

Figure D-9: WALLPAPER GROUP *pgg* (*p*2*gg*).

With two mirrors a primitive rectangular cell has symmetry group *p*2*mm*. A centered one is that for *cmm*,

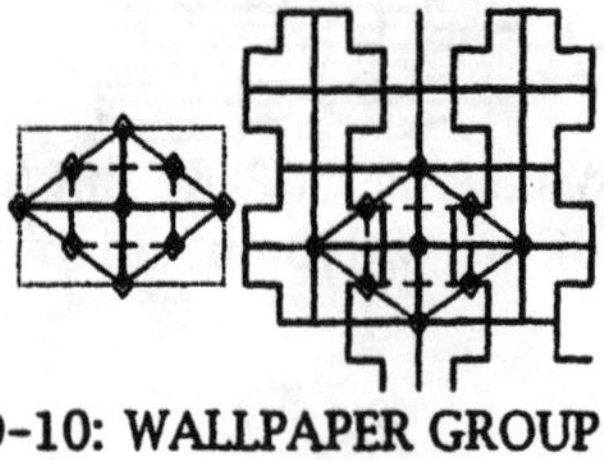

Figure D-10: WALLPAPER GROUP *cmm* (*c*2*mm*).

The pattern in the lattice cell (indicated by solid lines) repeated gives the entire pattern. This is also true of the centered cell, shown with dotted lines, but that shows more of the symmetry than does the lattice cell.

A square has the most symmetry, so to remove the mirror symmetry the motif has to be altered,

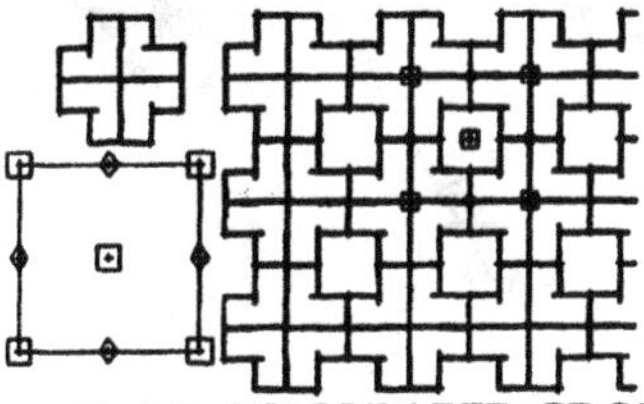

Figure D-11: WALLPAPER GROUP *p*4 (*p*4).

Mirrors give complete symmetry if there are no nonprimitive elements,

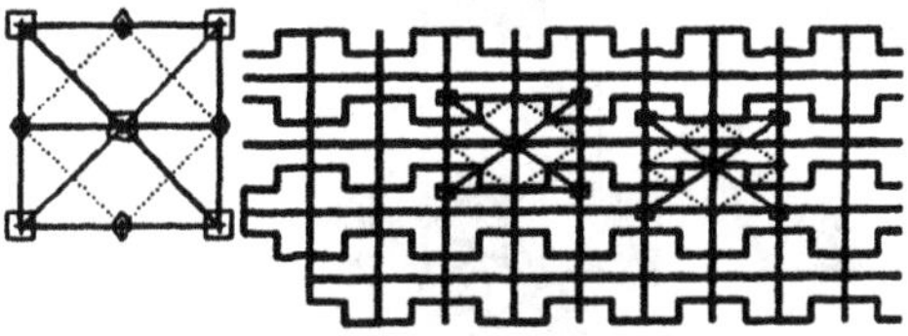

Figure D-12: WALLPAPER GROUP *p*4*m* (*p*4*mm*).

There can also be a glide,

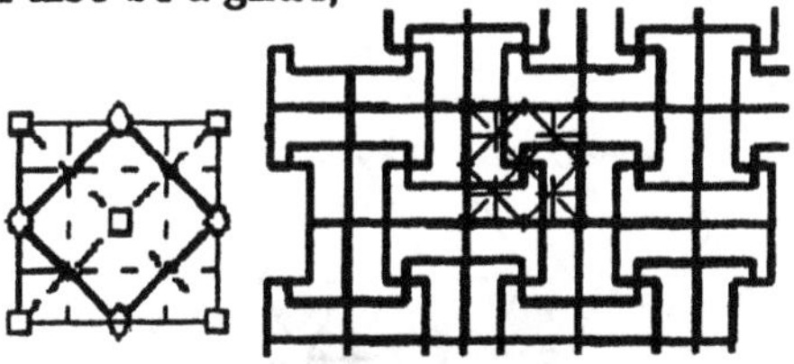

Figure D-13: WALLPAPER GROUP *p*4*g* (*p*4*gm*).

With three-fold symmetry a different motif is needed, this shown in stages of its construction, with a pattern (at the top left) placed within the lattice,

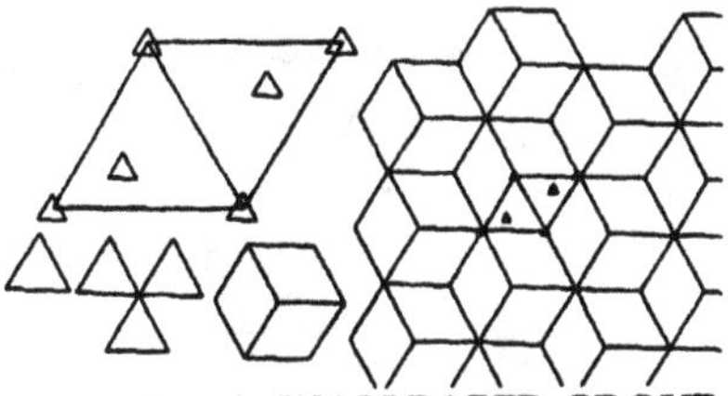

Figure D-14: WALLPAPER GROUP *p*3 (*p*3).

Adding mirrors gives

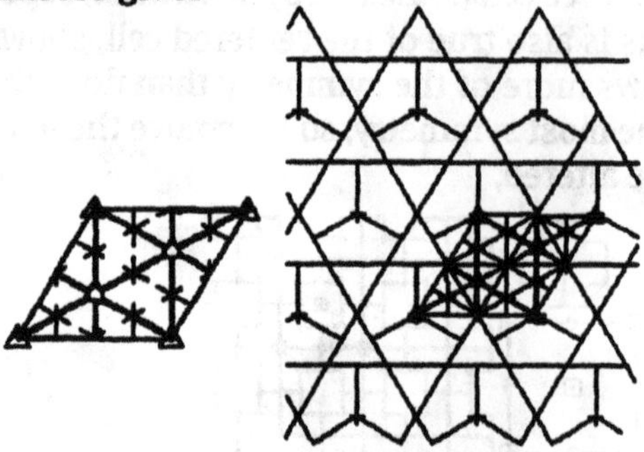

Figure D–15: WALLPAPER GROUP *p*3*m*1 (*p*3*m*1).

This has all 3-fold centers on mirrors. If that were not the case, then the group would be different, but its symbol would be almost the same,

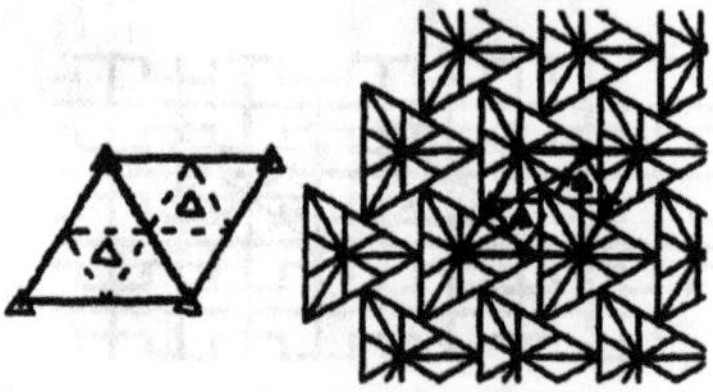

Figure D–16: WALLPAPER GROUP *p*31*m* (*p*31*m*).

This is an example with glides that do not go through any significant points in the object.

Hexagonal symmetry is exemplified by

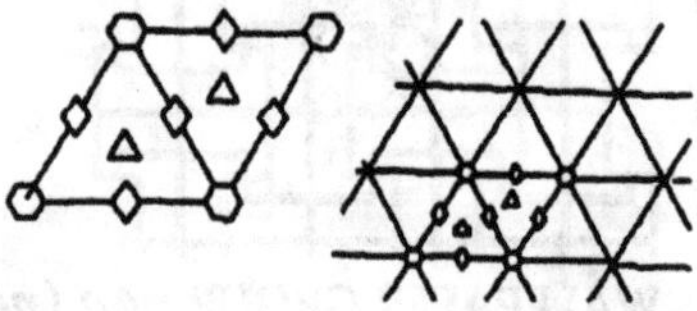

Figure D–17: WALLPAPER GROUP *p*6 (*p*6),

and with mirrors and glides

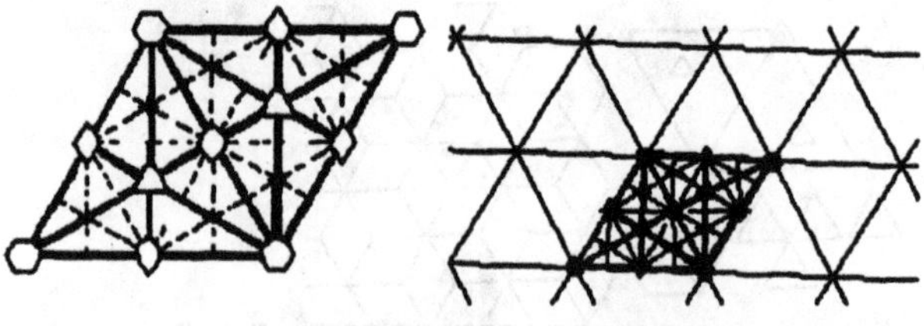

Figure D–18: WALLPAPER GROUP *p*6*m* (*p*6*mm*).

Appendix E

Point Group Character Tables

Here we give the character tables for the point groups [Burns (1977), p. 379; Lomont (1961), p. 78; Murnaghan (1963), p. 346; Wilson, Decius and Cross (1980), p. 323; Wooster (1967), p. 304].

E.1 DENOTING THE REPRESENTATIONS

Symbols for the representations are needed. There are three notations, Mulliken (chemical), Bethe and Bouckaert-Smoluchowski-Wigner (BSW) [Burns (1977), p. 379; Tsukerblatt (1994), p. 92]. There is a general convention for choosing the names [Wilson, Decius and Cross (1980), p. 322] in the first column. One-dimensional representations are given the letters A or B, two-dimensional ones E, three-dimensional representations τ (or F), and four and five get G and H respectively. In the D_n groups representations symmetric with respect to operation C_n are labeled A, those antisymmetric labeled B. The subscripts 1 and 2 denote representations symmetric, and antisymmetric, under a two-fold axis perpendicular to the principle axis, or a vertical plane, whichever is relevant. Representations symmetric under inversion are labeled by g (German for *gerade* — even), those antisymmetric by u (*ungerade* — odd). If there is a horizontal reflection as a generator, representations symmetric under σ_h are given a prime, those antisymmetric, a double prime. The relationship of the notations is [Burns (1977), p. 390]:

Table E.1-1: CONVENTIONS FOR THE REPRESENTATIONS

O_h M	B	BSW	D_{4h} M	B	BSW
A_{1g}	Γ_1^+	Γ_1	A_{1g}	Γ_1^+	M_1
A_{2g}	Γ_2^+	Γ_2	A_{2g}	Γ_2^+	M_2
E_g	Γ_3^+	Γ_{12}	B_{1g}	Γ_3^+	M_3
T_{1g}	Γ_4^+	Γ'_{15}	B_{2g}	Γ_4^+	M_4
T_{2g}	Γ_5^+	Γ'_{25}	E_g	Γ_5^+	M_5
A_{1u}	Γ_1^-	Γ'_1	A_{1u}	Γ_1^-	M'_1
A_{2u}	Γ_2^-	Γ'_2	A_{2u}	Γ_2^-	M'_2
E_u	Γ_3^-	Γ'_{12}	B_{1u}	Γ_3^-	M'_3
T_{1u}	Γ_4^-	Γ_{15}	B_{2u}	Γ_4^-	M'_4
T_{2u}	Γ_5^+	Γ_{25}	E_u	Γ_5^-	M'_5.

E.2 THE CHARACTER TABLES

The groups are labeled by both Schoenflies and international symbols. The top row of each table gives one element from each class with the number (not to be taken as multiplication) of class elements as a coefficient. The representations are listed on the left, and under each class element the character for that class. A star indicates complex conjugate. On the right, for each representation, is a set of functions of the coordinates that form basis vectors, or the corresponding functions for pseudo-vectors (which can be taken as angular momentum operators), whichever is relevant. One-dimensional representations are paired if their characters are complex conjugates of each other. Should they be degenerate then the pair can be regarded as a two-dimensional representation, with real characters.

Table E.2-1: THE CHARACTERS OF THE POINT GROUPS

C_1 (1)

	E
A	1

S_2 ($\bar{1}$)

	E	I		
A_g	1	1	R_x, R_x, R_x	$x^2, y^2, z^2, xy, xz, yz$
A_u	1	−1	x, y, z	

C_{1h} (m)

	E	σ_h		
A'	1	1	x, y, R_z	x^2, y^2, z^2, xy
A''	1	−1	z, R_x, R_y	xz, yz

C_2 (2)

	E	C_2		
A	1	1	z, R_z	x^2, y^2, z^2, xy
B	1	−1	x, y, R_x, R_y	xz, yz

C_3 (3)

	E	C_3	C_3^2	$\varepsilon = exp(\frac{2\pi i}{3})$	
A	1	1	1	z, R_z	$x^2 + y^2, z^2$
E	{1	ε	ε^*	$x + iy, R_x + iR_y$	$(x^2 - y^2, xy)$
	{1	ε^*	ε	$x - iy, R_x - iR_y$	(xz, yz)

C_4 (4)

	E	C_4	C_4^2	C_4^3		
A	1	1	1	1	z, R_z	$x^2 + y^2, z^2$
B	1	−1	1	−1		$x^2 - y^2, xy$
E	{1	i	−1	$-i$	$x + iy, R_x + iR_y$	(xz, yz)
	{1	$-i$	−1	i	$x - iy, R_x - iR_y$	

C_5

	E	C_5	C_5^2	C_5^3	C_5^4	$\varepsilon = exp(\frac{2\pi i}{5})$	
A	1	1	1	1	1	z, R_z	x^2+y^2, z^2
E_1	{1	ε	ε^2	ε^{2*}	ε^*	$x+iy, R_x+iR_y$	(xz, yz)
	{1	ε^*	ε^{2*}	ε^*	ε	$x-iy, R_x-iR_y$	
E_2	{1	ε^2	ε^*	ε	ε^{2*}		(x^2-y^2, xy)
	{1	ε^{2*}	ε	ε^*	ε^2		

C_6 (6)

	E	C_6	C_3	C_2	C_3^2	C_6^5	$\varepsilon = exp(\frac{2\pi i}{6})$	
A	1	1	1	1	1	1	z, R_z	x^2+y^2, z^2
B	1	−1	1	−1	1	−1		
E_1	{1	ε	$-\varepsilon^*$	−1	$-\varepsilon$	ε^*	$x+iy, R_x+iR_y$	(xz, yz)
	{1	ε^*	$-\varepsilon$	−1	$-\varepsilon^*$	ε	$x-iy, R_x-iR_y$	
E_2	{1	$-\varepsilon^*$	$-\varepsilon$	1	$-\varepsilon^*$	$-\varepsilon$		
	{1	$-\varepsilon$	$-\varepsilon^*$	1	$-\varepsilon$	$-\varepsilon^*$		x^2-y^2, xy

C_7

	E	C_7	C_7^2	C_7^3	C_7^4	C_7^5	C_7^6	$\varepsilon = exp(\frac{2\pi i}{7})$	
A	1	1	1	1	1	1	1	z, R_z	x^2+y^2, z^2
E_1	{1	ε	ε^2	ε^3	ε^{3*}	ε^{2*}	ε^*	$x+iy, R_x+iR_y$	(xz, yz)
	{1	ε^*	ε^{2*}	ε^{3*}	ε^3	ε^2	ε	$x-iy, R_x-iR_y$	
E_2	{1	ε^2	ε^{3*}	ε^*	ε	ε^3	ε^{2*}		
	{1	ε^{2*}	ε^3	ε	ε^*	ε^{3*}	ε^2		x^2-y^2, xy
E_3	{1	ε^3	ε^*	ε^2	ε^{2*}	ε	ε^{3*}		
	{1	ε^{3*}	ε	ε^{2*}	ε^2	ε^*	ε^3		

C_8

	E	C_8	C_4	C_8^3	C_2	C_8^5	C_4^3	C_8^7	$\varepsilon = exp(\frac{2\pi i}{8})$	
A	1	1	1	1	1	1	1	1	z, R_z	x^2+y^2, z^2
B	1	−1	1	−1	1	−1	1	−1		
E_1	{1	ε	i	$-\varepsilon^*$	−1	$-\varepsilon$	$-i$	ε^*	$x+iy, R_x+iR_y$	(xz, yz)
	{1	ε^*	$-i$	$-\varepsilon$	−1	$-\varepsilon^*$	i	ε	$x-iy, R_x-iR_y$	
E_2	{1	i	−1	$-i$	1	i	−1	$-i$		
	{1	$-i$	−1	i	1	$-i$	−1	i		x^2-y^2, xy
E_3	{1	$-\varepsilon$	i	ε^*	−1	ε	$-i$	$-\varepsilon^*$		
	{1	$-\varepsilon^*$	$-i$	ε	−1	ε^*	i	$-\varepsilon$		

C_{2v} (mm2)

	E	C_2	$\sigma_v(xz)$	$\sigma_v'(yz)$		
A_1	1	1	1	1	z	x^2, y^2, z^2
A_2	1	1	−1	−1	R_z	xy
B_1	1	−1	1	−1	x, R_y	xz
B_2	1	−1	−1	1	y, R_x	yz

C_{3v} (3m)

	E	$2C_3$	$3\sigma_v$		
A_1	1	1	1	z	x^2+y^2, z^2
A_2	1	1	−1	R_z	
E	2	−1	0	$(x, y), (R_x, R_y)$	$(x^2-y^2, xy), (xz, yz)$

C_{4v} (4mm)

	E	$2C_4$	C_2	$2\sigma_v$	$2\sigma_d$		
A_1	1	1	1	1	1	z	x^2+y^2, z^2
A_2	1	1	1	−1	−1	R_z	
B_1	1	−1	1	1	−1		x^2-y^2
B_2	1	−1	1	−1	1		xy
E	2	0	−2	0	0	$(x,y),(R_x,R_y)$	(xz,yz)

C_{5v}

	E	$2C_5$	$2C_5^2$	$5\sigma_v$	$\alpha = cos(\frac{2\pi}{5}),\ \beta = cos(\frac{4\pi}{5})$	
A_1	1	1	1	1	z	x^2+y^2, z^2
A_2	1	1	1	−1	R_z	
E_1	2	α	β	0	$(x,y),(R_x,R_y)$	(xz,yz)
E_2	2	β	α	0		(x^2-y^2,xy)

C_{6v} (6mm)

	E	$2C_6$	$2C_3$	C_2	$3\sigma_v$	$3\sigma_d$		
A_1	1	1	1	1	1	1	z	x^2+y^2, z^2
A_2	1	1	1	1	−1	−1	R_z	
B_1	1	−1	1	−1	1	−1		
B_2	1	−1	1	−1	−1	1		
E_1	2	1	−1	−2	0	0	$(x,y),(R_x,R_y)$	(xz,yz)
E_2	2	−1	−1	2	0	0		(x^2-y^2,xy)

C_{2h} $(\frac{2}{m})$

	E	C_2	I	σ_h		
A_g	1	1	1	1	R_z	x^2, y^2, z^2, xy
B_g	1	−1	1	−1	R_x, R_y	xz, yz
A_g	1	1	−1	−1	z	
B_g	1	−1	−1	1	x, y	

C_{3h} $(\bar{6})$

	E	C_3	$C_3{}^3$	σ_h	S_3	$S_3{}^5$	$\varepsilon = exp(\frac{2\pi i}{3})$	
A'	1	1	1	1	1	1	R_z	$x^2 + y^2, z^2$
E'	{1	ε	ε^*	1	ε	ε^*	$x + iy$	$(x^2 - y^2, xy)$
	{1	ε^*	ε	1	ε^*	ε	$x - iy$	
A''	1	1	1	−1	−1	−1	z	
E''	{1	ε	ε^*	−1	$-\varepsilon$	$-\varepsilon^*$	$R_x + iR_y$	(xz, yz)
	{1	ε^*	ε	−1	$-\varepsilon^*$	$-\varepsilon$	$R_x - iR_y$	

C_{4h} $(\frac{4}{m})$

	E	C_4	C_2	$C_4{}^3$	I	$S_4{}^3$	σ_h	S_4		
A_g	1	1	1	1	1	1	1	1	R_z	x^2+y^2, z^2
B_g	1	−1	1	−1	1	−1	1	−1		x^2-y^2, xy
E_g	{1	i	−1	$-i$	1	i	−1	$-i$	R_x+iR_y	(xz, yz)
	{1	$-i$	−1	i	1	$-i$	−1	i	R_x-iR_y	
A_u	1	1	1	1	−1	−1	−1	−1	z	
B_u	1	−1	1	−1	−1	1	−1	1		
E_u	{1	i	−1	$-i$	−1	$-i$	1	i	$x+iy$	
	{1	$-i$	−1	i	−1	i	1	$-i$	$x-iy$	

D_2 (222)

	E	C_2	$C_2(y)$	$C_2(x)$		
A	1	1	1	1		x^2, y^2, z^2
B_1	1	1	−1	−1	z, R_z	xy
B_2	1	−1	1	−1	y, R_y	xz
B_3	1	−1	−1	1	x, R_x	yz

D_3 (32)

	E	$2C_3$	$3C_2'$		
A_1	1	1	1		x^2+y^2, z^2
A_2	1	1	−1	z, R_z	
E	2	−1	0	$(x, y), (R_y, R_y)$	$(x^2-y^2, xy), (xz, yz)$

D_4 (422)

	E	$2C_4$	$C_2 = C_4^2$	$2C_2'$	$2C_2''$		
A_1	1	1	1	1	1		x^2+y^2, z^2
A_2	1	1	1	−1	−1	z, R_z	
B_1	1	−1	1	1	−1		x^2-y^2
B_2	1	−1	1	−1	1		xy
E	2	0	−2	0	0	$(x,y), (R_x, R_y)$	(xz, yz)

D_5

	E	$2C_5$	$2C_5^2$	$5C_2'$	$\alpha = cos(\frac{2\pi}{5})$,	$\beta = cos(\frac{4\pi}{5})$
A_1	1	1	1	1		x^2+y^2, z^2
A_2	1	1	1	−1	z, R_z	
E_1	2	α	β	0	$(x,y), (R_x, R_y)$	(xz, yz)
E_2	2	β	α	0		(x^2-y^2, xy)

D_6 (622)

	E	$2C_6$	$2C_3$	C_2	$3C_2'$	$3C_2''$		
A_1	1	1	1	1	1	1		x^2+y^2, z^2
A_2	1	1	1	1	−1	−1	z, R_z	
B_1	1	−1	1	−1	1	−1		
B_2	1	−1	1	−1	−1	1		
E_1	2	1	−1	−2	0	0	$(x,y), (R_x, R_y)$	(xz, yz)
E_2	2	−1	−1	2	0	0		(x^2-y^2, xy)

D_{2d} $(\bar{4}2m)$

	E	$2S_4$	C_2	$2C_2'$	$2\sigma_d$		
A_1	1	1	1	1	1		x^2+y^2, z^2
A_2	1	1	1	−1	−1	R_z	
B_1	1	−1	1	1	−1		x^2-y^2
B_2	1	−1	1	−1	1	z	xy
E	2	0	−2	0	0	$(x,y),(R_x,R_y)$	(xz,yz)

D_{3d} $(\bar{3}\frac{2}{m})$

	E	$2C_3$	$3C_2'$	I	$2S_6$	$3\sigma_d$		
A_{1g}	1	1	1	1	1	1		x^2+y^2, z^2
A_{2g}	1	1	−1	1	1	−1	R_z	
E_g	2	−1	0	2	−1	0	(R_x,R_y)	$(x^2-y^2,xy),(xz,yz)$
A_{1u}	1	1	1	−1	−1	−1		
A_{2u}	1	1	−1	−1	−1	1	z	
E_u	2	−1	0	−2	1	0	(x,y)	

D_{4d}

	E	$2S_8$	$2C_4$	$2S_8'$	C_2	$4C_2'$	$4\sigma_d$		
A_1	1	1	1	1	1	1	1		x^2+y^2, z^2
A_2	1	1	1	1	1	−1	−1	R_z	
B_1	1	−1	1	−1	1	1	−1		
B_2	1	−1	1	−1	1	−1	1	z	
E_1	2	$\sqrt{2}$	0	$-\sqrt{2}$	−2	0	0	(x, y)	
E_2	2	0	−2	0	2	0	0		(x^2-y^2, xy)
E_3	2	$-\sqrt{2}$	0	$\sqrt{2}$	−2	0	0	(R_x, R_y)	(xz, yz)

D_{5d}

	E	$2C_5$	$2C_5^2$	$5C_2'$	I	$2S_{10}^3$	$2S_{10}$	$5\sigma_d$	$\alpha = 2cos(\frac{2\pi}{5})$,	$\beta = 2cos(\frac{4\pi}{5})$
A_{1g}	1	1	1	1	1	1	1	1		x^2+y^2, z^2
A_{2g}	1	1	1	−1	1	1	1	−1	R_z	
E_{1g}	2	α	β	0	2	α	β	0	(R_x, R_y)	(xz, yz)
E_{2g}	2	β	α	0	2	β	α	0		x^2-y^2, xy
A_{1u}	1	1	1	1	−1	−1	−1	−1		
A_{2u}	1	1	1	−1	−1	−1	−1	1	z	
E_{1u}	2	α	β	0	−2	$-\alpha$	$-\beta$	0	(x, y)	
E_{2u}	2	β	α	0	−2	$-\beta$	$-\alpha$	0		

D_{6d}

	E	$2S_{12}$	$2C_6$	$2S_4$	$2C_3$	$2S_{12}^5$	C_2	$6C_2'$	$6\sigma_d$		
A_1	1	1	1	1	1	1	1	1	1		x^2+y^2, z^2
A_2	1	1	1	1	1	1	1	−1	−1	R_z	
B_1	1	−1	1	−1	1	−1	1	1	−1		
B_2	1	−1	1	−1	1	−1	1	−1	1	z	
E_1	2	$\sqrt{3}$	1	0	−1	$-\sqrt{3}$	−2	0	0	(x, y)	
E_2	2	1	−1	−2	−1	1	2	0	0		(x^2-y^2, xy)
E_3	2	0	−2	0	2	0	−2	0	0		
E_4	2	−1	−1	2	−1	−1	2	0	0		
E_5	2	$-\sqrt{3}$	1	0	−1	$\sqrt{3}$	−2	0	0	(R_x, R_y)	(xz, yz)

D_{2h} $(\frac{222}{mmm})$

	E	C_2	$C_2(y)$	$C_2(x)$	I	$\sigma(xy)$	$\sigma(xz)$	$\sigma(yz)$		
A_g	1	1	1	1	1	1	1	1		x^2, y^2, z^2
B_{1g}	1	1	−1	−1	1	1	−1	−1	R_z	xy
B_{2g}	1	−1	1	−1	1	−1	1	−1	R_y	xz
B_{3g}	1	−1	−1	1	1	−1	−1	1	R_x	yz
A_u	1	1	1	1	−1	−1	−1	−1		
B_{1u}	1	1	−1	−1	−1	−1	1	1	z	
B_{2u}	1	−1	1	−1	−1	1	−1	1	y	
B_{3u}	1	−1	−1	1	−1	1	1	−1	x	

D_{3h} ($\bar{6}m2$)

	E	$2C_3$	$3C_2$	σ_h	$2S_3$	$3\sigma_v$		
A_1'	1	1	1	1	1	1		x^2+y^2, z^2
A_2'	1	1	−1	1	1	−1	R_z	
E'	2	−1	0	2	−1	0	(x, y)	(x^2-y^2, xy)
A_1''	1	1	1	−1	−1	−1		
A_2''	1	1	−1	−1	−1	1	z	
E''	2	−1	0	−2	1	0	(R_x, R_y)	(xz, yz)

D_{4h} ($\frac{422}{mmm}$)

	E	$2C_4$	C_2	$2C_2'$	$2C_2''$	I	$2S_4$	σ_h	$2\sigma_v$	$2\sigma_d$		
A_{1g}	1	1	1	1	1	1	1	1	1	1		x^2+y^2, z^2
A_{2g}	1	1	1	−1	−1	1	1	1	−1	−1	R_z	
B_{1g}	1	−1	1	1	−1	1	−1	1	1	−1		x^2-y^2
B_{2g}	1	−1	1	−1	1	1	−1	1	−1	1		xy
E_g	2	0	−2	0	0	2	0	−2	0	0	(R_x, R_y)	(xz, yz)
A_{1u}	1	1	1	1	1	−1	−1	−1	−1	−1		
A_{2u}	1	1	1	−1	−1	−1	−1	−1	1	1	z	
B_{1u}	1	−1	1	1	−1	−1	1	−1	−1	1		
B_{2u}	1	−1	1	−1	1	−1	1	−1	1	−1		
E_u	2	0	−2	0	0	−2	0	2	0	0	(x, y)	

D_{5h}

	E	$2C_5$	$2C_5^2$	$5C_2'$	σ_h	$2S_5$	$2S_5^3$	$5\sigma_v$	$\alpha = 2\cos\frac{2\pi i}{5}$,	$\beta = 2\cos\frac{4\pi i}{5}$
A_1'	1	1	1	1	1	1	1	1		x^2+y^2, z^2
A_2'	1	1	1	−1	1	1	1	−1	R_z	
E_1'	2	α	β	0	2	α	β	0	(x, y)	
E_2'	2	β	α	0	2	β	α	0		x^2-y^2, xy
A_1''	1	1	1	1	−1	−1	−1	−1		
A_2''	1	1	1	−1	−1	−1	−1	1	z	
E_1''	2	α	β	0	−2	$-\alpha$	$-\beta$	0	(R_x, R_y)	(xz, yz)
E_2''	2	β	α	0	−2	$-\beta$	$-\alpha$	0		

D_{6h} $\left(\frac{622}{mmm}\right)$

	E	$2C_6$	$2C_3$	C_2	$3C_2'$	$3C_2''$	I	$2S_3$	$2S_6$	$\sigma_h(xy)$	$3\sigma_d$	$3\sigma_v$		
A_{1g}	1	1	1	1	1	1	1	1	1	1	1	1		x^2+y^2, z^2
A_{2g}	1	1	1	1	−1	−1	1	1	1	1	−1	−1	R_z	
B_{1g}	1	−1	1	−1	1	−1	1	−1	1	−1	1	−1		
B_{2g}	1	−1	1	−1	−1	1	1	−1	1	−1	−1	1		
E_{1g}	2	1	−1	−2	0	0	2	1	−1	−2	0	0	(R_x, R_y)	(xz, yz)
E_{2g}	2	−1	−1	2	0	0	2	−1	−1	2	0	0		(x^2-y^2, xy)
A_{1u}	1	1	1	1	1	1	−1	−1	−1	−1	−1	−1		
A_{2u}	1	1	1	1	−1	−1	−1	−1	−1	−1	1	1	z	
B_{1u}	1	−1	1	−1	1	−1	−1	1	−1	1	−1	1		
B_{2u}	1	−1	1	−1	−1	1	−1	1	−1	1	1	−1		
E_{1u}	2	1	−1	−2	0	0	−2	−1	1	2	0	0	(x, y)	
E_{2u}	2	−1	−1	2	0	0	−2	1	1	−2	0	0		

S_4 ($\bar{4}$)

	E	S_4	C_2	$S_4{}^3$		
A	1	1	1	1	R_z	x^2+y^2, z^2
B	1	−1	1	−1	z	x^2-y^2, xy
E	{1	i	−1	$-i$	$x+iy, R_x+iR_y$	(xz, yz)
	{1	$-i$	−1	i	$x-iy, R_x-iR_y$	

S_6 ($\bar{3}$)

	E	C_3	$C_3{}^2$	I	$S_6{}^5$	S_6	$\varepsilon = exp(\frac{2\pi i}{3})$	
A_g	1	1	1	1	1	1	R_z	x^2+y^2, z^2
E_g	{1	ε	ε^*	1	ε	ε^*	R_x+iR_y	$(x^2-y^2, xy), (xz, yz)$
	{1	ε^*	ε	1	ε^*	ε	R_x-iR_y	
A_u	1	1	1	−1	−1	−1	z	
E_u	{1	ε	ε^*	−1	$-\varepsilon$	$-\varepsilon^*$	$x+iy$	
	{1	ε^*	ε	−1	$-\varepsilon^*$	$-\varepsilon$	$x-iy$	

S_8

	E	S_8	C_4	$S_8{}^3$	C_2	$S_8{}^5$	$C_4{}^3$	$S_8{}^7$	$\varepsilon = exp(\frac{2\pi i}{8})$	
A	1	1	1	1	1	1	1	1	R_z	x^2+y^2, z^2
B	1	−1	1	−1	1	−1	1	−1	z	
E_1	{1	ε	i	$-\varepsilon^*$	−1	$-\varepsilon$	$-i$	ε^*	$x+iy$	
	{1	ε^*	$-i$	$-\varepsilon$	−1	$-\varepsilon^*$	i	ε	$x-iy$	
E_2	{1	i	−1	$-i$	1	i	−1	$-i$		x^2-y^2, xy
	{1	$-i$	−1	i	1	$-i$	−1	i		
E_3	{1	$-\varepsilon$	i	ε^*	−1	ε	$-i$	$-\varepsilon^*$	R_x+iR_y	(xz, yz)
	{1	$-\varepsilon^*$	$-i$	ε	−1	ε^*	i	$-\varepsilon$	R_x-iR_y	

Cubic Groups

T (23)

	E	$4C_3$	$4C_3^2$	$3C_2$	$\varepsilon = exp(\frac{2\pi i}{3})$	
A	1	1	1	1		$x^2+y^2+z^2$
E	{1	ε	ε^*	1		$(2z^2-x^2-y^2, x^2-y^2)$
	{1	ε^*	ε	1		
τ	3	0	0	−1	$(x,y,z), (R_x,R_y,R_z)$	(xy,xz,yz)

T_h $(\frac{2}{m}\bar{3})$

	E	$4C_3$	$4C_3^2$	$3C_2$	I	$4S_6^5$	$4S_6$	$3\sigma_h$	$\varepsilon = exp(\frac{2\pi i}{3})$	
A_g	1	1	1	1	1	1	1	1		$x^2+y^2+z^2$
E_g	{1	ε	ε^*	1	1	ε	ε^*	1		$(2z^2-x^2-y^2, x^2-y^2)$
	{1	ε^*	ε	1	1	ε^*	ε	1		
τ_g	3	0	0	−1	3	0	0	−1	(R_x,R_y,R_z)	(xy,xz,yz)
A_u	1	1	1	1	−1	−1	−1	−1		
E_u	{1	ε	ε^*	1	−1	$-\varepsilon$	$-\varepsilon^*$	−1		
	{1	ε^*	ε	1	−1	$-\varepsilon^*$	$-\varepsilon$	−1		
τ_u	3	0	0	−1	−3	0	0	1	(x,y,z)	

T_d ($\bar{4}3m$)

	E	$8C_3$	$6\sigma_d$	$6S_4$	$3C_2$		
O (432)							
	E	$8C_3$	$6C_2'$	$6C_4$	$3C_2$		($C_2 = C_4^2$)
A_1	1	1	1	1	1		$x^2+y^2+z^2$
A_2	1	1	−1	−1	1		
E	2	−1	0	0	2		$(2z^2-x^2-y^2, x^2-y^2)$
τ_1	3	0	−1	1	−1	$(x,y,z), (R_x,R_y,R_z)$	
τ_2	3	0	1	−1	−1	(x,y,z) in T_d	(xy,xz,yz)

O_h ($\frac{4}{m}\bar{3}\frac{2}{m}$)

	E	$8C_3$	$6C_2$	$6C_4$	$3C_2$	I	$6S_4$	$8S_6$	$3\sigma_h$	$6\sigma_d$		
A_{1g}	1	1	1	1	1	1	1	1	1	1		$x^2+y^2+z^2$
A_{2g}	1	1	−1	−1	1	1	−1	1	1	−1		
E_g	2	−1	0	0	2	2	0	−1	2	0		$(2z^2-x^2-y^2, x^2-y^2)$
τ_{1g}	3	0	−1	1	−1	3	1	0	−1	−1	(R_x,R_y,R_z)	
τ_{2g}	3	0	1	−1	−1	3	−1	0	−1	1		(xz,yz,xy)
A_{1u}	1	1	1	1	1	−1	−1	−1	−1	−1		
A_{2u}	1	1	−1	−1	1	−1	1	−1	−1	1		
E_u	2	−1	0	0	2	−2	0	1	−2	0		
τ_{1u}	3	0	−1	1	−1	−3	−1	0	1	1	(x,y,z)	
τ_{2u}	3	0	1	−1	−1	−3	1	0	1	−1		

Icosahedral Group

I_h

	E	$12C_5$	$12C_5^2$	$20C_3$	$15C_2$	I	$12S_{10}^3$	$12S_{10}$	$20S_6$	15σ	$\alpha = \frac{1}{2}(1+\sqrt{5})$	$\beta = \frac{1}{2}(1-\sqrt{5})$
A_g	1	1	1	1	1	1	1	1	1	1		$x^2+y^2+z^2$
A_{1g}	3	α	β	0	−1	3	α	β	0	−1	(R_x, R_y, R_z)	
τ_{2g}	3	β	α	0	−1	3	β	α	0	−1		
G_g	4	−1	−1	1	0	4	−1	−1	1	0		
H_g	5	0	0	−1	1	5	0	0	−1	1		u, v, xz, yz, xy
A_u	1	1	1	1	1	−1	−1	−1	−1	−1		
τ_{1u}	3	α	β	0	−1	−3	$-\alpha$	$-\beta$	0	1	(x, y, z)	
τ_{2u}	3	β	α	0	−1	−3	$-\beta$	$-\alpha$	0	1		
G_u	4	−1	−1	1	0	−4	1	1	−1	0		
H_u	5	0	0	−1	1	−5	0	0	1	−1		

$u = 2z^2 - x^2 - y^2, v = x^2 - y^2$

Linear Groups

$C_{\infty v}$

	E	$2C_\infty(\phi)$	$2C_\infty(2\phi)$	...	$\infty\sigma_v$		
$A_1(\Sigma^+)$	1	1	1	...	1	z	x^2+y^2, z^2
$A_2(\Sigma^-)$	1	1	1	...	−1	R_z	
$E_1(\Pi)$	2	$2cos(\phi)$	$2cos(2\phi)$	...	0	$(x, y), (R_x, R_y)$	(xz, yz)
$E_1(\Delta)$	2	$2cos(2\phi)$	$2cos(4\phi)$	...	0		(x^2-y^2, xy)
...	...						
...	...						
E_n	2	$2cos(n\phi)$	$2cos(2n\phi)$	...	0		

$D_{\infty h}$

	E	$2C_\infty(\phi)$	...	$\infty\sigma_v$	I	$2S_\infty(\phi)$	...	$\infty C_2'$		
$A_{1g}(\Sigma_g^+)$	1	1	...	1	1	1	...	1		x^2+y^2, z^2
$A_{2g}(\Sigma_g^-)$	1	1	...	−1	1	1	...	−1	R_z	
$E_{1g}(\Pi_g)$	2	$2\cos(\phi)$	...	0	2	$-2\cos(\phi)$	...	0	(R_x, R_y)	(xz, yz)
$E_{2g}(\Delta_g)$	2	$2\cos(2\phi)$	...	0	2	$2\cos(2\phi)$	...	0		(x^2-y^2, xy)
...		...	...							
E_{ng}	2	$2\cos(n\phi)$	...	0	2	$2(-1)^n\cos(n\phi)$	...	0		
...		...	...							
$A_{1u}(\Sigma_u^+)$	1	1	...	1	−1	−1	...	−1	z	
$A_{2u}(\Sigma_u^-)$	1	1	...	−1	−1	−1	...	1		
$E_{1u}(\Pi_u)$	2	$2\cos(\phi)$	...	0	−2	$2\cos(\phi)$	...	0	(x, y)	
$E_{2u}(\Delta_u)$	2	$2\cos(2\phi)$	...	0	−2	$-2\cos(2\phi)$	...	0		
...		...	...							
E_{nu}	2	$2\cos(n\phi)$	...	0	−2	$2(-1)^{n+1}\cos(n\phi)$	...	0		
...		...	...							

References

Altmann, Simon L. (1977), Induced Representations in Crystals and Molecules: Point, space and nonrigid molecule groups (London: Academic Press).

Altmann, Simon L. (1992), Icons and Symmetries (Oxford: Clarendon Press).

Altmann, Simon L. (1994), Band Theory of Solids: An Introduction from the Point of View of Symmetry (Oxford: Clarendon Press).

Altmann, S. L. and Bradley, C. J. (1965), Lattice Harmonics II. Hexagonal Closed-Packed Lattice, *Rev. Mod. Phys.* 37, #1, 33-45.

Arfken, George (1970), Mathematical Methods for Physicists, second ed. (New York: Academic Press).

Aris, Rutherford (1989), Vectors, Tensors and the Basic Equations of Fluid Mechanics (New York: Dover Publications).

Armstrong, M. A. (1988), Groups and Symmetry (New York: Springer-Verlag).

Audsley, W. and G. (1968), Designs and Patterns from Historic Ornament (New York: Dover Publications).

Barut, A. O. and Raczka R. (1986), Theory of Group Representations and Applications (Singapore: World Scientific Publishing Co.).

Bhagavantam, S. and Venkatarayudu, T. (1951), Theory of Groups and its Application to Physical Problems, second ed. (Waltair, India: Andhra University).

Birman, Joseph L. (1974), Theory of Space Groups and Lattice Dynamics (New York: Springer-Verlag).

Bishop, David M. (1993), Group Theory and Chemistry (New York: Dover Publications).

Blichfeldt, H. F. (1917), Finite Collineation Groups (Chicago: University of Chicago Press).

Blumenthal, Leonard M. (1980), A Modern View of Geometry (New York: Dover Publications).

Bohm, David (1951), Quantum Theory (New York: Prentice Hall).

Borchardt-Ott, Walter (1993), Crystallography (Berlin: Springer-Verlag).

Bradley, C. J. and Cracknell, A. P. (1972), The Mathematical Theory of Symmetry in Solids: Representation Theory for Point Groups and Space Groups (Oxford: Clarendon Press).

Bravais, A. (1969), On the Systems Formed by Points Regularly Distributed on a Plane or in Space (New York, American Crystallographic Association).

Brown, Harold, Rolf Bülow, Joachim Neubüser, Hans Wondratschek, Hans Zassenhaus (1978), Crystallographic Groups of Four-Dimensional Space (New York: John Wiley and Sons, Inc.).

Brown, Ronald F. (1975), Organic Chemistry (Belmont, CA: Wadsworth Publishing Co.).

Bruce, A. D. and Cowley, R. A. (1981), Structural Phase Transitions (London: Taylor & Francis, Ltd.).

Burn, R. P. (1991), Groups, A Path to Geometry (Cambridge: Cambridge University Press).

Burns, Gerald (1977), Introduction to Group Theory with Applications (New York: Academic Press).

Burns, Gerald and Glazer, A. M. (1978), Space Groups for Solid State Scientists (New York: Academic Press).

Burns, Gerald and Glazer, A. M. (1990), Space Groups for Solid State Scientists (San Diego: Academic Press).

Burrow, Martin (1993), Representation Theory of Finite Groups (New York: Dover Publications).

Califano, S. (1976), Vibrational States (London: John Wiley and Sons, Inc.).

Ceulemans, A. and Fowler, P. W. (1995), Symmetry Extensions of Euler's Theorem for Polyhedral, Toroidal and Benzenoid Molecules, *J. Chem. Soc. Faraday Trans.* **91**, 3089-3093.

Chen, Jin-Quan (1989), Group Representation Theory for Physicists (Singapore: World Scientific Publishing Co.).

Coleman, A. J. (1968), Induced and Subduced Representations in Loebl (1968), p. 57-118.

Cornwell, J. F. (1969), Group Theory and Electronic Energy Bands in Solids (Amsterdam - London: North-Holland Publishing Company).

Cornwell, J. F. (1984), Group Theory in Physics (London: Academic Press).

Cotton, F. Albert (1990), Chemical Applications of Group Theory (New York: John Wiley and Sons, Inc.).

Courant, R. and Hilbert, D. (1955), Methods of Mathematical Physics, vol. I (New York: Interscience Publishers, Inc.).

Coxeter, H. S. M. (1973) Regular Polytopes (New York: Dover Publications).

Cracknell, A. P. (1975), Group Theory in Solid-State Physics (London: Taylor & Francis, Ltd.).

D'Avennes, Prisse (1978), Arabic Art in Color (New York: Dover Publications).

Davies, B. L. and Dirl, R. (1994a), Space Group Structures: Integrated Software for IBM-compatible PCs, in Lulek, Florek and Walcerz (1994), p. 457-464.

Davies, B. L. and Dirl, R. (1994b), Space Group Representations: Integrated Software for IBM-compatible PCs, in Lulek, Florek and Walcerz (1994), p. 465-468.

de Groot, S. R. and Mazur, P. (1984), Non-Equilibrium Thermodynamics (New York: Dover Publications).

Dixon, John D. (1973), Problems in Group Theory (New York: Dover Publications).

Duffey, George H. (1992), Applied Group Theory for Physicists and Chemists (Englewood Cliffs, NJ: Prentice Hall).

Elliott, J. P. and Dawber, P. G. (1987), Symmetry in Physics, vol. 1, principles and simple applications (Houndsmills: Macmillan Publishers).

Elliott, J. P. and Dawber, P. G. (1986), Symmetry in Physics, vol. 2, further applications (New York: Oxford University Press).

Evarestov, R. A. and Smirnov, V. P. (1993), Site Symmetry in Crystals: Theory and Applications (Berlin: Springer-Verlag).

Falicov, L. M. (1966), Group Theory and its Physical Applications (Chicago: The University of Chicago Press).

Farmer, David W. (1996), Groups and Symmetry, A guide to discovering mathematics (Providence, RI: American Mathematical Society).

Fässler, A. and Stiefel, E. (1992), Group Theoretical Methods and Their Applications (Cambridge, MA: Birkhauser Boston).

Fedorov, E. S. (1971), Symmetry of Crystals (New York, American Crystallographic Association).

Fejes Toth, L. (1964), Regular Figures (Oxford: Pergamon Press).

Field, Michael and Golubitsky, Martin (1992), Symmetry in Chaos, A Search for Pattern in Mathematics, Art and Nature (Oxford: Oxford University Press).

Florek, Wojciech (1994), Magnetic Translation Groups as Group Extensions, *Rep. Math. Phys.*, 34, #1, 81-95.

Florek, Wojciech (1998), The role of a form of vector potential — normalization of the antisymmetric gauge, *J. Math. Phys.* **39**, #2, 739-748.

Ford, Lester R. (1972), Automorphic Functions (New York: Chelsea Publishing Co.).

Fulton, William and Harris, Joe (1991), Representation Theory (New York: Springer-Verlag).

Gasson, Peter C. (1989), Geometry of Spatial Forms (Chichester, West Sussex, Ellis Horwood Limited).
Gazeau, Jean Pierre (1997), From Periodic Crystals to Aperiodic Crystals, in Gieres, it et al (1997), p. 79-106.
Gel'fand, I. M. (1989), Lectures on Linear Algebra (New York: Dover Publications).
Ghyka, Matila (1977), The Geometry of Art and Life (New York: Dover Publications).
Gieres, F., M. Kibler, C. Lucchesi, O. Piguet (1997), Symmetries in Physics (Paris: Editions Frontières).
Gillon, Edmund V., Jr. (1969), Geometric Designs and Ornament (New York: Dover Publications).
Goldstein, Herbert (1953), Classical Mechanics (Cambridge, MA: Addison-Wesley Publishing Company).
Grafton, Carol Belanger (1992), Decorative Tile Designs (New York: Dover Publications).
Grossman, Israel and Magnus, Wilhelm (1992), Groups and their Graphs (Washington, DC: The Mathematical Association of America, New Mathematical Library).
Haase, Rolf (1990), Thermodynamics of Irreversible Processes (New York: Dover Publications).
Hahn, Theo, Ed. (1987), International Tables for Crystallography. Vol. A, Space-Group Symmetry. 4th ed. Published for the International Union of Crystallography (Norwell, MA, Kluwer Academic Publishers).
Hahn, Theo, Ed. (1989), International Tables for Crystallography. Vol. A, Space-Group Symmetry. Brief teaching edition. Published for the International Union of Crystallography (Norwell, MA, Kluwer Academic Publishers).
Hall, Marshall, Jr. (1959), The Theory of Groups (New York: Macmillan).
Hamermesh, M. (1962), Group Theory and its Application to Physical Problems (Reading MA: Addison-Wesley).
Hargittai, Istvan and Hargittai, Magdolna (1987), Symmetry through the Eyes of a Chemist (New York: VCH Publishers).
Harris, Daniel C. and Bertolucci, Michael D. (1989) Symmetry and Spectroscopy: An Introduction to Vibrational and Electronic Spectroscopy (New York: Dover Publications).
Harter, W. G. (1993), Principles of Symmetry, Dynamics and Spectroscopy (New York: John Wiley and Sons, Inc.).
Heine, Volker (1993), Group Theory in Quantum Mechanics (New York: Dover Publications).
Herring, Conyers (1942), Character Tables for Two Space Groups, *J. Franklin Institute* 233, 525-543.

Hessemer, F. M. (1990), Historic Designs and Patterns in Color from Arabic and Italian Sources (New York: Dover Publications).

Hilton, Harold (1963), Mathematical Crystallography and the Theory of Groups of Movement (New York: Dover Publications).

Hilton, Peter and Pederson, Jean (1996), The Euler Characteristic and Polya's Dream, *Am. Math. Monthly*, 103, 121-131.

Hochstrasser, Robin M. (1966), Molecular Aspects of Symmetry (New York: W. A. Benjamin, Inc.).

Holden, Alan (1991), Shapes, Space, and Symmetry (New York: Dover Publications).

Hornung, Clarence P. (1975), Allover Patterns for Designers and Craftsmen (New York: Dover Publications).

Inui, T., Tanabe, Y. and Onodera, Y. (1990), Group Theory and its Applications in Physics (Berlin: Springer-Verlag).

James, Gordon and Liebeck, Martin (1993), Representations and Characters of Groups (Cambridge: Cambridge University Press).

Janner, A. and Ascher, E. (1969a), Bravais Classes of Two-Dimensional Relativistic Lattices, *Physika* 45, 33-66.

Janner, A. and Ascher, E. (1969b), Relativistic Crystallographic Point Groups in Two Dimensions, *Physika* 45, 67-85.

Janner, A. and Ascher, E. (1971), Crystallographic Concepts for Inhomogeneous Subgroups of the Poincaré Group, *Physika* 54, 77-93.

Jansen, Laurens and Boon, Michael (1967), Theory of Finite Groups. Applications in Physics (Amsterdam: North-Holland Publishing Co.).

Janssen, T. (1973), Crystallographic Groups (Amsterdam: North-Holland Publishing Co.).

Jaric, Marko V., ed. (1989), Introduction to the Mathematics of Quasicrystals (Boston: Academic Press).

Jaswon, M. A. (1965), An Introduction to Mathematical Crystallography (New York: American elsevier Publishing Company, Inc.).

Jaswon, M. A. and Rose, M. A. (1983), Crystal Symmetry: Theory of Colour Crystallography (Chichester: Ellis Horwood Limited; New York: John Wiley and Sons, Inc.).

Jerphagnon, J., D. Chemla and R. Bonneville (1978), The description of the physical properties of condensed matter using irreducible tensors, *Advances in Physics* 27, #4, p. 609-650.

Johnston, Bernard L., and Richman, Fred (1997), Numbers and Symmetry: An Introduce to Algebra (Boca Raton, FL, CRC Press, Inc.).

Jones, H. (1975), The Theory of Brillouin Zones and Electronic States in Crystals Amsterdam - Oxford: North-Holland Publishing Company).

Jones, H. F. (1990), Groups, Representations and Physics (Bristol: Adam Hilger, IOP Publishing).

Jones, Owen (1987), The Grammer of Ornament (New York: Dover Publications).

Jones, William and March, Norman H. (1985), Theoretical Solid State Physics; Vol. 1 Perfect lattices in equilibrium; Vol. 2 Non-equilibrium and disorder (New York: Dover Publications).

Joshi, A. W. (1982), Elements of Group Theory for Physicists, third ed. (New York: John Wiley and Sons, Inc.).

Joshua, S. J. (1991), Symmetry Principles and Magnetic Symmetry in Solid State Physics (Bristol: Adam Hilger).

Kettle, S. F. A. (1995), Symmetry and Structure, readable group theory for chemists (Chichester: John Wiley and Sons, Inc.).

Kibler, M. (1997), Symmetries in Nuclear, Atomic and Molecular Spectroscopies, in Gieres, it et al (1997), p. 33-62.

Kocinski, Jerzy (1983), Theory of Symmetry Changes at Continuous Phase Transitions (New York: Elsevier Science Publishing Company, Inc.).

Kocinski, Jerzy (1990), Commensurate and Incommensurate Phase Transitions (New York: Elsevier Science Publishing Company, Inc.).

Koster, G. F. (1957), Space Groups and Their Representations, *Solid State Physics* (F. Seitz, D. Turnbull, eds.) 5, 173-256.

Kupersztych, J. (1976), Is There a Link between Gauge Invariance, Relativistic Invariance and Electron Spin?, *Nuovo Cimento* **31B**, #1, 1-11.

Kurosh, A. G. (1960a,b), The Theory of Groups, vol. 1,2 (New York: Chelsea Publishing Co.).

Ladd, M. F. C. (1989), Symmetry in Molecules and Crystals (Chichester: Ellis Horwood Limited; New York: John Wiley and Sons, Inc.).

Landau, L. D and Lifshitz, E. M. (1967), Statistical Physics (Oxford: Pergamon Press).

Lavenda, Bernard H. (1993), Thermodynamics of Irreversible Processes (New York: Dover Publications).

Lax, Melvin (1974), Symmetry Principles in Solid State and Molecular Physics (New York: John Wiley and Sons, Inc.).

Ledermann, Walter (1987), Introduction to Group Characters, second ed. (Cambridge: Cambridge University Press).

Leech, J. W. and D. J. Newman (1969), How to use Groups (London: Methuen and Co. Ltd).

Lifshitz, Ron (1997), Theory of Color Symmetry for Periodic and Quasiperiodic Crystals, *Rev. Mod. Phys.* 69, #4, 1181-1218.

Lines, L. (1965), Solid Geometry (New York: Dover Publications).

Lockwood, E. H. and Macmillan, R. H. (1978), Geometric Symmetry (Cambridge: Cambridge University Press).

Loebl, Ernest M. ed. (1968), Group Theory and its Applications (New York: Academic Press).

Lomont, J. S. (1961), Applications of Finite Groups (New York: Academic Press).

Lovett, D. R. (1990), Tensor Properties of Crystals (Bristol: Adam Hilger, IOP Publishing).

Ludwig, W. and Falter, C. (1988), Symmetries in Physics, Group Theory Applied to Physical Problems (Berlin: Springer-Verlag).

Lulek, T., W. Florek and S. Walcerz, eds., (1994), Symmetry and Structural Properties of Condensed Matter, Proceedings Third International School on Theoretical Physics (Singapore: World Scientific Publishing Co.).

Lyndon, Roger C. (1989), Groups and Geometry, London Mathematical Society, Lectures Notes Series 101 (Cambridge: Cambridge University Press).

Lyubarskii, G. Ya. (1960), The Application of Group Theory in Physics (Oxford: Pergamon Press).

Mackay, A. L. and Pawley, G. S. (1963), Bravais Lattices in Four-Dimensional Space, *Acta. Cryst.* **16**, 11-19.

Margenau, Henry and Murphey, George Mosely (1955), The Mathematics of Physics and Chemistry (New York: D. Van Nostrand Co., Inc.).

Martin, George E. (1987), Transformation Geometry, An Introduction to Symmetry (New York: Springer-Verlag).

McMullen, P. (1980), Convex Bodies which Tile Space by Translation, *Mathematika* **27**, 113-121.

Megaw, Helen D. (1973), Crystal Structures: A Working Approach (Philadelphia: W. B. Saunders Company).

Menten, Theodore (1975), Japanese Border Designs (New York: Dover Publications).

Merzbacher, Eugen (1967), Quantum Mechanics (New York: John Wiley and Sons, Inc.).

Meserve, Bruce E. (1983), Fundamental Concepts of Geometry (New York: Dover Publications).

Miller, Willard, Jr. (1972), Symmetry Groups and their Applications (New York: Academic Press).

Mirman, R. (1969), Coherent Superposition of Charge States, *Phys. Rev.* **186**, #5, 1380-1383.

Mirman, R. (1995a), Group Theory: An Intuitive Approach (Singapore: World Scientific Publishing Co.).

Mirman, R. (1995b), Group Theoretical Foundations of Quantum Mechanics (Commack, NY: Nova Science Publishers, Inc.).

Mirman, R. (1995c), Massless Representations of the Poincaré Group, electromagnetism, gravitation, quantum mechanics, geometry (Commack, NY: Nova Science Publishers, Inc.).

Murnaghan, Francis D. (1963), The Theory of Group Representations (New York: Dover Publications).

Neumann, Peter M., Gabrielle A. Stoy and Edward C. Thompson (1994), Groups and Geometry (Oxford: Oxford University Press).

Nikulin, V. V. and Shafarevich, I. R. (1987), Geometries and Groups (Berlin: Springer-Verlag).

Nowick, Arthur S. (1995), Crystal Properties Via Group Theory (Cambridge: Cambridge University Press).

Nye, J. F. (1992), Physical Properties of Crystals; Their Representation by Tensors and Matrices (Oxford: Clarendon Press).

Pauling, Linus (1988), General Chemistry (New York: Dover Publications).

Pauling, Linus and Wilson, E. Bright, Jr., (1935), Introduction to Quantum Mechanics, with applications to chemistry (New York: McGraw-Hill Book Co.).

Quilichini, M. and Janssen, T. (1997), Phonon Excitations in Quasicrystals, *Rev. Mod. Phys.* **69**, #1, 277-312.

Robinson, Derek J. S. (1993), A Course in the Theory of Groups (New York: Springer-Verlag).

Rosen, Joe (1977), Symmetry Discovered, Concepts and Applications in Nature and Science (Cambridge: Cambridge University Press).

Rosen, Joe (1981), Resource letter SP-2: Symmetry and group theory in physics, *Am. J. Phys.*, **49**, #4, 304-319.

Rotman, Joseph J. (1995), An Introduction to the Theory of Groups (New York: Springer-Verlag).

Sands, Donald E. (1993), Introduction to Crystallography (New York: Dover Publications).

Schattschneider, Doris (1978), The Plane Symmetry Groups: Their Recognition and Notation, *Am. Math. Monthly* **85**, #6, 439-450.

Schattschneider, Doris (1990), Visions of Symmetry: Notebooks, Periodic Drawings, and Related Work of M. C. Escher (New York: W. H. Freeman and Co.).

Schensted, Irene Verona (1976), A Course on the Application of Group Theory to Quantum Mechanics (Peaks Island, ME 04108: NEO Press).

Schiff, L. I. (1955), Quantum Mechanics, second ed. (New York: McGraw-Hill Book Co.).

Schwarzenberger, R. L. E. (1980), N-Dimensional Crystallography (London, Pitman Publishing Limited).

Senechal, Marjorie (1988), Color Symmetry, *Comput. Math. Applic.*, **16**, 545 - 553.

Senechal, Marjorie (1989), A Brief Introduction to Tilings in Jaric (1989), *p. 1-51.*

Senechal, Marjorie (1990), Crystalline Symmetries, An Informal Mathematical Introduction (Bristol: Adam Hilger, IOP Publishing).
Senechal, Marjorie (1995), Quasicrystals and Geometry (Cambridge: Cambridge University Press).
Shubnikov, A. V. and Belov, N. V. (1964), Colored Symmetry (New York: Pergamon Press).
Shubnikov, A. V. and Koptsik, V. A. (1974), Symmetry in Science and Art (New York: Plenum Press).
Simakoff, N. (1993), Islamic Designs in Color (New York: Dover Publications).
Simon, Barry (1996), Representations of Finite and Compact Groups (Providence, RI: American Mathematical Society).
Slater, J. C. (1965), Space Groups and Wave-Function Symmetry in Crystals, *Rev. Mod. Phys.* 37, #1, 68-83.
Sternberg, S. (1994), Group Theory and Physics (Cambridge: Cambridge University Press).
Streitwolf, Hans-Waldemar (1971), Group Theory in Solid-State Physics (London: Macdonald & Co.).
Tinkham, Michael (1964), Group Theory and Quantum Mechanics (New York: McGraw-Hill Book Co.).
Tolédano, Jean-Claude and Tolédano, Pierre (1987), The Landau Theory of Phase Transitions (Singapore: World Scientific Publishing Co.).
Tsukerblatt, Boris S., (1994), Group Theory in Chemistry and Spectroscopy (London: Academic Press).
Weyl, Hermann (1931), The Theory of Groups and Quantum Mechanics (New York: Dover Publications).
Weyl, Hermann (1989), Symmetry (Princeton, NJ: Princeton University Press).
Wherrett, Brian S. (1986), Group Theory for Atoms, Molecules and Solids (Englewood Cliffs, NJ: Prentice Hall).
Wigner, E. P. (1959), Group Theory, and its Application to Quantum Mechanics of Atomic Spectra (New York: Academic Press).
Wilson, E. Bright, Jr., J. C. Decius and Paul C. Cross (1980), Molecular Vibrations (New York: Dover Publications).
Wood, Elizabeth A. (1977), Crystals and Light, An Introduction to Optical Crystallography (New York: Dover Publications).
Wooster, W. A. (1973), Tensors and Group Theory for the Physical Properties of Crystals (Oxford: Clarendon Press).
Yale, Paul B. (1988), Geometry and Symmetry (New York: Dover Publications).
Zak, J., A. Casher, M. Glück, and Y. Gur (1969), The Irreducible Representations of Space Groups (New York: W. A. Benjamin, Inc.).

Index

Definitions are listed first, in bold; other references are italicized.

www.ingramcontent.com/pod-product-compliance
Lightning Source LLC
Chambersburg PA
CBHW050632190326
41458CB00008B/2240
* 9 7 8 9 8 1 0 2 3 7 3 2 5 *